ULTRAFAST INFRARED VIBRATIONAL SPECTROSCOPY

ULTRAFAST INFRARED VIBRATIONAL SPECTROSCOPY

Edited by **MICHAEL D. FAYER**

CRC Press
Taylor & Francis Group
Boca Raton London New York

CRC Press is an imprint of the
Taylor & Francis Group, an **informa** business

CRC Press
Taylor & Francis Group
6000 Broken Sound Parkway NW, Suite 300
Boca Raton, FL 33487-2742

First issued in paperback 2019

ISBN-13: 978-1-4665-1013-5 (hbk)
ISBN-13: 978-0-367-38030-4 (pbk)

Library of Congress Cataloging-in-Publication Data

Ultrafast infrared vibrational spectroscopy / editor, Michael D. Fayer.
p. cm.
Includes bibliographical references and index.
ISBN 978-1-4665-1013-5 (hardcover : alk. paper)
1. Infrared spectroscopy. 2. Vibrational spectra. I. Fayer, Michael D.

QD96.I5U485 2013
535.8'42--dc23 2012038081

Visit the Taylor & Francis Web site at
http://www.taylorandfrancis.com

and the CRC Press Web site at
http://www.crcpress.com

Contents

Preface

The dynamical timescales associated with molecular motions are very short. Molecules can range in size from water to proteins and DNA. While proteins and DNA are not small molecules, they are composed of amino acids and nucleotides. These building blocks are in essence small molecules, and the properties and dynamics of large biological molecules can depend on the motions and dynamical interactions of the small-molecule-sized subunits that, like their small molecule brethren, are inherently very fast. Ultrafast spectroscopy has provided tools that make it possible to study very fast motions of molecular systems on the timescales on which they actually occur. Initially, ultrafast laser spectroscopy was limited to the visible and ultraviolet spectral regimes because of the available laser technology. For many decades, UV/Vis ultrafast experiments have provided a wealth of information by exciting and probing molecular electronic-excited states.

Over approximately the last decade, sources of ultrafast infrared (IR) pulses have become readily available. The IR region of the spectrum corresponds to energies associated with molecular vibrations. The vibrational modes are the mechanical degrees of freedom of molecules. Time-independent linear IR absorption spectra can provide a vast amount of structural information about molecules. The mid-IR is often referred to as the fingerprint region because vibrational absorption spectra in this region are sensitive to molecular structure, conformation, and environment. In organic chemistry, vibrational spectra are often used to verify the presence of particular substituents and to determine the conformations. Vibrational spectra are sensitive to small differences in structure. For example, a system of long alkyl chains bound to a gold surface that is all trans has a slightly different spectrum than the same system with gauche defects.

The advent of laser-based sources of ultrafast IR pulses has extended the study of very fast molecular dynamics to the observation of processes manifested through their effects on the vibrations of molecules. In addition, nonlinear IR spectroscopic techniques make it possible to examine intra- and intermolecular interactions and how such interactions evolve on very fast timescales, but also in some instances on very slow timescales. To see the efficacy of using IR experiments to study molecular dynamics and interactions, it is useful to compare and contrast NMR, IR, and UV/Vis experiments. NMR spectra for molecules in liquids have very sharp lines that are exquisitely sensitive to structure. The sharp features of one-dimensional (1D) NMR spectra are augmented by two-dimensional (2D) and multidimensional methods and complex pulse sequences that make it possible to pull apart details of very complicated spectra of large molecules, including proteins. However, NMR inherently operates on slow timescales. Direct measurements of dynamics are generally limited to milliseconds and much longer. Using line shape analysis, NMR can provide information on fast timescales, but these are highly model dependent.

In contrast to the exceedingly sharp features of NMR spectra of molecules in liquids, 1D UV/Vis absorption spectra of chromophores in liquids generally are very broad and almost featureless. Very limited structural information is available from a 1D spectrum. Ultrafast UV/Vis spectroscopy does permit direct measurements of ultrafast processes that occur on excited-state potential surfaces, such as photoinduced electron transfer or the trans to cis isomerization of stilbene. However, with the exception of a few very specialized chromophore systems, particularly photosynthetic chromophore complexes, UV/Vis 2D methods provide limited improvement in the measurement of the details of structural dynamics and inter- and intramolecular interactions.

NMR and UV/Vis spectroscopies are in some sense-limiting extremes. NMR has very sharp spectral features that with multidimensional methods and pulse sequences give a great deal of structural information but can directly investigate dynamics only on slow timescales. UV/Vis spectra are very broad and mainly featureless but can provide dynamical information of ultrafast timescales usually with limited structural detail.

Vibrational spectroscopy bridges the gap. Vibrational 1D absorption spectra, in most cases, have relatively sharp lines that can be identified with specific molecular functional groups. As mentioned above, vibrational spectra are quite sensitive to structure and conformation although not nearly as sensitive as NMR. Vibrational experiments can be performed on ultrafast timescales. The typical wavelength of a time domain vibrational experiment is about 10 times longer than a UV/Vis experiment, which means that vibrational experiments cannot have the time resolution of UV/Vis experiments. However, time-dependent IR experiments are routinely conducted with subhundred femtosecond time resolution, which provides sufficient time resolution to study most molecular processes. The inverse of the vibrational linewidth ultimately sets the time resolution necessary for a dynamical experiment. Typical vibrational lines have linewidths of less than a few tens of wave numbers (the hydroxyl stretch of water is an important exception), which sets the fastest dynamical timescale at about a picosecond. The time resolution of mid-IR experiments is many orders of magnitude faster than can be achieved with NMR. In addition, the relatively sharp spectral features associated with vibrational spectra make them amenable to time-dependent 2D experiments that can extract a great deal of information that is not accessible by either NMR or UV/Vis methods. Even for physical problems that inherently involve electronic-excited states, combined time-dependent UV/Vis–IR experiments can provide information not readily ascertainable in straight UV/Vis experiments. For example, a UV/Vis excitation pulse can be used to initiate an electron transfer process while subsequent IR probing can observe changes in the vibrational spectrum to investigate the structural changes.

This book contains chapters that discuss experimental and theoretical topics that reflect the latest accomplishments and understanding of ultrafast IR vibrational spectroscopy. Many of the experiments employ 2D spectroscopy to obtain dynamics and structure. Each chapter provides background, details of methods, and explication of a topic of current research interest. Some of the chapters treat experiments that involve only IR pulse sequences, but others use a combination of visible and IR pulses. Experimental and theoretical studies cover topics as diverse as the dynamics of water and the dynamics and structure of biological molecules. Methods include vibrational echo chemical exchange spectroscopy, IR-Raman spectroscopy, time-resolved sum frequency generation, and 2D IR spectroscopy. In both experiments and theory, use is made of the polarization of the IR light pulses. Overall, this book contains material that will be of interest to people new to the field, experts in the field, and individuals who want to gain an understanding of particular methods and research topics.

Editor

Michael D. Fayer is the David Mulvane Ehrsam and Edward Curtis Franklin Professor of Chemistry at Stanford University. He grew up in Los Angeles, California, where he attended the public schools. He went to both undergraduate and graduate schools at the University of California at Berkeley. He received his PhD in chemistry in 1974. Fayer began his academic career in the same year as an assistant professor of chemistry at Stanford at the age of 26.

For many years, Fayer has been a pioneer in the development and application of ultrafast nonlinear laser techniques for the study of complex molecular systems ranging from solids at liquid helium temperature to flames. In large part due to his work, ultrafast nonlinear and coherent spectroscopic techniques such as transient gratings, photon echoes, and infrared vibrational echoes have become powerful techniques for studying the fast molecular processes, intermolecular interactions, and structure in complex molecular systems. His work has had a profound impact on modern physical chemistry, biophysics, and materials science, and his methods and approaches to the examination of problems involving dynamics and interactions in molecular systems have spread worldwide. Michael Fayer is a member of the National Academy of Sciences of the United States of America, and the American Academy of Arts and Sciences. He has won a number of national and international awards, including the Arthur L. Schawlow Prize in Laser Science conferred by the American Physical Society, the Ellis R. Lippincott Award given by the Optical Society of America, the E. Bright Wilson Award for Spectroscopy bestowed by the American Chemical Society, and the Earl K. Plyler Prize for Molecular Spectroscopy awarded by the American Physical Society.

Contributors

John B. Asbury
The Pennsylvania State University
University Park, Pennsylvania

Carlos R. Baiz
Massachusetts Institute of Technology
Cambridge, Massachusetts

H. J. Bakker
FOM Institute for Atomic and
Molecular Physics
Amsterdam, the Netherlands

M. Bonn
Max Planck Institute for Polymer
Research
Mainz, Germany

L. E. Buchanan
Department of Chemistry
University of Wisconsin
Madison, Wisconsin

Minhaeng Cho
Department of Chemistry
Korea University
Seoul, Korea

J. J. de Pablo
Department of Chemistry
University of Wisconsin
Madison, Wisconsin

Dana D. Dlott
School of Chemical Sciences
University of Illinois at
Urbana-Champaign
Urbana, Illinois

Benjamin Doughty
Department of Chemistry
Columbia University
New York, New York

E. B. Dunkelberger
Department of Chemistry
University of Wisconsin
Madison, Wisconsin

Kenneth B. Eisenthal
Department of Chemistry
Columbia University
New York, New York

Thomas Elsaesser
Max-Born-Institut für Nichtlineare Optik
und Kurzzeitspektroskopie
Berlin, Germany

Cyril Falvo
Institut des Sciences
Moléculaires d'Orsay
Université Paris Sud
Orsay, France

Michael D. Fayer
Department of Chemistry
Stanford University
Stanford, California

Robin M. Hochstrasser
Department of Chemistry
University of Pennsylvania
Philadelphia, Pennsylvania

James T. Hynes
Chemistry Department
Ecole Normale Supérieure
Paris, France
and
Department of Chemistry and Biochemistry
University of Colorado
Colorado

Daniel G. Kuroda
Department of Chemistry
University of Pennsylvania
Philadelphia, Pennsylvania

Kyung-Won Kwak
Department of Chemistry
Chung-Ang University
Seoul, Korea

Damien Laage
Chemistry Department
Ecole Normale Supérieure
Paris, France

Shaul Mukamel
Department of Chemistry
University of California, Irvine
Irvine, California

Kaoru Ohta
Molecular Photoscience Research Center
Kobe University
Kobe, Japan

Kwang-Hee Park
Department of Chemistry
Korea University
Seoul, Korea

Brandt C. Pein
School of Chemical Sciences
University of Illinois at Urbana-Champaign
Urbana, Illinois

Yi Rao
Department of Chemistry
Columbia University
New York, New York

Mike Reppert
Massachusetts Institute of Technology
Cambridge, Massachusetts

Igor V. Rubtsov
Department of Chemistry
Tulane University
New Orleans, Louisiana

Shinji Saito
Institute for Molecular Science
National Institutes of National Science
Okazaki, Japan

František Šanda
Faculty of Mathematics and Physics
Charles University
Prague, Czech Republic

J. L. Skinner
Department of Chemistry
University of Wisconsin
Madison, Wisconsin

Jumpei Tayama
Molecular Photoscience Research Center
Kobe University
Kobe, Japan

Andrei Tokmakoff
Massachusetts Institute of Technology
Cambridge, Massachusetts

Keisuke Tominaga
Molecular Photoscience Research Center
Kobe University
Kobe, Japan

Nicholas J. Turro
Department of Chemistry
Columbia University
New York, New York

L. Wang
Department of Chemistry
University of Wisconsin
Madison, Wisconsin

M. T. Zanni
Department of Chemistry
University of Wisconsin
Madison, Wisconsin

1 Vibrational Echo Chemical Exchange Spectroscopy

Michael D. Fayer

CONTENTS

1.1 INTRODUCTION

A wide variety of chemical and biological systems involve components or states that are in thermal equilibrium. A system can be composed of two or more chemical species that are constantly interconverting, one to another, without the concentrations of the species changing. The interconversion occurs on the ground electronic state potential surface. Call two species in equilibria A and B. The essence of equilibrium is that the forward rate is equal to the backward rate. That is, the rate of As turning into Bs is equal to the rate of Bs turning into As. By measuring the concentrations of the species, the equilibrium constant can be determined. However, knowing the equilibrium constant provides no information on the chemical dynamics.

In the past, for fast processes, the determination of the system's dynamics under thermal equilibrium conditions, that is, how fast the species are changing from one to the other, was very difficult. The interconversion of species under equilibrium conditions is often referred to as chemical exchange. For slow processes, it is possible to measure chemical exchange directly using multidimensional nuclear magnetic resonance (NMR) methods [1,2]. For example, the chemical exchange in a 1:1 mixture of $SnCl_4$ and $SnBr_4$ was measured using ^{119}Sn 2D NMR [2]. The system has five species in thermal equilibrium.

$$SnCl_4 \rightleftarrows SnCl_3Br \rightleftarrows SnCl_2Br_2 \rightleftarrows SnCl_1Br_3 \rightleftarrows SnBr_4$$

At short time, there are five peaks on the diagonal of the 2D spectrum. As time increases, off-diagonal peaks grow in on the tens of millisecond timescale. The growth of the off-diagonal peaks provides a direct determination of the chemical exchange kinetics. Such direct measurements can be extended into the many microsecond time ranges but not faster.

When 2D NMR methods are used to measure, for example, isomerization around a carbon–carbon single bond, systems must be studied at low temperature with bulky groups that make the barrier to isomerization very high. The low temperature and high barrier are necessary to slow the isomerization process down to NMR timescales [1]. Very fast kinetics for a molecule with a low barrier in a room-temperature solution, occurring on a picosecond timescale [3], cannot be deduced from the low-temperature high-barrier measurements because the rate constant is not a simple function of the temperature and the barrier height [4].

Very fast dynamics of a great many systems have been measured using ultrafast UV/Vis spectroscopy. A system that has been studied in great detail is the trans-to-cis isomerization of stilbene [5]. In the ground electronic state, the barrier for isomerization is so high that isomerization does not occur in a room-temperature liquid. In the ultrafast optical measurements, *trans*-stilbene is pumped with UV light to the first-excited singlet state (S_1). In S_1, the barrier for isomerization is low. The trans-to-cis isomerization occurs in ~70 ps [5]. While such experiments have taught us a great deal about fast chemical processes, they do not study the dynamics on the ground electronic state surface under thermal equilibrium conditions.

Now, ultrafast two-dimensional (2D) IR vibrational echo chemical exchange spectroscopy has made it possible to observe very fast chemical exchange processes in real time under thermal equilibrium conditions [6]. The 2D IR chemical exchange experiments are akin to 2D NMR experiments but operate on timescales many orders of magnitude faster than the equivalent NMR experiments. They have the ultrafast time resolution of UV/Vis ultrafast experiments but do not involve electronic-excited states. As will be shown in detail below, the excitation of a vibration is such a mild perturbation of the molecular system that it does not change the thermal equilibrium kinetics.

Both ultrafast 2D IR vibrational echo techniques and multidimensional NMR involve pulse sequences that generate and then read out the coherent evolution of excitations of a system of molecules (vibrations for IR and nuclear spins for NMR). The molecules are induced to "oscillate" in vibrational states or spin states, all at the same time and initially with the same phase by the first pulse in the sequence. However, the frequencies of the oscillations depend on the molecular environment. The effect of the first pulse and the manipulation of the phase relationships among the vibrational oscillators by the following pulses in the sequence is an important common feature that 2D vibrational echo spectroscopy has with 2D NMR. Subsequent pulses in the 2D IR vibrational echo pulse sequence generate observable signals that are sensitive to change in environments of individual molecules during the experiment, even if the total populations in two distinct environments (the observables in linear spectroscopy) do not change. The critical difference between the 2D IR and NMR variants is that the IR pulse sequence acts on timescales six or more orders of magnitude faster than multidimensional NMR experiments.

Since the first one-dimensional vibrational echo experiments were performed in 1993 on liquids and glasses [7] and soon after on proteins [8], ultrafast 2D IR vibrational echo-based experiments have been applied to the study of the intramolecular coupling of vibrational modes, the structure and dynamics of proteins, the hydrogen bond dynamics of water, and the dynamics of other liquids by observing the positions of peaks or the change with time in the shape of a band in the 2D spectra [9–31]. The change in shape of a band is caused by spectral diffusion, that is, the sampling of the many local structures that give rise to the inhomogeneously broadened IR absorption line. 2D IR spectroscopy has been extremely successful in understanding spectral diffusion in bulk water [14,32–37] and other hydrogen bonding systems [12,15,38], as well as in proteins and other biological molecules [21–23,39–44]. Useful methods have been developed for analyzing the change in the band shape caused by spectral diffusion and extracting the underlying dynamical information [45,46]. Spectral diffusion is related to but distinct from chemical exchange. Spectral diffusion reflects a single species or a system in a particular state experiencing the time evolution of its environment. In contrast, chemical exchange is the interconversion of one species or state into another. In the NMR chemical exchange example involving $SnCl_4$ and $SnBr_4$ mentioned above, distinct molecules are undergoing chemical reactions that change one into the other, resulting in new off-diagonal peaks growing in the spectrum.

In the 2D IR vibrational echo chemical exchange spectroscopy experiments described here, chemical exchange causes new peaks in the spectrum to appear and grow, yielding the rate of chemical exchange. First, the 2D IR chemical exchange method will be introduced. Solute–solvent complex formation and dissociation will be used as the initial example. Phenol forms π hydrogen bonding complexes with benzene in a solution with carbon tetrachloride (CCl_4). Two species exist in solution, the phenol–benzene complex and free phenol, that is, phenol that is not hydrogen bonded to a benzene. This system is used to demonstrate the method and to describe the data analysis and the extraction of the chemical exchange rate [6]. In addition, experiments show that the measured dynamics are not affected by the excitation of the hydroxyl stretch, which is the vibration studied in the experiments [6]. Data are presented for 13 solute–solvent complexes, and the relationship between the chemical exchange rates and the enthalpies of formation of the complexes is revealed [47,48].

Proteins, enzymes, and other large biological molecules are dynamic structures that sample many conformations under thermal equilibrium conditions. The structural fluctuations of a biological molecule are essential for their function. The types of structural evolution that a protein undergoes can be divided roughly into two types. One is the sampling of closely related structures within a relatively deep well that gives rise to a particular structural substate. Near the bottom of the well, there is a rough surface with many minima. Each minimum is associated with a variation in the structure of the substate, but these different structures do not change some fundamental characteristic of the substate. In the IR absorption spectrum of a vibrational probe, these different structures within a substate cause the inhomogeneous broadening of a single absorption line associated with the substate. Sampling of the structures causes spectral diffusion, which can be measured with 2D IR vibrational echo spectroscopy [17,19,21–24,28,49–51]. In some proteins, the IR absorption spectrum of a vibrational probe gives rise to two or more absorption lines. An important example is the protein myoglobin with CO bound at the active site as the vibrational probe (Mb–CO), which displays several CO absorption peaks reflecting a number of distinct substates [52–55]. The 2D IR chemical exchange method is applied to the direct observation of the interconversion between two well-defined structural substates of two mutants of Mb–CO. The results demonstrate that the interconversion of a protein from one well-defined substate to another can occur on a timescale of less than 100 ps. The results are compared to detailed molecular dynamics (MD) simulations. While the simulations are able to do a good job of reproducing the line shapes and spectral diffusion, they produce substate exchange times that are far too slow.

The 2D IR chemical exchange method is then applied to the isomerization around the carbon–carbon single bond of a substituted ethane molecule [3]. Ethane itself cannot be studied because the rotation of ethane's methyl groups around the carbon–carbon bond leaves the system with the identical structure. To use the chemical exchange method, the two (or more) species or states that are interconverting must have a vibration that has a distinct frequency in each state that changes upon interconversion. Since the rotational isomer states of ethane are structurally identical, a disubstituted ethane is studied that has a vibration with a different frequency in the trans and gauche conformations of the molecule. The results are compared to simulations of a similar molecule using a simple transition state theory approach and an estimate of the ethane isomerization time is obtained.

Another important problem that has been addressed using 2D IR vibrational echo chemical exchange spectroscopy is the dynamics of water hydrogen bonded to an ion [56]. A wide variety of important systems, from biology to geology, involve water, but not pure water. Frequently, there are ions that interact with the water and influence water's hydrogen bond dynamics. The ions may be due to dissolved salts or charged amino acids on the surface of a protein. Water molecules hydrogen bonded to ions and water molecules hydrogen bonded to other water molecules are in equilibrium and will undergo chemical exchange. The question is how fast can a water molecule hydrogen bonded to an ion exchange to become a water molecule hydrogen bonded to another water molecule and vice versa. The question can be answered with the appropriate 2D IR chemical exchange experiment in which water bound to water has a distinct spectrum from water bound to ions [56–58].

1.2 2D IR VIBRATIONAL ECHOES AND THE CHEMICAL EXCHANGE EXPERIMENT

As discussed above, the challenge is to directly observe fast timescale thermal equilibrium chemical events without changing the system's equilibrium behavior. Ultrafast 2D IR vibrational echo chemical exchange spectroscopy makes it possible to directly measure the rate of interconversion of chemical species under thermal equilibrium conditions. The method is an application of the 2D IR vibrational echo experiment, details of which have been described previously [33,59–61]. Here, the method will be briefly outlined.

Figure 1.1 is a schematic of the experiment and the pulse sequence. There are three excitation pulses, ~60 fs in duration produced with a Ti:sapphire regenerative amplifier pumped optical parametric amplifier (OPA). The pulse sequence is shown at the bottom of the figure. The OPA is tuned to the frequency of the vibrational mode under investigation. The IR pulses have sufficient bandwidth to span the spectral region of interest. The times between pulses 1 and 2 and between pulses 2 and 3 are called τ and T_w, respectively. The vibrational echo signal radiates from the sample at a time $\leq\tau$ after the third pulse in a unique direction. The vibrational echo signals are recorded by scanning τ at fixed T_w. The signal is spatially and temporally overlapped with a local oscillator for heterodyne detection, and the combined pulse is dispersed by a monochromator onto an IR array detector. Heterodyne detection provides both amplitude and phase information. If the vibrational echo is detected as it emerges from the sample, there is no phase information. The experiment produces a 2D spectrum in the frequency domain. The measurements are made in the time domain. To go from the time domain to the frequency domain requires two Fourier transforms. To perform a Fourier transform requires both amplitude and phase information. The local oscillator is another IR fs pulse derived from the same initial pulse that is split to make the excitation pulses. When it overlaps with the vibrational echo wave packet, interference occurs. As the time τ is scanned, the vibrational echo wave packet electric field moves in time across the fixed local oscillator wave packet electric field. The result is a temporal interferogram that provides the necessary phase information. An example of a temporal interferogram is shown in the lower right portion of Figure 1.1.

As shown in Figure 1.1, the combined vibrational echo/local oscillator pulse is passed through a monochromator used as a spectrograph and recorded by an IR array detector. The array detector permits the measurement of 32 wavelengths simultaneously, reducing the data acquisition time by

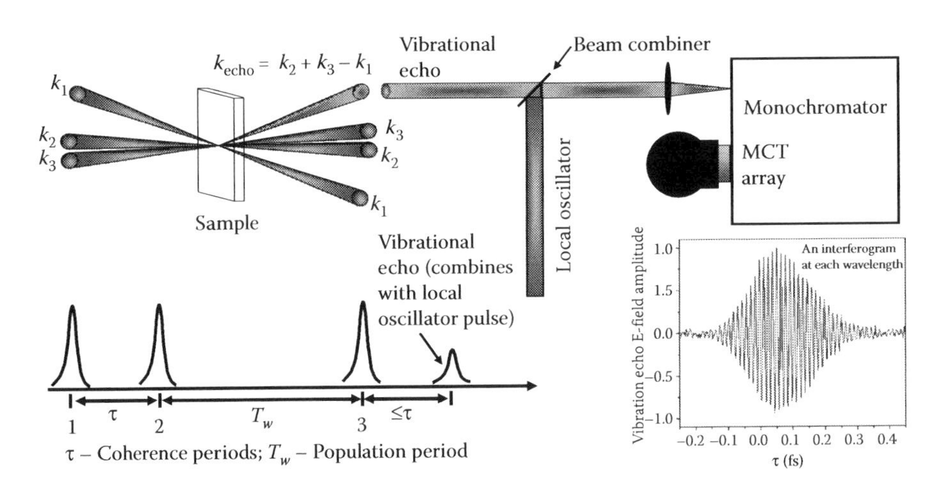

FIGURE 1.1 A schematic of the 2D IR vibrational echo experiment. Also shown are the vibrational echo pulse sequence and an example of a temporal interferogram, which is the result of heterodyne detection of the vibrational echo.

a factor of 32. Taking the spectrum of the heterodyne-detected vibrational echo signal performs one of the two Fourier transforms and gives the ω_m axis (vertical axis, *m* for monochromator) in the 2D IR spectra. When τ is scanned, a temporal interferogram is obtained at each ω_m. The temporal interferograms are numerically Fourier transformed to give the other axis, the ω_τ axis (horizontal axis, τ because the interferograms are obtained by scanning τ). 2D IR spectra are obtained for a range of T_ws. The information is contained in the changes in the 2D spectra as T_w is increased from short to long time.

Figure 1.2 shows a very simple example of the nature of the data. Figure 1.2a shows the spectrum of the hydroxyl stretching mode of phenol–OD. The H in the hydroxyl group has been replaced by a D. The IR absorption spectrum shows a single peak. The structure of the molecule is shown as an inset. When τ is scanned, the time of the output of the vibrational echo electric field, which is the signal $S(\tau)$, changes relative to the local oscillator electric field, L, which is fixed in time. The detector measures the absolute value squared of the sum of the electric fields.

$$\left|L + S(\tau)\right|^2 = L^2 + 2LS(\tau) + S^2 \tag{1.1}$$

The result is three terms. L is much larger than S. L^2 is constant in time. S^2 is very small and not detected relative to the much larger cross term, $2LS(\tau)$. In addition, S^2 is a slowly varying envelop. The $2LS(\tau)$ term has high-frequency oscillations. Windowing in the Fourier transform removes any vestiges of the S^2 term. As τ is scanned, the temporal interferogram is created (see Figure 1.1). There is one such interferogram for each monochromator frequency, ω_m. ω_m is the vertical axis of the 2D IR spectrum shown in the middle panel of Figure 1.2. This axis is obtained from the frequency measurements performed by the monochromator. The horizontal axis, the ω_τ axis, is obtained by numerical Fourier transformation of an interferogram like the one shown in Figure 1.1.

Figure 1.2b displays the 2D IR vibrational echo spectrum at $T_w = 16$ ps. It has two peaks. The peak on the diagonal (the dashed line) is positive going. The peak off-diagonal is negative going. The ω_τ axis is the frequency of the first interaction of the molecules with the radiation field (first pulse). This first interaction is represented by the dashed arrow on the left side of the energy level diagram in Figure 1.2c. The first pulse makes a coherent superposition state of the vibrational ground state (0) and the first vibrationally excited state [1]. The second pulse, represented by the solid arrow in the energy level diagram, transfers the vibrational coherence to frequency-dependent population changes in the 0 and 1 levels. Phase information is stored in the populations. The third pulse between 0 and 1 (dashed arrow between 0 and 1) again produces a coherent superposition state (a coherence), which gives rise to the vibrational echo emission (wavy arrow) at the frequency the molecules interact with the third pulse. The ω_m axis is the axis of the vibrational echo emission. The 0–1 vibrational peak is on the diagonal because the first interaction of the molecules with the first radiation field (ω_τ axis) is at the same frequency as the last interaction and vibrational echo emission (ω_m axis).

The off-diagonal 1–2 peak in the spectrum in Figure 1.2b is shifted along the ω_m axis by the vibrational anharmonicity (unequal spacing of the vibrational energy levels as shown in the energy level diagram in Figure 1.2c). The first two interactions are the same as discussed above. They produce population in the 1 level. Because the bandwidth of the laser is large, it is greater than the anharmonic shift of the levels. So the third interaction can produce a coherence between the 1 and 2 levels, which is represented by the dashed arrow connecting the 1 and 2 levels. The 1–2 coherence gives rise to vibrational echo emission at the 1–2 transition frequency (wavy arrow). Because the first interaction (ω_τ axis) is at the 0–1 transition frequency and the third interaction and vibrational echo emission (ω_m axis) is at the shifted 1–2 transition frequency, the 1–2 peak appears off-diagonal. For what follows, the difference between diagonal and off-diagonal peaks is very important. *If the first and last interactions are at the same frequency, a peak will be on the diagonal. If the first and last interactions are at different frequencies, a peak will be off-diagonal.*

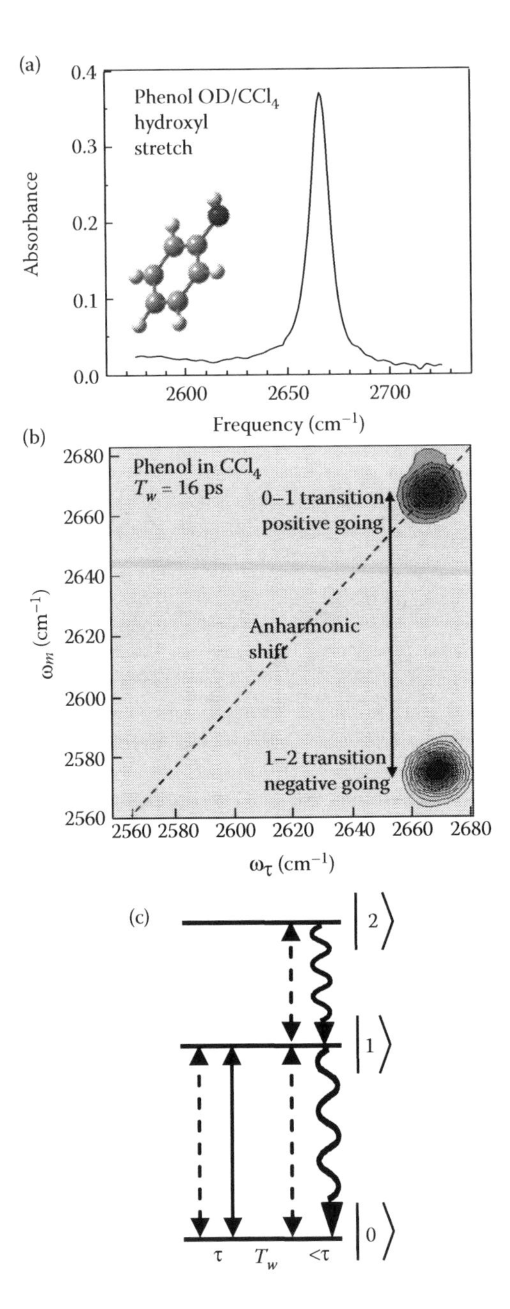

FIGURE 1.2 (a) The FT-IR absorption spectrum of the OD hydroxyl stretch of phenol–OD in CCl_4 solution. (b) A 2D IR spectrum of the phenol OD stretch in CCl_4 at $T_w = 16$ ps. The peak on the diagonal (dashed line) arises from the 0–1 vibrational transition and is positive going. The peak below the diagonal comes from vibrational echo emission at the 1–2 transition frequency, and is negative going. The separation between the two peaks on the ω_m axis is the vibrational anharmonicity. (c) An illustration of the quantum pathways that give rise to the two bands in the 2D IR spectrum in B. A dash double-headed arrow represents a coherent superposition of the two states. A solid double-headed arrow shows population changes created in the two states. The wavy arrow indicates vibrational echo emission arising from transitions between the two states.

A 2D IR vibrational echo chemical exchange experiment works in the following manner. The first laser pulse "labels" the initial structures of the species by establishing their initial frequencies, the ω_τ axis. The second pulse ends the first time period τ and starts the reaction time period T_w during which the labeled species undergo chemical exchange, that is, one type of species interconverts into the other. The third pulse ends the population period of length T_w and begins a third period of length $\leq\tau$, which ends with the emission of the vibrational echo pulse of frequency ω_m. The vibrational echo signal reads out information about the final structures of all labeled species by their frequencies, ω_m. During the period T_w between pulses 2 and 3, chemical exchange occurs. The exchange causes new off-diagonal peaks to grow in as T_w is increased. The growth of the off-diagonal peaks in the 2D IR spectra with increasing T_w is used to obtain the chemical exchange time constants.

Figure 1.3 illustrates the influence of chemical exchange between two species A and B that are in chemical equilibrium. As are constantly turning into Bs, and Bs are constantly turning into As, but the net numbers of As and Bs are not changing. The upper third of the figure shows what happens at very short time, a time short compared to any chemical exchange, that is, interconversion between As and Bs. In the 0–1 portion of the spectrum, there will be two peaks on the diagonal, one for species A at ω_A and one for species B at ω_B. The middle portion of the figures shows what happens at a long time, a time long compared to the time for chemical exchange. The energy level diagrams are like the one in Figure 1.2 except that there are two species and two frequencies. During the T_w period between pulses 2 and 3, As turn into Bs and Bs turn into As. When As turn into Bs, the first interaction is at ω_A but the last interaction and the vibrational echo emission is at ω_B. Therefore, an off-diagonal peak will be created in the lower right portion of the schematic spectrum. When Bs

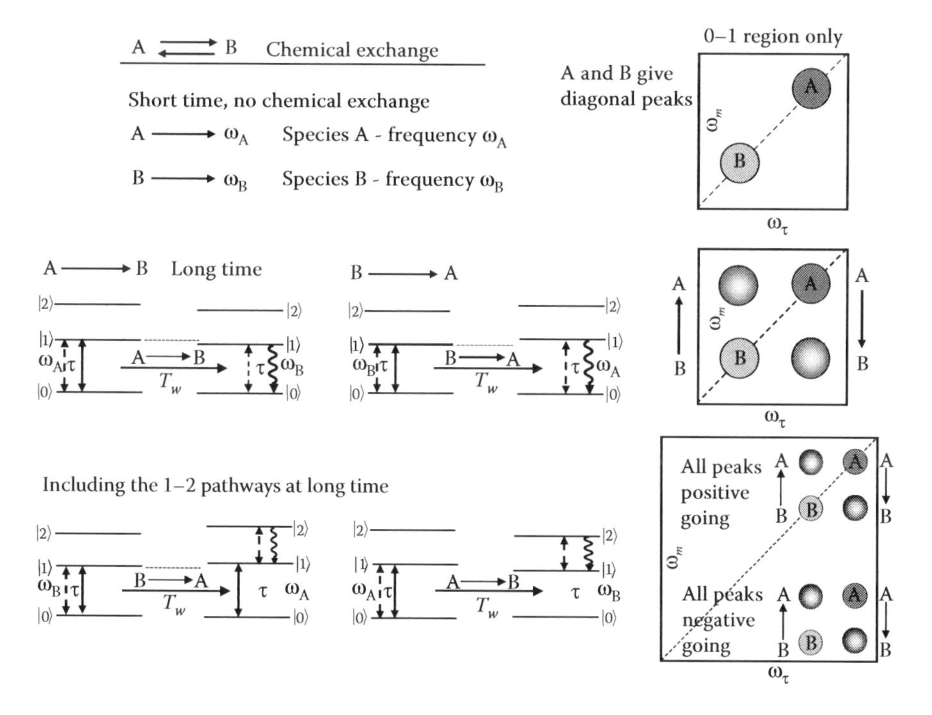

FIGURE 1.3 Two species, A and B, are in equilibrium. The top portion shows a 2D IR spectrum at short time when no chemical exchange has occurred. In the 0–1 region of the spectrum, there are two peaks on the diagonal. The middle portion shows the spectrum at long time. Off-diagonal peaks have grown in because of chemical exchange. The quantum pathways are shown. The bottom portion shows the long time spectrum including the 0–1 and 1–2 regions. The quantum pathways that give rise to the 1–2 chemical exchange peaks are shown.

turn into As, the first interaction is at ω_B but the last interaction and the vibrational echo emission is at ω_A. Therefore, an off-diagonal peak will be created in the upper left portion of the schematic spectrum. The important point is that chemical exchange will cause off-diagonal peaks to grow in as T_w goes from a short time to a long time. The rate of growth of the off-diagonal peaks can be used to directly determine the rate of interconversion of As and Bs under equilibrium conditions. The bottom portion of Figure 1.3 shows the nature of the spectrum at long time when chemical exchange has occurred and the 1–2 portion of the spectrum is included. Chemical exchange causes peaks to form in both the 0–1 and 1–2 regions of the spectrum. As discussed below, analyzing both the 0–1 and 1–2 regions of the chemical exchange 2D IR spectra demonstrates that the thermal equilibrium chemical exchange rates are not influenced by the vibrational excitation.

1.3 SOLUTE–SOLVENT COMPLEXES

To introduce the 2D IR chemical exchange experiments, the first systems to be presented are solute–solvent complexes. Solvents play an important role in chemistry by influencing the reactivity of dissolved solutes [62]. Specific intermolecular interactions, such as hydrogen bonding, can lead to structurally characterized solute–solvent complexes that are constantly forming and dissociating on very short timescales under thermal equilibrium conditions [63]. For most organic and other types of nonaqueous solutions, the solute–solvent complexes form and dissociate on subnanosecond timescales [6,47,48] that cannot be measured by NMR and other methods. The dynamics of transient solute–solvent complexes play an important role in the physical and chemical properties of a solute–solvent system by affecting reaction rates, reaction mechanisms, and product ratios [62].

A simple treatment of a solute in a solvent describes the solvent as a homogeneous continuum. The properties of the solvent are characterized by the solvent's dielectric constant. A more detailed description of a solute in a solvent considers both the solute and the solvent as molecular species. Generally, they are treated as spheres with some form of potential to account for the intermolecular interactions between the solute molecule with the solvent molecules, and the solvent molecules with each other. The interactions give rise to solvent shells around the solute and a radial distribution function, which gives the probability of finding a solvent molecule some distance from the solute. However, this description does not account for the shapes of molecules and the anisotropic intermolecular interactions between a solute and a solvent molecule.

The anisotropic intermolecular interactions between solute and solvent molecules can lead to the formation of a well-defined complex. The interactions are weak, comparable to the thermal energy, k_BT. Therefore, the complexes are short lived. There is equilibrium between the solute–solvent complex and the free solute, that is, a solute that is not complexed. Complexes are continually dissociating to form free solute molecules, and free solute molecules are constantly associating with solvent molecules to form complexes. Because the system is in thermal equilibrium, the numbers of complexes and free solute molecules are time independent.

The upper left portion of Figure 1.4 shows the IR absorption spectrum of the hydroxyl stretch of phenol–OD (OH hydroxyl replaced with OD) in the mixed solvent of benzene and CCl_4 (29 mol% benzene, 71 mol% CCl_4). Phenol–OD is used to shift the hydroxyl stretch absorption below the C–H stretch frequencies. The mixed solvent of benzene/CCl_4 is used to shift the equilibrium so that the two peaks are of similar amplitude. The structures of free phenol and the phenol–benzene complex are shown in the figure to the upper right [6,64]. The absorption spectrum shows that the two species coexist, but the spectrum cannot yield information on the time dependence of the dissociation and formation of the complexes.

The lower left portion of Figure 1.4 shows a 2D IR vibrational echo spectrum taken at a T_w short compared to the chemical exchange, 200 fs. The two bands on the diagonal (dashed line) arise from the two peaks in the absorption spectrum. As discussed above, the positive going bands on the diagonal arise from vibrational echo emission at the 0–1 vibrational transition frequency. The two negative going bands below the diagonal arise from vibrational echo emission at the 1–2 transition

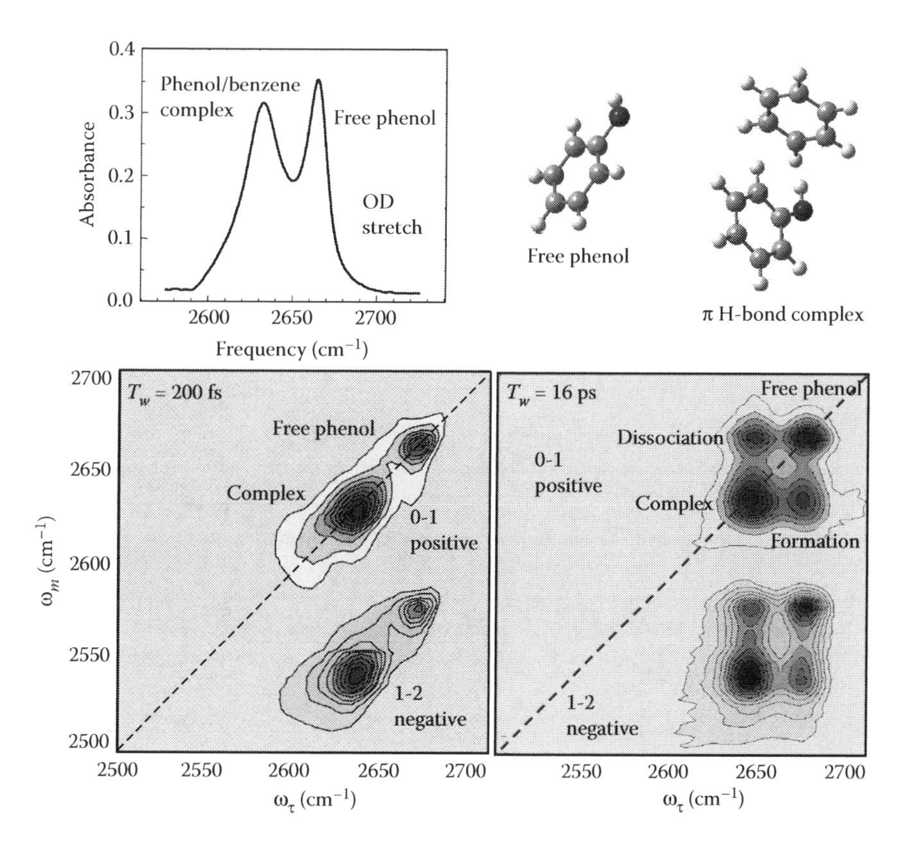

FIGURE 1.4 The upper left is an FT-IR absorption spectrum of OD stretch of phenol–OD in the mixed solvent benzene/CCl_4. Phenol forms π hydrogen bonds with benzene. The two peaks in the spectrum are from the phenol–benzene complex and free phenol (phenol not in a complex). To the right of the spectrum are the structures of the free phenol and the complex. The bottom left is the 2D IR spectrum at short time (200 fs), prior to chemical exchange. The bottom right is the 2D IR spectrum at long time (16 ps), after significant chemical exchange has occurred. The additional peaks in the spectrum arise from chemical exchange.

frequency. These 1–2 peaks are shifted along the ω_m axis by the vibrational anharmonicity of the OD hydroxyl stretching potential. The ω_τ axis is the frequency of the first interaction with the radiation field (first pulse). The bottom right side of Figure 1.4 shows the 2D IR spectrum at a time that is sufficiently long (16 ps) for a substantial amount of chemical exchange to have occurred. Off-diagonal peaks have grown in. Consider the peak-labeled dissociation in the 0–1 portion of the spectrum. At the time of the first pulse, species that are complexes have frequency $\omega_\tau = 2630$ cm^{-1}. Complexes that dissociate during the interval $T_w = 16$ ps become free and give rise to vibrational echo emission at frequency $\omega_m = 2665$ cm^{-1}. Therefore, the peak will be off-diagonal and to the upper left of the diagonal. This peak corresponds to dissociation. The peak-labeled formation arises in the same manner. Free species form complexes and give rise to the other off-diagonal peak. The peaks on the diagonal correspond to species that have not changed their character. The important point is that the time-dependent growth of the off-diagonal peaks yields direct information on the rate of chemical exchange. The data are analogous to the schematic illustration of chemical exchange shown in Figure 1.3.

Figure 1.5 shows three-dimensional representations of the 0–1 region of the 2D IR spectrum for a number of T_ws. As T_w increases, off-diagonal peaks grow in. At 200 fs, no off-diagonal peaks are visible. By 3 ps, they are becoming apparent; at 7 ps, they are well developed; and at 14 ps, the off-diagonal peaks are very prominent. From simple inspection, it is clear that the chemical exchange time is greater than a picosecond and on the order of a number of picoseconds.

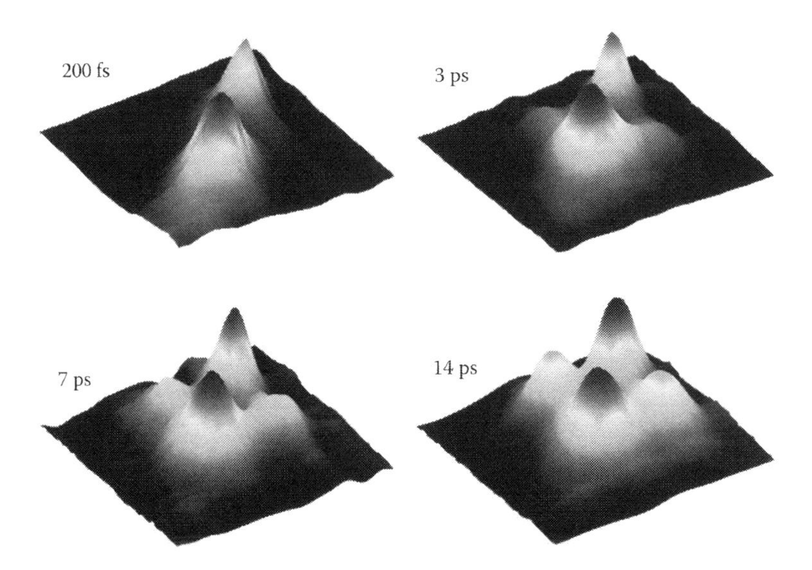

FIGURE 1.5 2D IR chemical exchange spectra for the phenol–benzene complex and free phenol system at four times. The off-diagonal chemical exchange peaks grow in with increasing time.

To analyze the data in detail, it is necessary to take into account the other dynamics that influence 2D IR vibrational echo chemical exchange spectra [6,60]. The vibrational excited states decay to the ground state with lifetimes, T_1^i. The free and complex species undergo orientational relaxation, with orientational relaxation time constants, τ_r^i. The various processes are summarized by the following kinetic scheme:

$$\xleftarrow[\text{decay}]{\tau_r^C, T_1^C} C \underset{k_f}{\overset{k_d = 1/\tau_d}{\rightleftarrows}} f \xrightarrow[\text{decay}]{\tau_r^f, T_1^f} \tag{1.2}$$

C stands for complex and f for the free phenol. k_f is the rate constant for complex formation, and k_d is the rate constant for complex dissociation. τ_d is the dissociation time constant. The T_1 processes cause decay of all of the peaks to zero. The orientational relaxation (τ_r) causes all of the peaks to be reduced in amplitude, but not decay to zero. The chemical exchange causes the diagonal peaks to decay and the off-diagonal peaks to grow.

Another contribution to the time dependence of the 2D IR spectra is spectral diffusion, which was discussed briefly in Section 1.1. Spectral diffusion is caused by thermally induced molecular motions that produce fluctuations in the transition frequency of a species [14,33,45,46,65,66]. Spectral diffusion changes the shape and amplitude of the peaks, but it does not change the peak volumes [6,60].

In fitting the data, all of the necessary input parameters except for the chemical exchange rate can be measured independently. The T_1s and τ_rs are measured with IR polarization and wavelength-selective pump-probe spectroscopy. The time-dependent peak volumes are used rather than the peak amplitudes to eliminate the contribution from spectral diffusion [6,60]. In addition, it is necessary to know the equilibrium constant and the ratio of the complex and free OD stretch transition dipoles. These are obtained with IR absorption spectroscopy [6,47]. Therefore, the time-dependent chemical exchange data can be fit with a single adjustable parameter, the complex dissociation time, τ_d, which is the inverse of the dissociation rate constant. The single parameter τ_d can be used because the system is in equilibrium [6], and therefore, the rate of formation is equal to the rate of dissociation.

The results of fitting the chemical exchange data for phenol–OD in the mixed benzene–CCl_4 solvent are shown in Figure 1.6. The data (symbols) are shown only for the 0–1 portion of the 2D

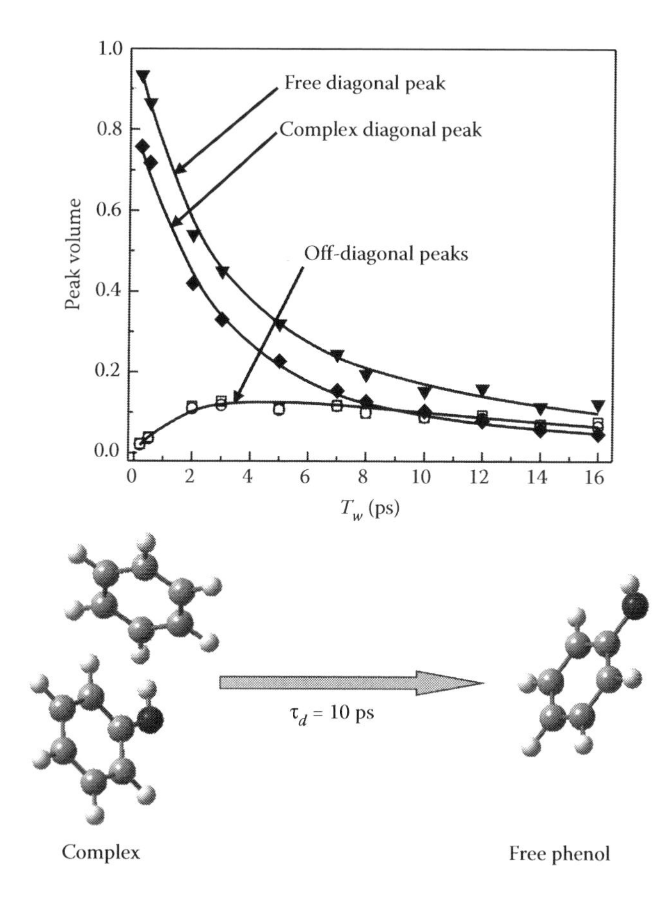

FIGURE 1.6 Data (symbols) for the phenol–benzene complex and free phenol system taken from spectra like those shown in Figure 1.5, which are for the 0–1 portion of the spectra. The curves through the data are from a single adjustable parameter fit that yield $\tau_d = 10$ ps, the time for the complex to dissociate under thermal equilibrium conditions.

IR spectra. The 1–2 region will be discussed below to show that the excitation of the hydroxyl stretch does not change the chemical exchange kinetics and therefore the equilibrium. There are four peaks, two diagonal and two off-diagonal. The diagonal peaks decay because of the lifetime orientational relaxation and chemical exchange. The off-diagonal peaks grow in initially because of chemical exchange, which competes with the lifetime and orientational relaxation. Ultimately, the off-diagonal peaks decay because of the vibrational lifetime decay. For the system in equilibrium, the rate of complex dissociation is equal to the rate of complex formation. Therefore, the off-diagonal peaks grow in at the same rate, which is clear from the data [6]. The curves through the data points are obtained from *the fit using the single adjustable parameter*, τ_d. The fit is very good. The fit yields the phenol–benzene complex dissociation time, $\tau_d = 10$ ps [6,47].

By comparing the 2D IR chemical exchange data taken in the 0–1 and 1–2 spectral regions, it is possible to show that the vibrational excitation does not affect the dissociation and formation rates, and therefore does not affect the thermal equilibrium of the system [6]. In Figures 1.2 and 1.3, simple diagrams were used to discuss the nature of the 2D IR experiment and the influence of chemical exchange. Here, it is necessary to expand on the description of the 2D IR experiment to see how the comparison of the 0–1 and 1–2 data can be used to address the effect of vibrational excitation. The 0–1 transition has two sets of quantum pathways that give rise to the signal. These are illustrated in Figures 1.7a and 1.7b for the complex dissociation ($C \rightarrow f$). Everything is analogous for complex formation. In Figure 1.7a, the first pulse generates a coherence between the 0 and 1 levels

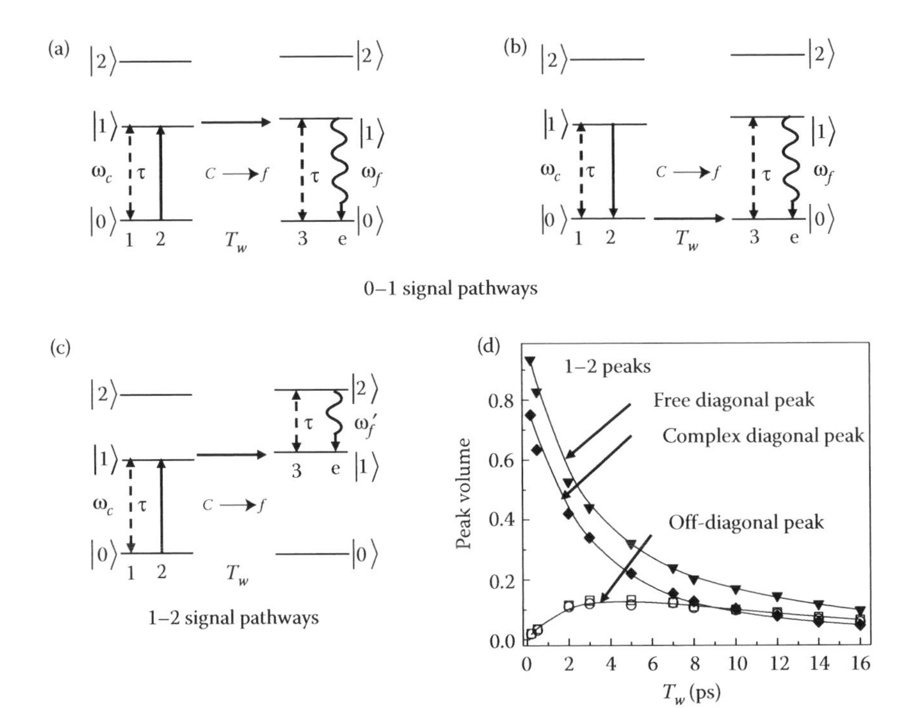

FIGURE 1.7 (a) Quantum pathway that contributes to the 0–1 portion of the 2D IR spectrum. The vibration is in the 1 level during the chemical exchange. (b) Quantum pathway that contributes to the 0–1 portion of the 2D IR spectrum. The vibration is in the 0 level during the chemical exchange. (c) Quantum pathway that gives the 1–2 portion of the 2D IR spectrum. The vibration is in the 1 level during the chemical exchange. (d) The chemical exchange data (symbols) from the 1–2 portion of the 2D IR spectrum. The curves through the data are calculated using the parameters obtained from the fit shown in Figure 1.6 using the 0–1 portion of the 2D IR spectrum.

(dash double-headed arrow) at ω_c. The second pulse produces a population in the vibrationally excited state, the 1 level (single-headed arrow going up from 0 to 1). The system then evolves in the vibrationally excited state for T_w and chemical exchange occurs. The third pulse makes a 0–1 coherence (dashed double-headed arrow) and the vibrational echo (wavy arrow) is emitted at ω_f. Therefore, chemical exchange occurs in the vibrational excited state. Figure 1.7b shows a different quantum pathway. Everything is the same except the second pulse produces a population in the ground vibrational state (single-headed arrow going down from 1 to 0). The system evolves for T_w and chemical exchange occurs in the ground vibrational state. The important point is that the 0–1 signal comes half from chemical exchange occurring with vibrational excitation and half from chemical exchange occurring with no vibrational excitation.

Figure 1.7c shows the quantum pathways that give rise to the 1–2 portion of the 2D IR spectrum. The first two interactions with the radiation fields are identical to those shown in Figure 1.7a. Chemical exchange occurs with vibrational excitation. However, the third radiation field interaction produces a 1–2 coherence (dashed double-headed arrow) followed by vibrational echo emission (wavy arrow) at the 1–2 transition frequency (ω_f'). For the 1–2 pathway, the signal only comes with chemical exchange occurring with the vibration excited. So, the 0–1 data come half from ground state exchange and half from excited state exchange, while the 1–2 data come only from excited state exchange. If being in the excited state changed the exchange rate, then the two sets of data would show different kinetics. Figure 1.7d shows the 1–2 chemical exchange data (symbols). The curves through the data were calculated using the parameters obtained from the 0–1 fits (Figure 1.6). The curves are not fits to the data, but just a reproduction of the 0–1 curves. As can be

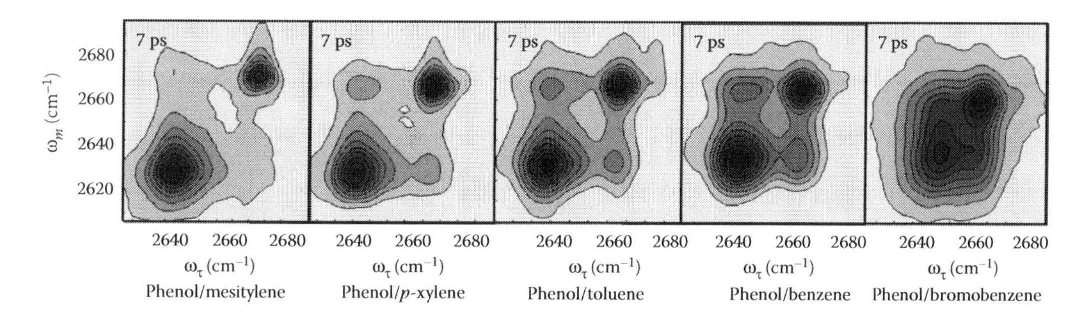

FIGURE 1.8 2D IR chemical exchange spectra all at the same $T_w = 7$ ps for a number of complexes of phenol with substituted benzenes. Going from left to right, there is an increasing amount of chemical exchange at 7 ps as shown by larger off-diagonal peaks. The increase in chemical exchange correlates with a decrease in the enthalpy of formation (ΔH^0) of the complexes. See Table 1.1 for complex dissociation times and enthalpies of formation.

seen from Figure 1.7d, the curves go right through the data. Therefore, within experimental error, vibrational excitation does not change the thermal equilibrium kinetics.

Figure 1.8 shows 2D IR chemical exchange spectra for five solute–solvent complexes between phenol and different substituted benzenes/CCl_4 solvents all at $T_w = 7$ ps [47]. The phenol complex partners, going from left to right in Figure 1.8, are mesitylene, *p*-xylene, toluene, benzene, and bromobenzene. For mesitylene (left panel), at 7 ps, the off-diagonal chemical exchange peaks are just beginning to appear. For *p*-xylene, at the same time, the off-diagonal peaks are somewhat more prominent. For toluene, they are becoming well developed. For benzene, the off-diagonal peaks are even more prominent. In the bromobenzene spectrum, the off-diagonal peaks are so large that they have merged with the diagonal peaks to give a characteristic "square" shape. Progressing from left to right across the figure, it is clear that the extent of chemical exchange at 7 ps increases. Mesitylene has three methyl groups. Methyls donate electron density to the benzene ring, which increases the strength of the π hydrogen bond with phenol. The increased hydrogen bond strength increases the dissociation time, τ_d. Therefore, the dissociation rate constant is smaller, and the overall chemical exchange rate is slower. *p*-Xylene has two methyls. There is less electron donation to the ring, making the π hydrogen bond weaker, which increases the chemical exchange rate. Toluene with one methyl group has an even weaker π hydrogen bond, and benzene with no methyl groups is weaker yet. The bromo group of bromobenzene withdraws electron density from the ring, making the π hydrogen bond very weak, and the exchange rate is very fast.

The trend between hydrogen bond strength and the chemical exchange rate can be quantified by comparing the dissociation time for the five complexes shown in Figure 1.8 with the enthalpies of formation of the complexes (ΔH^0). The enthalpies of formation were determined by measuring the temperature dependences of the complex-free phenol equilibrium constants. The results are shown in Table 1.1. For example, mesitylene has $\Delta H^0 = -2.45$ kcal/mol and $\tau_d = 32$ ps, while for bromobenzene $\Delta H^0 = -1.21$ kcal/mol and $\tau_d = 5$ ps.

Thirteen binary solute–solvent complexes have been studied with 2D IR vibrational echo chemical exchange spectroscopy [47,48]. Eight of them have phenol or substituted phenols as the solutes

TABLE 1.1
Complex Enthalpies of Formation and Dissociation Times

Phenol with	Mesitylene	*p*-Xylene	Toluene	Benzene	Bromobenzene
ΔH^0 (kcal/mol)	−2.45	−2.23	−1.98	−1.67	−1.21
τ_d (ps)	32	24	15	10	5

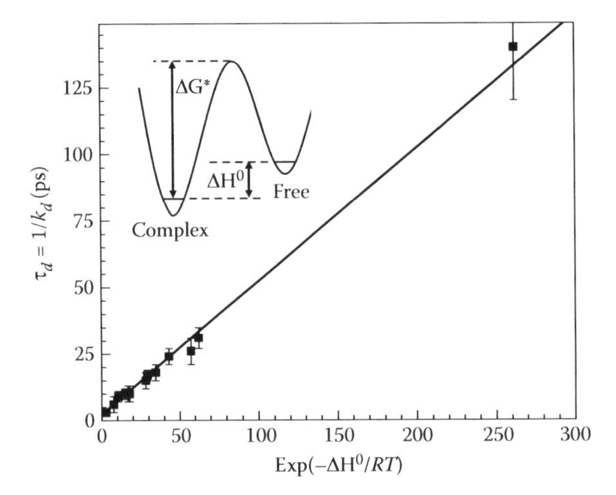

FIGURE 1.9 A plot showing the correlation between complex dissociation times and the enthalpies of complex formation, ΔH^0. The inset shows a schematic of the free energy curve. ΔG^* is the activation free energy.

[47], and the other five have triethylsilanol–OD (TES) as the solute [48]. All have solvents of substituted benzene–CCl_4 mixed solvents except for one which is TES in acetonitrile–CCl_4. Figure 1.9 displays the dissociation times, $\tau_d = 1/k_d$, for the 13 complexes [47,48]. The data are plotted versus $\exp(-\Delta H^0/RT)$, where ΔH^0 is the enthalpy of formation of the complexes. Over a range of dissociation times from ~4 to ~140 ps and ΔH^0 values ranging from −0.6 to −3.3 kcal/mol, the experimental points fall on a line. Transition state theory [67] states that k_d depends on the activation free energy, ΔG^*, not on the enthalpy of formation, ΔH^0 (see inset in Figure 1.9). However, if the activation enthalpy is proportional to the enthalpy of formation, $\Delta H^* \propto \Delta H^0$, and the activation entropy ΔS^* is essentially a constant, then the behavior displayed in Figure 1.9 is obtained. The results shown in Figure 1.9 demonstrate that the enthalpy of formation of a solute–solvent complex can be used as a guide for its dissociation time.

1.4 PROTEIN SUBSTATE INTERCONVERSION

Proteins are complex molecules that undergo constant structural dynamics at ambient temperatures. The structural fluctuations are central to their biological function. Molecules such as myoglobin and hemoglobin bind oxygen, carbon monoxide, and other small ligands at the active heme sites of these proteins. There is no channel from the exterior of the protein to the active site. Rather, the O_2 and CO move through the protein in what is effectively diffusion enabled by structural fluctuations. Enzymes are proteins that bind small molecular substrates in pockets in close proximity to the active site. The enzyme will perform chemistry on the substrate. Generally, there is a channel to bring the substrate to the pocket next to the active site. However, structural fluctuations are necessary to permit the substrate to reach the pocket and are in part responsible for selectivity in binding substrates [23]. In addition, structural fluctuations are required to bring the active site–substrate to the transition state for the chemical reaction. The study of the dynamics of biological molecules undergoing structural changes on a wide range of timescales, from femtoseconds to seconds, is a topic of great interest because the biological functions of proteins and enzymes are directly related to time-dependent structural changes, which occur over this wide range of times [68–78].

2D IR vibrational echo spectroscopy can measure protein structural evolution on timescales ranging from 100 fs to 100 ps [21,42,43,49,79]. Applications of 2D IR spectroscopy to biological molecules include studies of protein structure, dynamics, folding, and unfolding [8,16–23,43,49].

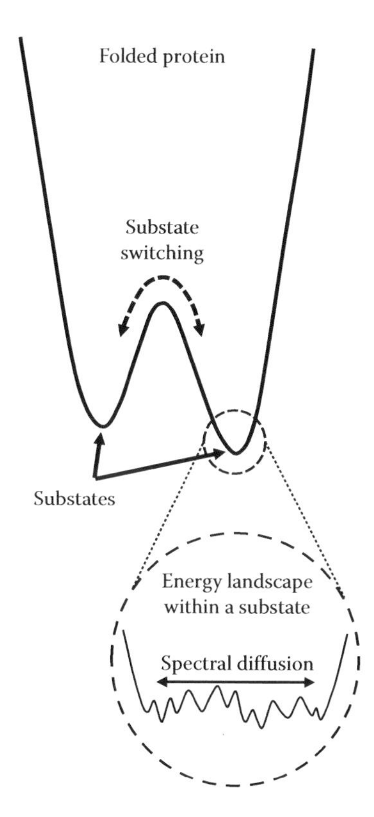

FIGURE 1.10 Top—a schematic illustration of a free energy curve that is part of the folded proteins energy landscape. Substate switching takes the structure from one minimum to the other causing chemical exchange in the 2D IR spectrum. Bottom—a schematic showing a blowup of the energy landscape around one of the minima. Transitions among the shallow minima produce spectral diffusion.

In the context of 2D IR vibrational echo spectroscopy, the structural dynamics of a folded protein can be roughly divided into spectral diffusion and substate switching (see Figure 1.10). Multiple substates can be reflected by more than one peak in the IR absorption spectrum of a vibrational probe. When there is a single peak in the IR absorption spectrum, there are still variations in substructure and dynamic sampling of these structures (see Figure 1.10). The IR spectra of vibrational probes in proteins are inhomogeneously broadened. The inhomogeneous broadening reflects a wide range of different structures. Within a relatively deep minimum associated with the folded structure, there is an energy landscape, that is, a rough potential surface with many minima associated with distinct variations on the folded structure. Interconversion among these minima causes spectral diffusion that can be measured with 2D IR vibrational echo experiments, as was discussed very briefly above. On some timescale, all of the distinct conformations that give rise to the inhomogeneously broadened IR absorption line will be sampled, at which point spectral diffusion is complete.

In some cases, the vibrational probe will produce IR absorption spectra with two or more peaks [23,49,52–55,80–82]. The distinct peaks show that the probe is interacting with significantly different protein structures, called protein substates, which go beyond the variations in conformation that give rise to inhomogeneous broadening of a single line. The substates on the energy landscape are shown schematically in Figure 1.10 as relatively deep wells separated by a significant barrier. It is important to note that distinct substates will only be manifested in the vibrational probe spectrum if the different structures have a significant influence on the vibrational frequency. Different substates that produce only small changes in the probe vibrational frequency will be part of the inhomogeneous broadening.

The problem of multiple substates has been studied extensively for the protein myoglobin with the ligand CO bound at the active site (Mb–CO) [20,43,80,81,83–86]. The infrared spectrum of the heme-ligated CO stretching mode of Mb shows two major absorption bands, denoted A_1 (1945 cm^{-1}) and A_3 (1932 cm^{-1}) [55]. Mb–CO interconverts between these two conformational substates under thermal equilibrium conditions. The distal histidine, His64, plays a prominent role in determining the conformational substates of Mb.

Here, the results of chemical exchange studies of two Mb mutants, L29I (leucine 29 replaced by an isoleucine) [43,86] and T67R/S92D (threonine 67 replaced by arginine and serine 92 replaced by aspartic acid) [20,86], are discussed. In wild-type Mb–CO, the A_3 band is a relatively small shoulder on the low-frequency side of the A_1 band. In the two mutants, the amplitude of the A_3 is increased relative to the A_1 band, which aids in the chemical exchange measurements.

The CO absorption spectrum of L29I–CO is shown in Figure 1.11. The lower frequency of A_3 compared to A_1 reflects a closer proximity of the protonated epsilon nitrogen of the imidazole side group of the distal histidine (H64) to the CO in A_3 (see discussion below) [80,87,88]. Each A substate exhibits a distinct ligand-binding rate [76,83]. Therefore, the peaks in the FT-IR spectrum of Mb–CO, L29I–CO, and T67R/S92D–CO reflect functionally distinct conformational substates.

2D IR spectra of CO bound to L29I at several T_ws are shown in Figure 1.12. The bands in the upper part of each panel are positive going and correspond to the 0–1 vibrational transitions. The bands in the lower part of each panel are negative going and arise from vibrational echo emission at the 1–2 transition frequencies. For $T_w = 0.5$ ps, there are only two peaks on the diagonal (dashed line), which arise from the 0–1 transition and the corresponding 1–2 peaks are observed. The diagonal peaks arise from the A_1 and A_3 bands in the FT-IR spectrum shown in Figure 1.11. The off-diagonal chemical exchange peaks grow in as T_w increases. By $T_w = 48$ ps, the off-diagonal peaks are readily apparent. The band to the upper left in the $T_w = 48$ ps panel is strong. Because the anharmonicity is not large, the negative going 1–2 diagonal band partially overlaps the positive going off-diagonal chemical exchange peak to the lower right of the two 0–1 diagonal peaks, reducing its amplitude.

The experimental diagonal and off-diagonal peak volumes for the 0–1 region of the spectra are plotted in Figure 1.13 [43]. In the figure, the off-diagonal peak volume values are multiplied by a factor of three. Because the orientational relaxation of a large protein like Mb is so slow, only the vibrational lifetimes were included in the kinetic data analysis. Both the positive (0–1) and negative (1–2) peak volumes were fit [43] using the same fitting procedure that was employed for the solute–solvent complex experiments. The substate switching time, τ_s, is the single adjustable parameter.

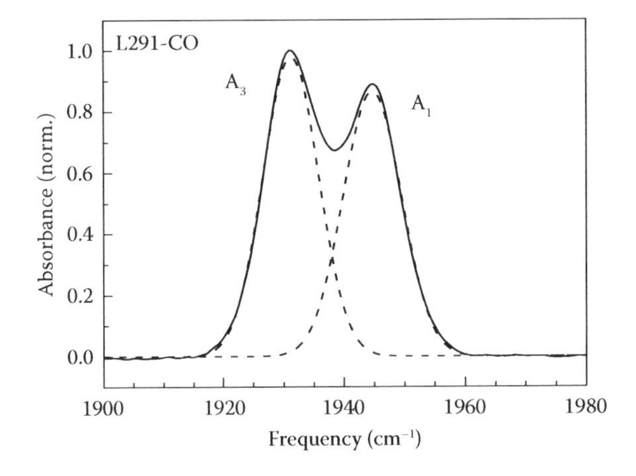

FIGURE 1.11 An FT-IR absorption spectrum (solid curve) of the CO stretching mode of CO bound to the heme of a mutant of myoglobin, L29I–CO. The dashed curves are Gaussian fits to the two peaks.

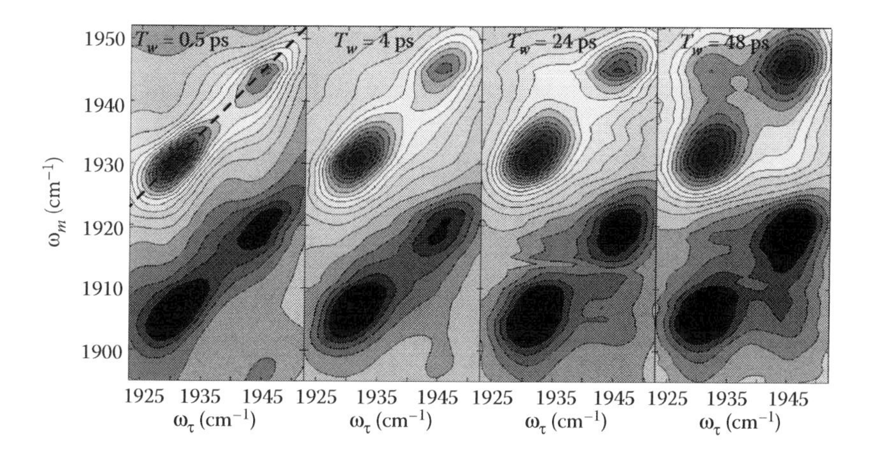

FIGURE 1.12 2D IR chemical exchange spectra of the CO stretching mode of L29I–CO at several T_ws. As T_w increases, off-diagonal peaks grow in. The off-diagonal peak in the upper left of the T_w = 48 ps spectrum is apparent.

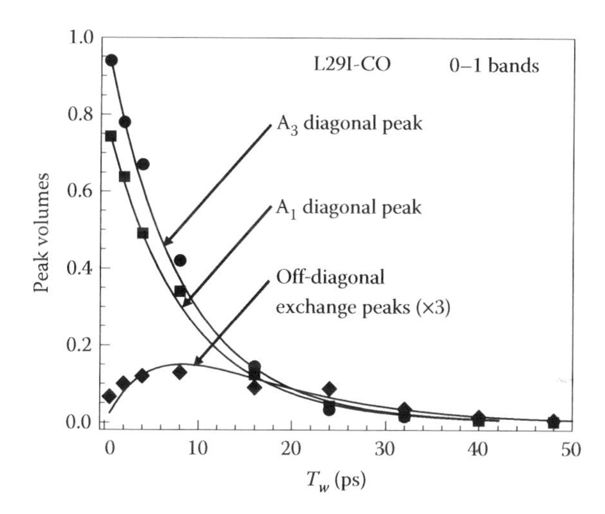

FIGURE 1.13 Chemical exchange data (symbols) showing protein substate switching for the myoglobin mutant, L29I–CO. The solid curves are the one adjustable parameter fit to the data that yields the substate switching time of 47 ps.

Fitting the three curves simultaneously yields $\tau_s = 47 \pm 8$ ps. This is the time for the A_1 and A_3 substates to interconvert under thermal equilibrium conditions.

Figure 1.14 displays the 2D IR chemical exchange data for the myoglobin mutant T67R/S92D. These data were fit in the same manner as the L29I–CO data with τ_s as the single adjustable parameter. The fit yields the solid curves through the data points. The results of the fit yield $\tau_s = 76 \pm 10$ ps.

The immediate cause of the difference in the frequency of the A_3 and A_1 bands is the position of the distal histidine, H64, as shown in Figure 1.15a. For the A_3, the protonated ε nitrogen of the imidazole side group of H64 is pointed toward the CO. The A_1 state has the protonated ε nitrogen pointing away from the CO [80,81]. The side group rotates by ~60°. However, the two substates involve more than the rotation of the imidazole side group. Changes in the configuration of the E helix (see Figure 1.15b) cause the distal histidine's imidazole side group to move relative to the CO [55,87].

That significant structural change occurs in the A_1–A_3 interconversion is supported by x-ray experiments. The high-resolution crystal structure of Mb–CO that contains two conformations has

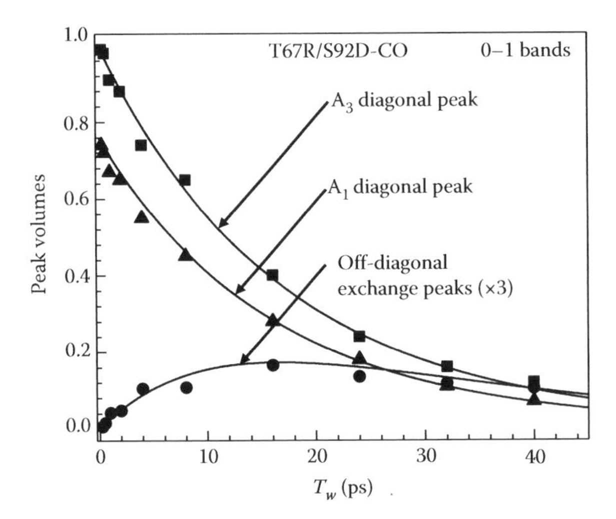

FIGURE 1.14 Chemical exchange data (symbols) showing protein substate switching for the myoglobin mutant, T76R/S92D–CO. The solid curves are the one adjustable parameter fit to the data that yields the substate switching time of 76 ps.

enabled modeling of the structure of A_1 and A_3 substates [87,89]. Although the distal histidine plays a critical role in determining the substates of Mb, structural comparison between the A_1 and A_3 substates shows that the A_3 substate contains an additional cavity, Xe3, and another transient cavity found in simulations [89]. Xe is used as a probe to identify the locations of cavities in proteins [90,91]. In Mb crystals, four Xe atoms (Xe1, Xe2, Xe3, and Xe4) occupy cavities, which may be involved in gas ligand migration [89,90]. The Xe3 site, which is near the surface and far from the iron atom, involves Trp7 and is located between helices E and H [90] (see Figure 1.15b). The existence of Xe3 in the A_3 substate but not in the A_1 substate demonstrates that the difference in the substates is significantly more than the rotation of the imidazole side group of the distal histidine.

NMR techniques probe protein motions in the microsecond, millisecond, and longer timescales [92]. Conformational change studied by NMR involves large structural transformations that occur, for example, in enzyme catalytic processes [72]. Such structural changes require the reconfiguration of many amino acids, helices, and various protein structures. The substate switching discussed here reports on a single elementary structural change. A slow response to a perturbation results from structural fluctuation sampling that produces elementary steps, which in turn combine to produce major restructuring.

Using ultrafast 2D IR vibrational echo chemical exchange spectroscopy, we have measured the time dependence of a single elementary step, the A_1–A_3 substate switching in two mutants of the protein myoglobin. The time constants for the substate switching are 47 and 76 ps. Attempts to simulate the substate switching in these two mutants have resulted in limited success [20,86]. The substate switching times from the simulations are too long.

The ability of proteins to undergo conformational switching is central to protein function. When an enzyme binds a substrate, the protein conformation will change [93]. On the path of protein folding, a protein will sample many conformations as it progresses toward the native folded structure [94]. Proteins can undergo large global conformational changes, which occur on long timescales, milliseconds to seconds. However, these large slow conformational changes involve a vast number of more local elementary conformational steps. The experimental determination of the timescale of elementary conformational steps is a long-standing problem that has now been successfully addressed using ultrafast two-dimensional infrared vibrational echo chemical exchange spectroscopy.

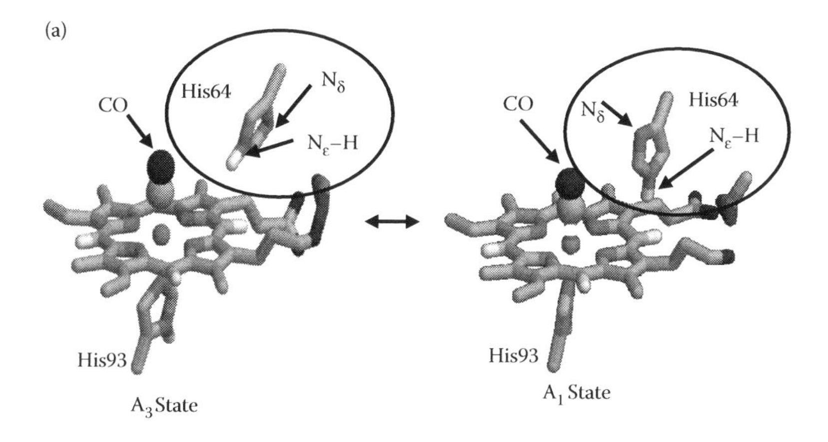

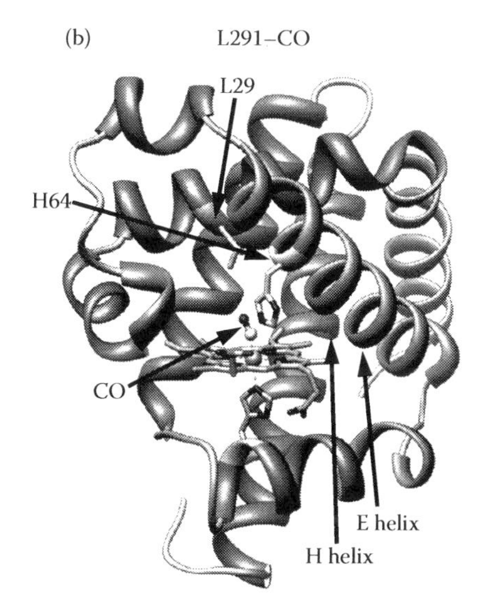

FIGURE 1.15 (a) Configurations of the distal histidine (His64) relative to the heme–CO in myoglobin–CO that are responsible for the two peaks in the FT-IR spectrum shown in Figure 1.11. (b) The myoglobin mutant, L29I, structure showing the location of key features. The change in structure illustrated in (a) is caused by a shift in the position of the E helix.

1.5 ISOMERIZATION AROUND A CARBON–CARBON SINGLE BOND

Orientational isomerization around a carbon–carbon single bond is central to organic chemistry. Such isomerization enables complex organic molecules to rapidly interconvert from one conformer to another. Such structural changes are important in fields from polymer materials science to biology.

Ethane and its derivatives are textbook examples of molecules that undergo isomerization by rotating around the carbon–carbon single bond [95]. Ethane isomerization is illustrated in Figure 1.16a. One hydrogen atom on each carbon atom is labeled as 1 and 2. Ethane starts in a staggered form, passes through the eclipsed transition state, and then ends in a new staggered state. As can be seen in the figure, the position of hydrogen 1 relative to hydrogen 2 has changed. In ethane, the transition from one staggered state to another leaves ethane structurally identical. Therefore, there is no change in the frequency of a vibration, which is needed to perform a chemical exchange spectroscopy experiment. Figure 1.16b shows a 1,2-disubstituted ethane, with two different substituents labeled A and B. Such a molecule has two distinct staggered conformations, gauche and trans, that

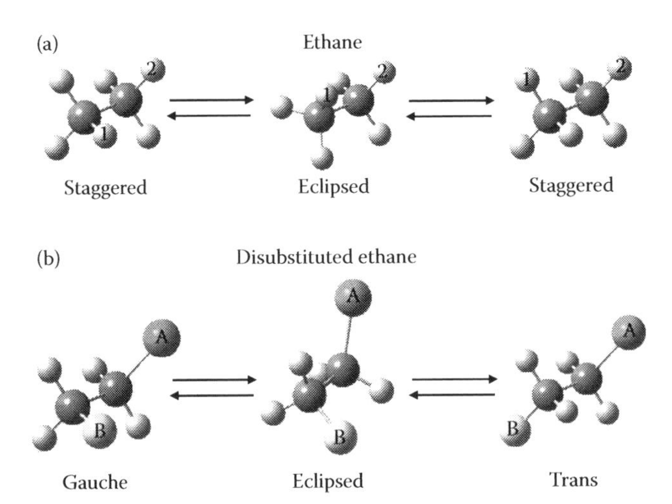

FIGURE 1.16 (a) Orientational isomerization around the carbon–carbon single bond of ethane. The eclipsed configuration is the transition state. (b) Orientational gauche–trans isomerization around the carbon–carbon single bond for a substituted ethane. The eclipsed form is the transition state.

have distinguishing characteristics because of the relative positions of the two substituents [95]. If the change in the relative positions of the two substituents changes a vibrational frequency of the molecule, then it is possible to do a 2D IR chemical exchange experiment to measure the isomerization rate.

The trans–gauche isomerization of 1,2-disubstituted ethane derivatives, for example, *n*-butane, is one of the simplest cases of a first-order chemical reaction. This type of isomerization has served as a basic model for modern chemical reaction kinetic theory and MD simulation studies in condensed phases [96–101]. In spite of extensive theoretical investigation, until recently, no corresponding kinetic experiments had been performed to test the results [3]. The experimental difficulty was due to the low-rotational energy barrier of the *n*-butane (~3.3 kcal/mol), and other simple 1,2-disubstituted ethane derivatives [102]. According to theoretical studies [96–101], the isomerization timescale ($1/k$, k is the rate constant) is in the range of 10–100 ps in room-temperature liquids.

Ultrafast 2D IR vibrational echo chemical exchange experiments were used to measure the trans–gauche isomerization about the carbon–carbon single bond of 1-fluoro-2-isocyanato-ethane (FICE) in CCl_4 solvent at 298 K [3]. The isocyanate group (N=C=O) was used as the vibrational probe. The structures of FICE in the gauche, eclipsed, and trans conformations are shown in Figure 1.17a. The eclipsed form is the transition state. The fluoro group is closer to the isocyanate in the gauche conformer than it is in the trans form. This difference changes the frequency of the isocyanate antisymmetric stretching mode. The IR absorption spectrum of FICE is shown in Figure 1.17b. The structures and their assignment to the peaks in the spectrum were obtained from electronic structure calculations using density functional theory (DFT) [103] at the B3LYP level and 6-31+G(d,p) basis set for the isolated molecules. The calculation also gave the barrier height for the eclipsed transition state as 3.3 kcal/mol [3].

Figure 1.18a shows the IR absorption spectrum over a wider frequency range. In addition to the two peaks shown in Figure 1.17b, there is another smaller unassigned combination band or overtone absorption at 2230 cm^{-1} [3]. This unassigned band is coupled by anharmonic terms in the molecular potential to the antisymmetric isocyanate stretching mode. The coupling produces additional off-diagonal peaks in the spectrum [3,104]. The nature of these additional off-diagonal peaks was investigated in detail by observing the 2D IR vibrational echo spectra of 1-bromo-2-isocyanato-ethane [3]. The 2D IR spectrum of the bromo compound is shown in the bottom portion of Figure 1.18b. The bromo group is so bulky that it raises the barrier height to the point where no

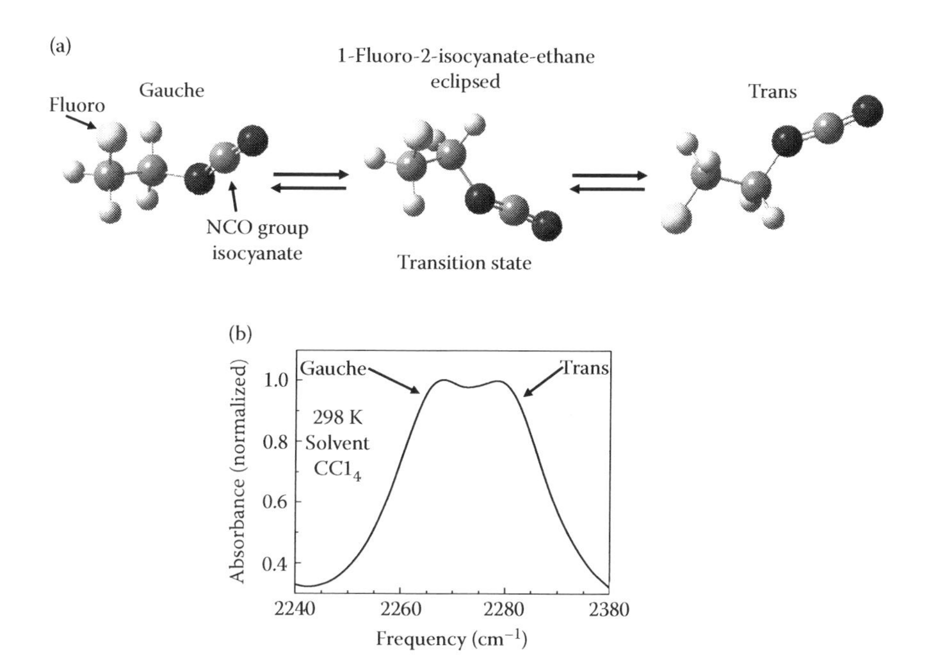

FIGURE 1.17 (a) The structure of 1-fluoro-2-isocyanate-ethane (FICE) showing the gauche, eclipsed, and trans conformations. The chemical exchange experiments were conducted on the antisymmetric stretching mode of the isocyanate (N=C=O) group. (b) The FT-IR spectrum of the antisymmetric stretch of the isocyanate group of FICE. The gauche and trans conformations produce different absorption frequencies.

gauche–trans isomerization occurs on the experimental timescale of several hundred picoseconds. As can be seen in the 2D IR spectrum in Figure 1.18b, the results of the experiments on the bromo compound show that there is a negative going off-diagonal band that overlaps the off-diagonal position where one of the positive going chemical exchange peaks will grow in through isomerization in the fluoro compound. The presence of this additional negative going peak is included in the data analysis and does not hinder the extraction of the isomerization rate from time-dependent growth of the off-diagonal bands.

Figure 1.19a displays 2D IR spectra of FICE in a CCl_4 solution at room temperature taken at four T_ws. The 200 fs panel corresponds to a short T_w at which negligible isomerization has occurred. The two peaks representing the gauche and trans conformers are on the diagonal. These are labeled in the upper left-hand panel. For a long time ($T_w = 25$ ps), isomerization has proceeded to a substantial degree. The obvious change is the additional peak that has appeared at the upper left of the $T_w = 25$ ps panel. This peak arises from gauche-to-trans isomerization. There is a corresponding peak to the lower right that is generated by trans-to-gauche isomerization, but it is reduced in amplitude by the negative going peak discussed briefly above [3,104]. Figure 1.19b shows calculated spectra that include the isomerization kinetics, the other off-diagonal peaks, and the vibrational dynamics other than the isomerization (vibrational lifetimes and orientational relaxation). Comparing Figures 1.19a and 1.19b shows that the data can be reproduced well by the calculations [3].

Figure 1.20 shows the T_w-dependent data for the diagonal and off-diagonal peaks. As in the solute–solvent complex experiments, all of the necessary input parameters are known except for the isomerization time, $\tau_{iso} = 1/k_{GT} = 1/k_{TG}$ [3]. In the analysis, the gauche-to-trans rate constant was taken to be equal to the trans-to-gauche rate constant. Within experimental error, the difference in the two rate constants could not be discerned. The solid curves through the data are the result of the fit with the single adjustable parameter, τ_{iso}. The isomerization time is $\tau_{iso} = 43 \pm 10$ ps. The error bars arise from the uncertainty in the parameters that go into the calculations. This is the first

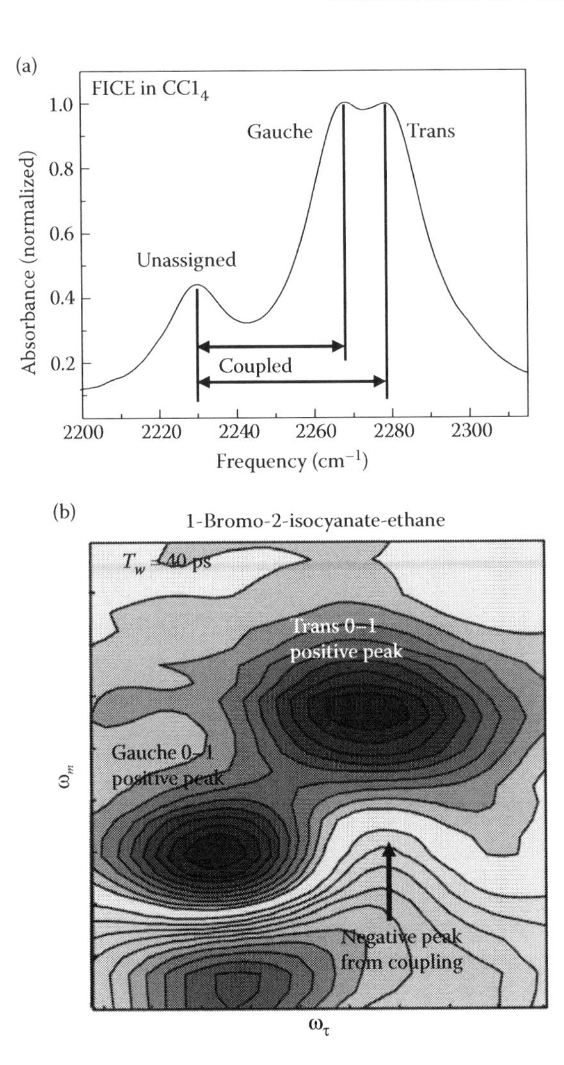

FIGURE 1.18 (a) The FT-IR spectrum of the antisymmetric stretch of the isocyanate group of 1-fluoro-2-isocyanate-ethane (FICE) shown over a wider frequency range than in Figure 1.17b. The unassigned peak is coupled to the antisymmetric stretch in both the gauche and trans forms of FICE. The coupling produces extra peaks in the 2D IR spectrum. (b) The 2D IR spectrum of 1-bromo-2-isocyanate-ethane. The large bromo group prevents gauche–trans isomerization on the experimental time scale. The 2D IR spectrum of the bromo compound allows one of the additional peaks that arises from coupling to the unassigned peak to be quantified.

determination of isomerization around a carbon–carbon single bond in a room-temperature liquid for a system with a typically low barrier height.

Employing the experimental results for FICE and transition state theory, it is possible to approximately calculate the gauche–trans isomerization rate of *n*-butane and the rotational isomerization rate of ethane under the same conditions used in this study, that is, a CCl_4 solution at 298 K [4]. It was assumed that the prefactors are the same because the transition states and the barrier heights are quite similar for the three systems [3]. The barrier heights were calculated with DFT calculations on all the systems using the same method (the B3LYP level and 6-31+ G(d,p) basis set). With the zero point energy correction, the trans-to-gauche isomerization of *n*-butane has a barrier of 3.3 kcal/mol. The barrier for ethane was calculated to be 2.5 kcal/mol [3], which is smaller than the result of more extensive calculations, 2.9 kcal/mol [105,106]. The 2.5 kcal/mol value was used for ethane so that all of the barriers were obtained with the same method, which should result in some cancellation of errors.

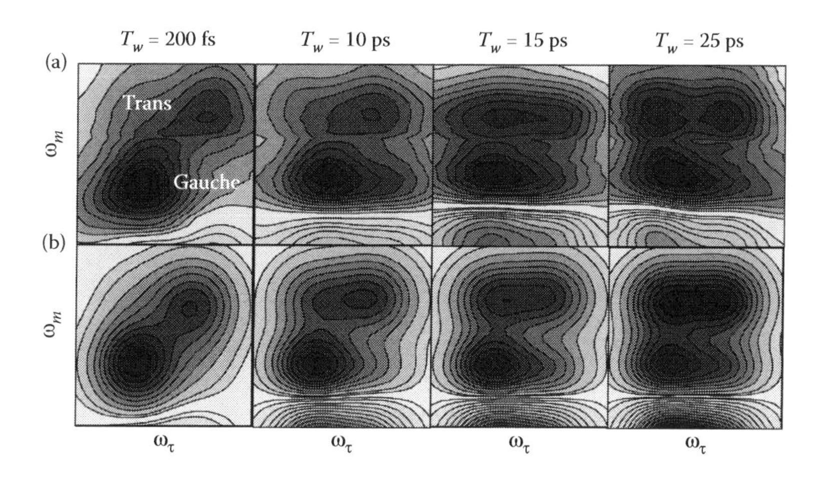

FIGURE 1.19 (a) 2D IR chemical exchange spectra at several T_ws of 1-fluoro-2-isocyanate-ethane. As T_w increases, off-diagonal peaks grow in. (b) The calculated 2D IR chemical exchange spectrum that includes the additional peaks that arise from the coupling of the isocyanate antisymmetric stretching mode to the unassigned peak (see Figure 1.18).

Using the calculated barriers for FICE and for n-butane yielded 43 ps for the n-butane trans-to-gauche isomerization time constant ($1/k_{TG}$). Rosenberg, Berne, and Chandler reported a 43 ps time constant for this process (in CCl_4 at 300 K) from MD simulations [99]. Other MD simulations gave isomerization rates in liquid n-butane at slightly lower temperatures: 52 ps (292 K) [100], 57 ps (292 K) [98], 50 ps (273 K) [98], and 61 ps (<292 K) [101]. All of these values are quite close to the value based on the experimental measurements on FICE. This comparison is the first experimental confirmation of the MD simulations. In the same manner, the isomerization time constant for ethane is found to be ~12 ps. The isomerization time obtained for ethane can be improved by better electronic structure calculations on FICE and calculations for both FICE and ethane that include the CCl_4 in determining the barriers. Thus, this first experimental determination of the time constant for the orientational isomerization about a carbon–carbon single bond for a disubstituted ethane

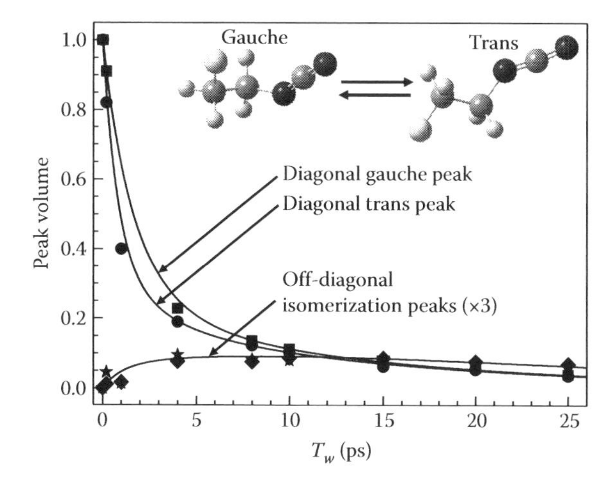

FIGURE 1.20 2D IR chemical exchange data (symbols) and the single adjustable parameter fit (solid curves) for the gauche–trans isomerization of 1-fluoro-2-isocyanate-ethane. The data analysis yields the time for isomerization around the carbon–carbon single bond of 43 ps.

with a low barrier in solution at room temperature also provides the first experimentally based value for the ethane isomerization time constant.

1.6 ION–WATER HYDROGEN BOND EXCHANGE

Water plays important roles in chemistry, biology, geology, and materials science. The properties of water are dominated by its hydrogen bonding network. Water and methane are about the same size and molecular weight; yet, methane is a gas at room temperature and does not liquefy until very low temperature (–162°C). Water is liquid at room temperature because it makes up to four hydrogen bonds with approximately tetrahedral geometry. However, the hydrogen bonding network is not static. Hydrogen bonds are constantly rearranging in a concerted manner. Hydrogen bonds are continually breaking and forming. The water's ability to rapidly rearrange its hydrogen bonding network (~2 ps) [14,33] enables it to solvate ions, transport protons, and accommodate protein folding.

In general, water does not occur as a pure bulk liquid. Rather it interacts with interfaces [107,108], proteins, and other larger molecules [109]. Frequently, water is interacting with charged species, such as the sulfonate groups that line the channels in Nafion and other polyelectrolyte fuel cell membranes [110], charged amino acids on the surfaces of proteins, and ions in solution [12,56,57,111]. An important question is to what extent do ions in aqueous solution influence the dynamics of hydrogen bond rearrangement. By selecting an appropriate anion, it is possible to use 2D IR chemical exchange spectroscopy to directly measure the rate of hydrogen bond switching between a water hydroxyl bound to an anion and the water hydroxyl bound to the oxygen of another water molecule [56,57].

Figure 1.21a shows a cartoon of the exchange process for an anion in water. On the left is a water hydroxyl bound to the anion. This is labeled *ha* for hydroxyl—anion. On the right, the water molecule has switched, and its hydroxyl is now hydrogen bonded to the oxygen of another water molecule. This is labeled *hw* for hydroxyl—water. These are in equilibrium, and there will be constant exchange between these configurations. The cartoon does not depict the nature of the actual process, which is discussed further below. To apply the 2D IR chemical exchange method, the two species, *ha* and *hw*, must have distinguishable spectra. In the experiments, we use a low concentration of HOD in H_2O, and make measurement on the OD stretch. This is necessary to eliminate vibrational excitation transport [112,113] that will occur in either pure H_2O or D_2O. MD simulations indicate that a dilute amount of HOD does not perturb the structure and properties of H_2O and that the OD stretch reports on the dynamics of water [114].

Many aqueous salt solutions, such as NaCl or NaBr, do not show a distinct peak associated with water bound to the anion [12,111]. Rather, the *ha* spectrum is not shifted substantially from that *hw* spectrum. The two spectra overlap to give a single broad band. However, aqueous solutions of sodium tetrafluoroborate, $NaBF_4$, yield two peaks in the IR absorption spectrum [56]. Figure 1.21b shows a spectrum of the OD stretching region of HOD in solutions of H_2O and $NaBF_4$. As the concentration of $NaBF_4$ is increased from pure water to 5.5 M $NaBF_4$, the *ha* band for the OD hydroxyl bound to the BF_4^- anion grows in. Figure 1.21c shows the spectrum of the 5.5 M solution. The narrow *ha* band is centered at 2650 cm^{-1}. While the *ha* band is evident, it is a shoulder on the blue side of the very broad *hw* band. The observation of these distinct bands in the IR absorption spectrum makes it possible to apply the chemical exchange method to the problem of hydrogen bond switching between the anion and the oxygen of another water molecule.

Figure 1.22a shows the 2D IR vibrational echo spectrum of the OD stretch of HOD in an aqueous solution containing 5.5 M $NaBF_4$ at very short $T_w = 200$ fs. The peaks labeled + are positive going and correspond to vibrational echo signals from the 0–1 transitions, and the peaks labeled – are negative going and correspond to vibrational echo emission from the 1–2 transitions. The 0–1 peaks are on the diagonal (labeled ha_{01} and hw_{01}), while the corresponding 1–2 peaks (labeled ha_{12} and hw_{12}) are shifted to lower frequency along the ω_m axis by the vibrational anharmonicity. From the absorption spectrum in Figure 1.21c, it can be seen that the *ha* peak is much narrower than the *hw* peak.

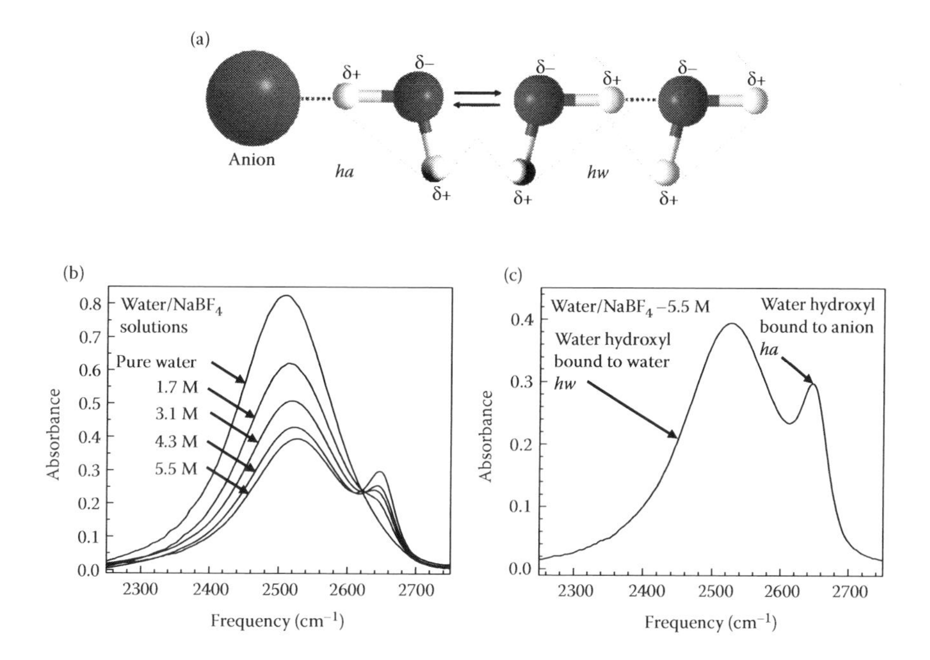

FIGURE 1.21 (a) A cartoon showing the switching of a water molecule hydrogen bonded to an anion (*ha*) to the water hydrogen bonded to the oxygen of another water molecule (*hw*). (b) The spectrum of the OD stretch of dilute HOD in water with various concentrations of sodium tetrafluoroborate ($NaBF_4$). As the concentration of $NaBF_4$ is increased, the peak at ~2650 cm^{-1} grows in. This peak arises from the OD hydroxyl hydrogen bonded to the tetrafluoroborate anion (*ha*). (c) The OD stretch spectrum of 5.5 M $NaBF_4$ in water with dilute HOD. The broad band is from OD hydroxyls hydrogen bonded to the oxygens of other water molecules (*hw*). The narrow band at ~2650 cm^{-1} is from the OD hydroxyl hydrogen bonded to the tetrafluoroborate anion (*ha*).

Then, in the 2D IR, the *ha* band is much narrower in both dimensions than the *hw* band. The anharmonicity of the *ha* transition is significantly smaller than that of the *hw* band. In addition, in 5.5 M $NaBF_4$ solution, the vibrational lifetime of *hw* (OD hydroxyl hydrogen bonded to another water oxygen) is $\tau_{hw} = 2.2 \pm 0.1$ ps, and the vibrational lifetime of *ha* (OD hydroxyl bound to a tetrafluoroborate anion) is $\tau_{ha} = 9.4 \pm 1$ ps [56]. The combination of different 2D peak sizes, different anharmonicities, and different lifetimes makes the time evolution of the 2D spectrum appear strange, but it can still be readily analyzed to obtain the chemical exchange dynamics [56].

Again, looking at the short-time spectrum (200 fs) in Figure 1.22a, consider where the additional chemical exchange peaks will grow in at long time. There will be a positive going chemical exchange peak to the left of the ha_{01} and above the hw_{01} peak. This chemical exchange band arises from *hw*s, becoming *ha*s. It will not overlap with other bands. However, the corresponding band for *ha*s becoming *hw*s will be below the ha_{01} and to the right of hw_{01} peak. This positive going band will be right on top of the negative going ha_{12} peak. Because the lifetime τ_{hw} is much shorter than τ_{ha}, the off-diagonal chemical exchange peaks are much lower in amplitude than the ha_{12} band. The result is that the negative ha_{12} band will dominate. The chemical exchange bands arising from the 1–2 transitions produce major changes in the appearance of the 2D spectrum. The negative going band arising from *hw*s going to *ha*s will appear right on top of the positive going diagonal band hw_{01}. The result is that this chemical exchange band eats out the middle of the diagonal hw_{01} band. From the short-time spectrum and the absorption spectrum, we know where all of the bands will appear. Therefore, we can analyze the data.

Figure 1.22b shows the spectrum at long time, $T_w = 4$ ps. Note that the wavelength ranges in Figure 1.22b have been reduced. The features discussed above are clear. The positive chemical exchange peak for *hw* going to *ha* appears in a nonoverlapping region of the spectrum. The corresponding

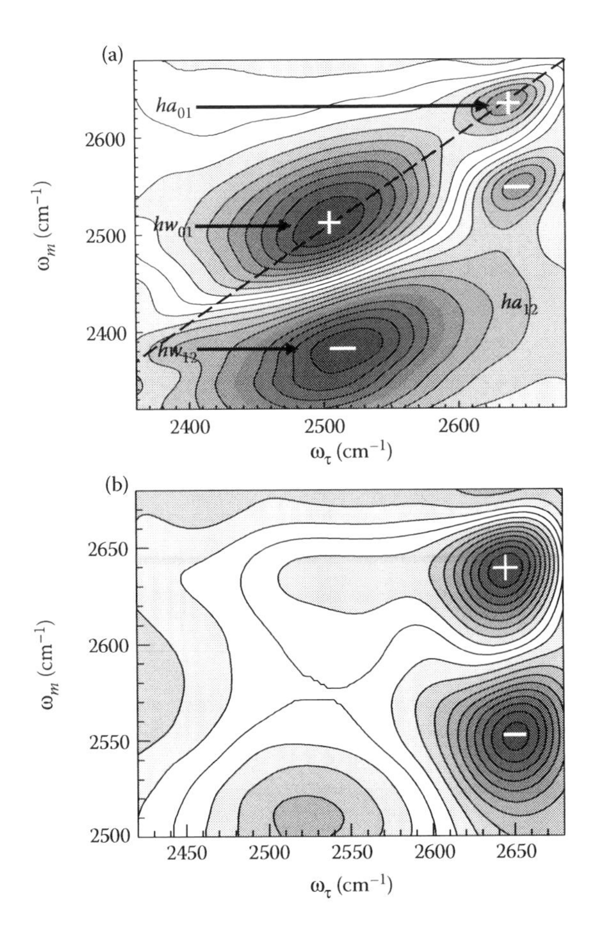

FIGURE 1.22 2D IR spectra of the OD hydroxyl stretch of HOD in 5.5 M solution of $NaBF_4$. (a) Short $T_w = 200$. *ha* peaks are from the OD hydrogen bonded to the tetrafluoroborate anion and *hw* peaks are for the OD hydrogen bonded to the oxygens of other water molecules. (b) Long $T_w = 4$ ps. (Note the change in the frequency axes.) Hydrogen bond switching between *ha* and *hw* causes off-diagonal chemical exchange peaks to grow in. The overlap of positive and negative peaks results in the structure of the spectrum.

positive peak arising from *ha* going to *hw* is overwhelmed by the negative ha_{12} band. In addition, the middle of the positive hw_{01} band has been almost negated by the negative 1–2 chemical exchange band from *hw* going to *ha*.

Analysis of the T_w-dependent changes in spectra like those shown in Figures 1.22a and 1.22b yields the chemical exchange rate [56]. The vibrational lifetimes and the orientational relaxation rates are included in the analysis and provide additional information and support of the analysis of the 2D spectra [56]. Figure 1.23 shows the results of the data analysis. The solid lines through the data points are the calculated curves. The results yield the chemical exchange rates. The time for water hydroxyls bound to anions to exchange and become water hydroxyls bound to other water oxygens is 7 ± 1 ps. This value is very close to that measured subsequently for the anion ClO_4^- using 2D IR chemical exchange spectroscopy [57,58].

The 7 ps hydrogen bond switching time cannot be compared indirectly to the information from bulk water [14,32,33,107,115–118]. The spectrum of the OD stretch of dilute HOD in bulk H_2O is a single very broad band that reflects the large range of hydrogen bond strengths in liquid water. 2D IR vibrational echo measurements of spectral diffusion show that there are dynamics on a variety of timescales. The slowest component of the spectral diffusion is 1.8 ps and is associated with the complete randomization of the hydrogen bond network [14,33]. Another observable is the orientational

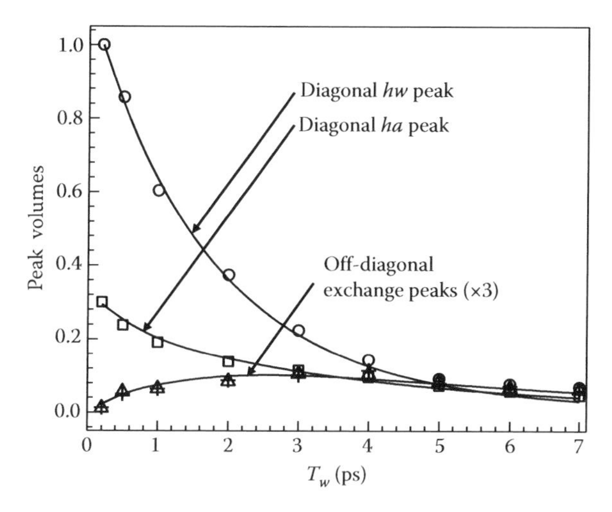

FIGURE 1.23 2D IR chemical exchange data (symbols) for the hydrogen bond switching between water bound to the tetrafluoroborate anion (*ha*) and water bound to the oxygen of other water molecules (*hw*). The solid curves are the fit to the data that yield the chemical exchange time (hydrogen bond switching time) of 7 ps.

relaxation time. Orientational relaxation in water can be described in terms of a jump reorientation model [119,120]. Water orientational relaxation is a concerted process. It involves the dissociation and formation of a number of hydrogen bonds in a collective manner. The orientational relaxation of HOD in H_2O is 2.6 ps. The long time component of the spectral diffusion and orientational relaxation involves similar but not identical processes. The two numbers, 1.8 and 2.6 ps, cannot be directly compared because they arise from different observables that measure different time correlation functions. Nonetheless, it is safe to say that the timescale for hydrogen bond rearrangement in pure bulk water is ~2 ps.

The ~2 ps hydrogen bond rearrangement time for pure water can be compared to the 7 ps time for a water bound to a tetrafluoroborate anion to switch to being bound to a water oxygen in the concentrated aqueous solution of $NaBF_4$. Hydrogen bonding to the anion slows down the hydrogen bond rearrangement but only by a factor of 3 or 4. The magnitude of this increase in the hydrogen bond rearrangement time for water bound to ions is likely to carry over to other systems, such as water interacting with charged amino acids on the surface of a protein or water interacting with the sulfonate head groups that line the channels in many polyelectrolyte fuel cell membranes. The effect of ions on water's hydrogen bond rearrangement is substantial, but water still undergoes rapid structural reorganization in the presence of ions.

1.7 CONCLUDING REMARKS

Here, the basics of 2D IR vibrational echo chemical exchange spectroscopy have been described. The method was illustrated with applications to four problems, solute–solvent complex formation and dissociation, proteins substate switching, gauche–trans isomerization around a carbon–carbon single bond, and water hydrogen bond switching between the water hydroxyl bound to an anion and the hydroxyl bound to the oxygen of another water molecule. In each case, the measurements are made under thermal equilibrium conditions in the electronic ground state.

In all of the systems discussed above, there were only two species or states. 2D IR chemical exchange spectroscopy has also been applied to systems with three distinct species. In one system, a solute molecule is observed to migrate from one binding site to another on a solvent molecule without the complex dissociation [121]. The system consists of a free species, phenol, and phenol bound to two distinct π regions of diphenylacetylene. In another system, chloroform forms weak hydrogen bonds with dimethylsulfoxide (DMSO) and acetone in a mixed solvent of DMSO–acetone, and the switching between the two solvent species is observed and quantified [122]. Again, there are three species, free chloroform and chloroform bound to either acetone or DMSO. Thus, the 2D IR chemical exchange experiment is not limited to equilibrium interconversion between two species.

2D IR chemical exchange spectroscopy can be applied to many types of systems, but there are limitations to the application of the method. Three conditions must be met for the method to work. First, there must be at least one IR active mode that has a distinct frequency for each species undergoing exchange. Second, the concentrations of all the equilibrated species must be high enough for detection. Third, the exchange rate must be comparable to or shorter than the vibrational lifetime of the vibration that is being used as the probe. It is important to note that the vibrational mode that is used as the probe need not be directly involved in the exchange process, as for example, the isocyanate group was employed to study the gauche–trans isomerization around the carbon–carbon single bond of FICE. It is only necessary that the exchange causes the mode frequency to change.

ACKNOWLEDGMENTS

I would like to thank the Air Force Office of Scientific Research, the National Institutes of Health, the Department of Energy, and the National Science Foundation for their support over many years of various aspects of this work. In addition, I would like to thank the many graduate students, post-doctoral students, and collaborators whose contributions made this work possible.

REFERENCES

1. Jackman LM and Cotton FA. 1975. *Dynamic Nuclear Magnetic Resonance Spectroscopy* (Academic Press, New York).
2. Perrin CL and Dwyer TJ. 1990. Application of two-dimensional NMR to kinetics of chemical exchange. *Chem. Rev.* 90:935–967.
3. Zheng J, Kwak K, Xie J, and Fayer MD. 2006. Ultrafast carbon-carbon single bond rotational isomerization in room temperature solution. *Science* 313:1951–1955.
4. Levine IN. 1978. *Physical Chemistry* (McGraw-Hill Book Company, New York).
5. Todd DC and Fleming GR. 1993. Cis-stilbene isomerization—Temperature-dependence and the role of mechanical friction. *J. Chem. Phys.* 98:269–279.
6. Zheng J, Kwak K, Asbury JB, Chen X, Piletic IR, and Fayer MD. 2005. Ultrafast dynamics of solute-solvent complexation observed at thermal equilibrium in real time. *Science* 309:1338–1343.
7. Zimdars D, Tokmakoff A, Chen S, Greenfield SR, Fayer MD, Smith TI, and Schwettman HA. 1993. Picosecond infrared vibrational photon echoes in a liquid and glass using a free electron laser. *Phys. Rev. Lett.* 70:2718–2721.
8. Rella CW, Kwok A, Rector KD, Hill JR, Schwettmann HA, Dlott DD, and Fayer MD. 1996. Vibrational echo studies of protein dynamics. *Phys. Rev. Lett.* 77:1648–1651.
9. Demirdoven N, Khalil M, Golonzka O, and Tokmakoff A. 2001. Correlation effects in the two-dimensional vibrational spectroscopy of coupled vibrations. *J. Phys. Chem. A* 105:8030.
10. Kim Y and Hochstrasser RM. 2005. Dynamics of amide-I modes of the alanine dipeptide in D_2O. *J. Phys. Chem. B* 109:6884–6891.
11. Asbury JB, Steinel T, and Fayer MD. 2003. Using ultrafast infrared multidimensional correlation spectroscopy to aid in vibrational spectral peak assignments. *Chem. Phys. Lett.* 381:139–146.
12. Park S and Fayer MD. 2007. Hydrogen bond dynamics in aqueous NaBr solutions. *Proc. Nat. Acad. Sci. U.S.A.* 104:16731–16738.
13. Steinel T, Asbury JB, Corcelli SA, Lawrence CP, Skinner JL, and Fayer MD. 2004. Water dynamics: Dependence on local structure probed with vibrational echo correlation spectroscopy. *Chem. Phys. Lett.* 386:295–300.
14. Asbury JB, Steinel T, Kwak K, Corcelli SA, Lawrence CP, Skinner JL, and Fayer MD. 2004. Dynamics of water probed with vibrational echo correlation spectroscopy. *J. Chem. Phys.* 121:12431–12446.
15. Asbury JB, Steinel T, Stromberg C, Gaffney KJ, Piletic IR, and Fayer MD. 2003. Hydrogen bond breaking probed with multidimensional stimulated vibrational echo correlation spectroscopy. *J. Chem. Phys.* 119:12981–12997.

16. Ganim Z, Chung HS, Smith AW, Deflores LP, Jones KC, and Tokmakoff A. 2008. Amide I two-dimensional infrared spectroscopy of proteins. *Acc. Chem. Res.* 41:432–441.
17. Bandaria JN, Dutta S, Nydegger MW, Rock W, Kohen A, and Cheatum CM. 2010. Characterizing the dynamics of functionally relevant complexes of formate dehydrogenase. *Proc. Nat. Acad. Sci. U.S.A.* 107:17974–17979.
18. Middleton CT, Woys AM, Mukherjee SS, and Zanni MT. 2010. Residue-specific structural kinetics of proteins through the union of isotope labeling, mid-IR pulse shaping, and coherent 2D IR spectroscopy. *Methods* 52:12–22.
19. Kim YS and Hochstrasser RM. 2009. Applications of 2D IR spectroscopy to peptides, proteins, and hydrogen-bond dynamics. *J. Phys. Chem. B* 113:8231–8251.
20. Bagchi S, Nebgen BT, Loring RF, and Fayer MD. 2010. Dynamics of a myoglobin mutant enzyme: 2D IR vibrational echo experiments and simulations. *J. Am. Chem. Soc.* 132:18367–18376.
21. Chung JK, Thielges MC, Bowman SJ, Bren KL, and Fayer MD. 2011. Temperature dependent equilibrium native to unfolded protein dynamics and properties observed with IR absorption and 2D IR vibrational echo experiments. *J. Am. Chem. Soc.* 133:6681–6691.
22. Chung JK, Thielges MC, and Fayer MD. 2011. Dynamics of the folded and unfolded villin headpiece (HP35) measured with ultrafast 2D IR vibrational echo spectroscopy. *Proc. Nat. Acad. Sci. U.S.A.* 108: 3578–3583.
23. Thielges MC, Chung JK, and Fayer MD. 2011. Protein dynamics in cytochrome P450 molecular recognition and substrate specificity using 2D IR vibrational echo spectroscopy. *J. Am. Chem. Soc.* 133:3995–4004.
24. Thielges MC, Axup JY, Wong D, Lee H, Chung JK, Schultz PG, and Fayer MD. 2011. Two-dimensional IR spectroscopy of protein dynamics using two vibrational labels: A site-specific genetically encoded unnatural amino acid and an active site ligand. *J. Chem. Phys. B* 115:11294–11304.
25. Zanni MT, Asplund MC, and Hochstrasser RM. 2001. Two-dimensional heterodyned and stimulated infrared photon echoes of *N*-methylacetamide-D. *J. Chem. Phys.* 114:4579.
26. Khalil M, Demirdoven N, and Tokmakoff A. 2003. Coherent 2D IR spectroscopy: Molecular structure and dynamics in solution. *J. Phys. Chem. A.* 107:5258–5279.
27. Chung HS, Khalil M, and Tokmakoff A. 2004. Nonlinear infrared spectroscopy of protein conformational change during thermal unfolding. *J. Phys. Chem. B* 108:15332–15342.
28. Mukherjee P, Krummel AT, Fulmer EC, Kass I, Arkin IT, and Zanni MT. 2004. Site-specific vibrational dynamics of the CD3 zeta membrane peptide using heterodyned two-dimensional infrared photon echo spectroscopy. *J. Chem. Phys.* 120:10215–10224.
29. Fulmer EC, Ding F, and Zanni MT. 2005. Heterodyned fifth-order 2D IR spectroscopy of the azide ion in an ionic glass. *J. Chem. Phys.* 122:034302(034312).
30. DeCamp MF, DeFlores L, McCracken JM, Tokmakoff A, Kwac K, and Cho M. 2005. Amide I vibrational dynamics of *N*-methylacetamide in polar solvents: The role of electrostatic interactions. *J. Phys. Chem. B* 109:11016–11026.
31. Golonzka O, Khalil M, Demirdoven N, and Tokmakoff A. 2001. Coupling and orientation between anharmonic vibrations characterized with two-dimensional infrared vibrational echo spectroscopy. *J. Chem. Phys.* 115:10814–10828.
32. Nicodemus RA, Ramasesha K, Roberts ST, and Tokmakoff A. 2010. Hydrogen bond rearrangements in water probed with temperature-dependent 2D IR. *J. Phys. Chem. Lett.* 1:1068–1072.
33. Asbury JB, Steinel T, Stromberg C, Corcelli SA, Lawrence CP, Skinner JL, and Fayer MD. 2004. Water dynamics: Vibrational echo correlation spectroscopy and comparison to molecular dynamics simulations. *J. Phys. Chem. A* 108:1107–1119.
34. Fecko CJ, Loparo JJ, Roberts ST, and Tokmakoff A. 2005. Local hydrogen bonding dynamics and collective reorganization in water: Ultrafast infrared spectroscopy of HOD/D_2O. *J. Chem. Phys.* 122: 054506-054518.
35. Cowan ML, Bruner BD, Huse N, Dwyer JR, Chugh B, Nibbering ETJ, Elsaesser T, and Miller RJD. 2005. Ultrafast memory loss and energy redistribution in the hydrogen bond network of liquid H_2O. *Nature* 434:199–202.
36. Loparo JJ, Roberts ST, and Tokmakoff A. 2006. Multidimensional infrared spectroscopy of water. I. Vibrational dynamics in two-dimensional IR line shapes. *J. Chem. Phys.* 125:194521.
37. Loparo JJ, Roberts ST, and Tokmakoff A. 2006. Multidimensional infrared spectroscopy of water. II. Hydrogen bond switching dynamics. *J. Chem. Phys.* 125:194522.
38. Roberts ST, Ramasesha K, Petersen PB, Mandal A, and Tokmakoff A. 2011. Proton transfer in concentrated aqueous hydroxide visualized using ultrafast infrared spectroscopy. *J. Phys. Chem. A* 115:3957–3972.

39. Ganim Z, Jones KC, and Tokmakoff A. 2010. Insulin dimer dissociation and unfolding revealed by amide I two-dimensional infrared spectroscopy. *Phys. Chem. Chem. Phys.* 12:3579–3588.
40. Finkelstein IJ, Zheng J, Ishikawa H, Kim S, Kwak K, and Fayer MD. 2007. Probing dynamics of complex molecular systems with ultrafast 2D IR vibrational echo spectroscopy. *Phys. Chem. Chem. Phys.* 9: 1533–1549.
41. DeFlores LP, Ganim Z, Ackley SF, Chung HS, and Tokmakoff A. 2006. The anharmonic vibrational potential and relaxation pathways of the amide I and Ii modes of *N*-methylacetamide. *J. Phys. Chem. B* 110:18973–18980.
42. Mukherjee P, Kass I, Arkin IT, and Zanni MT. 2006. Picosecond dynamics of a membrane protein revealed by 2D IR. *Proc. Nat. Acad. Sci. U.S.A.* 103:3528–3533.
43. Ishikawa H, Kwak K, Chung JK, Kim S, and Fayer MD. 2008. Direct observation of fast protein conformational switching. *Proc. Nat. Acad. Sci. U.S.A.* 105:8619–8624.
44. Tucker MJ, Gai XS, Fenlon EE, Brewer SH, and Hochstrasser RM. 2011. 2D IR photon echo of azido-probes for biomolecular dynamics. *Phys. Chem. Chem. Phys.* 13:2237–2241.
45. Kwak K, Park S, Finkelstein IJ, and Fayer MD. 2007. Frequency-frequency correlation functions and apodization in 2D IR vibrational echo spectroscopy, a new approach. *J. Chem. Phys.* 127:124503.
46. Kwak K, Rosenfeld DE, and Fayer MD. 2008. Taking apart the two-dimensional infrared vibrational echo spectra: More information and elimination of distortions. *J. Chem. Phys.* 128:204505.
47. Zheng JR and Fayer MD. 2007. Hydrogen bond lifetimes and energetics for solute/solvent complexes studied with 2D IR vibrational echo spectroscopy. *J. Am. Chem. Soc.* 129:4328–4335.
48. Zheng J and Fayer MD. 2008. Solute-solvent complex kinetics and thermodynamics probed by 2D IR vibrational echo chemical exchange spectroscopy. *J. Phys. Chem. B* 112:10221–10227.
49. Finkelstein IJ, Ishikawa H, Kim S, Massari AM, and Fayer MD. 2007. Substrate binding and protein conformational dynamics measured via 2D IR vibrational echo spectroscopy. *Proc. Nat. Acad. Sci. U.S.A.* 104:2637–2642.
50. Finkelstein IJ, Massari AM, and Fayer MD. 2007. Viscosity dependent protein dynamics. *Biophys. J.* 92: 3652–3662.
51. Urbanek DC, Vorobyev DY, Serrano AL, Gai F, and Hochstrasser RM. 2010. The two-dimensional vibrational echo of a nitrile probe of the Villin Hp35 protein. *J. Phys. Chem. Lett.* 1:3311–3315.
52. Makinen MW, Houtchens RA, and Caughey WS. 1979. Structure of carboxymyoglobin in crystals and in solution. *Proc. Nat. Acad. Sci. U.S.A.* 76:6042–6046.
53. Caughey WS, Shimada H, Choc MC, and Tucker MP. 1981. Dynamic protein structures: Infrared evidence for four discrete rapidly interconverting conformers at the carbon monoxide binding site of bovine heart myoglobin. *Proc. Nat. Acad. Sci. U.S.A.* 78:2903–2907.
54. Anderton CL, Hester RE, and Moore JN. 1997. A chemometric analysis of the resonance Raman spectra of mutant carbonmonoxy-myoglobins reveals the effects of polarity. *Biochim. Biophys. Acta* 1338: 107–120.
55. Li TS, Quillin ML, Phillips GN, Jr., and Olson JS. 1994. Structural determinants of the stretching frequency of Co bound to myoglobin. *Biochemistry* 33:1433–1446.
56. Moilanen DE, Wong D, Rosenfeld DE, Fenn EE, and Fayer MD. 2009. Ion-water hydrogen-bond switching observed with 2D IR vibrational echo chemical exchange spectroscopy. *Proc. Nat. Acad. Sci. U.S.A.* 106:375–380.
57. Park S, Odelius M, and J. GK. 2009. Ultrafast dynamics of hydrogen bond exchange in aqueous ionic solutions. *J. Phys. Chem. B* 113:7825–7835.
58. Ji M, Odelius M, and Gaffney KJ. 2010. Large angular jump mechanism observed for hydrogen bond exchange in aqueous perchlorate solution. *Science* 328:1003–1005.
59. Zheng J, Kwak K, Chen X, Asbury JB, and Fayer MD. 2006. Formation and dissociation of intra-intermolecular hydrogen bonded solute-solvent complexes: Chemical exchange 2D IR vibrational echo spectroscopy. *J. Am. Chem. Soc.* 128:2977–2987.
60. Kwak K, Zheng J, Cang H, and Fayer MD. 2006. Ultrafast 2D IR vibrational echo chemical exchange experiments and theory. *J. Phys. Chem. B* 110:19998–20013.
61. Park S, Kwak K, and Fayer MD. 2007. Ultrafast 2D IR vibrational echo spectroscopy: A probe of molecular dynamics. *Laser Phys. Lett.* 4:704–718.
62. Reichardt C. 2003. *Solvents and Solvent Effects in Organic Chemistry* (Wiley-VCH, Weinheim).
63. Vinogradov SN and Linnell RH. 1971. *Hydrogen Bonding* (Van Nostrand Reinhold Company, New York).
64. Kwac K, Lee C, Jung Y, Han J, Kwak K, Zheng JR, Fayer MD, and Cho M. 2006. Phenol-benzene complexation dynamics: Quantum chemistry calculation, molecular dynamics simulations, and two dimensional IR spectroscopy. *J. Chem. Phys.* 125:244508.

65. Mukamel S. 2000. Multidimensional femtosecond correlation spectroscopies of electronic and vibrational excitations. *Ann. Rev. Phys. Chem.* 51:691–729.
66. Mukamel S. 1995. *Principles of Nonlinear Optical Spectroscopy* (Oxford University Press, New York).
67. Chang R. 2000. *Physical Chemistry for the Chemical and Biological Sciences* (University Science Books, Sausalito), p 1018.
68. Eisenmesser EZ, Bosco DA, Akke M, and Kern D. 2002. Enzyme dynamics during catalysis. *Science* 295:1520–1523.
69. Henzler-Wildman KA, Lei M, Thai V, Kerns SJ, Karplus M, and Kern D. 2007. A hierarchy of timescales in protein dynamics is linked to enzyme catalysis. *Nature* 450:913–916.
70. Hammes-Schiffer S and Benkovic SJ. 2006. Relating protein motion to catalysis. *Annu. Rev. Biochem.* 75:519–541.
71. Erzberger JP and Berger JM. 2006. Evolutionary relationships and structural mechanisms of Aaa plus proteins. *Annu. Rev. Biophys. Biomol. Struct.* 35, 93–114.
72. Boehr DD, Dyson HJ, and Wright PE. 2006. An NMR perspective on enzyme dynamics. *Chem. Rev.* 106:3055–3079.
73. Hill SE, Bandaria JN, Fox M, Vanderah E, Kohen A, and Cheatum CM. 2009. Exploring the molecular origins of protein dynamics in the active site of human carbonic anhydrase Ii. *J. Phys. Chem. B* 113: 11505–11510.
74. Campbell BF, Chance MR, and Friedman JM. 1987. Linkage of functional and structural heterogeneity in proteins: Dynamic hole burning in carboxymyoglobin. *Science* 238:373–376.
75. Hong MK, Braunstein D, Cowen BR, Frauenfelder H, Iben IET, Mourant JR, Ormos P et al. 1990. Conformational substates and motions in myoglobin. *Biophys. J.* 58:429–436.
76. Frauenfelder H, Sligar SG, and Wolynes PG. 1991. The energy landscapes and motions of proteins. *Science* 254:1598–1603.
77. Frauenfelder H, McMahon BH, Austin RH, Chu K, and Groves JT. 2001. The role of structure, energy landscape, dynamics, and allostery in the enzymatic function of myoglobin. *Proc. Nat. Acad. Sci. U.S.A.* 98:2370–2374.
78. Andrews BK, Romo T, Clarage JB, Pettitt BM, and Phillips GN, Jr. 1998. Characterizing global substates of myoglobin. *Struct. Fold. Des.* 6:587–594.
79. Ghosh A, Qiu J, DeGrado WF, and Hochstrasser RM. 2011. Tidal Surge in the M2 proton channel, sensed by 2D IR spectroscopy. *Proc. Nat. Acad. Sci. U.S.A.* 108:6115–6120.
80. Merchant KA, Noid WG, Akiyama R, Finkelstein I, Goun A, McClain BL, Loring RF, and Fayer MD. 2003. Myoglobin-Co substate structures and dynamics: Multidimensional vibrational echoes and molecular dynamics simulations. *J. Am. Chem. Soc.* 125:13804–13818.
81. Merchant KA, Noid WG, Thompson DE, Akiyama R, Loring RF, and Fayer MD. 2003. Structural assignments and dynamics of the a substates of Mb-CO: Spectrally resolved vibrational echo experiments and molecular dynamics simulations. *J. Phys. Chem. B* 107:4–7.
82. Ishikawa H, Finkelstein IJ, Kim S, Kwak K, Chung JK, Wakasugi K, Massari AM, and Fayer MD. 2007. Neuroglobin dynamics observed with ultrafast 2D IR vibrational echo spectroscopy. *Proc. Nat. Acad. Sci. U.S.A.* 104:16116–16121.
83. Ansari A, Berendzen J, Braunstein D, Cowen BR, Frauenfelder H, Hong MK, Iben IET, Johnson JB, Ormos P, Sauke TB, Scholl R, Schulte A, Steinbach PJ, Vittitow J, and Young RD. 1987. Rebinding and relaxation in the myoglobin pocket. *Biophys. Chem.* 26:337–355.
84. Tian WD, Sage, JT, Champion, PM. 1993. Investigation of ligand association and dissociation rates in the "open" and "closed" states of myoglobin. *J. Mol. Biol.* 233:155–166.
85. Müller JD, McMahon BH, Chen EYT, Sligar SG, and Nienhaus GU. 1999. Connection between the taxonomic substates of protonation of histidines 64 and 97 in carbonmonoxy myoglobin. *Biophys. J.* 77: 1036–1051.
86. Bagchi S, Thorpe DG, Thorpe IF, Voth GA, and Fayer MD. 2010. Conformational switching between protein substates studied with 2D IR vibrational echo spectroscopy and molecular dynamics simulations. *J. Phys. Chem. B* 114:17187–17193.
87. Vojtechovsky J, Chu K, Berendzen J, Sweet RM, and Schlichting I. 1999. Crystal structures of myoglobin-ligand complexes at near atomic resolution. *Biophys. J.* 77:2153–2174.
88. Johnson JB, Lamb DC, Frauenfelder H, Müller JD, McMahon B, Nienhaus GU, and Young RD. 1996. Ligand binding to heme proteins. 6. Interconversion of taxonomic substates in carbonmonoxymyoglobin. *Biophys. J.* 71:1563–1573.
89. Teeter M. 2004. Myoglobin cavities provide interior ligand pathway. *Protein Sci.* 13:313–318.

90. Tilton RFJ, Kuntz IDJ, and Petsko GA. 1984. Cavities in proteins: Structure of a metmyoglobin-xenon complex solved to 1.9 Å. *Biochemistry* 23:2849–2857.
91. Doukov TI, Blasiak LC, Seravalli J, Ragsdale SW, and Drennan CL. 2008. Xenon in and at the end of the tunnel of bifunctional carbon monoxide dehydrogenase/acetyl-CoA synthase. *Biochemistry* 47: 3474–3483.
92. Palmer AGR. 1997. Probing molecular motion by NMR. *Curr. Opin. Struct. Biol.* 7:732–737.
93. Schnell JR, Dyson HJ, and Wright PE. 2004. Structure, dynamics, and catalytic function of dihydrofolate reductase. *Ann. Rev. Biophys. Biomol. Struct.* 33:119–140.
94. Oliveberg M and Wolynes PG. 2005. The experimental survey of protein-folding energy landscapes. *Q. Rev. Biophys.* 38:245–288.
95. March J. 1985. *Advanced Organic Chemistry* (John Wiley and Sons, New York); 3rd Ed, p 1346.
96. Chandler D. 1978 Statistical mechanics of isomerization dynamics in liquids and the transition state approximation. *J. Chem. Phys.* 68:2959–2970.
97. Weber TA. 1978 Simulation of *N*-butane using a skeletal alkane model. *J. Chem. Phys.* 69:2347–2354.
98. Brown D and Clarke JHR. 1990 A direct method of studying reaction rates by equilibrium molecular dynamics: Application to the kinetics of isomerization in liquid *N*-butane. *J. Chem. Phys.* 92:3062–3073.
99. Rosenberg RO, Berne BJ, and Chandler D. 1980 Isomerization dynamics in liquids by molecular dynamics. *Chem. Phys. Lett.* 75:162–168.
100. Edberg R, Evans DJ, and Morris GP. 1987. Conformational kinetics in liquid butane by nonequilibrium molecular dynamics. *J. Chem. Phys.* 87:5700–5708.
101. Ramirez J and Laso M. 2001. Conformational kinetics in liquid *N*-butane by transition path sampling. *J. Chem. Phys.* 115:7285–7292.
102. Streitwieser A and Taft RW. 1968. *Progress in Physical Organic Chemistry* (John Wiley and Sons, New York).
103. Parr RG and Yang W. 1989. *Density Functional Theory of Atoms and Molecules* (Oxford University Press, New York).
104. Khalil M, Demirdoven N, and Tokmakoff A. 2004. Vibrational coherence transfer characterized with Fourier-transform 2D IR spectroscopy. *J. Chem. Phys.* 121:362–373.
105. Pophristic V and Goodman L. 2001. Hyperconjugation not steric repulsion leads to the staggered structure of ethane. *Nature* 411:565–568.
106. Bickelhaupt FM and Baerends EJ. 2003. The case for steric repulsion causing the staggered conformation of ethane. *Angew. Chem., Int. Ed.* 42:4183–4188.
107. Moilanen DE, Fenn EE, Wong D, and Fayer MD. 2009. Water dynamics in large and small reverse micelles: From two ensembles to collective behavior. *J. Chem. Phys.* 131:014704.
108. Moilanen DE, Fenn EE, Wong D, and Fayer MD. 2009. Water dynamics in AOT lamellar structures and reverse micelles: Geometry and length scales vs. surface interactions. *J. Am. Chem. Soc.* 131:8318–8328.
109. Fenn EE, Moilanen DE, Levinger NE, and Fayer MD. 2009. Water dynamics and interactions in water-polyether binary mixtures. *J. Am. Chem. Soc.* 131:5530–5539.
110. Moilanen DE, Piletic IR, and Fayer MD. 2007. Water dynamics in Nafion fuel cell membranes: The effects of confinement and structural changes on the hydrogen bonding network. *J. Phys. Chem. C* 111: 8884–8891.
111. Smith JD, Saykally RJ, and Geissler PL. 2007. The effect of dissolved halide anions on hydrogen bonding in liquid water. *J. Am. Chem. Soc.* 129:13847–13856.
112. Woutersen S and Bakker HJ. 1999. Resonant intermolecular transfer of vibrational energy in liquid water. *Nature* 402:507–509.
113. Gaffney KJ, Piletic IR, and Fayer MD. 2003. Orientational relaxation and vibrational excitation transfer in methanol–carbon tetrachloride solutions. *J. Chem. Phys.* 118:2270–2278.
114. Corcelli S, Lawrence CP, and Skinner JL. 2004. Combined electronic structure/molecular dynamics approach for ultrafast infrared spectroscopy of dilute HOD in liquid H_2O and D_2O. *J. Chem. Phys.* 120: 8107.
115. Rezus YLA and Bakker HJ. 2006. Orientational dynamics of isotopically diluted H_2O and D_2O. *J. Chem. Phys.* 125:144512.
116. Moilanen DE, Fenn EE, Lin YS, Skinner JL, Bagchi B, and Fayer MD. 2008. Water inertial reorientation: Hydrogen bond strength and the angular potential. *Proc. Nat. Acad. Sci. U.S.A.* 105:5295–5300.
117. Schmidt JR, Roberts ST, Loparo JJ, Tokmakoff A, Fayer MD, and Skinner JL. 2007. Are water simulation models consistent with steady-state and ultrafast vibrational spectroscopy experiments? *Chem. Phys.* 341:143–157.
118. Fecko CJ, Eaves JD, Loparo JJ, Tokmakoff A, and Geissler PL. 2003. Ultrafast hydrogen-bond dynamics in the infrared spectroscopy of water. *Science* 301:1698–1702.

119. Laage D and Hynes JT. 2008. On the molecular mechanism of water reorientation. *J. Phys. Chem. B* 112: 14230–14242.
120. Laage D and Hynes JT (2006) A molecular jump mechanism of water reorientation. *Science* 311:832–835.
121. Rosenfeld DE, Kwak K, Gengeliczki Z, and Fayer MD. 2010. Hydrogen bond migration between molecular sites observed with ultrafast 2D IR chemical exchange spectroscopy. *J. Phys. Chem. B* 114: 2383–2389.
122. Kwak K, Rosenfeld DE, Chung JK, and Fayer MD. 2008. Solute-solvent complex switching dynamics of chloroform between acetone and dimethylsulfoxide-two-dimensional IR chemical exchange spectroscopy. *J. Phys. Chem. B* 112:13906–13915.

2 Ultrafast Vibrational Dynamics of Hydrogen-Bonded Dimers and Base Pairs

Thomas Elsaesser

CONTENTS

2.1 INTRODUCTION

Hydrogen bonding, a fundamental noncovalent interaction, plays a decisive role for the structure and the physical and chemical properties of a wide range of molecular systems [1–3]. In the most elementary structural motif, hydrogen bonding consists of an attractive interaction between a hydrogen donor group X–H, where X stands for O, N, or F, and an electronegative acceptor atom Y (Y = O, N, F, Cl). The binding energy is substantially smaller than that of covalent bonds and lies between 4 and 50 kJ/mol. The distance between the two heavy atoms X and Y, the hydrogen bond length, depends on the strength of the attractive interaction and covers a broad range from approximately 0.25 nm for strong hydrogen bonds up to 0.35 nm for weak bonds. In general, different microscopic forces contribute to hydrogen bonding such as attractive Coulomb interactions as well as van der Waals and dispersion forces.

In the condensed phase, hydrogen bonds are subject to interactions with their densely packed environment that can undergo structural dynamics on a multitude of timescales. This fact together with the moderate bond strength results in structural dynamics and fluctuations of hydrogen bonds, most elementary events being their breaking and reformation. In hydrogen-bonded liquids such as water, the fastest structural fluctuations occur in the sub-100 fs range while breaking and reformation of hydrogen bonds occur as random processes on a picosecond timescale [4–7]. In contrast, replication processes of hydrogen-bonded biomolecular structure involve a hierarchy of protein-initiated and enzyme-controlled steps on much slower timescales.

Linear vibrational spectroscopy in the frequency domain has been an important tool of hydrogen bond research since its beginning, mainly due to the fact that vibrational spectra provide insight into local molecular geometries and interactions. Most of such studies have focused on the X–H stretching vibration of the hydrogen donor group, which shifts to lower frequency upon formation of a hydrogen bond [1]. This shift reflects the softening of the X–H stretching potential by the attractive interaction with the hydrogen acceptor atom and is frequently connected with an enhancement of the (diagonal) anharmonicity. The absolute value of this red-shift has been considered a measure of hydrogen bond strength, and empirical correlations between the red-shift and the binding energy or hydrogen bond length have been derived, mainly for hydrogen bonds in solids (see Ref. [8] and references therein). Though frequently considered to be of general validity, such correlations can be applied for a particular type of hydrogen bonds only and neglect both other degrees of freedom, for example, the angle between the hydrogen-bonded groups, and structural fluctuations [9,10].

Other vibrational modes are affected by the formation of a hydrogen bond as well. The X–H bending mode and fingerprint modes involving elongations of the acceptor atom Y undergo limited frequency shifts of up to several tens of wavenumbers (cm^{-1}), in the case of the X–H bending mode to higher frequency. Moreover, the new so-called hydrogen bond modes arise that involve motions of the heavy atoms X and Y, which are now coupled through the comparably weak hydrogen bonding interactions. As a result, the frequency of such hydrogen bond modes is in the range below some 500 cm^{-1}.

Upon formation of a hydrogen bond, the strength of the X–H stretching bands is enhanced due to changes in the electronic charge distribution. Moreover, the stationary lineshapes of X–H stretch infrared and Raman bands undergo pronounced changes. For OH stretching bands of weak to medium-strong hydrogen bonds (binding energy $\leq$35 kJ/mol), one typically observes a spectral broadening by up to a factor of 10 and/or a highly complex substructure of the spectral envelopes. Such lineshapes reflect the coupling of the OH stretching oscillators to other intra- and intermolecular modes of the molecular system and—in the liquid phase—the influence of the fluctuating environment on the vibrational excitation. The latter gives rise to processes such as spectral diffusion of the vibrational transition frequency, which—vice versa—maps structural fluctuations [4,5,11,12]. The existence of different types of couplings of similar strength and the fact that steady-state vibrational spectra represent a time average over the system's dynamics make the quantitative analysis of linear vibrational spectra highly challenging. In spite of extensive theoretical work, linear vibrational spectroscopy has not allowed for deciphering the spectral envelopes in a quantitative way.

Nonlinear infrared techniques in the femtosecond time domain overcome such limitations of linear vibrational spectroscopy [4–6,13,14]. First, they allow for mapping microscopic dynamics of hydrogen bonds on the intrinsic timescale of vibrational excitations and nuclear motions. In recent years, a large variety of hydrogen-bonded systems has been investigated by femtosecond pump-probe and photon-echo methods in the mid-infrared spectral range from 500 to 4000 cm^{-1}. Second, two-dimensional (2D) infrared spectroscopy based on either the narrow-band pump/broadband probe technique or the heterodyne-detected photon echoes [13,15–17] gives specific insight into vibrational coupling schemes of hydrogen bonds and into spectral diffusion and other processes in fluctuating hydrogen-bonded structures.

This chapter focuses on hydrogen-bonded dimers and nucleobase pairs in a liquid environment and on base pairs in DNA helices. Complex O–H and N–H stretching lineshapes are analyzed by combining femtosecond pump-probe and 2D infrared spectroscopy with theoretical calculations, in some cases allowing for a quantitative description also of the linear infrared spectra. Different types of couplings such as Fermi resonances and couplings to low-frequency modes are separated and analyzed quantitatively. Coherent low-frequency motions of hydrogen bonds are resolved in time and their frequency spectra are determined. The kinetics and pathways of vibrational relaxation are briefly discussed. Interactions of DNA with its fluctuating aqueous environment, the hydration shell that plays an important role for DNA's structural and functional properties, as well as the properties of the water shell are addressed in experiments covering distinctly different hydration levels.

The chapter is organized as follows. In Section 2.2, basic vibrational couplings and system–bath interactions and their relevance for vibrational lineshapes are introduced, followed by a brief discussion of the relevant timescales and an account of the experimental methods applied. Section 2.3 addresses OH/OD stretching dynamics and coherent low-frequency motions of cyclic acetic acid dimers and NH stretching dynamics in nucleobase pairs in solution. Section 2.4 is devoted to base pairs in DNA and ultrafast hydration and energy dissipation processes for which both DNA vibrations and the OH stretching mode of the hydrating water serve as a probe. Conclusions and prospects are presented in Section 2.5.

2.2 VIBRATIONAL PROBES OF HYDROGEN BOND STRUCTURE AND DYNAMICS

2.2.1 LINEAR INFRARED SPECTRA AND VIBRATIONAL COUPLINGS

In this section, prototypical OH and NH stretching absorption spectra of hydrogen-bonded dimers and nucleobase pairs are introduced and the basic underlying couplings discussed. Femtosecond nonlinear infrared experiments on such systems will be presented in the later sections of this chapter.

2.2.1.1 Hydrogen-Bonded Dimers

In the liquid phase, the stretching vibration of a free OH and NH group displays infrared absorption as a single band with a respective maximum at approximately 3600 and 3500 cm^{-1}. The spectral width depends on the particular system and has values between 20 and 70 cm^{-1}. Upon formation of a hydrogen bond, the stretching bands undergo a spectral red-shift, a change of the spectral envelope, and a change of the overall absorption strength. This behavior is borne out by the OH stretching band of cyclic acetic acid dimers in Figure 2.1a and the NH stretching band of 7-azaindole dimers [18] in Figure 2.1b. The narrow weak absorption lines observed at high frequency are due to free OH and NH groups. The strong enhancement of absorption of the hydrogen-bonded groups compared to the free ones is due to changes of electronic structure and the enhanced electronic polarizability. To explain the strongly broadened and highly complex lineshapes, a number of coupling mechanisms have been invoked, which are schematically illustrated in Figure 2.2 for the cyclic acetic acid dimer, an important model system [19–24].

The cyclic acetic acid dimer has a planar structure and displays C_{2h} symmetry with respect to a rotation axis centered at the dimer and oriented perpendicular to the dimer plane (Figure 2.2a). For a description of couplings schemes, it is convenient to replace the local vibrational coordinates $q_{i,1}$ and $q_{i,2}$ of monomers 1 and 2 by cyclic coordinates $q_{i,g,u} = (q_{i,1} \pm q_{i,2})/\sqrt{2}$, which show a gerade (g) or ungerade (u) symmetry with respect to the rotation axis. While illustrated in Figure 2.2a for the OH stretching vibrations ($i = 1$), a similar choice of coordinates is applicable for the OH bending, the carbonyl stretching, and other fingerprint vibrations.

Without any coupling between the two OH oscillators, the infrared-active OH stretching fundamental is due to the dipole-allowed transition between the v = 0 and 1 states in the $q_{1,u}$ coordinate (Figure 2.2b). In the $q_{1,g}$ coordinate, the v = 0 to 1 transition is Raman allowed and occurs at the same frequency. An excitonic coupling V_0 between the two *local* OH stretching oscillators with coordinates $q_{1,1}$ and $q_{1,2}$ results in a splitting of the v = 1 states in the $q_{1,u}$ and $q_{1,g}$ coordinates by $2V_0$. The infrared- ($\Delta v > 0$ for $V_0 > 0$) and the Raman-active ($\Delta v < 0$ for $V_0 > 0$) OH stretching transitions are spectrally shifted by Δv, as schematically illustrated for the infrared transition in Figure 2.2c. In addition to excitonic coupling, Fermi resonances between the v = 1 state in the $q_{1,u}$ coordinate and overtone and combination tone of fingerprint modes may occur. This type of coupling leads

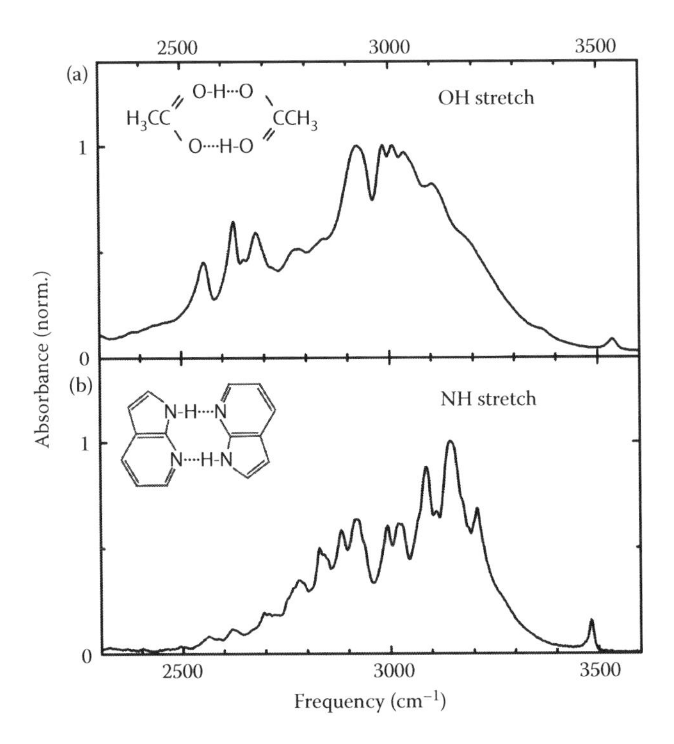

FIGURE 2.1 (a) OH stretching absorption band of cyclic acetic acid dimers dissolved in CCl_4 (sample temperature 298 K, concentration 0.8 M). The strong structured band extending from 2400 to 3500 cm^{-1} is due to the predominant cyclic dimers forming two O–H ··· O hydrogen bonds (inset) while the narrow weak absorption band above 3500 cm^{-1} originates from the small fraction of free OH groups of monomeric and chain-like species. (b) NH stretching absorption band of the 7-azaindole dimer dissolved in CCl_4 (sample temperature 298 K, concentration 0.35 M). The narrow NH stretching band at 3480 cm^{-1} is due to monomers. Inset: dimer structure with two N–H ··· N hydrogen bonds.

to a splitting of the v = 1 states and two dipole-allowed transitions as shown in Figures 2.2b and 2.2c. For a multitude of Fermi resonances, the single OH stretching absorption line breaks up into a pattern consisting of several lines. In many cases, there exists an additional anharmonic coupling between the OH stretching mode and hydrogen-bond modes, which involve motions of the heavy atoms of the hydrogen donor and acceptor groups. The force constant of such vibrations is determined by the hydrogen bond strength and the resulting vibrational frequencies are much lower than that of the OH stretch vibrations. This fact allows for a separation of timescales and the application of an adiabatic approximation, which is equivalent to the Born–Oppenheimer treatment of vibronic transitions [19]. In this picture, the high-frequency OH stretching mode defines a potential for the low-frequency hydrogen bond mode, which is treated as a quantized oscillator (Figure 2.2b, right panel). The anharmonic coupling is manifested in an origin shift of the potentials for different quantum states v_{OH} of the high-frequency mode, giving rise to low-frequency progressions in the OH stretching absorption (Figure 2.2c). The frequency of hydrogen bond modes is close to the thermal energy $kT = 200\ cm^{-1}$ at $T = 300$ K (k is the Boltzmann constant). Thus, excited states of the hydrogen bond modes are thermally populated and give rise to progression lines at frequencies lower than the OH stretching frequency.

In general, all types of couplings are of similar strength and, thus, occur in parallel. For instance, the presence of both excitonic coupling and coupling to low-frequency modes leads to different infrared-allowed progressions in the $q_{1,u}$ and $q_{1,g}$ coordinates [21,22]. This complexity makes it very difficult to decipher the overall infrared absorption bands. Nevertheless, the scheme outlined here has been applied to analyze the linear OH stretching absorption band of cyclic carboxylic acid dimers.

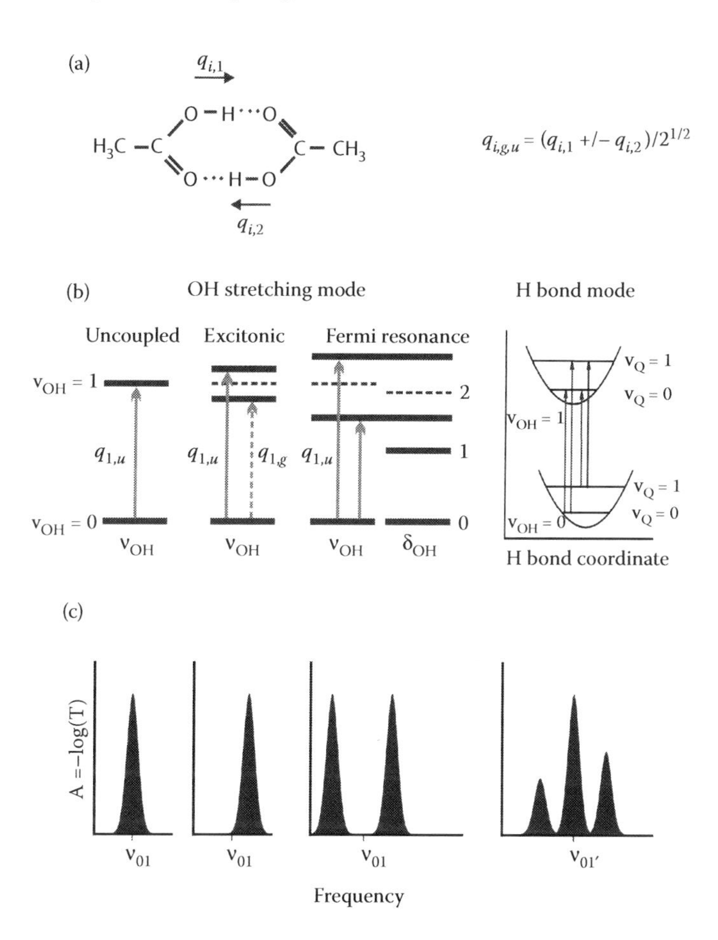

FIGURE 2.2 (a) Structure of cyclic acetic acid dimer of C_{2h} symmetry and definition of cyclic vibrational coordinates $q_{i,g,u}(q_{i,1},q_{i,2}$: local vibrational coordinates). (b) Schematic level schemes of the v = 0 to 1 OH stretching transition for different vibrational couplings: uncoupled dipole-allowed transition (solid arrow, $q_{1,u}$ coordinate), splitting of the $v_{OH} = 1$ state due to excitonic coupling and dipole-allowed transition in the $q_{1,u}$ coordinate (dashed arrow: dipole forbidden transition in the $q_{1,g}$ coordinate), transitions for a Fermi resonance of the OH stretching and bending modes (dashed lines: uncoupled v = 1/v = 2 states of the OH stretching/bending mode), and potential energy diagram illustrating the anharmonic coupling of the OH stretching and hydrogen bond mode. (c) Schematic infrared absorption lines for the uncoupled oscillator and the three coupling cases. Excitonic coupling leads to an up-shifted single line whereas Fermi resonances give rise to a pair of lines. On each of the latter, sidebands due to coupling of the hydrogen bond mode occur.

Such treatments were, however, either incomplete, that is, did not include all couplings, and/or inconclusive as different sets of coupling parameters give very similar fits to experimental spectra. Thus, a measurement of vibrational couplings by nonlinear vibrational spectroscopy combined with a theoretical analysis is mandatory for reaching a quantitative understanding.

2.2.1.2 Nucleobase Pairs

Nucleobases form hydrogen-bonded pairs that represent key building blocks of double-stranded DNA [25]. The guanine–cytosine base pair in Watson–Crick geometry (inset of Figure 2.3) is held together by two N–H ··· O and one N–H ··· N hydrogen bonds whereas the adenine–thymine base pair (inset of Figure 2.4b) displays two somewhat weaker hydrogen bonds, a N–H ··· N and a N–H ··· O bond [26]. In the gas phase and in the liquid solution, that is, without the structural constraints set by the DNA backbone, a variety of hydrogen-bonded geometries of guanine–cytosine and adenine–thymine

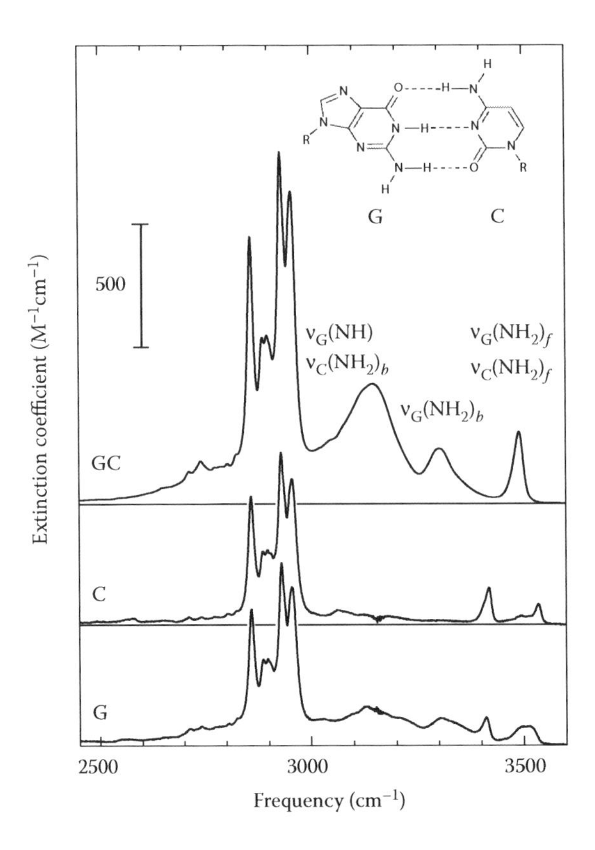

FIGURE 2.3 Infrared absorption spectra of guanine–cytosine (G–C) base pairs and C and G dissolved in $CDCl_3$ (from top to bottom). The sample concentrations were 50 mM (G–C pairs), 2 mM (C), and 2 mM (G). The bands above 3000 cm^{-1} are due to NH stretching vibrations while the strong narrow bands between 2800 and 3000 cm^{-1} are caused by CH stretching vibrations. In the GC dimer, there is a weak NH stretching component below and underneath the CH stretching bands. The assignment in the uppermost panel is explained in the text. The G and the C solutions exhibit narrow NH stretching bands of monomers as well as broad bands due to complexed G and C molecules.

pairs other than Watson–Crick pairing exist, representing an additional complication for the analysis of infrared absorption spectra [27]. However, appropriate chemical substitution of NH groups that are not involved in base pairing (N–R groups in the inset of Figure 2.3) can enforce Watson–Crick pairing of guanine–cytosine in solution. The infrared spectra of such a system consisting of 2′,3′,5′-*tert*-butyldimethylsilyl (TBDMS)-protected guanosine (G) and 3′,5′-TBDMS-protected deoxycytidine (C) [28] are presented in Figure 2.3, the uppermost panel showing the spectrum of the base pairs and the two lower panels the C and G spectra. All spectra show strong CH stretching absorption bands between 2850 and 3000 cm^{-1}, which will not be discussed in the following.

The spectra of the C monomer display the symmetric (s) and the asymmetric (as) NH_2 stretching band at 3418 and 3534 cm^{-1}, respectively. The NH_2 stretching bands of the G monomer are located at 3411 (s) and 3521 cm^{-1} (as) whereas the amino NH band is hidden under the broad absorption bands of GG dimers between 3000 and 3400 cm^{-1}. In Ref. [28], a frequency of 3340 cm^{-1} has been suggested for this vibration. Upon base pairing, the pattern of NH stretching bands changes substantially, now displaying three major components with maxima at 3145, 3303, and 3491 cm^{-1}. The two N–H · O hydrogen bonds result in a symmetry breaking at the two NH_2 groups, resulting in an effective "decoupling" into two local NH stretching modes, one of the free and the other of the hydrogen-bonded NH group. The assignments to particular NH stretching modes indicated in Figure 2.3 will be discussed in Section 2.3.2.

(a) 5′-TTA TAT ATA TAT ATA TAT ATA TT-3′
3′-AAT ATA TAT ATA TAT ATA TAT AA-5′

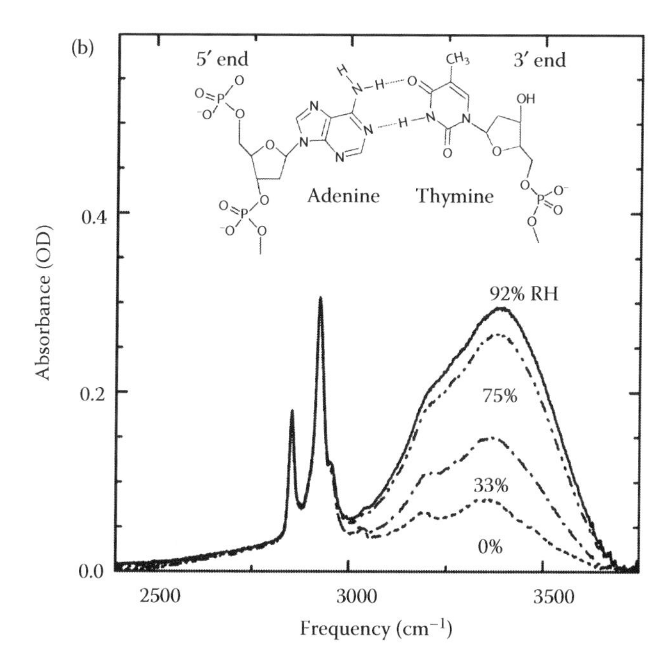

FIGURE 2.4 (a) Sequence of 23 alternating adenine–thymine (A–T) base pairs in artificial double-stranded DNA oligomers. (b) Infrared absorption spectra of a DNA thin-film sample (thickness approximately 10^{-3} cm) at hydration levels of 0% relative humidity (RH), 33% RH, 75% RH, and 92% RH. The spectrum for 0% RH displays the NH stretching bands of the A–T base pairs between 3000 and 3700 cm^{-1} whereas the narrow bands between 2600 and 3000 cm^{-1} are due to CH stretching oscillators of the DNA oligomers and the CTMA counterions. With increasing hydration level, the OH stretching vibrations of the surrounding water molecules contribute to the absorption between 3000 and 3700 cm^{-1}. Inset: molecular structure of an A–T base pair together with the sugar and phosphate groups of the backbone.

It should be noted that the spectral envelopes of the GC NH stretching bands are markedly different from that of NH stretching vibrations of the 7-azaindole dimer (Figure 2.1b), which has frequently been considered a model system for DNA base pairing [18,29]. The azaindole spectrum is centered at lower frequencies and shows a much larger spectral width and a complex substructure. The latter two features point to a more prominent role of Fermi resonances with overtone and combination tone of fingerprint vibrations, a behavior fully confirmed by recent theoretical calculations [30].

Watson–Crick pairing of nucleobases can be enforced in double-stranded DNA oligomers. Infrared spectra of DNA oligomers containing 23 alternating AT base pairs (sequence in Figure 2.4a) are shown in Figure 2.4b for different levels of DNA hydration [31]. A relative humidity (RH) of 0% corresponds to a residual water concentration of up to two water molecules per base pair whereas more than 20 water molecules per base pair are present at 92% RH corresponding to full hydration. At 0% RH, the broad infrared absorption between 3000 and 3600 cm^{-1} is mainly due to the NH stretching vibration of T and the NH_2 stretching vibrations of A. In contrast to the GC pairs in solution, there is no clear evidence for a decoupling into two local NH stretching modes of the NH_2 group. With increasing hydration, the overall band becomes stronger, now also containing a contribution of the OH stretch vibrations of the water shell. Such highly congested spectra do not allow for a separation of the NH and OH stretch contributions, an issue that will be discussed in Section 2.4.1.

In liquids and macromolecular structures at physiological temperatures, the fluctuating structure of the environment results in fluctuating forces exerted on dimers and base pairs. These forces are of a predominant electric character when the environment contains charged, that is, ionic groups and/or dipolar molecules such as water. The frequency spectrum of the fluctuating forces is determined by the vibrational, rotational/librational, and translational degrees of freedom of the environment and typically covers—in condensed phases—a very wide frequency interval. Fluctuating forces induce spectral diffusion and broadening of the vibrational transition lines [32]. Depending on the modulation strength and the fluctuation timescales, the spectra may vary between a distribution of transition frequencies corresponding to different hydrogen bond configurations (inhomogeneous broadening) or an averaged motionally narrowed transition (homogeneous broadening). Early theoretical work has treated such fluctuating system–bath interactions either by coupling the OH/NH stretching transition dipole directly to the local electric field induced by the solvent or via stochastically modulated low-frequency modes coupling to the OH/NH oscillator. An overview of the different classical and quantum approaches has been presented in Ref. [33]. More recent work on vibrational dephasing in aqueous systems has used empirical relations between the OH stretch frequency and the local electric field and calculated spectral diffusion processes with the help of classical MD simulations [4].

2.2.2 Ultrafast Vibrational Dynamics and Experimental Methods

Vibrational dynamics of neat hydrogen-bonded liquids and diluted hydrogen-bonded systems in a condensed-phase environment covers several orders of magnitude in time. The vibrational period of OH or NH stretching vibrations is of the order of 10 fs and may be considered a lower limit of the relevant range in time. Vibrational excitation of a molecular ensemble by a femtosecond infrared pulse initiates a sequence of relaxation processes through which the system eventually returns into an equilibrium state. The fact that different types of relaxation occur on similar overlapping timescales requires highly specific experimental probes to separate such processes in time and analyze the underlying microscopic interactions. Linear infrared absorption spectra reflect the different phenomena in a time-integrated way only and, in most cases, do not allow for deciphering molecular couplings in an unambiguous way.

Infrared excitation of an oscillator via a dipole-allowed transition generates a quantum-coherent superposition of the wavefunctions of the two optically coupled states, characterized by a well-defined quantum phase. In an ensemble of oscillators excited by a femtosecond infrared pulse, quantum coherence leads to a macroscopic coherent optical polarization at the infrared transition frequency. As time evolves, quantum coherence and, thus, the coherent infrared polarization decay by dephasing processes, which may originate from the interaction of the oscillators with their fluctuating liquid environment and from couplings among or relaxation processes of the oscillators themselves [4,32,33]. Vibrational dephasing and the related spectral diffusion of OH and NH stretching excitations cover a wide time interval with the fastest dynamics occurring in the sub-100 fs domain and slower components extending up to picoseconds. Their large anharmonicity makes hydrogen-bonded OH/NH stretching oscillators particularly sensitive to fluctuations of the environment [34]. Vice versa, mapping dephasing processes in a time-resolved way gives direct insight into dynamics of fluctuating structures.

Infrared excitation of a $v = 0$ to 1 transition of an oscillator populates the $v = 1$ state, which decays by population relaxation to other vibrational degrees of freedom. In general, both intramolecular anharmonic couplings between different oscillators of the molecular system and fluctuating forces of the environment, which modulate the energy positions of the coupled vibrational states, are essential for the pathways and timescale of relaxation [35]. Stretching vibrations of hydrogen-bonded OH and NH groups display $v = 1$ lifetimes, which are typically in the subpicosecond range, substantially shorter than the picosecond lifetimes of free OH and NH groups. The red-shift of OH/NH stretching frequencies upon hydrogen bonding brings the $v = 1$ state of the stretching oscillators closer to overtone and combination tone states of fingerprint vibrations, which couple to the stretching mode and through which population relaxation proceeds. In a number of systems, OH and NH bending vibrations have been identified as primary accepting modes [35,36]. In contrast to the stretching vibrations, the population decay of OH/NH bending modes and other fingerprint vibrations of hydrogen-bonded systems has been explored to a very limited extent only [37,38].

The excess energy released in any vibrational relaxation process is eventually redistributed in the manifold of low-frequency vibrations, librations, and other degrees of freedom. In the heated ensemble, structural changes such as a weakening, breaking, and/or reformation of hydrogen bonds may occur. While a selective probing of different low-frequency degrees of freedom is difficult, the overall kinetics of energy redistribution have been mapped via frequency shifts and changes of lineshapes of high-frequency modes to which low-frequency modes with a relaxation-induced excess population couple [39]. The concept of a so-called hot ground state describes the heated manifold of low-frequency vibrations by an elevated vibrational temperature, depending on the amount of excess energy initially supplied by the infrared excitation pulse [4,6,36,40]. While this picture has been widely used, the extent to which an equilibrium distribution of vibrational populations exists is unknown in most cases. The timescale on which the heated ground state is established ranges from a few up to tens of picoseconds. In liquids, cooling of the excited sample volume back to ambient temperature involves heat diffusion into unexcited parts of the sample volume, occurring in the micro- to millisecond time range.

A variety of methods of ultrafast nonlinear vibrational spectroscopy is required to map vibrational dynamics in a time-resolved way and separate the different relaxation phenomena. The results presented in this chapter are mainly based on femtosecond infrared spectroscopy in the frequency range from 1000 to 4000 cm^{-1} (wavelength range 10–2.5 μm). Both third-order pump-probe and photon-echo methods are applied with a time resolution of 50–100 fs. Heterodyne-detected photon echoes allow for deriving 2D infrared spectra of OH and NH stretching excitations. In the following, the experimental techniques are outlined briefly.

Femtosecond infrared pulses were generated by parametric frequency conversion of the output of an amplified Ti:sapphire laser working at a 1 kHz repetition rate and providing sub-100 fs pulses centered at a wavelength of 800 nm with an energy/pulse of typically 1 mJ [41]. The near-infrared component of a femtosecond white-light continuum generated by self-phase modulation in a 1-mm-thick sapphire plate is the signal seed pulse for parametric amplification in a nonlinear frequency converter. In this setup, a 4-mm-thick BBO crystal that is passed twice serves for signal amplification driven by pulses from the Ti:sapphire laser. The pulse energy is further enhanced by amplification in a subsequent 1-mm-thick BBO crystal. The amplified signal and idler pulses from the latter stage are then mixed in a 0.75-mm-thick $AgGaS_2$ or GaSe crystal to generate the difference frequency in the mid-infrared. The GaSe mixing crystal provides pulses tunable in the frequency range from 500 to 4000 cm^{-1} (wavelength range 20–2.5 μm) while a narrower range from 1000 to 4000 cm^{-1} (wavelength range 10–2.5 μm) is available from the $AgGaS_2$ crystal. Tuning is achieved by changing the frequencies of the input signal and idler pulses and adjusting the phase-matching angle for difference frequency mixing. The energy of the mid-infrared pulses is between 500 nJ and 8 μJ depending on the spectral position. In the range of the OH and NH stretching absorption, such sources in combination with a compensator of the linear chirp provide pulses of 50 fs duration and up to 8 μJ energy with extremely small intensity fluctuations of 0.2% (rms). For two-color experiments, two independent synchronized sources of this type were combined in a pump-probe setup. Third-order transient grating measurements with a spectrally resolved detection (TG-FROG) were performed for pulse characterization.

Single- and/or two-color pump-probe experiments were performed to study vibrational population dynamics and processes of energy redistribution and vibrational cooling. The samples were excited by a femtosecond pump pulse of up to 1 μJ energy and the resulting changes of vibrational absorption were measured as a function of pump-probe delay with probe pulses at least a factor of 30 weaker than the pump. After interaction with the sample, the transmitted probe pulse was spectrally dispersed and detected with a HgCdTe detector array (16, 32, or 64 elements, spectral resolution 2–6 cm^{-1}). To enhance the sensitivity of the experiments, a reference probe pulse is passed through an unexcited part of the sample and detected by a second identical array detector for normalizing the signals on a shot-to-shot basis.

The pump-probe experiments were performed with linearly polarized pulses. Measurements with parallel and perpendicular polarization of pump and probe pulses allowed for deriving the so-called isotropic (rotation-free) pump-probe signal $\Delta A_{iso} = (\Delta A_P + 2\Delta A_\perp)/3$, where $\Delta A_P = -\log(T_P(t_D)/T_{P0})\ T_P(t_D)$, T_{P0} is the sample transmission with and without excitation; t_D is the delay time] and $\Delta A_\perp$ is the absorbance change for parallel and perpendicular polarization, respectively. In addition, the pump-probe anisotropy $r(t_D) = (\Delta A_P - \Delta A_\perp)/(\Delta A_P + 2\Delta A_\perp)$ was derived from the data.

Femtosecond photon echoes were measured to study the dynamics of coherent vibrational polarizations and to determine 2D spectra of OH and NH stretching excitations. For three-pulse photon-echo studies in the widely applied boxcar geometry, three beams with wavevectors $\mathbf{k}_1$, $\mathbf{k}_2$, and $\mathbf{k}_3$ were focused onto the sample and the nonlinear signals diffracted into the directions $\mathbf{k}_3 + \mathbf{k}_2 - \mathbf{k}_1$ and $\mathbf{k}_3 - \mathbf{k}_2 + \mathbf{k}_1$ were measured as a function of both the coherence time τ, the delay between the two pulses generating a transient grating in the sample, and the population T, the delay between the second and the third pulse. Time-integrating InSb detectors were used for homodyne detection of the signals. In the heterodyne photon-echo experiments, two phase-locked pulse pairs were generated by reflection from a diffractive optic [12,42], the first two pulses ($\mathbf{k}_1$, $\mathbf{k}_2$) and the third and local oscillator pulses ($\mathbf{k}_3$, $\mathbf{k}_{LO}$), with a delay time t_{13} between them. After the diffractive optic, the coherence time $\tau = t_{12}$ is generated by delaying $\mathbf{k}_2$ with respect to pulse $\mathbf{k}_1$, and the local oscillator is delayed with respect to pulse $\mathbf{k}_3$. Locking of the relative phases of the beams was better than $\lambda/150$. The heterodyne detection scheme is based on spectral interferometry where spectral fringes are spectrally dispersed in a monochromator and detected with a HgCdTe array, giving the echo as a function of the detected frequency ν_3. The coherence time t_{12} is then scanned at constant population time T, and the signal is Fourier transformed along the $\tau = t_{12}$ dimension to produce the excitation frequency dimension ν_1. Measurements were performed with parallel linear polarizations of all four pulses (||||) and with perpendicular polarizations of pulses 3 and 4 relative to pulses 1 and 2 (||⊥⊥).

In both the pump-probe and the photon-echo experiments, the peak absorbance of the samples investigated was below A = 0.5 to avoid a distortion of the temporal pulse envelopes by propagation effects. The preparation of the different dimer and base pair samples and their characterization is described in the respective Sections 2.3.1.1, 2.3.2, and 2.4.1.

2.3 HYDROGEN-BONDED DIMERS AND NUCLEOBASE PAIRS IN SOLUTION

In this section, recent insight into ultrafast vibrational dynamics and the underlying interactions is discussed for hydrogen-bonded dimers and nucleobase pairs in the liquid phase. Prototype systems with well-defined hydrogen bond geometries were studied in aprotic and/or nonpolar solvents to exclude hydrogen bond formation with solvent molecules. On the femto- to picosecond timescales of the experiments, the dimer and base pair structures are stationary, that is, hydrogen bond breaking and other structure changes can be neglected. In contrast, the structure of the surrounding liquid undergoes stochastic structural fluctuations due to thermal excitation of low-frequency degrees of freedom.

2.3.1 OH Stretching Dynamics and Coherent Low-Frequency Motions in Carboxylic Acid Dimers

Carboxylic acid dimers represent model systems that have been studied extensively since many years and for which spectroscopic data have been reported in the gas, liquid, and solid phase. This experimental work has been complemented by theoretical calculations of vibrational spectra,

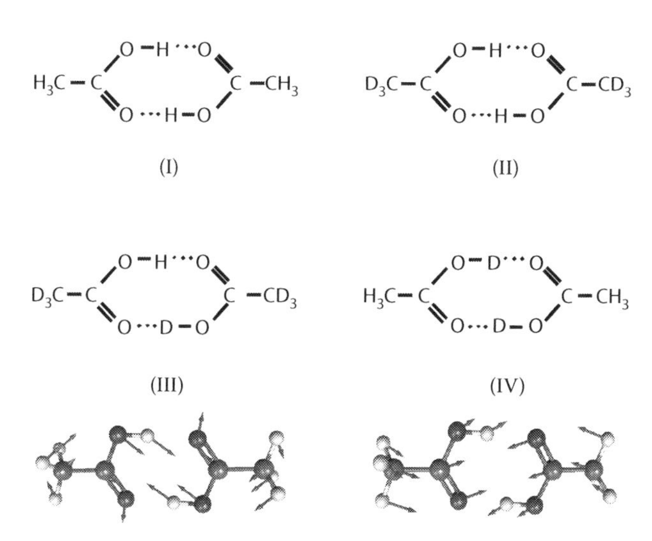

FIGURE 2.5 Structures of hydrogen-bonded cyclic dimers of acetic acid. (I) Acetic acid dimer, (II) methyl-deuterated acetic acid dimer, (III) mixed dimer containing one OH and one OD group, and (IV) dimer containing two OD groups. The bottom row shows the microscopic elongations of two Raman-active hydrogen bond vibrations, the dimer in-plane bending mode at 145 cm^{-1} (left) and the dimer in-plane stretching mode at 170 cm^{-1} (right).

mostly linear infrared absorption and Raman spectra, and by simulations of vibrational dynamics [19,20–24]. This section focuses on ultrafast vibrational dynamics of cyclic dimers of acetic acid (Figures 2.1, 2.2, and 2.5) in the aprotic solvent CCl_4. As shown in Figure 2.5, the OH and the methyl groups of acetic acid can be deuterated selectively, allowing for the preparation of dimers with two OH or OD groups as well as of so-called mixed dimers with one OH and one OD group. In the latter species, the C_{2h} symmetry is broken and, moreover, excitonic coupling between the OH and OD stretching oscillators is negligible because of the large energy mismatch of their v = 0 to 1 transitions.

2.3.1.1 Coherent Vibrational Dynamics of Cyclic Acetic Acid Dimers

Femtosecond excitation of the OH or OD stretching oscillators of cyclic acetic acid dimers generates different types of coherences. First, quantum coherences on the v = 0 to 1 transition give rise to a coherent macroscopic OH or OD stretching polarization. Second, coherences of anharmonically coupled low-frequency modes arise upon excitation with broadband pulses covering several transition lines of the low-frequency progression schematically shown in Figures 2.2b and 2.2c. The latter excitation scheme generates coherent wavepacket motions along low-frequency coordinates, which modulate the high-frequency OH stretching absorption and, thus, can be read out by mapping OH stretching dynamics. Both types of coherences were studied in a combination of femtosecond photon-echo and pump-probe experiments [43–46].

Cyclic dimers of CH_3COOH (species I, OH/OH) were prepared by dissolving acetic acid in CCl_4 with a concentration of 0.2 M. In this concentration range, the cyclic dimer represents the predominant species while monomers and chain-like aggregates play a minor role [47]. Mixed CD_3COOH/CD_3COOD (species III, OH/OD dimers) were prepared by dissolving CD_3COOH (c = 0.2 M) and CD_3COOD (c = 1.8 M) in CCl_3. The linear OH stretching absorption of the two types of dimers is shown in Figure 2.6a. The increase of absorption of the OH/OD dimers (dotted line) toward smaller frequencies is due to the onset of the OD stretching absorption band.

Three-pulse photon-echo experiments with homodyne detection of the signal give insight into the decay of the macroscopic OH stretching polarization. In Figure 2.7, the photon-echo signal from both types of dimers is plotted as a function of coherence time τ, the delay between the first and the

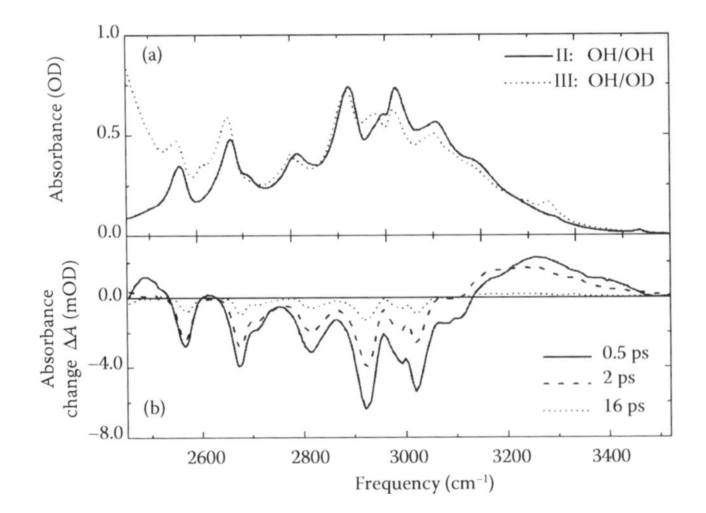

FIGURE 2.6 (a) OH stretching absorption band of cyclic acetic acid dimers containing two OH groups (solid line, species II of Figure 2.5) and one OH and OD group (dotted line, species III). The respective dimer concentration was c = 0.2 M in the aprotic solvent CCl_4. (b) Spectra of nonlinear OH stretch absorption of species II. The change of absorbance $\Delta A = -\log(T/T_0)$ is plotted as a function of probe frequency for three different delay times (T, T_0: sample transmission with and without excitation). The enhanced absorption at low frequencies is due to the v = 1 to 2 transition, while the enhanced absorption at high frequencies reflects the hot ground state formed after OH stretch relaxation. The negative absorption changes are due to ground state bleaching and stimulated emission on the v = 1 to 0 transition.

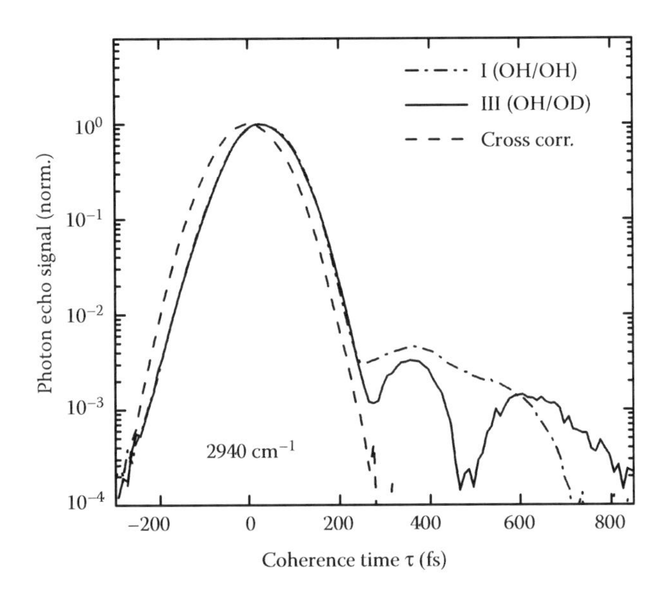

FIGURE 2.7 Three-pulse photon-echo signal from OH stretching excitations of acetic acid dimers dissolved in CCl_4 (concentration c = 0.2 M, species I: dash-dotted line, species III: solid line). The homodyne detected photon-echo signal generated with pulses centered at 2940 cm^{-1} is plotted as a function of coherence time τ. The data were recorded with population time T = 0.

second pulse. The measurements were performed with a population time T = 0 fs, that is, the third pulse from which the signal was derived overlapped in time with the second pulse. The data display a fast rise and decay of the signal over several orders of magnitude in intensity, followed by weak recurrences that are fully modulated for the mixed dimers. Measurements for different population times T show that the initial 30 fs peak shift of the photon-echo signal relative to delay zero (dashed line: cross-correlation of the pulses) decays to values below 10 fs on a 200 fs timescale, displaying weak oscillations (not shown, [43]).

The photon-echo measurements were complemented by spectrally resolved pump-probe experiments [45,46]. In Figure 2.6b, pump-probe spectra over the entire range of the OH stretching absorption of OH/OH dimers are plotted for three different delay times. The enhanced absorption at small probe frequencies is due to the v = 1 to 2 transition of the OH stretching oscillators and decays with the v = 1 lifetime of 200 fs [46]. The absorption decrease between 2500 and 3100 cm^{-1} is caused by bleaching of the v = 0 ground state and stimulated emission on the v = 1 to 0 transition. Up to the longest delay time of 16 ps, this part of the transient spectrum displays a peak structure very close to that of the linear absorption spectrum, pointing to a minor spectral diffusion of the different subcomponents. At picosecond delay times after the decay of the v = 1 population, the absorption decrease is entirely due to the persistent bleaching of the original v = 0 ground state. The enhanced absorption at high probe frequencies reflects the vibrationally hot ground state v′ = 0 of the dimers formed after the population decay of the OH stretching oscillators.

Pump-probe transients recorded at a fixed probe frequency are summarized in Figure 2.8a for OH/OH and OH/OD dimers and in Figure 2.8b for OD/OD dimers. On top of rate-like kinetics that reflect the population relaxation and energy redistribution processes, there are pronounced low-frequency oscillations that persist well into the picosecond time domain. The Fourier spectra of the oscillatory signals are shown in Figures 2.8d and 2.8e and display two major frequency components at 145 and 170 cm^{-1}. For the OD/OD dimers, a third weak frequency component around 50 cm^{-1} has been identified in pump-probe transients taken at a different spectral position [45].

Such results show that the OH stretching and the low-frequency coherences are characterized by distinctly different timescales of dephasing. The coherent polarization on the v = 0 to 1 transition of the OH stretching oscillators decays on a femtosecond timescale as is evident from the fast decay of the photon-echo signal in Figure 2.7. The absence of structural fluctuations of the dimers and their weak coupling to the aprotic nonpolar environment suggest a minor extent of spectral diffusion, a conclusion supported by the very small photon-echo peak shift and the time-independent line pattern in the pump-probe spectra (Figure 2.6b and Refs. [44,46]). This behavior points to a predominant homogeneous broadening of the individual lines, resulting in a free induction decay of the OH stretching polarization that is strongly influenced by the v = 1 population decay of the OH stretching mode. In contrast, the coherences generated in the anharmonically coupled low-frequency modes exist for much longer times and give rise to the recurrences of the photon-echo signal at late coherence times and to the oscillatory pump-probe signals, the latter extending into the picosecond time domain.

Low-frequency coherences are generated in both the v = 1 and the v = 0 state of the OH stretching oscillators, the latter via a Raman process, which is resonantly enhanced by the OH stretching transition dipole. The coherences in the v = 1 state are damped by the OH stretching population relaxation (lifetime 200 fs) while the ground state coherences exist up to a few picoseconds. Thus, the photon-echo experiments probe vibrational multilevel quantum coherences to which the v = 0 to 1 OH stretching transitions and a multitude of coherently coupled states of low-frequency modes contribute. The photon-echo signals have been analyzed by model calculations of the multilevel quantum coherence, which have been discussed in Ref. [43]. Such analysis suggests a dephasing time of the OH stretching excitations of 200 fs and two anharmonically coupled low-frequency modes at 50 and 150 cm^{-1} with a dephasing time $T_2 \geq 1$ ps.

In the pump-probe experiments where the first two interactions with the driving field occur simultaneously, one creates low-frequency coherences impulsively and reads them out by mapping

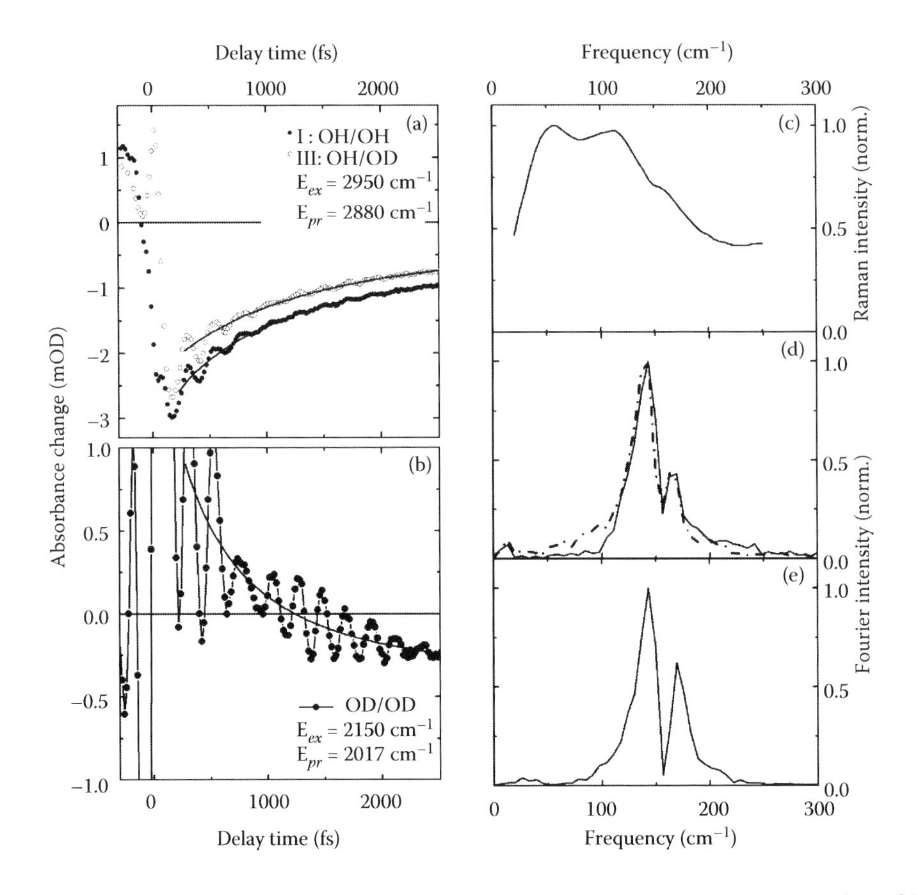

FIGURE 2.8 Time-resolved nonlinear OH stretching absorption of the acetic acid dimer species I (OH/OH) and III (OH/OD) after excitation by pulses centered at E_{ex} = 2950 cm^{-1}. The spectrally resolved absorbance change at E_{pr} = 2880 cm^{-1} is plotted as a function of pump-probe delay and displays low-frequency oscillations superimposed on a rate-like relaxation kinetics. (b) Nonlinear OD stretching absorption of species IV (OD/OD) measured with the pump and probe energies indicated. (c) Spontaneous Raman spectrum of low-frequency modes of acetic acid (taken from Ref. [48]). (d) Fourier spectra of the oscillatory component of the pump-probe signals in (a) (solid line: species I (OH/OH), dash-dotted line: species III (OH/OD)). (e) Fourier spectrum of the oscillatory component of the transient in (b).

the modulation of the OH stretching absorption with the time-delayed probe pulse. The Fourier spectra of the oscillatory pump-probe signals (Figures 2.8d and 2.8e) give more specific insight into the relevant low-frequency modes and allow—in combination with a vibrational mode analysis and *ab initio* calculations of vibrational couplings—an assignment of the relevant degrees of freedom. The Fourier spectra display two major components at 145 and 170 cm^{-1} and a much weaker component at 50 cm^{-1} (not shown). According to the selection rules for infrared transitions changing both the quantum numbers of the OH stretching and the low-frequency modes, only Raman-active low-frequency modes couple to the v = 0 to 1 OH stretching transition. The calculations that have been discussed in detail in Ref. [46] give four Raman-active modes of the dimers below 200 cm^{-1}, a CH_3/CD_3 torsion at 44/33 cm^{-1}, a dimer out-of-plane wagging mode at 118 cm^{-1}, the in-plane dimer bending mode at approximately 150 cm^{-1}, and the in-plane dimer stretching mode around 170 cm^{-1}. The calculated microscopic elongation patterns of the last two modes are shown in the bottom row of Figure 2.5. The generation of low-frequency wavepackets via high-frequency OH stretching excitation requires an anharmonic coupling between high- and low-frequency mode. The calculations give strong cubic couplings with an absolute value of approximately 100 cm^{-1} for the in-plane dimer bending and stretching modes while the coupling of the methyl torsion of 2 cm^{-1} is much smaller.

The coupling of the out-of-plane wagging to the OH stretching mode is negligible. Thus, the two prominent components in the Fourier spectra of Figures 2.8d and 2.8e are due to coherent motions in the in-plane dimer mode at 145 cm^{-1} and the in-plane stretching mode at 170 cm^{-1}, both modulating the dimer geometry. It is interesting to note that the spontaneous Raman spectrum of acetic acid (Figure 2.8c, Ref. [48]) does not allow for an unambiguous assignment of low-frequency modes, mainly because of the congested lineshape and the inherently large experimental uncertainty at low frequencies where the Rayleigh scattering wing has been subtracted. In contrast, the nonlinear femtosecond pump-probe experiments give the frequencies of the different dimer vibrations with high precision due to the underdamped character of the low-frequency dimer motions.

In summary, different types of quantum coherences occur after femtosecond OH stretching excitation of hydrogen-bonded carboxylic acid dimers due to the inherent multilevel character of anharmonically coupled high- and low-frequency vibrations. The results presented here give evidence for the pronounced anharmonic coupling of hydrogen donor stretching and low-frequency hydrogen bond vibrations, one of the basic coupling mechanisms contributing to the complex lineshapes of the infrared absorption spectra of hydrogen donor stretching modes (cf. Section 2.2). Similar cubic coupling strengths of the order of 100 cm^{-1} are found for OH/OH, OD/OD, and OH/OD dimers. The different timescales of dephasing of the high-frequency stretching modes and low-frequency hydrogen bond modes are a behavior found in a larger class of hydrogen-bonded systems with small structural fluctuations. In particular, 7-azaindole dimers [49] and intramolecular hydrogen bonds show pronounced low-frequency coherences [50].

2.3.1.2 Two-Dimensional Infrared Spectra of OH Stretching Excitations

The results presented in the previous section establish the interaction of OH stretching and low-frequency modes as one of the relevant coupling mechanisms. To get insight in the role of the other couplings introduced in Section 2.2 and to eventually understand the lineshape of the linear infrared absorption spectrum, 2D infrared spectra were derived from heterodyne detected three-pulse photon-echo measurements [44,51] and analyzed by a combination of *ab initio* density functional theory calculations and density matrix theory of the nonlinear response. In the following, OH stretching 2D spectra of OH/OH dimers of acetic acid are discussed.

In Figures 2.9b and 2.9c, 2D spectra are shown for population times of $T = 0$ and 400 fs. The real part of the nonlinear signal is plotted as a function of excitation and detection frequency [44]. The spectrum recorded at $T = 0$ fs exhibits positive amplitudes in the range of the $v = 0$ to 1 transitions above a detection frequency $\nu_3 = 2900$ cm^{-1} and negative amplitudes below. The negative components are caused by the partly overlapping $v = 1$ to 2 transitions that are anharmonically red-shifted. The latter contribution is absent in the 2D spectrum taken at a population time of $T = 400$ fs, due to the decay of the $v = 1$ state with a characteristic lifetime of 200 fs. In the central positive part of the spectra, strong diagonal peaks occur at 2920 and 2990 cm^{-1} together with their (off-diagonal) cross peaks. In addition, there is a variety of off-diagonal peaks over a broad range of detection frequencies, as is evident from cross sections through the spectra for constant excitation frequency. Cross section taken for $\nu_1 = 2920$ cm^{-1} along the detection frequency axis is shown in Figure 2.9a and display peaks at spectral positions very similar to those of the infrared absorption spectrum (thick solid line). The 2D spectra of mixed OH/OD dimers (not shown) exhibit a very similar peak pattern, pointing to a negligible role of excitonic OH/OH coupling. The positions and spectral shapes of both diagonal and off-diagonal peaks remain constant for longer population times T, suggesting a minor relevance of spectral diffusion processes on this subpicosecond timescale. This conclusion is in line with the pump-probe spectra discussed earlier.

The peaks located on the diagonal of the 2D spectra, that is, at identical excitation and detection frequencies, are due to interaction of the three pulses with a single transition and reflect the coarse structure of the linear absorption spectrum (Figure 2.9a). Off-diagonal peaks occur whenever excitation of a particular transition causes a signal on a transition at a different (detection) frequency. The off-diagonal peaks below $\nu_3 = 2900$ cm^{-1} originate from $v = 1$ to 2 transitions of

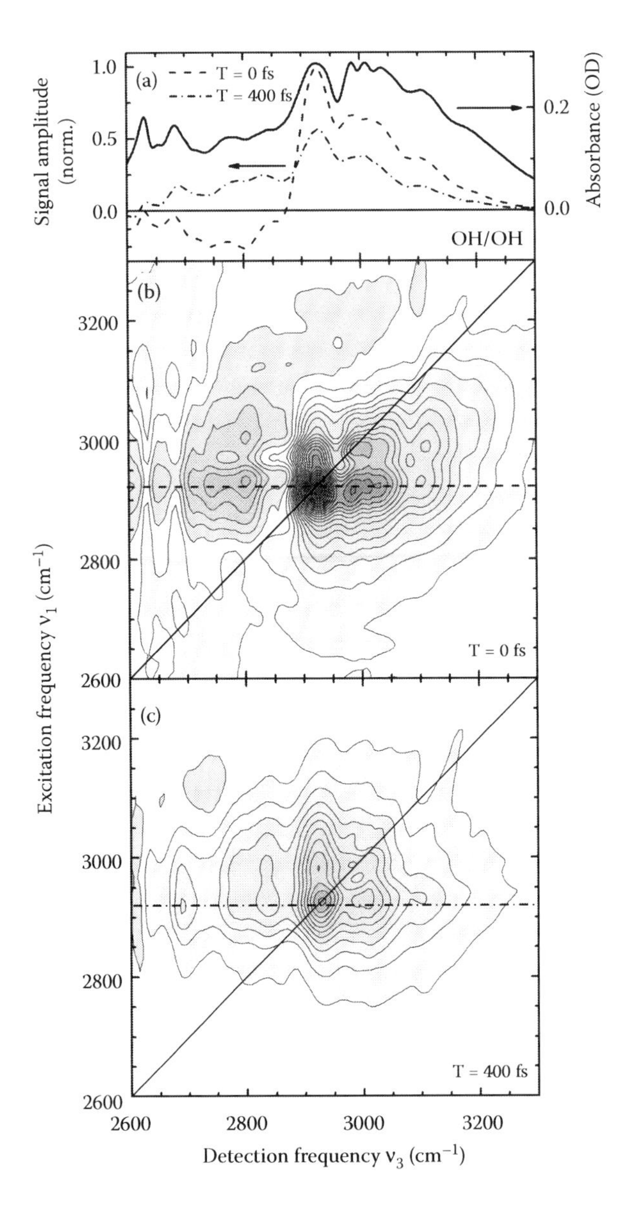

FIGURE 2.9 (a) OH stretching absorption band (thick solid line) and cross sections of the 2D spectra at an excitation frequency of 2920 cm^{-1} (dashed line: population time T = 0 fs, dash-dotted line: T = 400 fs) of the acetic acid dimer structure I (OH/OH) dissolved in CCl_4 (b). (b) 2D spectrum of acetic acid dimers (species I, OH/OH) measured at a population time of T = 0 fs. The real part of the signal is plotted as a function of excitation and detection frequency. The intensity change between neighboring contour lines is 5%. The cross section in (a) was taken along the dash-dotted line. (c) 2D spectrum at T = 400 fs.

the O–H stretching oscillators and will not be considered in the following. Fermi resonances of the v = 1 state of the OH stretching mode with combination tones of fingerprint modes cause cross peaks in the positive part of the spectrum (at $\nu_3 > 2900$ cm^{-1}), in parallel to cross peaks potentially arising from the coupling to low-frequency modes.

The highly complex 2D spectra were analyzed by theoretical calculations described in detail in Refs. [51,52]. In brief, density functional theory was applied to calculate vibrational eigenstates from a sixth-order anharmonic force field covering up to three-body interactions. Couplings of the high-frequency OH stretching normal mode to eight fingerprint normal modes, one half each being infrared- and Raman-active, and two Raman-active low-frequency modes result in an 11-dimensional vibrational Hamiltonian. The fingerprint vibrations are the C–O stretching, CH_3 wagging, O–H bending, and C=O stretching normal modes. The low-frequency normal modes are the in-plane bending and stretching hydrogen bond modes shown in Figure 2.5. 2D photon-echo signals in the phase matching direction $\mathbf{k}_{echo} = -\mathbf{k}_1 + \mathbf{k}_2 + \mathbf{k}_3$ were calculated with parallel linear polarization applying a sum-over-states formalism. For the calculation of 2D spectra, the three incident laser pulses were tuned to 2900 cm^{-1} assuming a rectangular electric field spectrum of ±400 cm^{-1} width for selecting resonant transitions. A homogeneous linewidth of $\Delta\nu = 36$ cm^{-1} was used for the individual subcomponents of the OH stretching band.

The theoretical 2D spectrum shown in Figure 2.10b was calculated from the ground-state Liouville space pathways only, that is, includes the nonlinear response function R_3 only [51]. This scenario corresponds to experimental 2D spectra recorded after the decay of the v = 1 states, such as the spectrum taken at a population time T = 400 fs (Figure 2.9c). The main features of the latter spectrum are well reproduced by the theoretical 2D spectrum. The calculation demonstrates that the main diagonal and cross peaks originate from Fermi resonances of the v = 1 OH stretching state (symmetry b_u) with combination tones made up of one quantum of an infrared-active fingerprint mode (symmetry b_u) and one quantum of a Raman-active mode (symmetry a_g). The strongest diagonal peak at 2920 cm^{-1} is due to the coupling of the OH stretching mode with a combination band of the Raman-active C–O and the infrared-active C=O stretching modes with a cubic coupling strength $\phi = -86$ cm^{-1}. This peak displays the largest relative amplitude of the OH stretching component. Other strong diagonal peaks and cross peaks between them occur at 2993 cm^{-1} (ν_{ag}C–O/ν_{bu}C=O, $\phi = 48$ cm^{-1}) and at 3022 cm^{-1} (γ_{bu} CH_3/ν_{ag}C=O, $\phi = 62$ cm^{-1}) [51]. In addition, prominent cross peaks exist for bands in the low-energy part of the spectrum, that is for the peak at 2555 cm^{-1}

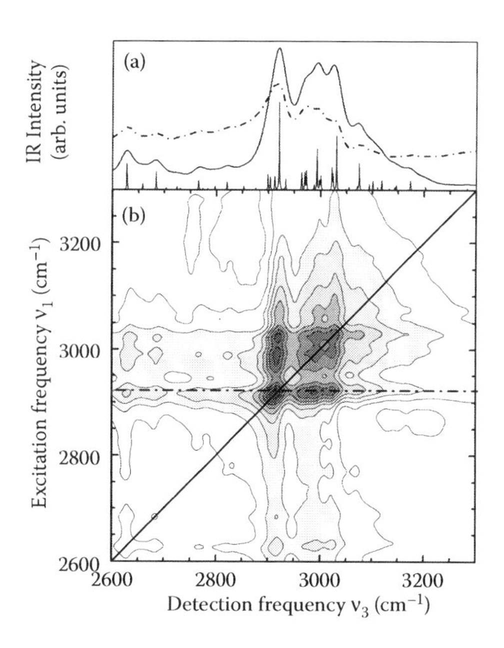

FIGURE 2.10 (a) Theoretical linear absorption spectra of the acetic acid dimer (species I, OH/OH) calculated with a homogeneous linewidth (FWHM) of $\Delta\nu = 1$ cm^{-1} (stick spectrum) and 36 cm^{-1} (solid line). The dash-dotted line shows a cross section through the calculated 2D spectrum of (b) at an excitation frequency $\nu_1 = 2920$ cm^{-1}. (b) Calculated OH stretching 2D spectrum (population time T = 400 fs) including v = 0 to 1 Liouville space pathways after the decay of the v = 1 population. The amplitude change between neighboring contour lines is 5%.

(ν_{ag} C–O/ν_{bu}C–O, $\phi = 150$ cm^{-1}, not shown), at 2627 cm^{-1} (ν_{ag} C–O/γ_{bu} CH_3, $\phi = -118$ cm^{-1}) and at 2684 cm^{-1} (ν_{ag} C–O/δ_{bu}OH, $\phi = -126$ cm^{-1}). A full account of such results has been presented in Ref. [51]. Compared to the prominent peaks caused by Fermi resonances, the peaks originating from the coupling to low-frequency hydrogen bond modes are much weaker and are not clearly discerned in the 2D spectra. This is mainly due to their substantially smaller transition dipoles.

The calculated molecular coupling strengths validated by the experimental 2D spectra represent an important input for simulating the linear OH stretching absorption spectrum of acetic acid dimers. Results of such simulations are shown in Figure 2.10a for different homogeneous line widths of the individual transitions [52]. Similar to the 2D spectrum, the main peaks of the spectral envelope are due to Fermi resonances. There are progressions in low-frequency modes starting at each of those peaks and contributing to the absorption in-between the main components. The comparably strong homogeneous broadening in the condensed phase and the small transition dipoles prevent that the individual progression lines occur as separate features in the linear spectrum. An in-depth comparison of the calculated spectra with gas phase data shows a quantitative agreement between theory and experiment [52], in contrast to earlier simplified calculations with an incomplete treatment of the different couplings.

In summary, 2D infrared spectra of OH stretching excitations of cyclic acetic acid dimers allow for deciphering the molecular couplings to combination tones of fingerprint modes and represent the key input for a quantitative theoretical modeling. Together with the photon-echo and pump-probe results discussed in Section 2.3.1.1, they demonstrate that the highly complex lineshapes of the 2D and the linear infrared spectra are determined by the interplay of different types of coupling mechanisms of similar strength. This work has clarified longstanding issues and controversies in the literature and generated a quantitative understanding of a coupling scheme that is expected to occur in other carboxylic acid dimers and related systems.

2.3.2 NH Stretching Excitations in Nucleobase Pairs in Solution

Hydrogen-bonded nucleobase pairs represent a key structural feature of double-stranded DNA helices that contain sequences of adenine–thymine and guanine–cytosine pairs. In B-DNA, the DNA structure prevalent under physiological conditions, the base pairs are arranged in the so-called Watson–Crick pairing geometries shown in the insets of Figures 2.3 and 2.4b [25]. Without the steric constraints set by the DNA backbone, nucleobases can form a variety of pairing geometries different from Watson–Crick, in both the gas and condensed phase. However, addition of appropriate side groups to the respective nucleobase structure is a method to enforce Watson–Crick pairing in individual base pairs in a liquid environment. This approach is complementary to studying base pairs in DNA helix structures where—as additional effects—inter-base pair couplings and changes of vibrational dynamics as well as energy dissipation pathways may arise. Liquid-phase studies of nucleobase pairs can also be performed in nonaqueous environments. In this section, results for TBDMS-protected guanosine–cytidine (G–C) base pairs [28] diluted in a $CHCl_3$ solution are presented.

The infrared absorption spectra of the G and C monomers and the G–C base pairs are shown in Figure 2.3. While the G and C monomers display symmetric and asymmetric NH_2 stretching bands (and G on top the amino NH stretching band), the NH stretching spectrum of the base pair exhibits a different pattern of lines above 3000 cm^{-1} and an additional shoulder between 2600 and 2900 cm^{-1}. An isolated narrow band occurs at 3491 cm^{-1} with a line width of some 30 cm^{-1}. Red-shifted bands of a much larger spectral width show maxima at 3303 and 3145 cm^{-1}. They are due to hydrogen-bonded NH groups in the G–C pairs.

A series of 2D infrared spectra recorded at different population times T is summarized in Figure 2.11 [53]. The absorptive 2D signal, that is, the sum of the rephasing and nonrephasing signal [54] is plotted as a function of excitation and detection frequency. Prominent diagonal peaks occur at the frequency positions $(\nu_1,\nu_3) = (3145,3145)$, (3303,3303), and (3491,3491) cm^{-1}. The peaks at (3145,3145) and (3491,3491) cm^{-1} display a more or less round shape while the peak at (3303,3303)

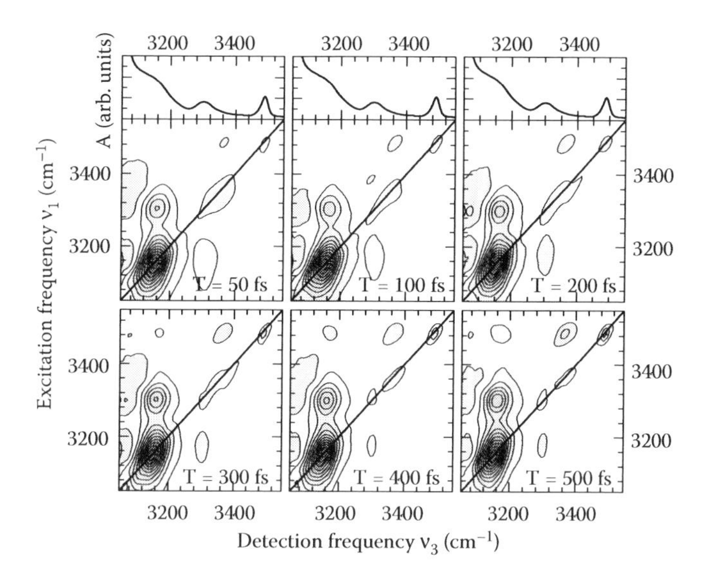

FIGURE 2.11 2D infrared spectra of NH stretching excitations of guanosine–cytidine base pairs in $CHCl_3$ solution (concentration c = 0.1 M). The absorptive 2D signal, the sum of the rephasing and nonrephasing signal, is plotted as a function of the excitation frequency ν_1 and the detection frequency ν_3. The contour lines were scaled for the maximum signal strength for each value of the population T. The intensity change between neighboring contour lines is 5%. The top row of panels shows the linear infrared absorption spectrum of the base pairs.

cm^{-1} is elongated along the diagonal. The spectral shapes of the diagonal peaks undergo minor changes as a function of population time T. This fact is also evident from cross sections of the 2D spectra along the diagonal, which are presented in Figure 2.12a. Cross peaks of positive sign occur at (3145,3303), (3303,3145), and (3491,3145) cm^{-1}. The negative cross peak at (3491,3360) cm^{-1} is due to the v = 1 to 2 transition of the oscillators having their v = 0 to 1 transition at (3491,3491) cm^{-1}.

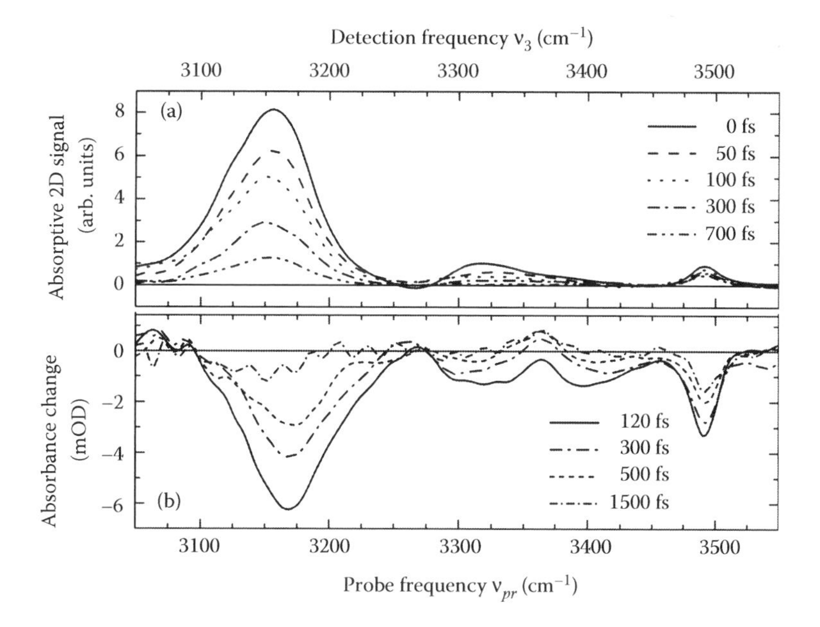

FIGURE 2.12 (a) Cross sections of the 2D spectra of Figure 2.11 along the diagonal $\nu_1 = \nu_3$ for different population times T. (b) Spectrally resolved pump-probe data for different pump-probe delays.

From the ν_3 position of the two peaks, one estimates a diagonal anharmonicity of approximately 130 cm^{-1}. With increasing population time T, the (3303,3145) cm^{-1} cross peak increases in strength relative to the diagonal peak at (3303,3303) cm^{-1}. The negative signals below a detection frequency $\nu_3 \approx 3100$ cm^{-1} are due to the v = 1 to 2 transitions of the different NH stretching oscillators and will not be analyzed in the following.

The intensity of the different diagonal and cross peaks was spectrally integrated to get insight into their time evolution. In Figure 2.13, such intensities are plotted as a function of population time T for (a) the three diagonal peaks and (b) selected cross peaks (symbols). The solid lines are mono-exponential fits to the intensity decays with the time constants given in Figure 2.13. The intensity ratio of the cross peak at (3303,3145) cm^{-1} and the diagonal peak at (3303,3303) cm^{-1} plotted as a function of population time T in Figure 2.13c rises with increasing T.

In addition to the 2D spectra of the GC pairs, pump-probe data were collected. Transient absorption spectra measured for different pump-probe delays are shown in Figure 2.12b. They display a pronounced decrease of absorption with maxima at 3140, 3305, and 3490 cm^{-1}, close to the maxima of

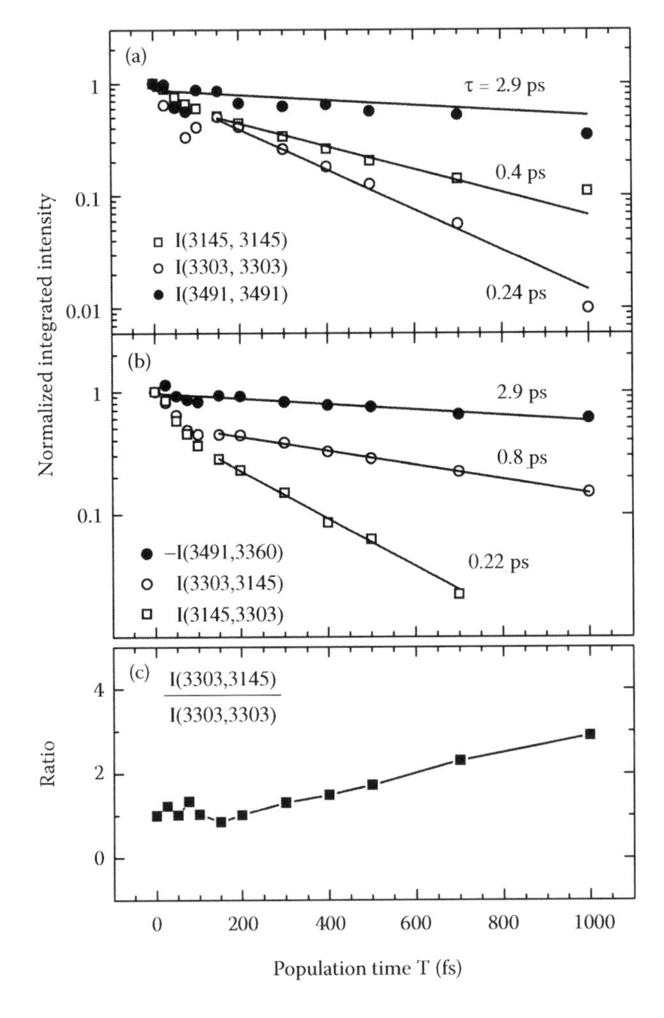

FIGURE 2.13 Spectrally integrated signal strengths of (a) 2D diagonal peaks at the frequency positions (ν_1, ν_3) indicated and (b) cross peaks as a function of population time T. The solid lines are single exponential fits with the respective time constants. (c) Intensity ratio between the cross peak at (3303,3145) cm^{-1} and the diagonal peak at (3303,3303) cm^{-1}. The solid line is to guide the eye.

the 2D signal in Figure 2.12a. Between such peaks, the pump-probe spectra show a slightly positive absorption change at 3360 cm^{-1}, which is absent in the diagonal 2D cross sections of Figure 2.12a. The pump-probe signal is equivalent to the 2D signal integrated over all excitation frequencies ν_1 and, thus, the cross peak at (3491,3360) cm^{-1} in the 2D spectra contributes to the pump-probe signal at $\nu_{pr} = 3360$ cm^{-1}. This component dominates in the integral and leads to the enhanced absorption at this spectral position. Time-resolved pump-probe transients were recorded at a variety of probe frequencies and give the following vibrational lifetimes of the different NH stretching contributions: $\nu_{pr} = 3145$ cm^{-1}: $\tau = 0.3(-0.1/+0.2)$ ps, and $\nu_{pr} = 3491$ cm^{-1}: $\tau = 2.9 \pm 0.7$ ps. The pump-probe kinetics measured around $\nu_{pr} = 3303$ cm^{-1} are highly complex and do not allow for the reliable extraction of a lifetime.

The base pairs contain a total of five NH units of which three are part of intermolecular hydrogen bonds and two are free (inset of Figure 2.3). On the two NH_2 groups, there is a mechanical coupling between the two NH stretching oscillators that can be estimated from the splitting $2|V|$ of the symmetric and asymmetric NH_2 stretching mode in the G and C monomers. The resulting values are $|V| = 58$ cm^{-1} for G and 55 cm^{-1} for C. Theoretical calculations and a comparison of the present system to other base pairs suggest that any couplings due to Coulomb and/or (resonant) dipole interactions between the five NH groups are small compared to the mechanical couplings. Upon formation of the GC pair, the NH stretching line pattern changes strongly compared to the monomers. Both the infrared absorption spectrum and the 2D spectra of the base pairs exhibit a contribution at $\nu_3 = 3491$ cm^{-1} with a comparably long vibrational lifetime of 2.9 ps. Relative to this band, the two other prominent 2D features at 3145 and 3303 cm^{-1} display a shift substantially larger than $2|V|$, the splitting between the symmetric and asymmetric NH_2 stretching bands of the G and C monomers. This fact is due to the strong detuning of the two local NH oscillators on the NH_2 groups by integrating one of them in a hydrogen bond. As a result, a picture of local NH modes of the free and the hydrogen-bonded NH stretching oscillators is appropriate. In this description, the band at $\nu_3 = 3491$ cm^{-1} is assigned to the fundamental stretching transition at $\nu_G(NH_2)_f = \nu_C(NH_2)_f$ of the free non-hydrogen-bonded NH groups of G and C.

Theoretical DFT calculations suggest that the three hydrogen bonds of the GC base pair have a slightly different strength, resulting in an N ⋯ O distance for the strongest $C(NH_2) \cdots G(O{=}C)$ bond of 0.273 nm, 0.287 nm for the $G(NH_2) \cdots C(O{=}C)$ bond, and 0.288 nm for the $G(NH) \cdots C(N)$ bond [55,56]. Such differences lead to different spectral positions of the NH stretching modes in the three hydrogen bonds. In line with the calculated hydrogen bond strengths of Ref. [55] and calculated patterns of NH stretching transition frequencies in isolated GC base pairs [56], we assign the NH stretching mode $\nu_G(NH_2)_b$ in the $G(NH_2) \cdots C(O{=}C)$ hydrogen bond to the absorption band at 3303 cm^{-1} and the NH stretching mode $\nu_C(NH_2)_b$ in the $C(NH_2) \cdots G(O{=}C)$ hydrogen bond to the band with a maximum at 3145 cm^{-1}. This band contains also the amine NH stretching mode $\nu_G(NH_2)_b$ in the $G(NH) \cdots C(N)$ bond. A separation of the latter two modes in the absorption and the 2D spectra requires an in-depth theoretical analysis of the different lineshapes, which has not been performed so far. Moreover, the NH stretching absorption band with maximum at 3145 cm^{-1} may extend well into the range of the CH stretching absorption, without a clear separation from the shoulder between 2600 and 2900 cm^{-1}. In this wide spectral range, one expects a mixing of the $v = 1$ states of the NH stretching modes with combination tones of fingerprint vibrations, similar to what has been observed in azaindole dimers (cf. Figure 2.1b, [30]) and—for OH stretching modes—in the acetic acid dimers.

In the 2D spectra, the shapes of the diagonal peaks do not change significantly for population times up to $T = 1$ ps, pointing to a limited role of spectral diffusion. Thus, the decay of the integrated intensity of the diagonal peaks (Figure 2.13a) can safely be assigned to the $v = 1$ population relaxation of the different NH stretching modes. Exponential fitting of the decay of the (3303,3303) cm^{-1} peak results in a 0.24 (−0.06/+0.08) ps time constant, while the (3145,3145) cm^{-1} diagonal peak shows a somewhat slower decay with a 0.4 (−0.10/+0.20) ps time constant. Within the experimental accuracy, this time constant agrees with the value of 0.3 ps derived from pump-probe data (not shown). Such decay times are much shorter than the 2.9 ps lifetime of the free NH stretching vibrations observed in the 2D peaks at (3491,3491) and (3491,3360) cm^{-1} and the pump-probe experiments. The lifetime shortening originates from enhanced anharmonic couplings with combination/overtone levels of fingerprint vibrations, often involving the NH bending mode, and a reduced energy mismatch between these combination/overtone levels and the down-shifted $v = 1$ states of the N–H stretching vibrations.

A comparison of the intensities of the diagonal and cross peaks shows clear differences in their time evolution as a function of T (Figures 2.13a and 2.13b). This fact demonstrates that there is excitation transfer between the different hydrogen-bonded NH stretching oscillators. The ratio of the intensities of the (3303,3145) cm^{-1} cross peak and the (3303,3303) cm^{-1} diagonal peak (Figure 2.13c) displays a pronounced increase with population time T, giving evidence of vibrational population transfer from the $\nu_G(NH_2)_b$ mode at 3303 cm^{-1} to the 3145 cm^{-1} modes. As a result, the measured decay rate of the 3303 cm^{-1} mode of $1/\tau' = (1/0.24)\ ps^{-1}$ represents the sum of the population transfer rate and the relaxation rate back to the $v = 0$ state. The ratio plotted in Figure 2.13c rises on a timescale of approximately 1 ps. This rise is determined by the population transfer and the relaxation of the 3303 and 3145 cm^{-1} modes, the latter decaying with a lifetime of 0.4 ps. An extraction of the transfer rate $1/\tau_{tr} = 1/\tau' - 1/\tau_1$ requires knowledge of the relaxation rate $1/\tau_1$ of the 3303 cm^{-1} mode, which is not available because of the strong spectral overlap of the different $v = 0$ to 1 and 1 to 2 transitions. Thus, one can derive only a lower limit of the transfer time of 0.24 ps and an upper limit of approximately 1.3 ps. A simple Fermi golden rule approach to estimate the corresponding vibrational coupling V gives values of $|V| = 2.3$–$12\ cm^{-1}$, much smaller than the mechanical coupling of the NH oscillators of the NH_2 groups. If through-space dipole–dipole coupling represents the predominant interaction mechanism, a transfer to the $\nu_G(NH)_b$ rather than the $\nu_C(NH_2)_b$ mode is favored because of the smaller spatial distance between the G(NH) and G(NH_2) groups.

In summary, the ultrafast NH stretching response of guanosine and cytidine bases in solution reveals a "decoupling" of the symmetric and asymmetric NH_2 stretching modes of G and C into local NH stretching vibrations upon formation of a base pair in Watson–Crick geometry. This effect is due to a detuning of the stretching vibrations of the hydrogen-bonded NH groups from those of the free NH groups, which is larger than the mechanical coupling of the two local oscillators. The hydrogen-bonded NH stretching oscillators, including the amine NH stretching mode of G, display a sub-500 fs lifetime of their $v = 1$ states while a 2.9 ps lifetime is found for the free NH stretching oscillators. The 2D spectra give evidence of a (sub)picosecond energy transfer from the hydrogen-bonded NH stretching oscillator on the NH_2 group of G to the other hydrogen-bonded oscillators at lower frequencies. The underlying coupling is substantially smaller than the mechanical coupling on the NH_2 groups. The infrared absorption spectrum of the GC base pairs shows a pronounced contribution between 2600 and 2900 cm^{-1}, well below the lowest NH stretching peak at 3145 cm^{-1}. This behavior points to a complex lineshape of the lowest NH stretching band, which may be influenced by Fermi resonances of the NH stretching modes with combination tones of fingerprint vibrations. More detailed theoretical work and 2D spectra in an extended spectral range are required to address this issue.

2.4 BASE PAIRS IN ARTIFICIAL DNA HELICES

In native and artificial DNA helices, a large number of functional units of the backbone and the base pairs contribute to the linear vibrational spectra that cover an extremely broad range from approximately 1 cm^{-1} for low-frequency modes delocalized in the macromolecular structure up to local NH stretching frequencies in the range around 3500 cm^{-1} [57–60]. Additional complexity arises from the interaction of DNA with an aqueous environment, the so-called hydration shell. The degree of hydration has a direct influence on the overall helix geometry and on particular local interactions between functional groups of DNA and water molecules [2]. Steady-state infrared and Raman spectroscopy has aimed at identifying the characteristic vibrational signatures of particular helix and hydration geometries and at unraveling couplings between different units of DNA such as neighboring base pairs and others. Substantial qualitative information and empirical rules have been derived from those spectra and time-averaged pictures of hydrated DNA structures have been developed [57]. In most cases, however, specific quantitative insight into couplings at the molecular level and into dynamics on the ultrafast timescale of molecular motions is lacking completely. So far, ultrafast vibrational spectroscopy has focused on the behavior of isolated bases and base pairs in solution ([53,61,62], cf. Section 2.3.2) and on vibrational couplings between C=O stretching, CN stretching, and CC ring modes of GC pairs in small model oligomers [63,64].

The surface of a DNA helix displays the so-called minor and major grooves of different width and depth [2]. At this surface, water molecules of the hydration shell interact with specific functional groups in the grooves such as the phosphate groups of the backbone and NH and C=O groups of bases [65–69]. The variety of local molecular geometries makes the first water layer highly heterogeneous from a structural point of view and distinctly different from the more distant water layers that are expected to be more similar to bulk water. The modification of water structure should affect the ultrafast dynamics of the water shell, that is, structural fluctuations, translational and orientational motions, vibrational relaxation, solvation processes, and eventually the hydrogen bond lifetimes. Numerous theoretical studies, in particular molecular dynamics simulations, have been performed to address hydration shell geometries and their femto- to picosecond dynamics [70–74]. At the DNA/water interface, average hydrogen bond lifetimes of approximately 10 ps, substantially longer than the few picoseconds in bulk water [7], and residence times of water molecules between 30 ps and several hundreds of picoseconds have been calculated, the long residence times being in qualitative agreement with NMR work. In addition to molecular trajectories, correlation functions of water dipoles and of hydrogen bond lifetimes between water and DNA have been generated, giving multiexponential decays with time constants between 0.5 and 200 ps [74]. Dipole correlations and hydrogen bonds in the minor groove are longer-lived than in the major groove while water–phosphate hydrogen bonds display an even shorter lifetime.

So far, femtosecond spectroscopy has addressed hydration shell dynamics of DNA and other macromolecules to a very limited extent only. In a first approach, electronic dipole excitations of a chromphore attached to or incorporated into a DNA helix structure are generated by femtosecond optical excitation and the induced reorientation of the polar surrounding, the dipole solvation process, is mapped by measuring the time-dependent Stokes shift of fluorescence from the chromophore [75–77]. Two-point frequency–time correlation functions (TCFs) have been derived from the transient Stokes shift. TCFs of different DNA/chromophore systems decay on a slower timescale than TCFs of the same chromophore in water and, in the DNA case, a highly complex multiexponential behavior of DNA has been observed, which covers orders of magnitude in time. Because of the long-range character of the underlying electrostatic (dipole) interactions, such TCFs represent the combined dynamics of the different constituents of the system, that is, the DNA structure, in particular its charged groups, the counterions of DNA, and the water shell. Nevertheless, the solvation results have been considered evidence for a slowing down of water reorientation dynamics at the surface of DNA and proteins. This conclusion has been challenged by recent NMR studies of proteins [77,78]. It should be noted that attaching an organic chromophore to DNA changes the structure of both the DNA helix and the hydration shell. An analysis of such geometries and their influence on hydration dynamics has been presented recently [79].

This section focuses on results from an extensive 2D infrared study of DNA oligomers in a wide range of hydration levels. As a model system, artificial double-stranded oligomers containing 23 alternating adenine–thymine base pairs are prepared in thin film samples the hydration of which can be varied in a controlled way. The frequency positions, couplings, and relaxation dynamics of the different NH stretching excitations of the base pairs are first analyzed at low hydration level. Experiments at higher hydration levels up to fully hydrated DNA oligomers allow for a separation of DNA and water shell dynamics, the latter being mapped via OH stretching excitations. In this way, a selective analysis of the dynamics of the water shell becomes possible. Time-dependent center line slopes [80] are derived from the 2D spectra at high hydration level and compared to calculated TCFs of bulk H_2O.

2.4.1 NH Stretching Excitations of Adenine–Thymine Base Pairs

The DNA oligomers studied here consist of a 5′-T(TA)$_{10}$-TT-3′ (A: adenine, T: thymine) strand and its complement (Figure 2.4a). To generate DNA films of high optical quality, the sodium counterions are replaced by the surfactant cetyltrimethylammonium (CTMA), forming complexes with DNA. DNA film samples of 5–30 μm thickness were prepared by a procedure described in detail in Refs. [81–83]. The complexes were cast on 0.5-μm-thick Si_3N_4 or 1-mm-thick CaF_2 substrates. From the size of the DNA/surfactant complexes, one estimates a DNA concentration of 1.5×10^{-2} M. The DNA samples were integrated into a home-built humidity cell, connected to a reservoir containing various agents to control the relative humidity (RH) in the cell and the DNA film. In the following, data are presented for humidity levels of 0% and 92% RH, which correspond to up to 2 and more than 20 water molecules per base pair. The water concentration in the film is $c \leq 0.57$ M for 0% RH and 5.7 M for 92% RH, the latter corresponding to full hydration of the DNA oligomers. The hydration level of the DNA films was verified by steady-state infrared measurements of the spectral position of the asymmetric PO_2^- stretching mode of the phosphate groups in the DNA backbone (cf. inset of Figure 2.4b) and by gravimetric measurements.

The DNA helix conformation depends on the hydration level [2]. X-ray diffraction has shown that DNA helices with alternating A–T base pairs exist in a B-like conformation at a humidity level above 70% RH [84]. Theoretical calculations [85] and infrared spectroscopy [86] have suggested that the B-helix conformation of DNA containing A–T base pairs prevails in an even wider humidity range. The frequency positions of phosphodiester backbone vibrations coupled to the sugar motions and the glycosidic bond torsion are sensitive probes of DNA conformation. In the samples studied here, two infrared bands at 835 and 890 cm^{-1} that are characteristic for the B-geometry are observed at 92% RH. Upon reducing the water content to 33% RH, these bands undergo minor shifts of 2–3 cm^{-1}, that is, the B-form prevails. In the whole range from 0% to 92% RH, infrared bands characteristic for the A-form of DNA at 805 and 860 cm^{-1} are absent.

The infrared absorption spectra of the DNA oligomers between 2400 and 3700 cm^{-1} are plotted in Figure 2.4 for different hydration levels. At 0% RH, the spectrum between 3000 and 3700 cm^{-1} is dominated by the NH stretching absorption of the base pairs while the OH stretching absorption of the water shell becomes predominant with increasing water content. Absorptive 2D infrared spectra, that is, the sum of the rephasing and nonrephasing signals of the DNA oligomers at 0% RH, are summarized in Figure 2.14 [87]. The spectra were measured with pulses centered at 3250 cm^{-1} and perpendicular linear polarizations of pulses 3 and 4 relative to pulses 1 and 2. The 2D signal normalized to the (electric field) spectra of pulse 3 and the local oscillator pulse 4 is plotted as a function of ν_1 and ν_3. The positive 2D signal due to the $v = 0$ to 1 transitions of the different NH stretching oscillators displays a pattern of partly overlapping components. This pattern consists of two strong diagonal peaks P1 with maximum at (3200,3200) cm^{-1} and P2 with maximum at (3350,3350) cm^{-1} as well as two pronounced cross peaks P3 at (3350,3200) cm^{-1} and P4 at (3200,3350) cm^{-1} (panel for $T = 100$ fs in Figure 2.14). The diagonal peak P1 is made up of two components, a strong, nearly homogeneously broadened component tilted relative to the diagonal and a second weaker component extended along the diagonal toward lower frequencies. The upper diagonal peak P2 is clearly elongated along the diagonal. The cross peak P3 is elongated parallel to the ν_1 axis pointing again to an essentially homogeneous lineshape while the cross peak P4 is essentially round. This peak pattern undergoes very minor changes up to the longest population time of $T = 500$ fs, as is confirmed by an analysis of the center lines of the different peaks [87]. As a function of T, the intensity of P3 relative to the diagonal peaks grows substantially. The negative signals dominating at detection frequencies $\nu_3 < 3100$ cm^{-1} are due to the $v = 1$ to 2 transitions of the different NH stretching oscillators.

In an independent series of experiments, 2D infrared spectra of DNA oligomers containing 23 nonalternating A–T pairs, that is, vertically stacked A and T bases linked to the respective backbone strand, were measured under the same experimental conditions. Within the experimental accuracy,

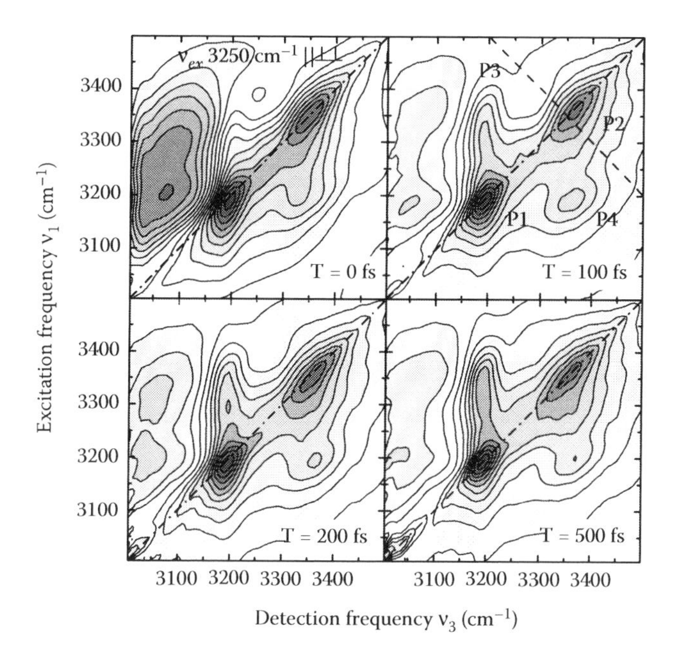

FIGURE 2.14 Absorptive 2D infrared spectra of DNA oligomers containing 23 alternating adenine–thymine (A–T) base pairs at 0% RH. The spectra were recorded at different population times T with pulses centered at 3250 cm^{-1} and perpendicular linear polarization of pulses 3 and 4 relative to pulses 1 and 2 (||⊥⊥). The four peaks P1 to P4 of the positive signal (T = 100 fs panel) and their surrounding are due to the fundamental NH stretching transitions of the A–T base pairs. The negative signals (contours at low detection frequencies) are due to the v = 1 to 2 transitions of the NH stretching oscillators. Each spectrum is normalized to its maximum positive signal component and contour lines correspond to 10% changes in amplitude. The dashed line in the T = 100 fs panel indicates the direction along which the antidiagonal cross sections in Figure 2.15c were taken.

such 2D spectra agree with the results of Figure 2.14. This fact demonstrates that inter-base pair couplings play a minor role for the ultrafast vibrational response studied here.

For a more detailed analysis of the 2D spectra of Figure 2.14, cuts along the diagonal $\nu_1 = \nu_3$ and along the antidiagonal going through the maximum of P2 at (3350,3350) cm^{-1} (dashed line in the T = 100 fs panel of Figure 2.14) are presented in Figures 2.15a and 2.15c for population times T up to 700 fs. The cuts along the diagonal (Figure 2.15a) consist of the broad components of P1 and P2 elongated along $\nu_1 = \nu_3$ and the additional narrow peak of P2 close to 3200 cm^{-1}. The antidiagonal cross sections in Figure 2.15c show the sharp edge of P2 toward low detection frequencies ν_3, which is superimposed by a comparably weak off-diagonal shoulder growing in over the T-period of 700 fs. Toward high values of ν_3, the profile displays a more gradual decrease, partly due to the overlap with the cross peak P4.

In Figures 2.16a and 2.16b, the spectrally integrated intensities of the diagonal peaks P1 and P2 and the cross peak P3 are plotted as a function of population time T (solid symbols). The diagonal peaks P1 and P2 show a partial decay within the first 500 fs, followed by a slower decrease. This behavior is very similar to kinetics of vibrational populations measured in spectrally and temporally resolved pump-probe experiments [83,88]. The pump-probe data reveal a complete decay of the slow component within 20 ps. The initial subpicosecond decay is due to the relaxation of the v = 1 population of the different NH stretching oscillators while the slower picosecond kinetics reflect vibrational redistribution among low-frequency modes and vibrational cooling within the DNA oligomers. The v = 1 relaxation of the NH stretching oscillators proceeds via combination and overtones of fingerprint modes close to the NH stretching v = 1 levels, as has been shown by

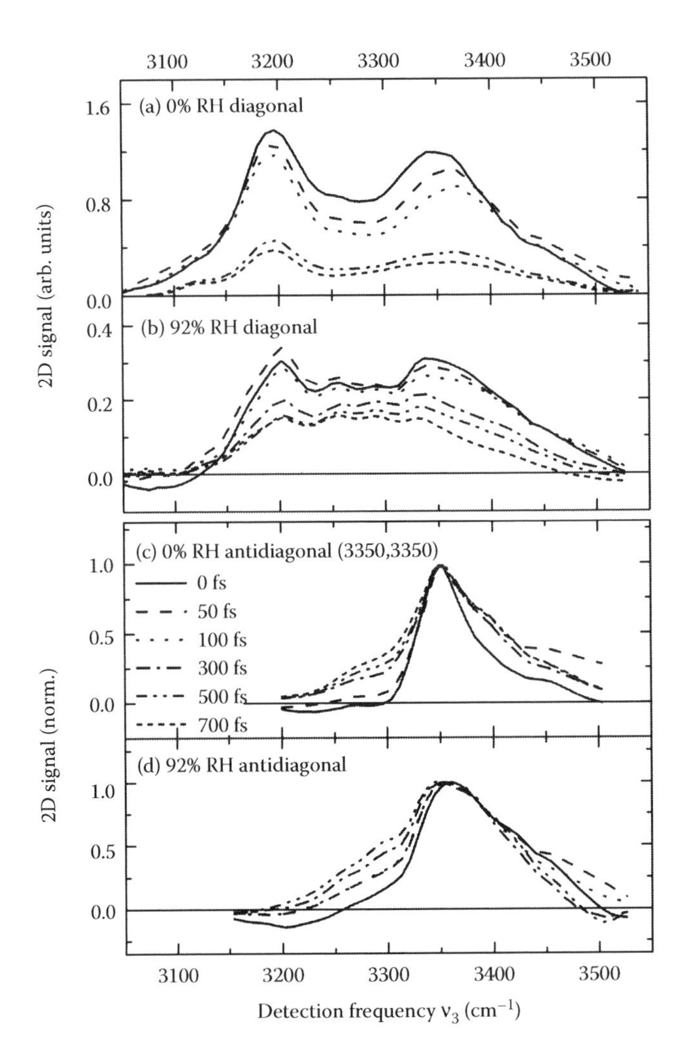

FIGURE 2.15 Cross sections of the 2D spectra recorded at 0% and 92% RH (cf. Figures 2.14 and 2.17) for different population times T. (a,b) Cross sections along the diagonal $\nu_1 = \nu_3$. (c,d) Cross sections along the antidiagonal through the frequency position (3350,3350) cm^{-1} (dashed lines in the T = 100 fs panels of Figures 2.14 and 2.17).

time-resolved anti-Stokes Raman scattering after femtosecond NH stretching excitation [89]. The intensity of the cross peak P3 (Figure 2.16b) exhibits a delayed rise and a slower gradual decay when compared to the diagonal peaks (Figure 20.16a). As a result, the intensity ratio of P3 relative to the diagonal peak P2 increases substantially with T as shown in Figure 2.16c.

For an assignment of the different components, the infrared absorption and the 2D spectra were compared to gas-phase spectra of A–T pairs and theoretical calculations. Details of this analysis have been discussed in Ref. [87]. It is important to note that—in contrast to the behavior of GC base pairs in solution (Section 2.3.2)—both the infrared absorption spectra and the 2D spectra do not display a significant decoupling of the two local NH oscillators on the NH_2 group of adenine, which would result in a stretching band of a free NH group around 3500 cm^{-1}. Thus, a description of the adenine NH_2 vibrations in terms of symmetric and asymmetric stretching modes represents a reasonable approximation for the A–T base pairs. In agreement with assignments based on previous pump-probe studies, the upper diagonal peak P2 in the 2D spectra of Figure 2.14 is attributed to the asymmetric NH_2 stretching mode of adenine. Its elongated shape points to an inhomogeneous

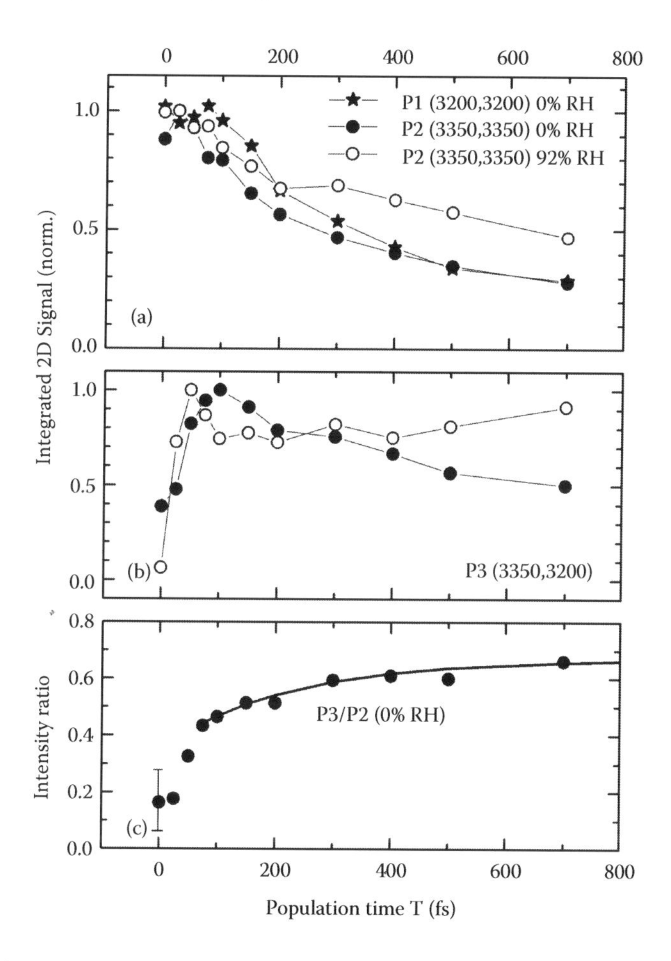

FIGURE 2.16 (a,b) Normalized spectrally integrated intensity of the diagonal peaks P1 and P2 and the cross peak P3 in the 2D spectra for 0% RH (solid symbols, cf. Figure 2.14). Intensities measured at 92% RH are shown as open symbols (cf. Figure 2.17). (c) Intensity ratio P3/P2 for 0% RH. The solid line is to guide the eye.

broadening, which is static on the timescale of the 2D experiments and may reflect structural disorder in the DNA oligomers. The 2D spectra and cross sections of P1 display two different features, a broad band elongated along the diagonal and a substantially narrower peak that is tilted toward the ν_1 axis. The broad elongated component resembles the upper diagonal peak and is assigned to the symmetric NH_2 stretching mode of adenine. Consequently, the narrow, nearly homogeneously broadened peak at 3200 cm^{-1} is assigned to the thymine NH stretching mode. This NH group is incorporated in a well-defined NH ··· N hydrogen bond inside the base pair and, thus, should be less affected by structural inhomogeneities. For excitation pulses centered at 3250 cm^{-1} and the very low water concentration in the DNA sample at 0% RH, OH stretching vibrations of the residual water molecules around 3500 cm^{-1} [83] make a negligible contribution to the 2D spectra.

The cross peaks P3 and P4 in the 2D spectra indicate an anharmonic coupling between the asymmetric NH_2 stretching mode and the modes around 3200 cm^{-1}. In general, anharmonic couplings are manifested in dispersive lineshapes of the cross peaks. Such features are impossible to discern in the present 2D spectra, due to the substantial overlap of the different diagonal and cross peaks and their comparably large spectral width of 70–100 cm^{-1}. The anharmonic couplings are—obviously—substantially smaller than the spectral width of the cross peaks, a conclusion supported by theoretical calculations. As shown in Figure 2.16c, the relative intensity of the cross peak P3 increases as a function of population time T. While the fast initial rise is strongly influenced by interferences and compensation effects with the partially overlapping v = 1 to 2 contribution in the 2D spectra, the subsequent slower rise with a time constant of the order of 500 fs is due to an energy transfer process from the asymmetric NH_2 stretching mode at $\nu_3 = 3350\ cm^{-1}$ to the modes at $\nu_3 = 3200\ cm^{-1}$. This slower increase of the intensity of P3 is also behind the rise of the shoulder in the antidiagonal cuts shown in Figure 2.15c. In the energy transfer process, the upper oscillators are deactivated and the (unshifted) v = 0 to 1 transition of the lower oscillators excited. The energy difference between the two excitations is accepted by the vibrational manifold of DNA. This downhill energy transfer enhances the positive P3 component relative to P2 whereas a potential negative, spectrally shifted P3 component decays by deactivation of the upper oscillators. The energy transfer process enhances the v = 1 population of the lower oscillators, leading to a P3 decay slower than that of P1 and P2 (Figures 2.16a and 2.16b). Taking the slower rise time of the P3 intensity of approximately 500 fs as a measure for the incoherent energy transfer time, a standard Fermi golden rule approach gives an absolute value of the coupling strength between the two oscillators of the order of 5 cm^{-1}. Couplings of the same order of magnitude between vibrational transition dipoles in base pairs have been found in Ref. [64] for fingerprint modes and in Ref. [88] for NH stretching modes. All such couplings are much smaller than the spectral widths of the different peaks in the NH stretching 2D spectra and cannot be derived by a lineshape analysis. If the present energy transfer process proceeds via through-space dipole–dipole coupling, a transfer from the asymmetric NH_2 stretching mode of adenine to the NH mode of thymine is favored because of the comparably small dipole separation of approximately 0.35 nm and the 30° angle between the transition dipoles, both being determined by the base pair geometry [2,27,56]. The orientation of the P3 envelope parallel to the ν_1 axis of the 2D spectra and the spectral width of P3 close to that of the thymine NH stretching component of the diagonal peak P1 supports this picture. In contrast, a predominant coupling between the asymmetric and symmetric NH_2 stretching modes can be ruled out because of the absence of quantum beats in spectrally resolved pump-probe data and the absence of signatures of such quantum coherences in the 2D spectra.

In summary, 2D spectra of DNA oligomers at a very low hydration level allow for deciphering the pattern of NH stretching vibrations of A–T base pairs and give insight into couplings between the different modes. Up to population times of the order of 1 ps, spectral diffusion plays a minor role in this system, resulting in essentially time-independent lineshapes of diagonal and cross peaks that display a spectral width of the order of 70–100 cm^{-1}. A comparison of 2D spectra of DNA oligomers with alternating and nonalternating A–T pairs suggests negligible inter-base pair couplings while coupling between NH stretching oscillators in a particular pair results in a subpicosecond energy transfer process between NH stretching modes.

2.4.2 NH and OH Stretching Dynamics in Hydrated DNA

The DNA thin-film sample technology applied here allows for a controlled step-wise increase of the hydration level up to 92% RH where more than 20 water molecules per base pair form a closed hydration shell around the DNA helices in B-geometry. In the following, 2D data for fully hydrated DNA are discussed; results for an intermediate hydration level of 33% RH have been presented in the supplement of Ref. [90].

Absorptive 2D infrared spectra of the DNA oligomers at 92% RH are shown in Figure 2.17 for six different population times T [90]. The measurements were performed under the same experimental conditions as for the data in Figure 2.14. In particular, the femtosecond pulses were centered at $\nu_{ex} = 3250\ cm^{-1}$ and the ||⊥⊥ polarization scheme was applied. The spectra in Figure 2.17 exhibit a pattern of NH stretch diagonal and cross peaks very similar to the spectra for 0% RH. At 92% RH, this pattern is preserved for the range of population times T up to 0.5 ps but complemented by an underlying spectrally broad contribution that originates from OH stretching excitations of the surrounding water shell. The 2D spectra in Figure 2.17 demonstrate a substantial reshaping of the latter component within the first 0.5 ps, going from a shape slightly elongated along the diagonal to

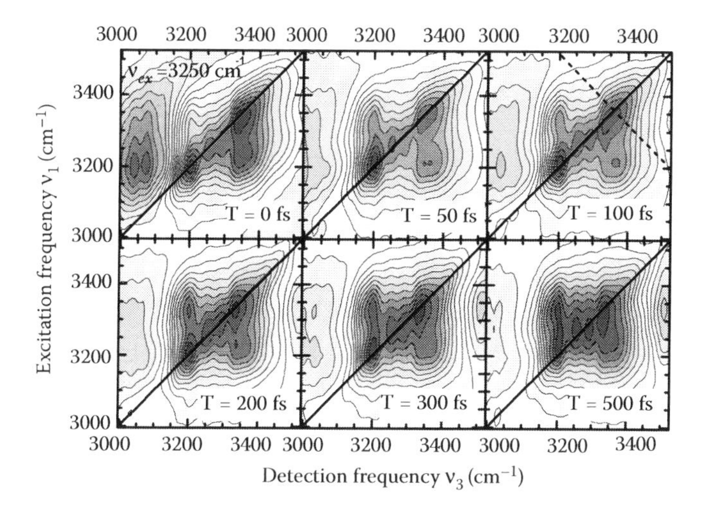

FIGURE 2.17 Absorptive 2D infrared spectra of DNA oligomers containing 23 alternating adenine–thymine (A–T) base pairs at 92% RH. The spectra were recorded at different population times T with pulses centered at 3250 cm^{-1} and perpendicular linear polarization of pulses 3 and 4 relative to pulses 1 and 2 (||⊥⊥). Each spectrum is normalized to its maximum positive signal component and contour lines correspond to 10% changes in amplitude. The contours below $\nu_3 = 3100$ cm^{-1} are negative signals due to v = 1 to 2 transitions. The negative signal at $\nu_3 > 3500$ cm^{-1} is caused by the hot ground state of the water shell. The dashed line in the T = 100 fs panel indicates the direction along which the antidiagonal cross sections in Figure 2.15d were taken.

an essentially round shape. The diagonal cuts of the 92% RH spectra shown in Figure 2.15b suggest a roughly additive behavior of the NH and OH stretch contributions with a spectral width of the NH bands similar to the 0% RH case. The decrease of intensity at high detection frequencies ν_3 with increasing T is mainly due to the buildup of the blue-shifted absorption band of the so-called hot water ground state that dominates at the highest detection frequencies shown in Figure 2.17. The hot ground state is formed by the v = 1 population decay of the OH stretching excitations and subsequent energy redistribution in the hydration shell (and the DNA oligomers). As a second measure of the linewidth of the diagonal NH stretching peaks, their frequency width $\Delta\nu_a$ along an antidiagonal going through the respective maxima at (3200,3200) and (3350,3350) cm^{-1} (Figures 2.15c and 2.15d) were estimated. The full width at half maximum displays a moderate increase by approximately 25% when going from 0% to 92% RH. The broader and stronger low-frequency shoulder in the cross sections of Figure 2.15d compared to Figure 2.15c reflects the additional spectral diffusion of the OH stretching component of the total signal.

The time-dependent intensity of the diagonal peak at (3350,3350) cm^{-1} in the 92% RH spectra is plotted in Figure 2.16a (open circles) and similar to the behavior at 0% RH (solid circles). The larger relative amplitude of the slower kinetics is due to the underlying water contribution. After an initial fast decrease, the intensity of the cross peak at (3350,3200) cm^{-1} (open circles in Figure 2.16b) displays a slow rise, which is due to the down-hill energy transfer from the NH stretch mode at 3350 cm^{-1} to the modes at 3200 cm^{-1} and a reshaping of the underlying OH stretching component of the water shell.

The moderate changes of the 2D NH stretching pattern with increasing hydration, the very similar time evolution of the peak intensities at the different hydration levels, and the "additivity" of NH and OH stretching contributions in the 2D spectra recorded at 92% RH point to a limited influence of the hydration shell on the properties and, in particular, the lineshapes of NH stretching excitations of the base pairs. In general, fluctuating Coulomb forces that originate from the ionic phosphate groups in the DNA backbone, the counterions of the DNA oligomers, and the water dipoles in the hydration shell give rise to vibrational dephasing reflected in 2D lineshapes.

The absence of a clear change with increasing water concentration suggests that the NH stretching lineshapes are governed by fluctuation mechanisms not involving the water shell, namely those originating from the DNA oligomers and/or the counterions. Taking the antidiagonal width $\Delta\nu_a$ of the NH stretching peaks of $\Delta\nu_a = 60–125\ cm^{-1}$ as a measure of the fluctuation-induced broadening, the timescale of the relevant fluctuations is clearly subpicosecond. The frequency spectrum of DNA fluctuations is governed by vibrational motions of the helix structure and the counterions while translational and rotational motions are much too slow to account for subpicosecond dephasing. Delocalized vibrations of the backbone and motions of the counterions relative to the DNA helix occur in the frequency range from 1 cm^{-1} up to 200 cm^{-1} [60,91,92] while bending and stretching motions of the phosphate ions have frequencies between 400 and 1300 cm^{-1}. At a temperature of 300 K, thermally excited modes below approximately 600 cm^{-1} dominate in the frequency spectrum of the fluctuations and contribute to the subpicosecond dephasing of the NH stretching excitations. It should be noted that—according to molecular dynamics simulations and NMR experiments—structural fluctuations of the first water layer at the DNA surface mainly occur in the picosecond time domain, too slow to make a significant contribution to the subpicosecond dephasing of the NH stretching modes.

The 2D spectra measured with pulses centered at $\nu_{ex} = 3250\ cm^{-1}$ display NH and OH stretching contributions of comparable amplitude. To get a better insight into the OH stretching component and to compare 2D spectra of the water shell of DNA with 2D spectra of bulk H_2O measured under the same conditions [93], a series of 2D spectra was recorded with pulses centered at a higher $\nu_{ex} = 3400\ cm^{-1}$, that is, at the maximum of the stationary OH stretching absorption band (Figure 2.4b). Such absorptive 2D spectra of DNA at 92% RH are shown in Figure 2.18. The 2D signal was corrected for the spectrum of the local oscillator only to facilitate a direct comparison with the published 2D spectra of bulk H_2O (Figure 1 in Ref. [93]). In Figure 2.18, the lineshape of

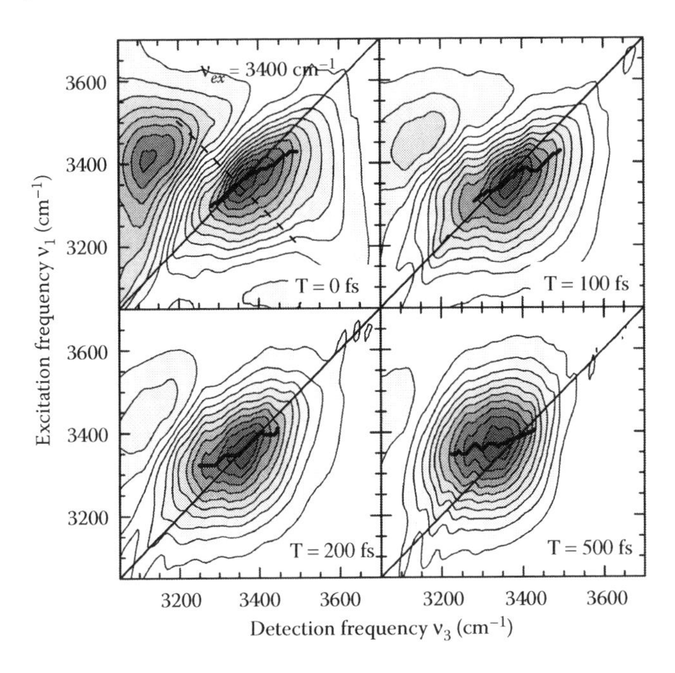

FIGURE 2.18 Absorptive 2D infrared spectra of hydrated DNA oligomers at 92% RH measured with femtosecond pulses centered at 3400 cm^{-1} (||⊥⊥ polarization scheme). The spectra are corrected for the spectrum of the local oscillator only (as spectra published in Ref. [93]). Each spectrum is normalized to its maximum positive signal component and contour lines correspond to 10% changes in amplitude. Thick solid lines: Center lines derived from cross sections along the excitation frequency axis ν_1. The dashed line in the T = 100 fs panel indicates the direction along which the antidiagonal cross sections in Figure 2.19 were taken.

the predominant OH stretching contribution (v = 0 to 1 transition) undergoes a transition from a diagonal to an essentially round envelope on a timescale of 500 fs. This reshaping directly reflects spectral diffusion in the water shell. A very similar behavior is evident from the 2D spectra of bulk H_2O [93], here, however, with an even stronger reshaping within the first 200 fs. In Figure 2.19, normalized spectral cross sections (symbols) along the antidiagonal going through the (3350,3350) cm^{-1} position (dashed line in the T = 0 fs panel of Figure 2.18) are shown for different population times T. Such cross sections undergo a substantial broadening toward low detection frequencies ν_3 within 500 fs. For comparison, cross sections along the same antidiagonal in the bulk H_2O 2D spectra of Ref. [93] are plotted (lines, cuts derived from 2D spectra for a sample temperature of 304 K). While the qualitative evolution as a function of population time T is very similar, the bulk H_2O 2D spectra broaden two to three times faster than that of the DNA hydration shell.

To quantify the timescale of spectral diffusion more accurately, the slope of center lines through the 2D spectra is determined for different population times T [80]. Center lines (thick black lines in Figure 2.18) connect frequency positions ($\nu_{max,1}$, ν_3) at which a spectral cross section taken for a particular ν_3 value along the ν_1 axis reaches its maximum. The change in the slope of center lines with increasing population time T is related to the frequency–time correlation function of the fluctuating ensemble. Such center-line analysis is meaningful for parts of 2D spectra in which the contributions of different optical transitions are clearly separated and measured lineshapes are undistorted by overlapping spectral envelopes. This condition is fulfilled for the central region of the positive 2D signals. In Figure 2.20, the center line slopes derived from the full 2D data set recorded with pulses centered at $\nu_{ex} = 3400$ cm^{-1} are plotted as a function of T (open circles). On a timescale of 500 fs, the slope decays from its initial value around 0.7 to 0.3. Applying the same procedure to the 2D spectra of bulk H_2O [93], one derives the solid squares. Here, the range of accessible population times T is more limited because of the short v = 1 lifetime of the OH stretching vibration of 200 fs and the subsequent formation of the hot ground state with a markedly different absorption band. The bulk H_2O data in Figure 2.20 display a strong decrease of the center line slope within the first 100 fs, which is much less pronounced in the data for the DNA hydration shell.

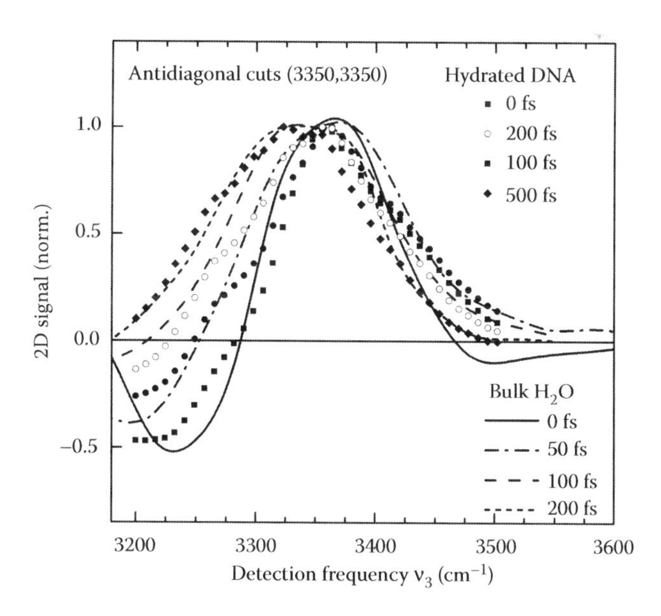

FIGURE 2.19 Antidiagonal cuts of the 2D spectra of neat H_2O (lines) as derived from the 2D spectra of Ref. [93] (sample temperature 304 K) and of the 2D spectra of hydrated DNA at 92% RH (symbols, cuts along dashed line in Figure 2.18). Spectral diffusion in the DNA case is slower than in H_2O.

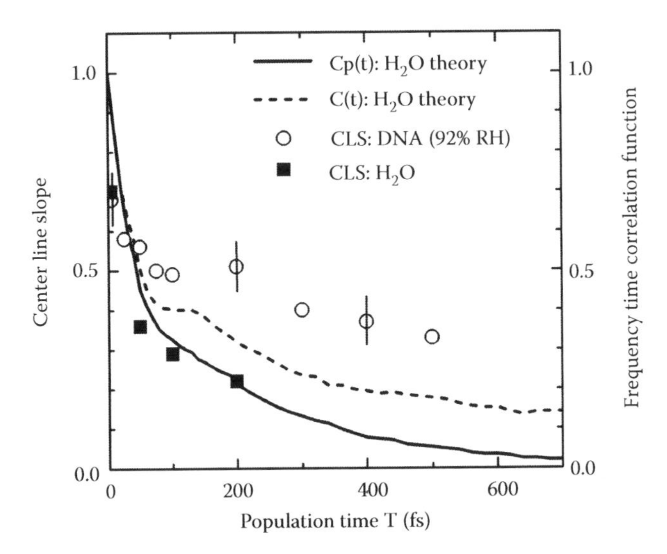

FIGURE 2.20 Comparison of center line slopes (CLS) derived from the 2D spectra of DNA oligomers at 92% RH (open circles, cf. Figure 2.18) and of bulk H_2O as derived from 2D spectra in Ref. [93] (solid squares). The lines represent calculated frequency–time correlation functions (TCFs) reported in Ref. [94]. The correlation function Cp(t) (solid line) was calculated from molecular dynamics simulations including resonant energy transfer between OH stretching oscillators of H_2O while this process was neglected in the calculation of C(t) (dashed line).

The solid line in Figure 2.20 represents the frequency–time correlation function Cp(t) of OH stretch excitations of bulk H_2O, as derived from molecular dynamics simulations [94]. These simulations include both fluctuating Coulomb forces due to structural fluctuations of the extended network of hydrogen-bonded water molecules and processes of resonant energy transfer, that is, the transfer of OH stretching excitations between different oscillators. The frequency–time correlation function at short times is in good agreement with the results for bulk H_2O (solid squares). At time zero, the measured center line slope is 0.7 compared to 1.0 in the calculation, an effect mainly due to time averaging over the 50–70 fs pulse duration in the experiment. In comparison to the calculated Cp(t) of bulk water, the center line slopes for hydrated DNA (open circles) exhibit a strongly reduced amplitude of the fast sub-100 fs component and a slower, more gradual decay between T = 70 fs and 500 fs. In particular, the pronounced decay of the calculated Cp(t) between 70 and 600 fs is absent. The slow decay found for hydrated DNA resembles the behavior of the correlation function C(t) of bulk water (dashed line), which has been calculated without including resonant energy transfer between OH stretching oscillators [94].

The results presented in Figures 2.18 and 2.19 and their analysis in Figure 2.20 provide specific insight into the dynamics of the hydration shell around the DNA oligomers. In contrast to the solvation studies with chromophores linked to DNA, the OH stretching excitations probe the water response in an undistorted hydration geometry and selectively. The observed nonlinear response is averaged over the entire water shell, including both the first highly heterogeneous layer of water molecules at the DNA surface and the outer layers. At a hydration level of 92% RH, the DNA oligomers are fully hydrated but the total water concentration in the sample of approximately 6 M is much smaller than in bulk H_2O (56 M). At such reduced concentration, resonant energy transfer between OH stretching oscillators [94–96] is significantly slowed down. Extrapolating from an experimental study of resonant energy transfer between diluted OH stretching oscillators (HOD in D_2O, Ref. [96]), one estimates picosecond energy transfer times within the water shell, which are longer than the subpicosecond vibrational lifetime of the OH stretching oscillators. As a result, resonant energy transfer is more or less suppressed and does not contribute to the decay of the center line slopes in the 2D spectra of hydrated DNA. This is particularly relevant for population times between 70 and 700 fs (Figure 2.20), where the decay of the calculated correlation function Cp(t) is strongly influenced by resonant energy transfer [94]. A second major difference between the fully hydrated DNA sample and bulk water is that a large fraction of water molecules (approximately 20 water molecules per base pair, including the backbone phosphate and sugar units) is located in the first hydration layer and attached to individual DNA groups by hydrogen bonds. Such local interactions slow down or even suppress orientational and translational water motions. Water molecules in the narrow minor groove of the helix are sterically hindered in their motions and hydrogen bond lifetimes are up to tens of picoseconds, as found in NMR measurements and MD simulations of correlation functions of hydrogen bond lifetimes [69,73,74]. The partial suppression of low-frequency water motions and the longer hydrogen bond lifetimes directly affect the frequency spectrum of the fluctuating Coulomb forces and result in a slowing down of the different decay components in the frequency–time correlation function of OH stretching excitations. The comparably small amplitude of the sub-100 fs component in the center line slopes of the hydrated DNA spectra (Figure 2.20) is attributed to this mechanism. On the other hand, both the time-dependent broadening of the 2D spectra (Figure 2.19) and the behavior of the center line slopes suggest a decay of the frequency–time correlation function on a timescale up to a few picoseconds. Thus, structural fluctuations in the water shell around DNA seem slowed down to a limited extent only but not by orders of magnitude in time. This conclusion is supported by recent theoretical simulations of water dynamics at DNA surfaces [97]. Such work gives decay times of the water frequency fluctuation correlation function of 0.4 and 2.7 ps.

In contrast to the moderate slowing down of structural dynamics of DNA's hydration shell, population relaxation of OH stretching excitation and the subsequent formation of a vibrationally hot ground state of the water shell proceed on timescales very similar to bulk H_2O. The latter processes have been studied in a series of femtosecond pump-probe experiments on OH stretching dynamics [83,88]. Moreover, the transient behavior of the asymmetric stretching vibration of phosphate groups in the DNA backbone, which is a highly sensitive probe of local hydration, reveals a highly efficient transfer of excess energy from DNA into the water shell, which represents the primary heat sink [98]. In the heated system, hydrogen bonds between water molecules and phosphate groups appear to be weakened and/or even broken.

In summary, the 2D infrared spectra of fully hydrated DNA allow for a separation of DNA's NH stretching response from the dynamics of the surrounding water shell. Similar NH stretching lineshapes are found in a very broad range of water concentration, suggesting a minor influence of water fluctuations on the time evolution of NH stretching excitations. The spectral widths of the different NH stretching peaks point to vibrational dephasing on the subpicosecond timescale, due to fluctuations of the DNA structure, including its counterions. For fully hydrated DNA, spectral diffusion of OH stretching excitations of the water shell is moderately slowed down in comparison to bulk H_2O. An analysis of center line slopes of 2D spectra of hydrated DNA and bulk H_2O shows a strong amplitude reduction of the sub-100 fs correlation decay in the hydration shell and a more gradual decay on the subpicosecond timescale. Both the minor role of resonant energy transfer between OH stretching oscillators in the water shell and the comparably rigid, highly heterogeneous structure of the first water layer interacting with the DNA surface are considered key mechanisms for the slowing down of the water response.

2.5 CONCLUSIONS AND PROSPECTS

The results presented in this chapter illustrate the rich variety of interactions and processes that are relevant for understanding the behavior of hydrogen-bonded dimers and nucleobase pairs in the ultrafast time domain. In contrast to linear infrared spectroscopy, femtosecond nonlinear pump-probe and photon-echo methods allow for a separation of processes and interactions via their characteristic timescales and particular signatures in the nonlinear absorption and, even more, 2D infrared spectra. In carboxylic acid dimers and nucleobase pairs, hydrogen-bonded systems with negligible fluctuations of their structure on the ultrafast timescale, different types of anharmonic couplings result in highly complex lineshapes of linear and nonlinear vibrational spectra. Key mechanisms are Fermi resonances originating from couplings to combination and overtones of fingerprint vibrations and anharmonic couplings to low-frequency hydrogen bond modes. The latter interaction allows for inducing and—in principle—steering low-frequency wavepacket motions along underdamped hydrogen bond degrees of freedom using tailored femtosecond infrared pulses. While the overall coupling scenario of OH stretching vibrations in carboxylic acid dimers is understood in great detail, the role of such couplings for NH stretching excitations in hydrogen-bonded nucleobase pairs is much less clear, in particular for NH stretching modes with a frequency in the range of combination and overtones. Here, 2D infrared studies in an extended frequency range and/or with deuterated samples should provide additional insight.

The study of nucleobase pairs in hydrated DNA oligomers reveals a surprisingly weak influence of the surrounding water shell on NH stretching dynamics, as is evident from pump-probe and 2D experiments in a very wide range of hydration levels. This behavior seems to be in line with the relatively rigid structure of the first water layer at the DNA surface. The latter also leads to a moderate slowing down of structural fluctuations in the water shell around DNA compared to bulk H_2O. The frequency–time correlation function of OH stretching excitations of the water shell demonstrates, however, a pronounced decay in the subpicosecond time domain, ruling out a slowing down of water dynamics by orders of magnitude.

While the work discussed here has focused on OH and NH stretching modes, extension of 2D techniques toward fingerprint modes, in particular of groups located in the DNA backbone, holds strong potential for unraveling DNA–water interactions with more specificity and, thus, in much greater detail. The symmetric and asymmetric phosphate stretching vibrations as well as a number of backbone modes between 500 and 1000 cm^{-1} are vibrational probes suitable for such investigations. Clarifying the interactions of DNA and other hydrogen-bonded biomolecular structures with their aqueous environment is also required for understanding the timescales and mechanisms of energy dissipation processes of electronically or vibrationally excited biomolecules at the molecular level.

ACKNOWLEDGMENTS

I would like to thank my present and former coworkers Satoshi Ashihara, Jens Dreyer, Jason Dwyer, Henk Fidder, Karsten Heyne, Nils Huse, Erik T. J. Nibbering, Łukasz Szyc, and Ming Yang for their important contributions to the work discussed in this chapter. I am strongly indebted to the collaborating groups of R. J. Dwayne Miller, Toronto and Hamburg, and Friedrich Temps, Kiel, and to Casey Hynes, Boulder, Jim Skinner, Madison, and Damien Laage, Paris, for discussions. Financial support by the European Research Council (FP7/2007-2013/ERC Grant Agreement No. 247051 "Ultradyne"), the Deutsche Forschungsgemeinschaft, and the Fonds der Chemischen Industrie is gratefully acknowledged.

REFERENCES

1. Schuster P, Zundel G, Sandorfy C (Eds.). 1976. *The Hydrogen Bond: Recent Developments in Theory and Experiments, Vol I–III*. North Holland, Amsterdam.
2. Saenger W. 1984. *Principles of Nucleic Acid Structure*. Springer Verlag, New York, Chapter 17.
3. Eisenberg D, Kauzmann W. 1969. *The Structure and Properties of Water*. Oxford University Press, New York.
4. Bakker HJ, Skinner JL. 2010. Vibrational spectroscopy as a probe of structure and dynamics in liquid water. *Chem Rev* 110:1498–1517.
5. Fayer MD. 2009. Dynamics of liquids, molecules, and proteins measured with ultrafast 2D IR vibrational echo chemical exchange spectroscopy. *Annu Rev Phys Chem* 60:21–38.

6. Nibbering ETJ, Elsaesser T. 2004. Ultrafast vibrational dynamics of hydrogen bonds in the condensed phase. *Chem Rev* 104:1887–1914.
7. Laage D, Hynes JT. 2006. A molecular jump mechanism of water reorientation. *Science* 311:832–835.
8. Mikenda W, Steinböck S. 1996. Stretching frequency vs bond distance correlation of hydrogen bonds in solid hydrates: A generalized correlation function. *J Mol Struct* 384:159–163.
9. Rey R, Møller KB, Hynes JT. 2002. Hydrogen bond dynamics in water and ultrafast infrared spectroscopy. *J Phys Chem A* 106:11993–11996.
10. Lawrence CP, Skinner JL. 2003. Vibrational spectroscopy of HOD in liquid D_2O. III. Spectral diffusion, and hydrogen-bonding and rotational dynamics. *J Chem Phys* 118:264–272.
11. Asbury JB, Steinel T, Kwak K, Corcelli SA, Lawrence CP, Skinner JL, Fayer MD. 2004. Dynamics of water probed with vibrational echo correlation spectroscopy. *J Chem Phys* 121:12431–12446.
12. Cowan ML, Bruner BD, Huse N, Dwyer JR, Chugh B, Nibbering ETJ, Elsaesser T, Miller RJD. 2005. Ultrafast memory loss and energy redistribution in the hydrogen bond network of liquid H_2O. *Nature* 434:199–202.
13. Fayer MD (Ed.). 2001. *Ultrafast Infrared and Raman Spectroscopy*. Marcel Dekker, New York.
14. Elsaesser T, Bakker HJ (Eds.). 2002. *Ultrafast Hydrogen Bonding Dynamics and Proton Transfer Processes in the Condensed Phase*. Kluwer, Dordrecht.
15. Hamm P, Lim M, Hochstrasser RM. 1998. Structure of the amide I band of peptides measured by femtosecond nonlinear infrared spectroscopy. *J Phys Chem B* 102:6123–6138.
16. Asplund MC, Zanni MT, Hochstrasser RM. 2000. Two-dimensional infrared spectroscopy of peptides by phase-controlled femtosecond vibrational photon echoes. *Proc Natl Acad Sci USA* 97:8219–8224.
17. Mukamel S. 2000. Multidimensional femtosecond correlation spectroscopies of electronic and vibrational excitations. *Annu Rev Phys Chem* 51:691–729.
18. Taylor CA, El-Bayoumi MA, Kasha M. 1969. Excited state two-proton tautomerism in hydrogen-bonded N-heterocyclic base pairs. *Proc Natl Acad Sci USA* 63:253–260.
19. Marechal Y, Witkowski A. 1968. Infrared spectra of H-bonded systems. *J Chem Phys* 48:3697–3705.
20. Marechal Y. 1987. IR spectra of carboxylic acids in the gas phase: A quantitative reinvestigation. *J Chem Phys* 87:6344–6353.
21. Chamma D, Henri-Rousseau O. 1999. IR theory of weak H-bonds: Davydov coupling, Fermi resonances and direct relaxations. I. Basis equations within the linear response theory. *Chem Phys* 248:53–70.
22. Chamma D, Henri-Rousseau O. 1999. IR theory of weak H-bonds: Davydov coupling, Fermi resonances and direct relaxations. II. General trends from numerical experiments. *Chem Phys* 248:71–89.
23. Florio GM, Zwier TS, Myshakin EM, Jordan KD, Sibert III EL. 2003. Theoretical modeling of the OH stretch infrared spectrum of carboxylic acid dimers based on first-principles anharmonic couplings. *J Chem Phys* 118:1735–1746.
24. Emmeluth C, Suhm MA, Luckhaus D. 2003. A monomers-in-dimers model for carboxylic acid dimers. *J Chem Phys* 118:2242–2255.
25. Watson JD, Crick FHC. 1953. Molecular structure of nucleic acids—A structure for deoxyribose nucleic acid. *Nature* 171:737–738.
26. Sponer J, Jurecka P, Hobza P. 2004. Accurate interaction energies of hydrogen-bonded nucleic acid base pairs. *J Am Chem Soc* 126:10142–10151.
27. Plutzer C, Hunig I, Kleinermanns K, Nir E, de Vries MS. 2003. Pairing of isolated nucleobases: Double resonance laser spectroscopy of adenine-thymine. *Chem Phys Chem* 8:838–842.
28. Schwalb NK, Michalak T, Temps F. 2009. Ultrashort fluorescence lifetimes of hydrogen-bonded base pairs of guanosine and cytidine in solution. *J Phys Chem B* 51:16365–16376.
29. Douhal A, Kim SK, Zewail AH. 1995. Femtosecond molecular dynamics of tautomerization in model base pairs. *Nature* 378:260–263.
30. Dreyer J. 2007. Unraveling the structure of hydrogen bond stretching mode infrared absorption bands: An anharmonic density functional theory study on 7-azaindole dimers. *J Chem Phys* 127:054309.
31. Szyc Ł, Yang M, Nibbering ETJ, Elsaesser T. 2010. Ultrafast vibrational dynamics and local interactions of hydrated DNA. *Angew Chem Int Ed* 49:3598–3610.
32. Oxtoby DW, Levesque D, Weis JJ. 1978. Molecular dynamics simulation of dephasing in liquid nitrogen. *J Chem Phys* 68:5528–5533.
33. Henri-Rousseau O, Blaise P. 1998. The infrared spectral density of weak hydrogen bonds within the linear response theory. *Adv Chem Phys* 103:1–186.
34. Stenger J, Madsen D, Hamm P, Nibbering ETJ, Elsaesser T. 2001. Ultrafast vibrational dephasing of liquid water. *Phys Rev Lett* 87:027401.

35. Rey R, Møller KB, Hynes JT. 2004. Ultrafast vibrational population dynamics of water and related systems: A theoretical perspective. *Chem Rev* 104:1915–1928.
36. Ashihara S, Huse N, Espagne A, Nibbering ETJ, Elsaesser T. 2007. Ultrafast structural dynamics of water induced by dissipation of vibrational energy. *J Phys Chem A* 111:743–746.
37. Ashihara S, Huse N, Espagne A, Nibbering ETJ, Elsaesser T. 2006. Vibrational couplings and ultrafast relaxation of the O-H bending mode in liquid H_2O. *Chem Phys Lett* 424:66–70.
38. Rey R, Ingrosso F, Elsaesser T, Hynes JT. 2009. Pathways for H_2O bend vibrational relaxation in liquid water. *J Phys Chem A* 113:8949–8962.
39. Hamm P, Ohline SM, Zinth W. 1997. Vibrational cooling after ultrafast photoisomerization of azobenzene measured by femtosecond infrared spectroscopy. *J Chem Phys* 106:519–529.
40. Lock AJ, Bakker HJ. 2002. Temperature dependence of vibrational relaxation in liquid H_2O. *J Chem Phys* 117:1708–1713.
41. Kaindl RA, Wurm M, Reimann K, Hamm P, Weiner AM, Woerner M. 2000. Generation, shaping, and characterization of intense femtosecond pulses tunable from 3 to 20 μm. *J Opt Soc Am B* 17:2086–2094.
42. Cowan ML, Ogilvie JP, Miller RJD. 2004. Two-dimensional spectroscopy using diffractive optics based phased-locked photon echoes. *Chem Phys Lett* 386:184–189.
43. Huse N, Heyne K, Dreyer J, Nibbering ETJ, Elsaesser T. 2003. Vibrational multilevel quantum coherence due to anharmonic couplings in intermolecular hydrogen bonds. *Phys Rev Lett* 91:197401.
44. Huse N, Bruner BD, Cowan ML, Dreyer J, Nibbering ETJ, Miller RJD, Elsaesser T. 2005. Anharmonic couplings underlying the ultrafast vibrational dynamics of hydrogen bonds in liquids. *Phys Rev Lett* 95:147402.
45. Heyne K, Huse N, Nibbering ETJ, Elsaesser T. 2003. Ultrafast coherent nuclear motions of hydrogen bonded carboxylic acid dimers. *Chem Phys Lett* 369:591–596.
46. Heyne K, Huse N, Dreyer J, Nibbering ETJ, Elsaesser T, Mukamel S. 2004. Coherent low-frequency motions of hydrogen bonded acetic acid dimers in the liquid phase. *J Chem Phys* 121:902–913.
47. Fujii Y, Yamada H, Mizuta M. 1988. Self-association of acetic acid in some organic solvents. *J Phys Chem* 92:6768–6772.
48. Faurskov Nielsen O, Lund PA. 1983. Intermolecular Raman active vibrations of hydrogen bonded acetic acid dimers in the liquid state. *J Chem Phys* 78:652–655.
49. Dwyer JR, Dreyer J, Nibbering ETJ, Elsaesser T. 2006. Ultrafast dynamics of vibrational N-H stretching excitations in the 7-azaindole dimer. *Chem Phys Lett* 432:146–151.
50. Stenger J, Madsen D, Dreyer J, Nibbering ETJ, Hamm P, Elsaesser T. 2001. Coherent response of hydrogen bonds in liquids probed by ultrafast vibrational spectroscopy. *J Phys Chem A* 105:2929–2932.
51. Dreyer J. 2005. Density functional theory simulations of two-dimensional infrared spectra for hydrogen-bonded acetic acid dimers. *Int J Quant Chem* 104:782–793.
52. Dreyer J. 2005. Hydrogen-bonded acetic acid dimers: Anharmonic coupling and linear infrared spectra studied with density-functional theory. *J Chem Phys* 122:184306.
53. Yang M, Szyc Ł, Röttger K, Fidder H, Nibbering ETJ, Elsaesser T, Temps F. 2011. Dynamics and couplings of NH stretching excitations of guanosine-cytidine base pairs in solution. *J Phys Chem B* 115:5484–5492.
54. Khalil M, Demirdoven N, Tokmakoff A. 2003. Coherent 2D IR spectroscopy: Molecular structure and dynamics in solution. *J Phys Chem A* 107:5258–5279.
55. Fonseca Guerra C, van der Wijst T, Bickelhaupt FM. 2006. Supramolecular switches based on the guanine-cytosine (GC) Watson-Crick pair: Effect of neutral and ionic substituents. *Chem Eur J* 12:3032–3042.
56. Wang GX, Ma XY, Wang JP. 2009. Anharmonic vibrational signatures of DNA bases and Watson-Crick base pairs. *Chin J Chem Phys* 22:563–570.
57. Falk M, Hartman KA, Lord RC. 1963. Hydration of deoxyribonucleic acid. II. An infrared study. *J Am Chem Soc* 85:387–391.
58. Tsuboi M. 1969. Application of infrared spectroscopy to structure studies of nucleic acids. *Appl Spectrosc Rev* 3:45–90.
59. Prescott B, Steinmetz W, Thomas Jr GJ. 1984. Characterization of DNA structures by laser Raman spectroscopy. *Biopolymers* 23:235–256.
60. Cocco S, Monasson R. 2000. Theoretical study of collective modes in DNA at ambient temperature. *J Chem Phys* 112:10017–10033.
61. Woutersen S, Cristalli G. 2004. Strong enhancement of vibrational relaxation by Watson-Crick base pairing. *J Chem Phys* 121:5381–5386.
62. Peng CS, Jones KC, Tokmakoff A. 2011. Anharmonic vibrational modes of nucleic acid bases revealed by 2D IR spectroscopy. *J Am Chem Soc* 133:15650–15660.

63. Krummel AT, Mukherjee P, Zanni MT. 2003. Inter- and intrastrand vibrational coupling in DNA studied with heterodyned 2D IR spectroscopy. *J Phys Chem B* 107:9165–9169.
64. Krummel AT, Zanni MT. 2006. DNA vibrational coupling revealed with two-dimensional infrared spectroscopy: Insight into why vibrational spectroscopy is sensitive to DNA structure. *J Phys Chem B* 110:13991–14000.
65. Kopka ML, Fratini AV, Drew HR, Dickerson RE. 1983. Ordered water structure around a B-DNA dodecamer—A quantitative study. *J Mol Biol* 163:129–146.
66. Schneider B, Berman HM. 1995. Hydration of the DNA bases is local. *Biophys J* 69:2661–2669.
67. Schneider B, Patel K, Berman HM. 1998. Hydration of the phosphate group in double-helical DNA. *Biophys J* 75:2422–2434.
68. Liepinsh E, Otting G, Wüthrich K. 1992. NMR observation of individual molecules of hydration water bound to DNA duplexes: Direct evidence for a spine of hydration water present in aqueous solution. *Nucl Acid Res* 20:6549–6553.
69. Halle B, Denisov VP. 1998. Water and monovalent ions in the minor groove of B-DNA oligonucleotides as seen by NMR. *Biopolymers* 48:210–233.
70. Beveridge DL, McConnell KJ. 2000. Nucleic acids: Theory and computer simulations, Y2K. *Curr Opinion Struct Biol* 10:182–196.
71. Pettitt BM, Makarov VA, Andrews BK. 1998. Protein hydration density: Theory, simulations and crystallography. *Curr Opinion Struct Biol* 8:218–221.
72. Feig M, Pettitt BM. 1999. Modeling high-resolution hydration patterns in correlation with DNA sequence and conformation. *J Mol Biol* 286:1075–1095.
73. Bonvin AMJJ, Sunnerhagen M, Otting G, van Gunsteren WF. 1998. Water molecules in DNA recognition II: A molecular dynamics view of the structure and hydration of the trp operator. *J Mol Biol* 282:859–873.
74. Pal S, Maiti PK, Bagchi B. 2006. Exploring DNA groove water dynamics through hydrogen bond lifetime and orientational relaxation. *J Chem Phys* 125:234903.
75. Pal SK, Zhao L, Zewail AH. 2003. Water at DNA surfaces: Ultrafast dynamics in minor groove recognition. *Proc Natl Acad Sci USA* 100:8113–8118.
76. Andreatta D, Lustres LP, Kovalenko SA, Ernsting NP, Murphy CJ, Coleman RS, Berg MA. 2005. Power-law solvation dynamics in DNA over six decades in time. *J Am Chem Soc* 127:7270–7271.
77. Zhong D, Pal SK, Zewail AH. 2011. Biological water: A critique. *Chem Phys Lett* 503:1–11.
78. Halle B, Nilsson L. 2009. Does the dynamic Stokes shift report on slow protein hydration dynamics? *J Phys Chem B* 113:8210–8213.
79. Furse KE, Corcelli S. 2010. Effects of an unnatural base pair replacement on the structure and dynamics of DNA and neighboring water and ions. *J Phys Chem B* 114:9934–9945.
80. Kwak K, Park S, Finkelstein IJ, Fayer MD. 2007. Frequency-frequency correlation functions and apodization in two-dimensional infrared vibrational echo spectroscopy: A new approach. *J Chem Phys* 127:124503.
81. Tanaka K, Okahata Y. 1996. A DNA-lipid complex in organic media and formation of an aligned cast film. *J Am Chem Soc* 118:10679–10683.
82. Yang C, Moses D, Heeger AJ. 2003. Base-pair stacking in oriented films of DNA-surfactant complex. *Adv Mater* 15:1364.
83. Dwyer JR, Szyc Ł, Nibbering ETJ, Elsaesser T. 2008. Ultrafast vibrational dynamics of adenine-thymine base pairs in DNA oligomers. *J Phys Chem B* 112:11194–11197.
84. Leslie AGW, Arnott S, Chandrasekaran R, Ratliff RL. 1980. Polymorphism of DNA double helices. *J Mol Biol* 143:49–72.
85. Mazur AK. 2005. Electrostatic polymer condensation and the A/B polymorphism in DNA: Sequence effects. *J Chem Theory Comput* 1:325–336.
86. Pilet J, Brahms J. 1973. Investigation of DNA structural changes by infrared spectroscopy. *J Biopolymers* 12:387–403.
87. Yang M, Szyc Ł, Elsaesser T. 2011. Femtosecond two-dimensional infrared spectroscopy of adenine-thymine base pairs in DNA oligomers. *J Phys Chem B* 115:1262–1267.
88. Szyc Ł, Dwyer JR, Nibbering ETJ, Elsaesser T. 2009. Ultrafast dynamics of NH and OH stretching excitations in hydrated DNA oligomers. *Chem Phys* 357:36–44.
89. Kozich V, Szyc Ł, Nibbering ETJ, Werncke W, Elsaesser, T. 2009. Ultrafast redistribution of vibrational energy after excitation of NH stretching modes in DNA oligomers. *Chem Phys Lett* 473:171–175.
90. Yang M, Szyc Ł, Elsaesser T. 2011. Decelerated water dynamics and vibrational couplings of hydrated DNA mapped by two-dimensional infrared spectroscopy. *J Phys Chem B* 115:13093–13100.

91. Urabe H, Hayashi H, Tominaga Y, Nishimura Y, Kubota K, Tsuboi M. 1985. Collective vibrational modes in molecular assembly of DNA and its application to biological systems—Low-frequency Raman spectroscopy. *J Chem Phys* 82:531–535.
92. Perepelytsya SM, Volkov SN. 2007. Counterion vibrations in the DNA low-frequency spectra. *Eur Phys J E* 24:261–269.
93. Kraemer D, Cowan ML, Paarmann A, Huse N, Nibbering ETJ, Elsaesser T, Miller RJD. 2008. Temperature dependence of the two-dimensional infrared spectrum of liquid H_2O. *Proc Natl Acad Sci USA* 105:437–442.
94. Jansen TLC, Auer BM, Yang M, Skinner JL. 2010. Two-dimensional infrared spectroscopy and ultrafast anisotropy decay of water. *J Chem Phys* 132:224503.
95. Paarmann A, Hayashi T, Mukamel S, Miller RJD. 2008. Probing inter-molecular couplings in liquid water with two-dimensional infrared photon echo spectroscopy. *J Chem Phys* 128:191103.
96. Woutersen S, Bakker HJ. 1999. Resonant intermolecular transfer of vibrational energy in liquid water. *Nature* 402:507–509.
97. Furse KE, Corcelli SA. 2008. The dynamics of water at DNA interfaces: Computational studies of Hoechst 33258 bound to DNA. *J Am Chem Soc* 130:13103–13109.
98. Szyc Ł, Yang M, Elsaesser T. 2010. Ultrafast energy exchange via water-phosphate interactions in hydrated DNA. *J Phys Chem B* 114:7951–7957.

3 Water Reorientation and Ultrafast Infrared Spectroscopy

Damien Laage and James T. Hynes

CONTENTS

3.1 OVERVIEW

The reorganization of water's hydrogen bond (HB) network by the breaking and making of HBs lies at the heart of many of the pure liquid's special features and of many aqueous media phenomena, including chemical reactions, ion transport, and protein activity, to name but a few. An important role in this reorganization is played by water molecule reorientation, which has been long described by very small angular displacement Debye rotational diffusion. A markedly contrasting picture has been recently proposed, based on simulation and analytic modeling: a sudden, large-amplitude jump mechanism, in which the reorienting water molecule rapidly exchanges HB partners in an activated process, which has all the hallmarks of a chemical reaction. In this chapter, we offer a recounting of the jump mechanism together with a discussion of its application to, and probing by, modern ultrafast infrared (IR) spectroscopy experiments in pure water and in aqueous solutions of ions, and of hydrophobic and amphiphilic solutes. Special emphasis is given to IR pump-probe anisotropy measurements and to the direct characterization of the jumps via pioneering two-dimensional IR (2D IR) spectroscopic measurements.

The organization of this chapter is as follows. In Section 3.2, we give an introduction to perspectives on water reorientation, both experimental and theoretical. Section 3.3 furnishes the jump model's description and its predictions for, via the extended jump model (EJM), and scrutiny by ultrafast IR spectroscopy. Section 3.4 is devoted to aqueous anionic solutions, the special issues that arise for those solutions, and the jump model's and EJM's predictions for and testing via IR anisotropy and 2D IR experiments. Amphiphilic solutions are examined both theoretically and experimentally in Section 3.5. Here, preliminary special attention is given to hydrophobic solutes and the interpretation of IR anisotropy experiments via the jump perspective and the transition state excluded volume (TSEV) factor, which is a second-tier theoretical EJM development. A culminating discussion is then given for amphiphilic solutes, emphasizing the different roles of their hydrophobic and hydrophilic portions in ultrafast IR anisotropy and 2D IR experiments; this involves a transition state hydrogen bond strength factor—an additional second-tier theoretical elaboration of the EJM—for the hydrophilic moieties of the amphiphiles. Finally, Section 3.6 gives a brief resumé and points to a few directions for further endeavor.

3.2 INTRODUCTION: WATER REORIENTATION

Among liquid water's many singular features is the great lability of its HB network, which constantly rearranges on a picosecond timescale by the breaking and forming of HBs [1–7]. This incessant reorganization dynamics has crucial roles to play in a wide span of fundamental chemical and biochemical processes, of which a partial listing would include S_N2 [8] and proton transfer reactions [9–11], proton transport [12,13], and protein activity [14,15].

The *modus operandi* of these HB network rearrangements involves in an essential fashion the reorientation of individual water molecules. Before we enter into a discussion of traditional images of this reorientation, we make an extensive discursion into the experimental methods by which this reorientation can be interrogated.

The simplest characterization of a water molecule's reorientation kinetics, from either an experimental or theoretical perspective, relies on the time-correlation function (tcf) of the molecular orientation. For a given body-fixed vector such as the water OH bond or the dipole moment, this function tracks how fast the memory of the initial orientation is lost. It is formally defined as the equilibrium average [16,17]

$$C_n(t) = \langle P_n[\mathbf{u}(0) \cdot \mathbf{u}(t)] \rangle, \tag{3.1}$$

where P_n is the Legendre polynomial of rank n and $\mathbf{u}(t)$ is the molecular orientation at time t. The two most commonly employed orientation tcfs are the first two functions C_1 and C_2. For the first, $P_1(x) = x$ and $C_1(t)$ thus follows the decay of $\cos(\theta)$, where θ is the angle between $\mathbf{u}(0)$ and $\mathbf{u}(t)$, while $P_2(x) = (3x^2 - 1)/2$ and $C_2(t)$ thus probes the decay of $\cos^2(\theta)$. Each of these tcfs decays from an initial value of 1 when the molecule has not yet reoriented to an asymptotic value of 0 corresponding to an isotropic distribution of orientations, reflecting a total memory loss. Before discussing the molecular information contained in these tcfs, we pause to survey the main techniques available to probe water reorientation dynamics.

We begin with the two experimental techniques traditionally employed to examine the water reorientational question: dielectric relaxation (DR) and nuclear magnetic resonance (NMR). These two techniques, which provide instructive comparisons with the IR spectroscopic techniques to be discussed below, measure rather different aspects of orientation dynamics. DR [18] actually probes the time correlation of the macroscopic, multiparticle, and electric polarization in the sample. Accordingly, the timescale for this collective variable is the Debye time, τ_D, which is the relaxation time of the first-order orientation tcf of the total dipole moment. This collective time τ_D can be related to the single-molecule reorientation time τ_1, but only through very approximate dielectric theory arguments [19]. In the case of water, the values τ_D and τ_1 differ by almost a factor of 2 [20,21].

In NMR, the measured longitudinal spin relaxation rate (in the extreme narrowing limit) is proportional to the orientation relaxation time, which is defined as the full time integral of the orientation tcf [22]. (As we will see, this fully time-integrated NMR aspect needs to be taken into account when comparisons with IR spectroscopic results are made.) Depending on the nuclei, different vectors and tensors can be followed. For experiments with ^1H, NMR probes the time integral of the $n = 2$ single-molecule tcf of the water HH vector. For ^2H (= D) and ^{17}O nuclei, each of which bears an electric quadrupole moment, NMR probes the tcf of the electric field gradient tensor [23]. For ^2H, this tensor is approximately uniaxial and NMR reports on the integral of the water OD vector $n = 2$ tcf. In contrast, the tensor for ^{17}O is not uniaxial and the reorientation tcf of the vector orthogonal to the water plane is only a very crude approximation to the measured tcf [23]. Overall, NMR is sensitive to the second-order ($n = 2$) orientation tcf $C_2(t)$ because both the dipolar and quadrupolar interactions involve $\cos^2(\theta)$ terms [22].

Currently, one of the most information-rich techniques to probe water reorientation dynamics is the more recently available ultrafast IR pump-probe spectroscopy, among whose attractive features is its possession of the necessary femtosecond time resolution to follow the very fast water reorientation motions. The pump pulse creates a population of vibrationally excited water molecules whose average orientation is aligned with the pump polarization. By probing the system with pulses polarized either parallel or perpendicular to the pump pulse and combining these signals, one recovers the anisotropy decay of this population, which is usually approximated to be 2/5 $C_2(t)$ [24,25]. Pump-probe spectroscopy is a third-order nonlinear technique that involves four pulse interactions, two with the pump and two with the probe [26], and it is thus sensitive to $\cos^2(\theta)$. Despite the great value of this technique, several limitations exist and need to be recognized from the beginning. First, since pump-probe spectroscopy follows the reorientation of a vibrationally excited water molecule, the anisotropy cannot be reliably measured for delays much longer than the vibrational population lifetime (i.e., at most 10 ps for water). Second, it has recently been shown that the connection between anisotropy and the orientation tcf is not as straightforward as is usually assumed; additional effects (e.g., non-Condon effects [25] and distributions of vibrational lifetimes in aqueous solution [27]) must be included for a quantitative comparison. We also note that a vibrationally excited water molecule's reorientation is probed, and the dynamics may differ somewhat from that of a water molecule in the ground vibrational state.

Finally, other techniques, including quasi-elastic neutron scattering (QENS) [23,28–31] and optical Kerr effect spectroscopy [32–35], have also been used to obtain some information about water reorientation dynamics.

Now that we are familiar with some of the experimental techniques to probe the reorientation of a water molecule, we turn to the issue of how this reorientation has been traditionally viewed. The most commonly invoked image by far of a water molecule's orientation motion is that of a very small step angular Brownian motion, that is, rotational diffusion [7]. This picture, introduced long ago by Debye [36], is so pervasive in the literature that it deserves some further explication. A diffusive picture requires for its validity a rapid loss of angular momentum memory before any significant angular motion occurs [16,17]. Accordingly, it is most plausible for large solutes in a viscous liquid solution, where slow reorientational motion is guaranteed. Despite the obvious size violation of this precept, one could well imagine that this picture could nonetheless apply for water reorientation by positing that the strong interaction of a water molecule with its hydrogen (H)-bonded neighbors could destroy, by a sort of orientational caging, any angular momentum memory before a significant angular displacement could be effected. Whether this is true or not would be expected to have been settled by experiment some time since, but the situation is in fact not so clear-cut, as is now discussed.

Since the rotational diffusion equation says that the time rate of change of the orientational distribution is the rotational diffusion constant D_r times the angular Laplacian of the distribution, the time dependence of the molecular tcf $C_n(t)$ is exponential [16,17]:

$$\frac{C_n(t)}{C_n(0)} = \exp\left(\frac{-t}{t_n}\right); \quad t_n = n(n+1)D_r, \tag{3.2}$$

the $n(n + 1)$ integer function here is also familiar from the quantum theory of, for example, diatomic rotational motion or the electron in the hydrogen atom, whose rotational kinetic energy operator is proportional to the angular Laplacian [37], appearing in the rotational diffusion equation [16,17]. The obvious question "What should be taken for D_r?" leads us to two important points concerning water's reorientational motion, now discussed.

The first point is that, for an H-bonded liquid such as water, there is apparently no useful and accurate simple formula available for D_r, and recourse is usually made to a macroscopic hydrodynamic approximation: the Einstein relation connects D_r inversely to the rotational friction constant [16,17] and the latter is proportional to the "solvent" shear viscosity η according to hydrodynamics, thereby producing the famous Debye–Stokes–Einstein relation [17]:

$$D_r = \frac{c}{(\eta R^3)}, \tag{3.3}$$

where c is a factor proportional to the temperature and dependent upon the hydrodynamic boundary condition [38] and R is the radius of the reorienting water molecule "solute." It must be immediately stressed that the validity of Equation 3.3 is an issue completely logically distinct from that of the validity of the time dependence Equation 3.2. Since the hydrodynamic description requires that the solute be large compared to the size of the solvent molecules, Equation 3.3's validity is clearly dubious, to say the least. Indeed, it can be readily criticized on a number of fundamental grounds [38,39], and concerns about its validity have been sometimes expressed [20,40–44]. Nonetheless, the combination of Equations 3.2 and 3.3 has sometimes had some success for water. Since this is no doubt assisted by the strong sensitivity of Equation 3.3 to the molecular radius R, which is often relegated to the status of a parameter, we must regard any such agreement for water as purely coincidental. The evidence of molecular dynamics (MD) simulations concerning the validity of the diffusive result Equation 3.2 itself, without recourse to any approximation for D_r, will be discussed presently.

The second point is that Equation 3.2 indicates that the *ratio* of different n orientation times is independent of D_r, in particular that

$$\frac{\tau_1}{\tau_2} = 3 \text{ (for rotational diffusion)}, \tag{3.4}$$

a happy circumstance that clearly has an experimental appeal and accordingly has often been examined. But here arises the drawback noted above that the experimental collective Debye time can only approximately provide the required molecular time τ_1. Nonetheless, Equation 3.4 is found to be approximately obeyed for water, although again with concerns voiced about the rotational diffusion description's validity [20,21,40–44]. The MD testimony on this score will be recounted below.

As we have described above, while the Debye rotational diffusion picture certainly has its merits, there is also room for doubt about its validity for water reorientation. We have in fact recently proposed a starkly different mechanism, in which a water molecule's reorientation mainly proceeds through quite sudden, and large-amplitude angular jumps [5,21]. In this new picture, large and abrupt jumps occur when a water hydroxyl (OH) group trades HB acceptors. In fact, one could regard water reorientation not only as the result of, but also a window on, such HB exchanges. It is also worth pointing out at this stage that the essentials of this mechanism have been shown to be robust vis-à-vis changes in the water force field and the definition of an HB; numerical values of jump times and amplitudes change somewhat, but the essential picture does not [21].

Subsequent simulation work has portrayed these angular jumps as a universal feature of liquid water; they were seen to occur not only in neat water (with both classical [5,21] and quantum [45–47] descriptions of the nuclear motions), but also in a considerable variety of aqueous environments, including aqueous solutions of anions [27,48] and of hydrophobic and amphiphilic solutes [49–51], at various aqueous interfaces [52–58] and in assorted biomolecular hydration layers [15,59–62]. This broad gamut of applicability lends strong support to the proposition that HB-exchanging jumps provide the elementary mechanism of aqueous HB network rearrangement.

In the remainder of this chapter, we will first recount some central features of the new picture and its detailed description via MD simulation and analytic modeling, highlighting those features that are particularly relevant in a spectroscopic context. We then turn to a focus on experimental ultrafast IR spectroscopic measurements, including rotational anisotropy experiments [63] and 2D IR spectroscopic experiments [4,64,65], with a twofold purpose: to consider the probing of the validity of this novel jump picture via these spectroscopies, and to illustrate how this picture can play a vigorous role in helping to interpret these often complex spectroscopic results. As we will see, it will often prove to be important in this endeavor to refer as well to NMR spectroscopic experiments [66,67] for a useful comparison for the time-resolved IR measurements. Our conclusion will be that the jump model and the equations related to that model provide a predictive theoretical framework to comprehend not only pure water dynamics but also how a solute alters the surrounding water's dynamics, as we will illustrate by recent ionic and amphiphilic aqueous solution examples. Of course, we do not pretend to furnish a complete discussion here, and refer the interested reader both to other reviews [7,68] for extensive further details and references and to the original papers [5,21] for discussion of, for example, the jump model details and some other approaches to and some historical aspects of reorientation issues.

3.3 JUMP REORIENTATION MECHANISM: BULK WATER

As we have noted in Section 3.2, MD simulations of water indicate the occurrence of large angular jumps when a water OH group trades HB acceptors [5,21]. Figure 3.1 depicts this exchange process. We immediately emphasize the key feature that this exchange has the character of an activated chemical reaction involving the reactant HB and the final HB, with a transition state (TS) involving a bifurcated HB. The first step in this reaction mechanism is the initial HB's elongation as the first hydration shell water recedes; at the same time, a new water oxygen acceptor approaches, in most cases (>75%) from the second hydration shell of the reorienting water molecule. Once the initial and final oxygen acceptors find themselves equidistant from the rotating water's oxygen, the water OH can suddenly

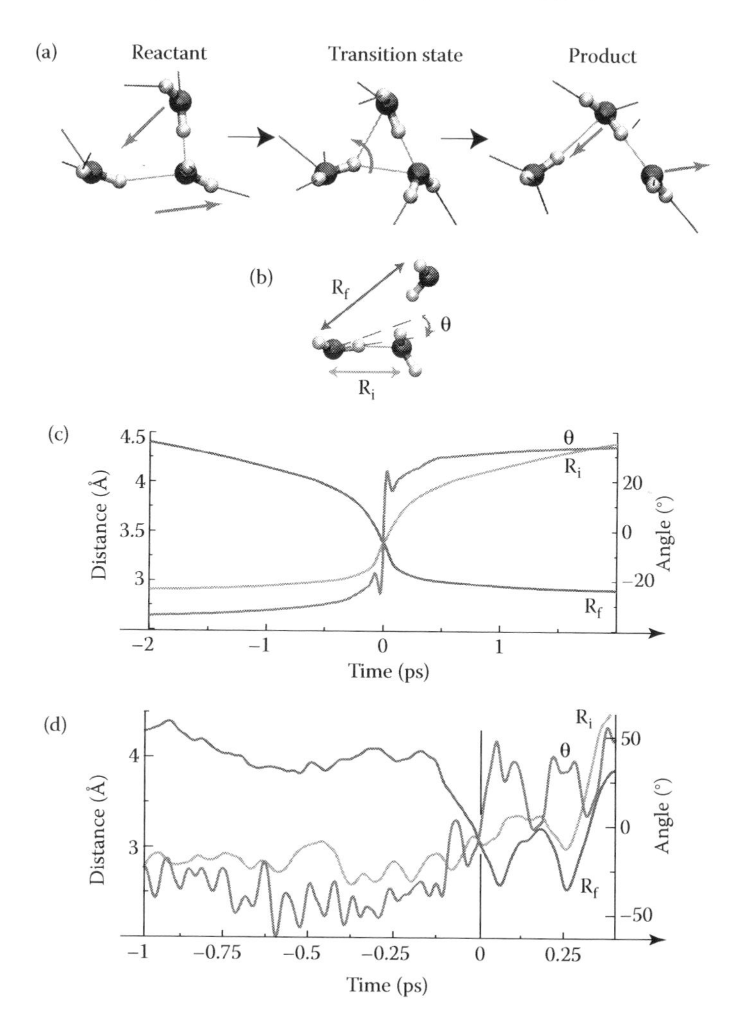

FIGURE 3.1 (a) Schematic jump reorientation mechanism (for a detailed description, see [5,21]). (b) Definition of the three key geometric parameters: distance to the initial acceptor and to the final acceptor, and angle with respect to the bisector plane. (c) Time dependence of these three parameters along the average jump path, where the system crosses the transition state at time 0. (d) Time dependence of these three parameters for one given jump event.

execute a large-amplitude angular jump from one acceptor to another. At the TS for this HB exchange reaction, the rotating water forms a symmetric bifurcated HB with its initial and final water acceptors [5,21] as noted above; this concerted feature of the reaction avoids the higher free energy cost of completely breaking the initial HB to form a "dangling" OH, since the new HB is partially formed while the original one is being broken [5,21]. The reaction concludes when the HB with the new partner eventually stabilizes, while the initial partner completes its departure. As is inherent in the concerted character we have just described, the jump mechanism's free energy barrier originates not only from the initial HB's elongation, but also from the new partner's penetration into the first shell (additional smaller contributions arise from HB fluctuations at the beginning and end of the process—which are responsible for the retreat and approach of the initial and final HB partners—and from the OH angular motion in the neighborhood of the TS) [21,69,70]. This mechanism has been found to be robust with respect to various changes in the water force fields and HB definitions employed [21].

It is important to stress that the mechanism displayed in Figure 3.1 should clearly be understood as an average, simplified, but nonetheless representative mechanism. The actual exchange paths are distributed around this typical mechanism; for example, there is a wide jump angle distribution found around the average amplitude of 68° [21], and as we noted above, not all final partners have their initial location in the second hydration shell of the reorienting water. Other aspects concerning departures from the average are discussed in Ref. [21]. But to date, the inclusion of such features has not been found necessary to describe both simulation and experimental results reasonably well, though naturally this may change in the future as water reorientation is probed more and more profoundly.

Since the jumps discussed above result in the reorientation of water molecules, their signature can be sought in the tcfs of the molecular orientation. As we discussed above, these dynamics can be probed by ultrafast IR spectroscopy [63], as well as more traditional NMR methods [66,67]. For a given body-fixed vector such as the water OH bond, this function reveals the rate of the loss of the memory of the initial orientation. Of most interest for the present chapter is the $n = 2$ orientational tcf defined above:

$$C_2(t) = \langle P_2[\mathbf{u}(0) \cdot \mathbf{u}(t)] \rangle, \tag{3.5}$$

with P_2 the second-order Legendre polynomial and $\mathbf{u}(t)$ the molecular orientation at time t, which we will usually take to be an OH (or OD) bond vector.* Ultrafast polarized pump-probe IR spectroscopy experiments measure the anisotropy decay, which is approximately proportional to $C_2(t)$ [25,63]. There are different time regimes relevant for this tcf (see Figure 3.2). Beyond an initial time interval (<200 fs) where water molecules partially reorient via fast librational motions [70,71], that is, rotations hindered by the HB network-imposed restoring torques, $C_2(t)$ decays exponentially with a characteristic time τ_2. The tcf $C_1(t)$ displays a similar behavior, with a longer characteristic time τ_1. Both of these times (as well as the Debye time τ_D) are in agreement with (the range of) experimental estimates [5,21], which lends credence to the simulation results.

Before proceeding to the theoretical description of the jump mechanism's prediction for the reorientation times, we consider what the simulated times [5,21] themselves reveal about the mechanism. Recall from Equation 3.4 that the rotational diffusion picture predicts the reorientation time ratio $\tau_1/\tau_2 = 3$. The MD results for various water potentials instead give a ratio ~2. This is a significant and revealing difference, particularly in view of the difficulty noted in Section 3.2 of an experimental estimate of τ_1 and thus the τ_1/τ_2 ratio. An even more striking indication of the failure of the diffusion description is the following: the rotational diffusion prediction Equation 3.2 for the time dependence of the tcfs can be tested via MD by separate simulation determination of the rotational diffusion constant D_r, which avoids having to appeal to the highly suspect Debye–Stokes–Einstein

* Most experimental 2D IR studies of water dynamics employ an isotopic mixture of dilute HOD in either H_2O or D_2O to avoid certain complications due to vibrational energy transfer (79,26). In this chapter, we ignore this and simply refer for example, to OH vibrations.

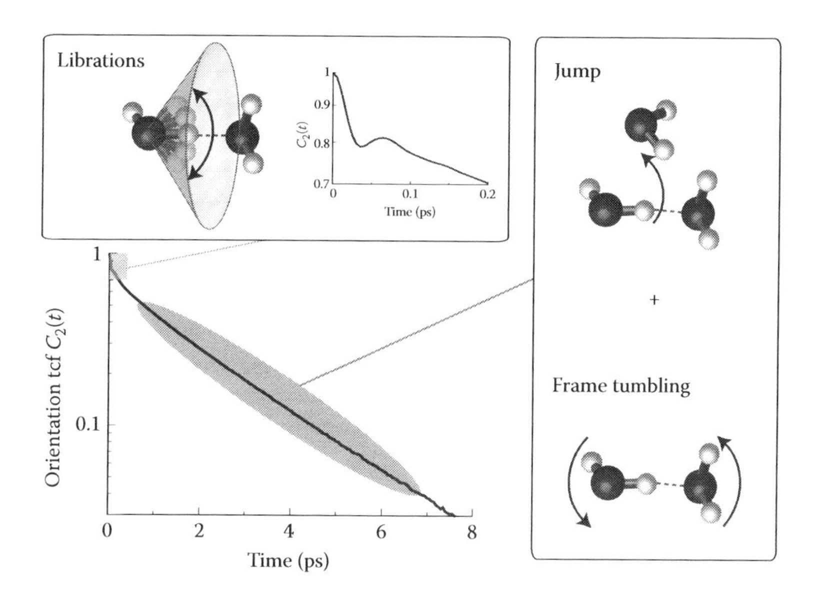

FIGURE 3.2 Bottom left panel: Second-rank orientation time-correlation function $C_2(t)$ (Equations 3.1 and 3.5) for a water OH bond at room temperature determined through molecular dynamics simulations [5,21]. For each time regime, the key reorientation mechanisms are indicated. In both the diffusive reorientation and jump model descriptions, the very short time librational regime (see top left panel) is not addressed. In the diffusive regime, this regime is ignored, and the subsequent exponential tail in the bottom left panel is assumed to be the entire time dependence, given by Equation 3.2. The extended jump picture addresses the exponential tail and describes its behavior in terms of the two mechanisms shown in the right panel, the major jump mechanism and the minority frame reorientation mechanism. The short time librational component is also included when the full time integral of the tcf is required. See the text.

relation Equation 3.3. This exercise shows that Equation 3.2 predicts decays for $C_1(t)$ and $C_2(t)$ that are four to six times more rapid than those of the MD tcfs [5,21].

We now return to our focus on τ_2 and address the theoretical description of this time from the jump perspective. We have shown [5,21] that the characteristic reorientation time resulting from a reorientation through large jumps can be well described via the jump model published by Ivanov in 1964 [72]. This model generalizes the diffusive angular Brownian motion picture to finite-amplitude jumps [5,21]. With the assumptions that the jumps have a constant amplitude $\Delta\theta$, are uncorrelated, and occur with a frequency of $1/\tau_{\text{jump}}$ around axes distributed isotropically, the (second-order) reorientation time is

$$\tau_2 = \tau_{\text{jump}} \left\{ 1 - \frac{1}{5} \frac{\sin(5\Delta\theta/2)}{\sin(\Delta\theta/2)} \right\}^{-1}. \tag{3.6}$$

The model however leaves unspecified two key features of the jumps, their amplitude $\Delta\theta$ and their frequency $1/\tau_{\text{jump}}$. The jump amplitude can be directly and straightforwardly determined from the simulations. The mechanism displayed in Figure 3.1 indicates the key identification of the jump frequency $1/\tau_{\text{jump}}$ as the forward rate constant for the reaction, which breaks an initial stable HB to form a new different stable HB [5,21]. It can be evaluated from the stable states picture theory for a rate constant [73], which is particularly well suited for low barrier reactions [74] (from simple point charge (SPC)/E water simulations at 300 K, the two jump parameters are $\Delta\theta = 68°$ and $\tau_{\text{jump}} = 3.3$ ps [21]).

However, this simple model incorrectly assumes that the molecular orientation remains fixed between jumps. In particular, this assumption implies that while a water OH bond retains the same HB acceptor, the OH direction remains frozen. But this clearly is not the case due to the intact HB axis reorientation through a tumbling motion of the local molecular frame of the hydrogen-bonded pair, whose inclusion is required (see Figure 3.2). With the assumption that the two contributions are statistically independent (in a fashion similar to analysis of multiple source reorientations in proteins [75,76]), the jump contribution can be combined with the frame component to produce the EJM [5,21], which gives an overall EJM reorientation time

$$\frac{1}{\tau_2^{\text{EJM}}} = \frac{1}{\tau_2^{\text{jump}}} + \frac{1}{\tau_2^{\text{frame}}}. \tag{3.7}$$

The structure of this result indicates that the faster of the two contributions will dominate the overall time; this will typically be the jump time, although on occasion the frame time can become an important contributor [48,61,77]. It is worth remarking parenthetically that the frame reorientation is a diffusive motion [21], and if it were dominant, then the Debye diffusion picture would hold. Possible examples of this will be discussed in the section on ionic solutions.

The EJM provides an excellent description of the water τ_2 reorientation times measured by both pump-probe IR anisotropy and NMR spectroscopies and calculated from simulations [5,21]. Despite this pleasant situation, it is possible that a diffusive model could still give rise to the experimental data. Thus, if the simulation data we have discussed is ignored, one could argue that the measured smooth exponential decay of $C_2(t)$ could result either from sharp large-amplitude jumps averaged over a great number of water molecules jumping at different times, or from infinitesimal reorientations executed by each water molecule. To definitively distinguish these quite disparate pictures via experiment, the ratio of the second-order time τ_2 with the first- or third-order times would be needed [5,21], but these latter times are currently not experimentally accessible. As we noted in Section 3.2, while the Debye DR time τ_D is a first-order reorientation time, it pertains to the *collective* reorientation of the total sample dipole, and due to strong static and dynamic correlations between individual water molecules, it is significantly longer than the first-order single-molecule reorientation time τ_1 of interest here [21]. Other direct and incisive experimental evidence is thus necessary to determine unambiguously the reorientation mechanism.

Some first experimental support of the jump model came from the analysis of QENS, discussed elsewhere [31]. But among the many experimental techniques probing water HB dynamics, including QENS [31], NMR spectroscopy [66], and ultrafast pump-probe IR spectroscopy [63], the most recent—and arguably most powerful—has been 2D IR spectroscopy, which accounts for its special emphasis here in the context of the issue of water reorientation in pure water, as well as in our subsequent discussions in this chapter of water reorientation in the neighborhood of assorted solutes.

Numerous excellent discussions of 2D IR spectroscopy are available [64,78–82]; our present discussion will address only those key relevant aspects necessary for water HB dynamics [4,6,64,82–84]. The most important feature of this technique for our purposes is that it follows the time fluctuations of vibrational frequencies and provides the correlation between the frequencies at two instants separated by a given time delay. Its application to water provides detailed HB dynamical information, since—as has been long recognized [85–87]—the water OH stretch vibration is a sensitive probe of hydrogen-bonding interaction. An OH engaged in an HB vibrates at a lower frequency than in the isolated water molecule: the OH bond is weakened by stabilizing interaction with the HB acceptor, inducing an average OH frequency red-shift, compared to the isolated case. For a water molecule immersed in pure water, a stronger than average HB with an HB accepting partner will result in an OH red-shift compared to the average OH frequency. Conversely, an OH that is more weakly hydrogen-bonded to an HB partner than on average exhibits a blue-shifted frequency compared to the average. The 2D IR probing of the OH frequency time evolution accordingly reports on HB network fluctuations and HB breaking and forming events. Accordingly, it has already been successfully exploited for water HB dynamics in an impressively broad gamut of contexts, from the bulk situation [3,4,6,69,84] of interest in this particular section to, for example, ionic aqueous solutions [64,82,83,88], confining environments [79,89], and biomolecular hydration [84], aspects of which we will discuss in separate sections.

That 2D IR spectroscopy has the potential to furnish insight on the presence and nature of angular jumps can be readily appreciated. Recall that the jump TS determined from MD simulations is a bifurcated HB structure in which the reorienting water OH donates two weak HBs to two acceptors (see Figure 3.1). This weakened HB structure results in an OH stretch vibrational frequency blue-shift [4–6,21]. The anticipated OH frequency changes during the jump thus suggest 2D IR spectroscopy as a promising tool for enquiry upon the presence of angular jumps in water, via the exploitation of its ability to selectively follow systems with given initial and final vibrational frequencies. As we will see, however, the situation is not as completely straightforward for liquid water itself as this rosy picture would suggest. In general, to characterize angular jumps, 2D IR often requires supplementary experimental information (e.g., polarization-resolved measurements) and its interpretation can be greatly aided by theory.

A first rather indirect support of the jump picture we have painted is to be found in a 2D IR study [4] performed in fact prior to the theoretical proposal of jumps [5]. Via a focus on the spectral relaxation of blue-shifted OH frequencies corresponding to very weak or broken HBs, this study first provided the evidence of the transient and unstable character of non-HB states, which very quickly (<200 fs) relax to form an HB [4]. While not itself suggesting or establishing a mechanism for HB exchange dynamics, this result is at least fully consistent with the jump mechanism in which HB acceptor exchanges occur through the concerted breaking and forming of HBs [5], in contrast with any sequential mechanism involving a long-lived broken and reorienting HB state, which is ruled out by the experiment.

In fact, most blue-shifted OHs do not even lie at the jump mechanism TS, since most of them do not execute a jump between HB acceptors, which is an activated, relatively infrequent event; the activation free energy barrier $\Delta G^{\ddagger}$ is estimated to be ~2 kcal/mol (which it is important to note is not the same thing as the activation energy [21]). Simulations indicate that most of the blue-shifted water OHs (~80%) only experience a transient HB break and quickly return to their initial HB acceptor and orientation without any jump, which would result in the formation of an HB with a new acceptor; the OHs jumping from one acceptor to another only represent a minor (~20%) part [21].

This implies, unfortunately, that water molecules close to the jump TS cannot be selectively excited using blue-shifted IR excitation. A further dispiriting fact is that the 2D IR measured frequency dynamics cannot distinguish the jump and libration contributions. This impotency is a result of the feature that the OH frequency dynamics is similar for a successful jump forming a new HB with a different water acceptor and for an unsuccessful jump (i.e., a large-amplitude libration) reforming an HB with the initial acceptor. This important feature was recently confirmed by the observed great similarity of the calculated 2D IR spectra of water with and without the jumps [21] (see Figure 3.3 and Ref. [21]).

Indeed, it is quite important to appreciate that fast transient HB breaking/reforming and slower HB jump exchanges provide very *different* contributions to the frequency dynamics (measured in 2D IR) and to the reorientation dynamics (monitored by the pump-probe IR anisotropy decay). Although fast transient HB breaks do not lead to a stable reorientation, they do cause most of the frequency dephasing [2,90]. In contrast, slower HB jump exchanges (whose rate constant is approximately four times smaller than that of the transient breakings at room temperature [5]) provide the key reorientation pathway; these jump exchanges also lead to frequency decorrelation, but the key point here is that by the time a jump occurs, most of the memory of the initial frequency has already been lost due to the transient HB breaks [2]. A further critical aspect is that for HB exchanges from one water acceptor to another, different, water acceptor, the product HB's frequency is of course just the same as that of the reactant HB, so the former's production will make no contribution to the frequency dephasing. All of these fundamental considerations will later prove to be crucial when we consider 2D IR spectroscopy for aqueous solutions of amphiphilic solutes in Section 3.5.

In view of the limitations just discussed, what is to be done? Happily, there is a way forward, and it is the following: to attain discrimination between successful jumps and failed jump attempts, the

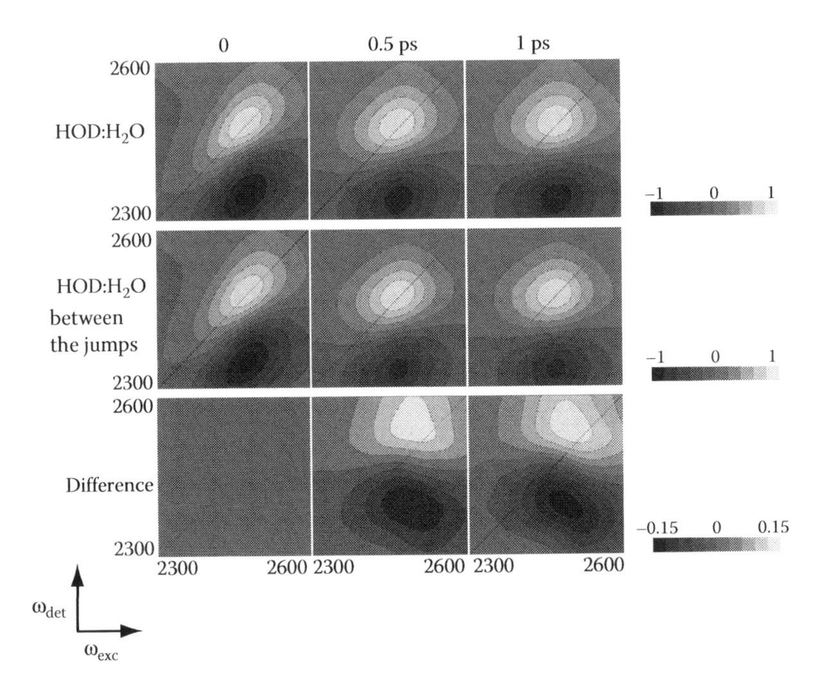

FIGURE 3.3 Bulk water 2D IR spectra calculated [50] for different population delays from a regular simulation (top panel) and from the same simulation during the time intervals between the jumps (middle panel). The difference between these two sets of spectra (bottom panel) provides an estimate of the jump contributions (see also [69]). Horizontal and vertical axes correspond to excitation and detection frequencies.

key criterion is that only the former lead to a stable, long-time reorientation, that is, the formation of a stable, product HB. Accordingly, specific evidence for angular jumps can only emerge from the *combined* study of spectral dynamics and orientation dynamics. This challenging combination is achievable through polarization-resolved 2D IR spectroscopy [69,88,91], as will be discussed in the next section. Such experiments provide a frequency-resolved extension of the conventional anisotropy decay $C_2(t)$: probing of the decaying correlation for each delay t, with the reorientation now given as a function of the initial and final water OH vibration frequencies. Accordingly, this technique allows, for example, the specific following of the reorientation rate of water OHs, which start and end in weakly bonded configurations, that is, with blue-shifted frequencies. Experimentally, such 2D anisotropy maps can be obtained either by polarization-resolved pump-probe [92] or by 2D IR methods [69,82,88,91].

A frequency-dependent version of the EJM (which we label as FD-EJM) is required to interpret such 2D anisotropy diagrams and to identify those spectral features that unambiguously reveal the presence of angular jumps [69]. The basic new feature of this extension describes the jump probability's moderate dependence on the initial HB strength, and thus on the water molecule's OH vibrational frequency. The free energy barrier $\Delta G^{\ddagger}$ to reach the jump TS had already been shown to be mostly due to the concerted elongation of the initial HB and penetration of the new partner within the rotating water's first shell [21]. The respective free energy costs of the elongation and of the new partner's approach were found to be similar [69]. These are nearly independent contributions to the jump activation barrier $\Delta G^{\ddagger}$, so that an exclusive elongation of the initial partner separation costs about $\Delta G^{\ddagger}/2$; a further $\Delta G^{\ddagger}/2$ activation must be provided by the new partner's approach. An initially blue-shifted frequency corresponds to a water OH that has already weakened its initial HB, but such a frequency does not imply that a new HB partner is available to effect a jump. This provides the explanation of the observations that the jump probability increases with vibrational frequency blue-shift, but that this increase is very moderate [21,69].

The FD-EJM predicts that water angular jumps should result in a faster anisotropy decrease for blue-shifted frequencies, provided that memory of the initial frequency is retained, that is, the decay time of the frequency tcf sets an important limiting timescale here [69]. Such accelerated reorientation has already been observed in frequency-resolved pump-probe spectroscopic studies [92] and it can be rigorously connected to the presence of jumps through the FD-EJM. Future polarization-resolved 2D IR measurements should give a more detailed picture. As of the present writing, however, a more telling investigation of the water angular jumps is available from studies of water dynamics in the hydration shells of solutes, which we take up next.

3.4 AQUEOUS SALT SOLUTIONS

We have presented in Section 3.2 a brief indication of the importance of water dynamics in ionic solutions in connection with transport and reactivity. We now commence the discussion of the jump picture and associated IR spectroscopic experiments for water dynamics in the hydration shells of anionic solutes. The particularly attractive feature of anions is that water OH groups will be hydrogen-bonded to the anions and the frequency shifts referenced to the bulk OH values can serve as a spectroscopic handle. In this section, we first consider these dynamics in the context of theory and experiment related to pump-probe experiments, and then turn to the situation for 2D IR spectroscopy.

Pioneering pump-probe ultrafast IR experiments on water dynamics in the hydration shells of anions [93–97] have provided a new—and it must be emphasized, time-dependent—spectroscopic window in this important area for transport and reactivity. These measurements complement traditional NMR spectroscopic results [67], which yield dynamical information at the more restricted level of a single relaxation time. An early MD study of the iodide anion hydration dynamics [98] first identified an important contribution to the time dependence inferred from these experiments arising from exchange of a water molecule between the anion hydration shell and the water bulk, a phenomenon detectable in simulations and experiments sensitive to frequency fluctuations, due to a significant frequency blue-shift of iodine hydration shell water compared to the bulk. In particular, it was shown that the time-evolving, nonequilibrium average OH frequency for a water molecule initially H-bonded to the anion exhibited a longer time tail not present in bulk water. When the simulation was altered to restrict the water to remain H-bonded for a period somewhat longer than the bulk water timescale, no such tail was observed. However, when the restriction was released, the tail present in the fully unrestricted calculations appeared, unambiguously identifying the tail as a signature of the anionic hydration shell–bulk exchange of the water molecule. Since past a fairly short transient period, these nonequilibrium results are equivalent to a measurement of the frequency tcf, a strictly analogous statement about the tail of that tcf applies.

We now turn to the theoretical [27] investigation of the experimental results [96,97] for the Cl^- anion in water. From a theoretical perspective, water reorientation in this anion's hydration shell presents a more complex situation than that of pure water, in the sense that it provides a paradigm heterogeneous situation. A water molecule's HB to the anion is weaker than its HB to another water, a perhaps surprising fact reflected in, for example, the blue-shift in aqueous chloride solutions [99]. An MD and EJM study of water reorientation dynamics in the Cl^- hydration shell [27] showed that again a jump mechanism applied, albeit with asymmetric features reflecting the disparity in the HB strengths mentioned above. It also examined the pump-probe anisotropy for this system, for which a timescale noticeably longer than the bulk water reorientation time was measured experimentally [95]. The original interpretation [100,101] of this longer timescale was that water in the Cl^- hydration shell reorients noticeably more slowly than in the bulk. In language commonly employed [102], this would portray Cl^- as a structure-making anion, a designation at odds with its NMR experimental depiction as a weakly structure-breaking anion [67,102], that is, inducing slightly faster hydration shell dynamics than in the bulk. An essential aspect of the experiment (and our subsequent discussion), which is key for the resolution of this disturbing discrepancy is that the hydration shell waters are only "visible" in the experiment due to the longer vibrational population lifetime for them compared to bulk water. Thus, after the bulk vibrationally excited water signal has decayed, the longer living vibrationally excited anionic hydration shell water signal can be monitored.

Turning now to the theoretical MD/EJM study [27], it was shown that the contribution from water molecules that remain hydrogen-bonded to Cl^- (i.e., which only reorient through the minority mechanism of slow frame tumbling and do not jump), is overemphasized by their longer water OH vibrational lifetime, a distortion that accompanies the very feature that makes these waters visible. As a consequence, the ultrafast IR spectroscopy measures a reorientation time similar to the frame reorientation time associated with an intact hydrogen-bonded anion–water pair (and is thus an excellent tool to examine the dynamics of these minority players), whereas the NMR measurement includes the dominant faster contribution of the jump mechanism, which is the main reorientation pathway for the hydration shell waters (a similar conclusion has been reached for the I^- ion in Ref. [48]).

The EJM is successful in accounting for the NMR time, although as was noted in Section 3.2, the NMR time involves a full time integral, which has a contribution from the short time libration motion of the hydration shell water, an aspect not addressed by the EJM, which focuses on the longer time exponential decay. Thus, it is the combination of this librational component and the EJM that is in agreement with the NMR results [27,48]. In this connection, it should also be remarked that the fact that there is a distribution of angular jump amplitudes, as noted in Section 3.3, has the following consequence. A better numerical value of the reorientation time τ_2, compared to simulation results, is obtained [48] if the average of the jump time in Equation 3.6 is used rather than inserting the average jump angle in that expression.

For completeness, we should note that it has been suggested [101] that the theoretical explanation given above is inconsistent with the long residence time of chloride's hydration layer water calculated by *ab initio* MD simulations [103]. However, this apparent inconsistency originated from an inadequate calculational procedure for water's residence time next to labile ions [74].

In the Cl^- example just discussed, the frame reorientation component has only a minor role in the unbiased hydration shell water reorientation, but has a major role to play in the vibrational lifetime-distorted ultrafast IR experimentally measured lifetime. Will the frame reorientation of an intact ion–water hydrogen-bonded pair ever become important or even dominant for hydration shell dynamics? From the EJM Equation 3.7, this will require τ_2^{frame} to be comparable to or much smaller than τ_2^{jump}, a situation favored by strong HBs. A recent simulation/EJM examination of the small, high-charge-density F^- case indicates that due to the exceptionally strong anion–water HB, the jump and frame contributions to the reorientation time are in fact comparable, representing an anomaly among the halides [48]. Unfortunately, the F^- system cannot be studied by pump-probe methods, since this anion's vibrational population lifetime is too short [94]. Systems where τ_2^{jump} is likely to dominate the hydration shell reorientation are multiply charged ions such as Mg^{2+}. Indeed, all multiply charged ions with very long water residence times [104] are excellent candidates for this dominance.

So far, our discussion of ionic solutions has focused on pump-probe spectroscopic results (and their NMR counterparts). These solutions also have a central, direct role to play in the experimental probing per se of the validity of the jump model, and here, 2D IR spectroscopy comes again to the fore. We begin by indicating why this is a powerful tool in this context.

As we have discussed in the previous section, the principal difficulty in distinguishing an actual jump from a large-amplitude libration in neat water arises from the symmetry between initial and final HB acceptors, which are both water oxygens. Accordingly, ions have been used to break this symmetry and furnish a direct characterization of the angular jumps through 2D IR experiments [82,83,88,105]. As indicated at this section's beginning, simulations indicate that in the presence of a salt, some jumps occur from an initial state where an HB is donated to an anion to a final state with an HB to a water molecule [27,48]. If the anion accepts very weak HBs [82,83,88,105], the vibrational frequency of the water OH-bonded to the anion is sufficiently blue-shifted to become separate from the broad distribution of vibrational frequencies observed for OH groups bonded to a water. An example of this is the I^- anion, where there is an ~60 cm^{-1} OH frequency blue-shift compared to bulk water [99]. In the linear IR spectrum, this results in the telltale signature of two distinct OH stretch bands (Figure 3.4).

The great strength of 2D IR spectroscopy here is the ability to extract dynamical information, and to reveal and measure the kinetics of the chemical exchange between these two water bonding states, hydrogen-bonded to an anion and to a water oxygen. Chemical exchange between two states usually manifests itself in 2D IR spectra through two signatures: the presence of two distinct diagonal peaks and the growth of off-diagonal peaks [81–83,88,105,106]. The latter peaks correspond to OH groups with the same type of HB acceptor when the system is first excited and after a delay T when the correlation is measured (Figure 3.4). As this delay T is increased, exchange between the two populations causes these two diagonal peaks to decrease, while off-diagonal peaks progressively grow. These off-diagonal peaks correspond to OH groups, which have different initial and final frequencies, indicating that they have undergone chemical exchange between the two states (Figure 3.4).

2D IR experiments on concentrated salt solutions (5.5 M $NaBF_4$ [83] and 6 M $NaClO_4$ [82,88,105]) measured the exchange time for a water OH to transit from an anion acceptor to a water acceptor, which approximately corresponds to the jump time τ_{jump} in the EJM terminology. (For this statement to be correct, it is required that the measured time be less than the inverse rate constant for the reverse reaction, so that there is no complication by this reverse reaction.) The experimental values found (7 ps [83] and 6 ps [88]) are consistent with the time determined from a simulation study of a different salt solution (3.6 ps in a 3 M NaCl solution), considering the different nature of the salt and—more importantly—the strong concentration dependence of the jump time, which dramatically increases for increasing salt concentrations [27].

These pioneering 2D IR experiments provided the first time-resolved measurement of water exchange kinetics between different HB acceptors. Despite this achievement and the consistency

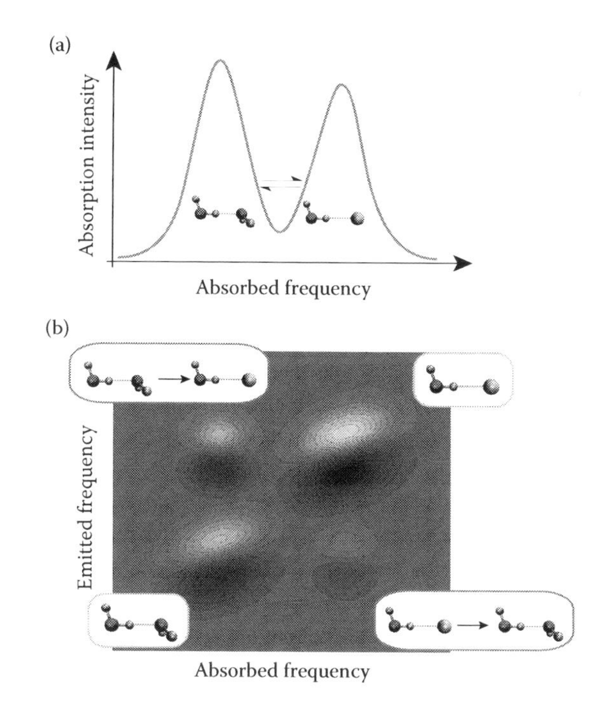

FIGURE 3.4 Schematic representation, for a concentrated ionic aqueous solution, of (a) the linear infrared spectrum and (b) a 2D IR spectrum after a few picoseconds, with growing off-diagonal peaks due to exchanges.

with a salt jump time noted above, there is a limitation. In particular, the experiments could not establish unambiguously that exchanges occur through large angular jumps, since they do not contain any direct geometric information on the exchange mechanism. Subsequent polarization-resolved 2D IR experiments have recently removed this lacuna and have given a quantitative measure of the reorientation associated with HB exchange [82,88]. These experiments fulfill the requirements, stressed near the end of Section 3.3, of combined probing of spectral and orientation dynamics. By following the same approach as for pump-probe IR anisotropy measurements, the comparison of 2D IR spectra acquired with parallel and perpendicular polarizations indicated that systems that exchange between the two states (off-diagonal peaks) experience a much larger reorientation than those that remain in the same state (diagonal peaks). This points toward a jump mechanism and indeed, analysis of the spectra via a kinetic model leads to an average jump angle of $49 \pm 4°$ [88], in qualitative agreement with the distributions of jump angles determined from MD simulations of various aqueous solutions, whose averages lie between 60° and 70° [5,15,21,27,48,49,52,61].

2D IR spectroscopy has thus shown itself to be an exquisite technique to probe water HB exchange kinetics, and unambiguously supports the existence of angular jumps. One could therefore hope to extend these measurements to other aqueous environments to assess how the jump mechanism is followed or altered, according to the different environments, and to compare with theoretical predictions. As always, however, there are some limitations. A broad applicability of 2D IR is constrained by a number of factors. Perhaps most important among these is that a large fraction of the water hydroxyls bonded to solute acceptors is necessary to obtain a detectable spectral peak. Experiments on very dilute solutions (<1 M) available for instance in NMR measurements [66] are thus still inaccessible, making direct experimental assessment of unambiguously molecular orientation dynamics free of concentration effects still to be achieved. To pursue this point, it is most unfortunate that water HB dynamics is extremely sensitive to the solute concentration, so that dilute solution behavior cannot be easily inferred from concentrated solution measurement. In addition and as we have emphasized, these 2D IR studies require the existence of spectrally distinct populations; this means that the vibrational frequency shift between the two states should exceed the frequency fluctuations within each state. Unfortunately, this condition is not met for most singly charged anions, for example, (most of the) halides, for which 2D IR cannot be used to directly follow HB exchanges, although all is not lost since information on spectral diffusion can still be garnered [107]. Nonetheless, one can look forward to further aqueous solution 2D IR experiments on ions satisfying the frequency shift requirement.

Before leaving the aqueous ionic solution topic, we remark that most of the simulations to date have been classical rather than quantum in nature. A situation where the latter will be required is for ions that can be regarded as electronically polarizable over charge localized valence bond (VB) states. For example, the inclusion of the nitrate ion's (electronically coupled) three VB states corresponding to three charge and bonding distributions over the NO_3^- framework [108] allows internal rearrangement of charge and chemical bonds as a water molecule reorients in the anion's hydration shell, either intramolecularly or intermolecularly [109]. One can anticipate that, just as was pointed out for vibrational relaxation dynamics [110], a good source of candidates here would be the ions identified by Pauling [111] as likely to involve multiple VB states of importance.

3.5 WATER NEXT TO AMPHIPHILIC SOLUTES

Of course, many, if not most solutes of interest are amphiphilic, possessed of both hydrophobic and hydrophilic portions. In this section, we will consider water dynamics about such solutes. Our discussion will begin with a fairly extended consideration of the hydrophobic limit, which, as we will see, involves the solute size and thus applies to any type of solute. We then turn to the full amphiphilic situation. We remark at the outset that from the theoretical side, treatment of these matters requires a second layer of theory beyond the first layer of the EJM, analytic formulations aimed at explicitly predicting the impact of hydrophobic and hydrophilic groups on the jump time. Another relevant remark to make before we begin our exposition is that we will need to deal with a certain amount of controversy along the way.

The famous iceberg model of Frank and Evans [112] explained the observed entropy decrease upon hydration of hydrophobic solutes by the structuring of water molecules into ice-like cages around hydrophobic groups. Although its simplicity has a certain appeal, this model has been seriously challenged [14,113], and despite an active search, the ice-like structures have never been evidenced experimentally [114]. However, the recent first time-resolved pump-probe IR anisotropy measurements of water reorientation in the hydration shell of amphiphilic molecules [115] were interpreted as showing that some water molecules are immobilized by hydrophobic groups, thus supporting the old iceberg concept from a dynamical, as opposed to structural, point of view. The observation of a large residual anisotropy after a delay (well beyond the 2.5 ps bulk water reorientation time) was assigned to the immobilization of four water OHs per methyl group [115]. It is important to note here, for later reference, that this immobilization is stated to be intrinsic and independent of the solute concentration [115]. According to these studies, the jump exchange of HB acceptors is suppressed for some hydration shell water molecules, preventing their reorientation [115]. This suppression is also proposed to strongly slow down related phenomena, including proton mobility [116] and OH vibrational dephasing [117–119].

As we will discuss below, an alternate interpretation of these experiments without the invocation of any immobilization has been advocated [49], but the aforementioned studies already raise several questions, which cast some doubt on the immobilization conclusion drawn. For example, only a fraction of the first shell water OHs are supposed to be immobilized [115], whereas the remaining shell waters would exhibit bulk-like dynamics; this stark contrast is at the very least intriguing. In addition, the interpretation of these experiments assumed the presence of a bulk water population even at concentrations close to the solute saturation condition [115], which does not seem realistic.

In any event, this proposed dynamic iceberg picture contrasts quite strongly with the conclusions of earlier NMR and DR experiments on a wide span of solutes, ranging from pure hydrophobes (e.g., xenon [120]) to amphiphiles (alcohols, methylated urea, and small peptides [66,121–126]). In all of these experimental studies, together with all MD simulations performed with a large variety of hydrophobic solutes, it was concluded that water reorientation is only moderately slowed in the hydration shell for dilute solutions: The retardation referenced to bulk water rarely exceeds a factor of 2.

The first theoretical EJM/simulation investigation of the hydrophobic hydration shell dynamics issue focused on aqueous solutions of increasing concentrations of the $(CH_3)_3N^+O^-$ trimethylamine-*N*-oxide (TMAO) molecule, also employed in the experiments (see Scheme 3.1). This molecule bears a hydrophobic portion of three methyl groups (and it should also be noted as a hydrophilic N^+O^- portion; more of that later). Other hydrophobic systems were examined as well, and will be mentioned below. A first point to make is that again the jump picture applies for hydrophobic solute hydration layer dynamics. In addition to the neat water and salt solution cases that we have already discussed, simulations indicate that water next to hydrophobic groups also reorients through large angular jumps [49]. (Here and until further notice, we are speaking of the dilute concentration case, in which reorientation about a given solute molecule is independent of any other solute. This limit is the most

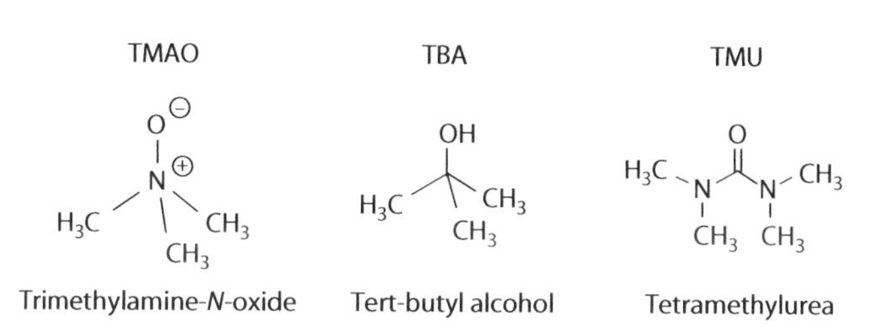

SCHEME 3.1 Chemical structures of the three studied amphiphilic molecules.

fundamental, and as we noted earlier, the water immobilization was also claimed to apply there.) The jump mechanism and amplitude observed in the simulations are almost identical to those found in bulk water, but the jump rate constant $1/\tau_{jump}$ is smaller [49]. Correspondingly, the jump time (and the τ_2 reorientation time) were found in the MD simulation to be ~1.4 times the bulk water values [49].

This jump rate reduction was explained through a TSEV effect for HB exchange (Figure 3.5a). This is the first of the second theoretical layer treatments that identify and characterize important physical effects in the jump time/rate within the EJM. The TSEV provided the first quantitative treatment of a hydrophobic group's influence on the surrounding water HB and reorientation dynamics. The TSEV theory is based upon TS theory for the HB exchange rate constant, exploiting the reaction character of the HB exchange. It assumes that the activation enthalpy component of the activation-free energy $\Delta G^{\ddagger}$ for the HB exchange rate is the same as in bulk water (an assumption reasonably well supported by simulation and experiment near room temperature [49,120]). This leaves the slowdown factor ρ_V

$$\rho_v = \frac{\tau_2^{\text{shell}}}{\tau_2^{\text{bulk}}} = \exp\left[\frac{\Delta\Delta S^{\ddagger}}{R}\right] \tag{3.8}$$

as an exponential of the difference $\Delta\Delta S^{\ddagger} = \Delta S^{\ddagger}_{bulk} - \Delta S^{\ddagger}_{shell}$ of the activation entropies for the exchange in bulk water and in the hydrophobic solute hydration shell. The further simplifying assumption that entropies can be approximated by the Boltzmann constant times the natural logarithm of the microcanonical number of states at the TS then reduces the problem to a geometrical one for a steric effect at the TS, as now explained.

This model determines how the approach of a new water partner necessary for the HB exchange is hindered by the hydrophobe's presence. In the bulk, that new HB partner can approach from an unrestricted range of angles at the TS. But for hydrophobic hydration shell water molecules, the TS angular range is restricted (i.e., there is a lower activation entropy) by the presence of the solute (Figure 3.5a). The TSEV slowdown factor in the jump time (relative to the bulk) ρ_V is then directly related to the local fraction of space occupied by the solute [49], that is, a steric effect. For simple solutes, this can be calculated analytically, and in any event can be computed via simulation for any solute and aqueous solvent situation. For usual convex hydrophobic groups (e.g., methyl, dimethyl, and trimethyl groups, whose radii range between 3 and 5 Å), the TSEV slowdown factor ρ_V is close

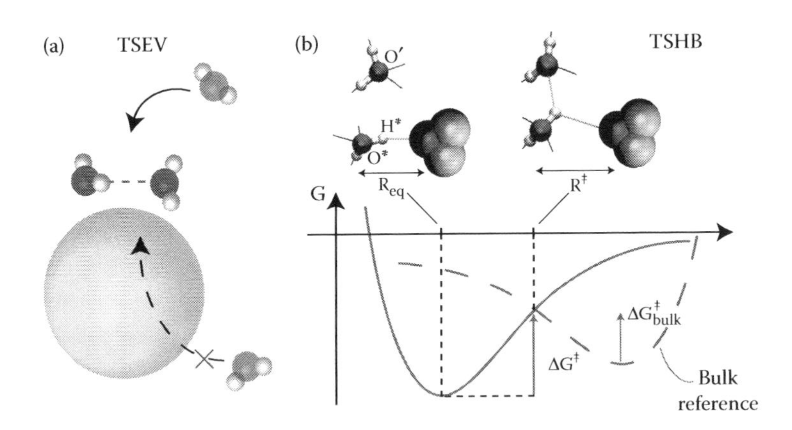

FIGURE 3.5 Schematic representations of the (a) TSEV and (b) TSHB effects of a solute on the water jump rate constant. The TSEV model describes the jump slowdown due to the solute that blocks the approach of some new HB acceptors. The TSHB model connects the free energy cost $\Delta G^{\ddagger}$—to stretch the initial HB from its equilibrium length R_{eq} to its transition state length $R^{\ddagger}$—to the acceleration or slowdown factor induced by the HB acceptor on the water jump dynamics.

to 1.4. This modest deceleration shows that a single hydrophobic group has a very limited effect on water HB dynamics. There is no immobilization. That there is such a moderate slowdown quantitatively agrees with dilute solution studies using a broad range of techniques, including NMR [66], DR [121], Kerr effect spectroscopy [33], light scattering [127], and MD simulations of a wide range of hydrophobic and hydrophobic group-bearing solutes (with both classical force fields [49,51,62] and first principles dynamics [128,129]).

We do note in passing that there is a tendency for the TSEV theory—which admittedly in its present form is a zeroth-order description designed to capture the essentials of the phenomenon—to slightly numerically overestimate the slowdown compared to MD simulation results [49,51]; this might be addressed in future by a more refined treatment of the activation entropy using modern liquid-state structural theories [130]. Further refinements will also include a molecular description of the consequences of different structural fluctuations in the bulk and hydrophobic shell environments on the temperature dependence of the TSEV factor [66], and an explicit consideration of the different jump activation enthalpies in the shell and in the bulk.

So far, we have restricted the discussion to the fundamental regime of low concentrations. What of the experimental results at higher concentrations [115,117–119,131]? For these higher concentrations, significant complications arise from aggregation [132], and especially from its consequences on the fraction of water molecules affected by the solutes and its unknown temperature dependence; we will thus mostly focus on TMAO, which does not aggregate in aqueous solution, even at high concentration [51]. The TSEV approach well predicts [49] the experimental results for high TMAO concentrations, at least to within the fairly wide scatter of the experimental points due, for example, to measurements taken far after the population lifetime of the vibrationally excited water molecules, which it should be recalled are the molecules probed. The TSEV explanation of the increased slowdown at higher concentration is that, in addition to the influence of the hydrophobic solute being trivially more apparent at higher concentration, the slowdown factor ρ_V itself increases. For example, this is due to the presence of other TMAO molecules in the neighborhood of any given TMAO, reducing the possibilities for the TS approach of the new HB partner water molecule required for a jump. At the highest concentrations, there is a further contribution to the slowdown of τ_2 arising from the increasing solution viscosity's deceleration of the diffusive frame reorientation factor in the EJM Equation 3.7.

Finally, what of the influence of the hydrophilic N^+O^- moiety of the TMAO molecule? In this case, the number of waters hydrogen-bonded to this portion is small compared to those in the neighborhood of TMAO's three contiguous methyl groups and their contribution to the average dynamics in the shell is small. But hydrophilic groups in amphiphilic molecules will not always be silent, as we will presently see.

To what extent can 2D IR spectroscopy provide an experimental characterization of the water jumps next to hydrophobic groups? Recent experiments [117–119] and calculations [50] have been performed on a series of small (amphiphilic) solutes containing hydrophobic methyl groups, among them TMAO. The computed spectra [50] (Figure 3.6) compare very well with experiment [117–119], which lends some confidence in the force fields employed. In dilute (1 mol/kg) solution, the 2D spectra are very similar to those of bulk water, due to the dominant bulk water population in the collected signal; there are simply too few solutes to have a visible effect. But in the concentrated (8 mol/kg) case, the spectral relaxation becomes much slower, and the frequency correlation persists over several picoseconds [50,117–119]. (Recall from Section 3.4 that it is just the OH frequency correlations to which 2D IR is sensitive.) This very slow spectral relaxation was interpreted by some authors [117–119] as again revealing a dramatic intrinsic effect of hydrophobic groups on water dynamics. In this view, hydrophobic groups would suppress HB acceptor jumps and "immobilize" the orientation of a fraction of the water molecules within their hydration layer, in line with previous controversial conclusions reached by IR pump-probe anisotropy experiments [115] on these solutes. But we argued above that this interpretation involving an immobilization of water molecules is not correct for hydrophobic groups and needs to be replaced by the TSEV perspective. We now show

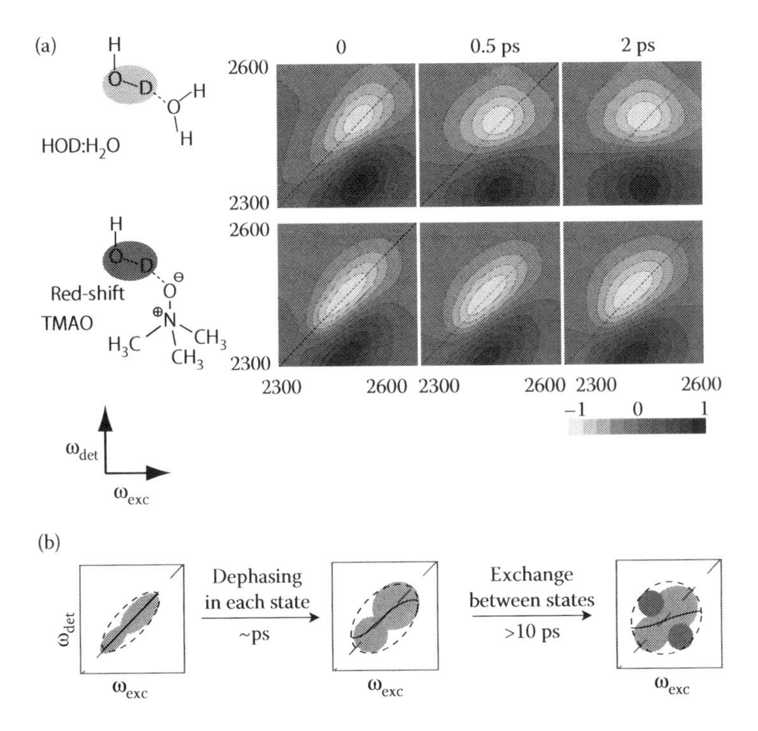

FIGURE 3.6 (a) Simulated 2D IR spectra at several delays for HOD:H_2O [50] and for a concentrated (8 mol/kg) aqueous solutions of TMAO in HOD:H_2O. (b) Schematic representation of the 2D IR spectral decays (omitting the negative 1–2 transition peak), first due to dephasing within each HB state and then due to exchange between HB acceptors.

that the 2D IR spectra can be probing the water dynamics of more than the hydrophobic portion of an amphiphilic molecule and that this disagreement can be resolved through a careful interpretation of 2D IR spectra. This effort leads to a picture fully consistent with the moderate hydrophobic effect predicted by the EJM/TSEV model, and in addition is consistent with the second theoretical layer of the EJM, introduced below, which is necessary to deal with hydrophilic groups.

To begin our exposition, some general considerations will be essential. The hydrophobic portion of these amphiphilic solutes can neither donate nor receive HBs; accordingly, the water molecules in their vicinity form HBs with other water molecules. In the hydration layer of hydrophobic groups, jumps therefore occur from one water oxygen to another. This is the same type of symmetric jump that we discussed in Section 3.3 for the neat water case. Recall that it was shown there that 2D IR was unable to discriminate between spectral dynamics due to transient HB breaking/reforming with a single HB acceptor and the spectral dynamics due to actual jumps between two different HB acceptors. How is it possible then that the 2D IR can be probing the presence or absence of water jump dynamics per se in the hydration of the hydrophobic portion of the solute? A telling indication that the slowdown in the spectral decay is in fact not due to suppressed jumps for waters in the hydrophobic hydration layer comes from the 2D IR spectra computed in the total absence of jumps (Figure 3.3); these are found to be very similar to the regular spectra including jumps. This similarity is a consequence of the dominant contribution made by transient HB breaks (unsuccessful jumps) to spectral dephasing [50]. The conclusion then is that very retarded 2D IR spectral decay for aqueous solutions of amphiphilic solutes is thus very unlikely to originate from suppressed jumps and water immobilization by hydrophobic groups. Another source of the slowdown should be sought.

As to be detailed presently, this slow spectral decay actually has an origin different from the hydrophobic groups [50]: it arises from the slow exchange of water molecules donating an HB to the other moiety of these solute molecules, which are amphiphilic [7,49,66,69,117–119] and—here is the key point—contain an HB accepting hydrophilic group. This hydrophilic slowdown source implies that just as for the ionic solutions we discussed in Section 3.4, two states are possible for each water OH, corresponding to two different HB acceptors: either a water oxygen or a hydrophilic headgroup (whose respective fractions are 76% and 24% at 8 m [69]). Depending on the HB strength with the hydrophilic head, these two populations may have distinct vibrational frequency dynamics.

The amphiphilic solutes considered in the experiments and simulations include in addition to TMAO with its $N^{+}O^{-}$ headgroup, the molecules $(CH_3)_2NCON(CH_3)_2$, tetramethylurea (TMU) with a carbonyl headgroup and $(CH_3)_3COH$, *tert*-butyl alcohol (TBA) with an OH headgroup (see Scheme 3.1). Recall that in our discussion of the pump-probe anisotropy experiments earlier in this section, we stated that the role of the TMAO hydrophilic group was negligible compared to that of the hydrophobic groups in view of the much larger number of water molecules associated with the latter (although not stated there, the same applies to TMU and TBA [50]). But since 2D IR is sensitive to frequency correlations, this type of argument will no longer apply if the hydrophilic groups are an important source of such correlations. At first glance, however, it is not immediately obvious that they are, when we recall from Section 3.4, the requirement of sufficient OH frequency shift compared to the bulk. Because the TMAO, TMU, and TBA polar hydrophilic heads differ, but are not enormously different from a water oxygen, the frequency shift between the OH populations respectively hydrogen-bonded to water and to the solute is at most 30 cm^{-1}, the value for TMAO, whose headgroup has a dipole moment about twice that of the water molecule [50,133]. But this is less than the width of the individual peaks, leading to a single broad band in the linear IR spectrum [50,117]. In the 2D IR spectra, the two frequency distributions are also too close to be resolved separately, and the characteristic signatures of exchange discussed in Section 3.4 for ionic solutions, including the presence of off-diagonal peaks, are not visible. But despite this nonpromising situation, the exchange between these two populations in fact has an important dynamical consequence: it can significantly slow down the spectral relaxation, as we now recount.

In these two-component systems, the spectral relaxation proceeds via two mechanisms. The first, faster contribution arises from transient HB breaking and making events [1,2,90,98], and is similar to the mechanism in neat water that we discussed in Section 3.3. These HB breaks occur without any change in the HB acceptor identity and so they do not lead to an exchange between the two populations. The resulting spectral relaxation time depends only moderately on the HB strength, and as a result is similar in the two states. After ~1 ps, the 2D IR spectra have decayed via this mechanism from the sum of two diagonally elongated bands to the sum of two round peaks (see the schematic representation in Figure 3.6b, which is supported by quantitative frequency tcf [50]). When these two peaks are offset slightly as is the case for the solute discussed, the resulting 2D IR spectra remain diagonally elongated [50,81], which explains the observed incomplete spectral relaxation. Attainment of a full spectral relaxation then requires a second, slower contribution coming from exchanges between the two peaks, that is, HB jumps between water oxygens and solute hydrophilic heads. The visibility of this exchange in the so-called center line slope (CLS) of the 2D IR spectrum [134] depends on the magnitude of this second component, which is (approximately) proportional to the frequency shift between the peaks [50]. For TMAO in particular, the quite polar oxygen headgroup is a sufficiently strong HB acceptor that it induces a marked red-shift ~30 cm^{-1} relative to the bulk, which gives a clearly visible contribution to the CLS. It is this additional contribution due to the slow exchange between the two types of HB acceptors that causes the marked slowdown in the spectral relaxation compared to the bulk. In bulk water, this contribution is absent because there is only a single type of HB acceptor, so that there is zero background for the hydrophilic group exchange contribution.

In drawing this lengthy section to a close, it is worthwhile to stress and expand upon the subtle effects induced by amphiphilic solutes in 2D IR spectra: the same chemical group can have radically different impacts on the water reorientation dynamics and on the water spectral dynamics and in the different experiments that probe those dynamics. While, at low concentration, a hydrophobic group slows down the surrounding water molecules' reorientation by a factor <2 (cf. the beginning portion of this section), it has a very limited effect on water spectral relaxation, since a new HB population is not produced to alter the OH frequency; spectral relaxation remains almost bulk-like, resulting from transient HB breaks largely unaffected by the hydrophobic group. Hydrophilic HB acceptor groups can also have very different impacts on the water HB exchange and thus reorientation dynamics. A hydrophilic group accepting an HB stronger than a water–water HB leads to a reorientation dynamics slowdown (which are potentially much more pronounced than that induced by hydrophobic groups [61]); conversely, weak HB acceptors lead to an accelerated water reorientation. Both of these effects are described via the other second layer theoretical component of the EJM, a transition state hydrogen-bond (TSHB) model [61] (Figure 3.5b). The TSHB model provides a quantitative connection between the free energy cost $\Delta\Delta G^{\ddagger}$ to stretch the initial HB to its TS length (compared to the bulk water case) and the acceleration or slowdown factor ρ_{HB} induced by the hydrophilic group on the reorientation dynamics of the HB donating water. This factor

$$\rho_{HB} = \exp\left[\frac{\Delta\Delta G^{\ddagger}}{k_B T}\right] \tag{3.9}$$

when combined with the TSEV factor ρ_V describes the slowdown (or acceleration) factor for an amphiphilic solute and has been successfully applied, for example, to a series of amino acids [61]. In contrast with this considerable span of possible effects on water reorientation and pump-probe anisotropy, all hydrophilic HB acceptors lead to a new HB type, and therefore—except when the two populations have superimposable frequency distributions—lead to a unidirectional effect in frequency correlation decay and 2D IR: an additional slow component in the water spectral relaxation associated with chemical exchanges between HB acceptors. This asymmetry can have some striking consequences: some weak HB acceptors can accelerate the water reorientation dynamics while retarding the spectral relaxation.

3.6 CONCLUDING REMARKS

Simulation, analytic modeling, and ultrafast IR spectroscopic experiments have led to some extensive and significant new insights for our understanding of water and aqueous solution HB dynamics in recent years. In our discussion in this chapter, we have shown that the presence of large and sudden angular jumps in water, absent in the tradition Debye rotational diffusion description and initially suggested through simulations and analytic models [5], has subsequently received clear support from pioneering 2D IR spectroscopy measurements [82,83,88,105]. These jumps appear to be a universal feature of liquid water and have been found in a wide gamut of environments, including, for example, the bulk, the interface with small hydrophilic groups (e.g., ions), hydrophobic and amphiphilic solutes, the interface with extended surfaces [52,53], and the hydration layer of biomolecules [15,60,61]. Predictive analytic models describe how different chemical moieties affect the jump dynamics and kinetics, and MD simulations and analytic modeling coupled with spectra calculations greatly assist in providing a molecular picture for pump-probe anisotropy experiments and complex and sometimes ambiguous 2D IR spectral patterns [50]. We can confidently anticipate that novel theoretical advances, in improving and extending the theory, and experimental developments, especially in polarization-resolved 2D IR [69,82,88], surface-sensitive 2D IR [135], and 3D-IR [136] spectroscopies, will furnish additional insight into the water's fascinating dynamics in its various regimes and habitats. Among those currently under theoretical investigation are supercooled water, water (and aqueous solutes) next to electrodes and at other interfaces, in the grooves of DNA and water in the course of chemical reaction dynamics.

ACKNOWLEDGMENTS

We thank Guillaume Stirnemann (present address: Department of Chemistry, Columbia University) and Fabio Sterpone (present address: Institut de Biologie Physico Chimique, Paris) for their important participation in much of the research discussed here. This work was supported in part by NSF grant CHE-1112564 (JTH) and by the P. G. de Gennes Foundation for Research.

REFERENCES

1. Rey R, Møller KB, and Hynes JT. 2002. Hydrogen bond dynamics in water and ultrafast infrared spectroscopy. *J Phys Chem A* 106:11993–11996.
2. Møller KB, Rey R, and Hynes JT. 2004. Hydrogen bond dynamics in water and ultrafast infrared spectroscopy: A theoretical study. *J Phys Chem A* 108:1275–1289.
3. Asbury JB et al. 2004. Dynamics of water probed with vibrational echo correlation spectroscopy. *J Chem Phys* 121:12431–12446.
4. Eaves JD, Loparo JJ, Fecko CJ, Roberts ST, Tokmakoff A, and Geissler PL. 2005. Hydrogen bonds in liquid water are broken only fleetingly. *Proc Natl Acad Sci USA* 102:13019–13022.
5. Laage D, and Hynes JT. 2006. A molecular jump mechanism of water reorientation. *Science* 311:832–835.
6. Roberts ST, Ramasesha K, and Tokmakoff A. 2009. Structural rearrangements in water viewed through two-dimensional infrared spectroscopy. *Acc Chem Res* 42:1239–1249.
7. Laage D, Stirnemann G, Sterpone F, Rey R, and Hynes JT. 2011. Reorientation and allied dynamics in water and aqueous solutions. *Annu Rev Phys Chem* 62:395–416.
8. Gertner BJ, Whitnell RM, Wilson KR, and Hynes JT. 1991. Activation to the transition state: Reactant and solvent energy flow for a model S_N2 reaction in water. *J Am Chem Soc* 113:74–87.
9. Ando K, and Hynes JT. 1995. HCl acid ionization in water: A theoretical molecular modeling. *J Mol Liq* 64:25–37.
10. Ando K, and Hynes JT. 1997. Molecular mechanism of HCl acid ionization in water: Ab initio potential energy surfaces and Monte Carlo simulations. *J Phys Chem B* 101:10464–10478.
11. Wang S, Bianco R, and Hynes JT. 2009. Depth-dependent dissociation of nitric acid at an aqueous surface: Car-Parrinello molecular dynamics. *J Phys Chem A* 113:1295–1307.
12. Marx D, Tuckerman ME, Hutter J, and Parrinello M. 1999. The nature of the hydrated excess proton in water. *Nature* 397:601–604.
13. Berkelbach TC, and Tuckerman ME. 2009. Concerted hydrogen-bond dynamics in the transport mechanism of the hydrated proton: A first-principles molecular dynamics study. *Phys Rev Lett* 103:238302.
14. Ball P. 2008. Water as an active constituent in cell biology. *Chem Rev* 108:74–108.
15. Sterpone F, Stirnemann G, and Laage D. 2012. Magnitude and molecular origin of water slowdown next to a protein. *J Am Chem Soc* 134:4116–4119.
16. Berne BJ, and Pecora R. 2000. *Dynamic Light Scattering: With Applications to Chemistry, Biology, and Physics* (Dover, Mineola, NY).
17. McQuarrie DA. 2000. *Statistical Mechanics* (University Science Books, Sausalito, CA).
18. Frölich H. 1958. *Theory of Dielectrics: Dielectric Constant and Dielectric Loss* (Clarendon Press, Oxford).
19. Madden P, and Kivelson D. 1984. A consistent molecular treatment of dielectric phenomena. *Adv Chem Phys* 56: 467–566.
20. van der Spoel D, van Maaren PJ, and Berendsen HJC. 1998. A systematic study of water models for molecular simulation: Derivation of water models optimized for use with a reaction field. *J Chem Phys* 108:10220–10230.
21. Laage D, and Hynes JT. 2008. On the molecular mechanism of water reorientation. *J Phys Chem B* 112:14230–14242.
22. Abragam A. 1961. *The Principles of Nuclear Magnetism* (Oxford University Press, USA).
23. Qvist J, Schober H, and Halle B. 2011. Structural dynamics of supercooled water from quasielastic neutron scattering and molecular simulations. *J Chem Phys* 134:144508.
24. Lipari G, and Szabo A. 1980. Effect of librational motion on fluorescence depolarization and nuclear magnetic resonance relaxation in macromolecules and membranes. *Biophys J* 30:489–506.
25. Lin YS, Pieniazek PA, Yang M, and Skinner JL. 2010. On the calculation of rotational anisotropy decay, as measured by ultrafast polarization-resolved vibrational pump-probe experiments. *J Chem Phys* 132:174505.

26. Mukamel S. 1999. *Principles of Nonlinear Optical Spectroscopy* (Oxford University Press, New York, USA).
27. Laage D, and Hynes JT. 2007. Reorientational dynamics of water molecules in anionic hydration shells. *Proc Natl Acad Sci USA* 104:11167–11172.
28. Teixeira J, Bellissent-Funel M, Chen SH, and Dianoux AJ. 1985. Experimental determination of the nature of diffusive motions of water molecules at low temperatures. *Phys Rev A* 31:1913–1917.
29. Russo D, Murarka RK, Hura G, Verschell E, Copley JRD, and Head-Gordon T. 2004. Evidence for anomalous hydration dynamics near a model hydrophobic peptide. *J Phys Chem B* 108:19885–19893.
30. Russo D, Hura G, and Head-Gordon T. 2004. Hydration dynamics near a model protein surface. *Biophys J* 86:1852–1862.
31. Laage D. 2009. Reinterpretation of the liquid water quasi-elastic neutron scattering spectra based on a nondiffusive jump reorientation mechanism. *J Phys Chem B* 113:2684–2687.
32. Hunt NT, Kattner L, Shanks RP, and Wynne K. 2007. The dynamics of water-protein interaction studied by ultrafast optical Kerr-effect spectroscopy. *J Am Chem Soc* 129:3168–3172.
33. Mazur K, Heisler IA, and Meech SR. 2011. THz spectra and dynamics of aqueous solutions studied by the ultrafast optical Kerr effect. *J Phys Chem B* 115:2563–2573.
34. Mazur K, Heisler IA, and Meech SR. 2010. Ultrafast dynamics and hydrogen-bond structure in aqueous solutions of model peptides. *J Phys Chem B* 114:10684–10691.
35. Mazur K, Heisler IA, and Meech SR. 2011. Water dynamics at protein interfaces: Ultrafast optical Kerr effect study. *J Phys Chem A* 116:2678–2685.
36. Debye PJW. 1929. *Polar Molecules* (The Chemical Catalog Company, New York).
37. Levine IN. 2008. *Quantum Chemistry* (Prentice Hall, Upper Saddle River, NJ).
38. Hynes JT, Kapral R, and Weinberg M. 1978. Molecular rotation and reorientation: Microscopic and hydrodynamic contributions. *J Chem Phys* 69:2725–2733.
39. Hynes JT, Kapral R, and Weinberg M. 1977. Microscopic boundary layer effects and rough sphere rotation. *J Chem Phys* 67:3256–3267.
40. Eisenberg D, and Kauzmann W. 2005. The structure and properties of water (Oxford University Press, Oxford, UK).
41. Bagchi B. 2005. Water dynamics in the hydration layer around proteins and micelles. *Chem Rev* 105:3197–219.
42. Rahman A, and Stillinger H. 1971. Molecular dynamics study of liquid water. *J Chem Phys* 55:3336–3359.
43. O'Reilly DE. 1974. Self-diffusion coefficients and rotational correlation times in polar liquids. VI. Water. *J Chem Phys* 60:1607–1618.
44. Winkler K, Lindner J, Bürsing H, and Vöhringer P. 2000. Ultrafast Raman-induced Kerr-effect of water: Single molecule versus collective motions. *J Chem Phys* 113:4674–4682.
45. Paesani F, Iuchi S, and Voth GA. 2007. Quantum effects in liquid water from an ab initio-based polarizable force field. *J Chem Phys* 127:074506.
46. Paesani F, Yoo S, Bakker HJ, and Xantheas SS. 2010. Nuclear quantum effects in the reorientation of water. *J Phys Chem Lett* 1:2316–2321.
47. Ono J, Hyeon-Deuk K, and Ando K. 2012. Semiquantal molecular dynamics simulations of hydrogen-bond dynamics in liquid water using spherical Gaussian wavepackets. *Int J Quantum Chem*, doi: 10.1002/qua.24146.
48. Boisson J, Stirnemann G, Laage D, and Hynes JT. 2011. Water reorientation dynamics in the first hydration shells of F^- and I^-. *Phys Chem Chem Phys* 13:19895–19901.
49. Laage D, Stirnemann G, and Hynes JT. 2009. Why water reorientation slows without iceberg formation around hydrophobic solutes. *J Phys Chem B* 113:2428–2435.
50. Stirnemann G, Hynes JT, and Laage D. 2010. Water hydrogen bond dynamics in aqueous solutions of amphiphiles. *J Phys Chem B* 114:3052–3059.
51. Stirnemann G, Sterpone F, and Laage D. 2011. Dynamics of water in concentrated solutions of amphiphiles: Key roles of local structure and aggregation. *J Phys Chem B* 115:3254–3262.
52. Stirnemann G, Rossky PJ, Hynes JT, and Laage D. 2010. Water reorientation, hydrogen-bond dynamics and 2D IR spectroscopy next to an extended hydrophobic surface. *Faraday Discuss* 146:263–281.
53. Stirnemann G, Castrillón SR, Hynes JT, Rossky PJ, Debenedetti PG, and Laage D. 2011. Non-monotonic dependence of water reorientation dynamics on surface hydrophilicity: Competing effects of the hydration structure and hydrogen-bond strength. *Phys Chem Chem Phys* 13:19911–19917.
54. Chowdhary J, and Ladanyi BM. 2009. Hydrogen bond dynamics at the water/hydrocarbon interface. *J Phys Chem B* 113:4045–4053.

55. Rosenfeld DE, and Schmuttenmaer CA. 2011. Dynamics of the water hydrogen bond network at ionic, nonionic, and hydrophobic interfaces in nanopores and reverse micelles. *J Phys Chem B* 115:1021–1031.
56. Laage D, and Thompson WH. 2012. Reorientation dynamics of nanoconfined water: Power-law decay, hydrogen-bond jumps, and test of a two-state model. *J Chem Phys* 136:044513.
57. Malani A, and Ayappa G. 2012. Relaxation and jump dynamics of water at the mica interface. *J Chem Phys* 136:194701.
58. Mukherjee B, Maiti PK, Dasgupta C, and Sood AK. 2009. Jump reorientation of water molecules confined in narrow carbon nanotubes. *J Phys Chem B* 113:10322–10330.
59. Jana B, Pal S, and Bagchi B. 2008. Hydrogen bond breaking mechanism and water reorientational dynamics in the hydration layer of lysozyme. *J Phys Chem B* 112:9112–9117.
60. Zhang Z, and Berkowitz ML. 2009. Orientational dynamics of water in phospholipid bilayers with different hydration levels. *J Phys Chem B* 113:7676–7680.
61. Sterpone F, Stirnemann G, Hynes JT, and Laage D. 2010. Water hydrogen-bond dynamics around amino acids: The key role of hydrophilic hydrogen-bond acceptor groups. *J Phys Chem B* 114:2083–2089.
62. Verde AV, and Campen RK. 2011. Disaccharide topology induces slowdown in local water dynamics. *J Phys Chem B* 115:7069–7084.
63. Bakker HJ, and Skinner JL. 2009. Vibrational spectroscopy as a probe of structure and dynamics in liquid water. *Chem Rev* 110:1498–1517.
64. Fayer MD, Moilanen DE, Wong D, Rosenfeld DE, Fenn EE, and Park S. 2009. Water dynamics in salt solutions studied with ultrafast two-dimensional infrared (2D IR) vibrational echo spectroscopy. *Acc Chem Res* 42:1210–1219.
65. Ji M, and Gaffney KJ. 2011. Orientational relaxation dynamics in aqueous ionic solution: Polarization-selective two-dimensional infrared study of angular jump-exchange dynamics in aqueous 6M $NaClO_4$. *J Chem Phys* 134:044516.
66. Qvist J, and Halle B. 2008. Thermal signature of hydrophobic hydration dynamics. *J Am Chem Soc* 130:10345–10353.
67. Endom L, Hertz HG, Thül B, and Zeidler MD. 1967. A microdynamic model of electrolyte solutions as derived from nuclear magnetic relaxation and self-diffusion data. *Ber Bunsenges Phys Chem* 71:1008–1031.
68. Laage D, Stirnemann G, Sterpone F, and Hynes JT. 2012. Water jump reorientation: From theoretical prediction to experimental observation. *Acc Chem Res* 45:53–62.
69. Stirnemann G, and Laage D. 2010. Direct evidence of angular jumps during water reorientation through two-dimensional infrared anisotropy. *J Phys Chem Lett* 1:1511–1516.
70. Laage D, and Hynes JT. 2006. Do more strongly hydrogen-bonded water molecules reorient more slowly? *Chem Phys Lett* 433:80–85.
71. Moilanen DE, Fenn EE, Lin YS, Skinner JL, Bagchi B, and Fayer MD. 2008. Water inertial reorientation: Hydrogen bond strength and the angular potential. *Proc Natl Acad Sci USA* 105:5295–300.
72. Ivanov EN. 1964. Theory of rotational Brownian motion. *Sov Phys JETP* 18:1041–1045.
73. Northrup SH, and Hynes JT. 1980. The stable states picture of chemical reactions. I. Formulation for rate constants and initial condition effects. *J Chem Phys* 73:2700–2714.
74. Laage D, and Hynes JT. 2008. On the residence time for water in a solute hydration shell: Application to aqueous halide solutions. *J Phys Chem B* 112:7697–7701.
75. Szabo A. 1984. Theory of fluorescence depolarization in macromolecules and membranes. *J Chem Phys* 81:150–167.
76. Tjandra N, Szabo A, and Bax A. 1996. Protein backbone dynamics and 15N chemical shift anisotropy from quantitative measurement of relaxation interference effects. *J Am Chem Soc* 118:6986–6991.
77. Vartia AA, Mitchell-Koch KR, Stirnemann G, Laage D, and Thompson WH. 2011. On the reorientation and hydrogen-bond dynamics of alcohols. *J Phys Chem B* 115:12173–12178.
78. Zheng J, Kwak K, and Fayer MD. 2007. Ultrafast 2D IR vibrational echo spectroscopy. *Acc Chem Res* 40:75–83.
79. Fayer MD. 2009. Dynamics of liquids, molecules, and proteins measured with ultrafast 2D IR vibrational echo chemical exchange spectroscopy. *Annu Rev Phys Chem* 60:21–38.
80. Hamm P, and Zanni M. 2011. *Concepts and Methods of 2D Infrared Spectroscopy* (Cambridge University Press, Cambridge, UK).
81. Kim YS, and Hochstrasser RM. 2009. Applications of 2D IR spectroscopy to peptides, proteins, and hydrogen-bond dynamics. *J Phys Chem B* 113:8231–8251.

82. Gaffney KJ, Ji M, Odelius M, Park S, and Sun Z. 2011. H-bond switching and ligand exchange dynamics in aqueous ionic solution. *Chem Phys Lett* 504:1–6.
83. Moilanen DE, Wong D, Rosenfeld DE, Fenn EE, and Fayer MD. 2009. Ion-water hydrogen-bond switching observed with 2D IR vibrational echo chemical exchange spectroscopy. *Proc Natl Acad Sci USA* 106:375–380.
84. Elsaesser T. 2009. Two-dimensional infrared spectroscopy of intermolecular hydrogen bonds in the condensed phase. *Acc Chem Res* 42:1220–1228.
85. Pimentel G, and McClellan A. 1960. *The Hydrogen Bond* (W. H. Freeman, San Francisco).
86. Mikenda W. 1986. Stretching frequency versus bond distance correlation of O–D (H) Y (Y = N, O, S, Se, Cl, Br, I) hydrogen bonds in solid hydrates. *J Mol Struct* 147:1–15.
87. Mikenda W, and Steinböck S. 1996. Stretching frequency vs. bond distance correlation of hydrogen bonds in solid hydrates: A generalized correlation function. *J Mol Struct* 384:159–163.
88. Ji M, Odelius M, and Gaffney KJ. 2010. Large angular jump mechanism observed for hydrogen bond exchange in aqueous perchlorate solution. *Science* 328:1003–1005.
89. Fenn EE, Wong DB, and Fayer MD. 2009. Water dynamics at neutral and ionic interfaces. *Proc Natl Acad Sci USA* 106:15243–15248.
90. Lawrence P, and Skinner L. 2003. Vibrational spectroscopy of HOD in liquid D_2O. III. Spectral diffusion, and hydrogen-bonding and rotational dynamics. *J Chem Phys* 118:264–272.
91. Ramasesha K, Roberts ST, Nicodemus RA, Mandal A, and Tokmakoff A. 2011. Ultrafast 2D IR anisotropy of water reveals reorientation during hydrogen-bond switching. *J Chem Phys* 135:054509.
92. Bakker HJ, Rezus YL, and Timmer RL. 2008. Molecular reorientation of liquid water studied with femtosecond midinfrared spectroscopy. *J Phys Chem A* 112:11523–11534.
93. Omta AW, Kropman MF, Woutersen S, and Bakker HJ. 2003. Influence of ions on the hydrogen-bond structure in liquid water. *J Chem Phys* 119:12457–12461.
94. Kropman MF, and Bakker HJ. 2004. Effect of ions on the vibrational relaxation of liquid water. *J Am Chem Soc* 126:9135–9141.
95. Kropman MF, Nienhuys HK, and Bakker HJ. 2002. Real-time measurement of the orientational dynamics of aqueous solvation shells in bulk liquid water. *Phys Rev Lett* 88:077601.
96. Omta AW, Kropman MF, Woutersen S, and Bakker HJ. 2003. Negligible effect of ions on the hydrogen-bond structure in liquid water. *Science* 301:347–349.
97. Kropman MF, and Bakker HJ. 2001. Dynamics of water molecules in aqueous solvation shells. *Science* 291:2118–2120.
98. Nigro B, Re S, Laage D, Rey R, and Hynes JT. 2006. On the ultrafast infrared spectroscopy of anion hydration shell hydrogen bond dynamics. *J Phys Chem A* 110:11237–11243.
99. Bergstroem PA, Lindgren J, and Kristiansson O. 1991. An IR study of the hydration of perchlorate, nitrate, iodide, bromide, chloride and sulfate anions in aqueous solution. *J Phys Chem* 95:8575–8580.
100. Bakker J, Kropman F, and Omta W. 2005. Effect of ions on the structure and dynamics of liquid water. *J Phys: Cond Matt* 17:S3215–S3224.
101. Bakker HJ. 2008. Structural dynamics of aqueous salt solutions. *Chem Rev* 108:1456–1473.
102. Marcus Y. 2009. Effect of ions on the structure of water: Structure making and breaking. *Chem Rev* 109:1346–1370.
103. Heuft M, and Meijer J. 2003. Density functional theory based molecular-dynamics study of aqueous chloride solvation. *J Chem Phys* 119:11788–11791.
104. Ohtaki H, and Radnai T. 1993. Structure and dynamics of hydrated ions. *Chem Rev* 93:1157–1204.
105. Park S, Odelius M, and Gaffney KJ. 2009. Ultrafast dynamics of hydrogen bond exchange in aqueous ionic solutions. *J Phys Chem B* 113:7825–7835.
106. Dlott DD. 2005. Chemistry. Ultrafast chemical exchange seen with 2D vibrational echoes. *Science* 309:1333–1334.
107. Park S, and Fayer MD. 2007. Hydrogen bond dynamics in aqueous NaBr solutions. *Proc Natl Acad Sci USA* 104:16731–16738.
108. Ramesh SG, Re S, Boisson J, and Hynes JT. 2010. Vibrational symmetry breaking of NO_3- in aqueous solution: NO asymmetric stretch frequency distribution and mean splitting. *J Phys Chem A* 114:1255–1269.
109. Boisson J. 2008. PhD Thesis, University of Paris, UPMC: Sur l'interaction eau/anion;? les caractères structurants et destructurants, la rupture de symétrie du nitrate.
110. Rey R, and Hynes JT. 2001. Coulomb force and intramolecular energy flow effects for vibrational energy transfer for small molecules in polar solvents. In *Ultrafast Infrared and Raman Spectroscopy* (Marcel Dekker, New York). M. D. Fayer, Editor.

111. Pauling L. 1960. *The Nature of the Chemical Bond and the Structure of Molecules and Crystals: An Introduction to Modern Structural Chemistry* (Cornell University Press, Ithaca, NY).
112. Frank HS, and Evans MW. 1945. Free volume and entropy in condensed systems III. Entropy in binary liquid mixtures; partial molal entropy in dilute solutions; structure and thermodynamics in aqueous electrolytes. *J Chem Phys* 13:507–532.
113. Blokzijl W, and Engberts JBFN. 1993. Hydrophobic effects. Opinions and facts. *Angew Chem Int Edit Engl* 32:1545–1579.
114. Buchanan P, Aldiwan N, Soper K, Creek L, and Koh A. 2005. Decreased structure on dissolving methane in water. *Chem Phys Lett* 415:89–93.
115. Rezus Y, and Bakker H. 2007. Observation of immobilized water molecules around hydrophobic groups. *Phys Rev Lett* 99:148301.
116. Bonn M et al. 2009. Suppression of proton mobility by hydrophobic hydration. *J Am Chem Soc* 131: 17070–17071.
117. Bakulin AA, Liang C, la Cour Jansen T, Wiersma DA, Bakker HJ, and Pshenichnikov MS. 2009. Hydrophobic solvation: A 2D IR spectroscopic inquest. *Acc Chem Res* 42:1229–1238.
118. Petersen C, Bakulin AA, Pavelyev VG, Pshenichnikov MS, and Bakker HJ. 2010. Femtosecond midinfrared study of aggregation behavior in aqueous solutions of amphiphilic molecules. *J Chem Phys* 133:164514.
119. Bakulin AA, Pshenichnikov MS, Bakker HJ, and Petersen C. 2011. Hydrophobic molecules slow down the hydrogen-bond dynamics of water. *J Phys Chem A* 115:1821–1829.
120. Weingärtner H, Haselmeier R, and Holz M. 1996. Effect of xenon upon the dynamical anomalies of supercooled water. A test of scaling-law behavior. *J Phys Chem* 100:1303–1308.
121. Hallenga K, Grigera JR, and Berendsen HJC. 1980. Influence of hydrophobic solutes on the dynamic behavior of water. *J Phys Chem* 84:2381–2390.
122. Okouchi S, Moto T, Ishihara Y, Numajiri H, and Uedaira H. 1996. Hydration of amines, diamines, polyamines and amides studied by NMR. *J. Chem. Soc. Faraday Trans* 92:1853–1857.
123. Ishihara Y, Okouchi S, and Uedaira H. 1997. Dynamics of hydration of alcohols and diols inaqueous solutions. *J. Chem. Soc. Faraday Trans* 93:3337–3342.
124. Okouchi S, Ashida T, Sakaguchi S, Tsuchida K, Ishihara Y, and Uedaira H. 2002. Dynamics of the hydration of halogenoalcohols in aqueous solution. *Bull Chem Soc Jpn* 75:59–63.
125. Okouchi S, Tsuchida K, Yoshida S, Ishihara Y, Ikeda S, and Uedaira H. 2005. Dynamics of the hydration of amino alcohols and diamines in aqueous solution. *Bull Chem Soc Jpn* 78:424–429.
126. Shimizu A, Fumino K, Yukiyasu K, and Taniguchi Y. 2000. NMR studies on dynamic behavior of water molecule in aqueous denaturant solutions at 25°C: Effects of guanidine hydrochloride, urea and alkylated ureas. *J Mol Liq* 85:269–278.
127. Lupi L et al. 2011. Hydrophobic hydration of tert-butyl alcohol studied by Brillouin light and inelastic ultraviolet scattering. *J Chem Phys* 134:055104.
128. Silvestrelli PL. 2009. Are there immobilized water molecules around hydrophobic groups? Aqueous solvation of methanol from first principles. *J Phys Chem B* 113:10728–10731.
129. Rossato L, Rossetto F, and Silvestrelli PL. 2012. Aqueous solvation of methane from first principles. *J Phys Chem B* 116:4552–4560.
130. Pratt LR, and Pohorille A. 2002. Hydrophobic effects and modeling of biophysical aqueous solution interfaces. *Chem Rev* 102:2671–2692.
131. Tielrooij KJ, Hunger J, Buchner R, Bonn M, and Bakker HJ. 2010. Influence of concentration and temperature on the dynamics of water in the hydrophobic hydration shell of tetramethylurea. *J Am Chem Soc* 132:15671–15678.
132. Almásy L, Len A, Székely NK, and Pleštil J. 2007. Solute aggregation in dilute aqueous solutions of tetramethylurea. *Fluid Phase Equilibr* 257:114–119.
133. Paul S, and Patey GN. 2006. Why *tert*-butyl alcohol associates in aqueous solution but trimethylamine-*N*-oxide does not. *J Phys Chem B* 110:10514–10518.
134. Kwak K, Rosenfeld DE, and Fayer MD. 2008. Taking apart the two-dimensional infrared vibrational echo spectra: More information and elimination of distortions. *J Chem Phys* 128:204505.
135. Zhang Z, Piatkowski L, Bakker HJ, and Bonn M. 2011. Communication: Interfacial water structure revealed by ultrafast two-dimensional surface vibrational spectroscopy. *J Chem Phys* 135:021101.
136. Garrett-Roe S, and Hamm P. 2009. What can we learn from three-dimensional infrared spectroscopy? *Acc Chem Res* 42:1412–1422.

4 Femtosecond Vibrational Spectroscopy of Aqueous Systems

H. J. Bakker and M. Bonn

CONTENTS

4.1 INTRODUCTION

Many of the anomalous properties of liquid water find their origin in the large number of directional hydrogen-bond interactions present in this liquid [1]. The extremely high density of hydrogen bonds leads to a strong cohesion between the water molecules with the result that water has an anomalously high freezing and melting point in view of its small molecular mass. The high density of hydrogen bonds also explains the high heat capacity of water: the breaking of the high density of hydrogen bonds requires an extremely large amount of energy. As a result, water is an ideal thermal regulator as the release or acceptance of large amounts of energy only leads to moderate changes of the temperature [2,3]. When water is frozen to ice, the water molecules form an extended spatial hydrogen-bonded network due to the strongly directional character of the hydrogen bonds and the bend shape of the water molecule. As a result, the specific volume increases at the phase transition to ice, which is probably the best known anomaly of water.

Water is also an excellent solvent due to its polarity, high dielectric constant, and its ability to form hydrogen bonds to other molecules and ions. It is one of the few solvents showing sufficiently strong interactions with ions to break up the strong Coulomb interactions of salts. The interaction with water also plays an essential role in the determination of the conformation of biomolecular systems. In aqueous solution, proteins tend to acquire a conformation in which their hydrophilic molecular groups are located on the outside having strong interaction with the surrounding water, while their hydrophobic molecular groups are located on the inside and are shielded from the water. A similar self-organizing process leads to the formation of bilipid membranes in which the polar head groups of the lipid molecules are interacting with water and the apolar tails are shielded from the water.

The structure and dynamics of liquid water and aqueous solutions have been studied with many experimental and theoretical techniques. The molecular-scale structure has been investigated with x-ray diffraction and neutron scattering. From these studies, it follows that the average oxygen–oxygen distance of the water molecules amounts to 2.8 Å. Unfortunately, these techniques do not give information on the position of the hydrogen atoms. It is quite generally believed that most water molecules arrange their hydroxyl groups in near-linear O–H···O hydrogen-bonded configurations. Most molecular dynamics simulations also indicate that liquid water possesses a high density of near-linear O–H···O hydrogen bonds, corresponding to an average of ~3.5 hydrogen bonds per water molecule [4–8]. However, this picture has recently been contested by x-ray absorption, x-ray emission, and x-ray Raman scattering studies that probe the electronic states of the oxygen atoms of the water molecules [9,10]. In these studies, it was claimed that most water molecules participate in two hydrogen bonds only (single donor, single acceptor) instead of four. This interpretation has later in turn been disputed [11,12].

The molecular-scale dynamics of the water molecules in liquid water and aqueous solutions have been investigated with several experimental techniques like nuclear magnetic resonance (NMR), dielectric relaxation spectroscopy, and nonlinear femtosecond vibrational spectroscopy. NMR can be used to measure the longitudinal proton spin relaxation time from which the average correlation time constants of the rotation over different molecular axes of the water molecule can be derived [13–18]. With dielectric relaxation, the polarization response of water to an oscillatory electric field is measured [19–22]. The frequency dependence of this response gives information on the reorientation time of the water molecules.

A shortcoming of NMR and dielectric relaxation spectroscopy is that only an average of the dynamics of the water molecules can be measured. NMR echo techniques are in principle capable of distinguishing different subsets of water molecules, but the interconversion between the different types of waters occurs on timescales (picoseconds) much shorter than the typical timescale of NMR experiments (microseconds). As a result, NMR probes the average dynamics of all water molecules, making it difficult to extract unambiguous information on the properties of different subensembles. This property of NMR is a problem in the study of aqueous solutions, for which it can be expected that the dynamics of the water molecules solvating molecules and ions differ from the dynamics of the bulk water molecules.

Nonlinear femtosecond vibrational spectroscopy is quite unique in that it can probe the dynamics of different water molecules on a timescale that is shorter than their interconversion. An essential feature of this technique is its capability to excite a molecular vibration of a subensemble of water molecules with femtosecond mid-infrared light pulses and to subsequently follow that the dynamics of these molecules with additional femtosecond mid-infrared pulses. Provided the dynamics are slower than the duration of the light pulses, it is thus possible to time resolve the dynamics. This technique has thus been applied successfully to study the molecular-scale dynamics [37–46] of water molecules. Over the last 15 years, nonlinear femtosecond vibrational spectroscopy has thus been applied successfully to study the molecular-scale dynamics of neat liquid water and different aqueous solutions. In this chapter, we present an overview of the different variations of the technique and the information that has been obtained to date.

4.2 NONLINEAR VIBRATIONAL SPECTROSCOPY

4.2.1 Generation of Femtosecond Mid-Infrared Pulses

Nonlinear vibrational spectroscopy of aqueous systems requires mid-infrared laser pulses that are resonant with the molecular vibrations of H_2O, HDO, and/or D_2O. The energy relaxation dynamics and molecular motions of water molecules take place on a subpicosecond to picosecond timescale. Hence, the time-resolved study of these dynamics requires mid-infrared pulses with a pulse duration on the order of 100 fs. In all reported femtosecond vibrational mid-infrared experiments of aqueous systems, the mid-infrared pulses were generated via a sequence of nonlinear frequency-conversion processes that are pumped by the intense femtosecond laser pulses delivered by Ti:sapphire multipass and/or regenerative amplifiers (800 nm, pulse duration 30–100 fs, pulse energy ≥1 mJ, repetition rate 1 kHz).

The infrared pulses required to study the O–H and O–D stretch vibrations are usually generated via an optical parametric amplification process in a KTP or $KNbO_3$ crystal that is pumped by a part of the 800 nm pulses of the Ti:sapphire laser [38–45]. The study of the O–H vibration(s) of H_2O or HDO requires pulses at 3 μm, and to generate pulses at this wavelength, the parametric amplification process has to be seeded by 1100 nm light. Similarly, the study of the O–D vibration(s) of D_2O or HDO requires infrared pulses at 4 μm, and to obtain these pulses, the parametric amplification has to be seeded with 1000 nm light. The 1000/1100 nm seed can be produced with different methods. For instance, a small part of the 800 nm pulse can be used to generate a white-light continuum [40,42,43]. The part of this continuum at 1100/1000 nm can then be parametrically amplified in a BBO (β-barium borate) crystal. This process is pumped by a pulse at 400 nm that is generated via second harmonic generation of part of the 800 nm pulse [40], or by part of the fundamental 800 nm pulse [42,43]. The amplified light at 1100/1000 nm is subsequently used as a seed in a final parametric amplification process in KTP [40] or $KNbO_3$ [42,43] to generate the desired pulses at 3 or 4 μm. The 1000/1100 nm seed pulses can also be obtained by first generating 1300/1400 nm (signal) and 2200/2000 nm (idler) via optical parametric amplification in a BBO crystal [38,39,44,45], followed by frequency doubling of the idler pulses in a second BBO crystal. These doubled idler pulses are then used to seed a final parametric amplification process in KTP [38,39,45] or $KNbO_3$ [44].

A completely different approach to generate 4 μm pulses is by generating ~1330 nm (signal) and ~2000 nm (idler) via optical parametric amplification in BBO, followed by difference frequency mixing of the signal and idler in a $AgGaS_2$ crystal [41,46]. Difference frequency mixing of signal and idler has also been used to generate pulses that are resonant with the H–O–H bending mode of H_2O (~6 μm) [47] or with the librations of the water molecules (~10 μm) [48].

The energy of the generated pulses typically varies between 2 [41] and 10 μJ [37–40,44,45]. In some studies, the used KTP and $KNbO_3$ crystals are quite short (1 mm) and the 800 nm pulses have a pulse duration of 30 [42,43] or 40 fs [41,46]. The resulting mid-infrared pulses are also short, having a pulse duration of ~50 fs [41–43,46]. The bandwidth of these pulses is ~400 cm^{-1}, thereby completely covering the O–H/O–D absorption band. In other studies, longer crystals (4–5 mm) and longer 800 nm pulses (~100 fs) have been used, leading to the generation of longer (~150 fs) mid-infrared pump pulses with an appreciably narrower spectral bandwidth of ~100 cm^{-1} [37–40,44,45].

4.2.2 Nonlinear Optical Response of the Molecular Vibrations of Water

In all nonlinear vibrational spectroscopic experiments on aqueous systems, one or two mid-infrared pulses are used to excite a significant fraction of one of the vibrations of the water molecule from the $v = 0$ ground state to the first excited vibrational state $v = 1$. As there are less molecules in the ground state, this excitation leads to a transient bleaching signal of the fundamental $v = 0 \rightarrow 1$ absorption. In addition, the occupation of the $v = 1$ state gives rise to $v = 1 \rightarrow 0$ stimulated emission. Both effects lead to an increased transmission of light at the $v = 0 \rightarrow 1$ transition frequency. In addition, the occupation of the $v = 1$ state gives rise to a new absorption. The excited molecules will absorb light at a frequency that corresponds to the excitation from $v = 1$ to the second excited vibrational state, $v = 2$. For most molecular vibrations, the frequency of the $v = 1 \rightarrow 2$ transition is lower than that of the fundamental $v = 0 \rightarrow 1$ transition. At this point, it should also be noted that in case the frequency of the $v = 1 \rightarrow 2$ transition would be the same as that of the $v = 0 \rightarrow 1$ transition, as is the case for the harmonic oscillator, all the absorption changes would cancel, which implies that there would be no nonlinear spectroscopic response. Hence, nonlinear vibrational spectroscopy relies on the *anharmonic* character of the molecular vibrations.

The water molecule possesses two stretch vibrations and one bending vibration. All three vibrations have been probed with nonlinear vibrational spectroscopy. In most experiments, isotopically diluted water is studied, probing either the OH stretch vibration of HDO dissolved in D_2O or the OD vibration of HDO dissolved in H_2O. The use of isotopically diluted water has advantages that heating effects are limited, and, that the signals are not affected by the strong intra- and intermolecular couplings of the OH/OD stretch vibrations of pure H_2O or D_2O [49,55]. In cases where the O–H stretch vibration of HDO:D_2O is studied, the concentration of HDO is ~1%. When the OD stretch vibration of HDO:H_2O is studied, the concentration of HDO is usually somewhat higher, ranging from 2.5% to 4%. In the latter experiments, a higher concentration has to be used, because the H_2O solvent has a nonnegligible absorption in the frequency region of the O–D stretch vibration.

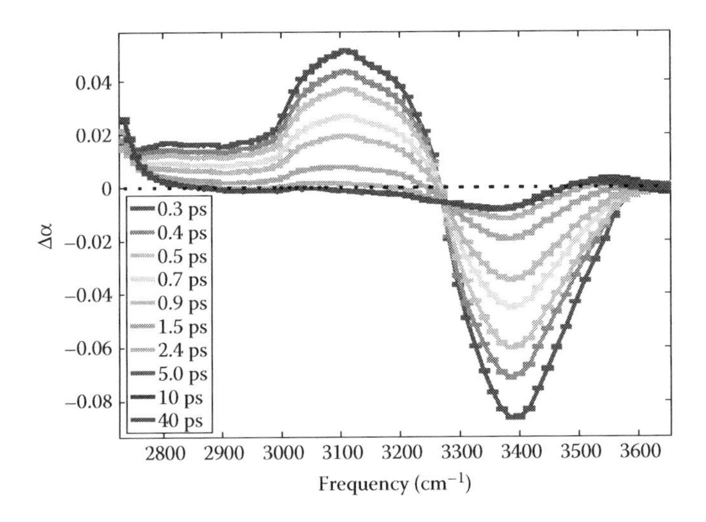

FIGURE 4.1 Transient spectra of the O–H stretch vibration of HDO molecules for a 4% solution of HDO in D_2O at different time delays after excitation with a pump pulse centered at 3400 cm^{-1}.

The excitation of the O–H stretch vibration of HDO dissolved in D_2O gives rise to a bleaching of the fundamental $v = 0 \rightarrow 1$ centered at 3400 cm^{-1} (2.94 μm) and a broad induced $v = 1 \rightarrow 2$ absorption band centered at 3180 cm^{-1} (3.4 μm). The excitation of the O–D stretch vibration of HDO dissolved in H_2O gives rise to a bleaching of the fundamental $v = 0 \rightarrow 1$ centered at 2500 cm^{-1} (4.0 μm) and a broad induced $v = 1 \rightarrow 2$ absorption band centered at 2350 cm^{-1} (4.3 μm). In Figure 4.1, the transient absorption signals at different delays after excitation of the O–H stretch absorption band of HDO:D_2O are presented. The transient spectra clearly show the decreased absorption due to the bleaching of the $v = 0 \rightarrow 1$ transition and the induced $v = 1 \rightarrow 2$ excited state absorption of the O–H vibration of HDO. In addition, a small rising signal is observed at frequencies below 2800 cm^{-1}. This signal represents the change in the blue wing of the absorption band of the O–D stretch vibrations of D_2O that results from the small rise in temperature that follows from the thermalization of the energy of the excitation pulse.

4.3 DYNAMICS OF NEAT LIQUID WATER

The transient absorption changes are the basis of different nonlinear vibrational spectroscopic techniques that provide information on the energy relaxation dynamics and molecular motions of the water molecules. In this section, the different techniques will be described together with the information that has been obtained with these techniques on the dynamics of neat liquid water.

4.3.1 Vibrational Energy Relaxation

The signals associated with the bleaching of the fundamental $v = 0 \rightarrow 1$ transition and the induced excited state $v = 1 \rightarrow 2$ absorption will both decay when the molecules relax back from the excited $v = 1$ state to the ground state. These relaxation dynamics are characterized by the population relaxation time T_1. The dynamics observed with a probe pulse at the $0 \rightarrow 1$ transition can be different from the dynamics at the $1 \rightarrow 2$ transition, in case the relaxation proceeds via an intermediate state for which the absorption spectrum differs from that of the molecules in thermal equilibrium. Hence, the comparison of the two signals provides information on the relaxation mechanism. Further information on the relaxation mechanism can be obtained by tuning the probing pulse to other molecular vibrations. If the originally excited state relaxes via transient excitation of the probed vibration, a delayed transient bleaching or induced absorption signal of this vibration will be observed.

The vibrational energy transfer and relaxation processes of the different isotopic variations of water have been studied with transient absorption spectroscopy [24,41–43,47–55] and with heterodyne-detected photon-echo spectroscopy [56,57]. The O–H stretch vibration of a dilute solution of HDO in D_2O was observed to relax with a time constant of 740 ± 30 fs at 295 K. For the complementary system, that is, the O–D stretch vibration of HDO dissolved in H_2O, the relaxation time was observed to be longer: 1.8 ± 0.2 ps [41,44].

The vibrational relaxation of the excited O–H and O–D vibrations of pure H_2O and D_2O is substantially faster. The excited O–H stretch vibrations of pure H_2O have a vibrational relaxation time constant T_1 of ~200 fs [51,56] and the excited O–D stretch vibrations of pure D_2O have a T_1 of ~400 fs [55]. These vibrational lifetimes would lead to absorption line widths of ~50 and ~25 cm^{-1} for H_2O and D_2O, respectively. However, the observed absorption line widths are ~400 and ~250 cm^{-1} for H_2O and D_2O, respectively, which means that the broadening of the line is dominated by pure dephasing, which contains both homogeneous dephasing and inhomogeneous dephasing contributions, as will be discussed in detail in Section 4.3.2.

The relaxation mechanism of the O–H stretch vibrations of H_2O has been studied in detail. The O–H stretch vibrations (3200–3500 cm^{-1}) are in resonance with a combination tone of two quanta of H–O–H the bending mode of H_2O (~3250 cm^{-1}). These quanta can be located on the same molecule, that is, corresponding to an overtone excitation [48,54], or on different H_2O molecules [53]. The strong coupling between a single excitation in the O–H stretch and two quanta in the H–O–H bending mode leads to a mixing (Fermi resonance) of the two states [54]. The excitation of the mixed states is observed to lead to a fast relaxation (T_1 = 200 fs) to the v = 1 state of the bending vibration, that in turn relaxes with a vibrational lifetime of ~170 fs [47,48,52,54]. The much longer lifetimes of the O–H and O–D stretch vibrations of HDO can be explained from the fact that for HDO neither the O–H (3400 cm^{-1}) nor the O–D (2500 cm^{-1}) are in resonance with the fundamental (1450 cm^{-1}) or overtone (2900 cm^{-1}) of the H–O–D bending mode.

For all isotopic variations of water, the vibrational lifetime shows an anomalous temperature dependence. The vibrational lifetime of the O–H stretch vibration is ~400 fs for HDO:D_2O ice, jumps to ~650 fs at the phase transition (275 K) to liquid HDO:D_2O, and then further increases to ~950 fs at 360 K. This temperature dependence can be described well [50] by a phenomenological expression that relates the frequency of the O–H stretch vibration to the vibrational predissociation time of hydrogen-bonded gas-phase complexes [58]. The good correspondence of the data with this description indicates that the hydrogen bond is one of the accepting modes of the vibrational energy. The observed anomalous temperature dependence can then be explained from the decrease of the effective anharmonic interaction between the hydrogen bonds and the O–H stretch vibration with increasing temperature. However, molecular dynamics simulations showed that the anomalous temperature dependence can also be explained from an increase in the energy gap between the excited O–H stretch vibration and its most likely accepting mode, which is the bending vibration of the HDO molecule [6]. With increasing temperature, the O–H stretch vibration shifts to higher frequencies while the frequency of the bending mode hardly changes and even slightly decreases. Therefore, the energy difference with both the fundamental (1450 cm^{-1}) and the overtone (~2900 cm^{-1}) increases with increasing temperature leading to a slowing down of the relaxation. For pure H_2O, the value of T_1 is observed to increase from 260 fs at 295 K to 350 fs at 343 K [51]. This increase can be explained from an increasing energy mismatch, in this case, between the excited O–H/O–D stretch vibration and the combination tone of two quanta in the bending mode vibration [51].

4.3.2 Hydrogen-Bond Dynamics

The frequency of the O–H/O–D stretch vibrations of the water molecule and the strength of the hydrogen-bond interaction are highly correlated. Water molecules with short and linear hydrogen bonds absorb at lower frequencies than water molecules with long and bent hydrogen bonds [59–61]. Owing to this correlation, the dynamics of the hydrogen bonds can be studied by monitoring the spectral diffusion of these vibrations.

4.3.2.1 Spectral Hole Burning

The first studies of the spectral diffusion of the O–H stretch vibration of HDO dissolved in D_2O were performed with spectral hole-burning spectroscopy [23–25,28,62]. Spectral hole burning is a variation of transient absorption spectroscopy in which the inhomogeneously broadened absorption band of a molecular vibration is excited with a mid-IR excitation pulse that has a smaller bandwidth than the width of the absorption band. The infrared pulse will only excite the molecules of which the vibration is resonant with the infrared light. Hence, only for part of the molecules the $v = 0 \rightarrow 1$ transition will be bleached. The absorption band will thus acquire a so-called spectral hole. When the frequencies of the excited vibration change, for instance, because the strength of the hydrogen-bond interactions change, the spectral hole will broaden and will eventually evolve into an overall bleaching of the complete absorption band.

The earliest spectral hole-burning studies of water were performed with infrared pulses with a pulse duration >1 ps. From these measurements, it was concluded that the broad inhomogeneous absorption band of the O–H stretch vibration would be composed of several subbands that would correspond to specific hydrogen-bonded structures of water molecules [23,62]. However, later studies employing pulses with pulse durations of ~100 fs did not confirm the existence of these subbands [24,25,28]. In fact, strong indications were found that the absorption band is the result of a broad, continuous distribution of different O–H stretch frequencies, thus reflecting a rather continuous variation of hydrogen-bond configurations. The dynamics of the spectral hole could be well described by a Gauss–Markov stochastic spectral diffusion process.

For a Gauss–Markov process, the distribution of frequencies of the oscillators has a Gaussian shape and the time-dependent instantaneous frequency at time t is determined solely by its value at time $t - \Delta t$, $\Delta t \rightarrow 0$ (no memory effect), so that the spectral diffusion is characterized by a two-point frequency–frequency correlation function (FFCF) $< \delta\omega(t)\delta\omega(0) >$, where $\delta\omega(t) = \omega(t) - < \omega >$. Thus, $\omega(t)$ is the time-dependent transition frequency of a given O–H/O–D oscillator, and $< \omega >$ is the average frequency of the ensemble. For a Gauss–Markov process, the FFCF decays exponentially:

$$< \delta\omega(t)\delta\omega(0) > = \Delta^2 e^{-|t|/\tau_c}, \tag{4.1}$$

where Δ^2 is the mean-squared frequency fluctuation and τ_c is the frequency–frequency correlation time.

Usually there are several spectral diffusion processes active with specific values of the correlation time constant $\tau_{c,i}$ and the spectral width of the distribution Δ_i. In case $\Delta_i\tau_{c,i} << 1$, the spectral modulation process is fast and in the homogeneous limit. For such a process, no spectral diffusion can be observed. The homogeneous pure dephasing time T_2^* of such a process is given by $1/(\Delta_i^2\tau_{c,i})$. Hence, the dephasing time constant becomes longer when the frequency fluctuations (characterized by $\tau_{c,i}$) become faster. At first sight, this result may seem counterintuitive, but one has to realize that in the case of rapid frequency fluctuations all oscillators acquire phase at approximately the same averaged rate, meaning that the oscillators get slower out of phase with each other than in case they would have kept their specific frequencies for a long time. This effect is often referred to as motional narrowing, and it has as a consequence that the broad Gaussian spectral distribution of oscillators with width Δ_i evolves into a much narrower Lorentzian line with width $\Delta_i^2\,\tau_{c,i}$.

In a spectral hole-burning experiment, the spectral diffusion processes for which $\Delta_i\tau_{c,i} << 1$ and the vibrational population relaxation together define the initial (homogeneous) width of the spectral hole. For the O–H stretch vibration of HDO:D_2O, the initial hole was observed to possess a relatively large width of ~150 cm^{-1}, which means that there is substantial homogeneous broadening [24]. This line width corresponds to a value of $T_{2,\text{hom}}$ of ~140 fs. It follows that the vibrational population relaxation with a time constant of ~740 fs constitutes only a minor contribution to the homogeneous line width. The homogeneous line width is likely associated with rapid variations in the angle and length of the local O–H···O hydrogen bonds.

The width of the spectral hole at different delays defines the so-called dynamic line width. The broadening of the spectral hole toward the linear absorption spectrum with a width of ~250 cm^{-1} was observed to take place on two timescales. The first fast process has a time constant of ~170 fs. This process was assigned to an adaption of the hydrogen bond upon the excitation to the $v = 1$ state [28]. The main component of the broadening dynamics of the spectral hole can be modeled well with a Gauss–Markov process with a relatively long time constant τ_c of ~900 fs [24,25,28]. For this latter process, $\Delta_i\tau_{c,i} >> 1$, meaning that this process is approaching the so-called inhomogeneous limit. The absorption band of the O–H stretch vibration of HDO:D_2O is thus a convolution of a homogeneous band (fast spectral diffusion processes + vibrational population relaxation) and an inhomogeneous contribution. The spectral diffusion process with a time constant τ_c of ~900 fs is likely associated with collective reorganizations of the hydrogen-bond network of liquid water.

4.3.2.2 Photon-Echo Peak Shift Spectroscopy

The spectral diffusion of HDO dissolved in D_2O has also been studied with different forms of photon-echo spectroscopy, employing shorter pulses than those used in the spectral hole-burning studies. In photon-echo spectroscopy, the $v = 1$ state of the vibrations is excited with two broad-band IR laser pulses with wave vectors $\mathbf{k}_1$ and $\mathbf{k}_2$ that are incident onto the sample at different angles. As a result of constructive and destructive interference between the two pulses, the amplitude of the resulting population of the $v = 1$ state is periodically modulated with wave vectors $\mathbf{k}_1 - \mathbf{k}_2$ and $\mathbf{k}_2 - \mathbf{k}_1$. Owing to the inhomogeneous broadening of the absorption band, there will be such modulated populations (population gratings) for each different frequency in the inhomogeneous distribution. The relative phases of these population gratings will change with increasing delay between the two excitation pulses, because of the difference in the rate of phase evolution of the oscillators in the time interval between the two excitation pulses. The population gratings are read out by a third pulse that has a direction $\mathbf{k}_3$. This third pulse generates a polarization in the directions $\mathbf{k}_3 + \mathbf{k}_2 - \mathbf{k}_1$ and $\mathbf{k}_3 + \mathbf{k}_1 - \mathbf{k}_2$. In the so-called rephasing direction, which is $\mathbf{k}_3 + \mathbf{k}_2 - \mathbf{k}_1$ if the pulse with wave vector $\mathbf{k}_1$ enters first, the phase accumulation is opposite to the phase accumulation between the first and second pulses. Hence, the polarizations generated by the third pulse get into phase again after a time delay that corresponds to the time difference between the first two excitation pulses. At that moment in time, all polarizations add up constructively, and a light pulse is emitted from the sample in the rephasing direction. Because the emission is delayed with respect to the entrance time of the third pulse, the emitted light pulse is denoted as a photon echo.

The nonlinear polarization that leads to the echo signals shows the same dependence on the excitation and probing electric fields as the nonlinear polarization that leads to the transient absorption signal: both techniques rely on the third-order nonlinear susceptibility of the probed vibration. The main difference is that the echo signals are generated in new directions whereas in transient absorption spectroscopy the nonlinear signal is generated in the direction of the probing pulse and leads to a change of the absorption of the probing pulse. Another difference is that in photon-echo spectroscopy the delay between the excitation field interactions can be varied, which can be used to get information on the dephasing dynamics.

In an echo-peak shift measurement, the delay τ_1 between the first two excitation pulses, that gives the maximum echo signal, is measured as a function of the so-called waiting time τ_2 between the excitation pulses and the third interrogation pulse. In the absence of spectral diffusion and for short waiting times τ_2, the maximum photon-echo signal is attained for times τ_1 that are of the order of the pulse duration, because this configuration leads to the largest amplitude of the population gratings. However, if there is spectral diffusion and the waiting time is comparable to the characteristic timescale of these processes, the optimal delay τ_1 for getting the maximum echo signal will become smaller. This can be understood as follows. For nonzero delay τ_1 between the excitation pulses, the different population (frequency) gratings have nonzero phase differences. The correlation of these phase differences with the frequency of the oscillator is an essential feature for the generation of the echo signal in the rephasing direction. However, because of spectral diffusion, the phase differences will become increasingly uncorrelated from their original excitation frequency with increasing waiting time τ_2. This correlation loss leads to a decay of the echo signal as it prevents the rephasing. If the excitation pulses enter the sample simultaneously ($\tau_1 = 0$), all population gratings are excited in phase, and the echo signal is generated directly with the entrance of the third pulse. In this case, changes in the frequencies of the oscillators during τ_2 will not affect their relative phases, as they are all zero. Therefore, for $\tau_1 = 0$, the polarizations generated by the third pulse always add up constructively, irrespective of the value of τ_2. Hence, for large waiting times τ_2 the maximum echo signal is obtained for zero or very small delays τ_1 between the first two excitation pulses. It can be shown that the dependence of the optimal value of τ_1 on the value of τ_2 directly represents the time dependence of the FFCF [63,64]. This optimal value of τ_1 as a function of waiting time τ_2 constitutes the so-called echo-peak signal.

The spectral diffusion of the O–H stretch vibration of HDO dissolved in D_2O has been measured with several echo-peak shift spectroscopic studies [27,30,43]. These studies confirmed the presence of a spectral fluctuation process with a time constant of ~1 ps, as was also observed with spectral hole burning. In one study, very slow additional dynamics were observed occurring on a timescale of 5–15 ps [27]. However, in a later study, it was argued that these dynamics likely result from an interference effect with signal generated from the D_2O solvent [65]. In another echo-peak shift study by Tokmakoff and coworkers, a slow spectral diffusion process with a time constant of 1.4 ps, and a fast spectral diffusion process with a time constant of ~50 fs were observed [30,43]. In addition, an increase in the FFCF extracted from experiment was observed at a waiting time of ~180 fs. This "recurrence" was assigned to an oscillation associated with the stretching vibration of the OH···O H-bond [4,5]. It follows from this observation that the OH···O H-bond between HDO and a D_2O molecule is underdamped.

4.3.2.3 Other Photon-Echo Experiments

In the first photon-echo experiment on water, only two pulses were used, and the second pulse served to both excite and probe the population gratings [26]. In this experiment, the time-integrated echo signal was measured as a function of the delay between the two excitation pulses. As there is no delay between a second and a third pulse, this experiment does not give information on the timescales of the slow (inhomogeneous) frequency fluctuation processes, but it does give information on the dephasing time of the O–H oscillators. Using certain assumptions, this dephasing time can be related to the timescale of the frequency fluctuations. In the delay time interval between the two pulses, the oscillators will get out of phase because they have a different central frequency (inhomogeneous broadening effect) but also because there are rapid frequency fluctuations (homogeneous broadening effect). The phase differences due to the differences in central frequency will rephase, and thus lead to an echo signal. Hence, the inhomogeneous broadening will not lead to a decay of the echo signal (in a two-pulse echo experiment). However, the loss in phase due to the homogeneous broadening effects leads to a decay of the polarization at each central oscillator frequency and thus to a decay of the amplitude of the corresponding population grating. This loss in amplitude is not restored in the generation of the echo. Hence, the generation of a two-pulse echo relies on the presence of inhomogeneous broadening, but its dependence on the delay between the two pulses gives information on the homogeneous broadening. For HDO:D_2O, it was thus found that $T_{2,\mathrm{hom}}$ ~132 fs, in case the absorption is dominantly inhomogeneously broadened [26]. This value agrees well with the value for $T_{2,\mathrm{hom}}$ of ~140 fs found in the spectral hole-burning studies [24,28].

In another two-pulse photon-echo experiment, the echo signal was time resolved by interfering this signal with a reference pulse (local oscillator) [29]. The delay of the maximum of the echo signal with respect to the second pulse was measured as a function of the delay between the first and second pulses. For a purely inhomogeneously broadened system, the delay of the echo signal would always correspond to the delay between the first and second pulses. In the presence of frequency fluctuations, the delay of the maximum of the echo signal will not exactly follow the increase of the delay between the two excitation pulses, because the frequency fluctuations prevent the full rephasing of the echo signal to occur. Hence, the maximum echo signal will occur at a shorter delay after the second pulse. From these experiments, it followed that the spectral modulation includes two processes with timescales of 130 and 900 fs. This observation agrees well with the results of the spectral hole-burning studies [24,25,28] and the echo-peak shift studies [27,30,43].

4.3.2.4 Two-Dimensional Vibrational Spectroscopy

Two-dimensional vibrational spectroscopy is another variation of third-order nonlinear infrared spectroscopy of molecular vibrations. The name two-dimensional vibrational spectroscopy refers to the fact that the measured signals are plotted in a contour plot as a function of the excitation frequency and the detection frequency. In the literature, two different implementations of this technique have been reported, double-resonance spectroscopy and heterodyne-detected photon-echo spectroscopy. Double-resonance spectroscopy is essentially a conventional spectral hole-burning experiment in which a narrow-band pump pulse is scanned through the broad (inhomogeneous) absorption band. This narrow-band pump is usually generated by passing a short, broadband pump pulse through a piezo-controlled Fabry–Perot etalon. With the etalon, both the central frequency and the bandwidth can be varied. The probe pulse can be generated in an independent manner, but can also be a fraction of the broadband input pulse. After its transmission through the sample, the broadband probe is spectrally dispersed and detected in a frequency-resolved manner. The main difference with a conventional spectral hole-burning experiment is formed by the adjustable Fabry–Perot filter used to control the pump spectrum.

2D IR photon-echo spectroscopy is a three-pulse photon-echo experiment in which the sample is excited by two identical, time-delayed infrared pulses. The associated spectrum of this excitation has a modulation frequency that is inversely proportional to the time delay. This modulated excitation spectrum burns multiple holes in the absorption line. Increasing the time separation between the pulses in the first pair leads to a finer frequency modulation of the holes and thus to a more severe smearing of the holes. The dependence of the signal on the excitation frequency is obtained by performing many experiments in which the delay between the two excitation pulses is varied. Fourier transform of the signals at different delays gives the dependence of the signal on the excitation frequency, thus defining one of the two frequency axes of the 2D spectrum. The third pulse enters the sample after a (waiting) time delay with respect to the second of pulse of the excitation pulse pair and generates the echo by being diffracted from the population gratings generated by the first two excitation pulses.

The detection frequency axis can be obtained in two different ways. In the first method, the echo signal is interfered with a fourth laser pulse that acts as a local oscillator in a heterodyning experiment. The frequency of the echo signal can then be determined in the time domain by scanning the time delay of the local oscillator and Fourier transforming thus obtained signal with respect to this time variable. The heterodyne-detected 2D IR photon-echo technique is schematically illustrated in Figure 4.2.

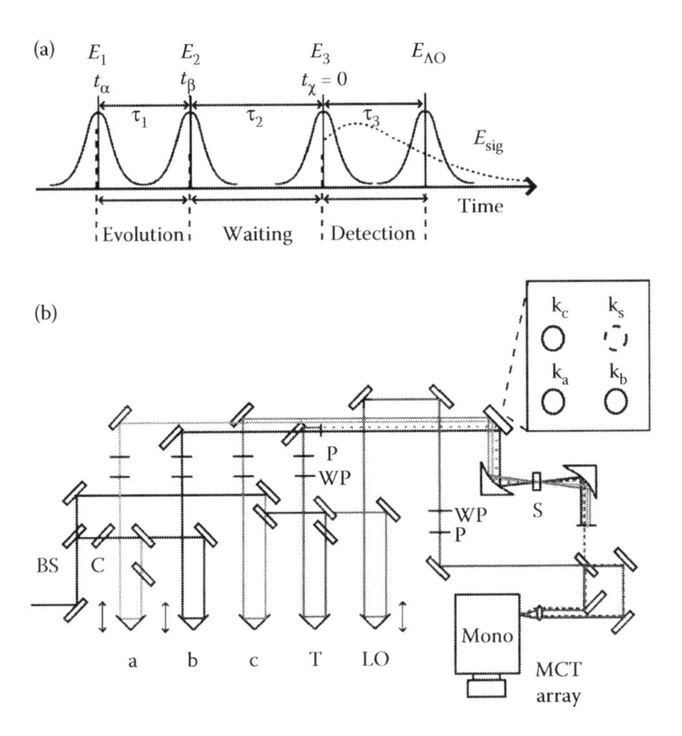

FIGURE 4.2 (a) Illustration of the excitation pulse ordering and time variables of a two-dimensional infrared (2D IR) heterodyne-detected photon-echo experiment. (b) Layout of the five beam interferometer showing the three excitation pulses a, b, and c, the tracer T, and the local oscillator field LO; BS, 50–50, 3-mm-thick CaF_2 thick beam splitter; C, 3-mm-thick CaF_2 compensation plate; WP, $\lambda/2$ tunable wave plate; P, wire grid CaF_2 polarizer; S, a 50 μm jet of HOD in D_2O; mono, monochromator. (Reprinted with permission from Loparo JJ, Roberts ST, Tokmakoff A. Multidimensonal infrared spectroscopy of water. I. Vibrational dynamics in 2D lineshapes. *J Chem Phys* 125: 194521. Copyright 2006, American Institute of Physics.)

The detection frequency axis can also be obtained by sending the echo signal through a spectrograph and by measuring the spectrally dispersed light with an infrared detector array. In the latter case, the echo signal can also be interfered with a local oscillator pulse to amplify the weak echo signal [31,32].

The two types of two-dimensional spectroscopic experiments in principle give the same information on the spectral diffusion of the excited vibration. The double-resonance technique has an advantage that it is relatively simple and that it does not require phase stability, but has a disadvantage that the measured spectral shapes are always convoluted with the prechosen bandwidth of the pump pulse. As a result, fine spectral details (narrow homogeneous lines) could be missed if the prechosen bandwidth of the excitation pulse is too large. The 2D IR photon-echo technique has an advantage that it automatically provides the optimal frequency resolution for studying the spectral dynamics. If the spectrum contains very narrow homogeneous lines, the signals obtained with long delay times between the excitation pulses will strongly contribute to the 2D spectrum, meaning that the 2D spectrum will show these fine spectral details. A disadvantage of this technique is that the outcome of the experiment is dependent on the phase difference between the two excitation pulses and the phase difference between the echo signal and the local oscillator field. This means that the setup must show a high mechanical stability. In Figure 4.3, 2D IR spectra of $HDO:D_2O$ are shown that are measured with the heterodyne-detected photon-echo technique [36].

If the waiting time between the excitation and the detection (τ_2 in Figure 4.3) is short, there will be a strong correlation between the frequency at which the vibration is excited and the frequency at which it is probed. In the 2D spectrum (left panel of Figure 4.3), this is seen as an elongated shape along the diagonal. With increasing delay, the 2D spectral shape will become increasingly circular due to spectral diffusion. Hence, the dynamics of the spectrum as a function of time delay between the excitation and the probing pulse give information on the structural dynamics of the water molecules. The FFCF derived from the 2D lineshapes shows fast (~100 fs) and slow (~1 ps) components, for both the O–H and the O–D vibration of HDO. For the O–H vibration, the dynamics were modeled with two exponential time constants with values of 60 fs and 1.4 ps [34,35], for the O–D vibration with three exponential time constants with values of 48 fs, 400 fs, and 1.4 ps [32]. The short time constant of 60/48 fs is likely associated with very fast local fluctuations in the angle and length of the O–D···O hydrogen bond, while the slower time constants of 400 fs and 1.4 ps are most probably associated with the collective reorganization of the water hydrogen-bond network.

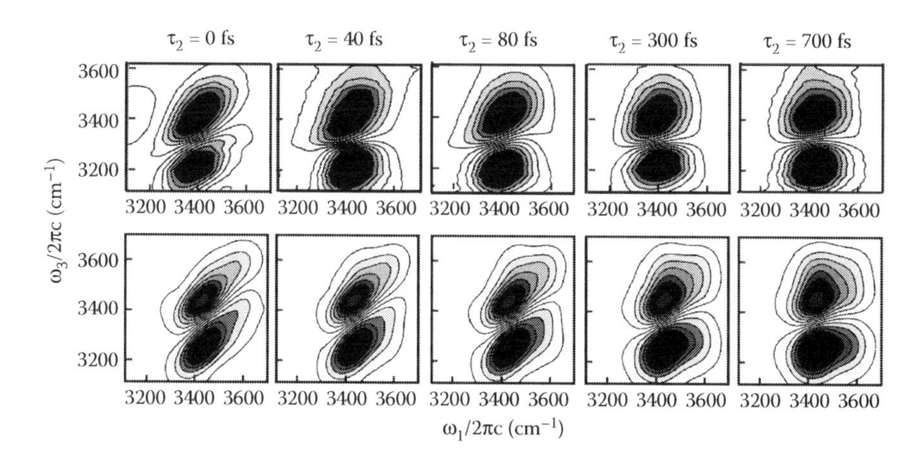

FIGURE 4.3 Experimental (top) and simulated (bottom) absorptive two-dimensional infrared heterodyne-detected photon-echo spectra of the OH stretch vibration of $HDO:D_2O$ for several waiting times τ_2. (Reprinted with permission from Loparo JJ, Roberts ST, Tokmakoff A. Multidimensional infrared spectroscopy of water. II. Hydrogen bond switching dynamics. *J Chem Phys* 125: 194522. Copyright 2006, American Institute of Physics.)

The cross section of the spectral contour of the 2D spectrum along the probing frequency (ω_3) for a fixed excitation frequency (ω_1) defines the dynamical line width. This line width broadens with increasing waiting time, closely following the dynamics of the FFCF [32]. The initial line width of the O–D stretch vibration of HDO:H_2O has a value of ~115 cm^{-1} (corresponding to $T_{2,hom}$ ~180 fs) [32], indicating that the O–D absorption band contains a significant homogeneous line broadening component. This observation is in line with the broad homogeneous line width of the O–H vibration of HDO:D_2O that was observed in spectral hole-burning studies [24,28], and in two-pulse photon-echo experiments [26]. An interesting result of the 2D measurements is that at short times (~100 fs) the dynamic line width depends on the excitation frequency, being wider on the blue side of the absorption band than in the center of this band [33,36]. This result indicates that water molecules on the blue side of the absorption band experience more rapid spectral diffusion. This finding was interpreted as resulting from the less constrained environments for water molecules with weaker H-bonds [33], and from the rapid decay of weakly (bifurcated) hydrogen-bonded structures [36].

Heterodyne-detected photon-echo spectroscopy has also been used to measure the 2D IR spectra of pure liquid H_2O [56,57]. These spectra are much more difficult to interpret than the spectra of isotopically diluted water because the vibrational lifetime of the O–H stretch vibrations of H_2O is very short (~200 fs), and because the vibrational relaxation leads to an overwhelmingly strong thermal effect on the 2D spectrum. The 2D IR spectrum of H_2O was found to show extremely fast spectral dynamics with a time constant of ~50 fs [56].

Pure liquid H_2O possesses an extremely high density of (near-)resonant O–H oscillators that are strongly coupled, meaning that the O–H will be delocalized over several H_2O molecules (see Section 4.3.3). The distribution of these delocalized vibrations over the local O–H vibrations is strongly dependent on the frequencies of the uncoupled O–H stretch vibrational frequency that in turn depend on the local hydrogen-bond strengths. Fluctuations of these hydrogen bonds will lead to a strong change of the character and frequency of the delocalized modes and thus to spectral diffusion.

The fast spectral dynamics of H_2O were thus interpreted to indicate that the hydrogen-bond network of water shows an ultrafast memory loss with a time constant of ~50 fs [56]. However, for H_2O, relatively small changes in the length of the hydrogen bond will already lead to large changes in the frequencies (and character) of the delocalized O–H stretch modes. This means that fast hydrogen-bond length fluctuations over a limited angle and/or length interval are already sufficient to get a full equilibration of the O–H stretch frequencies over the complete absorption band. As a result, the possible presence of further slower larger-amplitude hydrogen-bond fluctuations will not be discernible anymore, as the spectral diffusion is already complete. Based on the similarity in the strength and variation of the hydrogen bonds in H_2O and D_2O, it is very likely that H_2O contains a similar slow large-amplitude component with a time constant of ~1 ps in its hydrogen-bond dynamics as has been observed for isotopically diluted HDO:D_2O and HDO:D_2O.

4.3.3 Förster Vibrational Energy Transfer

In pure liquid H_2O, the average distance between the H_2O molecules is only ~3.1 Å, which leads to a very strong resonant dipole–dipole coupling of the O–H stretch vibrations located on different molecules. This coupling leads to a strong delocalization of the O–H stretch vibration over several H_2O. The dynamics of these delocalized vibrations (quantum interference and fluctuations in character) can be equivalently described as rapid Förster energy transfer between the local O–H stretch vibrations of the different water molecules. The rate of Förster energy transfer strongly depends on the distance between the donating and accepting resonance, and as such electronic Förster energy transfer is widely employed as a molecular ruler to determine the distance between different chromophores. For molecular vibrations, the transition dipole moment is much smaller than for electronic resonances and thus vibrational Förster energy transfer is not as widely studied and observed as electronic Förster energy transfer.

Vibrational Förster energy transfer has been studied for both liquid H_2O and D_2O by monitoring the dynamics of the anisotropy of the O–H and O–D stretch excitations, respectively [49,55,56]. If a vibration is excited with a linearly polarized infrared pulse, the resulting excitation probability is anisotropic showing a $\cos^2(\theta)$ dependence, where θ is the angle between the transition dipole of the excited vibration and the polarization direction of the excitation pulse. The decay of this anisotropy can be measured by rotating the polarization of the excitation pulse at 45° with respect to the polarization of the probe pulse. In transient absorption spectroscopy, the dynamics of the anisotropy are measured by choosing alternatingly the polarization components of the probe parallel and perpendicular to the pump polarization with a polarizer [37,38,40,42–46]. The measured absorption changes $\Delta\alpha_{\parallel}(\tau)$ and $\Delta\alpha_{\perp}(\tau)$ are used to construct the so-called rotational anisotropy:

$$R(\tau) = \frac{\Delta\alpha_{\parallel}(\tau) - \Delta\alpha_{\perp}(\tau)}{\Delta\alpha_{\parallel}(\tau) + 2\Delta\alpha_{\perp}(\tau)} = \frac{\Delta\alpha_{\parallel}(\tau) - \Delta\alpha_{\perp}(\tau)}{3\Delta\alpha_{iso}(\tau)}. \tag{4.2}$$

The denominator of Equation 4.2 is not affected by reorientation and resonant energy transfer and thus only represents isotropic dynamics like the vibrational relaxation. Hence, the isotropic effects are divided out and $R(\tau)$ is directly related to the angular correlation function [66]:

$$R(\tau) = \frac{2}{5}C_2(\tau) = \frac{2}{5} < P_2(\cos\theta(\tau)) >, \tag{4.3}$$

where $P_2(x)$ is the second Legendre polynomial and $\theta(t)$ is the angle between the vibrational transition dipole moment vectors at time 0 and time τ. The anisotropy will decay as a result of intra- and intermolecular (Förster) energy transfer between vibrations with different orientations, but also because of molecular reorientation (see Section 4.3.4). For bulk D_2O and bulk H_2O, it was observed that Förster energy transfer between the O–D/O–H stretch vibrations leads to a fast and complete decay of the anisotropy [49,55].

For liquid H_2O, the Förster energy transfer was observed to occur on a timescale <100 fs [49,56]. For the O–D vibrations of D_2O, the resonant energy transfer is observed to be two to three times slower, which can be explained from the smaller transition dipole moment of the O–D vibration in comparison to the O–H vibration [55]. The rapid resonant energy exchange implies that for pure H_2O/D_2O the O–H/O–D stretch excitations are strongly delocalized. When D_2O is mixed with H_2O, the average distance between the O–D vibrations increases, leading to a strong slowing down of the resonant energy transfer between the O–D vibrations, as illustrated in Figure 4.4. At low concentrations (<1%), the anisotropy decay only reflects the molecular reorientation of the probed O–H or O–D group of HDO, as will be discussed in Section 4.3.4.

Vibrational energy transfer will also occur for the hydroxyl vibrations of water molecules at interfaces. The dynamics of these water molecules can be probed with high surface selectivity using time-resolved surface sum-frequency generation (SFG) [67,68]. With this technique, the vibrations are first excited with a strong mid-infrared pulse. This excitation is probed by combining a second infrared pulse with a visible pulse to generate the sum frequency of the latter two pulses. This SFG process is highly surface specific as it is symmetry forbidden in centrosymmetric media, that is, the bulk liquid. The SFG process is resonantly enhanced when the infrared frequency is resonant with a molecular vibration at the interface. SFG can thus be used to measure the spectrum of molecular vibrations located at an interface. The excitation by the first strong mid-infrared pulse (which is by itself not surface specific) will lead to a bleaching of the absorption of the molecular vibrations, both in the bulk and at the surface. The bleaching of the absorption of the vibrations at the surface will lead to a transient decrease of the SFG signal. The SFG signal will recover when the excited vibrations at the surface relax.

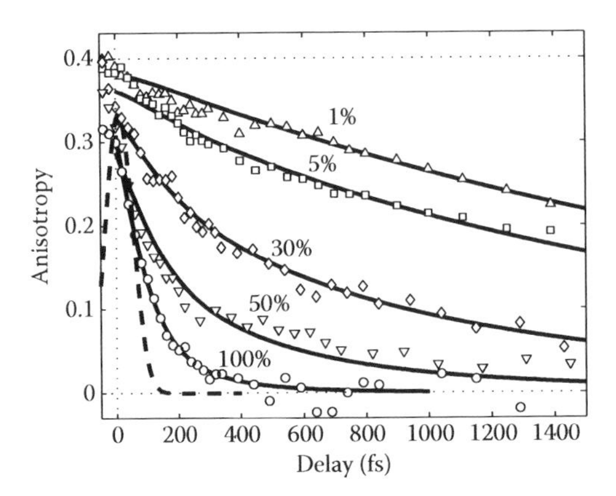

FIGURE 4.4 Anisotropy decay curves of the O–D stretch vibrations for solutions of different concentrations of D_2O in H_2O (100%—open circles; 50%—downward pointing triangles; 30%—diamonds; 5%—squares; 1%—upward pointing triangles). The anisotropy decays are measured with excitation and detection pulses centered at 2500 cm^{-1} with a pulse duration of 100 fs. The solid lines represent fits of the measured anisotropy decays to a model describing Förster energy transfer between the O–D vibrations. The dashed line represents a Gaussian profile fitted to the cross-correlated signal of the excitation and detection pulses. (From Piatkowski L, Eisenthal KB, Bakker HJ. 2009. Ultrafast intermolecular energy transfer in heavy water. *Phys Chem Chem Phys* 11: 9033–9038. Reprinted with permission from the Royal Chemical Society.)

Recently, the process of vibrational Förster energy transfer has been studied for the interface of liquid D_2O and air using two-dimensional surface sum-frequency generation (2D-SFG) [69,70], which is a new form of time-resolved SFG [71–73]. In this technique, the first infrared excitation pulse has a relatively narrow bandwidth and is tuned through the broad inhomogeneous absorption band of the O–D stretch vibration. The principle of the technique is illustrated in Figure 4.5.

In Figure 4.6, 2D-SFG spectra at different delays are shown. It is seen that for early delays, the 2D spectrum has an elliptical shape and a nonzero slope, showing the inhomogeneity of the O–D absorption band of hydrogen-bonded O–D groups at the D_2O interface. With increasing delay, the ellipticity rapidly vanishes and the slope decays to zero. This finding shows that the O–D vibrations at the surface are subject to rapid spectral diffusion [70]. This result agrees with the results of previous time-resolved SFG experiments of water-fused silica [67] and water–air interfaces [68]. The rapid spectral diffusion can be well explained from the occurrence of Förster energy transfer between the O–D vibrations of the D_2O molecules. The rate of this Förster transfer is lower than is observed for bulk liquid D_2O, which follows from the lower density of acceptors at the interface. For the topmost layer of D_2O molecules, the number density of available acceptors is expected to be approximately half the number density of D_2O molecules in the bulk.

The 2D-SFG data show a clear cross peak of the non-hydrogen-bonded O–D group sticking out of the surface (absorbing at 2750 cm^{-1}) and the O–D group located on the same D_2O molecule that is oriented toward the bulk (absorbing at 2550 cm^{-1}). This cross peak results from the energy transfer between the two O–D groups and has a time constant of 300 ± 60 fs for the transfer from the free O–D group to the hydrogen-bonded O–D group. This result shows that the two O–D groups located on the same molecule are strongly coupled, which agrees with the results of a recent study of the effects of isotopic substitution on the frequency of the free O–D/O–H group [74].

4.3.4 Molecular Reorientation

The anisotropy of the vibrational excitation not only decays as a result of Förster energy transfer, but also because of molecular reorientation. The two mechanisms can be separated by changing the isotopic composition. The rate of Förster energy transfer rapidly decreases with increasing distance

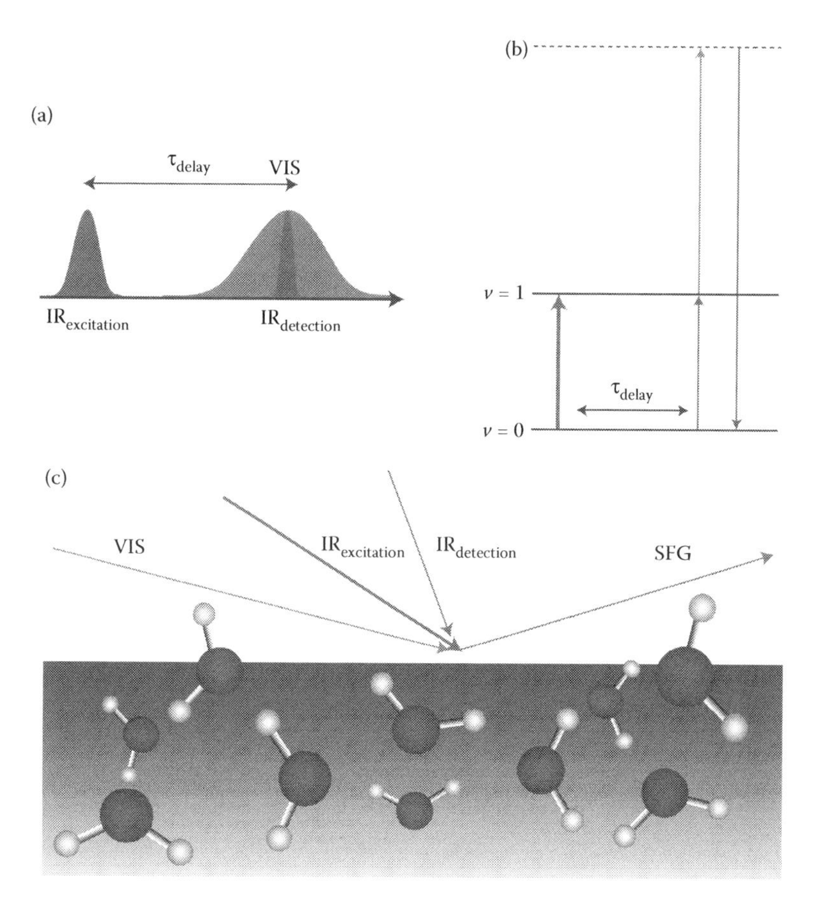

FIGURE 4.5 Experimental scheme for two-dimensional surface sum-frequency generation spectroscopy. (a–c) A tunable infrared pulse excites a subset of water molecules from the ground ($v = 0$) state to the first vibrationally excited ($v = 1$) state. The interfacial response is detected over a wide frequency range as a function of delay, using a pair of infrared and visible detection pulses (a) that generate light at their sum frequency at the interface (b,c). (From Zhang Z et al. 2011. *Nat Chem* 3: 888–893.)

between the donor and acceptor of the vibrational excitation. Hence, for a sufficiently dilute solution of HDO in D_2O, the anisotropy dynamics of the excitation of the O–H vibration is no longer affected by Förster energy transfer and only reflects the reorientation of the O–H groups of the HDO molecules. The reported femtosecond mid-infrared measurements of the orientational dynamics of water thus always involve studies of the anisotropy of the excitation of the O–H vibration of a dilute solution of HDO in D_2O or of the O–D vibration of a dilute solution of HDO in H_2O [37–46].

The two isotopic systems have their specific advantages and disadvantages. The O–H stretch vibration of HDO:D_2O is more strongly inhomogeneously broadened than the O–D vibration of HDO:H_2O [27,32], thus making it easier to resolve dynamical inhomogeneities. In addition, for HDO:D_2O, a lower HDO concentration can be used than for HDO:H_2O as the background signal of D_2O in the frequency region of the O–H stretch vibration is much smaller than that of the H_2O in the frequency region of the O–D vibration. On the other hand, studying the O–D vibration of HDO:H_2O has the advantage that the lifetime of the O–D vibration is more than two times longer than that of the O–H vibration, thereby allowing the measurement of the orientational dynamics of the O–D group over a significantly longer time interval.

The orientational dynamics of the O–H/O–D stretch vibrations of HDO dissolved in D_2O/H_2O have been studied with polarization-resolved transient absorption spectroscopy [37–41,43–46,75]

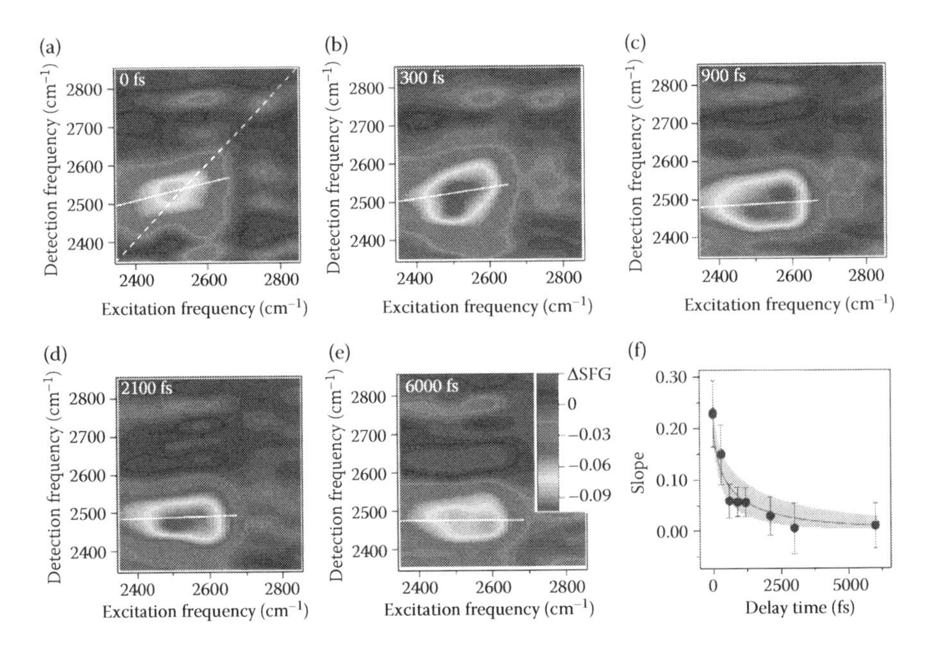

FIGURE 4.6 Two-dimensional surface sum-frequency generation spectra of the D_2O water/air interface at various delay times after the excitation. (a–e) the solid lines represent the IR frequency corresponding to the maximum SFG response as a function of the excitation frequency. The slopes of the solid lines are plotted as a function of delay in (f). The solid line in (f) is the result of a model calculation that accounts for the spectral diffusion by resonant Förster energy transfer. (From Zhang Z et al. 2011. *Nat Chem* 3: 888–893.)

and 2D IR photon-echo spectroscopy [76]. In the 2D IR photon-echo experiment, the anisotropy dynamics of the vibrational excitation are determined by choosing the polarization of the third pulse either parallel or perpendicular to the polarization of the two excitation pulses [76].

For both HDO:D_2O and HDO:H_2O, it is observed that the anisotropy shows a rapid partial decay in the first 100 fs after the excitation [42,43,46,77]. This decay is due to the librational (hindered rotational) motion of the O–H groups that keep the O–H···O hydrogen bond intact, and due to Förster resonant vibrational energy transfer. It may seem surprising that solutions that can be considered to be isotopically dilute, like a solution of 2.5% HDO in H_2O, still show a fast initial decay of the anisotropy as a result of Förster resonant vibrational energy transfer. However, even for solutions containing only a few percent of HDO, the probability of having two of the probed O–D/O–H vibrations close to each other is not negligible, and thus a small rapid decay of the anisotropy due to the energy transfer between these oscillators is observed. If the concentration of HDO dissolved in H_2O is reduced from 3% to 0.5%, the amplitude of the initial decay of the anisotropy becomes smaller, as illustrated in Figure 4.7.

Figure 4.7 shows the anisotropy after 100 fs. In absence of a fast decay, its value should be close to 0.4. Reducing the HDO concentration clearly reduces the deviation from 0.4. The experimental results obtained for a solution of 0.5% HDO in H_2O agree very well with the results of a calculation that describes the fast librational motion, that includes non-Condon effects (i.e., including the dependence of the absorption cross section on the frequency of the O–D stretch vibration), and that includes the contribution of the $v = 1 \rightarrow 2$ induced absorption [77]. The contribution of the librational motion to the initial anisotropy decay shows quite an interesting temperature dependence [46]. At a relatively high temperature of 65°C, the initial drop is more or less four times larger in the blue wing than in the red wing of the absorption band. This result can be explained from the fact that the O–D···O hydrogen bond that restricts the librational motion is much weaker for an O–D vibration absorbing in

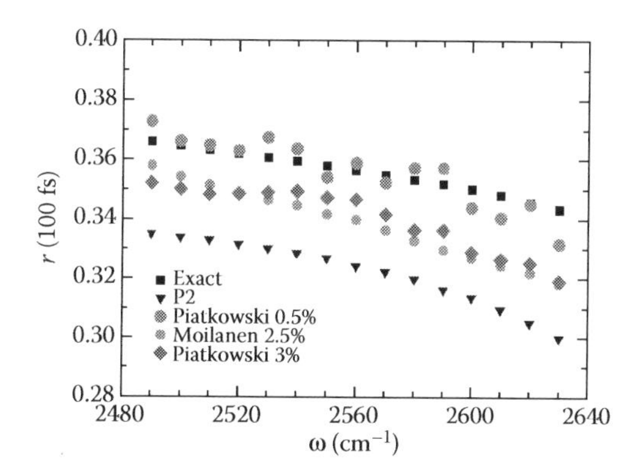

FIGURE 4.7 Experimental and calculated anisotropy of the O–D stretch vibration of HDO at 100 fs delay for solutions of HDO in H_2O of different concentration. (Reprinted with permission from Lin YS et al. On the calculation of rotational anisotropy decay, as measured by ultrafast polarization-resolved vibrational pump-probe experiments. *J Chem Phys* 132: 174505. Copyright 2010, American Institute of Physics.)

the blue wing than for an O–D vibration absorbing in the red wing of the absorption band. However, at 1°C, the initial drop of the anisotropy due to librations is nearly frequency independent. This last result indicates that just above the freezing point the librational motion is more collective [46].

At early delay times, the anisotropy dynamics depends on the excitation and detection frequency. This frequency dependence has been studied with transient absorption spectroscopy [75] and with 2D IR photon-echo spectroscopy [76], and gives information on the mechanism by which the water molecules reorient. In Figure 4.8, the anisotropy of the O–D stretch vibration of HDO:H_2O is shown as a function of detection frequency at five different delay times. In case the O–D stretch vibration is pumped close to its central frequency (upper panels), the anisotropy shows very little frequency dependence, even at early delay times. When the pump frequency is tuned to the blue wing of the absorption spectrum (lower panels of Figure 4.8), the anisotropy in the center and the red wing is significantly lower than 0.4, already at a delay of 0.2 ps. This low anisotropy shows that molecules that are excited in the blue wing reorient while changing their frequencies from the excited blue wing to the center and the red wing of the absorption band. This finding agrees with the molecular jump model for reorientation that was developed by Laage and Hynes based on molecular dynamics simulations [78].

In the model of Laage and Hynes, the reorientation of water proceeds through a transition state in which the reorienting hydroxyl group forms a weak bifurcated hydrogen bond with its old hydrogen-bonded partner water molecule and its new hydrogen-bonded partner water molecule. The transition to a single strong hydrogen bond with the new partner leads to a large and fast change of the frequency of the O–D vibration. The observed frequency dependence can be well explained if O–D vibrations absorbing in the blue wing of the absorption band are much closer to the bifurcated transition state than O–D vibrations absorbing in the center and the red wing of the absorption band. If this is the case, a significant part of O–D vibrations excited in the blue wing undergo a fast change in orientation and frequency, thus leading to a low anisotropy in the center and red wing of the absorption band at early delay times.

In a recent 2D IR photon-echo study [76], the measured transient absorption line shapes of the O–H stretch vibration of HDO:D_2O showed a similar frequency dependence as was observed for the O–D vibration of HDO:H_2O with transient absorption spectroscopy [75], that is, a lower anisotropy in the center of the absorption band at short delays after excitation in the blue wing of the band. However, the corresponding 2D IR *power* spectra, which comprise both the excitation-induced change in absorption and the excitation-induced change in refractive index, did not show

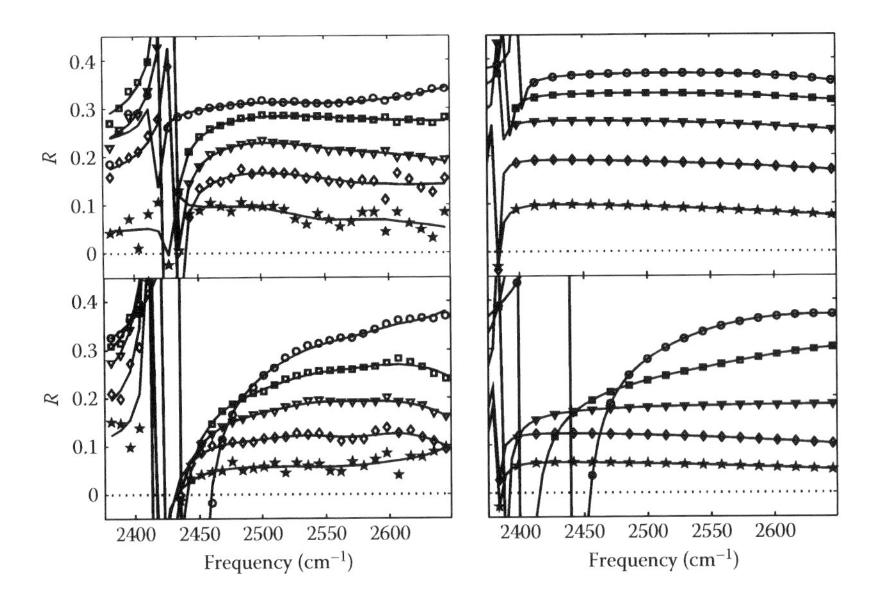

FIGURE 4.8 Anisotropy as a function of frequency at delays of 0.2 ps (circles), 0.5 ps (squares), 1 ps (triangles), 2 ps (diamonds), and 4 ps (stars). Shown are results obtained with a pump frequency of 2500 cm^{-1} (upper panels), and 2600 cm^{-1} (lower panels). The left panels show experimental results, the right panels present calculated results, based on a model that includes the large frequency modulation resulting from the transition to and through the bifurcated hydrogen-bond transition state for the molecular reorientation of water. (Reprinted with permission from Bakker HJ, Rezus YLA, Timmer RLA. Molecular reorientation of liquid water studied with femtosecond mid-infrared spectroscopy. *J Phys Chem A* 112: 11523–11534. Copyright 2008, American Chemical Society.)

this frequency dependence. In the 2D IR power spectra, the lowest initial anisotropy is obtained if the excitation and detection frequency are both in the blue wing of the absorption band [76].

The anisotropy dynamics become frequency independent at delay times >1 ps, showing that after this time the spectral equilibration is near complete. After this time, the anisotropy shows a single exponential decay. The time constant of this decay is observed to be somewhat shorter for the O–D vibration of HDO in H_2O (2.5 ± 0.2 ps) [44,46,79] than for the O–H vibration of HDO in D_2O (3 ± 0.3 ps). It may seem surprising that the orientational relaxation of the O–D group proceeds faster than that of the OH group, as its moment of inertia is almost twice as large. However, it should be realized that the orientational dynamics on longer timescales are not determined by the moment of inertia, but are governed by the relative motions of the water molecules, and, in particular, by the dynamics of H-bond breaking and reformation. A good measure for the translational mobility of the water molecules is provided by the value of the viscosity. Using the viscosities of H_2O and D_2O (0.9 and 1.1 mPa s, respectively), the reorientation times of the OH and OD are estimated to show a ratio of 0.8, which is very similar to the measured ratio of the reorientation time constants. The scaling with the viscosity indicates that the rate-limiting steps of the formation of the bifurcated transition state for reorientation are formed by translational molecular motions for which the rate is closely related to the viscosity [80].

The measured molecular reorientation times of 2.5/3 ps in H_2O/D_2O agree quite well with the results obtained with other techniques. NMR studies arrive at a reorientation time of the water molecule of 2.35–2.5 ps in liquid H_2O at 298 K [13,17,18] and 2.4–2.9 ps in liquid D_2O at 298 K [16,17]. In comparing the results of femtosecond pump-probe and NMR experiments with the results of dielectric relaxation studies and THz absorption, it should be realized that these techniques measure different orientational time-correlation functions. Femtosecond pump-probe and NMR experiments probe the second Legendre polynomial of the orientational correlation function ($\langle P_2(\cos\theta(\tau))\rangle$)

with time constant $\tau_{r,2}$, the decay time τ_D measured in dielectric relaxation and THz absorption spectroscopy is related to the time constant $\tau_{r,1}$ of the decay of the first Legendre polynomial of the orientational correlation function ($< P_1(\cos\theta(\tau)) >$).

In dielectric relaxation studies of liquid water, a main relaxation component with a time constant τ_D of 8.3 ps was found [19]. Similar values were found in THz spectroscopic studies of H_2O and D_2O [21,22]; at room temperature, the Debye times τ_D of the slow component were determined to be 8.5 ps for H_2O, and 10 ps for D_2O. To arrive at the time constant τ_1 of the decay of $< P_1(\cos\theta(\tau)) >$, the values of τ_D have to be corrected for collective effects. Using the correction proposed by Wallqvist and Berne [81], one arrives at values of τ_1 of ~7 and ~9 ps for H_2O and D_2O, respectively. The ratio between the first- and second-rank decay times $\tau_{r,1}$ and $\tau_{r,2}$ is determined by the nature of the reorientation mechanism. The rotation of water proceeds primarily via a jump mechanism in which the orientation angle of the O–H group changes by a value of ~60° [78]. The extended jump model for water predicts a ratio $\tau_{r,1}/\tau_{r,2}$ of ~2.5. The values of ~7/ ~9 ps for $\tau_{r,1}$ and 2.5/3 ps for $\tau_{r,2}$ observed for H_2O/D_2O are indeed close to this ratio.

The reorientation of water molecules specifically at interfaces has recently been measured at the air/water interface using time- and polarization-resolved IR pump/vibrational sum frequency probe measurements [82]. The approach is very similar to the bulk measurements: the pump pulse induces vibrational excitation for molecules oriented preferentially along the polarization axis of the pump field. The reorientation is subsequently followed with time-resolved SFG spectroscopy. From these experiments, the reorientation of the free O–H group, that is, the non-hydrogen-bonded O–H group sticking out of the water surface, was found to be three times faster than the reorientation of hydrogen-bonded O–H groups in bulk, due to the lower degree of hydrogen-bond coordination at the interface. At the same time, the vibrational relaxation is four times slower, which can be explained from the relatively weak intra- and intermolecular couplings of these O–H groups. The observed rapid reorientation of the free O–H groups is likely a general feature of water near extended hydrophobic surfaces, which, as will be shown in Section 4.6, is very different from the behavior of water molecules hydrating subnanometer-structured hydrophobic molecular groups.

4.4 AQUEOUS SALT SOLUTIONS

The addition of salts to water leads to significant changes of the structure and dynamics of the water hydrogen-bonded network. For many negatively charged anions, the interaction between water and the anion takes the form of a new type of hydrogen bond between the ion and the solvating water molecule [83,84]. These newly formed $O–H\cdots X^-$ hydrogen bonds are often directional in character [83,84], which means that the O–H bond and the $O\cdots X^-$ hydrogen-bond coordinates are collinear. This new type of hydrogen bond can be observed in the linear absorption spectrum. For instance, within the halogenic series (F^-, Cl^-, Br^-, I^-), the absorption spectrum of the O–H stretch vibration is observed to shift to higher frequencies, which indicates that the $O–H\cdots X^-$ hydrogen bond becomes longer and weaker in this series [59–61]. The dynamics of water molecules in salt solutions have been studied with transient absorption spectroscopy and with 2D IR spectroscopic techniques [85–91]. In all reported studies, a low-concentration solution of HDO in D_2O or HDO in H_2O is used as a solvent, to prevent the measurements being affected by resonant energy transfer among the O–H or O–D vibrations, respectively.

4.4.1 VIBRATIONAL ENERGY RELAXATION

For the O–H stretch vibration of HDO in solutions of NaCl, NaBr, and NaI in D_2O, a biexponential decay is observed of which the shorter time constant has a value of ~0.8 ps for all salt solutions [85–88]. The time constant of ~0.8 ps compares very well with the vibrational relaxation time constant of the O–H stretch vibration of pure HDO:D_2O [50]. The ~0.8 ps component originates from O–H groups that form O–H···O hydrogen bonds to other water molecules. This component includes the response of the HDO molecules in the first hydration shell of the Na^+ cations, because for these HDO molecules the O–H group will point away from the ion and forms an O–H···O hydrogen bond with a D_2O molecule [92–94]. A similar biexponential decay was observed for the O–D stretch vibration of HDO in solutions of NaBr in HDO:H_2O [89,90]. In this study, the fast component has a time constant of ~1.2 ps, which is somewhat faster than the value of T_1 of the O–D vibration of HDO:H_2O without salt added.

With increasing salt concentration, the amplitude of the fast component decreases and the amplitude of the slow relaxation component increases [86]. The time constant of the slow component increases within the halogenic series from Cl^- to Br^- to I^-. For solutions of 6 M NaCl, NaBr, and NaI, time constants of 2.6 ± 0.3, 3.1 ± 0.3, and 3.6 ± 0.3 ps are observed. The dependence of the vibrational lifetime of the slow component on the nature of the anion indicates that the O–H···X^- hydrogen bond (X^- = Cl^-, Br^-, I^-) plays an important role in the vibrational relaxation mechanism. The slow relaxation component is thus assigned to HDO molecules forming an O–H···X^- hydrogen bond to the anion. The increase in frequency of the O–H vibration of the O–H···X^- system in the halogenic series Cl^-, Br^-, and I^- indicates that the hydrogen bond becomes weaker [59–61]. A weaker hydrogen-bond interaction in turn leads to a decrease of the anharmonic interaction between the O–H stretch vibration and the hydrogen-bond mode [95], thus slowing down the vibrational relaxation.

For a solution of KF, no slow component is observed. The absorption spectrum of a KF solution is slightly red-shifted with respect to the spectrum of HDO:D_2O, which indicates that the O–H···F^- hydrogen bond is stronger than the O–H···O hydrogen bond. Hence, for O–H groups forming O–H···F^- hydrogen bonds to F^- ions, the vibrational lifetime will be relatively short, which explains the absence of a slow relaxation component for a solution of KF in HDO:D_2O [85,86].

In Figure 4.9, the excited-state absorption is shown as a function of delay time for solutions containing different concentrations of NaI. The delay time curves show a biexponential decay reflecting the vibrational relaxation of the O–H···O groups and the O–H···I groups. In addition, it is seen that the T_1 value of the O–H···I^- groups shows a significant dependence on concentration: T_1 increases from 2.4 ± 0.2 ps at 0.5 M to 4.7 ps at 10 M [88]. For the O–D stretch vibration of HDO in solutions of NaBr in HDO:H_2O, a similar trend is observed [89,90]. When the concentration of NaBr is increased from 1.5 to 6 M, the time constant of the slow relaxation component increases from 3.2 to 5.7 ps.

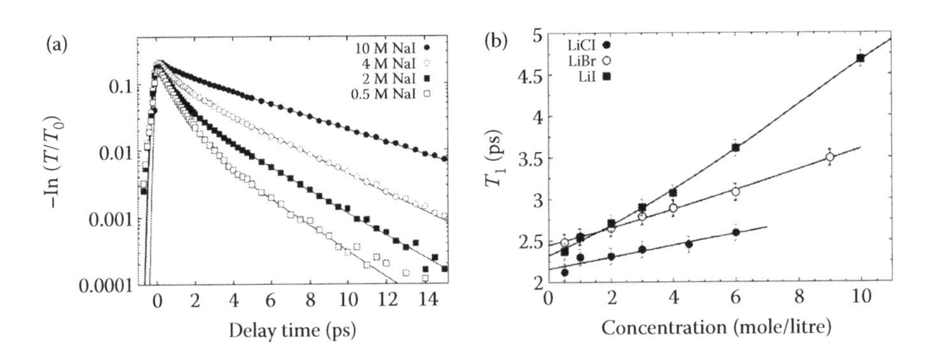

FIGURE 4.9 (a) Pump-probe traces of solutions of several concentrations of NaI in HDO:D_2O. The measurements were performed at a pump frequency of 3450 cm^{-1} and a probe frequency of 3200 cm^{-1}. The lines are biexponential fits to the data. (b) T_1 of the O–H stretch vibration of the O–H···X^- system (X^- = Cl^-, Br^- or I^-) as a function of concentration for three salts containing the same cation: Li, LiBr, and LiI. The lines are fits to the data. (Reprinted with permission from Kropman MF, Bakker HJ. Effect of ions on the vibrational relaxation of liquid water. *J Am Chem Soc* 126: 9135–9141. Copyright 2004, American Chemical Society.)

The dependence of T_1 of the O–H···X$^-$ system on concentration has been explained from a slowing down of the out-of-shell rotation with concentration [96]. The rotation of the O–H group out of the hydration shell involves the breaking of the O–H···X$^-$ hydrogen bond and the formation of a new O–H···O hydrogen bond to a neighboring water molecule. The latter system will show a rapid vibrational relaxation with a time constant of 0.8 ps. Hence, the out-of-shell rotation constitutes a relaxation channel for the O–H···X$^-$. As the out-of-shell rotation is expected to be much slower than the 0.8 ps relaxation, the out-of-shell rotation forms the rate-limiting step of this relaxation channel. With increasing salt concentration, the number density of water molecules decreases and the translational motions of the water molecules and ions will slow down. Hence, the out-of-shell rotation will also slow down with increasing concentration, as it requires a water molecule to approach the O–H···X$^-$ system to form a bifurcated hydrogen-bonded transition state [97]. If the difference in T_1 at low and high concentration would be completely due to the vanishing of the out-of-shell rotation, $\tau_{oos}(Cl^-) = 1/[(1/2.2) - (1/2.6)] = 11$ ps, $\tau_{oos}(Br^-) = 1/[(1/2.5) - (1/3.6)] = 8$ ps, and $\tau_{oos}(I^-) = 1/[(1/2.5) - (1/3.6)] = 6$ ps. These time constants decrease going from Cl$^-$ to Br$^-$ to I$^-$, which is consistent with the increasingly deformable character of the anion hydration shells in the halogenic series [96].

The vibrational lifetime of water molecules in the solvation shell of the anion is also influenced by the nature of the cation, in particular when the hydration shell is shared by the anion and the cation, thus forming Y$^+$ O–H···X$^-$ systems. The vibrational lifetime of HDO solvating Cl$^-$ and Br$^-$ is observed to decrease in the cationic series Na$^+$, Li$^+$, and Mg^{2+} [88]. The effect of cations on the vibrational lifetime can be explained from the electric field exerted by the cation. This electric field polarizes the O–H···O and O–H···X$^-$ hydrogen bonds of water molecules adjacent to the cation. With increasing electric field, the hydrogen bonds become stronger, which in turn leads to an increase of the anharmonic interaction with the O–H stretch vibration of an HDO molecule solvating the X$^-$ anion. Li$^+$ is smaller than Na$^+$, while Mg^{2+} has a similar size as Na$^+$, but possesses twice the charge. Therefore, the local electric field exerted by the cation increases going from Na$^+$ to Li$^+$ to Mg^{2+}, which agrees with the observed decrease of the vibrational lifetime of the anion hydration shells in this cation series.

4.4.2 Hydrogen-Bond Dynamics

The molecular motions of the water molecules hydrating specific anions like Cl$^-$, Br$^-$, I$^-$, BF_4^-, and ClO_4^- have been measured with two-color pump-probe spectroscopy [85–87], heterodyne-detected 2D photon-echo spectroscopy [89,90], and with spectral hole-burning spectroscopy [91]. As in the case of the studies of pure water, information on the fluctuations of the O–H···X$^-$ or O–D···X$^-$ hydrogen bond can be obtained from the spectral diffusion of the high-frequency O–H stretch or O–D vibration. A challenge in these studies is that the response of the water molecules hydrating the ions has to be distinguished from the response of the bulk-like water molecules.

The absorption spectrum of water hydrating Cl$^-$, Br$^-$, and I$^-$ is somewhat blue-shifted with respect to the spectrum of bulk liquid water, but the shift is small, meaning that the absorption bands of the separate O–D···X$^-$ and O–D···O components strongly overlap. Fortunately, O–H/O–D groups forming a hydrogen bond to Cl$^-$, Br$^-$, I$^-$, or ClO_4^- have a three to five times longer vibrational lifetime than O–H/O–D groups that form a hydrogen bond to an oxygen atom of another water molecule [85–91]. Hence, with increasing delay time, the measured signal will show an increasing bias to the dynamics of the hydroxyl groups that form a hydrogen bond to the anion. The spectral dynamics at later delay times will thus be dominated by the hydrogen-bond dynamics of the hydration shell.

In a recent spectral hole-burning study of solutions of NaBr and LiBr, the spectral dynamics at delays >3 ps were thus assigned to the O–H groups forming O–H···Br$^-$ hydrogen bonds. The FFCF derived from the spectral holes was observed to show a time constant of 4.3 ± 0.3 ps [91] (Figure 4.10). This time constant agrees very well with the time constant of 4.8 ± 0.6 ps of the

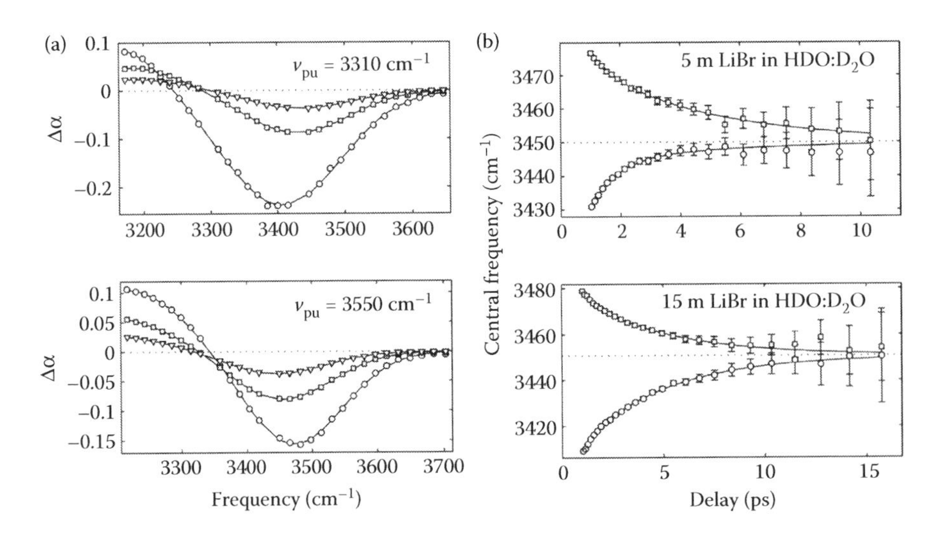

FIGURE 4.10 (a) Transient spectra at delay times of 0.5 ps (circles), 1 ps (squares), and 2 ps (triangles) measured at different delay times for a solution with a molality of 5 m LiBr in 2% HDO:D_2O. (b) Spectral diffusion shown by the position of the central frequency (first moment) of the bleach signal for solutions of 5 m LiBr (upper panel) and 15 m LiBr (lower panel) in 2% HDO:D_2O. Circles and squares denote excitation frequencies tuned to the red and blue wing of the absorption band, respectively. (Reprinted with permission from Timmer RLA, Bakker HJ. Hydrogen bond fluctuations of the hydration shell of the bromide anion. *J Phys Chem A* 113: 6104–6110. Copyright 2009, American Chemical Society.)

slowest spectral diffusion component observed for a 6 M NaBr solution with heterodyne-detected 2D photon-echo spectroscopy [89,90].

The hydrogen-bond dynamics of water molecules hydrating BF_4^- and ClO_4^- ions have been studied with 2D IR spectroscopy [98–100]. For these ions, the absorption spectrum of the O–H/O–D groups forming hydrogen bonds to the anion is quite well separated from the absorption spectrum of the other O–H/O–D groups in the solution. Hence, for these ions, the rate at which water is exchanged between the hydration shell and the bulk can be determined by measuring the exchange rate between the corresponding bands in the vibrational spectrum, as illustrated in Figure 4.11.

In Figure 4.11a, the 2D spectrum measured for a solution of $NaBF_4$ in HDO:H_2O is shown. The spectrum contains peaks corresponding to the $v = 0 \rightarrow 1$ and $v = 1 \rightarrow 2$ transition of O–D groups that are hydrogen-bonded to the oxygen atom of H_2O molecules, denoted as *hw*, and peaks corresponding to these transitions of O–D groups that are hydrogen-bonded to BF_4^-, denoted as *ha*. In addition, the spectrum contains so-called cross peaks. The presence of a cross peak indicates that a mode that is excited as *hw* gives a response at the frequency position of *ha* and vice versa. The amplitude of the cross peak will rise in the case of chemical exchange, that is, the switching of a hydrogen bond to BF_4^- ion to a hydrogen bond to H_2O or vice versa. The most evident chemical exchange peak is the $v = 0 \rightarrow 1$ (*hw* → *ha*) peak, labeled A. The spectrum contains two other exchange peaks. The $v = 1 \rightarrow 2$ (*hw* → *ha*) peak is labeled B. Because this peak represents an induced absorption, it has a negative sign. The last exchange peak in Figure 4.11a is the $v = 0 \rightarrow 1$ (*ha* → *hw*) peak, labeled C. Like peak A, it is going in the positive direction and is manifested as a reduction in the bottom portion of the $v = 1 \rightarrow 2$ *ha* peak going in the negative direction. The observed timescale of ~7 ps for a hydrogen bond to switch from BF_4^- to H_2O forms the out-of-shell rotation time of the hydration shell of the anion in dilute solutions [98]. For ClO_4^- ions, the switching time of an O–D group from the ion to a water molecule was observed to be 9 ps [99,100]. These time constants are similar to the out-of-shell rotation times that were estimated from the concentration dependence of the vibrational

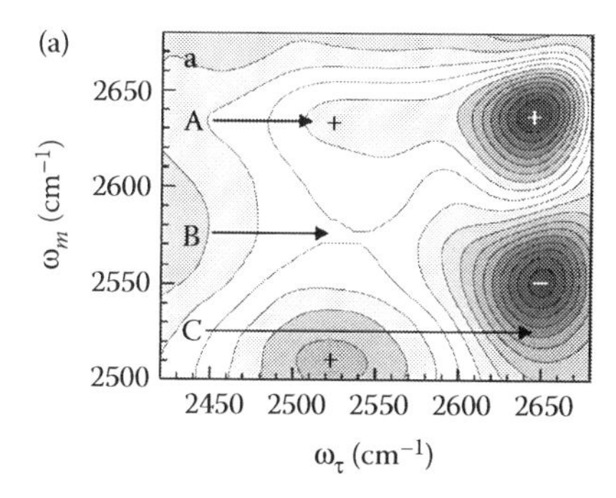

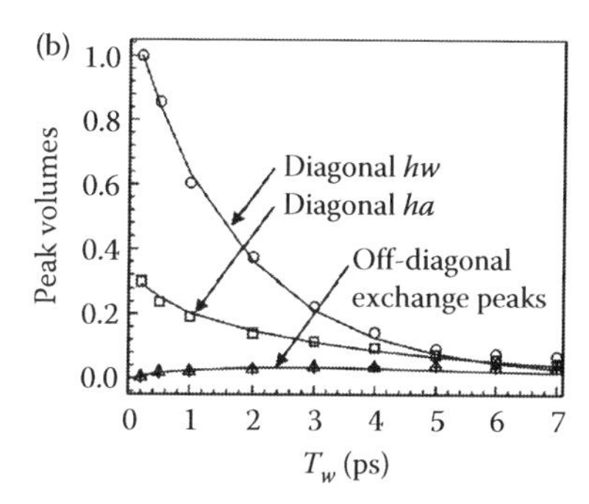

FIGURE 4.11 (a) The two-dimensional infrared vibrational echo spectrum of a solution with 7 H_2O per $NaBF_4$ at a waiting time T_w (= τ_2) of 4 ps. Peaks going in the positive direction are marked by +, and peaks going in the negative direction are marked by –. By 4 ps, additional peaks labeled A, B, and C have grown in because of chemical exchange. (b) Peak volumes as a function of the waiting time T_w(= τ_2) of the diagonal and chemical exchange peaks. *hw* denotes the O–D groups of HDO that are hydrogen-bonded to H_2O, *ha* denotes the O–D groups that are hydrogen-bonded to the BF_4^- anion. The solid curves are the result of a kinetic model. (From Moilanen DE et al. 2009. Ion-water hydrogen bond switching observed with 2D IR vibrational echo chemical exchange spectroscopy. *Proc Natl Acad Sci USA* 106: 375–380. Reprinted with permission from National Academy of Sciences of the United States of America.)

lifetime of O–H groups hydrogen-bonded to Cl^-, Br^-, I^- (see Section 4.4.1). The hydrogen-bond switching times of the anion hydration shells are thus 5–10 times longer than the slowest component of the hydrogen-bond dynamics of pure liquid water [32–36].

4.4.3 Molecular Reorientation

The orientational dynamics of water molecules in the hydration shells of different ions have been studied with polarization-resolved transient absorption spectroscopy [87,89,101,102]. Most of these studies addressed the dynamics of water hydrating the halogenic anions Cl^-, Br^-, and I^- ions. Again, a complication in these studies is that the spectral response of the water molecules hydrating the halogenic anions overlaps with that of the bulk water molecules. Hence, the measured anisotropy dynamics usually represent the dynamics of both the bulk water molecules and the hydration shell water molecules. Recently, it was shown that for all delay times it is possible to decompose the measured transient spectra in the spectrum of water hydroxyl groups forming O–H/O–D···O hydrogen bonds to the oxygen atom of another water molecule, and the spectrum of water hydroxyl groups forming O–H/O–D···X^- hydrogen bonds to the X^- anion ($X^- = Cl^-, Br^-, I^-$). This decomposition is greatly aided by the fact that the vibrational lifetime of the O–H/O–D stretch vibrations forming O–H/O–D···X^- hydrogen bonds is much longer than that of the other O–H/O–D stretch vibrations. Figure 4.12 shows the anisotropy dynamics of the different types of hydrogen-bonded O–D groups obtained from this decomposition. The anisotropy dynamics of the O–D···O component can be fitted well with a single exponential function with a time constant of ~2.5 ps, which means that the anisotropy dynamics of the O–D···O oscillators is very similar to the dynamics observed for neat HDO:H_2O [44]. The dynamics of the O–D···X^- is seen to follow a biexponential function with time constants of 2 ± 0.3 and 9 ± 2 ps. It is also seen that the amplitude of the fast component is somewhat larger for O–D···I^- than for O–D···Cl^-.

The fast component of the anisotropy decays shown in Figure 4.12 is likely due to a wobbling motion of the O–D group that keeps the O–D···X^- hydrogen bond with the anion intact. The wobbling motion will lead to a partial decay of the anisotropy, and the amplitude of this partial decay is determined by the angular cone of the wobbling motion. The I^- ion is much larger than the Cl^- ion and thus allows for a larger angular spread of the O–D group while keeping the hydrogen bond

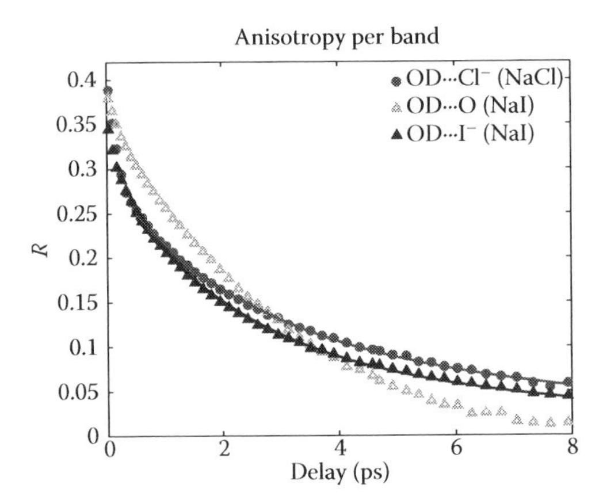

FIGURE 4.12 Anisotropy dynamics of the spectral components associated with O–D···O oscillators (open triangles) O–D···Cl^- oscillators (solid circles), and O–D···I^- oscillators (solid triangles). The anisotropy decay of the O–D···O oscillators can be fitted well to a single exponential decay with a time constant of 2.5 ps, and is thus very similar to the dynamics observed for neat HDO:H_2O. The anisotropy decay of the O–D···I^- and O–D···Cl^- oscillators can be fitted well to a biexponential function with time constants of 2 ± 0.3 and 9 ± 2 ps.

intact, which explains why the amplitude of the fast component is larger for O–D···I^- than for O–D···Cl^- (Figure 4.12). The wobbling motion will be induced by the translational and orientational motions of the H_2O molecules that surround the hydration shell. Therefore, the time constant of the wobbling motion can be expected to be similar to of the 2.5 ps associated with the anisotropy decay of the O–D···O component. The time constant of the fast decay of the anisotropy of 2 ± 0.3 ps agrees with this picture.

It should be realized that the rotation of O–D oscillators out of the hydration shell will not directly lead to a decay of the anisotropy of the O–D···X^- oscillators. Rotation out of the shell implies that the O–D···X^- hydrogen bond is exchanged for an O–D···O hydrogen bond. As a result, the O–D oscillator no longer contributes to the O–D···X^- spectral component. The rotation out of the hydration shell thus leads to a decay of the *amplitude* of the O–D···X^- spectral component, but it will not affect the *anisotropy* of this component. However, there is an equal number of O–D groups that were excited as O–D···O oscillators and rotate into the hydration shell to become O–D···X^- oscillators. These oscillators will decrease the anisotropy of the O–D···X^- spectral component, because the angle of rotation in the exchange between the bulk and the hydration shell is ~55° [78,97,103]. At longer delays, there will be very little O–D···O excited oscillators left. Hence, at later delay times (>3 ps), the switching from O–D···O to O–D···X^- will no longer contribute to the slow decay component of the anisotropy of the O–D···X^- component. At later delay times, the slow component will thus mainly represent the diffusion of the O–D group over the surface of the ion, and the reorientation of the complete hydration shell of the halogenic anion.

For a solution of 3 M NaCl in HDO:D_2O, it is observed that the reorientation time constant determined for delays >3 ps decreases from 9.6 ± 0.6 ps at 300 K to 4.2 ± 0.4 ps at 379 K [87]. In the same temperature interval, τ_{or} decreases from 12 ± 2 to 6 ± 1 ps for a solution of 3 M NaBr, and from 7.6 ± 1 to 2.6 ± 0.4 ps for a solution of 3 M NaI. The value of τ_{or} is smaller for I^- than for Br^- and Cl^-, which shows that the probed O–H group moves faster around I^- than around Br^- and Cl^-. This indicates that the hydration shell of the I^- anion is less structured and rigid than the hydration shells of Br^- and Cl^-, in agreement with the results from molecular dynamics simulations [104,105]. The temperature dependence of the reorientation can be modeled well with a Stokes–Einstein relation for orientational diffusion [87]. Following this relation, $\tau_{or}(T) \sim \eta(T)/T$, where $\eta(T)$ represents the temperature-dependent viscosity.

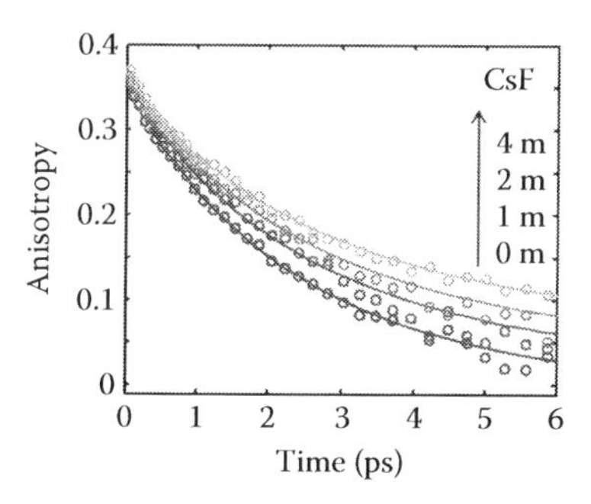

FIGURE 4.13 Anisotropy decay, obtained with pump/probe frequency centered at 2480 cm^{-1}, as a function of delay time for different concentrations of CsF. (Reprinted with permission from Tielrooij KJ et al. Anisotropic water reorientation around ions. *J Phys Chem B* 115: 12638–12647. Copyright 2011, American Chemical Society.)

The bias for the dynamics of the intact hydration shell at later delay times can be avoided by choosing anions for which the hydrating water molecules possess a vibrational lifetime similar to that of water in the bulk. Figure 4.13 shows the anisotropy decay as a function of delay time for CsF solutions of different concentration. These anisotropy decays represent the orientational dynamics of all water molecules since the O–D stretch mode of fluoride-bound water molecules have a similar absorption spectrum and vibrational lifetime as bulk and cation-bound water molecules. With increasing concentration, the anisotropy decay is observed to become slower. To quantify this effect, we describe the anisotropy decay with a biexponential function. The extracted slow water fraction is presented as a function of concentration in Figure 4.14. The linear slopes correspond to a hydration number of 9 for CsF.

The effect of ions on the dynamics of water can also be studied with gigahertz and terahertz dielectric relaxation spectroscopy. In this technique, the reorientation of the water dipoles in an applied external electric field is being probed. The addition of ions is often observed to lead to a decrease of the polarization response (depolarization). This depolarization results from the tight binding of water molecules to the ions with the result that these water molecules no longer contribute to the polarization response of the sample. The slow water fraction of CsF solutions as obtained with dielectric relaxation spectroscopy is shown in the right panel of Figure 4.14. From the slope, it follows that the hydration number of CsF is only 2.

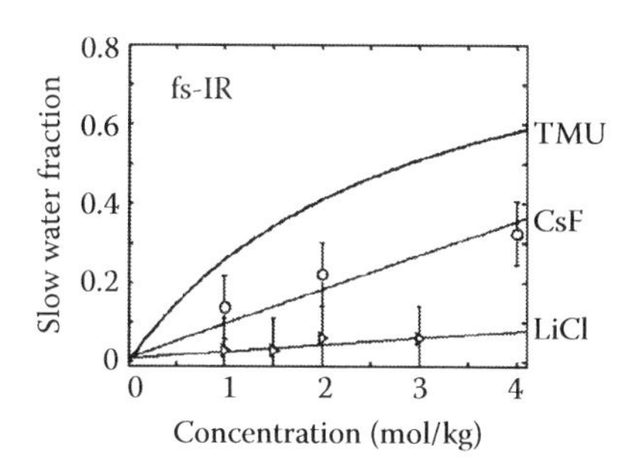

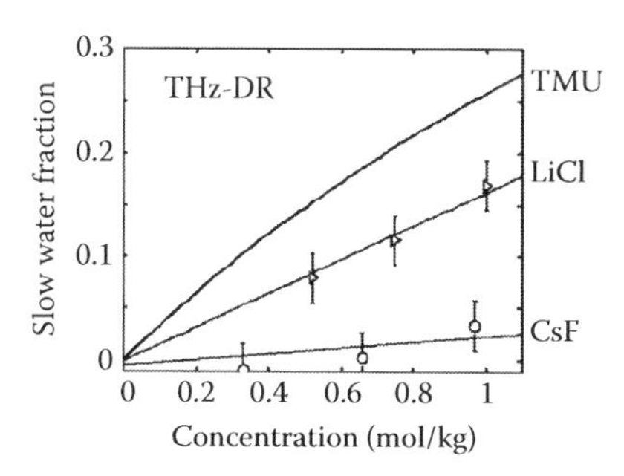

FIGURE 4.14 Comparison of the fractions of slow hydration shell water for LiCl, CsF, and TMU, as found by femtosecond infrared spectroscopy (left panel) and dielectric relaxation spectroscopy (right panel). Note that the two panels have different concentration ranges. (Reprinted with permission from Tielrooij KJ et al. Anisotropic water reorientation around ions. *J Phys Chem B* 115: 12638–12647. Copyright 2011, American Chemical Society.)

Figure 4.14 clearly shows that femtosecond IR spectroscopy arrives at a substantially larger slow water fraction for CsF solutions than dielectric relaxation measurements. This difference can be understood from the fact that femtosecond mid-IR measures the reorientation dynamics of the O–H/O–D groups of the water molecule, whereas dielectric relaxation measurements are sensitive to the reorientation dynamics of the water dipole (bisectrix of the water molecule). We thus find that Cs^+ and F^- ions mainly slow down the reorientation of the O–H/O–D groups, while leaving the reorientation of the water dipoles relatively unaffected. In Figure 4.14, the slow water fractions of CsF are compared with those measured for LiCl. In the femtosecond IR experiments on LiCl, the excitation and detection pulses were centered around 2480 cm^{-1}, and have limited spectral overlap with the blue-shifted O–D stretch modes of water molecules that are bound to chloride ions [102]. As a result, the observed results primarily represent the dynamics of the bulk-like and cation-bound water molecules. It is seen that the behavior is quite opposite to that of CsF: the femtosecond IR experiments find a rather low fraction of slow water corresponding to a hydration number of 2, whereas dielectric relaxation studies show a large hydration number of 8. It thus follows that Li^+ ions mainly slow down the reorientation of water dipoles and leave the reorientation of the O–H/O–D groups largely unaffected.

The effect of positive ions like Li^+ on the dynamics of water is thus mainly the result of the local electric field exerted by the ion, which fixes the water dipoles in a radial direction from the positive charge. The O–H groups of these water molecules remain relatively free to reorient in a propeller-like fashion around the fixed water dipole. The reorientation rate of these water molecules will be similar to that of neat liquid water. Earlier femtosecond IR measurements on $Mg(ClO_4)_2$ and $NaClO_4$ [106,107] also showed that even strongly hydrating cations like Mg^{2+} had very little effect on the dynamics of the hydroxyl groups of the water molecules. For CsF, the interaction is mainly due to the formation of directional hydrogen bonds between O–D groups and the negative charge of the F^- ion. Owing to this strong hydrogen-bond interaction, the reorientation of the OD groups pointing toward the F^- ion is strongly slowed down. However, the dipole (bisectrix) of the water molecule to which this O–D group belongs is hardly affected in its orientational mobility. Hence, the strongly hydrating F^- ion has little effect on the orientational mobility of the water dipoles. Hence, we arrive at a picture of semirigid hydration of ions: the reorientation of water molecules is only affected along a certain vector, whereas reorientation in other directions remains largely unaffected. The range of the semirigid hydration effect is restricted mainly to water molecules in the first solvation shell around the hydrated ions, as can be concluded from the hydration numbers of for Li^+ and F^-. Outside the first solvation shell, water molecules behave predominantly bulk-like. This result agrees with the findings of molecular dynamics simulations [108,109].

The above picture of semirigid hydration largely applies to cases where one of the ions is strongly hydrating and the other is weakly hydrating. A special situation arises for salt solutions for which both ions are strongly hydrating, as is for instance the case in $MgSO_4$ solutions. For $MgSO_4$ solutions, the hydration number is ~18 both in dielectric relaxation and in femtosecond IR measurements [101]. In view of this large hydration number, the corresponding hydration structures must extend well beyond the first hydration shells of the ions. The fact that the same number of affected water molecules is observed in dielectric relaxation and femtosecond IR demonstrates that these hydrating water molecules are affected along both the dipolar and the bond axes. The formation of such extended rigid hydration structures probably arise as a result of the combination of a strong fixing of the dipole vectors of the water molecules by the cation and a strong fixing of the direction of the hydroxyl groups by the anion. The ions thus cooperate in impeding the orientational mobility of water over relatively long ranges, and probably induce the formation of a locked hydrogen-bonded structure of water molecules in between the ions. This latter picture is in line with the observation of solvent-separated ion pairs for solutions of $MgSO_4$ [110].

4.5 WATER INTERACTING WITH PROTONS AND HYDROXIDE IONS

4.5.1 Protons

The proton (H^+) forms a very special ion in water as it is not a well-localized ion, like for instance Na^+ or Cl^-. Instead, the proton charge is delocalized in extended hydration structures of which the most well known are the so-called Zundel ($H_5O_2^+$) and Eigen ($H_9O_4^+$) structures. A consequence of this delocalized character is that protons in liquid water are not transferred by a conventional Stokes diffusion process but instead are transported by transferring the protonic charge between hydrogen atoms. The proton is thus conducted through liquid water via a mechanism that is denoted as Grotthuss conduction [111,112]. Recent *ab initio* molecular dynamics simulations showed that this conduction involves the rapid interchange of Eigen and Zundel hydration structures [113,114]. Recently, the mechanism of proton conduction was studied experimentally with femtosecond transient absorption spectroscopy [115]. In this study, it was observed that the excitation of the vibrations of the Eigen structure also leads to a quasi-instantaneous response of the vibrations of the Zundel structure (Figure 4.15), which implies that the interconversion between the hydration structures must take place on a timescale <50 fs, in agreement with the results of the *ab initio* molecular dynamics simulations. In addition, it was observed that the vibrations of the Eigen hydration structure absorbing at 2900 cm^{-1} show a very rapid vibrational relaxation with a time constant of 110 fs, as illustrated in Figure 4.15.

Owing to its short vibrational lifetime, the hydrated proton also functions as a sink of vibrational energy of nearby resonant vibrations. For solutions of HDO in HCl/H_2O, it was observed that the relaxation of the O–D stretch vibration of HDO shows an additional nonexponential decay that becomes faster with increasing proton concentration [116]. This additional decay finds its origin in vibrational Förster energy transfer from the O–D vibration at 2500 cm^{-1} to water molecules hydrating the proton. These water molecules have a very broad absorption spectrum ranging from 1000 to 3400 cm^{-1} and can thus resonantly take up the energy of a nearby excited O–D vibration. For the complementary system, that is, the O–H stretch vibration HDO molecules hydrating the deuteron, no acceleration of the vibrational relaxation was observed because the O–H vibration (~3400 cm^{-1}) is strongly out of resonance with the O–D stretch vibrations of D_2O molecules hydrating the deuteron (700–2500 cm^{-1}).

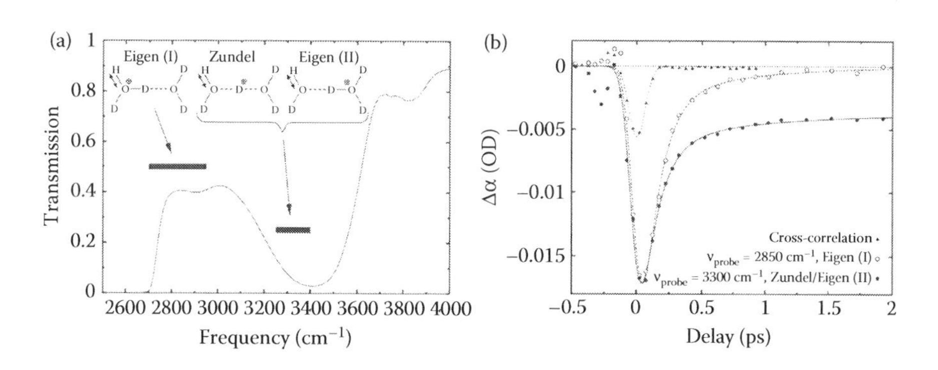

FIGURE 4.15 (a) Infrared spectrum of a 5 M solution of HCl:DCl in $HDO:D_2O$, with a H:D ratio of 1:20. The bars indicate the frequency regions of the O–H-stretching modes of the Eigen(I) and Zundel/Eigen(II) structures. (b) Absorption change as function of delay after resonant excitation (at 2935 cm^{-1}) of the O–H stretching mode of the Eigen structure. The absorption change is shown for two probing frequencies, one resonant with the Eigen I structure, and one resonant with the Zundel and Eigen II structures. The time constants are 120 fs and 0.7 ps for the dotted curve, and 130 fs and 0.8 ps for the solid curve. (From Woutersen S, Bakker HJ. 2006. *Phys Rev Lett* 96: 138305.)

Detailed information on the rate and mechanism of proton transfer can be obtained if the proton transfer can be triggered, for instance, by optical excitation of a photo-acid. In the last decades, many studies of aqueous proton transfer have been reported in which different pyrene photo-acids, like 8-hydroxy-1,3,6-pyrenetrisulfonic acid trisodium salt (HPTS), were used [117–120]. HPTS has a strong absorption near 400 nm and is thus easily excited using the second harmonic of the output of a Ti:sapphire laser at 800 nm. The excitation leads to an enhancement in the acidity of the molecule by a factor of 10^6. HPTS has been used to study the dynamics of acid dissociation [117–120] and acid–base reactions [121,122] using different time-resolved spectroscopic techniques, including transient absorption and time-resolved fluorescence spectroscopy. It was found that HPTS* dissociation in water (proton transfer to solvent) occurs with a time constant of 90/220 ps in H_2O/D_2O. When a stronger base than water is added at sufficient concentration, the proton transfer reaction speeds up because then direct proton transfer between the acid and the base becomes the dominant reaction pathway (rather than proton transfer to the water solvent) [121,122].

The proton transfer reaction from a photo-acid to water has also been studied with femtosecond mid-infrared laser pulses [123–134]. In this approach, the proton transfer is followed by probing the vibrational resonances of the photo-acid, the conjugate photo-base, the hydrated proton, and the accepting base. The responses of the photo-acid and its conjugate photo-base reflect the time the proton needs to leave the photo-acid, the response of the proton in water shows when the proton is taken up by the water in between the photo-acid and the accepting base, and finally the response of the conjugate acid of the accepting base signals when the proton arrives at the base. As a result, a complete picture of the proton transfer reaction can be obtained.

Most of the reported femtosecond mid-infrared studies have employed the photo-acid HPTS, but recently also other photo-acids like naphthol salts [133,134] have been used. In all studies, it is observed that the proton transfer reaction is highly nonexponential, as illustrated in Figure 4.16. The nonexponential behavior can be well explained from the presence of a (statistical) distribution of acid–base distances in solution [123–134]. The proton transfer will be fast if the nearest accepting base is close to the excited photo-acid, and will be slow in case the nearest base is separated by many water molecules. Hence, a distribution of reaction rates is observed.

The distribution of reaction rates for different photo-acid–base separations has been modeled in different ways. In one approach, the generation and reaction rate of each water-separated acid–base

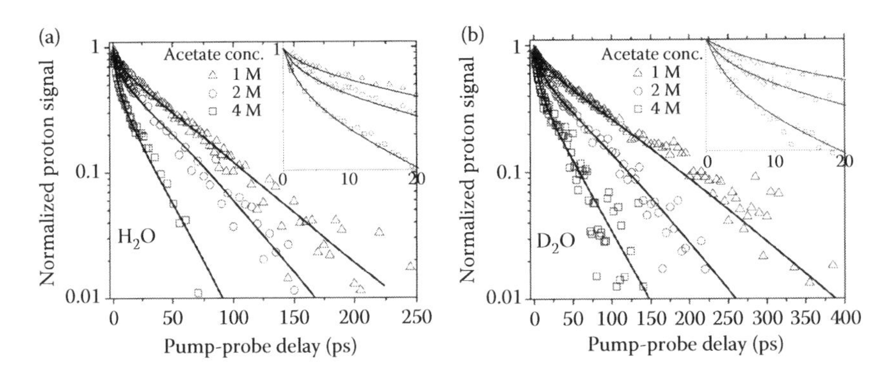

FIGURE 4.16 Response of the proton/deuteron vibrations as a function of delay for solutions of 10 mM of HPTS and 1, 2, and 4 M of acetate in H_2O (a) and D_2O (b). In the inset, the response measured in the first 20 ps is shown, illustrating the highly nonexponential character of the proton transfer. The solid lines are calculated curves using a conduction model in which the rate of transfer decreases by a constant factor for every additional water molecule in the short-living hydrogen-bonded water wire connecting the acid and the base. (Reprinted with permission from Siwick BJ, Cox MJ, Bakker HJ. Long-range proton transfer in aqueous acid-base reactions. *J Phys Chem B* 112: 378–389. Copyright 2008, American Chemical Society.)

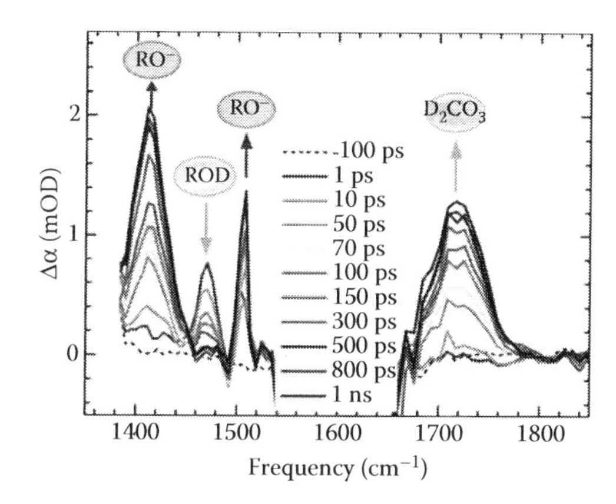

FIGURE 4.17 Transient infrared absorption spectra at different delays following excitation of the photo-acid 2-naphthol-6,8-disulfonate. The transient spectra show the marker modes of the excited photo-acid (ROD at 1472 cm^{-1}), the conjugate photo-base (RO^- at 1410 and 1510 cm^{-1}), and of D_2CO_3 at 1720 cm^{-1}. (From Adamczyk K et al. 2009. Real-time observation of carbonic acid formation in aqueous solution. *Science* 326: 1690–1694. Reprinted with permission of AAAS.)

complex was modeled independently from the other acid–base complexes [126,127], which results in a large number of independent rate constants. In another approach, the proton transfer dynamics were described with a model in which the proton transfer rate decreases by the same factor for every additional water molecule separating the acid from the base [128,129]. For the base acetate, it was found that at a base concentration of 1 M, most proton transfer events take place in reaction complexes in which the photo-acid and the acetate base are separated by two or three water molecules [129].

The kinetic isotope effect of the proton transfer to the base was found to be 1.5 [128,129], which is very similar to the kinetic isotope effect observed for the mobility of free hydrated protons/deuterons in H_2O/D_2O. This finding supports a mechanism of proton transfer from the photo-acid to the base via a Grotthuss-type conduction mechanism that is similar to the mechanism of proton transfer in neat liquid water. The proton conduction thus likely takes place via short-living wires of water molecules that for a very short time connect the acid with the base. At low concentrations and for weak bases, it is also possible that the proton is first transferred to water and later taken up (scavenged) by the base, or that the photo-acid and the base first diffuse in closer proximity before proton conduction occurs [130].

The transfer of a proton from HPTS to a base has also been used to study the properties of important transient acid species like the carbonic acid H_2CO_3 [133]. The proton release from HPTS and the temporary uptake by a bicarbonate base leads to the production of H_2CO_3, as is illustrated in Figure 4.17. It was found in this study that the produced carbonic acid is stable on timescales well beyond 1 nanosecond.

4.5.2 Hydroxide Ions

The properties of the hydroxide ion (OH^-) in water resemble those of the proton in water. Like the proton, the transfer of the hydroxide ion involves a Grotthuss conduction mechanism in which the charge of the hydroxide ion is being transferred among water molecules. However, the details of the hydration structures of the hydroxide ion and their interconversion dynamics are quite different from those of the proton in water [135].

The dynamical properties of the hydroxide ion in water have been studied with femtosecond transient absorption spectroscopy [136–138] and 2D vibrational spectroscopy [139,140]. The addition of OH^- ions to water leads to a small shoulder in the absorption spectrum of the O–H vibrations at ~3600 cm^{-1}. This shoulder has been assigned to the absorption of the OH^- ion. The high frequency of this absorption indicates that the donated hydrogen bond of the hydrogen atom of the OH^- ion is very weak, which can be explained from the fact that the hydrogen atom of OH^- is somewhat negatively charged. The addition of OH^- to water also leads to a broad-band absorption in the frequency region between 2800 and 3400 cm^{-1}. This broad-band absorption is associated with the O–H stretch vibrations of water molecules that donate strong hydrogen bonds to the negatively charged oxygen atom of the OH^- ion.

The dynamics of the broad-band absorption have been studied for the isotopically diluted system of HDO and OH^- in solutions of NaOD in D_2O. The HDO molecules hydrating OD^- are observed to show a very fast vibrational relaxation, for which time constants of ~160 fs [136] and ~110 fs [139] have been reported. This relaxation is substantially faster than the relaxation of the O–H stretch vibration of HDO in pure D_2O for which $T_1 = 740 \pm 30$ fs. The fast relaxation of the HDO molecules has been explained in different ways. In the transient absorption study, this component has been assigned to O–H groups of HDO molecules in the hydration shell of $O–D^-$ that are spectators to the deuteron transfer between D_2O and $O–D^-$ in a $D–O–D\cdots^-O–D$ system [136]. The transfer is accompanied by electron transfer from the moving deuterium atom to the O–D fragment that is left behind. This deuteron transfer forms an essential step in the conduction mechanism of the hydroxide ion through liquid water. As a result of the transfer, the vibrational frequency of the O–H group of an HDO molecule hydrogen-bonded to one of the oxygen atoms in the $D–O–D\cdots^-O–D$ system will be strongly modulated, leading to a fast vibrational relaxation. In the 2D vibrational spectroscopic study, the broad vibrational response and fast vibrational relaxation were assigned to the $v = 0 \rightarrow 2$ transition of the stretch vibration of the hydrogen atom in a nearly symmetric Zundel-type $D–O–H\cdots O–D^-$ system [139]. Hence, in the latter explanation, the observed relaxation is not associated with a spectator O–H group, but rather due to the fast vibrational relaxation of the transferring proton/deuteron itself.

The two-dimensional infrared spectra also show a rise of cross-peak signals between the O–H stretch vibrations of HDO and OH^-, which indicates the occurrence of exchange between the two species. This exchange also results from deuteron transfer, as illustrated in Figure 4.18. The transfer is found to take place on a relatively long timescale with a lower boundary of 3 ps [140].

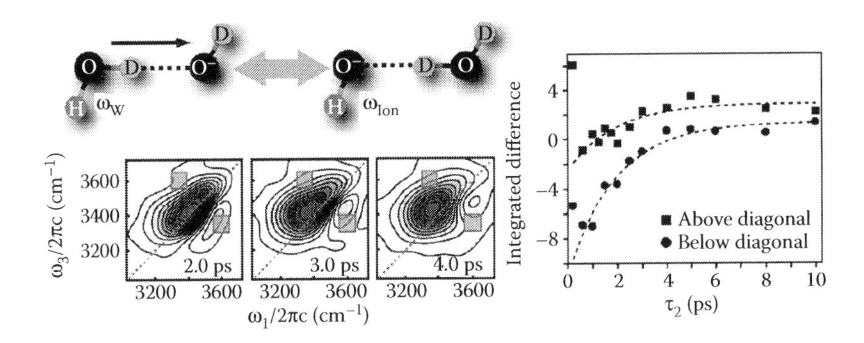

FIGURE 4.18 Two-dimensional infrared spectra in the region of the absorption O–H stretch vibration for a solution of HDO, OH^-, and 10.6 M NaOD in D_2O. The right-hand part of the figure shows the integrated difference in the off-diagonal regions highlighted by the shaded gray squares above and below the diagonal axis (ω_1 = 3300–3400 cm^{-1}, ω_3 = 3550–3600 cm^{-1} and its reflection about the diagonal). The cartoon in the top of the figure shows the likely origin of the observed rise of the cross-peak intensities: deuteron transfer between HDO and OD^-. (Reprinted with permission from Roberts ST et al. Proton transfer in concentrated aqueous hydroxide visualized using ultrafast infrared spectroscopy. *J Phys Chem A* 115: 3957–3972. Copyright 2011, American Chemical Society.)

Recently, the energy relaxation dynamics following the vibrational relaxation of the OH^- hydration complex were studied for a solution of NaOH in pure H_2O [137]. The rapid <200 fs relaxation of the O–H vibration of H_2O molecules hydrating the OH^- ion is observed to lead to a rapid local heating of the whole hydration complex. The hydration complex cools with a time constant that increases from 1.2 ps for a 0.5 M solution to 4.5 ps for a 10 M solution. The equilibration can be modeled well with a heat diffusion model from which the size of the hydration structure can be determined. The obtained value for the radius of the hydration complex of 0.36 ± 0.03 nm corresponds to a hydration shell of ~4.5 water molecules, in excellent agreement with the results of *ab initio* molecular dynamics calculations [135].

The dynamics of the water molecules outside the first hydration shell of the hydroxide ion have also been studied both for the O–H stretch vibration of HDO for a solution of NaOD in D_2O and for the O–D stretch vibration of HDO for a solution of NaOH in H_2O. For both cases, the vibrational relaxation is observed to become faster in the presence of the hydroxyl ion. The value of T_1 of the O–D stretching vibration of HDO molecules outside the first hydration shell of OH^- decreases from 1.7 ± 0.2 ps for neat water to 1.0 ± 0.2 ps for a solution of 5 M NaOH in HDO:H_2O [138]. For the O–H vibration of HDO molecules outside, the first hydration shell of OD^-, a similar acceleration is observed, from 750 ± 50 fs for neat HDO:D_2O to 600 ± 50 fs for a solution of 6 M NaOD in HDO:D_2O. The acceleration of the vibrational relaxation can be explained from fluctuations in the energy levels of the HDO molecules due to charge transfer events and charge fluctuations. The reorientation dynamics of water molecules outside the first hydration shell have the same time constant of 2.5 ± 0.2 ps as in bulk liquid water, indicating that there is no long range effect of the hydroxide ion on the hydrogen-bond structure of liquid water [138].

4.6 WATER INTERACTING WITH DISSOLVED MOLECULES

Water can strongly interact with other molecules in case the molecule possesses molecular groups with strong dipole moments or with a high polarizability. These interactions are generally denoted as hydrophilic. For molecules and molecular groups without a dipole moment or large polarizability, the interaction with water will be weak. Usually, it is energetically favorable to shield these hydrophobic groups from water, which results in a tendency of these groups to cluster. This hydrophobic driving force plays a crucial role in many self-organizing processes like the folding of proteins and the formation of bilayer membranes. Femtosecond mid-infrared spectroscopy is an ideal technique to obtain molecular-scale information on the interactions between water and hydrophilic and hydrophobic molecular groups. In this section, the results of studies of these interactions will be presented.

4.6.1 Hydrophilic Molecular Groups

The hydrophilic interactions between water and dissolved molecules often take the form of hydrogen bonds. An example is the urea molecule $OC(NH_2)_2$. Urea can form up to eight hydrogen bonds with water molecules. The vibrational energy relaxation and reorientation dynamics of water molecules in a solution of urea in HDO:H_2O have been studied with femtosecond transient absorption spectroscopy [141]. The O–D groups of HDO molecules that are hydrogen-bonded to urea were observed to have the same vibrational relaxation time constant as the O–D groups of HDO molecules that are hydrogen-bonded to H_2O molecules. In addition, the rate of orientational relaxation of the solvating water was observed to be practically the same as for pure liquid water. This finding agrees with results from NMR studies in which it was also found that urea has very little effect on the average molecular reorientation rate of water [142]. The surprisingly small effect of urea on the dynamics of water can be explained from the fact that urea fits quite well in the hydrogen-bond network of liquid water: the urea molecule replaces a dimer of water molecules, thus leaving the hydrogen-bond and orientational dynamics of the remaining water very similar to the corresponding dynamics of neat liquid water.

The dynamics of water molecules interacting with the polymer poly(ethylene)oxide (PEO) have been studied by probing the dynamics of the O–D stretch vibration of a solution of this polymer in HDO:H_2O [143]. PEO is an amphiphilic system containing ether oxygen atoms and hydrophobic aliphatic groups. For PEO solutions, the relaxation of the O–D stretch vibration becomes slower with decreasing water content. This can be well explained from the existence of two different water fractions: water molecules that form a hydrogen bond to other water molecules and those that form a hydrogen bond to the oxygen atoms of PEO. The slower vibrational relaxation is consistent with the fact that the O–D spectrum shows a clear blue shift with decreasing water content. This blue shift indicates that the hydrogen bonds become weaker, which in turn leads to a slower vibrational relaxation [58]. The orientational relaxation also shows two components. The water-bonded O–D groups show an orientational relaxation that is similar to pure HDO:H_2O, whereas the PEO-bonded water molecules are significantly slower, with time constants >15 ps. Another interpretation given in Ref. [143] is that the orientational dynamics are in fact uniform for all water molecules. In this picture, the fast component corresponds to reorientation of the water molecules within a certain restricted angular space, while the slow reorientation follows from the complete randomization of the orientation. This behavior can be described with a wobbling-in-a-cone model. Yet another interpretation would be that the slow reorientation component is associated with O–D groups hydrating the hydrophobic parts of PEO (see Section 4.6.2).

4.6.2 Hydrophobic Molecular Groups

The properties of hydrophobic molecular groups in water have been studied with several experimental techniques. In these studies, usually amphiphilic molecules are used because these molecules can be dissolved up to high concentrations. Clearly, a disadvantage of this approach is that the observations can also be (partly) due to water molecules interacting with the hydrophilic part of the molecule.

Naively, one would expect that the introduction of a weakly interacting hydrophobic solute in water leads to a break-up of strong water–water hydrogen bonds and an increase in the disorder of the water network. Hence, the dissolution of a hydrophobic solute is expected to lead to an increase of both the enthalpy ΔH and the excess entropy ΔS_{exc} (the entropy contribution other than the translational entropy). However, Frank and Evans observed that the dissolution of hydrophobic compounds is associated with *negative* changes in both the enthalpy and the excess entropy ($\Delta H < 0$, $\Delta S_{exc} < 0$) [144]. Both effects indicate that the introduction of hydrophobic groups leads to an increased structuring of the surrounding water molecules. These structures were denoted by Franks and Evans as hydrophobic icebergs [144].

In spite of the thermodynamic results, molecular scale studies did not find the structure of water surrounding hydrophobic groups to be very different from bulk liquid water. For instance, in neutron scattering studies the oxygen–oxygen distances of the solvating water molecules were observed to be similar as in bulk liquid water [145–148]. However, with respect to the water *dynamics*, techniques like NMR and dielectric relaxation do find differences between water hydrating hydrophobic groups and bulk liquid water [149–152]. Both methods show that the average orientational mobility of water molecules in solutions containing hydrophobic solutes is decreased. However, as these techniques only measure a response that is averaged over all water molecules, it is not clear whether the observations result from many water molecules that show slightly slower dynamics than bulk liquid water or rather from a small number of water molecules that show much slower dynamics.

Recently, the orientational dynamics of water molecules in the solvating hydrophobic groups has been studied by measuring the orientational dynamics of the O–D vibration of HDO for solutions of amphiphilic molecules like tetra-methyl-urea (TMU), tri-methyl-aminoxide (TMAO), *N*-methyl-amide (NMA), and proline in HDO:H_2O [153,154]. This method has as an advantage

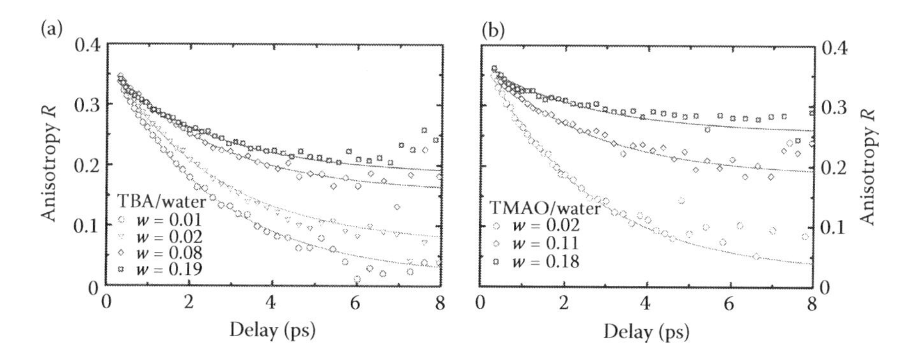

FIGURE 4.19 Anisotropy decays of the excitation of the O–D stretch vibration of HDO molecules for four different solutions of tertiary butyl alcohol (TBA) (a) and three different solutions of trimethylamine-*N*-oxide (TMAO) (b) in HDO:H_2O. (From Petersen C et al. 2010. *J Chem Phys* 133: 164514.)

over NMR and dielectric relaxation studies that the full orientational correlation function is measured.

In Figure 4.19, the anisotropy dynamics are shown as a function of delay for several concentrations of tertiary butyl alcohol (TBA) and trimethylamine-*N*-oxide (TMAO). For all solutions, the anisotropy dynamics are observed to be very different from the dynamics of bulk HDO:H_2O. The relaxation of the anisotropy shows a fast component with a time constant of ~2.5 ps and a much slower component with a time constant >10 ps. The amplitude of the slow component increases with the concentration of TBA and TMAO. At low concentrations, this increase is linear, and comparison of different solutes shows that the amplitude also scales with the number of methyl groups contained in the solute [153,154]. Hence, the slow component has been assigned to water molecules solvating the hydrophobic parts of the solute. The fast component is assigned to water molecules outside the hydrophobic hydration shell. It is thus found that the hydrophobic hydration shell shows much slower orientational dynamics than bulk liquid water.

Recent 2D IR studies showed that the water around the hydrophobic molecular groups also shows much slower spectral diffusion than bulk-like water [155,156]. This effect is illustrated in Figure 4.20. This figure shows that for amphiphilic solutes the elongation of the 2D spectrum along the diagonal is much longer-lived than for pure water. This result implies that the spectral diffusion is substantially slower for water molecules in solutions of TBA, TMAO, and TMU than for bulk liquid water. This slowing down of the spectral diffusion dynamics turns out to be strongly correlated with the slowing down of the orientational mobility of the water molecules [156], which indicates that these effects have a common origin in the influence of hydrophobic molecular groups on the hydrogen-bond dynamics of water. A similar correlation between spectral dynamics and reorientation was recently reported for solutions of NaBr [89]. It was observed that the spectral diffusion and the reorientation show a very similar slowing down with salt concentration, which indicates that both slowing down effects possess a common origin in the influence of the ions on the hydrogen-bond dynamics of water.

As was discussed in Section 4.3.4, the molecular reorientation of water involves a jump mechanism in which the hydrogen atom of the reorienting hydroxyl group forms a bifurcated hydrogen bond to the oxygen atoms of two nearby water molecules [36,75,78]. The frequency of the O–H stretch vibration in this transition state is strongly blue-shifted with respect to the O–H stretch vibration of linearly hydrogen-bonded O–H groups. Hence, the evolution from a linear hydrogen bond to the bifurcated transition state is accompanied by a large change in vibrational frequency and thus leads to strong spectral diffusion, irrespective of whether this evolution leads to a new hydrogen bond and thus reorientation (successful switch) or to the restoring of the original

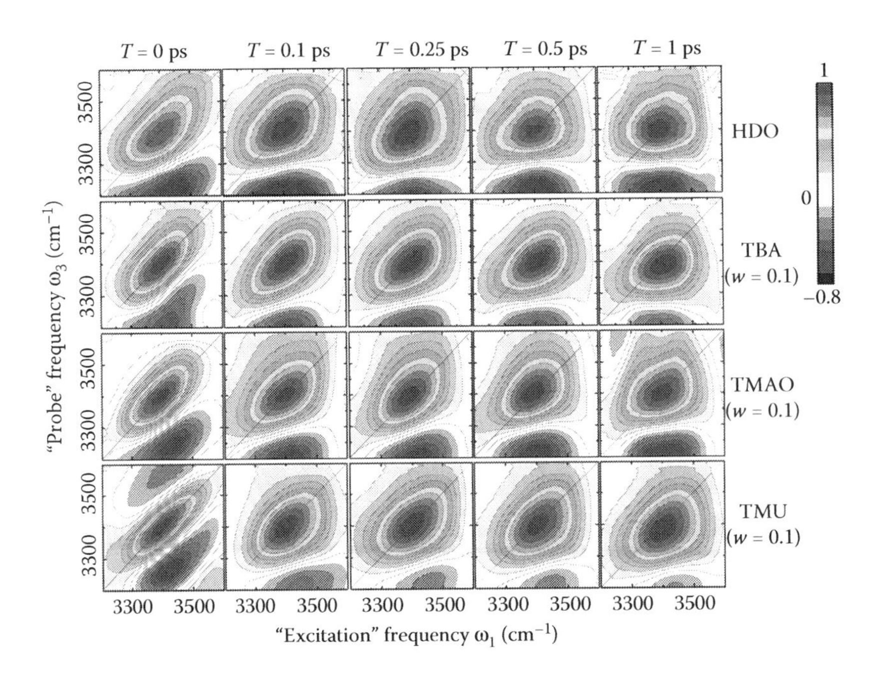

FIGURE 4.20 Two-dimensional infrared spectra of the O–H stretch vibration for pure HDO:D_2O (top left) and for a solution of 5 M trimethylamine-*N*-oxide (TMAO) in HDO:D_2O (lower left) at a waiting time of 2 ps. The right-hand side of the figure illustrates to which type of water molecules the two-dimensional infrared spectra correspond. (Reprinted with permission from Bakulin A et al. Hydrophobic molecules slow down the hydrogen-bond dynamics of water. *J Phys Chem A* 115: 1821–1829. Copyright 2011, American Chemical Society.)

hydrogen bond (unsuccessful switch). The correlated slowing down of the spectral diffusion and the reorientation indicates that water near hydrophobic groups can no longer evolve to the bifurcated hydrogen-bond structure. A possible explanation for this observation is that the hydrophobic solute methyl groups fill up the cavities of the hydrogen-bond network of water, thereby preventing the local collapses of the network that are required for the formation of bifurcated hydrogen bonds [157].

The reorientation and spectral diffusion dynamics of water in solutions of amphiphiles have also been studied with classical molecular dynamics simulations [158–160]. In these studies, a slowing down of the reorientation and spectral diffusion of the water molecules around the hydrophobic groups is found, but not to the extent as is observed in the femtosecond transient absorption and the 2D IR experiments. According to the MD simulations, the effects of TBA, TMAO, and TMU on the two-dimensional infrared spectrum of water are dominated by the hydrophilic groups of these solutes and not by the hydrophobic groups [159]. For instance, TMAO (with three methyl groups), which has a strongly polar NO group, was calculated to have the largest effect, even larger than TMU (with four methyl groups), and TBA (with three methyl groups), which has a hydroxyl group similar to the hydroxyl groups of water, was calculated to have a negligible effect on the spectral dynamics of water. This trend is quite different from what is observed in the experiments. In the 2D IR experiments, TBA and TMAO are observed to have similar effects on the water dynamics and TMU is observed to show the largest effect (Figures 4.20 and 4.21). These latter results indicate that the hydrophobic groups are more important in affecting the dynamics of water than the hydrophilic group. The relatively strong effect of hydrophobic groups on the dynamics of water agrees with the results of previous NMR [142,150,161,162] and dielectric relaxation studies [163,164]. In all these studies, it is found that the effect of the solute molecules on the

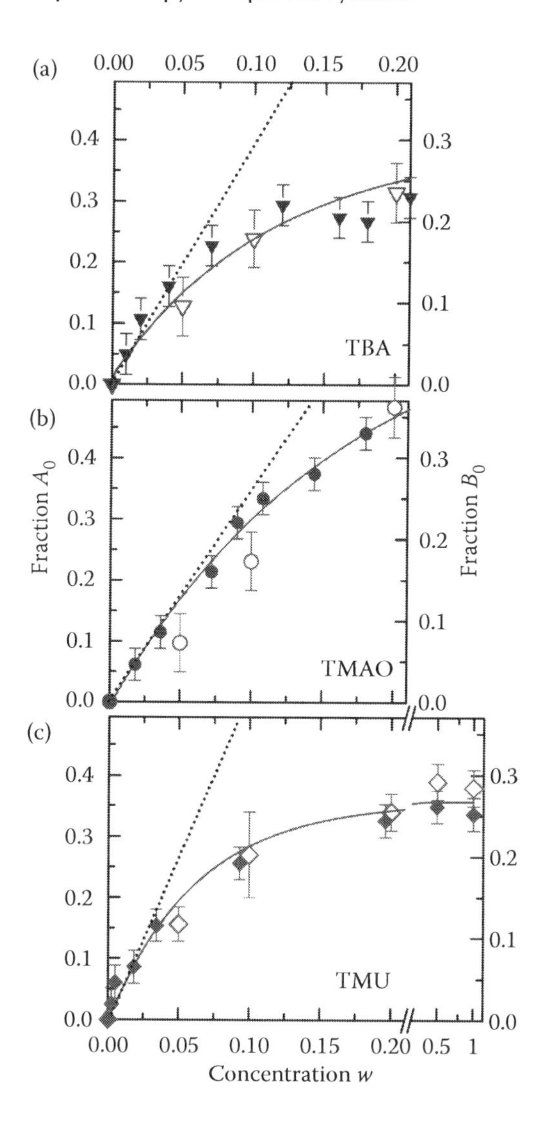

FIGURE 4.21 Fraction of water molecules showing slow spectral diffusion A_0 (left axis, open symbols) and fraction B_0 of slowly reorienting water molecules (right axis, filled symbols), as a function of the solute–water ratio w for solutions of tertiary butyl alcohol (TBA) (a), trimethylamine-*N*-oxide (TMAO) (b), and tetramethylurea (TMU) (c). The dotted lines represent linear fits to the low-concentration ($w < 0.05$) anisotropy data. (Reprinted with permission from Bakulin A et al. Hydrophobic molecules slow down the hydrogen-bond dynamics of water. *J Phys Chem A* 115: 1821–1829. Copyright 2011, American Chemical Society.)

water dynamics more or less scales with the size of the hydrophobic part of the molecule, indicating that the interactions with the hydrophilic group are relatively unimportant for the dynamics of the water molecules.

Figure 4.21 shows that at high concentrations the slow water fraction shows a saturation effect due to the sharing of the hydration shells of neighboring solute molecules. For TBA and TMU, the saturation effect is very strong, which can be explained from aggregation of the solute molecules. Beyond a certain TBA/TMU concentration, all additionally added TBA/TMU molecules are completely embedded within the aggregated TBA/TMU molecular clusters that are already present in the solution. As a result, the fraction of water showing slowed-down behavior no longer changes in proportion to the TBA/TMU concentration. The observation of clustering of TBA and TMU at high concentrations agrees with previous findings from NMR studies, SPC/E molecular dynamics simulations, and neutron scattering studies [142,160,165]. For TMAO, the saturation effects are much less strong, which indicates that TMAO does not aggregate, even at high concentrations [166].

The orientational relaxation rate of the water molecules hydrating the hydrophobic molecular groups strongly changes when the temperature is increased [157]. The reorientation time constant decreases from >10 ps at 2°C to ~2 ps at 65°C. The reorientation of the solvating water molecules is thus accelerated by a factor of ~5 over a temperature interval of only 40°C. The activation energy of the reorientation in the hydrophobic hydration shell is ~30 ± 3 kJ/mol, which is almost two times larger than the activation energy of molecular reorientation found for bulk water [21,157]. The observation of a ~2 times higher activation energy for water in the hydrophobic hydration shell agrees with the results of several NMR studies [150,161,167]. In an NMR study of the dynamics of water in supercooled aqueous solutions, the activation energy of water in the hydration shell of TMA and TMU was found to be ~20 kJ/mol at 300 K, only slightly higher than that of bulk liquid water [162]. However, the activation energy reported in this study also shows an extremely strong temperature dependence, and changes from −40 to −20 kJ/mol over the studied temperature interval of 255–300 K.

4.7 WATER IN NANOCONFINEMENT

4.7.1 Embedded Single Water Molecules

For dilute solutions of water in other solvents, the dissolved water molecules can become isolated from the other water molecules. The dynamics of single water molecules have thus been studied for water in acetone [168,169], acetonitrile [170], dimethyl sulfoxide (DMSO) [171], and *N,N*-dimethylacetamide (DMA) [172]. The isolated water molecules are observed to form one or two hydrogen bonds with their O–H groups to the solvent molecules. In the case of acetonitrile and DMA, the water molecule forms two hydrogen bonds, and the vibrational absorption spectrum of an isolated H_2O molecule shows distinct peaks corresponding to the symmetric and asymmetric O–H stretch vibrations [170,172]. In case only one of the O–H groups is hydrogen-bonded, again two peaks are observed, but now corresponding to the hydrogen-bonded O–H group and the nonbonded O–H group [168,169]. For single water molecules in acetone and in DMSO, both doubly and singly hydrogen-bonded configurations are observed [168,169,171].

The vibrational lifetime of the O–H stretch vibration of isolated water molecules is observed to be much longer than for bulk liquid water. For a single H_2O molecule, $T_1 = 8$ ps in acetonitrile [170], 6.3 ps in acetone [168,169], and 0.8 ps in DMA [172]. The red shift of the absorption spectrum increases in the same series, which thus implies that the vibrational lifetime strongly decreases with increasing hydrogen-bond strength, in line with what is observed for most hydrogen-bonded systems [58]. For all isolated H_2O molecules, it is observed that the excitation of the O–H stretch vibration rapidly equilibrates over the two O–H groups. In acetonitrile, this process takes only 0.2 ps, in DMA 0.8 ps, and in acetone 1.3 ps. In the latter solvent, the energy transfer between the two O–H groups is governed by the rate at which the H_2O molecule forms and breaks hydrogen bonds to the embedding acetone molecules, a process that is illustrated in Figure 4.22. In the dominant singly hydrogen-bonded structure, the O–H stretch frequency of the bonded group is far out of resonance with the frequency of the other, nonbonded O–H group. The transient formation of a hydrogen bond to the nonbonded O–H group brings the stretch frequency of this O–H group into resonance with that of the other O–H group, thereby enabling resonant energy transfer. The doubly hydrogen-bonded configuration thus forms the transition state for the energy transfer between the two O–H groups of the singly hydrogen-bonded configurations.

For water embedded by two DMA molecules, the orientational dynamics are observed to be strongly anisotropic. The orientational dynamics of this complex can be probed in detail by comparing the anisotropy dynamics of the O–H stretch vibration of HDO with the anisotropy dynamics of

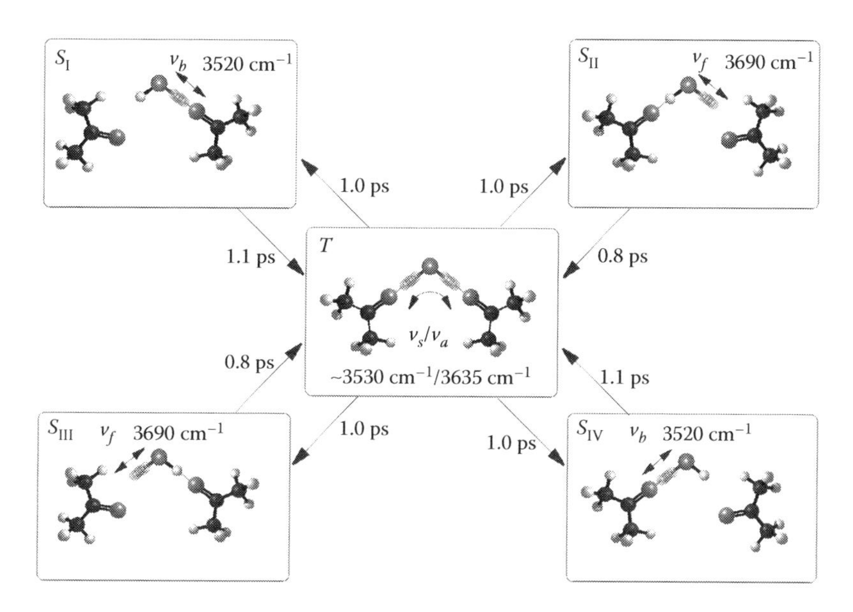

FIGURE 4.22 Schematic picture of the mechanism of energy transfer between the two O–H groups of an H_2O molecule that is hydrogen-bonded to acetone. The four structures S_{I-IV} differ in which O–H group is hydrogen-bonded and in which local O–H vibration (ν_b or ν_f) is excited. An arrow next to an O–H group denotes that the stretch vibration of that group is excited. (From Gilijamse JJ, Lock AJ, Bakker HJ. 2005. Dynamics of confined water molecules. *Proc Natl Acad Sci USA* 102: 3202–3207. Reprinted with permission from National Academy of Sciences of the United States of America.)

the symmetric and antisymmetric O–H stretch vibrations of H_2O [172]. These three modes have their transition dipole moment pointing in different directions, thus providing a complete picture of the reorientation dynamics. It is found that that the water molecule shows a hinging motion in between the two DMA molecules to which it is hydrogen-bonded, with a time constant of 0.5 ± 0.2 ps.

4.7.2 Reverse Micelles

Water is often found in nano-confined geometries, as is for instance the case in a biological cell. The presence of a nano-confining interface can have a profound effect on the structure and dynamics of water. A suitable model system for studying the dynamics of confined water is formed by reverse micelles [173,174]. Reverse micelles are nanometer-sized water droplets that form in a three-component mixture of water, apolar solvent, and certain surfactants. The surfactants are molecules with a polar or charged part that favorably interacts with the water nanodroplet, and an apolar part that interacts with the apolar solvent surrounding the droplet. The size of the droplet can often be varied by changing the ratio of water and surfactant, conventionally denoted by the parameter $w_0 = [H_2O]/[surfactant]$.

Most studies on reverse micelles have been performed using the anionic sodium bis(2-ethylhexyl) sulfosuccinate (AOT) as the surfactant. Water-AOT solutions in apolar solvents lead to the formation of reverse micelles of which the size can be varied over a wide range and that are quite monodisperse (>85%). The absorption spectrum of the O–H stretch vibrations of $HDO:D_2O$ contained in AOT reverse micelles can be well described as a sum of an interfacial and a core absorption spectrum, where the interfacial spectrum is blue-shifted by ~90 cm^{-1} in comparison with the absorption spectrum of the O–H vibrations in bulk $HDO:D_2O$ [175,176]. For the O–D absorption spectrum of $HDO:H_2O$ micelles, the same behavior is observed: the absorption spectrum of the O–D vibration of HDO molecules at the interface is ~60 cm^{-1} blue-shifted with respect to the absorption spectrum of the O–D vibration of HDO in bulk $HDO:H_2O$ [79,177].

The water molecules at the interface with the AOT surfactant show a significantly slower vibrational relaxation than the water molecules in the core of the reverse micelle. The relaxation time constant T_1 of the O–H stretch vibration is ~2.8 ps for HDO bound to the interface, and ~1 ps for the O–H stretch vibration of HDO in the core of the reverse micelle. With increasing reverse micelle size, the latter time constant approaches the relaxation time constant of 740 fs of the O–H vibration of bulk HDO:D_2O [175]. For the O–D vibration of HDO:H_2O, similar behavior is observed: T_1 of the O–D stretch vibration is ~4.3 ps for HDO at the interface and ~1.8 ps for HDO in the core of the micelle, the latter value being similar to the value of T_1 of the O–D stretch vibration of HDO in bulk HDO:H_2O solution [79,177,178]. The longer relaxation time of the interfacial water can be explained from the fact that the hydrogen bond between the O–H/O–D group of HDO and the sulfonate group (SO_3^-) of the AOT surfactant molecule is weaker than the hydrogen bond between two water molecules.

For pure H_2O micelles, the relaxation is observed to be significantly faster and to slow down significantly with decreasing reverse micelle size. In one study, it was found that the reverse micelles show quite uniform energy relaxation dynamics, which can be explained from the rapid resonant Förster vibrational energy transfer among the O–H stretch vibrations [179]. The value of T_1 is observed to decrease from ~1 ps for $w_0 = 2$ to 400 fs for $w_0 = 10$. In another study, the observed dynamics are fit to a biexponential decay where the short time constant with a value of 270 fs is assigned to the vibrational relaxation of the O–H stretch vibrations of H_2O molecule in the core of the reverse micelle, and the longer time constant with a value of 850 fs is assigned to the relaxation of the O–H stretch vibration of H_2O at the interface [176]. This latter explanation is supported by the observation that the anisotropy decays show a very slow component, that is likely associated with the water molecules at the AOT interface. This observation indicates that at least part of the water molecules at the interface does not exchange vibrational energy with the water molecules in the core of the micelle on a picosecond timescale.

The molecular reorientation dynamics of water in reverse micelles is observed to be strongly nonexponential [79,175–177]. This nonexponential character has been explained from the presence of interfacial and core water in the AOT micelles [175,176]. The interfacial water is observed to show a much longer reorientation time constant $\tau_{or,i} > 15$ ps than the water in the core of the micelle ($\tau_{or,c} \sim 3$ ps). This explanation is supported by the fact that the fractions of slow and fast water show the same dependence on the micelle size as the fractions of interfacial and core water that are derived from the spectrally resolved vibrational relaxation data [175]. The strong differences in reorientation and vibrational relaxation time constants of the interfacial and the core water can lead to quite anomalous dynamics for the measured anisotropy, as illustrated in Figure 4.23. In this figure, it is seen that with increasing delay, the anisotropy rises. This rise can be explained from the fact that with increasing delay the signal becomes increasingly dominated by the interfacial water, which has a much longer lifetime and also shows the slower reorientation dynamics.

In a recent study [177,178], it is found that for large micelles ($w_0 > 10$), the observed orientational dynamics form the sum of the behavior of interfacial water and core water, in agreement with the findings of Ref. [175], but that for small micelles ($w_0 \leq 5$), the orientational dynamics would behave in a more uniform manner. For small reverse micelles, the fast reorientation is thus assigned to the reorientation of the water molecules in a restricted angular space, and the slow reorientation to the complete randomization of the orientation. This behavior can be described with a wobbling-in-a-cone model [79]. Recent molecular dynamics simulations show that the orientational dynamics gradually become faster with increasing distance from the interface [180]. Hence, it is to be expected that there exists a distribution of subensembles that each show somewhat different orientational dynamics. The descriptions of the water reorientation dynamics of small micelles as two distinct water components [175] and as uniform nonexponential wobbling-in-a-cone dynamics [79,177,178] are probably both quite good approximations of a distribution of reorientation rates. For large micelles, the description with two components works better, as for large micelles the distribution will be highly bimodal as it will consist of slowly relaxing and reorienting interfacial water on one end and quickly relaxing and reorienting core water on the other end [175,177,178].

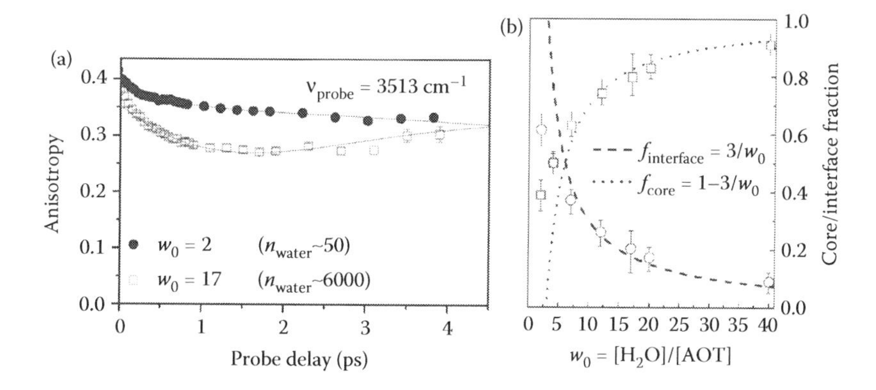

FIGURE 4.23 (a) Comparison of the orientational relaxation for two different reverse micelle sizes ($w_0 = 2,17$) at a single probe wavelength of 3510 cm^{-1}. (b) The relative core and interfacial fractions, which are obtained by spectrally integrating the positive (bleaching) part of the spectrum for each of the two components. (Dokter AM, Woutersen S, Bakker HJ. 2006. Inhomogeneous dynamics in confined water nanodroplets. *Proc Natl Acad Sci USA* 103: 15355–15358. Reprinted with permission from National Academy of Sciences of the United States of America.)

In the absence of an apolar solvent, mixtures of AOT and water will form lamellar structures [181]. The water molecules in these lamellar structures show very similar dynamics as observed for the AOT reverse micelles. The only difference is that the fraction of slow water appears to be somewhat larger for the lamellar structures than for reverse micelles with the same water to surfactant ratio w [181]. This finding indicates that water molecules penetrate somewhat more strongly into the AOT layer for lamellar AOT–water structures than for reverse micelles [181].

Reverse micelles can also be produced with cationic surfactants like cetyl trimethylammonium bromide (CTAB) [182]. The observed water dynamics for these systems is strongly dominated by the water molecules that form hydrogen bonds to the negative counterion of the surfactant, like for instance Br^-. An interesting aspect is that the dynamics of water interacting with these ions can be studied at extremely high concentrations, with the advantage that the contribution to the signal of the bulk-like water (water with hydrogen bonds to an oxygen of another water molecule) becomes negligible. For small CTAB micelles for which the concentration of Br^- is extremely high, it is thus observed that the O–H group that is hydrogen-bonded to the Br^- anion not only shows a very slow spectral diffusion and reorientation components, as was observed in bulk solutions of bromide salts [89–91], but also a fast component in its reorientation dynamics [182]. This fast component is due to the wobbling motion of the O–D group that keeps the O–H···Br^- hydrogen bond with the Br^- anion intact. Similar wobbling components are observed for the hydration shells of Cl^- and I^-, as was discussed in Section 4.4.3 and illustrated in Figure 4.12.

4.7.3 Nanochannels

The properties of water in nanochannels are highly relevant for the transfer of ions (protons) through biological membranes and the membranes of fuel cells. Hydrogen-bond rearrangement is an important parameter for proton transfer in aqueous media, and thus it can be expected that the performance of polymer electrolyte membranes like Nafion will be strongly dependent on the dynamical properties of the water inside the membranes.

The properties of water in Nafion membranes have been studied with polarization-resolved transient absorption spectroscopy [183]. In this study, the Nafion nanochannels were hydrated with a dilute solution of HDO in H_2O. The vibrational relaxation of the O–D vibration of the HDO molecules was observed to be strongly frequency dependent and indicated the presence of two distinct ensembles of water molecules in the Nafion nanochannels. The orientational mobility of the water molecules was observed to depend strongly on the hydration level of the Nafion membranes. At low hydration levels, the water molecules showed a very low orientational mobility, indicating that the water molecules are probably hydrogen-bonded to one or more sulfonate oxygens with no other mobile water molecules nearby to provide new hydrogen-bond partners. With increasing hydration, the reorientation becomes much faster, indicating an increasing ability for hydrogen-bond network rearrangements in the nanochannels to occur [183].

The proton transfer in a sodium-substituted Nafion membrane has been studied by probing the proton release from the photo-acid HPTS embedded in the membrane [184]. For low hydration levels, the proton diffusion constant is observed to be dramatically reduced in comparison to bulk liquid water. This indicates that the low water content and low water mobility prevent the structural hydrogen-bond rearrangements that are required to conduct the proton charge away from HPTS. The results were compared with the proton transfer dynamics of HPTS to water in AOT reverse micelles at similar hydration levels. From this comparison, it was concluded that the water pools in sodium-substituted Nafion membranes are quite similar to those of AOT reverse micelles [184].

4.7.4 Membranes and DNA

The interactions with water are key to the self-organized formation of bi(phospho)lipid membranes. In water, the (phospho)lipids organize in such a way that they form bilipid layers with the hydrophobic tails inside and the hydrophilic head groups outside interacting with water. The dynamics of water hydrating several model membrane like dimyristoyl-phosphatidylcholine (DMPC) [185,186], 1-palmitoyl-2-linoleyl phosphatidylcholine (PLPC) [187], and dilauroyl phosphatidylcholine (DLPC) [188] have been studied with transient absorption spectroscopy [185,188] and 2D IR spectroscopy [187,186]. For all studied model membranes, the linear infrared absorption spectrum of the water molecules interacting with the phospholipid headgroups is red-shifted with respect to the bulk water spectrum. This observation indicates that water molecules form quite strong hydrogen bonds with the head groups of the phospholipid molecules, in particular with phosphate.

In all studied hydrated membrane systems, different types of water molecules were distinguished. The 2D vibrational spectroscopic studies of Refs. [187,186] showed the presence of doubly hydrogen-bonded, singly hydrogen-bonded, and free water molecules. For the singly hydrogen-bonded species, one of the O–H groups is (strongly) hydrogen-bonded to the phospholipid while the other O–H group is dangling. For a PLPC:water ratio of 2:1, the doubly hydrogen-bonded, singly hydrogen-bonded, and free water molecules show contributions of 10%, 40%, and 50%, respectively [187]. In the study of Ref. [188], the nonexponential decay of the transient absorption signals were explained from the presence of two different water molecules: water molecules that are bound to the phosphate group and water molecules that are associated with the choline group of the lipid.

For all three studied lipid systems, the vibrational relaxation of the interacting water is observed to be quite strongly frequency dependent [185–188]. For DMPC and PLPC, the dynamics of the O–H/O–D stretch vibration of pure H_2O/D_2O were investigated, while for DLPC, the dynamics of the O–D stretch vibration of isotopically diluted HDO in H_2O were studied. This difference leads to an interesting difference in the frequency dependence of the vibrational relaxation rate. For DLPC, the relaxation slows down with increasing frequency of the O–D stretch vibration of HDO, as is usually observed for hydrogen-bonded O–H groups [188]. However, for pure H_2O/D_2O interacting with DMPC and PLPC, the relaxation becomes faster with increasing frequency of the O–H/O–D stretch vibration of H_2O/D_2O [185–187]. This anomalous frequency dependence of the relaxation of water bonded to DMPC and PLPC is explained from the presence of a down-hill intramolecular energy transfer process with a time constant of ~600 fs within singly hydrogen-bonded H_2O/D_2O molecules. In this intramolecular transfer process, the energy of the high-frequency dangling O–H/O–D group is transferred to the low-frequency hydrogen-bonded O–H/O–D group, thus accelerating the observed decay in the blue wing of the absorption band.

The vibrational lifetime of lipid-bound water has also been measured with time-resolved SFG spectroscopy [67,68]. These pump-probe experiments can be performed on lipid monolayers at the water–air interface [189–191], which has the advantage that the lipids are in their equilibrium hydration state. In the experiments, excitation by a first strong mid-infrared pulse leads to excitation of a significant fraction of O–H groups to the first excited state, and a concomitant reduction in SFG intensity. The SFG signal recovery reflects the vibrational relaxation of the excited O–H vibrations at the surface. While in initial experiments, the frequency dependence of the vibrational lifetime was attributed to a highly inhomogeneous distribution of water at lipid monolayer interfaces [190], later experiments [191] revealed that a two-component description sufficed to describe the observed dynamics. Specifically, a distinction could be made between bulk-like water at the water–lipid interface, and lipid-bound water. The latter water molecules showed very fast vibrational relaxation, and no significant energy exchange with the bulk water phase.

The interaction with water plays an important role in the conformational dynamics of DNA. The properties of water hydrating DNA have been studied with transient absorption spectroscopy [192–194] and 2D IR spectroscopy [194,195]. At low hydration levels, H_2O molecules interacting with DNA are likely hydrogen-bonded to the ionic phosphate groups of the DNA backbone, giving rise to an O–H stretch absorption band near 3500 cm^{-1}. These water molecules show a vibrational relaxation time constant T_1 of ~500 fs [192,194], which is substantially longer than the value of T_1 of ~200 fs observed for the O–H stretch vibrational absorption of pure H_2O [49,51,56]. The water molecules that are hydrogen-bonded to the ionic phosphate groups also show a high, nondecaying anisotropy after vibrational excitation, which indicates that the molecules neither rotate nor show resonant energy transfer on the picosecond timescale [192,194]. This latter finding implies that the two O–H oscillators of the H_2O molecule are decoupled, probably because only one of the oscillators is engaged in a hydrogen bond to the DNA molecule.

At higher hydration levels, a much broader water band is observed. In this case, excitation near 3500 cm^{-1} leads to a similar population relaxation time of 500 fs as is observed at low hydration levels. However, at lower frequencies, a significantly faster population relaxation time of ~250 fs is observed, similar to the relaxation rate of the O–H stretch vibrations in bulk liquid H_2O [192,194]. The rapid vibrational relaxation of these water molecules suggest that these molecules are primarily surrounded by other water molecules. The O–H stretch vibrations of these water molecules show spectral diffusion on a 500 fs timescale [195], which is much slower than is observed for pure liquid water [56]. This reduction in the rate of spectral diffusion shows that the confinement of water in the grooves of DNA leads to a strong reduction of the resonant vibrational energy transfer and a suppression of the low-frequency water motions [195].

The excitation of the O–H stretch vibration of water molecules around fully hydrated DNA induces a response of the DNA phosphate groups that increases with the relaxation of the O–H stretching vibrations [193]. This response includes a rearrangement of the hydration shell and a reduction of the average number of phosphate–water hydrogen bonds. In turn, excitation of the phosphate groups of DNA leads to energy transfer to the hydration water on a timescale similar to or even faster than the vibrational lifetime of 340 fs of the excited (anti-symmetric) phosphate vibration. The water shells around the phosphates thus serve as a primary heat sink accepting vibrational excess energy from DNA on a femtosecond timescale [193].

4.8 CONCLUSIONS AND OUTLOOK

The development of techniques to generate intense femtosecond mid-infrared laser pulses has led to a new approach to the study of the structure and dynamics of water and aqueous solutions. With these pulses, it became possible to measure the molecular motions and energy dynamics of water molecules with nonlinear time-resolved vibrational spectroscopic techniques like transient absorption spectroscopy, vibrational photon-echo spectroscopy, and two-dimensional infrared vibrational spectroscopy.

For different isotopic variations of pure liquid water, it was found that the excited molecular stretching and bending vibrations of the water molecule relax on a timescale ranging from 0.2 to 2 ps. The hydrogen-bond dynamics and the molecular reorientation were observed to show a few different timescales associated with distinctly different molecular motions. On a timescale <100 fs, the hydrogen bond between water molecules remains intact, but does show rapid variations in angle (librations) and in length (translations). Both types of fluctuations lead to a rapid spectral diffusion of the O–H stretch vibrations of the water molecule. On a timescale of 1–3 ps, the water molecules break their hydrogen bonds and rotate. These motions are associated with a collective reorganization of the hydrogen-bond network. The time constants of the energy dynamics and the molecular motions of the different isotopic variations of neat liquid water are summarized in Table 4.1.

Femtosecond mid-infrared spectroscopic techniques also enable the probing of the dynamics of subensembles of the water molecules, which is especially advantageous for the study of aqueous solutions that are intrinsically heterogeneous. Moreover, these techniques allow the probing of the dynamics on timescales that are short compared to the exchange time of water between the bulk liquid and the hydration shells.

It was found that the hydrogen-bond dynamics and molecular rotation of water molecules in the hydration shells of ions possess a component that is much slower than is observed in bulk liquid water. A similar slowing down is observed for water near hydrophobic molecular groups and for water close to the surface of reverse micelles. This slowing down is observed even in case the hydrogen bond to the solute is significantly weaker than the average hydrogen bond between two water molecules. Examples are the hydration shells of BF_4^- and ClO_4^- ions and the shell water of AOT reverse micelles.

The slowing down can best be explained from steric effects that lead to a kinetic inhibition of the collective reorganization processes that are active in bulk liquid water. For instance, molecular rotation in bulk liquid water proceeds via the transient formation of a bifurcated donated hydrogen bond. This bifurcated hydrogen bond presents an optimal low-energy transition state configuration for the rotation, but its formation requires a significant repositioning of several nearby water molecules. In hydration shells and near surfaces, much less water molecules are available and the

TABLE 4.1
Time Constants of the Dynamical Processes of Water Obtained by Probing the Stretch Vibrations of the Different Isotopes of Water with Femtosecond Vibrational Spectroscopy

	OH of H_2O	OD of D_2O	OH of HDO:D_2O	OD of HDO:H_2O
Vibrational lifetime T_1	0.23 ± 0.03 ps [51,56]	0.40 ± 0.03 ps [55]	0.74 ± 0.03 ps [50,43]	1.8 ± 0.2 ps [41,44]
Resonant energy transfer	0.08 ± 0.03 ps [49,56]	0.20 ± 0.05 ps [55]	∞	∞
Dephasing time $T_{2,hom}$	–	–	0.14 ± 0.03 ps [24,26]	0.18 ps [32]
H-bond dynamics	0.05 ± 0.02 ps [56,57]/–	–/–	0.1 ± 0.05[29,30]/1.0 ± 0.3 ps [24,30]	0.048/0.4/1.8 ps [32]
Reorientation	–	–	< 0.1 ps [43]/3.0 ± 0.2 ps [38,43]	< 0.1 ps [46]/ 2.6 ± 0.2 ps [44,46]

Note: The reported time constants of the resonant energy transfer process are approximations of the early time decay rates, as this process is strongly nonexponential. For pure H_2O and D_2O, the molecular reorientation cannot be measured because the (anisotropy) dynamics are overwhelmed by the resonant energy transfer process.

formation of such an energetically favorable transition state is strongly hampered, thus explaining the slowing down of the molecular reorientation. An important conclusion is thus that the dynamics of a water molecule are much more determined by the structure of its local environment than by the strength of its hydrogen bonds.

Femtosecond mid-infrared spectroscopy techniques have also been used to study the role of water molecules in the transfer of protons and hydroxide ions through aqueous media. Clear evidence was found that water molecules actively participate in this transfer by conducting the charge of the proton/hydroxide ions over short-living chains of hydrogen-bonded water molecules in a Grotthuss-type conduction mechanism. It was also found that the transfer of a proton from an acid to a base does not require the mutual diffusion of the reactants to come into close contact. Instead, the proton can be transferred via water molecules connecting the acid and the base. The water molecules thus form a short-living conduction wire for the proton.

In the near future, femtosecond mid-infrared spectroscopies will be further developed and applied to study the role of water in more complex systems. For instance, it is to be expected that femtosecond mid-infrared spectroscopy will be used to study the structure and dynamics of water at the surface of and inside biological membranes and proteins. Future femtosecond mid-infrared spectroscopic studies will thus hopefully provide new insights in the way water molecules participate in the self-organization and chemical reactivity of complex biological systems.

REFERENCES

1. Chaplin M. 2012. Water structure and science: Anomalous properties of water. http://www.lsbu.ac.uk/water/anomalies.html.
2. Ball P. 1999. *Life's Matrix: A Biography of Water* (Farrar, Straus, and Giroux: New York).
3. Ball P. 2008. Water: Water—an enduring mystery. *Nature* 452: 291–292.
4. Rey R, Moller KB, Hynes JT. 2002. Hydrogen bond dynamics in water and ultrafast infrared spectroscopy. *J Phys Chem A* 106: 11993–11996.
5. Lawrence CP, Skinner JL. 2002. Vibrational spectroscopy of HOD in liquid D_2O. II. Infrared line shapes and vibrational Stokes shift. *J Chem Phys* 117: 8847–8854.
6. Lawrence CP, Skinner JL. 2003. Vibrational spectroscopy of HOD in liquid D_2O. III. Spectral diffusion, and hydrogen-bonding and rotational dynamics. *J Chem Phys* 118: 264–272.
7. Corcelli SA, Lawrence CP, Skinner JL. 2004. Combined electronic structure/molecular dynamics approach for ultrafast infrared spectroscopy of dilute HOD in liquid H_2O and D_2O. *J Chem Phys* 120: 8107–8117.
8. Auer B, Kumar R, Schmidt JR, Skinner JL. 2007. Hydrogen bonding and Raman, IR, and 2D IR spectroscopy of dilute HOD in liquid D_2O. *Proc Natl Acad Sci USA* 104: 14215–14220.
9. Wernet P, Nordlund D, Bergmann U, Cavalleri M, Odelius M, Ogasawara H, Naslund LA et al. 2004. The structure of the first coordination shell in liquid water. *Science* 304: 995–999.
10. Nilsson A, Wernet P, Nordlund D, Bergmann U, Cavalleri M, Odelius M, Ogasawara H. et al. 2005. Comment on "Energetics of hydrogen bond network rearrangements in liquid water." *Science* 308: 793–793a.
11. Smith JD, Cappa CD, Wilson KR, Messer BM, Cohen RC, Saykally RJ. 2004. Energetics of hydrogen bond network rearrangements in liquid water. *Science* 306: 851–853.
12. Smith JD, Cappa CD, Wilson KR, Messer BM, Cohen RC, Saykally RJ. 2005. Response to comment on "Energetics of hydrogen bond network rearrangements in liquid water." *Science* 308: 793.
13. Smith DWG, Powles JG. 1966. Proton spin-lattice relaxation in liquid water and liquid ammonia. *Mol Phys* 10: 451–463.
14. Godralla BC, Zeidler MD. 1986. Molecular dynamics in the system water-dimethylsulphoxide. *Mol Phys* 59: 817–828.
15. Hardy EH, Zygar A, Zeidler MD, Holz M, Sacher FD. 2001. Isotope effect on the translational and rotational motion in liquid water and ammonia. *J Chem Phys* 114: 3174–3181.
16. Ropp J, Lawrence C, Farrar TC, Skinner JL. 2001. Rotational motion in liquid water is anisotropic: A nuclear magnetic resonance and molecular dynamics simulation study. *J Am Chem Soc* 123: 8047–8052.

17. Jonas J, DeFries T, Wilbur DJ. 1976. Molecular motions in compressed liquid water. *J Chem Phys* 65: 582–588.
18. Lang E, Lüdemann HD. 1977. Pressure and temperature dependence of the longitudinal proton relaxation times in supercooled water to −87°C and 2500 bar. *J Chem Phys* 67: 718–723.
19. Barthel J, Bachhuber K, Buchner R, Hetzenauer H. 1990. Dielectric spectra of some common solvents in the microwave region. Water and lower alcohols. *Chem Phys Lett* 165: 369–373.
20. Kindt JT, Schmuttenmaer CA. 1996. Far-infrared dielectric properties of polar liquids probed by femtosecond terahertz pulse spectroscopy. *J Phys Chem* 100: 10373–10379.
21. Roenne C, Thrane L, Astrand PO, Wallqvist A, Mikkelsen KV, Keiding SR. 1997. Investigation of the temperature dependence of dielectric relaxation in liquid water by THz reflection spectroscopy and molecular dynamics simulation. *J Chem Phys* 107: 5319–5331.
22. Roenne C, Astrand PO, Keiding SR. 1999. THz spectroscopy of liquid H_2O and D_2O. *Phys Rev Lett* 82: 2888–2891.
23. Laenen R, Rauscher C, Laubereau A. 1998. Dynamics of local substructures in water observed by ultrafast infrared hole burning. *Phys Rev Lett* 80: 2622–2625.
24. Gale GM, Gallot G, Hache F, Lascoux N, Bratos S, Leicknam JC. 1999. Femtosecond dynamics of hydrogen bonds in liquid water: A real time study. *Phys Rev Lett* 82: 1068–1071.
25. Woutersen S, Bakker HJ. 1999. Hydrogen bond in liquid water as a Brownian oscillator. *Phys Rev Lett* 83: 2077–2080.
26. Stenger J, Madsen D, Hamm P, Nibbering ETJ, Elsaesser T. 2001. Ultrafast vibrational dephasing of liquid water. *Phys Rev Lett* 87: 027401.
27. Stenger J, Madsen D, Hamm P, Nibbering ETJ, Elsaesser T. 2002. A photon echo peak shift study of liquid water. *J Phys Chem A* 106: 2341–2350.
28. Bakker HJ, Nienhuys HK, Gallot G, Lascoux N, Gale GM, Leicknam JC, Bratos S. 2002. Transient absorption of vibrationally excited water. *J Chem Phys* 116: 2592–2598.
29. Yeremenko S, Pshenichnikov MS, Wiersma DA. 2003. Hydrogen-bond dynamics in water explored by heterodyne-detected photon echo. *Chem Phys Lett* 369: 107–113.
30. Fecko CJ, Eaves JD, Loparo JJ, Tokmakoff A, Geissler PL. 2003. Ultrafast hydrogen-bond dynamics in the infrared spectroscopy of water. *Science* 301: 1698–1702.
31. Asbury JB, Steinel T, Stromberg C, Corcelli SA, Lawrence CP, Skinner JL, Fayer MD. 2004. Water dynamics: Vibrational echo correlation spectroscopy and comparison to molecular dynamics simulations. *J Phys Chem A* 108: 1107–1119.
32. Asbury JB, Steinel T, Kwak K, Corcelli SA, Lawrence CP, Skinner JL, Fayer MD. 2004. Dynamics of water probed with vibrational echo correlation spectroscopy. *J Chem Phys* 121: 12431–12446.
33. Steinel T, Asbury JB, Corcelli SA, Lawrence CP, Skinner JL, Fayer MD. 2004. Water dynamics: Dependence on local structure probed with vibrational echo correlation spectroscopy. *Chem Phys Lett* 386: 295–300.
34. Eaves JD, Loparo JJ, Fecko CJ, Roberts ST, Tokmakoff A, Geissler PL. 2005. Hydrogen bonds in liquid water are broken only fleetingly. *Proc Natl Acad Sci USA* 102: 13019–13022.
35. Loparo JJ, Roberts ST, Tokmakoff A. 2006. Multidimensonal infrared spectroscopy of water. I. Vibrational dynamics in 2D lineshapes. *J Chem Phys* 125:194521.
36. Loparo JJ, Roberts ST, Tokmakoff A. 2006. Multidimensional infrared spectroscopy of water. II. Hydrogen bond switching dynamics. *J Chem Phys* 125: 194522.
37. Woutersen S, Emmerichs U, Bakker HJ. 1997. Femtosecond mid-IR pump-probe spectroscopy of liquid water: Evidence for a two-component structure. *Science* 278: 658–660.
38. Nienhuys HK, van Santen RA, Bakker HJ. 2000. Orientational relaxation of liquid water molecules as an activated process. *J Chem Phys* 112: 8487–8494.
39. Bakker HJ, Woutersen S, Nienhuys HK. 2000. Reorientational motion and hydrogen-bond stretching dynamics in liquid water. *Chem Phys* 258: 233–245.
40. Gallot G, Bratos S, Pommeret S, Lascoux N, Leicknam JC, Kozinski M, Amir W, Gale GM. 2002. Coupling between molecular rotations and OH···O motions in liquid water: Theory and experiment. *J Chem Phys* 117: 11301–11309.
41. Steinel T, Asbury JB, Zheng J, Fayer MD. 2004. Watching hydrogen bonds break: A transient absorption study of water. *J Phys Chem A* 108: 10957–10964.
42. Loparo JJ, Fecko CJ, Eaves JD, Roberts ST, Tokmakoff A. 2004. Reorientational and configurational fluctuations in water observed on molecular length scales. *Phys Rev B* 70: 180201.
43. Fecko CJ, Loparo JJ, Roberts ST, Tokmakoff A. 2005. Local hydrogen bonding dynamics and collective reorganization in water: Ultrafast IR spectroscopy of HOD/D_2O. *J Chem Phys* 122: 054506.

44. Rezus YLA, Bakker HJ. 2005. On the orientational relaxation of HDO in liquid water. *J Chem Phys* 123: 114502.
45. Rezus YLA, Bakker HJ. 2006. Orientational dynamics of isotopically diluted H_2O and D_2O. *J Chem Phys* 125: 144512.
46. Moilanen DE, Fenn EE, Lin YS, Skinner JL, Bagchi B, Fayer MD. 2008. Water inertial reorientation: Hydrogen bond strength and the angular potential. *Proc Natl Acad Sci USA* 105: 5295–5300.
47. Bodis P, Larsen OFA, Woutersen S. 2005. Vibrational relaxation of the bending mode of HDO in liquid D_2O. *J Phys Chem A* 109: 5303–5306.
48. Ashihara S, Huse N, Espagne A, Nibbering ETJ, Elsaesser T. 2007. Ultrafast structural dynamics of water induced by dissipation of vibrational energy. *J Phys Chem A* 111: 743–746.
49. Woutersen S, Bakker HJ. 1999. Resonant intermolecular transfer of vibrational energy in liquid water. *Nature* 402: 507–509.
50. Woutersen S, Emmerichs U, Nienhuys HK, Bakker HJ. 1998. Anomalous temperature dependence of vibrational lifetimes in water and ice. *Phys Rev Lett* 81: 1106–1109.
51. Lock AJ, Bakker HJ. 2002. Temperature dependence of vibrational relaxation in liquid H_2O. *J Chem Phys* 117: 1708–1713.
52. Huse N, Ashihara S, Nibbering ETJ, Elsaesser T. 2005. Ultrafast vibrational relaxation of O-H bending and librational excitations in liquid H_2O. *Chem Phys Lett* 404: 389–393.
53. Lindner J, Vöhringer P, Pshenichnikov MS, Cringus D, Wiersma DA, Mostovoy M. 2006. Vibrational relaxation of pure liquid water. *Chem Phys Lett* 421: 329–333.
54. Ashihara S, Huse N, Espagne A, Nibbering ETJ, Elsaesser T. 2006. Vibrational couplings and ultrafast relaxation of the O-H bending mode in liquid H_2O. *Chem Phys Lett* 424: 66–70.
55. Piatkowski L, Eisenthal KB, Bakker HJ. 2009. Ultrafast intermolecular energy transfer in heavy water. *Phys Chem Chem Phys* 11: 9033–9038.
56. Cowan ML, Bruner BD, Huse N, Dwyer JR, Chugh B, Nibbering ETJ, Elsaesser T, Miller RJD. 2005. Ultrafast memory loss and energy redistribution in the hydrogen bond network of liquid H_2O. *Nature* 434: 199–202.
57. Paarmann A, Hayashi T, Mukamel S, Miller RJD. 2008. Probing intermolecular couplings in liquid water with two-dimensional infrared photon echo spectroscopy. *J Chem Phys* 128: 191103.
58. Miller RE. 1988. The vibrational spectroscopy and dynamics of weakly bound neutral complexes. *Science* 240: 447–453.
59. Novak A. 1974. Hydrogen bonding in solids. Correlation of spectroscopic and crystallographic data. *Struct Bonding (Berlin)* 18: 177–216.
60. Mikenda W. 1986. Stretching frequency versus bond distance correlation of O–D(H)···Y hydrogen bonds in solid hydrates. *J Mol Struct* 147: 1–15.
61. Mikenda W, Steinböck S. 1996. Stretching frequency vs bond distance correlation of hydrogen bonds in solid hydrates: A generalized correlation function *J Mol Struct* 384: 159–163.
62. Graener H, Seifert G, Laubereau A. 1991. New spectroscopy of water using tunable picosecond pulses in the infrared. *Phys Rev Lett* 66: 2092–2095.
63. Cho M, Yu JY, Joo T, Nagasawa Y, Passino SA, Fleming GR. 1996. The integrated photon echo and solvation dynamics. *J Phys Chem* 100: 11944–11953.
64. Piryatinski A, Skinner JL. 2002. Determining vibrational solvation-correlation functions from three-pulse infrared photon echoes. *J Phys Chem B* 106: 8055–8063.
65. Yeremenko S, Pshenichnikov NS, Wiersma DA. 2006. Interference effects in IR photon echo spectroscopy of liquid water. *Phys Rev A* 73: 021804.
66. Lipari G, Szabo A. 1980. Effect of librational motion on fluorescence depolarization and nuclear magnetic resonance relaxation in macromolecules and membranes. *Biophys J* 30: 489–506.
67. McGuire JA, Shen YR. 2006. Ultrafast vibrational dynamics at water interfaces. *Science* 313: 1945–1948.
68. Smits M, Ghosh A, Sterrer M, Müller M, Bonn M. 2007. Ultrafast vibrational energy transfer between surface and bulk water at the air-water interface. *Phys Rev Lett* 98: 098302.
69. Zhang Z, Piatkowski L, Bakker HJ, Bonn M. 2011. Interfacial water structure revealed by ultrafast two-dimensional surface vibrational spectroscopy. *J Chem Phys* 135: 021101.
70. Zhang Z, Piatkowski L, Bakker HJ, Bonn M. 2011. Ultrafast vibrational energy transfer at the water/air interface revealed by two-dimensional surface vibrational spectroscopy. *Nat Chem* 3: 888–893.
71. Bredenbeck J, Ghosh A, Smits M, Bonn M. 2008. Ultrafast two dimensional-infrared spectroscopy of a molecular monolayer. *J Am Chem Soc* 130: 2152–2153.

72. Bredenbeck J, Ghosh A, Nienhuys HK, Bonn M. 2009. Interface-specific ultrafast two-dimensional vibrational spectroscopy. *Acc Chem Res* 42: 1332–1342.
73. Xiong W, Laaser JE, Mehlenbacher RD, Zanni MT. 2011. Adding a dimension to the infrared spectra of interfaces using heterodyne detected 2D sum-frequency generation (HD 2D SFG) spectroscopy. *Proc Natl Acad Sci USA* 108: 20902–20907.
74. Stiopkin IV, Weeraman C, Pieniazek PA, Shalhout FY, Skinner JL, Benderskii AV. 2011. Hydrogen bonding at the water surface revealed by isotopic dilution spectroscopy. *Nature* 474: 192–195.
75. Bakker HJ, Rezus YLA, Timmer RLA. 2008. Molecular reorientation of liquid water studied with femtosecond mid-infrared spectroscopy. *J Phys Chem A* 112: 11523–11534.
76. Ramasesha K, Roberts ST, Nicodemus RA, Mandal A, Tokmakoff A. 2011. Ultrafast 2D IR anisotropy of water reveals reorientation during hydrogen-bond switching. *J Chem Phys* 135: 054509.
77. Lin YS, Pieniazek PA, Yang M, Skinner JL. 2010. On the calculation of rotational anisotropy decay, as measured by ultrafast polarization-resolved vibrational pump-probe experiments. *J Chem Phys* 132: 174505.
78. Laage D, Hynes JT. 2006. A molecular jump mechanism of water reorientation. *Science* 311: 832–835.
79. Piletic IR, Moilanen DE, Spry DB, Levinger NE, Fayer MD. 2006. Testing the core/shell model of nanoconfined water in reverse micelles using linear and nonlinear IR spectroscopy. *J Phys Chem A* 110: 4985–4999.
80. Laage D, Hynes JT. 2008. On the molecular mechanism of water reorientation. *J Phys Chem B* 112: 14230–14242.
81. Wallqvist A, Berne BJ. 1993. Effective potentials for liquid water using polarizable and nonpolarizable models. *J Phys Chem* 97: 13841–13851.
82. Hsieh CS, Campen RK, Verde ACV, Bolhuis P, Nienhuys HK, Bonn M. 2011. Ultrafast reorientation of dangling OH groups at the air-water interface using femtosecond vibrational spectroscopy. *Phys Rev Lett* 107: 116102.
83. Walrafen GE. 1962. Raman spectral studies of the effects of electrolytes on water. *J Chem Phys* 36: 1035–1042.
84. Bergström PA, Lindgren J. 1991. An IR study of the hydration of ClO_4^-, NO_3^-, I^-, Br^-, Cl^-, and SO_4^{2-} anions in aqueous solution. *J Phys Chem* 95: 8575–8580.
85. Kropman MF, Bakker HJ. 2001. Dynamics of water molecules in aqueous solvation shells. *Science* 291: 2118–2120.
86. Kropman MF, Bakker HJ. 2001. Femtosecond mid-infrared spectroscopy of aqueous solvation shells. *J Chem Phys* 115: 8942–8948.
87. Kropman MF, Nienhuys HK, Bakker HJ. 2002. Real-time measurement of the orientational dynamics of aqueous solvation shells in bulk liquid water. *Phys Rev Lett* 88: 77601.
88. Kropman MF, Bakker HJ. 2004. Effect of ions on the vibrational relaxation of liquid water. *J Am Chem Soc* 126: 9135–9141.
89. Park S, Fayer MD. 2007. Hydrogen bond dynamics in aqueous NaBr solutions. *Proc Natl Acad Sci USA* 104: 16731–16738.
90. Park S, Moilanen DE, Fayer MD. 2008. Water dynamics—the effects of ions and nanoconfinement. *J Phys Chem B* 112:5279–5290.
91. Timmer RLA, Bakker HJ. 2009. Hydrogen bond fluctuations of the hydration shell of the bromide anion. *J Phys Chem A* 113: 6104–6110.
92. Hashimoto K, Morokuma K. 1994. *Ab-initio* theoretical study of surface and interior structures of the $Na(H_2O)_4$ cluster and its cation. *Chem Phys Lett* 223: 423–430.
93. Asada T, Nishimoto K. 1995. Monte-Carlo simulations of $M^+ Cl^- (H_2O)_n$ (M = Li, Na) clusters and the dissolving mechanism of ion-pairs in water. *Chem Phys Lett* 232: 518–523.
94. Ramaniah LM, Bernasconi M, Parrinello M. 1998. Density-functional study of hydration of sodium in water clusters. *J Chem Phys* 109: 6839–6843.
95. Staib A, Hynes JT. 1993. Vibrational predissociation in hydrogen-bonded OH···O complexes via OH stretch OO stretch energy-transfer. *Chem Phys Lett* 204: 197–205.
96. Bakker HJ. 2008. Structural dynamics of aqueous salt solutions. *Chem Rev* 108: 1456–1473.
97. Laage D, Hynes JT. 2007. Reorientational dynamics of water molecules in anionic hydration shells. *Proc Natl Acad Sci USA* 104: 11167–11172.
98. Moilanen DE, Wong DB, Rosenfeld DE, Fenn EE, Fayer MD. 2009. Ion-water hydrogen bond switching observed with 2D IR vibrational echo chemical exchange spectroscopy. *Proc Natl Acad Sci USA* 106: 375–380.

99. Park S, Odelius M, Gaffney KJ. 2009. Ultrafast dynamics of hydrogen bond exchange in aqueous ionic solutions. *J Phys Chem B* 113: 7825–7835.
100. Ji M, Odelius M, Gaffney KJ. 2010. Large angular jump mechanism observed for hydrogen bond exchange in aqueous perchlorate solution. *Science* 328: 1003–1005.
101. Tielrooij KJ, Garcia-Araez N, Bonn M, Bakker HJ. 2011. Cooperativity in ion hydration. *Science* 328: 1006–1009.
102. Tielrooij KJ, van der Post ST, Hunger J, Bonn M, Bakker HJ. 2011. Anisotropic water reorientation around ions. *J Phys Chem B* 115: 12638–12647.
103. Laage D, Hynes JT. 2008. On the residence time for water in a solute hydration shell: Application to aqueous halide solutions. *J Phys Chem B* 112: 7697–7701.
104. Lee SH, Rasaiah JC. 1996. Molecular dynamics simulation of ion mobility. 2. Alkali metal and halide ions using the SPC/E model for water at 25°C. *J Phys Chem* 100: 1420–1425.
105. Heuft JM, Meijer EJ. 2005. Density functional theory based molecular-dynamics study of aqueous iodide solvation. *J Chem Phys* 123: 094506.
106. Omta AW, Kropman MF, Woutersen S, Bakker HJ. 2003. Negligible effect of ions on the hydrogen-bond structure in liquid water. *Science* 273: 347–349.
107. Omta AW, Kropman MF, Woutersen S, Bakker HJ. 2003. Influence of ions on the hydrogen-bond structure in liquid water. *J Chem Phys* 119: 12457–12461.
108. Guardia E, Laria D, Marti J. 2006. Hydrogen bond structure and dynamics in aqueous electrolytes at ambient and supercritical conditions. *J Phys Chem B* 110: 6332–6338.
109. Lin YS, Auer BM, Skinner JL. 2009. Water structure, dynamics, and vibrational spectroscopy in sodium bromide solutions. *J Chem Phys* 131: 144511.
110. Buchner R, Chen T, Hefter G. 2004. Complexity in simple electrolyte solutions: Ion pairing in $MgSO_4$ (aq). *J Phys Chem B* 108: 2365–2375.
111. de Grotthuss CJT. 1806. Sur la décomposition de l'eau et des corps qu'elle tient en dissolution à l'aide de l'électricité galvanique. *Annales de Chimie* 58: 54–73.
112. Agmon N. 1995. The Grotthuss mechanism. *Chem Phys Lett* 244: 456–462.
113. Marx D, Tuckerman ME, Hutter J, Parrinello M. 1999. The nature of the hydrated excess proton in water. *Nature* 397: 601–604.
114. Schmitt UW, Voth GA. 1999. The computer simulation of proton transport in water. *J Chem Phys* 111: 9361–9381.
115. Woutersen S, Bakker HJ. 2006. Ultrafast vibrational and structural dynamics of the proton in liquid water. *Phys Rev Lett* 96: 138305.
116. Timmer RLA, Tielrooij KJ, Bakker HJ. 2010. Vibrational Förster transfer to hydrated protons. *J Chem Phys* 132: 194504.
117. Pines E, Huppert D, Agmon N. 1998. Geminate recombination in excited-state proton-transfer reactions: Numerical solution of the Debye-Smoluchowski equation with backreaction and comparison with experimental results. *J Chem Phys* 88: 5620–5630.
118. Tran-Thi TH, Gustavsson T, Prayer C, Pommeret S, Hynes JT. 2000. Primary ultrafast events preceding the photoinduced proton transfer from pyranine to water. *Chem Phys Lett* 329: 421–430.
119. Spry DB, Goun A, Fayer MD. 2007. Deprotonation dynamics and stokes shift of pyranine (HPTS). *J Phys Chem A*, 230–237.
120. Spry DB, Fayer MD. 2008. Charge redistribution and photoacidity: Neutral versus cationic photoacids. *J Chem Phys* 128: 084508.
121. Pines E, Manes BZ, Land MJ, Fleming GR. 1997. Direct measurement of intrinsic proton transfer rates in diffusion-controlled reactions. *Chem Phys Lett* 281: 413–420.
122. Genosar L, Cohen B, Huppert D. 2000. Ultrafast direct photoacid-base reaction. *J Phys Chem A* 104: 6689–6698.
123. Rini M, Magnes BZ, Pines E, Nibbering ETJ. 2003. Real-time observation of bimodal proton transfer in acid-base pairs in water. *Science* 301: 349–352.
124. Rini M, Pines D, Magnes BZ, Pines E, Nibbering ETJ. 2004. Bimodal proton transfer in acid-base reactions in water. *J Chem Phys* 121: 9593–9610.
125. Mohammed OF, Pines D, Dreyer J, Pines E, Nibbering ETJ. 2005. Sequential proton transfer through water bridges in acid-base reactions. *Science* 310: 83–86.
126. Mohammed OF, Pines D, Nibbering ETJ, Pines E. 2007. Base-induced solvent switches in acid-base reactions *Angew Chem Intern Ed* 46: 1458–1461.
127. Mohammed OF, Pines D, Pines E, Nibbering ETJ. 2007. Aqueous bimolecular proton transfer in acid-base neutralization. *Chem Phys* 341: 240–257.

128. Siwick BJ, Bakker HJ. 2007. On the role of water in intermolecular proton-transfer reactions. *J Am Chem Soc* 129: 13412–13420.
129. Siwick BJ, Cox MJ, Bakker HJ. 2008. Long-range proton transfer in aqueous acid-base reactions. *J Phys Chem B* 112: 378–389.
130. Cox MJ, Bakker HJ. 2008. Parallel proton transfer pathways in aqueous acid-base reactions. *J Chem Phys* 128: 174501.
131. Cox MJ, Timmer RLA, Bakker HJ, Park S, Agmon N. 2009. Distance-dependent proton transfer along water wires connecting acid-base pairs. *J Phys Chem A* 113: 6599–6606.
132. Cox MJ, Siwick BJ, Bakker HJ. 2009. Influence of ions on aqueous acid-base reactions. *Chemphyschem* 10: 236–244.
133. Adamczyk K, Premont-Schwarz M, Pines D, Pines E, Nibbering ETJ. 2009. Real-time observation of carbonic acid formation in aqueous solution. *Science* 326: 1690–1694.
134. Cox MJ, Bakker HJ. 2010. Femtosecond study of the deuteron-transfer dynamics of naphtol salts in water. *J Phys Chem A* 114: 10523–10530.
135. Marx D, Chandra A, Tuckerman ME. 2010. Aqueous basic solutions: Hydroxide solvation, structural diffusion, and comparison to the hydrated proton. *Chem Rev* 110: 2174–2216.
136. Nienhuys HK, Lock AJ, van Santen RA, Bakker HJ. 2002. Dynamics of water molecules in an alkaline environment. *J Chem Phys* 117: 8021–8029.
137. Liu L, Hunger J, Bakker HJ. 2011. Energy relaxation dynamics of the hydration complex of hydroxide. *J Phys Chem A* 115: 14593–14598.
138. Hunger J, Liu L, Tielrooij KJ, Bonn M, Bakker HJ. 2011. Vibrational and orientational dynamics of water in aqueous hydroxide solutions. *J Chem Phys* 135: 124517.
139. Roberts ST, Petersen PB, Ramasesha K, Tokmakoff A, Ufimtsev IS, Martinez TJ. 2009. Observation of a Zundel-like transition state during proton transfer in aqueous hydroxide solutions. *Proc Natl Acad Sci USA* 106: 15154–15159.
140. Roberts ST, Ramasesha K, Petersen PB, Mandal A, Tokmakoff A. 2011. Proton transfer in concentrated aqueous hydroxide visualized using ultrafast infrared spectroscopy. *J Phys Chem A* 115: 3957–3972.
141. Rezus YLA, Bakker HJ. 2006. Effect of urea on the structural dynamics of water. *Proc Natl Acad Sci USA* 103: 18417–18420.
142. Shimizu A, Fumino K, Yukiyasu K, Taniguchi Y. 2000. NMR studies on dynamic behavior of water molecule in aqueous denaturant solutions at 25°C: Effects of guanidine hydrochloride, urea and alkylated ureas. *J Mol Liq* 85: 269–278.
143. Fenn EE, Moilanen DE, Levinger NE, Fayer MD. 2009. Water dynamics and interactions in water-polyether binary mixtures. *J Am Chem Soc* 131: 5530–5539.
144. Frank HS, Evans MW. 1945. Free volume and entropy in condensed systems III. Entropy in binary liquid mixtures; partial molal entropy in dilute solutions; structure and thermodynamics in aqueous electrolytes. *J Chem Phys* 13: 507–532.
145. Soper AK, Finney JL. Hydration of methanol in aqueous solution. *Phys Rev Lett* 71: 4346–4349.
146. Turner J, Soper AK. 1994. The effect of apolar solutes on water structure: Alcohols and tetraalkylammonium ions. *J Chem Phys* 101: 6116–6125.
147. Dixit S, Crain J, Poon WCK, Finney JL, Soper AK. 2002. Molecular segregation observed in a concentrated alcohol-water solution. *Nature* 416: 829–832.
148. Buchanan P, Aldiwan N, Soper AK, Creek JL, Koh CA. 2005. Decreased structure on dissolving methane in water. *Chem Phys Lett* 415: 89–93.
149. Haselmaier R, Holz M, Marbach W, Weingartner H. 1995. Water dynamics near a dissolved noble gas—First direct experimental evidence for a retardation effect. *J Phys Chem* 99: 2243–2246.
150. Ishihara Y, Okouchi S, Uedaira H. 1997. Dynamics of hydration of alcohols and diols in aqueous solutions. *J Chem Soc Faraday Trans* 93: 3337–3342.
151. Kaatze U, Gerke U, Pottel R. 1986. Dielectric relaxation in aqueous solutions of urea and some of its derivatives. *J Phys Chem* 90: 5464–5469.
152. Wachter W, Buchner R, Hefter G. 2006. Hydration of tetraphenylphosphonium and tetraphenylborate ions by dielectric relaxation spectroscopy. *J Phys Chem B* 110: 5147–5154.
153. Rezus YLA, Bakker HJ. 2007. Observation of immobilized water molecules around hydrophobic groups. *Phys Rev Lett* 99: 148301–148304.
154. Rezus YLA, Bakker HJ. 2008. Strong slowing down of water reorientation in mixtures of water and tetramethylurea. *J Phys Chem A* 112: 2355–2361.
155. Bakulin A, Liang C, Jansen TL, Wiersma DA, Bakker HJ, Pshenichnikov MS. 2009. Hydrophobic solvation: A 2D IR spectroscopic inquest. *Acc Chem Res* 42: 1229–1238.

156. Bakulin A, Pshenichnikov MS, Bakker HJ, Petersen C. 2011. Hydrophobic molecules slow down the hydrogen-bond dynamics of water. *J Phys Chem A* 115: 1821–1829.
157. Petersen C, Bakker HJ. 2009. Strong temperature dependence of water reorientation in hydrophobic hydration shells. *J Chem Phys* 130: 214511.
158. Laage D, Stirnemann G, Hynes JT. 2009. Why water reorientation slows without iceberg formation around hydrophobic solutes. *J Phys Chem B* 113: 2428–2435.
159. Stirnemann G, Hynes JT, Laage D. 2010. Water hydrogen bond dynamics in aqueous solutions of amphiphiles. *J Phys Chem B* 114: 3052–3059.
160. Stirnemann G, Sterpone F, Laage D. 2011. Dynamics of water in concentrated solutions of amphiphiles: Key roles of local structure and aggregation. *J Phys Chem B* 115: 3254–3262.
161. Fumino K, Yukiyasu K, Shimizu A, Taniguchi Y. 1998. NMR studies on dynamic behavior of water molecules in tetraalkylammonium bromide-D_2O solutions at 5–25°C. *J Mol Liq* 75: 1–12.
162. Qvist J, Halle B. 2008. Thermal signature of hydrophobic hydration dynamics. *J Am Chem Soc* 130: 10345–10353.
163. Schrödle S, Buchner R, Kunz W. 2004. Effect of the chain length on the inter- and intramolecular dynamics of liquid oligo(ethylene glycol)s. *J Phys Chem B* 108: 6281–6287.
164. Tielrooij KJ, Hunger J, Buchner R, Bonn M, Bakker HJ. 2010. Influence of concentration and temperature on the dynamics of water in the hydrophobic hydration shell of tetramethylurea. *J Am Chem Soc* 134: 15671–15678.
165. Bowron DT, Soper AK, Finney JL. 2001. Temperature dependence of the structure of a 0.06 mole fraction tertiary butanol-water solution. *J Chem Phys* 114: 6203–6219.
166. Petersen C, Bakulin AA, Pavelyev VG, Pshenichnikov MS, Bakker HJ. 2010. Femtosecond mid-infrared study of aggregation behavior in aqueous solutions of amphiphilic molecules *J Chem Phys* 133: 164514.
167. Yoshida K, Ibuki K, Ueno M. 1998. Pressure and temperature effects on ^{2}H spin-lattice relaxation times and ^{1}H chemical shifts in tert-butyl alcohol- and urea-D_2O solutions. *J Chem Phys* 108: 1360–1367.
168. Gilijamse JJ, Lock AJ, Bakker HJ. 2005. Dynamics of confined water molecules. *Proc Natl Acad Sci USA* 102: 3202–3207.
169. Bakker HJ, Gilijamse JJ, Lock AJ. 2005. Energy transfer in single hydrogen-bonded water molecules *Chem Phys Chem* 6: 1146–1156.
170. Cringus D, Jansen TIC, Pshenichnikov MS, Wiersma DA. 2007. Ultrafast anisotropy dynamics of water molecules dissolved in acetonitrile. *J Chem Phys* 127: 084507.
171. Wulf A, Ludwig R. 2006. Structure and dynamics of water confined in dimethyl sulfoxide. *Chem Phys Chem* 7: 266–272.
172. Timmer RLA, Bakker HJ. 2007. Water as a molecular hinge in amidelike structures. *J Chem Phys* 126: 154507.
173. Levinger NE. 2002. Water in confinement. *Science* 298: 1722–1723.
174. Deak JC, Pang Y, Sechler TD, Wang Z, Dlott DD. 2004. Vibrational energy transfer across a reverse micelle surfactant layer. *Science* 306: 473–476.
175. Dokter AM, Woutersen S, Bakker HJ. 2006. Inhomogeneous dynamics in confined water nanodroplets. *Proc Natl Acad Sci USA* 103: 15355–15358.
176. Cringus D, Bakulin A, Lindner J, Vöhringer P, Pshenichnikov MS, Wiersma DA. 2007. Ultrafast energy transfer in water-AOT reverse micelles. *J Phys Chem B* 111: 14193–14207.
177. Moilanen DE, Fenn EE, Wong DB, Fayer MD. 2009. Water dynamics in large and small reverse micelles: From two ensembles to collective behavior. *J Chem Phys* 131: 014704.
178. Moilanen DE, Fenn EE, Wong DB, Fayer MD. 2009. Water dynamics at the interface in AOT reverse micelles. *J Phys Chem B* 113: 8560–8568.
179. Dokter AM, Woutersen S, Bakker HJ. 2005. Anomalous slowing down of the vibrational relaxation of liquid water upon nanoscale confinement. *Phys Rev Lett* 94: 178301.
180. Pieniazek PA, Lin YS, Chowdhary J, Ladanyi BM, Skinner JL. 2009. Vibrational spectroscopy and dynamics of water confined inside reverse micelles. *J Phys Chem B* 113: 15017–15028.
181. Moilanen DE, Fenn EE, Wong D, Fayer MD. 2009. Geometry and nanolength scales versus interface interactions: Water dynamics in AOT lamellar structures and reverse micelles *J Am Chem Soc* 113: 8318–8328.
182. Dokter AM, Woutersen S, Bakker HJ. 2007. Ultrafast dynamics of water in cationic micelles. *J Chem Phys* 126: 124507.
183. Moilanen DE, Piletic IR, Fayer MD. 2007. Water dynamics in Nafion fuel cell membranes: The effects of confinement and structural changes on the hydrogen bond network. *J Phys Chem C* 111: 8884–8891.
184. Spry DB, Goun A, Glusac K, Moilanen DE, Fayer MD. 2007. Proton transport and the water environment in nafion fuel cell membranes and AOT reverse micelles. *J Am Chem Soc* 129: 8122–8130.

185. Volkov VV, Palmer DJ, Righini R. 2007. Heterogeneity of water at the phospholipid membrane interface. *J Phys Chem B* 111: 1377–1383.
186. Volkov VV, Takaoka Y, Righini R. 2009. What are the sites water occupies at the interface of a phospholipid membrane? *J Phys Chem B* 113: 4119–4124.
187. Volkov VV, Palmer DJ, Righini R. 2007. Distinct water species confined at the interface of a phospholipid membrane. *Phys Rev Lett* 99: 078302.
188. Zhao W, Moilanen DE, Fenn EE, Fayer MD. 2008. Water at the surfaces of aligned phospholipid multibilayer model membranes probed with ultrafast vibrational spectroscopy. *J Am Chem Soc* 130: 13927–13927.
189. Ghosh A, Campen RK, Sovago M, Bonn M. 2009. Structure and dynamics of interfacial water in model lung surfactants. *Faraday Disc* 141: 145–159.
190. Ghosh A, Smits M, Bredenbeck J, Bonn M. 2007. Membrane-bound water is energetically decoupled from nearby bulk water: An ultrafast surface-specific investigation. *J Am Chem Soc* 129: 9608–9609.
191. Bonn M, Bakker HJ, Ghosh A, Yamamoto S, Sovago M, Campen RK. 2011. Structural inhomogeneity of interfacial water at lipid monolayers revealed by surface-specific vibrational pump-probe spectroscopy. *J Am Chem Soc* 132: 14971–14978.
192. Szyc L, Dwyer JR, Nibbering ETJ, Elsaesser T. 2009. Ultrafast dynamics of N-H and O-H stretching excitations in hydrated DNA oligomers. *Chem Phys Lett* 357: 36–44.
193. Szyc L, Yang M, Elsaesser T. 2010. Ultrafast energy exchange via water-phosphate interactions in hydrated DNA. *J Phys Chem B* 114: 7951–7957.
194. Szyc L, Yang M, Nibbering ETJ, Elsaesser T. 2010. Ultrafast vibrational dynamics and local interactions of hydrated DNA. *Angew Chem Intern Ed* 49: 3598–3610.
195. Yang M, Szyc L, Elsaesser T. 2011. Decelerated water dynamics and vibrational couplings of hydrated DNA mapped by two-dimensional infrared spectroscopy. *J Phys Chem B* 115: 13093–13100.

5 Solvation Dynamics of Vibrational States in Hydrogen-Bonding Solvents

Vibrational Frequency Fluctuations Studied by the Three-Pulse Infrared Photon Echo Method

Kaoru Ohta, Jumpei Tayama, Shinji Saito, and Keisuke Tominaga

CONTENTS

5.1 INTRODUCTION

Solute–solvent interaction plays an important role in chemical reaction dynamics and in many relaxation processes in the condensed phase [1]. Vibrational transition is a good probe for local interactions between solute and solvent molecules in the condensed phase because the frequency and magnitude of transition dipole moments of molecular vibrations depend strongly on the structure and charge distribution of molecules [2]. Therefore, time-resolved infrared (IR) spectroscopy provides us information on the conformational changes of solute molecules in solution and the dynamics of the surrounding environments [3–6]. In particular, nonlinear IR spectroscopy, such as IR photon echo and two-dimensional (2D) IR spectroscopy, is shown to be a powerful tool for studying the dynamical responses of the solvation structure.

One of the key observations in nonlinear IR spectroscopy is vibrational frequency fluctuations, $\Delta\omega(t) = \omega(t) - \omega_{ave}$, where $\omega(t)$ is the time-dependent vibrational frequency at time t, and ω_{ave} is its average value. The fluctuations are characterized by its time-correlation function (TCF), $\langle\Delta\omega(t)\ \Delta\omega(0)\rangle$, which reflects the amplitude of solute–solvent interactions and the timescales of the solvation

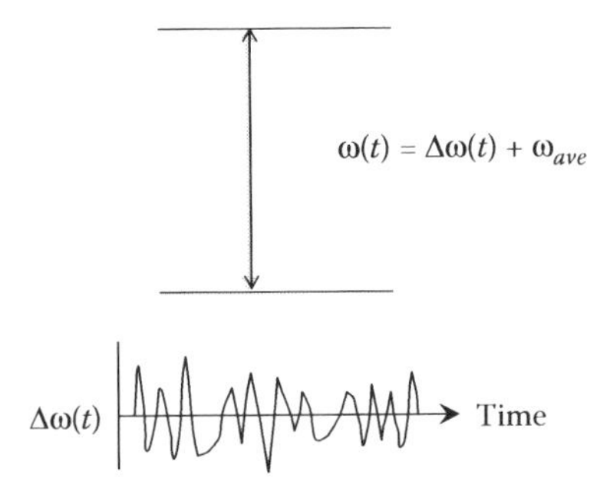

FIGURE 5.1 A schematic illustration of the frequency fluctuation.

dynamics [3,7]. In Figure 5.1, a schematic illustration of the frequency fluctuation is shown. Many research groups have focused on investigating the hydrogen-bonding dynamics of water and small alcohols by probing the vibrational dynamics of the OH stretching mode [7–11]. Furthermore, nonlinear IR spectroscopy has also been applied to the study of the structure and dynamics of peptides and small proteins by probing amide I vibrations [12–18].

In previous studies, the vibrational dynamics of small ions in polar solvents serves as a model system for studying the effect of the charge distributions and the normal coordinates on their vibrational dynamics. Until now, a number of groups, including us, investigated the vibrational population relaxation, orientational relaxation, and spectral diffusion process of azide (N_3^-), cyanate (OCN^-), and thiocyanate (SCN^-) by using ultrafast nonlinear IR spectroscopy [19–27]. Hochstrasser and coworkers performed seminal works in the early 1990s [19–21]. Hamm et al. have used three-pulse IR photon echo methods to investigate the spectral diffusion process of the antisymmetric stretching mode of azide in water [3]. Furthermore, we have studied the spectral diffusion processes of the antisymmetric stretching modes of OCN^- and SCN^- in various solvents, and the CN stretching modes for metal complexes in water by using three-pulse IR photon echo methods [28–32]. We have recently reported a detailed analysis of the temperature dependence of the vibrational dynamics of azide in water [33]. Furthermore, azide and thiocyanate groups can be incorporated into peptides and proteins, which are very useful to investigate site-specific fluctuation of biomolecules [34–38]. Detailed insight into the sensitivity of these vibrational modes to solvation structure and dynamics in simple solution should be very useful to understand the mechanism of site-specific fluctuation of the proteins and membranes.

Hydrogen-bonding liquids are particularly important solvents because of their ability to form hydrogen bonds with solvent molecules and solute molecules. They can often serve as good solvents for chemical reactions. In these solvents, hydrogen-bonding networks are formed, which continuously repeats the formation and breaking of the hydrogen bonds as well as structural rearrangement of the network. Such fluctuations of the networks may cause large effects in the vibrational and electronic states of a solute molecule. Among the hydrogen-bonding solvents, liquid water forms three-dimensional hydrogen-bonding network because of its nearly tetrahedral structure with two lone pairs and two OH bonds. This three-dimensional structure is the source of various peculiar behaviors of water. Furthermore, hydrogen-bonding dynamics in water is essential for various physiological reactions in life.

In this chapter, we summarize the three-pulse IR photon echo studies to elucidate spectral diffusion processes of the vibrational modes of ions such as three atomic ions and metal complexes in hydrogen-bonding solvents. These studies have clarified the microscopic interactions and molecular dynamics (MD) of solute ions in hydrogen-bonding solvents. Especially, we focus on the role of the dynamics of the hydrogen bond in the vibrational frequency fluctuations.

5.2 THEORY: THREE-PULSE INFRARED PHOTON ECHO

IR photon echo is one of the time-domain nonlinear techniques based on the third-order nonlinearity of the optical polarization. In our experimental setup, a mid-IR pulse has a pulse width of about

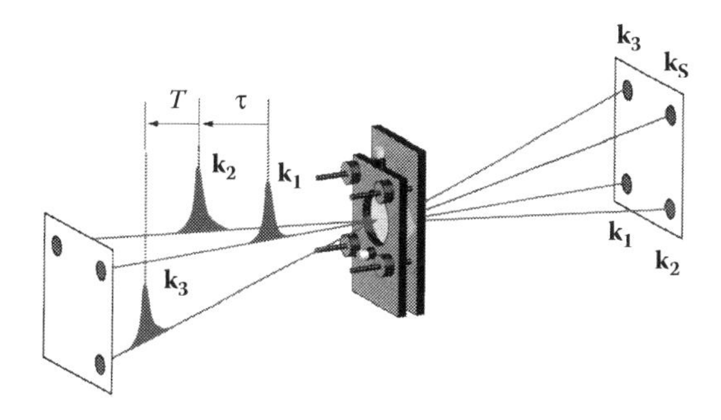

FIGURE 5.2 An illustration of the pulse geometry for the three-pulse photon echo experiment.

140–160 fs, a bandwidth of 120–130 cm^{-1}, and a pulse energy of 3–4 μJ/pulse at a repetition rate of 1 kHz [26,28–33,39–43]. The IR beam is split into three and focused at the sample in a box-car geometry (Figure 5.2). The photon echo signal is detected in the phase-matched direction of $-\mathbf{k}_1 + \mathbf{k}_2 + \mathbf{k}_3$, where $\mathbf{k}_1$, $\mathbf{k}_2$, and $\mathbf{k}_3$ are the wavevectors of the first, second, and third pulses, respectively. The time interval is defined as τ and T, where τ is the time interval between $\mathbf{k}_1$ and $\mathbf{k}_2$ beams while T is the time interval between $\mathbf{k}_2$ and $\mathbf{k}_3$ beams for $\tau > 0$ or between $\mathbf{k}_1$ and $\mathbf{k}_3$ beams for $\tau < 0$.

IR photon echo and transient grating methods are time-domain nonlinear techniques based on the third-order nonlinearity of the optical polarization [44]. A third-order nonlinear optical signal is expressed by integrating the polarization with respect to t, which is given as follows:

$$I(\tau,T) = \int_0^{\infty} \left|P(t,T,\tau)\right|^2 \mathrm{d}t, \tag{5.1}$$

where P is the third-order nonlinear polarization. Third-order polarization is expressed by the convolution of the response function, $R(t_1, t_2, t_3)$, with the electric fields of the laser pulses [44]

$$\begin{aligned} P_i^{(3)}(t,T,\tau) = {} & \int_0^{\infty} \mathrm{d}t_3 \int_0^{\infty} \mathrm{d}t_2 \int_0^{\infty} \mathrm{d}t_1 R_{ijkl}(t_3,t_2,t_1) \\ & \times E_{3j}(t-t_3) E_{2k}(t+T-t_3-t_2) E_{1l}(t+T+\tau-t_3-t_2-t_1). \end{aligned} \tag{5.2}$$

Here, i, j, k, and l are the laboratory frame orientational indices. We assume that the orientational motion is not coupled to the vibrational dynamics. This means that the response function can be described by the product of an isotropic vibrational response function and an orientational response function as follows [45]:

$$R_{ijkl}(t_3,t_2,t_1) = Y_{ijkl}(t_3,t_2,t_1) G(t_3,t_2,t_1), \tag{5.3}$$

where $Y_{ijkl}(t_3, t_2, t_1)$ is the orientational contribution to the response function and $G(t_3, t_2, t_1)$ is the isotropic part of the vibrational response function. The orientational response function depends on the polarization of the input electric fields. Assuming that the molecule is regarded as a spherical rotor, the orientational contribution of the polarization with the parallel and perpendicular polarizations is given by, respectively [45]

$$Y_{ZZZZ}(t_3,t_2,t_1) = \frac{1}{9}\exp(-2Dt_3)\left[1+\frac{4}{5}\exp(-6Dt_2)\right]\exp(-2Dt_1) \tag{5.4}$$

$$Y_{YYZZ}(t_3,t_2,t_1) = \frac{1}{9}\exp(-2Dt_3)\left[1-\frac{2}{5}\exp(-6Dt_2)\right]\exp(-2Dt_1), \tag{5.5}$$

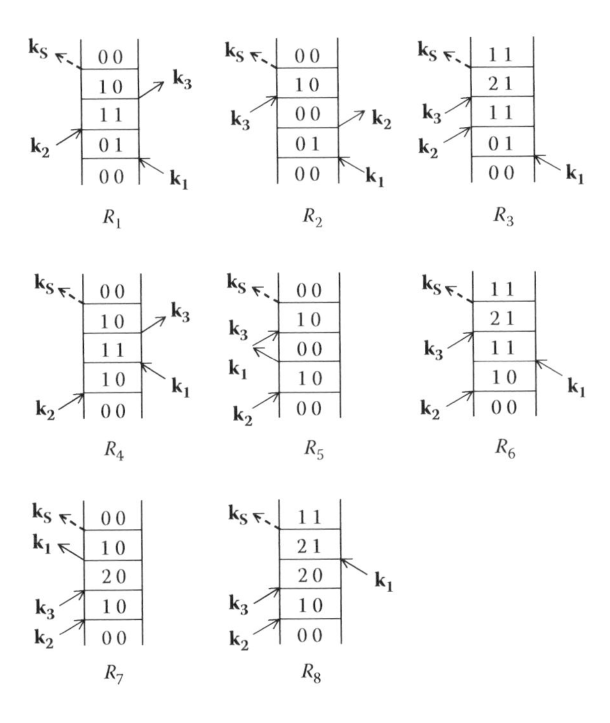

FIGURE 5.3 Double-side Feynman diagrams that contribute to the photon echo signals.

where D is the diffusion constant of the orientational motion and the orientational time is defined as

$$\tau_r = \frac{1}{6D}. \tag{5.6}$$

A detailed description of the isotropic part of the response functions has been given in other papers [3,44]. Briefly, interaction of the system with the electric fields is expressed by Feynman diagrams. The response function is a sum of contributions from each of Feynman paths as shown in Figure 5.3. The first three diagrams contribute to the photon echo signal when the delay time τ is positive. These are the so-called rephasing diagrams, which give photon echo signals. The rest of the diagrams contribute to the signal when the delay time τ is negative. These are the so-called nonrephasing diagrams, which give free induction decays. The response functions, R_1, R_2, R_4, and R_5 involve only the coherence of the $v = 0$–1 transition propagating during the delay times τ and t. The response functions, R_3 and R_6, involve the coherence of the $v = 0$–1 transition propagating during τ and the coherence of the $v = 1$–2 transition propagating during t. R_7 and R_8 contribute to the signal only when T is close to zero.

For the transient grating measurements, we set τ to zero. The time dependence of the transient grating signals at the parallel and perpendicular polarizations of the pump and probe pulses in the impulsive limit is given by, respectively

$$I_{TG,parallel}\left(T\right) = I_0\left[N\left(T\right)\right]^2\left[1 + \frac{4}{5}\exp\left(-6DT\right)\right]^2 \tag{5.7}$$

$$I_{TG,perpendicular}\left(T\right) = I_0\left[N\left(T\right)\right]^2\left[1 - \frac{2}{5}\exp\left(-6DT\right)\right]^2, \tag{5.8}$$

where $N(T)$ is the population of the $v = 1$ state at time T. As is well known, the transient grating signals at the magic angle condition contain only population dynamics, eliminating the orientational contribution to the signal. The anisotropy of the transient grating signals can be calculated from the following equation:

$$r(T) = \frac{\sqrt{I_{TG,parallel}(T)} - \sqrt{I_{TG,perpendicular}(T)}}{\sqrt{I_{TG,parallel}(T)} + 2\sqrt{I_{TG,perpendicular}(T)}}. \quad (5.9)$$

For the transient absorption measurements, IR pulse was split into pump and probe pulses by a CaF_2 wedged window. The pump and probe pulses were focused into the sample by a parabolic mirror. The probe pulse was dispersed in a monochromator and was then detected with a liquid N_2-cooled InSb detector or mercury cadmium telluride (MCT) array detector. The polarization of the probe pulse with respect to the pump pulse is controlled by two wire grid polarizers. One is placed before the sample and the other is after the sample. Similar to the transient grating measurements, the anisotropy decay is given by the following equations:

$$r(T) = \frac{I_{PP,parallel}(T) - I_{PP,perpendicular}(T)}{I_{PP,parallel}(T) + 2I_{PP,perpendicular}(T)}, \quad (5.10)$$

where $I_{PP,\,parallel}(T)$ and $I_{PP,\,perpendicular}(T)$ are the pump-probe signals at the parallel and perpendicular polarizations of the pump and probe pulses, which are proportional to orientational response functions, Y_{ZZZZ} and Y_{YYZZ}, respectively.

For the photon echo measurements, we measure the nonlinear optical signals by scanning the delay time τ (coherence time) at a certain T (population time) [3]. As shown in the previous studies, the three-pulse photon echo signals provide information on the temporal evolution of the inhomogeneous distribution of the vibrational transition frequency. In the presence of the inhomogeneity, the integrated intensity becomes larger for $\tau > 0$ than that for $\tau < 0.9$. As delay time T increases, spectral diffusion destroys the static inhomogeneity in the distribution of the vibrational frequencies. When there is no inhomogeneity in the system or the inhomogeneity is washed out completely due to the spectral diffusion, rephasing and nonrephasing diagrams contribute equally to the signals, and the photon echo signal becomes symmetric with respect to $\tau = 0$ fs. Therefore, the asymmetry of the photon echo signals is a sensitive probe of the degree of the inhomogeneity in the distribution of the vibrational frequencies [3].

The distribution of the vibrational frequency fluctuations is characterized by the TCF of the vibrational frequency fluctuations

$$M(t) = \langle \Delta\omega_{01}(t)\Delta\omega_{01}(0)\rangle, \quad (5.11)$$

where $\Delta\omega_{01}(t)$ is the shift of the vibrational frequency at time t from the average value. We calculate third-order nonlinear response functions from the line-broadening function, $g(t)$

$$g(t) = \int_0^t dt_1 \int_0^{t_1} dt_2 \langle \Delta\omega_{01}(t_2)\Delta\omega_{01}(0)\rangle. \quad (5.12)$$

Here, we assume that the anharmonicity fluctuation is small so that $\Delta\omega_{12}(t)$ is equal to $\Delta\omega_{01}(t)$. In the following simulation, we take into account an effect of the finite pulse width, the population relaxation, and orientational relaxation. The contribution of each diagram R_1–R_8 depends on the time ordering of the three pulses and all of the diagrams are included in the calculation.

As shown later, the population relaxation occurs in a nonexponential fashion. We have to incorporate the nonexponential population kinetics in the response function. According to Lim and

Hochstrasser [46], we use the following term in the population decay from the $v = 1$ state in place of a single exponential decay function

$$P_{11}(t) = \exp\left[-\int_0^t k_1(t')\mathrm{d}t'\right], \tag{5.13}$$

where $k_1(t)$ is the time-dependent population decay rate from the $v = 1$ state. The population decay factors of the 0–1 coherence and 1–2 coherence during the time τ and t are given by

$$P_{01}(t) = \exp\left[-\frac{1}{2}\int_0^t k_1(t')\mathrm{d}t'\right] \tag{5.14}$$

$$P_{12}(t) = \exp\left[-\frac{1}{2}\int_0^t \{k_1(t') + k_2(t')\}\mathrm{d}t'\right], \tag{5.15}$$

where $k_2(t)$ is the time-dependent population decay rate from the $v = 2$ state. We assume that the population relaxation time from the $v = 2$ state is a half of that from the $v = 1$ state. This assumption is based on the harmonic approximation for the transition dipole moment, that is, $2\mu_{10}^2 = \mu_{21}^2$. The transition rate is given by the Fermi's golden rule and is linear in the vibrational quantum number for the harmonic oscillator [47].

5.3 POPULATION RELAXATION AND ORIENTATIONAL RELAXATION

In this chapter, we focus on vibrational frequency fluctuation of solute molecules in hydrogen-bonding solvents. However, as we mentioned in the previous section, we need information on population relaxation and orientational relaxation of the vibrational state to obtain the TCF of the vibrational frequency fluctuation from analysis of the photon echo signal. These dynamical quantities are investigated by pump-probe or transient grating techniques in the IR region, which are also third-order nonlinear techniques. The pump-probe method is a heterodyne detection technique, and the population relaxation between the $v = 0$ and $v = 1$ states and between the $v = 1$ and $v = 2$ states can be differentiated. On the other hand, the transient grating method, which is based on homodyne detection, does not observe these dynamics separately if we do not perform the frequency-resolved measurement. In Table 5.1, we summarize the results of the population relaxation and orientational relaxation for the methanol and aqueous solutions at room temperature. FT-IR spectra of the antisymmetric stretching modes of SCN^- and N_3^- in some hydrogen-bonding solvents are shown in Figure 5.4 as examples.

Here, we briefly mention interesting observations in Table 5.1. We studied vibrational dynamics of the triply degenerate T_{1u} mode of $[Fe(CN)_6]^{4-}$ [29,30] and $[Ru(CN)_6]^{4-}$ [32]. For the ruthenium complex, the decay time constants of the fast and slow components at the magic angle condition are 0.7 and 23.0 ps, respectively. The initial anisotropy starts near 0.4 and decays with a time constant of 2.6 ps. We interpreted that the fast decay of the anisotropy is due to symmetry breaking of the complex. These metal complexes have an O_h symmetry, and the T_{1u} mode of the CN stretching is triply degenerate. However, coupling with other intramolecular modes and/or fluctuations of the surrounding solvents break the triple degeneracy, yielding three modes with small energy splitting. The symmetry breaking causes the anisotropy decay of the T_{1u} mode through two different mechanisms. One mechanism is population transfer and/or dephasing among the three states of the T_{1u} mode of the CN stretching mode [48,49]. The asymmetric stretching modes of the three pairs of cyanide ligands are directed along the three axes of the orthogonal coordinate system, x, y, and z. Excitation with the polarized light creates a certain superposition state among the three different states. Fluctuations of

TABLE 5.1
Parameters of Absorption Spectra of the Probe Molecule and Population Relaxation and Orientational Relaxation Times at Room Temperature

Solute	Solvent	V_{max}/cm^{-1}	ΔV/cm^{-1} (FWHM)	Population Relaxation Time T_1/ps	Orientational Relaxation Time T_R/ps	References
OCN^-	CH_3OH	2161	20	2.9	6.6	[28]
SCN^-	CH_3OH	2062	45	11.0	8.8	[28]
SCN^-	D_2O	2063	35	18.3	4.7	[29]
SCN^-	FA	2059	28	24.4	8.0	[90]
SCN^-	NMF	2056	32	27.9	5.7	[90]
N_3^-	D_2O	2043	18	2.3	7.1	[91]
N_3^-	H_2O	2048	25	0.8 ± 0.1[d]	1.3 ± 0.3[d]	[42]
N_3^-	CH_3OH	2044	22	3.0	11.5	[90]
N_3^-	NMF	2028	23	5.5	10.6	[90]
$[Fe(CN)_6]^{4-}$	D_2O	2036	16	0.70 (17%), 23.0 (83%)	2.6	[30]
$[Fe(CN)_6]^{4-}$	H_2O	2037	16	0.60 (20%), 3.7 (80%)	2.0	[30]
$[Ru(CN)_6]^{4-}$	D_2O	2045	14	0.8 ± 0.1 (39%), 20.8 ± 1.3 (61%)[b]	3.1 ± 0.4	[32]
$[Fe(CN)_5(NO)]^{3-}$[a]	D_2O	1935	16	7.3	16	[54]
$[Fe(CN)_5(NO)]^{3-}$	H_2O	1936	15	22	20	[54]
N_3^-	RM[c]	2037	28	1.4 ± 0.2[d]	7.6 ± 2[d]	[40]

Note: FA, formamide; NMF, *N*-methylformamide.
[a] The probe vibrational mode is the NO stretching mode.
[b] Obtained from the isotropic component of the pump-probe signal for the $v = 2\text{–}1$ transition.
[c] RM denotes reverse micelle. See the text for details.
[d] Taken from Ref. [22].

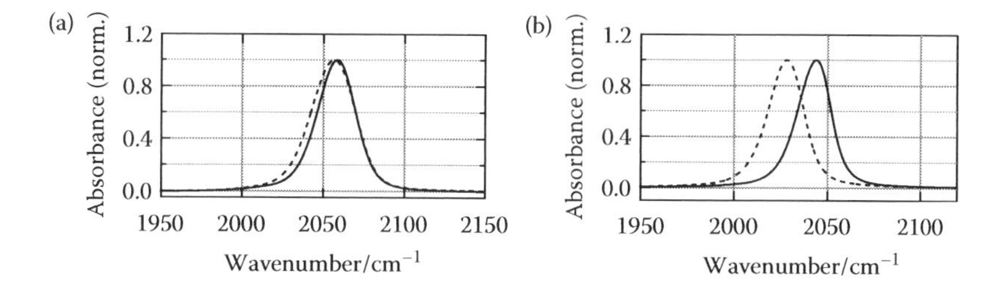

FIGURE 5.4 FT-IR spectra of (a) the antisymmetric stretching mode of SCN^- in FA (solid line) and NMF (dashed line), and (b) the antisymmetric stretching mode of N_3^- in methanol (solid line) and NMF (dashed line).

the solvent environment and other intramolecular modes with a certain symmetry cause changes of both energy separation and anharmonic coupling interactions among the three levels [49]. Such interactions with the T_{1u} modes induce the transition from one of the states of the T_{1u} modes to the others or pure dephasing. Thus, the direction of the transition dipole changes as a function of time because the composition of the population distribution along the three different axes changes, causing the depolarization of the transient grating signals. Therefore, the time evolution of the superposition states of the triply degenerate T_{1u} mode of the CN stretching mode is regarded as the orientational relaxation of the initial excited dipoles.

We observed the fast-decaying component (~0.70 ps) in the transient grating signals at the magic angle condition for $[Fe(CN)_6]^{4-}$. In this case, the fast-decaying component can be attributed to the population equilibration between the T_{1u} mode and the Raman active E_g and A_{1g} modes since the energy difference between the IR active mode and Raman active mode is small [29,50]. Frequency differences between the T_{1u} and E_g modes, and between the T_{1u} and A_{1g} modes for $[Fe(CN)_6]^{4-}$ in D_2O are 21 and 58 cm^{-1}, respectively. For $[Ru(CN)_6]^{4-}$, it was reported that these are 23 and 63 cm^{-1}, respectively. Therefore, we can consider that the population transfer to a higher energy level is feasible for both metal complexes, which is assisted by the low-frequency motions of the solvent. Using a femtosecond IR pulse, we could obtain time constants of this energy transfer to be 0.8 and 0.7 ps for $[Ru(CN)_6]^{4-}$ and $[Fe(CN)_6]^{4-}$, respectively. These time constants are quite similar with each other. This is because the energy differences between the IR active mode and Raman active modes are similar for the two metal complexes. Furthermore, the coupling strength and the low-frequency motions of the solvent, which are important for the population transfer, should be the same for the two complexes.

The slowly decaying component of the isotropic part of the pump-probe signal is assigned to the vibrational population relaxation of the $v = 1$ state from the T_{1u} mode of the CN stretching mode. The population of the initial excited vibrational mode will be equilibrated with the other CN stretching modes of different symmetries as pointed out earlier. It can also relax by transferring energy to a combination of the lower-frequency internal vibrations of the solute molecules and solvent phonon vibrational modes. Given the vibrational frequency of the solute and solvents, at least two other vibrations and a phonon of the solvent have to be excited in the course of the vibrational relaxation of the T_{1u} mode. The frequency of the bending mode of D_2O is around 1210 cm^{-1}. Therefore, the initial excitation of the T_{1u} mode may transfer energy to the stretching and bending motion of the cyano–metal complexes, the bending motion of D_2O and a solvent phonon that makes up for the energy mismatch. Although the frequency differences of the low-frequency modes can be observed between the $[Ru(CN)_6]^{4-}$ and $[Fe(CN)_6]^{4-}$ in Table 5.2, the fact that the population

TABLE 5.2
Parameters for the Time-Correlation Function of the Frequency Fluctuations

Molecule	Solvent	Δ_1 (ps^{-1})	T_1 (ps)	T_2^* (ps)	Δ_2 (ps^{-1})	T_2 (ps)	Δ_∞ (ps^{-1})	References
OCN^-	CH_3OH	1.3	0.12	4.9	1.6	4.5	0.55	[28]
SCN^-	CH_3OH	2.6	0.09	1.6	3.6	4.1	0.1	[28]
SCN^-	D_2O	4.3	0.08	0.7	2.7	1.3	0.0	[29]
SCN^-	FA	2.8	0.09	1.4	1.8	4.7	0.6	[90]
SCN^-	NMF	2.75	0.09	1.5	2.55	5.4	0.3	[90]
N_3^-	D_2O	2.6	0.08	1.8	1.4	1.3	0.3	[3]
N_3^-	H_2O	4.0	0.08	0.8	1.0	1.2	0.2	[42]
N_3^-	CH_3OH	3.1	0.09	1.2	1.25	3.5	0.55	[90]
N_3^-	NMF	3.0	0.09	1.2	1.45	3.8	0.75	[90]
$[Fe(CN)_6]^{4-}$	D_2O	2.8	0.08	1.6	1.15	1.5	0.0	[30]
$[Fe(CN)_6]^{4-}$	H_2O	2.95	0.08	1.4	1.0	1.4	0.0	[30]
$[Ru(CN)_6]^{4-}$	D_2O	3.0	0.08	1.4	0.8	1.4	0.1	[32]
$[Fe(CN)_5(NO)]^{3-}$ [a]	D_2O	3.0	0.09	1.2	1.3	1.0	0.2	[54]
$[Fe(CN)_5(NO)]^{3-}$	H_2O	2.6	0.09	1.6	1.3	1.0	0.2	[54]
N_3^-	RM [b]	3.3	0.08	1.1	1.2	1.2	1.0	[40]

Note: FA, formamide; NMF, *N*-methylformamide.

[a] The probe vibrational mode is the NO stretching mode.

[b] RM denotes reverse micelle. See the text for details.

relaxation times are similar for the two complexes indicate that the population relaxation is not affected by such differences.

5.4 SPECTRAL DIFFUSION IN HYDROGEN-BONDING SOLVENTS

Figure 5.5 shows the three-pulse photon echo signals of SCN^- in formamide (FA) as a function of coherence time (τ). In Figure 5.6, the photon echo signals are displayed along the two axes, coherence time (τ) and population time (T). At population times T earlier than 5 ps, the center of mass of the photon echo signal is located at around 300 fs, indicating the presence of an inhomogeneous distribution of the vibrational frequency. The peak of the echo signal shifts toward zero as the population time T increases. This behavior shows that the local environment of each oscillator is inhomogeneously distributed and evolves on a timescale of a few picoseconds. To characterize the degree of asymmetry of the decay of the photon echo signal along the τ-axis, the first moment of the photon echo signal is calculated as a function of T (Figure 5.6b). The first moment is defined as

$$FM(T) = \frac{\int_{-\infty}^{\infty} \mathrm{d}\tau\tau I(\tau,T)}{\int_{-\infty}^{\infty} \mathrm{d}\tau I(\tau,T)}, \tag{5.16}$$

where $I(\tau,T)$ is the intensity of the experimentally observed photon echo signals. As the delay time T increases, frequencies are sampled throughout the inhomogeneously broadened line, which makes the first moment decay to zero. Therefore, the first moment of the photon echo signal is a sensitive measure for the degree of "transient inhomogeneity" in the distribution of the transition frequencies, and the timescale of the decay is approximately proportional to that of the TCF of the frequency fluctuation [51–53]. The first moment of the three-pulse IR photon echo signals can be directly obtained from the experimental data without complex numerical simulations, and the first moment is a useful quantity for surveying the inhomogeneous fluctuation time.

We quantitatively evaluated the parameters for the TCF of the vibrational transition frequency fluctuation by simulating the temporal profile of the three-pulse IR photon echo signals and IR absorption spectrum simultaneously. The detailed procedure of the simulation of the IR photon echo signal and IR absorption spectrum has been described elsewhere [28,29]. From the measurements of the three-pulse photon echo signals as a function of delay times τ and T, we can obtain information

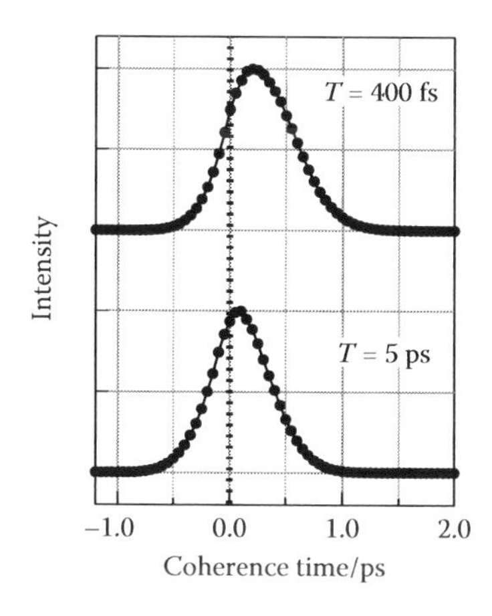

FIGURE 5.5 Three-pulse photon echo signals as a function of coherence time (τ) of SCN^- in FA at $T = 400$ fs and at $T = 5$ ps.

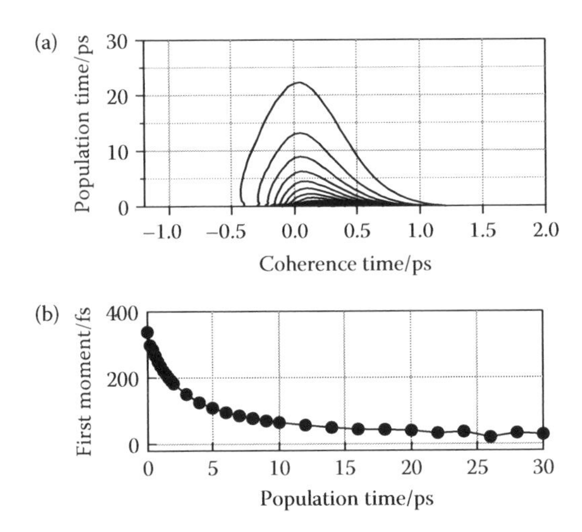

FIGURE 5.6 (a) Three-pulse photon echo signals plotted with respect to the delay times τ and T for SCN^- in FA. (b) The first moment of the photon echo signals from the experimental data (solid line with closed circles).

on the correlation function of the vibrational frequency fluctuation. The correlation function of the vibrational frequency fluctuation is assumed to be a sum of exponentials plus a constant

$$M(t) = \sum_{i=1}^{2} \Delta_i^2 \exp(-t/\tau_i) + \Delta_\infty^2. \tag{5.17}$$

This biexponential function works quite well to reproduce both the photon echo signal and the absorption spectrum. We calculated the temporal profile of the photon echo signal at each delay time T according to Equation 5.1, taking into account the pulse width and the time ordering of the laser pulses, and then the first moment of the photon echo signal was calculated based on Equation 5.15. We also calculated a linear absorption spectrum along with the experimental absorption spectrum. The absorption spectrum is calculated from the following equation [3,44]:

$$I(\omega) = 2\,\mathrm{Re}\int_{-\infty}^{\infty} \exp\left[-i(\omega - \omega_{01})t\right]\exp\left[-g(t) - 2Dt\right]P_{01}(t)\,dt, \tag{5.18}$$

where $P_{01}(t)$ is the population decay from the $v = 1$ state. It can be seen that a good agreement between the experimental results and the simulation is obtained.

In Table 5.2, we summarize the obtained parameters of the vibrational frequency fluctuations for the hydrogen-bonding solvents. The ultrafast component is in a rapid modulation limit ($\tau_1 \times \Delta_1 < 1$), so that the only dephasing time ($T_2^* = 1/\tau_1\Delta_1^2$) can be obtained with a good accuracy, and the uncertainty of the amplitude Δ_1 and the time constant τ_1 may be ±50%. This timescale is similar to that observed in the solvation dynamics of the electronic state, which was assigned due to inertial motion of the solvent. Regarding the slow and constant component, the amplitudes of the TCF are determined within errors of a few percent. On the other hand, the uncertainty of the time constant τ_2 is around ±20%.

From Table 5.2, we can find that the time constant for the slow component τ_2 does not significantly depend on the solute. In contrast, the amplitude for the slow component depends on both the solvent and the solute. This is especially true for water, both H_2O and D_2O. The value of τ_2 ranges from 1.0 to 1.5 ps, although the value of Δ_2 is from 0.8 to 2.7 ps^{-1} depending on the solute molecule. For the other solvents, the value of τ_2 is several picoseconds. Since the intensity of the photon echo signal decays in parallel with the population relaxation T_1, a system with short T_1 such as N_3^-/CH_3OH or N_3^-/*N*-methylformamide (NMF), precise determination of the τ_2 value from the analysis of the photon echo signal is rather difficult for these systems if τ_2 is similar to or longer than T_1.

Before discussing the molecular picture of the frequency fluctuation of the solute, we briefly point out a couple of interesting results of the three-pulse photon echo experiments.

5.4.1 Effect of Mode Degeneracy on the Frequency Fluctuation

The timescales of the correlation function for the vibrational frequency fluctuation for $[Fe(CN)_6]^{4-}$ in D_2O are similar to those for N_3^- and SCN^- in D_2O. These results show the absence of the difference between the frequency fluctuations in the degenerate and nondegenerate systems. Fayer and coworkers showed that the temperature dependence on the pure dephasing of $Rh(CO)_2(C_5H_7O_2)$ is different from that of $W(CO)_6$ in the same solvent [49]. They proposed that this difference results from the difference in mode degeneracy; the dephasing of CO in $W(CO)_6$ is caused by lowering the symmetry due to local fluctuations in the solvent, which is not seen in $Rh(CO)_2(C_5H_7O_2)$. Even though the vibrational dephasing could be caused by both the energy gap fluctuation between the $v = 0$ and $v = 1$ states for the T_{1u} mode and the fluctuation of the energy splitting among the triply degenerate T_{1u} modes, our results suggest that the timescale of the spectral diffusion process for $[Fe(CN)_6]^{4-}$ is controlled by the same mechanism as that for SCN^-, not by the time-dependent anisotropic solute–solvent interaction. To confirm this point, we measured the photon echo signal of the NO stretching mode of nitroprusside $[Fe(CN)_5(NO)]^{2-}$ in both H_2O and D_2O [54]. As can be seen in Table 5.2, the τ_2 value is almost the same as the other cases, which confirms that the symmetry breaking is not an additional source of the spectral diffusion process, which is important for the homogeneous vibrational dephasing.

5.4.2 Quasi-Static Component

For some systems such as $[Ru(CN)_6]^{4-}/D_2O$, the first moment $M(T)$ indicates a slowly decaying component over a very long time ($T > 10$ ps), although the signal-to-noise ratio is getting poor in this time region. Therefore, we have also performed simulation by assuming a triexponential function without a quasi-static term for the TCF of the frequency fluctuation [32],

$$M(t) = \sum_{i=1}^{3} \Delta_i^2 \exp(-t/\tau_i). \tag{5.19}$$

The obtained parameters are the following: $\Delta_1 = 3.0\ \text{ps}^{-1}$, $\tau_1 = 0.08$ ps, $\Delta_2 = 0.7\ \text{ps}^{-1}$, $\tau_2 = 1.4$ ps, $\Delta_3 = 0.1\ \text{ps}^{-1}$, and $\tau_3 = 10$ ps. Because of the weak intensity of the signal in the longer time-delay region, it is difficult to determine the Δ_3 and τ_3 values precisely. To investigate such a weak signal, the 2D IR may be more suitable because it is based on the heterodyne detection technique. We also found that the inclusion of the third component does not influence the simulation results of the faster component such as τ_2.

It is worth noting that the quasi-static component of the TCF of the frequency fluctuation, Δ_∞, is observed in the case of azide in D_2O. The long timescale components corresponding to the quasi-static component of the TCF, Δ_∞, were observed in optical Kerr effect [55] and 2D IR [56–58] experiments using aqueous salt solutions. These components may be caused by the presence of cooperative rotational motion of azide with a solvation shell, which would be consistent with the long residence times for water molecules in the solvation shell of ions [59,60]. Nevertheless, we suggest that Na^+ and/or N_3^- ions do not alter the dynamics of water because the concentration of the solution in this study is low in comparison with the above experiments using aqueous salt solutions.

5.4.3 Reverse Micelle

A reverse micelle (RM), a water-in-oil microemulsion, is one of the well-known model systems to provide the confined space with a well-defined size of nanometer dimension. Since various properties such as viscosity or dielectric constant is different from those of bulk water, it is interesting to see the vibrational frequency fluctuation of solute molecules in RM. We chose azide as a probe and nonionic RM, nonylphenol poly(oxyethylene)$_7$ [40]. The system is chosen as the same one as Owrutsky and coworkers studied for the vibrational dynamics to compare confinement effects on the population relaxation, orientational relaxation, and transition frequency fluctuations of the vibrational mode [22,23].

The ultrafast and picosecond components in the RM have similar amplitudes and timescales to those in the bulk water. On the other hand, the contribution of the static component is much larger in the RM. The picosecond component observed in the RM has the same timescale (1.2 ps) as that in the bulk water. The amplitudes of the picosecond component are 1.2 and 1.0 ps^{-1} for the RM and bulk water. This clearly indicates that the N_3^- antisymmetric stretching mode in the RM is perturbed on a picosecond timescale by the same interaction and dynamics as in the bulk water.

The existence of the static component shows that slow dynamics with a timescale longer than, at least, 1.2 ps occurs in the RM. For the present RM case, it is difficult to ascribe whether the observed static component reflects dynamics slower than a few picoseconds or essentially static inhomogeneity in the water pool. Recent time-resolved studies in the RMs have shown longer time constants for dynamical processes such as solvation dynamics [61–65] and dielectric relaxation [66]. The TCF of the vibrational fluctuation could have such a slower component. The previous study on the population and orientational relaxation in the RM suggests that the azide ion may exist in somewhat uniform environment of the water pool, implying a small inhomogeneous contribution.

5.5 MOLECULAR PICTURE OF VIBRATIONAL FREQUENCY FLUCTUATION

The MD of water is a key to understand the molecular picture of dynamics of the hydrogen-bonding networks in water that have been extensively investigated by ultrafast IR spectroscopy in conjunction with MD simulation [7,10,67–75]. Tokmakoff and coworkers analyzed a 2D line shape of the OH stretching mode of HOD in D_2O. The dynamics show a rapid 60 fs decay, an underdamped oscillation on a 130 fs timescale induced by hydrogen bond stretching, and a long time-decay constant of 1.4 ps [74]. MD simulation has yielded the relationship between the frequency TCF and the time evolution of the liquid structure. A comparison between experimental and computational results indicates that the slow component of the correlation function is due to structural reorganization, including the collective rearrangement of the hydrogen-bonding network. Moreover, 2D IR experiments on the OD stretching mode of HOD in H_2O by Fayer and coworkers extracted TCF with decay components of 48 fs, 400 fs, and 1.4 ps [9]. They showed that MD simulation that employs a polarizable water model predicts TCF, which is very close to the experimental results. They concluded that the slow component in TCF is associated with hydrogen bond equilibration and the making and breaking of hydrogen bonds. The comparable spectral diffusion of the HOD in D_2O and H_2O systems shows that the hydrogen bond dynamics of D_2O and H_2O are very similar [76].

We also observed the small isotope effect on the timescale of the slow component in the TCF for the ionic solutes in water. Furthermore, the timescales of the longest component in the spectral diffusion of the HOD in D_2O and H_2O systems are similar to those for the aqueous solution systems. Because of these similarities between the pure water and the aqueous solutions, the slow component in the frequency TCF for the aqueous solutions can be considered to originate from the same mechanism as that for the OH or OD stretching mode of the pure water system.

Theoretical studies on the vibrational frequency fluctuations have been conducted for aqueous solution systems by employing the MD simulations. Results of the simulation on CN^- and N_3^-

in water and methanol indicated that the timescale of the hydrogen bond dynamics in methanol is about three times longer than that in water [77]. This was consistent with our experimental observations. However, the MD simulation study also showed that the hydrogen-bonding dynamics is very sensitive to the charge distribution of the ion. Since the hydrogen bond dynamics is a short-range electrostatic interaction between the ion and solvent, it is readily understandable that a small change in the charge of the solute affects the timescale of making and breaking of the hydrogen bond. Although we do not have information on the charge distributions of the ions and an effect of their differences on the hydrogen bond dynamics in a quantitative level, the experimental results show that the dynamics of the correlation function are mostly affected by properties that are characteristic to the solvent.

Skinner and coworkers investigated the vibrational frequency fluctuation of N_3^- in D_2O by MD simulations [78]. They calculated the TCF of the electric field from surrounding solvents along the antisymmetric stretching mode of N_3^-, vibrational frequency fluctuation of the mode, and solute–solvent hydrogen bond number fluctuation. They also calculated the TCF of the hydrogen bond number fluctuation $\langle \Delta n(t) \Delta n(0) \rangle$, where $\Delta n(t) = n(t) - \langle n \rangle$ is the fluctuation of the number of hydrogen bonds from equilibrium. They found that the hydrogen bond TCF decays slightly faster at longer time ($t > 0.5$ ps) than the TCF of the frequency fluctuation. They also calculated the projection of the electric field along the molecular axis of azide due to the surrounding water molecules. They showed that the decay of the TCF of the electric field fluctuation is very similar to that of the TFC of the vibrational frequency fluctuation. This is consistent with the findings by Tokmakoff and coworkers for HOD/D_2O system, where they concluded that at long times ($t > 200$ fs), the relaxation originates from large-scale cooperative reorganization and not from specific molecular motions such as hydrogen bond making and breaking [79]. Skinner and coworkers also pointed out that these two viewpoints, hydrogen bond versus electric field fluctuations, are not really mutually exclusive because the field is dominated by the nearest neighbors, the ones involved in making and breaking hydrogen bonds [78]. The importance of the electric field fluctuations in the computed vibrational frequency fluctuations for both cases, N_3^-/D_2O and HOD/D_2O systems, is consistent with our observation that the long time decay in the TCF for the solutes in aqueous solutions is similar to that of HOD/D_2O system.

From the above-mentioned discussion, the vibrational frequency fluctuation of the ions in aqueous solutions can be viewed as follows; the hydrogen bond between the ionic probe and water molecule influences the vibrational transition frequency of the intramolecular mode of the probe. However, hydrogen bond breaking and making process between the solute and solvent is not a major source, which induces such a modulation on the vibrational mode of the ion. Rather, more collective dynamics of water around the ion, which is characteristic to water itself, is important for the vibrational fluctuation. The fluctuation of the local electric field produced by water molecules around the ion may be a source of the vibrational fluctuation, as concluded by Skinner and coworkers for their theoretical work. In this case, the electrostatic interaction between the ion probe and adjacent water molecule may make a relatively strong hydrogen bond, and this hydrogen bond fluctuates without breaking and making, which is controlled by structural rearrangement of the hydrogen bond network. In other words, the probe molecule "feels" the fluctuation of the hydrogen bond network of water through the hydrogen bond between the probe and water. This picture is schematically illustrated in Figure 5.7. This microscopic picture explains a couple of experimental observations such as the insensitivity of the time constant of the slow component τ_2 to the solute and the similarity of τ_2 to the time constant of the frequency fluctuation of pure water. This picture also suggests that the presence of the ion does not significantly affect the dynamics of the hydrogen-bonding network around the ion probe at least for the picosecond timescale, since the picosecond component of the TCF nearly depends on the probe ion.

For the other hydrogen-bonding solvents such as methanol, NMF, or FA, the time constants of the spectral diffusion τ_2 of the ion probe are several picoseconds, which are longer than that of

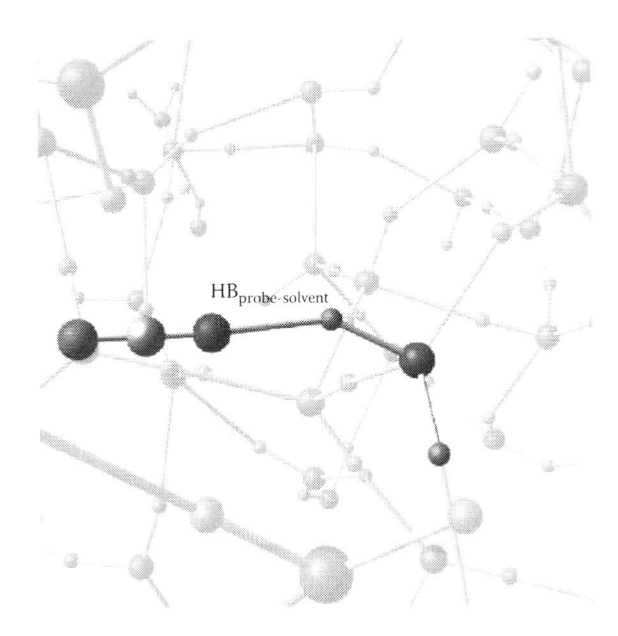

FIGURE 5.7 A schematic illustration of the vibrational frequency fluctuation of solute molecule in water.

the aqueous solutions. It is interesting to note that the timescale of the TCF of pure FA is different from that of the solute vibrational mode in FA. Recent 2D IR studies of N–H stretching vibrations in FA showed that the TCF of the vibrational frequency fluctuation can be expressed by the sum of exponentials with time constants of 0.24, 0.8, and 11 ps [80]. The time constants of the slowest-decaying components that we obtained for SCN^- in FA are about one-half that obtained by monitoring the N–H stretching vibrations. This result suggests that the slow-decaying component of the TCF of the antisymmetric mode of the three atomic ions originates from a mechanism different from that in the aqueous solution. By making and breaking hydrogen bonds between solute ions and solvent molecules, the spectra associated with different hydrogen-bonded complexes are exchanged with each other and merged into one band. This exchange is conceivably an origin of slow-decaying components of the TCF, and the decay of the TCF corresponds to the timescales of this exchange. Further investigation using 2D IR spectroscopy will provide us with useful information to identify the underlying structures of the IR absorption spectra of the antisymmetric stretching modes of SCN^- and to resolve the formation and dissociation of solute–solvent hydrogen-bonded complexes in protic solvents.

5.6 SOLVATION DYNAMICS STUDIED BY FLUORESCENCE DYNAMIC STOKES SHIFT

Finally, we briefly mention about the frequency fluctuation of the electronic state. So far, we have been discussing the TCF of the fluctuation of the transition frequency of the vibrational mode of solute molecules. This quantity corresponds to fluctuations of the vibrational state in the thermal equilibrium state. Within a context of linear response theory, the TCF of the fluctuation in the equilibrium state is equal to the normalized response function that characterizes relaxation of the nonequilibrium state. For example, for the electronic state of the solute in polar solvents, dynamic fluorescence Stokes shift experiments have been used to study solvation dynamics [81,82]. In these measurements, a probe molecule such as coumarin 153 is excited to the electronically excited state. The charge distribution of a probe molecule is changed instantaneously upon photoexcitation. One can monitor the dynamic response of solvents by measuring the time dependence of emission frequency, which is called the fluorescence dynamic Stokes shift. The relaxation is characterized by the response function

$$C(t) = \frac{\nu(t) - \nu(\infty)}{\nu(0) - \nu(\infty)}, \tag{5.20}$$

where $\nu(t)$, $\nu(0)$, and $\nu(\infty)$ are the peak wavenumbers of the time-dependent fluorescence spectra at t, 0, and ∞, respectively. The equality between $M(t)$ (in Equation 5.10) and $C(t)$ is based on the following concepts: in a thermal equilibrium state, the transition frequency continuously fluctuates around its mean value. On the other hand, after the excitation to the excited state, the transition frequency continuously changes corresponding to the relaxation from the initial nonequilibrium configuration on the excited state. The change in the nonequilibrium transition energy can be described as the TCF of transition frequency with a linear response approximation.

Solvation dynamics of water have been studied by means of femtosecond fluorescence up-conversion method by several groups. Barbara and coworkers reported that the water solvation dynamics is characterized by a biexponential function with time constants of 0.16 and 1.2 ps [83]. They used the anionic coumarin dye molecule, 7-(dimethylamino)coumarin-4-acetate ion. They also used coumarin 343 as a probe to investigate temperature dependence of solvation dynamics in water [84]. Later, Fleming and coworkers observed a sub-100 fs component with an experimental setup of improved time resolution, which is assigned to the inertial motion of the solvent [85]. The striking finding is that this ultrafast component is a dominant contribution in the total Stokes shift. They also used three-pulse photon echo peak shift method, which is based on the third-order nonlinear effect, to investigate solvation dynamics in water [86]. Quite recently, Ernsting and coworkers chose *N*-methyl-6-oxyquinolone as a probe and reported temperature dependence of solvation time [87]. Above 20°C, solvation dynamics in H_2O and D_2O are identical, whereas, below that temperature, the dynamics in D_2O slows down. It is interesting to note that the response function of solvation dynamics $C(t)$ has a picosecond component whose timescale is similar to that of $M(t)$, TCF of the vibrational mode of solute in aqueous solutions.

The solvation dynamics of a polar solute in methanol has been extensively investigated by the time-dependent dynamic Stokes shift [88,89]. These studies showed that solvation dynamics occurs on multiple timescales. The 100 fs component results from the inertial solvation dynamics. The slow picosecond-decaying component is ascribed to the diffusive type of the solvent motion. For example, Horng et al. obtained 3.2 and 15.3 ps decay components of the Stokes shift function for coumarin 153 in methanol with the fluorescence Stokes shift measurements as well as a faster subpicosecond component [88]. It was also shown that the solvation dynamics in methanol takes place slower than that in water for both inertial and diffusive types of the solvent motion. Comparing the solvation dynamics for the electronic transition and vibrational transition in water and methanol, there seems to be a correlation of timescales of the slowly decaying components, although the coupling strength to solvent is different in two orders of magnitudes for the two cases. It is expected that the relative importance of dynamics due to short-range interaction such as the hydrogen-bonding dynamics or the local electric field fluctuation against dynamics due to the long-range interaction in the solvation correlation function can be different for the electronic and vibrational transitions. It will be interesting to study how the solvation dynamics is correlated between the electronic transition and the vibrational transition.

5.7 SUMMARY

In this chapter, we summarized the results of the three-pulse IR photon echo experiments to investigate the vibrational frequency fluctuations of the ionic probes in hydrogen-bonding solvents such as water, methanol, FA, and NMF. The TCF is modeled by a biexponential function with a quasi-static term; the time constants of the two components are sub-100 fs and picoseconds. For the aqueous solutions, the insensitivity of the time constant of the picosecond timescale and the similarity to the picosecond component observed for the liquid water case suggests that the vibrational frequency fluctuation is mostly influenced by collective dynamics of water surrounding the probe ion rather than the breaking and making process of the hydrogen bond between the ion and solvent. For methanol, FA, or NMF, to investigate the mechanisms of the vibrational frequency fluctuations of the solute ions in a molecular level, it is desirable to accumulate knowledge on dynamics of neat liquids of these solvents by, for example, 2D IR spectroscopy. Moreover, further studies, including MD simulations, will be required to clarify a more detailed picture of the vibrational frequency fluctuations in hydrogen-bonding solvents.

REFERENCES

1. Stratt RM and Maroncelli M. 1996. Nonreactive dynamics in solution: The emerging molecular view of solvation dynamics and vibrational relaxation. *J Phys Chem USA* 100(31):12981–12996.
2. Fayer MD. 2000. *Ultrafast Infrared and Raman Spectroscopy* (Marcel Dekker, New York).
3. Hamm P, Lim M, and Hochstrasser RM. 1998. Non-Markovian dynamics of the vibrations of ions in water from femtosecond infrared three-pulse photon echoes. *Phys Rev Lett* 81(24):5326–5329.
4. Hamm P, Lim MH, and Hochstrasser RM. 1998. Structure of the amide I band of peptides measured by femtosecond nonlinear-infrared spectroscopy. *J Phys Chem B* 102(31):6123–6138.
5. Nibbering ETJ and Elsaesser T. 2004. Ultrafast vibrational dynamics of hydrogen bonds in the condensed phase. *Chem Rev* 104(4):1887–1914.
6. Hamm P and Zanni MT. 2011. *Concepts and Methods of 2D Infrared Spectroscopy* (Cambridge University Press, Cambridge).
7. Stenger J, Madsen D, Hamm P, Nibbering ETJ, and Elsaesser T. 2002. A photon echo peak shift study of liquid water. *J Phys Chem A* 106(10):2341–2350.
8. Asbury JB et al. 2003. Ultrafast heterodyne detected infrared multidimensional vibrational stimulated echo studies of hydrogen bond dynamics. *Chem Phys Lett* 374(3–4):362–371.
9. Asbury JB et al. 2004. Dynamics of water probed with vibrational echo correlation spectroscopy. *J Chem Phys* 121(24):12431–12446.
10. Fecko CJ, Eaves JD, Loparo JJ, Tokmakoff A, and Geissler PL. 2003. Ultrafast hydrogen-bond dynamics in the infrared spectroscopy of water. *Science* 301(5640):1698–1702.
11. Loparo JJ, Roberts ST, and Tokmakoff A. 2006. Multidimensional infrared spectroscopy of water. I. Vibrational dynamics in two-dimensional IR line shapes. *J Chem Phys* 125(19):194521.
12. DeCamp MF et al. 2005. Amide I vibrational dynamics of *N*-methylacetamide in polar solvents: The role of electrostatic interactions. *J Phys Chem B* 109(21):11016–11026.
13. Woutersen S and Hamm P. 2000. Structure determination of trialanine in water using polarization sensitive two-dimensional vibrational spectroscopy. *J Phys Chem B* 104(47):11316–11320.
14. Kim YS, Wang JP, and Hochstrasser RM. 2005. Two-dimensional infrared spectroscopy of the alanine dipeptide in aqueous solution. *J Phys Chem B* 109(15):7511–7521.
15. Maekawa H, Toniolo C, Broxterman QB, and Ge NH. 2007. Two-dimensional infrared spectral signatures of 3(10)- and alpha-helical peptides. *J Phys Chem B* 111(12):3222–3235.
16. Demirdoven N et al. 2004. Two-dimensional infrared spectroscopy of antiparallel beta-sheet secondary structure. *J Am Chem Soc* 126(25):7981–7990.
17. Mukherjee P, Kass I, Arkin I, and Zanni MT. 2006. Picosecond dynamics of a membrane protein revealed by 2D IR. *Proc Natl Acad Sci USA* 103(10):3528–3533.
18. Ganim Z et al. 2008. Amide I two-dimensional infrared spectroscopy of proteins. *Acc Chem Res* 41(3):432–441.
19. Owrutsky JC, Kim YR, Li M, Sarisky MJ, and Hochstrasser RM. 1991. Determination of the vibrational energy relaxation time of the azide ion in protic solvents by two-color transient infrared spectroscopy. *Chem Phys Lett* 184(5–6):368–374.
20. Li M et al. 1993. Vibrational and rotational relaxation times of solvated molecular ions. *J Chem Phys* 98(7):5499–5507.
21. Owrutsky JC, Raftery D, and Hochstrasser RM. 1994. Vibrational relaxation dynamics in solutions. *Annu Rev Phys Chem* 45:519–555.

22. Zhong Q, Baronavski AP, and Owrutsky JC. 2003. Reorientation and vibrational energy relaxation of pseudohalide ions confined in reverse micelle water pools. *J Chem Phys* 119(17):9171–9177.
23. Zhong Q, Baronavski AP, and Owrutsky JC. 2003. Vibrational energy relaxation of aqueous azide ion confined in reverse micelles. *J Chem Phys* 118(15):7074–7080.
24. Lenchenkov V, She CX, and Lian TQ. 2006. Vibrational relaxation of CN stretch of pseudo-halide anions (OCN-, SCN-, and SeCN-) in polar solvents. *J Phys Chem B* 110(40):19990–19997.
25. Sando GM, Dahl K, and Owrutsky JC. 2007. Vibrational spectroscopy and dynamics of azide ion in ionic liquid and dimethyl sulfoxide water mixtures. *J Phys Chem B* 111(18):4901–4909.
26. Ohta K and Tominaga K. 2006. Vibrational population relaxation of thiocyanate ion in polar solvents studied by ultrafast infrared spectroscopy. *Chem Phys Lett* 429(1–3):136–140.
27. Dahl K, Sando GM, Fox DM, Sutto TE, and Owrutsky JC. 2005. Vibrational spectroscopy and dynamics of small anions in ionic liquid solutions. *J Chem Phys* 123(8):084504.
28. Ohta K, Maekawa H, Saito S, and Tominaga K. 2003. Probing the spectral diffusion of vibrational transitions of OCN- and SCN- in methanol by three-pulse infrared photon echo spectroscopy. *J Phys Chem A* 107(30):5643–5649.
29. Ohta K, Maekawa H, and Tominaga K. 2004. Vibrational population relaxation and dephasing dynamics of Fe(CN)(6)(4-) in D_2O with third-order nonlinear infrared spectroscopy. *J Phys Chem A* 108(8):1333–1341.
30. Ohta K, Maekawa H, and Tominaga K. 2004. Vibrational population relaxation and dephasing dynamics Fe(CN)(4-)(6) in water: Deuterium isotope effect of solvents. *Chem Phys Lett* 386(1–3):32–37.
31. Ohta K and Tominaga K. 2005. Dynamical interactions between solute and solvent studied by three-pulse photon echo method. *B Chem Soc Jpn* 78(9):1581–1594.
32. Tayama J, Banno M, Ohta K, and Tominaga K. 2010. Vibrational dynamics of the CN stretching mode of [Ru(CN)(6)](4-) in D(2)O studied by nonlinear infrared spectroscopy. *Sci China Phys Mech* 53(6):1013–1019.
33. Tayama J et al. 2010. Temperature dependence of vibrational frequency fluctuation of N(3)(-) in D(2)O. *J Chem Phys* 133(1):014505.
34. Fafarman AT and Boxer SG. 2010. Nitrile bonds as infrared probes of electrostatics in ribonuclease S. *J Phys Chem B* 114(42):13536–13544.
35. Fafarman AT, Sigala PA, Herschlag D, and Boxer SG. 2010. Decomposition of vibrational shifts of nitriles into electrostatic and hydrogen-bonding effects. *J Am Chem Soc* 132(37):12811–12813.
36. Lindquist BA, Furse KE, and Corcelli SA. 2009. Nitrile groups as vibrational probes of biomolecular structure and dynamics: An overview. *Phys Chem Chem Phys* 11(37):8119–8132.
37. Taskent-Sezgin H et al. 2010. Azidohomoalanine: A conformationally sensitive IR probe of protein folding, protein structure, and electrostatics. *Angew Chem Int Edit* 49(41):7473–7475.
38. Waegele MM, Culik RM, and Gai F. 2011. Site-specific spectroscopic reporters of the local electric field, hydration, structure, and dynamics of biomolecules. *J Phys Chem Lett* 2(20):2598–2609.
39. Maekawa H, Ohta K, and Tominaga K. 2004. Vibrational population relaxation of the $-N=C=N-$ anti-symmetric stretching mode of carbodiimide studied by the infrared transient grating method. *J Phys Chem A* 108(44):9484–9491.
40. Maekawa H, Ohta K, and Tominaga K. 2004. Spectral diffusion of the anti-symmetric stretching mode of azide ion in a reverse micelle studied by infrared three-pulse photon echo method. *Phys Chem Chem Phys* 6(16):4074–4077.
41. Maekawa H, Ohta K, and Tominaga K. 2004. Vibrational dynamics of the OH stretching mode of water in reverse micelles studied by infrared nonlinear spectroscopy. *Mater Res Soc Symp P* 790:73–83.
42. Maekawa H, Ohta K, and Tominaga K. 2005. Vibrational dynamics in liquids studied by non-linear infrared spectroscopy. *Res Chem Intermediat* 31(7–8):703–716.
43. Ohta K and Tominaga K. 2007. Vibrational population relaxation of hydrogen-bonded phenol complexes in solution: Investigation by ultrafast infrared pump-probe spectroscopy. *Chem Phys* 341(1–3):310–319.
44. Mukamel S. 1995. *Principles of Nonlinear Optical Spectroscopy* (Oxford University, New York).
45. Tokmakoff A. 1996. Orientational correlation functions and polarization selectivity for nonlinear spectroscopy of isotropic media.1. Third order. *J Chem Phys* 105(1):1–12.
46. Lim M and Hochstrasser RM. 2001. Unusual vibrational dynamics of the acetic acid dimer. *J Chem Phys* 115(16):7629–7643.
47. Fourkas JT, Kawashima H, and Nelson KA. 1995. Theory of nonlinear-optical experiments with harmonic-oscillators. *J Chem Phys* 103(11):4393–4407.

48. Tokmakoff A and Fayer MD. 1995. Homogeneous vibrational dynamics and inhomogeneous broadening in glass-forming liquids—Infrared photon-echo experiments from room-temperature to 10 K. *J Chem Phys* 103(8):2810–2826.
49. Rector KD and Fayer MD. 1998. Vibrational dephasing mechanisms in liquids and glasses: Vibrational echo experiments. *J Chem Phys* 108(5):1794–1803.
50. Tokmakoff A, Sauter B, Kwok AS, and Fayer MD. 1994. Phonon-induced scattering between vibrations and multiphoton vibrational up-pumping in liquid solution. *Chem Phys Lett* 221(5–6):412–418.
51. Cho MH et al. 1996. The integrated photon echo and solvation dynamics. *J Phys Chem USA* 100(29):11944–11953.
52. deBoeij WP, Pshenichnikov MS, and Wiersma DA. 1996. On the relation between the echo-peak shift and Brownian-oscillator correlation function. *Chem Phys Lett* 253(1–2):53–60.
53. Fleming GR and Cho MH. 1996. Chromophore-solvent dynamics. *Annu Rev Phys Chem* 47:109–134.
54. Tayama J, Ohta K, and Tominaga K. 2012. Vibrational transition frequency fluctuation of the NO stretching mode of sodium nitroprusside in aqueous solutions. *Chem Lett* 41(4):366–368.
55. Turton DA, Hunger J, Hefter G, Buchner R, and Wynne K. 2008. Glasslike behavior in aqueous electrolyte solutions. *J Chem Phys* 128(16):161102.
56. Park S and Fayer MD. 2007. Hydrogen bond dynamics in aqueous NaBr solutions. *Proc Natl Acad Sci USA* 104(43):16731–16738.
57. Ishikawa H et al. 2007. Neuroglobin dynamics observed with ultrafast 2D IR vibrational echo spectroscopy. *Proc Natl Acad Sci USA* 104(41):16116–16121.
58. Bonner GM, Ridley AR, Ibrahim SK, Pickett CJ, and Hunt NT. 2010. Probing the effect of the solution environment on the vibrational dynamics of an enzyme model system with ultrafast 2D IR spectroscopy. *Faraday Discuss* 145:429–442.
59. Ohtaki H and Radnai T. 1993. Structure and dynamics of hydrated ions. *Chem Rev* 93(3):1157–1204.
60. Marcus Y. 2009. Effect of ions on the structure of water: Structure making and breaking. *Chem Rev* 109(3):1346–1370.
61. Nandi N, Bhattacharyya K, and Bagchi B. 2000. Dielectric relaxation and solvation dynamics of water in complex chemical and biological systems. *Chem Rev* 100(6):2013–2045.
62. Bhattacharyya K. 2003. Solvation dynamics and proton transfer in supramolecular assemblies. *Acc Chem Res* 36(2):95–101.
63. Mandal D, Datta A, Pal SK, and Bhattacharyya K. 1998. Solvation dynamics of 4-aminophthalimide in water-in-oil microemulsion of Triton X-100 in mixed solvents. *J Phys Chem B* 102(45):9070–9073.
64. Riter RE, Willard DM, and Levinger NE. 1998. Water immobilization at surfactant interfaces in reverse micelles. *J Phys Chem B* 102(15):2705–2714.
65. Satoh T, Okuno H, Tominaga K, and Bhattacharyya K. 2004. Excitation wavelength dependence of solvation dynamics in a water pool of a reversed micelle. *Chem Lett* 33(9):1090–1091.
66. Fioretto D, Freda M, Mannaioli S, Onori G, and Santucci A. 1999. Infrared and dielectric study of Ca(AOT)(2) reverse micelles. *J Phys Chem B* 103(14):2631–2635.
67. Moilanen DE et al. 2008. Water inertial reorientation: Hydrogen bond strength and the angular potential. *Proc Natl Acad Sci USA* 105(14):5295–5300.
68. Asbury JB et al. 2003. Hydrogen bond dynamics probed with ultrafast infrared heterodyne-detected multidimensional vibrational stimulated echoes. *Phys Rev Lett* 91(23):237402.
69. Stenger J, Madsen D, Hamm P, Nibbering ETJ, and Elsaesser T. 2001. Ultrafast vibrational dephasing of liquid water. *Phys Rev Lett* 87(2):027401.
70. Asbury JB et al. 2003. Hydrogen bond breaking probed with multidimensional stimulated vibrational echo correlation spectroscopy. *J Chem Phys* 119(24):12981–12997.
71. Bakker HJ and Skinner JL. 2010. Vibrational spectroscopy as a probe of structure and dynamics in liquid water. *Chem Rev* 110(3):1498–1517.
72. Loparo JJ, Roberts ST, and Tokmakoff A. 2006. Multidimensional infrared spectroscopy of water. I. Vibrational dynamics in two-dimensional IR line shapes. *J Chem Phys* 125(19):194521.
73. Asbury JB et al. 2004. Water dynamics: Vibrational echo correlation spectroscopy and comparison to molecular dynamics simulations. *J Phys Chem A* 108(7):1107–1119.
74. Fecko CJ, Loparo JJ, Roberts ST, and Tokmakoff A. 2005. Local hydrogen bonding dynamics and collective reorganization in water: Ultrafast infrared spectroscopy of HOD/D_2O. *J Chem Phys* 122(5):054506.
75. Steinel T et al. 2004. Water dynamics: Dependence on local structure probed with vibrational echo correlation spectroscopy. *Chem Phys Lett* 386(4–6):295–300.
76. Kraemer D et al. 2008. Temperature dependence of the two-dimensional infrared spectrum of liquid H2O. *Proc Natl Acad Sci USA* 105(2):437–442.

77. Ferrario M, Klein ML, and Mcdonald IR. 1993. Dynamical behavior of the azide ion in protic solvents. *Chem Phys Lett* 213(5–6):537–540.
78. Li SZ, Schmidt JR, Piryatinski A, Lawrence CP, and Skinner JL. 2006. Vibrational spectral diffusion of azide in water. *J Phys Chem B* 110(38):18933–18938.
79. Eaves JD, Tokmakoff A, and Geissler PL. 2005. Electric field fluctuations drive vibrational dephasing in water. *J Phys Chem A* 109(42):9424–9436.
80. Park J, Ha JH, and Hochstrasser RM. 2004. Multidimensional infrared spectroscopy of the N-H bond motions in formamide. *J Chem Phys* 121(15):7281–7292.
81. Barbara, PF and Jarzeba, W. 2007. Ultrafast photochemical intramolecular charge and excited state solvation, in *Advances in Photochemistry*, Volume 15, eds. D. H. Volman, G. S. Hammond, and K. Gollnick (John Wiley & Sons, Inc., Hoboken, NJ).
82. Maroncelli M. 1993. The dynamics of solvation in polar liquids. *J Mol Liq* 57:1–37.
83. Jarzeba W, Walker GC, Johnson AE, Kahlow MA, and Barbara PF. 1988. Femtosecond microscopic solvation dynamics of aqueous-solutions. *J Phys Chem USA* 92(25):7039–7041.
84. Barbara PF, Walker GC, Kang TJ, and Jarzeba W. 1990. Ultrafast experiments on electron-transfer. *Proc Soc PhotoOpt Ins* 1209:18–31.
85. Jimenez R, Fleming GR, Kumar PV, and Maroncelli M. 1994. Femtosecond solvation dynamics of water. *Nature* 369(6480):471–473.
86. Lang MJ, Jordanides XJ, Song X, and Fleming GR. 1999. Aqueous solvation dynamics studied by photon echo spectroscopy. *J Chem Phys* 110(12):5884–5892.
87. Sajadi M, Weinberger M, Wagenknecht HA, and Ernsting NP. 2011. Polar solvation dynamics in water and methanol: Search for molecularity. *Phys Chem Chem Phys* 13(39):17768–17774.
88. Horng ML, Gardecki JA, Papazyan A, and Maroncelli M. 1995. Subpicosecond measurements of polar solvation dynamics—Coumarin-153 revisited. *J Phys Chem USA* 99(48):17311–17337.
89. Rosenthal SJ, Jimenez R, Fleming GR, Kumar PV, and Maroncelli M. 1994. Solvation dynamics in methanol—Experimental and molecular-dynamics simulation studies. *J Mol Liq* 60(1–3):25–56.
90. Ohta K, Tayama J, and Tominaga K. 2012. Ultrafast vibrational dynamics of SCN^- and N_3^- in polar solvents studied by nonlinear infrared spectroscopy. *Phys Chem Chem Phys* 14:10455–10465.
91. Li M et al. 1993. Vibrational and rotational relaxation-times of solvated molecular ions. *J Chem Phys* 98(7):5499–5507.

6 Polarization Anisotropy Effects for Degenerate Vibrational Levels

Daniel G. Kuroda and Robin M. Hochstrasser

CONTENTS

6.1 INTRODUCTION

When molecular ions in the gas phase have high-enough symmetry to support degenerate vibrational states, one can expect the effect of solvation to be particularly apparent in as much as motions of the solvent as can be expected to lower the ion symmetry and remove these degeneracies. This perturbation may give rise to splittings of the degenerate vibrational states if the average symmetry is lowered in the solvent. But the splittings may only be transient while leaving the average structure with apparently degenerate states. These possibilities must be considered on an individual basis.

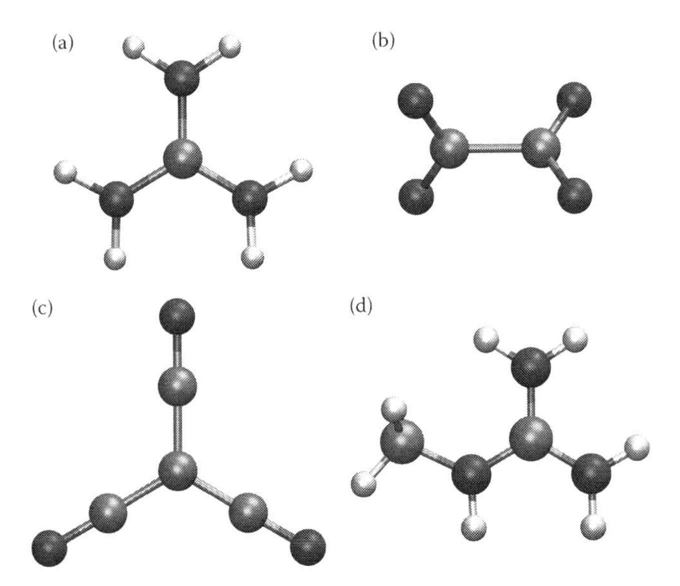

FIGURE 6.1 Symmetric ions: (a) guanidinium cation, (b) tricyanomethanide anion (TCM), (c) oxalate dianion, and (d) methyl guanidinium cation.

In either case, it is not obvious what is the composition of the nearly degenerate normal modes of the molecule. The splitting or lack of splitting, and the properties of the resulting eigenmodes can provide direct information of the solvent shell structure and dynamics that may not be so distinctive in similar experiments on lower symmetry ions. Thus, this chapter focuses on what are the important parameters of the structural and hydration dynamics of highly symmetric molecular ions that pertain to their vibrational dynamics and structure. We describe the special properties of highly symmetric ions and how their interactions with water are seen by linear, vibrational pump-probe, and echo infrared (IR) methods applied to their degenerate modes.

Focusing on small molecular ions that have degenerate states, there have been many theoretical studies on tetratomic ions typified by NO_3^- and CO_3^- that have predicted vibrational spectra and solvent-induced modifications of the symmetry [1–3]. At the time of preparation of this chapter, there are no reports on the vibrational dynamics of these ions as would be measured by two-dimensional infrared (2D IR) although there are x-ray [4], neutron diffraction [5], and Raman studies [6] that indicate symmetry breaking of these ions in water. However, there are nonlinear 2D IR experiments that provide interesting new views of the dynamics of near degenerate states of some relatively small molecular ions (Figure 6.1), such as guanidinium cation [7,8], $C(NH_2)_3^+$, and its perturbed variants methyl guanidinium cation and arginine [9], the hexatomic oxalate double-negative anion [10], $[O_2C\text{–}CO_2]^{2-}$, and tricyanomethanide anion [11], $C(CN)_3^-$. Although there are extensive experimental [12,13] and theoretical [14,15] studies of the highly symmetric triatomic azide ion in water, no dynamics experiments on its degenerate bending mode have been reported so far.

6.2 DEGENERATE VIBRATIONAL MODES

The usual way of describing degenerate vibrational modes of molecules involves recognizing that if the force constant matrix has two identical eigenvalues, the part of the harmonic vibrational Hamiltonian (in units of the zero-point energy of the mode) that describes them will have the form

$$\begin{aligned} H &= H_a + H_b \\ &= -\left(\frac{\partial^2}{\partial y_a^2} + \frac{\partial^2}{\partial y_b^2} \right) + y_a^2 + y_b^2 \end{aligned} \tag{6.1}$$

where the y's are dimensionless normal mode displacements. The eigenstates of H_a and H_b on the first line are the usual harmonic oscillators that have the same frequencies, so that the two oscillator systems consist of products of occupations of those modes, which will be referred to as $|V_aV_b\rangle$. However, the full Schrodinger equation for the 2D harmonic oscillator on the second line of Equation 6.1 can be solved (such as by conversion to polar coordinates, illustrating that the axes a and b are not unique) and it has solutions that are eigenfunctions of the vibrational angular momentum, which will be referred to as $|V,l\rangle$, where V is the vibrational quantum number and $l\hbar$ is the associated vibrational angular momentum, with $l = \pm V, \pm(V-2), \ldots, \pm 1$, or 0. We could add another kinetic energy $\partial^2/\partial y_c^2$ and coordinate y_c to Equation 6.1 and thereby describe a three-dimensional oscillator. In a realistic situation, these levels are coupled to the overall rotational states through the Coriolis interaction. In a full calculation that considers the total angular momentum, all the degenerate modes would need to be brought in. For the present discussion, we only consider the properties of the $V = 0$, $V = 1$, and $V = 2$ levels of one of the degenerate modes in a group of molecules. Then, there are very simple relationships between the two types of solutions $|V,l\rangle$ and $|V_aV_b\rangle$, and we give the relevant ones here:

$$
\begin{aligned}
|0,0\rangle &= |00\rangle \\
|1,\pm 1\rangle &= \frac{-1}{\sqrt{2}}\left(|10\rangle \pm i|01\rangle\right) \\
|2,0\rangle &= \frac{-1}{\sqrt{2}}\left(|20\rangle + |02\rangle\right) \\
|2,\pm 2\rangle &= \frac{1}{\sqrt{2}}\left\{\frac{1}{\sqrt{2}}\left(|20\rangle - |02\rangle\right) \pm i|11\rangle\right\}
\end{aligned}
\tag{6.2}
$$

This formulation of the eigenvectors makes it obvious that the angular momentum in the $V = 2$ states is quenched by any significant anharmonic coupling in a nonrotating molecule and so they do not need to be considered explicitly. For example, the diagonal anharmonicity of the local mode, which is known in all relevant cases to be significant, is taken into account by an effective Hamiltonian of the type:

$$
H_A = -\Delta/2\left\{|10\rangle\langle 10| + |01\rangle\langle 01|\right\} - 3\Delta/2\left\{|20\rangle\langle 20| + |02\rangle\langle 02|\right\} - \delta|11\rangle\langle 11| \tag{6.3}
$$

This operator couples $|2, +2\rangle$ with $|2, -2\rangle$ and shifts the antisymmetric linear combination of these angular momentum eigenstates to a lower frequency by Δ. This antisymmetric combination remains degenerate at $-\Delta$ with the state $|2,0\rangle$. These new $V = 2$ states are real and have no preference for right or left circularly polarized light in transitions with nondegenerate states. So it appears that other routes would be needed for such experiments in the dipole approximation, as initially concluded in our earlier work on polarization effects in four wave experiments [16]. The degenerate pair at $-\Delta$ may be equally well described as the pair $|20\rangle,|02\rangle$, which cannot be split by anharmonic terms in the Hamiltonian, which turns out to be the most convenient and commonly used basis in simple modeling of the 2D harmonic oscillator applications, including interpretations of the 2D IR spectra of slightly anharmonic oscillators [7,9,10,17,18]. However, the anharmonic coupling H_A does not quench the angular momentum in the $V = 1$ states that in a small magnetic field can form the states $|1, \pm 1\rangle$. This implies that the $V = 1$ transitions are individually excitable by left and right circularly polarized light competing with the relaxation of the angular momentum from the coupling to overall rotation of the molecule [19]. In the following discussion, the coherences in the superposition states of $|V_i,V_j\rangle$ and $|V_k,V_l\rangle$ will appear in the signals and they are labeled as $\rho_{ij,kl}$.

6.3 POLARIZATION PROPERTIES OF THE PUMP-PROBE SPECTROSCOPY

In the absence of external fields, the real degenerate vibrational states $|a\rangle$ and $|s\rangle$ of the $V = 1$ level might in principle be chosen as any orthogonal linear combinations of $|10\rangle$ and $|01\rangle$:

$$\begin{aligned} |s\rangle &= \cos\theta|10\rangle + \sin\theta|01\rangle \\ |a\rangle &= -\sin\theta|10\rangle + \cos\theta|01\rangle \end{aligned} \tag{6.4}$$

where θ is any angle. The transition dipoles $\langle 00|\vec{\mu}|10\rangle$ and $\langle 00|\vec{\mu}|01\rangle$ must be perpendicular to form a basis for the twofold degenerate representation of the mode so that the transition dipoles $\langle 00|\vec{\mu}|s\rangle$ and $\langle 00|\vec{\mu}|a\rangle$ are also perpendicular. This raises a subtle difference between degenerate states and exciton states with small coupling. The symmetric and antisymmetric exciton states of a dimer have transition dipoles that are perpendicular but generally they have different dipole lengths because the site transition dipoles are not perpendicular. In that case, the limiting anisotropy (see following discussion) would not be 0.1, as shown below, but would exhibit a dependence on the angle between the site dipoles with the value $r = 0.1$ being a special case of the angle being $\pi/2$ [10].

In the usual way of representing Liouville pathways associated with 2D IR and other 2D spectroscopies, each pathway involves four electric dipole transition matrix elements. Thus, in the case of real transition dipoles, each pathway is characterized by a product of four cosines corresponding to the projections of the transition dipoles onto the laboratory axes chosen as the directions of the polarization of the driving fields. As in our earlier paper on polarization of 2D IR signals [20], we use $a,b,\ldots$ to represent the laboratory axes so that the most general path has an orientational factor $\langle i_a(0)j_b(0)k_c(T)l_d(T)\rangle \equiv \langle i_a j_b k_c l_d\rangle$, where the parentheses indicate averaging over an isotropic distribution of system transition dipoles, and a spectral shape factor for the pathway $S_{ijkl}(\omega)$. The signals are determined by summing over all possible pathways. In this notation, $(ijkl)$ are the directions of the dipoles of the time-ordered transitions responsible for attaining the next step in the pathway, so this set is the description of the pathway needed to compute optically anisotropic properties. While the $(ijkl)$ sequence is time ordered, the four pulse centers could have any time ordering, though we will assume i and j to be driven by the first pulse in a pump-probe experiment [21].

The pump-probe signals of molecules with degenerate states are very sensitive to the anharmonic coupling, which determines the distribution of states at the $V = 2$ level. Normally, the signal at the bleach and stimulated emission frequency is described by a four-level system as shown in Figure 6.2, where the upper state near $2\omega_0$ is the combination mode $|11\rangle$ at $2\omega_0 - \delta$, where ω_0 is the fundamental frequency. The overtone levels $|02\rangle$ and $|20\rangle$ are generally shifted significantly from twice the vibrational frequency by the diagonal anharmonicity Δ. This result is very evident from the 2D IR spectra of the various systems that have been studied such as amide-I modes [22], nitriles [23], and Gdm+ degenerate modes [7]. As we have discussed previously, the anisotropy found for this four-level system is often close to 0.4 [7]. The expected result is found by adding the pathways for stimulated emission, coherence, and bleaching neglecting any interstate dynamics and assuming that all the transition dipoles are given as the values for harmonic oscillators, in the following manner:

$$r(\omega,T) = \frac{\sum_{\text{pathways}} \left[\langle i_a j_a k_a l_a\rangle - \langle i_a j_a k_b l_b\rangle\right] S_{ijkl}(\omega) e^{-T/T_1}}{\sum_{\text{pathways}} \left[\langle i_a j_a k_a l_a\rangle + 2\langle i_a j_a k_b l_b\rangle\right] S_{ijkl}(\omega) e^{-T/T_1}} \tag{6.5}$$

where $a \perp b$, the molecular indices are the transition dipole directions from the ground state to mode 1 ($|10\rangle$) or mode 2 ($|01\rangle$) of the degenerate pair, and $S_{ijkl}(\omega)$ is the spectral shape factor for the pathway. The angle brackets involve an average over an isotropic angular distribution of molecular axes.

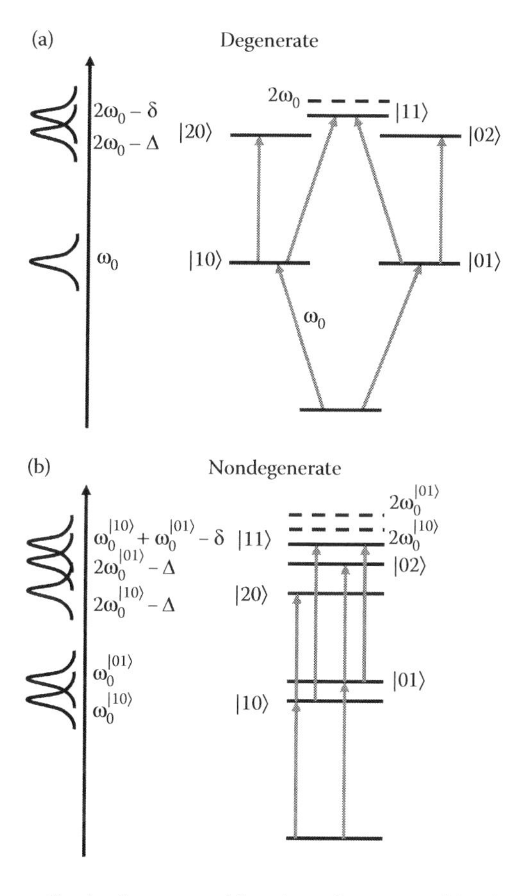

FIGURE 6.2 Energy diagram for the degenerate (a) and nondegenerate (b) pair of normal modes.

The vibrational relaxation is T_1, considered to be the same for both one-quantum levels involved and will be canceled in all expressions that follow. The diagrams are written out explicitly in the following:

$$r(\omega) = \frac{\left(F_{\parallel} - F_{\perp}\right)S_F\left(\omega_0 - \omega\right) + \left(C_{\parallel} - C_{\perp}\right)S_C\left(\omega_0 - \delta - \omega\right)}{\left(F_{\parallel} + 2F_{\perp}\right)S_F\left(\omega_0 - \omega\right) + \left(C_{\parallel} + 2C_{\perp}\right)S_C\left(\omega_0 - \delta - \omega\right)} \tag{6.6}$$

where $S_F(\omega_0 - \omega)$ and $S_C(\omega_0 - \delta - \omega)$ are the spectral shape factors (see Section 6.4 for more detail on these factors) for the fundamental transitions and the transitions from the fundamentals to the combination modes, which peak respectively at $\omega = \omega_0$ and $\omega = \omega_0 - \delta$, where δ is the off-diagonal anharmonicity. In effect, F means that the detected coherences involve the $|00\rangle$ and mode 1 ($|01\rangle$) or mode 2 ($|10\rangle$), whereas C means these coherences involve the one-quantum modes with the combination mode $|11\rangle$. The orientational parts of the signals for F and C, written out for each of the signed Liouville pathways involved in the pump-probe signals, are given by

$$\begin{aligned} F_{\parallel} - F_{\perp} &= 2\left[\left\langle 1_a1_a1_a1_a\right\rangle - \left\langle 1_a1_a1_b1_b\right\rangle\right] + \left\langle 1_a2_a1_a2_a\right\rangle - \left\langle 1_a2_a1_b2_b\right\rangle + \left\langle 1_a1_a2_a2_a\right\rangle - \left\langle 1_a1_a2_b2_b\right\rangle \\ C_{\parallel} - C_{\perp} &= -\left[\left\langle 1_a2_a1_a2_a\right\rangle - \left\langle 1_a2_a1_b2_b\right\rangle\right] - \left[\left\langle 1_a1_a2_a2_a\right\rangle - \left\langle 1_a1_a2_b2_b\right\rangle\right] \end{aligned} \tag{6.7}$$

The orientational averages can be carried in a standard manner [20] assuming fixed orientations to yield

$$F_{\parallel} - F_{\perp} = \frac{3}{10}; \quad F_{\parallel} + 2F_{\perp} = 1$$
$$C_{\parallel} - C_{\perp} = -\frac{1}{30}; \quad C_{\parallel} + 2C_{\perp} = -\frac{1}{3} \tag{6.8}$$

In the absence of anharmonic mixing of the two modes that would cause a shift of the combination state $|11\rangle$ away from the frequency $2\omega_0$, and under the assumption that the frequency–frequency correlation functions and T_1 relaxations are the same for all diagrams involving the fundamentals, as frequently appears to be approximately the case, we shall see that the anisotropy must be close to 0.4 for systems with molecular axes that are not rotating. It should be noted that it is assumed for now that the transition dipole from a $V = 1$ state to the combination state is equal to that of the fundamental, so that the transition dipoles cancel in the numerator and denominator of the anisotropy expressions. It is also assumed that there is no significant variation of transition dipole with frequency in the inhomogeneous distribution of frequencies. Under these conditions, the anisotropy is given by

$$r(\omega) = \frac{0.9S_F(\omega_0 - \omega) - 0.1S_C(\omega_0 - \delta - \omega)}{3S_F(\omega_0 - \omega) - S_C(\omega_0 - \delta - \omega)} \tag{6.9}$$

As an example, if both the C and F terms had Gaussian profiles, we would have

$$S_F(\omega_0 - \omega) = 1/\sigma_F\sqrt{2}e^{-(\omega_F-\omega)^2/2\sigma_F^2}; \quad S_C(\omega_0 - \delta - \omega) = 1/\sigma_c\sqrt{2}e^{-(\omega_F-\delta-\omega)^2/2\sigma_C^2} \tag{6.10}$$

Then one can see directly (Figure 6.3) that the anisotropy of the separated combination band would be 0.1 while that of the separated F band would be 0.3. If the spectral factors are equal, the anisotropy is fixed at 0.4. This is the condition where the off-diagonal anharmonicity δ can be neglected and when the dynamical parameters composing the linewidths are very similar for the two types of transition. However, the experimental line shapes are not generally Gaussian, as is shown later, but the principles are nevertheless the same.

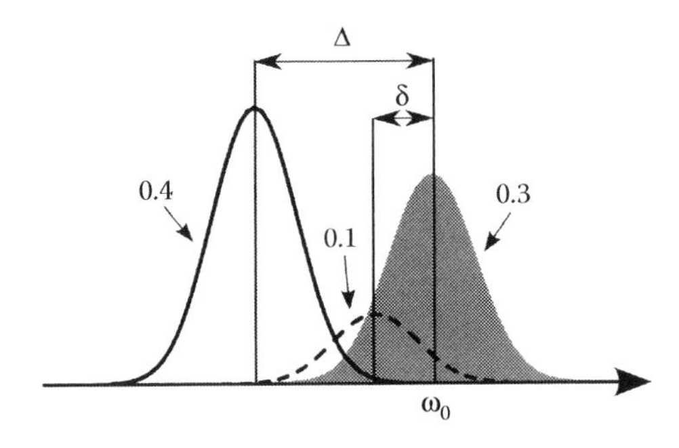

FIGURE 6.3 Spectral components of the anisotropy at $T = 0$ of transitions to a twofold degenerate level. The signal corresponding to the bleach and the stimulated absorption (gray solid band) are separated from the new absorption signal (black line band) by the anharmonicity Δ and from the combination mode signal (dashed line band) by the off-diagonal anharmonicity δ. The numbers correspond to the anisotropy values observed for each of the signals at $T = 0$ and calculated according to the text.

Clearly, the anisotropy of a transition to a degenerate vibrational level of a set of fixed molecules in space may vary significantly with frequency and may be found anywhere between 0.1 and 0.4 depending on the anharmonicity and the observation frequency. When the observation is made at $\omega = \omega_0$, the anisotropy decreases smoothly from 0.4 to 0.3 as δ increases to exceed the linewidth of the transitions. These results were discussed and used to fit the data from numerical simulations in our paper on Gdm+ [7] in which the actual form of the correlation function and measured dynamical parameters were used in the computation of the spectral factors and their contribution to the anisotropy for all the Liouville path diagrams. As is seen, there is really no specific value expected for the pump-probe anisotropy of degenerate vibrational states such as 0.4, the value found for single oscillators. Furthermore, we find no values exceeding 0.4 as had previously been observed for the fluorescence from degenerate electronic transitions [24] and discussed by Jonas et al. [25,26].

There are only a few experimental examples of pump-probe or 2D IR vibrational anisotropy of sufficiently high-symmetry systems where doubly degenerate states are examined. Vibrational pump-probe anisotropy measurements include those on the triply degenerate states of high-symmetry molecules such as ferric and ferro-hexacyanide [27] and tungsten hexacarbonyl [28–30], which have been shown to have initial anisotropies of approximately 0.4 and 0.25–0.4, respectively. Also, the pump-probe and vibrational photon echo anisotropy of the degenerate CN stretching mode of Gdm+, was just slightly <0.4 (or significantly more than 0.3 depending on your point of view) as expected from the model given above if δ is small. The conditions for Gdm+ were that the 1/e width of the degenerate transition is 11 cm^{-1} while the off-diagonal anharmonicity is 1–2 cm^{-1}. We recently completed measurements of the pump-probe anisotropy and photon echo tensors of another threefold symmetric nitrile, $C(CN)_3^-$ [11], which has a degenerate CN stretching mode in the 2200 cm^{-1} range. In this example, the early time anisotropy is 0.29 in the bleach region, which is in agreement with the concepts of the foregoing model with off-diagonal anharmonicity comparable with half the linewidth. Another recent example of a small molecular ion concerned oxalate that has D_{2d} symmetry and a degenerate CO_2^- stretching mode in the 1600 cm^{-1} range. Here, the anisotropy was not measurably different from 0.4, again in agreement with the model described above for the case of negligible anharmonic coupling between the modes. In this example, discussed in more detail below, the coupling between the two CO_2^- groups must vanish in D_{2d} symmetry, which can explain why the off-diagonal anharmonicity is small.

The formulas given above only have utility at very near-zero pump-probe delay ($T = 0$) because in reality, the components of degenerate states may exchange population very rapidly, especially in fast hydrogen-bonding solvents such as water or D_2O, where the forces exerted on the modes are large enough to significantly mix them. The vibrational mode mixing causes major ultrafast changes in the anisotropy, which then requires a completely different model from that given above except, as stated, at nearly zero delay time between the pump and the probe. Also, the model does not include contributions from diagrams that would arise if the pulses were actually overlapping in time such as $\langle 1_a(T)2_b(0)1_b(0)2_a(T)\rangle$. The overall motion also degrades the anisotropy with increasing delay time. There could be circumstances where the spectra have lines that are sharp enough compared with the mixed-mode anharmonic shift, δ, that the bleach and combination band signals occur in different spectral regions. In that event, the anisotropy in the bleaching region would be 3/10 as shown above and rather more similar to expectations recently emphasized for electronic transitions [26]. However, this is a less likely situation in the condensed phase where vibrational linewidths are often larger than or comparable with the off-diagonal anharmonicity. In the next section, we consider the changes that occur when interstate dynamics and exchange can occur. When dealing with degenerate states, this should be the expected situation for modes involving groups that are sensitive to the locations and movements of solvent molecules that can then couple the modes on vibrational motion timescales. In that case, the anisotropy becomes time dependent and it is more challenging to find a general solution to the transition dipole reorientation dynamics of states that are on average nearly degenerate.

6.4 GENERAL STATEMENTS REGARDING THE LONG TIME ANISOTROPY

We have seen that the initial values of the anisotropy of degenerate states of threefold symmetric systems are dependent on the anharmonicity and generally would lie close to 0.4 but not <0.3. A more general statement can be made about the limiting anisotropy for the equilibrium distribution of symmetric molecules fixed in space. As T increases, so does the probability that the pulse created population, say $\rho_{10,10}$ will transfer to $\rho_{01,01}$ until finally these two populations will equilibrate. Similarly, we could assume that the created coherence $\rho_{10,01}$will equalize with $\rho_{01,10}$. Thus, with the same caveats as described above, the factors of Equation 6.8 at significantly large T would become

$$\begin{aligned} F_{\parallel} - F_{\perp} &= \frac{1}{10}; \quad F_{\parallel} + 2F_{\perp} = 1 \\ C_{\parallel} - C_{\perp} &= -\frac{1}{30}; \quad C_{\parallel} + 2C_{\perp} = -\frac{1}{3} \end{aligned} \tag{6.11}$$

It is notable that the C terms do not change as the quasi-equilibrium state is achieved: under the foregoing assumptions, the angular averages have the same value and the number of pathways remains the same whether equilibration occurred or not, assuming that the transition dipoles for $|01\rangle \rightarrow |11\rangle$ and $|10\rangle \rightarrow |11\rangle$ equal those for $|00\rangle \rightarrow |01\rangle$ and $|00\rangle \rightarrow |10\rangle$. Following from Equation 6.11 is the prediction from Equation 6.5 that the anisotropy has the limiting value $r(\infty) = 0.1$, *which is now independent of the anharmonicity*. This is analogous to a well-known result in the fluorescence spectroscopy of circular absorbers when the excitations become equally distributed between the two emitters [31]. The overall conclusion is that the pump-probe anisotropy of degenerate vibrational transitions is expected to drop from a value near 0.4, but greater than 0.3, to a value of 0.1 as the waiting time increases. Note that this is an idealized result based on the stated assumptions.

6.5 RELATIONS BETWEEN PUMP-PROBE AND 2D IR ANISOTROPY OF DEGENERATE LEVELS

Pump-probe experiments in which the first two field interactions both arise from delta pump pulses involve a coherence time of zero. On the contrary, 2D IR signals do not have this time constraint because the first two interactions are scanned in time (usually labeled τ) and the detection time, no longer dictated by the time response of a detector, is instead scanned and heterodyne-detected at each value of t. The 2D IR spectrum plots the coherence frequency ω_τ versus the detection frequency ω_t. The polarization possibilities are substantially increased compared with pump probe because now the four lab polarizations in the term $\langle i_a j_b k_c l_d \rangle$ in the signal can be chosen independently [20]. Furthermore, the terms involving both component states of the near degeneracy are spectrally separated into cross peaks between the diagonal signals from pathways involving only one of the modes. These factors influence not only the methods of polarization analysis but also the spectral shapes. As a result, the 2D IR spectra of nearly degenerate levels have unique time-dependent shapes that are formed as the energy transfer between the component states takes place.

6.6 WAITING TIME DEPENDENCE OF THE ANISOTROPY

The waiting time dependence, introduced in some of the above expressions, has been dealt with by many previous investigators. Hochstrasser presented a useful table of angular coefficients required for each of the Liouville paths in which the relevant matrix of $(i_a j_b k_c l_d)$ is given with the choice of *ijkl* as the columns and *abcd* as the rows [20]. The anisotropic character of many processes involving vibrational dynamics can be expressed in terms of these tensor elements. The spectral signal for pump-probe experiments can be written as a T-dependent orientational part times a spectral factor such as we have assumed in the above. The interaction with solvent molecules (water in the present example) causes transitions to occur between the two component states of the degenerate level. The specific mechanisms of such effects are not obvious *a priori*. If interactions such as hydrogen bonding occur, the solvent, water or D_2O molecules, exerts forces that cause the molecule to deviate from its threefold symmetry so that the two degenerate modes will be coupled and energy will flow from one to the other as described above. In the case of Gdm+, oxalate, and the recently studied $C(CN)_3^-$, the anisotropy drops to the limiting value 0.1 on a time subpicosecond scale. This energy flow can be represented by a kinetic rate process occurring during T. The polarization of the pulses creates the initial conditions for the resulting kinetics. The molecules responsible for the signal contribution represented by $\langle 1_a 1_a 2_b(T) 2_b(T)\rangle$ might be in dynamic equilibrium with $\langle 1_a 1_a 1_b(T) 1_b(T)\rangle$ during the waiting time. Thus, although the first two pulses created a population of one component of the degenerate level, labeled 1, the system may have transferred to 2 before interacting with the third pulse to generate the signal. Moreover, each molecule that is initiated in a state 1 will only follow the signals described above if it is found to be in state 1 at the time of the interaction with the third pulse. Therefore, it is seen that the signal becomes spread over many more pathways as a result of the energy transfer. The pathways of the signals are written out explicitly in Figure 6.4 (note the different notation for the states used in this figure) with the symbolism that a horizontal dashed line implies that an energy or coherence transferred state is detected by the probe pulse. The symbols P_{11} and P_{12} underneath the diagram implies that the diagram contributes to the signal only if after starting in state 1 it is found in state 1 when probed (P_{11}) or after starting in state 1 it is found in state 2 when probed (P_{12}). The coherences between the component states of the degenerate level also equilibrate during the waiting time. Similarly, a coherence created by the second pulse may also transform to its conjugate during T. These diagrams are extremely helpful in quickly recalling which pathways contribute to the pump-probe or 2D IR signals (see, e.g., Ref. [21] where these diagrams are described in detail). The coherences and populations are read horizontally at each time step moving upwards in the diagram. For example, in pathway C_1 (Figure 6.4), the system goes through the sequence $\rho_{01} \xrightarrow{\tau} \rho_{21} \rightarrow \xrightarrow{T} \rho_{1+2,1} \xrightarrow{t} \omega_{1+2,1}$, where the last coherence indicates the emission frequency of $\omega_0 - \delta$. Considering all the diagrams, the emitting frequency could be $\omega_{10} = \omega_0 = \omega_{20}$, $\omega_{1+2,1} = \omega_0 - \delta$, or $\omega_{1+1,1} = \omega_0 - \Delta$. The pump-probe signals have no delay between the first two pulses ($\tau = 0$).

The Redfield theory presents one approach to the two-state vibrational coherence dynamics (see, e.g., Ref. [32]). For nearly degenerate vibrational modes, the Redfield matrix elements coupling the populations and coherences created by the interaction of the second pulse are zero. When the populations and coherences can be treated independently, the separate master equations are

$$\frac{\mathrm{d}}{\mathrm{d}T}\begin{bmatrix}\rho_{11}\\ \rho_{22}\end{bmatrix} = \begin{bmatrix}-k_{et} & k_{et}\\ k_{et} & -k_{et}\end{bmatrix}\begin{bmatrix}\rho_{11}\\ \rho_{22}\end{bmatrix} \tag{6.12}$$

and

$$\frac{\mathrm{d}}{\mathrm{d}T}\begin{bmatrix}\rho_{12}\\ \rho_{21}\end{bmatrix} = \begin{bmatrix}-k_{et} + i\omega_{12} - \gamma & k_{et}\\ k_{et} & -k_{et} + i\omega_{12} - \gamma\end{bmatrix}\begin{bmatrix}\rho_{12}\\ \rho_{21}\end{bmatrix} \tag{6.13}$$

The energy transfer rates are exactly equal for the two degenerate modes ($\omega_{12} = 0$) and are denoted as k_{et} while γ is an empirical, smaller, coherence dephasing rate. In principle, the energy transfer could be calculated exactly from

$$k_{et} = \frac{1}{\hbar^2}\int_{-\infty}^{\infty} \mathrm{d}t\, \mathrm{e}^{i\omega_{12}t}\left\langle V_{12}(t)V_{21}(0)\right\rangle \tag{6.14}$$

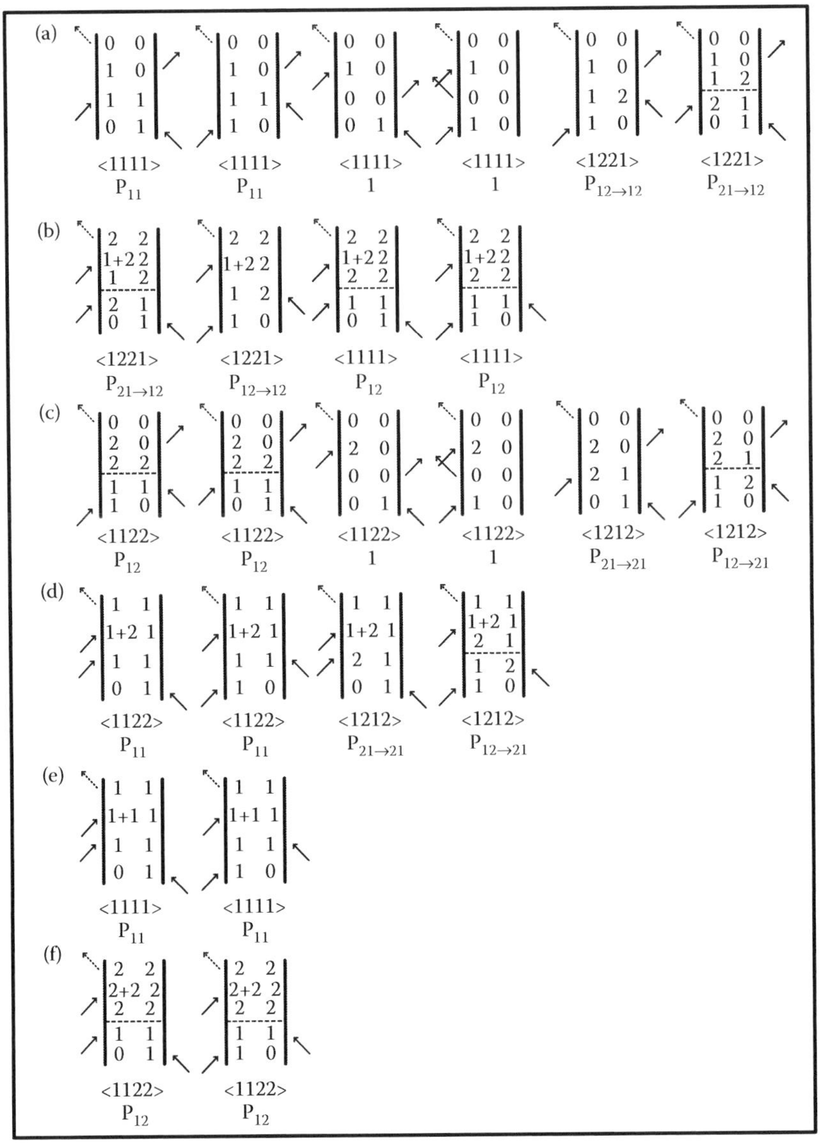

(a) <1111> P_{11} <1111> P_{11} <1111> 1 <1111> 1 <1221> $P_{12\to12}$ <1221> $P_{21\to12}$

(b) <1221> $P_{21\to12}$ <1221> $P_{12\to12}$ <1111> P_{12} <1111> P_{12}

(c) <1122> P_{12} <1122> P_{12} <1122> 1 <1122> 1 <1212> $P_{21\to21}$ <1212> $P_{12\to21}$

(d) <1122> P_{11} <1122> P_{11} <1212> $P_{21\to21}$ <1212> $P_{12\to21}$

(e) <1111> P_{11} <1111> P_{11}

(f) <1122> P_{12} <1122> P_{12}

FIGURE 6.4 Liouville pathways for two coupled oscillators for the echo signal emitted in the $-\mathbf{k}_1 + \mathbf{k}_2 + \mathbf{k}_3$ direction. Horizontal dashed lines imply a spontaneous coherence or population transfer in the waiting time period. The diagrams are arranged according to their detection frequency on the ω_t axis. (a) $\omega_t = \omega_1$, (b) $\omega_t = \omega_1 - \delta$, (c) $\omega_t = \omega_2$, (d) $\omega_t = \omega_2 - \delta$, (e) $\omega_t = \omega_1 - \Delta$, and (f) $\omega_t = \omega_2 - \Delta$, where Δ is the diagonal and δ the mixed-mode anharmonicity. In this figure, a different convention is used for the eigenstates to save space. Thus, 1, 2, 1 + 2, 1 + 1, and 2 + 2 replace $|10\rangle$, $|01\rangle$, $|11\rangle$, $|20\rangle$, and $|02\rangle$, respectively (see also Figure 6.5). (Extracted from Ghosh A, Tucker MJ, and Hochstrasser RM. 2011. *J Phys Chem A* 115(34):9731–9738.)

where the angle brackets represent a trace over the solvent (bath) coordinates and $V_{12}(t)$ is a matrix element of the system–bath interaction [33] between the two components of the degenerate level. The energy transfer rate can be thought of as a *degenerate vibrational energy relaxation*. A zero-frequency relaxation depends directly on the full variance of the forces acting on the mode. To couple the two molecular states, one must know the changes in one of the modes (say, mode 2) that are brought about by fluctuations in the force exerted on the other mode (mode 1) by the solvent.

This part of the potential corresponds to the term $\left[(\partial/\partial Q_2)(\partial V(t)/\partial Q_1)\right]_0 Q_1Q_2 \equiv \lambda_{21}(t)Q_1Q_2$ and the equal term with 1 and 2 interchanged. Thus, if $\omega_{21} \approx 0$, it is the time integral of the correlation function $\langle\lambda_{12}(t)\lambda_{21}(0)\rangle$ that will determine the transfer rate. For the simple molecules discussed here, it is possible to compute reasonable values of k_{et} from classical simulations. The present discussion is qualitative and describes a mechanism that might be considered as analogous to the quantitative treatment of the relaxation between asymmetric and symmetric stretches of azide ions [14]. In practice, for a meaningful computation of the rate k_{et}, one would have to take into account that the solvent forces act on all the coordinates of the solute, not just on the two normal modes of interest. In any event, each pathway involving the evolution of a population during T has an associated conditional probability P that energy transfer between the modes will occur during the T evolution. The probability that if a molecule is excited into a population state i at $T = 0$, it will be found in the state j at time t is written as $P_{ij}(t)$, where $i,j = 1,2$. The conditional probability factors can be obtained from Equation 6.1 and they are

$$P_{11}(t) = \left(1 + e^{-2k_{et}T}\right)/2$$
$$P_{12}(t) = \left(1 - e^{-2k_{et}T}\right)/2 \tag{6.15}$$

Diagrams that involve the ground-state bleaching during T are assumed to have a T-independent conditional probability distribution because if a molecule is in the ground state at $T = 0$, then it will still be in the ground state at all $T > 0$ unless it interacts with an electromagnetic field. Of course, the $V = 1$ populations of both states 1 and 2 are decreasing and the ground-state population is growing with time constant T_1, assumed equal for both excited states. The pathways that involve the evolution of a coherence during T also need to be considered. The density operator of the subensemble of the system in which the interaction with the second pulse has created a coherence can be calculated from Equation 6.13 as

$$\rho^{(12)}(T) = C_{12\to12}(T)|1\rangle\langle2| + C_{12\to21}(T)|2\rangle\langle1| \tag{6.16}$$

where the coefficients are given by

$$C_{12\to12}(T) = e^{-(k_{et}+\gamma)T}\left[\cos(\Omega T) - \frac{i\omega_{12}}{\Omega}\sin(\Omega T)\right] = P_{12\to12} - i\frac{\omega_{12}}{k_{et}}P_{12\to21} \tag{6.17}$$

$$C_{12\to21}(T) = P_{12\to21} \tag{6.18}$$

with $\Omega = \sqrt{\omega_{12}^2 - k_{et}^2}$. Given that the energy transfer rates are measured to be in the range of 0.5 ps [7,10,11], we see that these coefficients will become oscillatory only when the splitting of the degeneracy exceeds ca. 10 cm^{-1}. The coherence diagrams used to calculate the anisotropy are also given in Figure 6.4.

6.7 EFFECT OF THE TRANSIENT ABSORPTION

So far we considered only the signal in the frequency range of the overlapping bleach, stimulated emission, and the combination band. Of course there is also a signal at $\omega_0 - \Delta$. The anisotropy in this case is determined by only one Liouville pathway having relative orientational factor $S_D(\omega_0 - \omega - \Delta)$ $\langle 1_a 1_a 1_b 1_b\rangle$ so that its anisotropy is $(1/5 - 1/15)(1/5 + 2/15) = 0.4$. When the diagonal anharmonicity is sufficiently large, the transient absorption can be treated separately but the anisotropy becomes more complex to interpret when linewidths become comparable with the diagonal anharmonicity. Then, Equation 6.3 can be replaced with

$$r(\omega) = \frac{0.9S_F(\omega_0 - \omega) - 0.1S_C(\omega_0 - \delta - \omega) - 0.4S_D(\omega_0 - \Delta - \omega)}{3S_F(\omega_0 - \omega) - S_C(\omega_0 - \delta - \omega) - S_D(\omega_0 - \Delta - \omega)} \tag{6.19}$$

Note that the anisotropy is 0.4 when the anharmonicity vanishes completely in comparison with the linewidth. This situation would also prevail in a broad band pump-broadband probe experiment without any frequency selection of the probe beam if the pulses had sufficient bandwidth to accommodate all the transitions. In the presence of energy equilibration between the two vibrational components, the limiting anisotropy (large values of T) of the transient absorption component is again 0.1.

It is of some interest that the transient absorption anisotropy is independent of whether the intermediate states are $|1, \pm 1\rangle$ or $|10\rangle$, $|01\rangle$, so this signal does not appear to be useful as a probe of the vibrational angular momentum relaxation.

6.8 TRIPLY DEGENERATE STATE ANISOTROPY

As mentioned earlier, there have been some experiments on the anisotropy of triply degenerate states [28–30]. By means of the methods just outlined, it is straightforward to predict a model result for the anisotropy of threefold degenerate vibrations and the effects of anharmonicity in solution-phase IR experiments. For this application, the pathway diagrams on Figure 6.4 should be extended by adding one pathway that includes a state 3 for each of those containing state 2 (i.e., where the index 3 replaces the index 2), while the remainder stays the same. Here, we defined the states 1, 2, and 3 as $|100\rangle$, $|010\rangle$, and $|001\rangle$. In the two-quantum regime, we have the six states composed of three combinations $|110\rangle$, and $|101\rangle$ and $|011\rangle$ three overtone states $|200\rangle$, $|020\rangle$, and $|002\rangle$. The transition dipoles from 0 to 1, 2, and 3 are perpendicular. In this way, all pathways starting with an interaction with state 1 are included. Those starting with state 2 or 3 would just give the same result. The anisotropy is found to be

$$r_3(\omega) = \frac{S_F(\omega_0 - \omega) - 0.2S_C(\omega_0 - \delta - \omega) + 0.8S_D(\omega_0 - \Delta - \omega)}{4S_F(\omega_0 - \omega) - 2S_C(\omega_0 - \delta - \omega) + 2S_D(\omega_0 - \Delta - \omega)} \tag{6.20}$$

It is easily seen from this equation that the anisotropy of the bleach plus stimulated emission is 0.25, but becomes 0.4 when the off-diagonal anharmonicity becomes very small in which case S_F and S_C are both fully included. The combination mode signal has an anisotropy of 0.1, similar to the twofold degenerate case (Equation 6.19). The new absorption also has an anisotropy of 0.4 (Equation 6.19). Therefore, the situation is analogous to that of the twofold degeneracy: the anisotropy in the bleach, stimulated emission, and combination band region might be found with any value between 0.25 and 0.4 depending on the anharmonicity. If broadband excitation was used that encompassed all the possible transitions, or if the anharmonicity was vanishingly small, we would expect an anisotropy of 0.4. In the case of a threefold degeneracy, the limiting value of the anisotropy is zero arising from the value $\frac{1}{3}(0.4 - 0.2 - 0.2)$ for three equilibrated perpendicularly polarized states. When the spectral transitions are narrow and the resolution is high enough to resolve component states, then the anisotropy should be computed for each of the symmetry species involved in the $A \rightarrow T$ and $T \rightarrow A + E + T_2$ dipole transitions.

6.9 SEPARATION OF THE ANISOTROPY AND THE SPECTRUM

To this point, we have assumed that the pump-probe signal can be written as a product of a spectrum S and a transition dipole ensemble average $\langle \ldots \rangle$. The pump-probe signal arises from the third-order responses described by Mukamel [34]. For a single absorber, there are the three responses to consider: the bleach, the stimulated emission, and the excited-state absorption. The signal is the real part of half-Fourier transforms of these responses and, apart from constants, all have the form

$$I_{aabb}(T,\omega) = \mathrm{Re}\int_0^\infty \mathrm{d}t \left\langle \hat{a}\cdot\mu_i(0)\hat{a}\cdot\mu_j(0)\hat{b}\cdot\mu_k(T)\hat{b}\cdot\mu_l(T+t)\mathrm{e}^{i\omega t+i\int_T^{T+t}\omega_{mn}(\tau)\mathrm{d}\tau - t/2T_1}\right\rangle \tag{6.21}$$

where $\omega_{mn}(t)$ is the fluctuating vibrational frequency associated with the emitting coherence. This definition leads directly to the equations we have used if the product of the dipole projections is statistically independent of the exponential term and most importantly that the detection time range of t, determined by the frequency fluctuations and the vibrational relaxation, is small enough that we can assume $T+t \approx T$ in the orientational portion. The latter condition implies that the pump-probe delay is larger than the inverse bandwidth of the vibrational transition. We have also assumed in the simple foregoing relationships that the dipole moment magnitudes are independent of time. We will present an example later for the oxalate dianion where this condition is certainly not met. However, with these assumptions and sufficiently far from $T=0$, the signal can be written as

$$I_{aabb}(T,\omega) = \left\langle i_a j_a k_b(T) l_b(T)\right\rangle S(\omega_{mn}-\omega) \tag{6.22}$$

where the spectrum can now be defined as

$$S(\omega_{nm}-\omega) = \mathrm{e}^{-T/T_1}\,\mathrm{Re}\int_0^\infty \mathrm{d}t\left\langle \mathrm{e}^{i\omega t+i\int_T^{T+t}\omega_{mn}(\tau)\mathrm{d}\tau - t/2T_1}\right\rangle \tag{6.23}$$

In many cases, the ensemble-averaged exponential can be transformed by means of a second-order cumulant expansion [34] into common analytic spectral forms that depend on the vibrational frequency–frequency correlation function [21,35].

6.10 EFFECTS OF COHERENCE TRANSFER

In addition to anharmonic coupling, energy transfer, and exchange, anisotropy might be affected by coherence transfer. This possibility has been considered some time ago [36]. Even if the pump pulse is assumed to be a delta function, the finite width of the probe can introduce the possibility that coherence transfers among the 12 coherences are composed of superpositions of one and two quantum states. The vibrational energy transfer times between degenerate states is a few hundred femtoseconds, so we have questioned [7] whether coherence transfer rates might also be fast enough to act during the detection time.

The essential point about coherence transfer in a pump-probe experiment on the degenerate level is that it results in some shifting of the combination band signal from $\omega_0-\delta$ to a signal at $\omega_0-\Delta$ and some doubly excited-state signal shifts from $\omega_0-\Delta$ to $\omega_0-\delta$. As a further explanation of this point, recall that the signal in the vibrational pump-probe experiment arises from the coherence created by the probe pulse. For example, a group of molecules having a population density matrix $\rho_{10,10}$ can interact with a probe pulse to create a coherence $\rho_{11,10}$, resulting in emission at frequency

$\omega_0 - \delta$. Coherence transfer from $\rho_{11,10}$ could yield a coherence $\rho_{02,01}$ that radiates at the frequency $\omega_0 - \Delta$. The anisotropy becomes modified if some of these coherence transfer steps occur quickly enough. Equation 6.4 becomes

$$\begin{aligned} C_{\parallel} - C_{\perp} &= -\frac{1}{30} + p\sqrt{2}\left[\langle 1_a 1_a 1_b 1_b\rangle - \langle 1_a 1_a 1_a 1_a\rangle\right] \\ C_{\parallel} + 2C_{\perp} &= -\frac{1}{3} + p\sqrt{2}\left[\langle 1_a 1_a 1_a 1_a\rangle + 2\langle 1_a 1_a 1_b 1_b\rangle\right] \end{aligned} \tag{6.24}$$

The use of a harmonic approximation results in the $V = 1 \rightarrow V = 2$ transition dipole being $\sqrt{2}$ times that of $V = 0 \rightarrow V = 1$. There is a net increase in the contribution of the combination band to the bleaching/stimulated emission signal, which, with some assumptions, results in a modified and relatively simple formula for $r(\omega)$:

$$r(\omega) \approx \frac{0.9S_F(\omega_0 - \omega) - 0.1(1 + 4\sqrt{2}p)S_C(\omega_0 - \delta - \omega)}{3S_F(\omega_0 - \omega) - (1 + \sqrt{2}p)S_C(\omega_0 - \delta - \omega)} \tag{6.25}$$

where p is the small "integrated probability" that a coherence transfer will occur between pathways identified in Equation 6.24 that yield the signal. The result (Equation 6.25) assumes that the rates of coherence transfer are equal for each of the permitted pathways. We will see later that if the rate of coherence transfer is in the range of few hundred femtoseconds, then the effect on the probe signal may be up to ca. 10%. Since the coherence transfer is occurring during the detection time, t, its effect is small unless it occurs faster than the free decay of the probe signal and the pulse width. Most coherence transfer diagrams are eliminated from the observed pump-probe signal by the angle average over the isotropic distribution or by the rotating wave approximation. For example, there is no coherence transfer signal from the pathways that utilize only the fundamental frequencies (the F terms) because they all vanish by orientation averaging, by energy mismatches, or by the vanishingly small transition dipoles. The net increase in the combination band signal arises from the imbalance of outgoing and incoming transfers. Thus, it is found that transfers from coherence $\rho_{11,10}$ to $\rho_{02,01}$ can occur without change in the orientational factor of the term, while transfers into $\rho_{11,10}$ can occur from $\rho_{20,10}$ introducing a $\langle 1111\rangle$ character into the combination band signal that is absent from $C_{\parallel} - C_{\perp}$ term of Equation 6.11. The p-factor has no effect on the anisotropy near ω_0 when δ becomes very large, since this condition implies that there is no effect of the combination band on the bleaching and stimulated emission signals. In other cases, there might be a more significant effect. These effects would be directly measured if a heterodyned transient grating configuration was used instead of pump probe because then the detection time t can be experimentally scanned and the anisotropy can be measured for each value of t. The quantity p in that case would be directly related to the conditional probability that if the particular diagram is contributing to the signal at time zero, it will not contribute at time t, because it will have possibly transferred its coherence at $\omega_0 - \delta$ to a new frequency $\omega_0 - \Delta$ that is out of range of the signal acquisition at the stimulated emission frequency ω_0.

As a concrete example of how coherence transfer might influence anisotropy, we consider the transient absorption signal at $\omega_0 - \Delta$. Transfers can occur across the dashed lines shown for each of the contributing pathways in Figure 6.5. Each path contributes to the signal in proportion to its number of occurrences, its amplitude from a harmonic approximation and its angular factor. We assume that the transfers on lines 2 and 3 of Figure 6.5 have coherence transfer rates of k_2 and k_3, respectively with each having exponential growths $P_i(t) = 1 - \exp[-k_i t]$, where $i = 2,3$, so that the anisotropy at t becomes

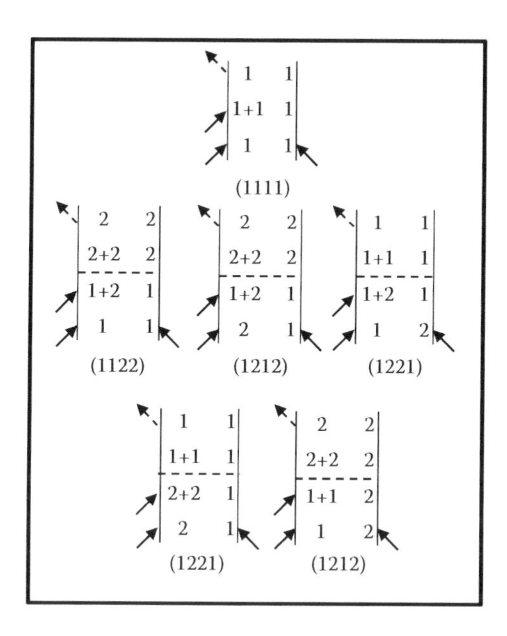

FIGURE 6.5 Liouville pathways for two coupled oscillators for the pump-probe transient absorption signal. Horizontal dashed lines imply a spontaneous coherence transfer in the t time period.

$$r(t) = \frac{4\sqrt{2} + P_2(t) + 3P_3(t)}{10\sqrt{2} + 5P_2(t)} \tag{6.26}$$

The numerator and denominator of Equation 6.26 form the anisotropic portion of signals in the t period, which are measured in 2D IR heterodyned echo or transient grating experiments. In a pump-probe experiment, which is self-heterodyned, the measurement is the t-integral of the product of the probe electric field times the convolution of the probe field with the anisotropic signal fields. Therefore, the anisotropy for delta pulse excitation depends on the coherence transfer rates as in Equation 6.26 and the probe field envelope. This effect always reduces the anisotropy as illustrated in Figure 6.6 for various pulse widths. The pathways in lines 2 and three are fourth- and

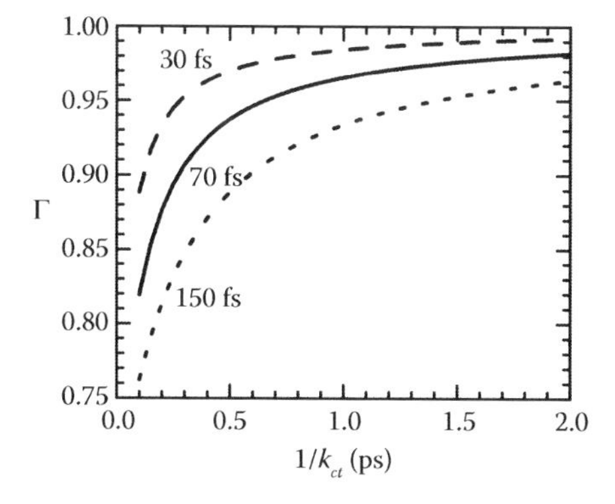

FIGURE 6.6 Effect of the coherence transfer in the pump-probe anisotropy. The coefficient Γ is calculated as the ratio of the anisotropy with and without coherence transfer and presented as a function of the coherence transfer rates, k_{ct}.

sixth order, respectively, in the expansion of the system bath interaction so at least k_3 should be extremely small.

6.11 EXAMPLES OF THE ANISOTROPY OF DEGENERATE VIBRATIONAL TRANSITIONS

In this section, we summarize results for a few examples from this laboratory on the anisotropy of high-symmetry molecules having a degenerate vibrational mode. Each example brings a somewhat different set of characteristics that cover what we expect are typical of many other systems. The common feature of all examples is that the component modes of the doubly degenerate pair are strongly coupled by interaction with the solvent, which in all cases considered here is water (or D_2O), and the effective coupling is presumably hydrogen (or deuterium) bonding. In all cases, the splitting of the degeneracy is similar to or less than the vibrational linewidth so that even in aqueous media, the ions have approximately threefold symmetry on average.

6.11.1 Guanidimium Ion

Calculations of gas-phase Gdm+ predict a D_3 symmetry ion with the C and N atoms on the plane perpendicular to a threefold symmetry axis [7,8]. The H–N–H bond angle is 120° and the plane of each of the NH_2 groups makes an angle ~12° with the CN_3 plane (Figure 6.1). The normal mode analysis for the free ion demonstrates that there is a degenerate mode consisting of mainly a CN_3 stretch and NH_2 scissors motion at a frequency near 1600 cm^{-1} that has a large dipole derivative in the plane of the CN_3 atoms. The carbon atom of Gdm+ has a positive charge that contributes to repelling water hydrogen atoms from the region immediately above and below the CN_3 plane. So, the strong hydrogen bonds in this case involve the NH_2 groups and water molecules approaching on the periphery of the ion. The pump-probe anisotropy of Gdm+ (Figure 6.7) is best fit by the multiexponential form with the fast component being caused by transfer between the two degenerate components and the slow one by overall rotational diffusion. The time separation between these processes gives structural meaning to the amplitudes. The fast process is ca. 0.4 ps whereas the rotational diffusion is typically longer than 5 ps. This difference allows extrapolation to the so-called limiting anisotropy, which as we have stated is predicted to be 0.10. In the case of Gdm+, experiments were performed for the bleaching and transient absorption regions of five different D_2O/glycerol mixtures. For these 10 measurements, the limiting anisotropy was found to be 0.1 ± 0.02. In higher-viscosity solvents such as glycerol/water mixtures, the degeneracy of the Gdm+ ion C–N transition is distinctly split into two components indicating that the molecules are distorted from 3-fold symmetry. Even in this latter case, the vibrational excitation becomes averaged over the two states on a subpicosecond timescale.

6.11.2 Oxalate Dianion

The oxalate dianion presents a particularly interesting symmetry arising from two apposed carboxylate groups [10]. The symmetry in the gas phase is D_{2d} and the ion has three degenerate modes. The symmetric and antisymmetric pairs of stretch modes of the carboxylate groups of oxalate are near 1400–1500 cm^{-1} and 1500–1700 cm^{-1}. The asymmetric stretch vibration of carboxylate is near 1575 cm^{-1}. In oxalate, the frequencies of these modes depend on whether or not the carboxylate planes are parallel. If they are parallel (dihedral angle zero, D_{2h} symmetry), then only one of the two asymmetric stretch components is IR allowed [10]. Otherwise, there are two allowed transitions that, when the dihedral angle is 90°, become the degenerate E mode of the D_{2d} symmetry. The gas-phase structure has a dihedral angle of 90° and this is the most likely structure of oxalate in water [10].

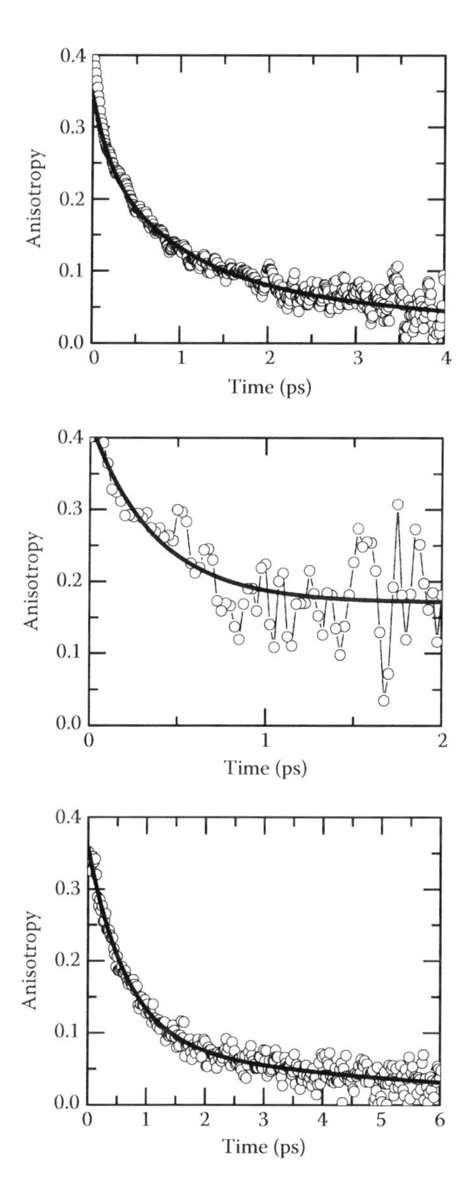

FIGURE 6.7 Experimental anisotropy of various ions. Anisotropy signal (open circles) and exponential fit (black line) for guanidinium (a), oxalate (b), and tricyanomethanide (c). The signal-to-noise in the case of oxalate is degraded by the ultrafast relaxation, 0.3 ps of the carboxylate vibration.

The hydration shells of the two carboxylates of oxalate are close enough (1.5 Å) that they must share D_2O molecules or at least each bonded solvent molecule will be sensitive to both negative ion centers. The oxalate dianion is a very special case since the degeneracy of the mode is split by changes in the carboxylate dihedral angle and it is only degenerate at the mean angle of 90°. Furthermore, the transition dipoles strongly depend on the twist angle. The carboxylate dihedral angle must adjust to near 90° to minimize the Coulombic repulsion, whereas the best fit with the water network is not necessarily at the 90° conformation. Therefore, we obtained the potential of mean force for changes in the dihedral angle. This potential is sufficiently shallow that at room temperature the root mean squared angular variation is 30°. On this basis, we explained the 2D IR spectra of oxalate in terms of transitions occurring to a pair of states undergoing significant angular fluctuations about the degenerate configuration. We will say more later about how these measurements can be interpreted by quantum dynamical modeling. In addition, oxalate, like the other symmetric ions, shows a decay of anisotropy in its pump-probe signal (Figure 6.7) on subpicosecond timescale. This ultrafast dipole reorientation dynamics is related to the population exchange between excitonic states and it has been related to fast fluctuations of the carboxylate frequencies (see Section 6.12).

6.11.3 Tricyanomethanide Ion

Tricyanomethanide ion is a threefold symmetric anion composed of three cyano groups attached to a central carbon [11] (Figure 6.1). The isolated TCM ion has three CN groups arranged in D_{3h} symmetry, which generates three IR modes in the CN stretch region. Owing to symmetry considerations, two of these three modes are degenerate. The degenerate pair corresponds to a pair of asymmetric stretch transitions (A_1 and A_2 modes) [11]. The remaining mode is a symmetric stretch (S mode) [11]. The degenerate mode is IR active and located at 2172 cm^{-1} while the other mode is not IR active and has been observed through Raman spectroscopy at 2225 cm^{-1}. Unlike Gdm+ ion, the central carbon atom has a negative charge that can act as a hydrogen bond acceptor. So in this case, the ion contains two chemically different sites that can undergo hydrogen bonding: one being the central carbon and the other the CN groups. While it is expected that the CN groups will interact with the peripheral water molecules, the central carbon will only interact with those water molecules approaching from the top or bottom of the molecule. As in the case of Gdm+, the TCM ion has an anisotropic pump-probe signal (Figure 6.7) that is well described by the multiexponential function where the ultrafast component produced is by population transfer between the degenerate states and the picosecond component is caused by rotational diffusion. Moreover, TCM also shows the typical limiting anisotropy of 0.10 because of the difference between the timescale of population transfer and rotational diffusion. However, unlike Gdm+ where it is believed that the interaction of the NH_2 groups with water molecules induces the population transfers, in TCM, the exchange may also be assisted by water forming hydrogen bonds with the center carbon of the ion [11]. In addition, the population transfer mechanism is supported by the waiting time evolution of the 2D IR spectra in which the analysis of the cross-peak growth provides a very similar population transfer rate as the one observed in the pump-probe anisotropy.

In our study of the TCM-vibrational dynamics, a new approach has been developed to describe the frequencies of degenerate modes [11]. It consists of directly evaluating the frequencies for each of the instantaneous ion configurations observed in a molecular dynamics simulation. Because of the symmetry of the ion, the vibrational transitions of interest are nearly degenerate over the whole trajectory. Hence, the transition frequencies do not have a fixed and simple assignment to a particular vibrational mode. To identify and follow the frequencies of each of the degenerate transitions over the course of the MD trajectory, we computed a similarity factor that compares the displacement vectors for each of the instantaneous modes in two consecutive MD snapshots. Our procedure allows one to instantaneously tag each mode of a degenerate pair and their corresponding frequencies during the trajectory. It provides the tools to evaluate experimental observables such as the frequency–frequency correlation function of nearly degenerate vibrational transitions and the anisotropy.

Another strategy to evaluate the vibrational dynamics is by normal mode analysis where the idea is to represent the trajectory of ion Cartesian displacement coordinates, freed from overall translational and rotational motions, as linear combinations of the displacement vectors of normal mode eigenvectors. The normal mode coordinates contain the time evolution of the vibrational modes, which readily can be related to observables such as anisotropy decays or population transfer rates. In our study of tricyanomethanide ion, we have used the normal mode analysis to investigate the vibrational dynamics of the degenerate transition. This methodology produces a model for the energy transfer between the degenerate transitions of the TCM ion and hence the anisotropy, which is in good agreement with the experiments [11].

6.11.4 Symmetry Similarity of Protonated Arginine and Gdm+

Isolated Gdm+ has D_3 symmetry, with the C and three N atoms being on the plane perpendicular to the C_3 axis [7,8]. In aqueous solution, the threefold symmetry is slightly perturbed and the degeneracy of the mode at ~1600 cm^{-1} is split by a few wavenumbers. As we have seen, the D_2O induces picosecond timescale energy transfer between these nearly degenerate modes by a mechanism discussed above. The side chain of arginine (Arg) consists of a guanidyl group (Figure 6.1), which differs from the symmetric Gdm+ system only by virtue of its slightly broken symmetry caused by alkyl substitution of one of the six hydrogen atoms of the N–H groups (see Figure 6.1). The two CN_3 stretching modes of guanidinium portion of arginine are not exactly degenerate, but the amine substitution has only a small effect on the degenerate normal mode and the fast energy transfer between the component states is not significantly altered in Arg+ compared with Gdm+. The 2D IR spectroscopy permits direct observation through the anisotropy of the ultrafast energy transfer; therefore, this process was proposed to serve as a unique spectral signature of Arg+ even in complex environments [9]. Quantum computations showed the effect of a methyl perturbation on the Gdm+ degeneracy: the degenerate level is split by ~20 cm^{-1} in MeGdm+, which is close to the experimental value for Arg–dipeptide [9]. This splitting is much smaller than k_BT and apparently has little effect on the energy transport time that is reported as 500 fs. However, it was noted that symmetry lowering causes the transition dipoles undergoing interchanges to show small deviations from being perpendicular. Moreover, the existence of the splitting makes possible direct measurements with anisotropic 2D IR of the equilibrium dynamics involving the two nearly degenerate components.

The perturbation of the alkyl group is also sufficient to introduce an asymmetry into the transition dipole magnitudes. So, the pathways always go with a factor of either μ_1^4 or $\mu_1^2\mu_2^2$. Assuming that the spectral shape factors are the same for all the pathways, and δ is very small, we can write out an approximation to the ratio of the cross peak to the diagonal 2D IR signals in 2D IR spectrum. In the same notation as used above, with the transition dipole factors rationalized and equal spectral factors, the ratio in 2D IR of the cross-peak signal S_{12} to a diagonal peak S_{22} of the split degeneracy is written directly from the diagrams in Figure 6.3 by noting that the cross peaks involve the diagrams having a different component state (1 or 2) in the coherence and detection time intervals:

$$\frac{S_{12}}{S_{22}} \approx \frac{2(P_{12}+1)\langle 1_a1_a2_b2_b\rangle + P_{1212}\langle 1_a2_a1_b2_b\rangle - 2P_{11}\langle 1_a1_a2_b2_b\rangle - P_{1212}\langle 1_a2_a1_b2_b\rangle}{2\left(\frac{\mu_2^2}{\mu_1^2}\right)(P_{11}+1)\langle 1_a1_a1_b1_b\rangle + P_{1212}\langle 1_a2_a2_b1_b\rangle - 2\left(\frac{\mu_2^2}{\mu_1^2}\right)P_{12}\langle 1_a1_a1_b1_b\rangle - P_{1212}\langle 1_a2_a2_b1_b\rangle}$$

$$= \frac{P_{12}\langle 1_a1_a2_b2_b\rangle}{(\mu_2^2/\mu_1^2)P_{11}\langle 1_a1_a1_b1_b\rangle} \tag{6.27}$$

where $P_{1212} = P_{12\to12} + P_{12\to21}$.

The 2D IR spectrum of MeGdm+ is shown in Figure 6.8 as a function of waiting time with the growth of the cross peak due to energy transfer highlighted. The results for alkyl Gdm+ suggest that D_2O could enable the energy transfer between the two component states without altering the transition dipole directions in the plane perpendicular to the threefold axis, since the computations show that the methyl substitution already establishes principal dielectric axes.

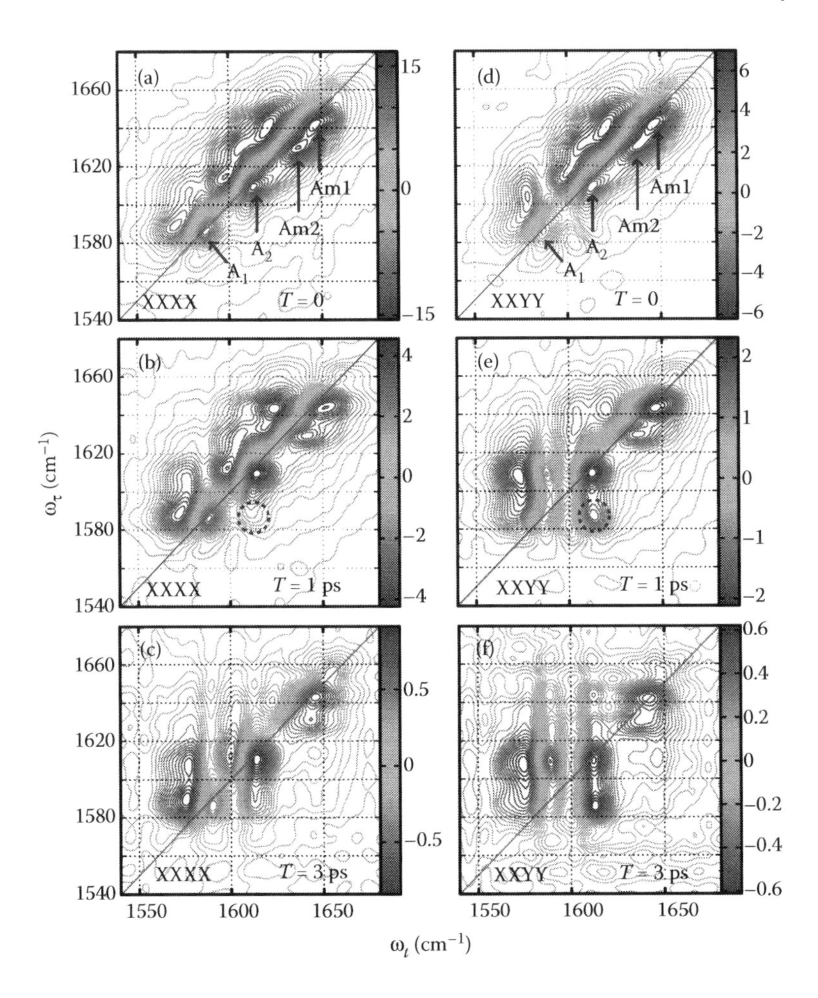

FIGURE 6.8 Absorptive 2D IR spectra of arginine dipeptide in D_2O for waiting times 0, 1, and 3 ps. The spectra for the XXXX polarization scheme are plotted on the left column, and those for the XXYY polarization scheme are plotted on the right column. The two amide-I modes are labeled Am1 and Am2. The cross-peak region is highlighted by dotted circles. (Adapted from Ghosh A, Tucker MJ, and Hochstrasser RM. 2011. *J Phys Chem A* 115(34):9731–9738.)

6.12 MOLECULAR DESCRIPTIONS OF THE ANISOTROPY

In symmetric ions such as guanidinium, tricyanomethanide, and oxalate, there are various molecular pictures of how the ultrafast decay of the anisotropy occurs.

The first approach has already been discussed and involves the dipole reorientation caused by energy transfer between the two degenerate states where their transition dipoles to the ground state are perpendicular (Figure 6.9a). This process can be conceptualized as the influence on one of the modes of a force acting on the other as described earlier. The second possibility involves the pseudorotation mechanism (Figure 6.9b). In this case, the system is conceptualized as ion and solvent forming a long-lived and stable conformation in which the solvent is arranged asymmetrically around the ion. This model gives rise to a nondegenerate system of states produced by the nonsymmetric water shell of the ion. The loss of anisotropy following excitation of the level pair is then produced by the jump of the solvent between *identical* structures that are rotated by 120° in the molecular frame. In these two apparently different mechanisms, the anisotropy can be evaluated as the contributions from ions that did and those that did not undergo change in transition dipole

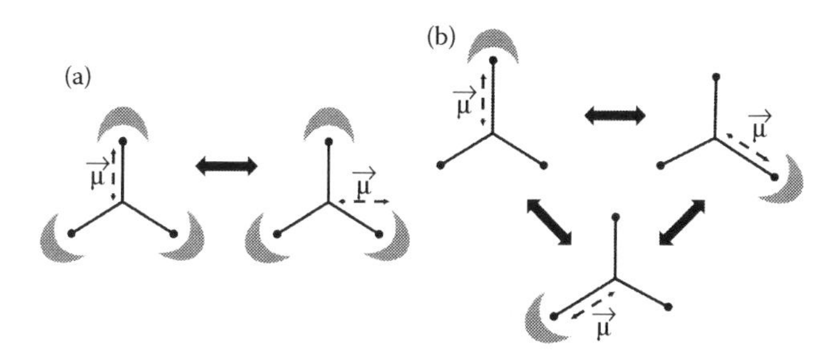

FIGURE 6.9 Pseudorotation and superposition of states-of-states mechanism. In the superposition of states mechanism (a), the jumps of the dipole direction are only 90° and in the pseudorotation mechanism (b), the transition dipole direction makes jumps of 120°.

moment direction prior to the arrival of the probe pulse. The (*aabb*) tensor component of the polarized signal at some instant *t* is therefore proportional to an ensemble average:

$$I_{aabb}(t) = \left\langle P_{ii}(t)\mu_{0i}^2(0)\mu_{0i}^2(t)(i)_a^2(i)_b^2(T) + \sum_{i\neq j} P_{ij}(t)\mu_{0i}^2(0)\mu_{0j}^2(t)(i)_a^2(j)_b^2(T) \right\rangle \tag{6.28}$$

where $P_{ii}(t)$ is the probability that if a molecule is pumped to the state $|i\rangle$, it will be in state $|i\rangle$ after time *t* and $P_{ij}(t)$ as the probability that it has transferred to state $|j\rangle$ and the sum represents all the other possible transition dipole directions that the system can acquire. When there are only a few states involved in the transfer, the kinetics of population transfer between them is appropriately represented by simple conditional kinetic factors such as those for two- and three-state jump kinetics. In the case of population transfer between the degenerate transitions, these factors are as given in Equation 6.15 where the transition dipole jumps by an angle $\pi/2$. If three equivalent sites are involved in the transfer, such as in the pseudorotation of a molecule having a threefold axis, the kinetic factors are given by

$$\begin{aligned} P_{11}(t) = P_{22}(t) = P_{33}(t) = 1/3(1 + 2e^{-3kt}) \\ P_{12}(t) = P_{13}(t) = P_{23}(t) = 1/3(1 - e^{-3kt}) \end{aligned} \tag{6.29}$$

where *k* is the site-to-site jump rate coefficient. The transition dipole $\hat{\mu}_{0i}$ of one transition is perpendicular to $\hat{\mu}_{0j}$ at all times because both are normal modes of the ion. Using that the Cartesian tensors are $\langle 1_a^2 2_b^2(t)\rangle = 2/15$, $\langle 1_a^2 1_a^2(t)\rangle = 1/5$, and $\langle 1_a^2 1_b^2(t)\rangle = 1/15 = \langle 1_a^2 2_a^2(t)\rangle$, one can derive the numerator of the anisotropy for the population transfer mechanism as

$$I_{aaaa} - I_{aabb} = \frac{e^{-6Dt}}{3}\left(\frac{2}{5}\langle \mu_1^2(0)\mu_1^2(t)\rangle_\theta P_{11}(t) - \frac{1}{5}\langle \mu_1^2(0)\mu_2^2(t)\rangle_\theta P_{12}(t)\right) \tag{6.30}$$

Similarly, for the pseudorotation, I_{aabb}^P, the numerator of Equation 6.5 is

$$I_{aaaa}^P - I_{aabb}^P = \frac{e^{-6Dt}}{3}\left\{\begin{aligned} &\frac{2}{5}\langle \mu_1^2(0)\mu_1^2(t)\rangle_\theta P_{11}(t) \\ &-\frac{1}{20}\left(\langle \mu_1^2(0)\mu_2^2(t)\rangle_\theta P_{12}(t) + \langle \mu_1^2(0)\mu_3^2(t)\rangle_\theta P_{13}(t)\right) \end{aligned}\right\} \tag{6.31}$$

Finally, the anisotropy can be expressed as

$$r(t) = e^{-6Dt} \frac{\left\{0.4\left\langle\mu_1^2(0)\mu_1^2(t)\right\rangle_\theta P_{11}(t) - 0.2\left\langle\mu_1^2(0)\mu_2^2(t)\right\rangle_\theta P_{i12}(t)\right\}}{\left\{\left\langle\mu_1^2(0)\mu_1^2(t)\right\rangle_\theta P_{11}(t) + \left\langle\mu_1^2(0)\mu_2^2(t)\right\rangle_\theta P_{12}(t)\right\}} \quad (6.32)$$

and

$$r_p(t) = e^{-6Dt} \frac{\left\{0.4\left\langle\mu_1^2(0)\mu_1^2(t)\right\rangle_\theta P_{11}(t) - 0.05\left(\left\langle\mu_1^2(0)\mu_2^2(t)\right\rangle_\theta P_{12}(t) + \left\langle\mu_1^2(0)\mu_3^2(t)\right\rangle_\theta P_{13}(t)\right)\right\}}{\left\{\left\langle\mu_1^2(0)\mu_1^2(t)\right\rangle_\theta P_{11}(t) + \left\langle\mu_1^2(0)\mu_2^2(t)\right\rangle_\theta P_{12}(t) + \left\langle\mu_1^2(0)\mu_3^2(t)\right\rangle_\theta P_{13}(t)\right\}} \quad (6.33)$$

If the transition dipole were to be constant or to undergo very slow correlation decay and be equal for all the transitions, the anisotropies could be simplified to the usual forms:

$$r(t) = e^{-6Dt}\left[0.4P_{11}(t) - 0.2P_{12}(t)\right] \quad (6.34)$$

and

$$r(t) = e^{-6Dt}\left[0.4P_{11}(t) - 0.05P_{12}(t) - 0.05P_{13}(t)\right] \quad (6.35)$$

From these two results, one can see that the limiting values for both mechanisms (omitting the rotational diffusion) are the same, $r(0) = 0.4$ and $r(\infty) = 0.1$. Therefore, the anisotropy measurement provides timescale of the dynamics, but its results do not distinguish the proposed microscopic mechanisms for the transition dipole reorientations.

However, the 2D IR spectrum can resolve this ambiguity. The population transfer causes the 2D IR spectrum to have cross peaks, which is the case of energy transfer between the components of a split degeneracy. Whereas the pseudorotation alters the anisotropy of the diagonal peaks, it does not produce any cross peaks. The system jumps between *identical* but rotated states, which involve Liouville pathways that are indistinguishable. A simulation of the 2D IR spectrum (Figure 6.10) of

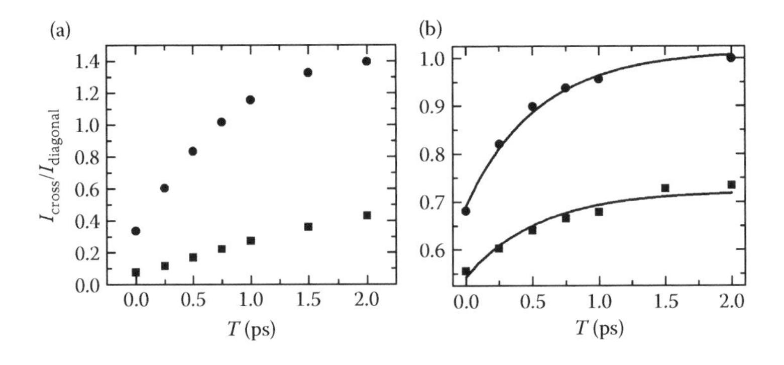

FIGURE 6.10 Simulated and experimental time dependence of the ratios of 2D IR cross- and diagonal-peak intensities for DGdm+ in 59% D-glycerol/D_2O. (a) Simulated 2D IR spectra for XXXX (circles) and XXYY (squares) polarizations with the assumption of population exchange between component states. (b) The ratio of the amplitudes between the cross peak at (1601 cm^{-1},1585 cm^{-1}), and the diagonal peak at (1601 cm^{-1},1601 cm^{-1}). The polarizations are XXXX (squares) and XXYY (circles). The data refer to the 2D IR absorptive spectra. Single exponential fits are shown as solid curves. (Adapted from Vorobyev DY et al. 2009. *J Phys Chem B* 113(46):15382–15391.)

a pair of oscillators well separated in frequency shows up the effect most clearly for a pair of vibrational transitions undergoing energy transfer.

In the two presented examples, the dipole magnitudes are considered constant, which validate Equations 6.34 and 6.35. However, a really interesting situation arises when dipole magnitudes in Equation 6.32 are changing significantly with time. In the case of the oxalate dianion, the correlation function $\langle \mu_i^2(0)\mu_i^2(t)\rangle_\theta$ arising from the fluctuations in the angle θ between the two coupled carboxylate transitions is the cause of the time dependence of the anisotropy. In the next section, we summarize a typical numerical approach to such situations.

6.13 THEORETICAL MODELING

One of the main challenges when describing symmetric ions is how the site properties of the ion relate to the experimental observables such as frequency correlation times, anisotropy decays, and so on. One possible way of linking site modes and normal modes is by modeling the symmetric ions as systems of coupled oscillators in which each site is represented by a single oscillator. For example, in the case where the ion can be described as two coupled oscillators (Figure 6.11), the vibrational Hamiltonian containing the ground state and the one and two-quanta transitions for the local sites are represented by

$$H = \begin{bmatrix} 0 & & & & & \\ & \omega_1(t) & \beta(t) & & & \\ & \beta(t) & \omega_2(t) & & & \\ & & & 2\omega_1(t)-\Delta & 0 & \sqrt{2}\beta(t) \\ & & & 0 & 2\omega_2(t)-\Delta & \sqrt{2}\beta(t) \\ & & & \sqrt{2}\beta(t) & \sqrt{2}\beta(t) & \omega_1(t)+\omega_2(t)-\delta \end{bmatrix} \tag{6.36}$$

where $\omega_i(\mathrm{t}) = \omega_{10} + \delta\omega_i(t)$ are the frequencies of the sites (ω_{10}) perturbed by the solvent ($\delta\omega_i(t)$), and in the absence of coupling, $\beta(t)$ is the coupling strength, and Δ and δ are the anharmonicities of the local modes and the mixed mode, respectively. The factor $\sqrt{2}$ in the coupling constant in the two-quanta part of Hamiltonian arises from the harmonic approximation. Since it is assumed that there

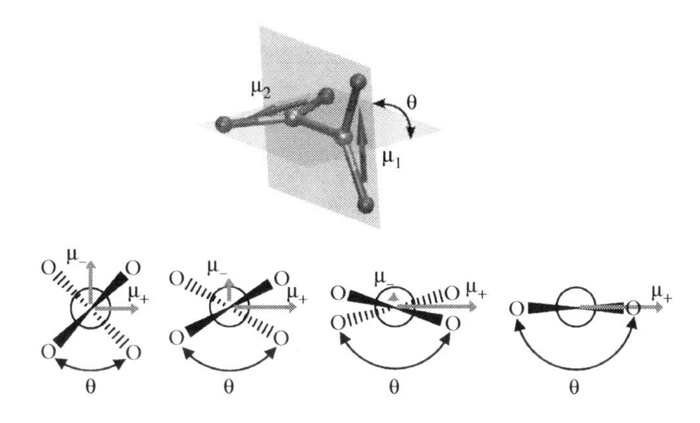

FIGURE 6.11 Angle dependence of the transition dipole magnitude of the two exciton transitions (|+⟩ and |–⟩) in the oxalate dianion. The dihedral angle between the two carboxylate planes is defined by the angle θ.

is no coupling between the one- and two-quanta manifold, this Hamiltonian can be solved in parts. The one-quanta manifold of the Hamiltonian has an analytical solution in the form of

$$H = \begin{bmatrix} \frac{\omega_1(t)+\omega_2(t)}{2} - \Omega(t) & \\ & \frac{\omega_1(t)+\omega_2(t)}{2} + \Omega(t) \end{bmatrix} \tag{6.37}$$

where $\Omega(t) = \sqrt{\left((\omega_1(t)-\omega_2(t))^2 + (2\beta(t))^2\right)}\Big/2$ and the eigenvectors

$$\begin{aligned} |+\rangle(t) &= \cos\frac{\Theta(t)}{2}|10\rangle + \sin\frac{\Theta(t)}{2}|01\rangle \\ |-\rangle(t) &= -\sin\frac{\Theta(t)}{2}|10\rangle + \cos\frac{\Theta(t)}{2}|01\rangle \end{aligned} \tag{6.38}$$

where $\tan\Theta(t) = 2|\beta(t)|/(\omega_1(t)-\omega_2(t))$ and as shown in Figure 6.12, $\beta(t) = |\beta(t)|$.

The two-quanta manifold of the Hamiltonian has an analytical solution, but the eigenvalues cannot be expressed by a simple equation. However, the two-quanta eigenstates can be written symbolically as

$$\begin{aligned} |S_+\rangle(t) &= C_{20}^{S_+}(t)|20\rangle + C_{02}^{S_+}(t)|02\rangle + C_{11}^{S_+}(t)|11\rangle \\ |S_-\rangle(t) &= C_{20}^{S_-}(t)|20\rangle + C_{02}^{S_-}(t)|02\rangle + C_{11}^{S_-}(t)|11\rangle \\ |a\rangle(t) &= C_{20}^{a}(t)|20\rangle + C_{02}^{a}(t)|02\rangle + C_{11}^{a}(t)|11\rangle \end{aligned} \tag{6.39}$$

Note that this representation reproduces the experimental vibrational spectrum since the transition dipole moment operator in the site basis set can be described as

$$\mu = \begin{bmatrix} & \mu_1 & \mu_2 \\ \mu_1 & & \\ \mu_2 & & \end{bmatrix} \tag{6.40}$$

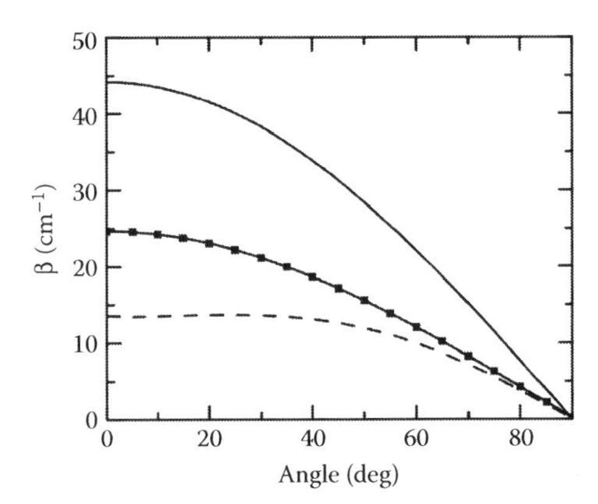

FIGURE 6.12 Coupling between the asymmetric stretches of the two carboxylates of the oxalate dianion as a function of the dihedral angle for different approximations for separated systems. The solid, solid with filled squares, and dashed lines represent the coupling constant predicted by TDC, TCC, and DFT, respectively. (Extracted from Kuroda DG and Hochstrasser RM. 2011. *J Chem Phys* 135(20): 044501.)

where μ_i is the transition dipole vector of the ith site. The transition dipole operator of the normal mode basis set is calculated by applying the same unitary transformation that diagonalizes the vibrational Hamiltonian. Thus, the transition dipole magnitudes are

$$|\mu_{0+}(t)|^2 = |\langle + | \mu | 0 \rangle|^2 = 1 + \sin\Theta(t)\cos\theta_{ij}(t) \quad (6.41)$$

and

$$|\mu_{0-}(t)|^2 = |\langle - | \mu | 0 \rangle|^2 = 1 - \sin\Theta(t)\cos\theta_{ij}(t) \quad (6.42)$$

where $\theta_{ij}(t)$ is the angle between the transition dipoles in the local site basis set.

If all molecular observables such as site frequency fluctuations and coupling constant are known, it is possible to solve the instantaneous Hamiltonian and calculate the linear IR absorption and 2D IR spectrum for the ion. Thus, we focus the next part of this section obtaining these molecular variables from molecular dynamics simulations and/or *ab initio* computations.

6.13.1 Evaluations of the Coupling Constants

One of the great advantages of 2D IR is its ability to determine coupling between different modes through the existence of cross peaks. In some high-symmetry molecules, a reasonable zero-order description for the vibrational states is the sum of vibrational Hamiltonians corresponding to each separated systems or sites for the excitations as in Frenkel exciton theory. The coupling between these site modes can be evaluated by using transition dipole coupling (TDC), transition charge coupling (TCC), or *ab initio* theoretical calculations. Such is the case for the coupling between the carboxylate groups of the oxalate dianion where each carboxylate has transition dipoles parallel to the line joining the two carboxylate oxygen atoms.

The TDC model is the simplest in that the sites only interact via a dipole–dipole approximation to the Coulomb potential of the form

$$V(r_{ij}) = \frac{1}{4\pi\varepsilon_0 r_{ij}^3}\left[\vec{\mu}_i \cdot \vec{\mu}_j - \frac{3(\vec{\mu}_i \cdot \vec{r}_{ij})(\vec{\mu}_j \cdot \vec{r}_{ij})}{r_{ij}^2}\right] \quad (6.43)$$

where $\vec{\mu}_i$ is the transition dipole of the ith site and $\vec{r}_{ij}$ is the vector joining them. This model is reasonable when the distance between the separated system dipoles is much larger than the dipole magnitudes, $|\vec{\mu}_i| \ll |\vec{r}_{ij}|$. In TCC, the coupling constant is modeled as the change in the energy due to the presence of two oscillators and includes the effect of the charge flux occurring during the normal mode displacement [37]. The TCC model takes into account not only the dipole–dipole interaction, but also other multipole interactions such as the dipole–quadrupole but omits through bond interactions. The charges and charge derivatives are obtained from *ab initio* calculations on the most computationally convenient molecule containing the site, for example, formate for the carboxylate group, and cyanide for nitriles. Of course, the coupling constant can be obtained from direct *ab initio* calculations.

The oxalate dianion provides an example where the coupling constants obtained by the different approaches have been compared [10]. Oxalate presents a very particular situation in which the coupling constant varies with the dihedral angle between the two carboxylate planes. Moreover, the two carboxylates are only separated by 1.5 Å, which makes inappropriate the use of certain simple models. As expected, the predicted angular interaction between carboxylate groups of oxalate, either by TDC or by TCC, yields larger couplings than the density functional theory (DFT) computation (Figure 6.12). The TDC appears to significantly overestimate the coupling by almost a factor of four. While TCC shows a better agreement with DFT calculations, it still gives a significantly larger coupling constant compared with the DFT computation by almost a factor of two. The mismatch between either TDC or TCC and DFT arises from the nonnegligible through bond interaction between the carboxylate groups of oxalate since the sites have a significant overlap in their electronic densities. However, all the coupling models tend toward zero coupling as the dihedral angle goes to $\pi/2$ and the symmetry tends to D_{2d}.

6.13.2 Frequency Fluctuations

The fluctuations of excitation frequencies of certain vibrators can be calculated using frequency maps in conjunction with classical molecular dynamics simulations. Frequency maps are based on an empirical correlation between the electric field exerted by the solvent onto the atoms of an oscillator, and its vibrational frequency, as has been shown by Cho and coworkers [38]. The frequency of the vibrational mode is modeled as

$$\omega = \omega_0 + \sum_{i\alpha} C_{i\alpha} E_{i\alpha}(r_i) \tag{6.44}$$

where ω_0 is the site frequency in vacuum, $\alpha = \{x,y,z\}$, E_i is the electrostatic field or its first or second derivative in a given atom of the site, and $C_{i\alpha}$ are the electrostatic correlation parameters obtained from the *ab initio* calculations. Many different maps have been developed to describe the IR absorption frequencies and bandwidths of common IR probes such as nitriles, azide, carboxylate, and the amide-I mode [39–52]. In our work on oxalate, we used the frequency maps of Falvo et al. [43] that demanded the evaluation of the electric field at 18 points on the each of the carboxylate ion components and have been successfully used to describe carboxylate groups of other molecular systems [43,53].

6.13.3 Modeling of the Vibrational Dynamics

Owing to the dynamics of the ion sites and solvent molecules in the hydration shell, the vibrational Hamiltonian of the ion–solvent complex is time dependent (Equation 6.36). Valuable formulas with parameters given in terms of auto- and cross-correlation functions for general two-level dynamics were already reported by Silbey and Wertheimer [32]. However, in some cases, perturbation theory is not sufficient and the full time-dependent Schrodinger equation must be solved. We will give one example of such a computation in this section. The solution of the two site time-dependent Hamiltonian leads to a system of differential equations of the form

$$\begin{aligned} i\frac{dc_1}{dt} &= \omega_1(t)c_1 + \beta(t)c_2 \\ i\frac{dc_2}{dt} &= \omega_2(t)c_2 + \beta(t)c_1 \end{aligned} \tag{6.45}$$

where $\omega_i(t)$ are the instantaneous frequencies obtained from Equation 6.44, $\beta(t)$ is the instantaneous coupling constant, and c's are the coefficients of the vibrational wave functions in the site basis set. The system of differential equations obtained from the time-dependent Schrodinger equation can be solved with a standard algorithm. The coefficients provide the instantaneous populations of the sites and the instantaneous interstate coherences. However, to compute the observables seen in experiments such as transient gratings and 2D IR, the statistical average over the ensemble must be computed. Since it is computationally impossible to simulate an ensemble of molecules, the molecular dynamics simulation is done for a long time window. Thus, the density matrix can be computed by averaging the instantaneous values $c_m^*(t)c_n(t)$ over a large number of shorter time windows chosen from the trajectory, which is equivalent to generating an ensemble of values [54]. Such a procedure yields the statistical density matrix elements

$$\rho_{nm}(t) = \left\langle c_m^*(t)c_n(t)\right\rangle \tag{6.46}$$

This density matrix allows us to calculate mean values of properties from traces of system operators over the density matrix (Equation 6.46).

The time dependence of the density matrix elements provides the information necessary to describe the vibrational dynamics of the system, for example, $\rho_{11}(t)$ and $\rho_{12}(t)$ represent the time evolution of the population and coherences of a system of coupled oscillators interacting with a bath. The density matrix elements can be related to experimental observables. For example, the time evolution of the population can be used to understand the population transfer mechanism in ions with two nearly degenerate transitions, such as the one presented in Section 6.10.

In our study of oxalate [10], we have used the density matrix evolution to obtain the characteristic times and anisotropies of population transfer and coherence transfer in both the site and exciton basis set. Moreover, the comparison between the population transfer and anisotropy decays allows us to model the process occurring in oxalate in terms of simple site parameters such as frequency fluctuations and coupling constants.

6.13.4 Linear Absorption Spectrum

The linear absorption spectrum is given by the Fourier transform of the dipole time correlation function as

$$I(\omega) \sim \int_{-\infty}^{\infty} dt e^{-i\omega t} \left\langle \mu_a(t)\mu_a(0)\right\rangle \tag{6.47}$$

where μ_a is a-component of the dipole operator. In the semiclassical limit, the transition rate from the ground to the first excited state of a particular vibrational mode is

$$I(\omega) \sim \int_{-\infty}^{\infty} e^{-i\omega t} \left\langle \hat{1}_a(0)\hat{1}_a(t)\mu_{01}(0)\mu_{10}(t)\exp\left[i\int_0^t \omega_{10}(\tau)d\tau\right]\right\rangle dt \tag{6.48}$$

where $\hat{1}_a$ is the unit transition dipole projected into the laboratory axis, μ_{10} is the magnitude of the transition dipole, and ω_{10} is the time-dependent transition frequency. Assuming that the rotational diffusion is not correlated to the changes in the magnitude of the transition dipole and adding the lifetime as an empirical factor, the linear IR spectrum expression becomes

$$I(\omega) \sim \int_{-\infty}^{\infty} e^{-i\omega t}e^{-t/T_1} \left\langle \hat{1}_a(0)\hat{1}_a(t)\right\rangle\left\langle \mu_{10}(0)\mu_{10}(t)\exp\left[i\int_0^t \omega_{10}(\tau)d\tau\right]\right\rangle dt \tag{6.49}$$

where T_1 is the vibrational lifetime of the state. In this last equation, the first average represents the orientational dynamics (considered to be that of a sphere) of the system, which is

$$\langle \hat{1}_a(0)\hat{1}_a(t)\rangle = \exp(-2Dt) \tag{6.50}$$

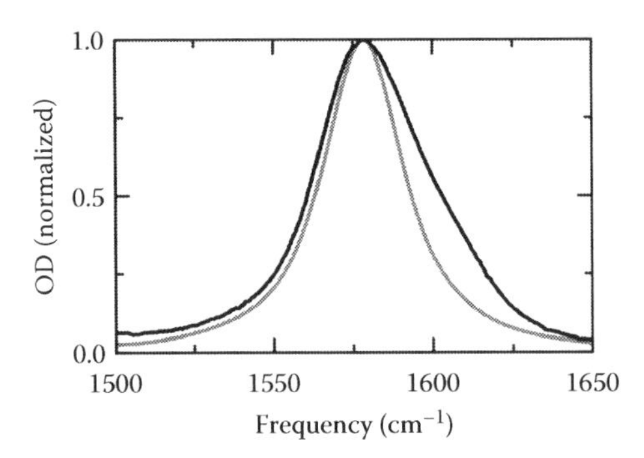

FIGURE 6.13 Experimental and simulated linear IR spectra of oxalate. The black line is the experimental absorption line shape and the gray line is predicted by the model discussed in the text relating to Equation 6.36. (Extracted from Kuroda DG and Hochstrasser RM. 2011. *J Chem Phys* 135(20): 044501.)

The result obtained in Equation 6.49 can be highly simplified if the transition dipole magnitudes are nearly constant in time. However, some cases exist where the fluctuations of frequencies are not Gaussian and the transition dipoles are undergoing large-amplitude fluctuations. Such is the case with the oxalate dianion because the time dependence of the dihedral angles implies that the transition dipole magnitudes must be considered.

In the computation of the oxalate IR spectrum (Figure 6.13), Equation 6.49 has to be expanded to take into account the two excitonic states ($|+\rangle$ and $|-\rangle$) present in the oxalate dianion. Because the two excitonic states have perpendicular transition dipoles, the linear IR spectrum is the sum over the two exciton states as

$$I(\omega) \sim \int_{-\infty}^{\infty} e^{-i\omega t} e^{-t/T_1} \sum_{j=|+\rangle,|-\rangle} \left\langle \hat{j}_a(0)\hat{j}_a(t) \right\rangle \left\langle \mu_{j0}(0)\mu_{j0}(t)\exp\left[i\int_0^t \omega_{j0}(\tau)d\tau \right] \right\rangle dt \tag{6.51}$$

Note that no explicit term of population exchange has been included in Equation 6.51 because exchange of populations is already taken into consideration when the instantaneous frequencies of the $|+\rangle$ and $|-\rangle$ states are calculated from the analytical solutions of the instantaneous Hamiltonian (Equation 6.36). The result obtained by numerical calculation of the absorption spectrum of oxalate depicted in Figure 6.13 shows that the linear IR spectrum has a slightly asymmetric shape. The presence of an asymmetry in the linear spectrum indicates that the transition dipole magnitude fluctuations might play a significant role in the description of the linear IR spectrum of coupled systems such as small symmetric ions.

6.13.5 2D IR Absorption Spectrum

In the 2D IR spectrum, the signal $S(\omega_\tau,T,\omega_t)$ arises from the interaction of three IR pulses with the system. This field is generated by an impulsively induced macroscopic polarization and is given by

$$S(\omega_\tau,T,\omega_t) = \text{Re}\left[\sum_{i=1}^{20} S_i(\omega_\tau,T,\omega_t) \right] \tag{6.52}$$

where

$$S_i(\omega_\tau, T, \omega_t) = \int_0^\infty dt_1 \int_0^\infty dt_3 \exp(i\omega_t t \mp \omega_\tau \tau) R_i(\tau, T, t) \tag{6.53}$$

The signs of the exponentials define the rephasing (+) or nonrephasing (–) contributions to the signal and $R_i(\tau,T,t)$ are the corresponding response functions that are given by the different Liouville-space pathways. In the case of an ion composed of two excitonic states, the number of possible response functions (Liouville-space pathways) that contribute to the 2D IR spectrum is 20 (Figure 6.14).

In a case such as oxalate, where the exciton has a time-dependent interaction, the response functions can be used without invoking the Condon approximation. However, it is most likely that the transition dipole magnitudes are not correlated with their direction, which implies a separation of the internal and the overall motion. For oxalate, the much slower rotational diffusion is not correlated with internal motions, such as the ultrafast dihedral angle changes. Thus, under

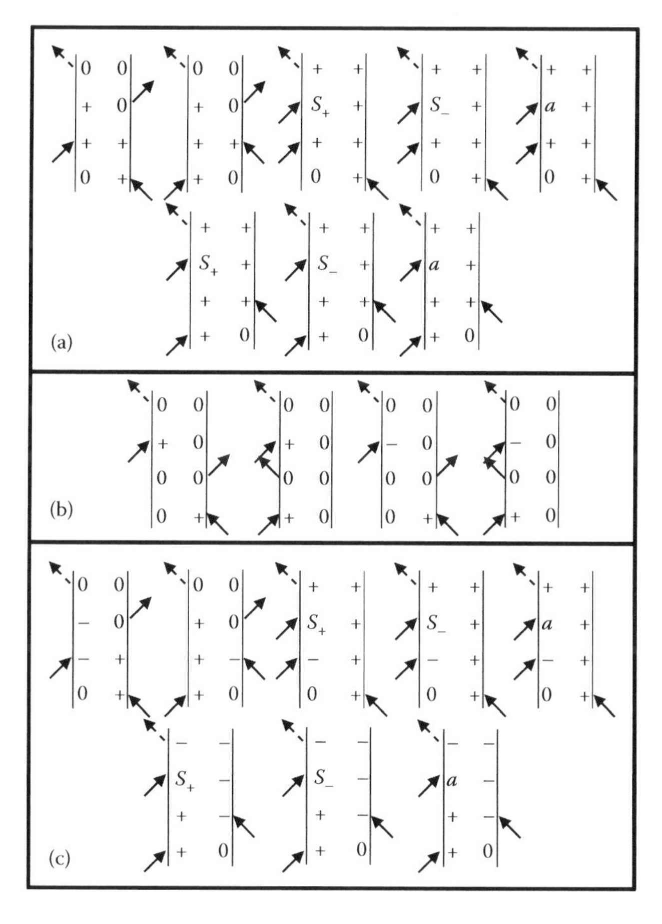

FIGURE 6.14 Liouville path diagrams for exciton transitions contributing to the photon echo signal emitted in the $-\mathbf{k}_1 + \mathbf{k}_2 + \mathbf{k}_3$ direction. Diagrams are ordered according to their contributions to: (a) new absorption, (b) bleach, and (c) those involving coherences in the T time. The complete set of diagrams needed for simulations includes those with indices + and – interchanged.

this assumption, the response function can be simplified. For example, the response $R_i(\tau,T,t)$ in Figure 6.14 is

$$R_1(\tau,T,t) = \left\langle \hat{+}_a(0)\hat{+}_b(\tau)\hat{+}_c(\tau+T)\hat{+}_d(\tau+T+t) \right\rangle$$
$$\times \left\langle \mu_{+0}(0)\exp\left[i\int_0^{\tau} \omega_{+0}(\Gamma)d\Gamma \right]\mu_{+0}(\tau)\mu_{+0}(\tau+T)\exp\left[-i\int_{\tau+T}^{\tau+T+t} \omega_{+0}(\Gamma)d\Gamma \right]\mu_{+0}(\tau+T+t) \right\rangle \tag{6.54}$$

where $\hat{+}_a$ is the unit vector of the *plus* state transition dipole projected onto the a-axis, μ_{+0} is the magnitude of the transition dipole, and ω_{+0} is the time-dependent transition frequency. The first average of Equation 6.54 is the orientational factor that has been described earlier in this chapter.

The presented methodology allows one to calculate the 2D IR spectrum even when the commonly used assumption of a Condon approximation cannot be applied.

6.14 EXCITONIC ANISOTROPIES

The oxalate dianion with its two coupled carboxylates presents some unique conditions in its anisotropic responses. The $V = 1$ state is formed from two interacting carboxylates so they are effectively exciton states where the coupling vanishes identically at D_{2d} symmetry (see Figure 6.11). Naturally, it is possible to excite coherent superposition of these states so in principle the anisotropy should contain all the coherence diagrams as well as those involving intermediate population states as described earlier through Equation 6.5. An excitonic model is used to describe the dynamic coupling between the two carboxylate asymmetric stretches. Each of the carboxylate transition dipoles is directed parallel to a line connecting the two oxygen atoms of either carboxylate and they are represented as vectors $\mu_{01}^{(1)} = \mu_{01}\hat{1}$ and $\mu_{01}^{(2)} = \mu_{01}\hat{2}$ located at the centers of mass of the carboxylate groups where μ_{01} is the transition dipole *magnitude* for the $V = 0$ to $V = 1$ transition of a single carboxylate asymmetric stretch mode. The unit vectors $\hat{\pm} = (\hat{1}(t) \pm \hat{2}(t))/\sqrt{2}$ are *independent of time*, neglecting overall rotation, because the changes in the dihedral angle only result in changes of the magnitudes of the transition dipoles (Figure 6.11) to the delocalized states that are given at time t as

$$|\mu_{\pm 0}(t)| = \mu_{01}\sqrt{(1 \pm \cos\theta(t))} \tag{6.55}$$

The computation described in more detail below shows that the off-diagonal elements of this exciton density matrix decay very rapidly. The very rapid tens of femtoseconds relaxation of the angle distribution contributes to the decay of these coherences that have completely gone in the computation by ca. 100 fs. Therefore, the exciton coherences are not required to be included in the anisotropy expressions. The parallel and perpendicular signals of the pump probe are then calculated from $I_{aabb}(T)$ given by Equation 6.28. We use the same notation as before for the projection of the pathway state (±) onto a laboratory axis $(\pm)_a$. The result is the same as in Equation 6.32 where the motions of the dihedral angle and the overall motion are assumed to be statistically independent. The angle brackets around the dipole amplitudes imply averaging of the *magnitudes* of the transition moments over the dihedral angle θ distribution; these terms, typified by $\langle \mu_{0+}^2(0)\mu_{0+}^2(t) \rangle_\theta$ in Equation 6.32, depend on the time variation of the magnitudes due to fluctuations in the dihedral angle. Were it not for the transition dipole magnitudes that vary with fluctuations in the dihedral angle, the anisotropy would decay from 0.4 to 0.1 with time constant $1/k$ as is expressed in Equation 6.34. This result also comes directly from adding all Liouville path diagrams (see Figure 6.3) that contribute to the pump-probe signal where a pair of states is undergoing equilibration. The result incorporates an assumption that the off-diagonal anharmonicity is zero, such that the combination band of the two exciton transitions exactly equals twice the fundamental frequency: this eliminates the contributions of the coherent excitations of exciton states from the signal. Also, to cancel these diagrams, it is required that the fluctuations in the exciton frequencies are negligible, which is consistent with the model result of there being only small coupling fluctuations [32]. The most important feature of this model is that population exchange between $|+\rangle$ and $|-\rangle$ (P_{11} and P_{12} in Equation 6.32) will cause a fast drop in the anisotropy, regardless of any orientational dynamics. Furthermore, the fluctuations in frequency of the site states must be responsible for the decay of the anisotropy because the coupling between the two carboxylates is very small. In this limit, the rate expression 6.3 becomes

$$k_{et} = \frac{1}{\hbar^2}\int_{-\infty}^{\infty} dt \left\langle V_{+-}(t)V_{-+}(0)\right\rangle \approx \int_{-\infty}^{\infty} dt \left\langle \delta\omega_{10}(t)\delta\omega_{10}(0)\right\rangle \tag{6.56}$$

Introducing the results from Equation 6.57 to Equation 6.32 leads to an expression for the anisotropy of the form

$$r(t) = e^{-6Dt}\,\frac{0.4P_{++}(t) - 0.2\alpha P_{+-}(t)}{P_{++}(t) + \alpha P_{+-}(t)} \tag{6.57}$$

The parameter α depends on the dihedral angle dynamics which is determined by the potential function and the temperature and is given by

$$\alpha(t) = \frac{\left\langle \cos^2\left(\theta(0)/2\right)\sin^2\left(\theta(t)/2\right)\right\rangle}{\left\langle \cos^2\left(\theta(0)/2\right)\cos^2\left(\theta(t)/2\right)\right\rangle} \tag{6.58}$$

where $\theta(t)$ is the dihedral angle reached by time t from the initial value of $\theta(0)$ at $t = 0$. This α parameter varies from 0 to 1, limiting the second term of Equation 6.57 to values from 0.1 to 0.4. Thus, the anisotropy is affected by the presence of fluctuation in the transition dipole magnitude. However, if the parameter α grows very quickly to unity compared with anisotropy changes, the effect of the time-dependent transition dipole magnitude in the anisotropy is diminished. For oxalate [10], α decays within 100–200 fs. It can be seen from this analysis that the model distinguishes between whether the system is initially excited to a localized or a delocalized state and it provides a framework for interpreting the time dependence of the anisotropy in terms of loss of the coherence between the sites.

6.15 SURVEY OF POLARIZATION EFFECTS IN DEGENERATE VIBRATIONAL TRANSITIONS

Some relatively small molecular ions have high-enough symmetry to have degenerate vibrational modes. We gave examples of the threefold symmetric ions: the guanidinium cation, the tricyanomethanide anion, and the D_{2d} oxalate dianion. These ions have degenerate levels in IR spectral regions accessible to modern pump-probe and 2D IR spectroscopies. In water, the ions are strongly hydrogen bonded and the degenerate levels provide sensitive probes of the solvent structures. The solvent dynamics causes fluctuations from the high-symmetry configurations, but it is also observed that the degeneracies may be split on average. These results indicate that the water prefers to bind to asymmetric ion structures so that the distorted ion structure is the one with lowest free energy. However, the splitting of the degeneracy can be extremely small, even much smaller than the spectral linewidth as in the case of tricyanomethanide such that 2D IR spectral shape analysis is needed to establish the splitting.

The IR polarization of such transitions is extremely revealing of the dynamic processes. We discussed the origins of the pump-probe anisotropy for such degenerate levels and characterized the important role of anharmonic coupling on the IR polarization properties. We showed that the anisotropy is not a canonical 0.4 but may lie between 0.3 and 0.4 with the latter being the rule for the transient absorption signal. The canonical 0.4 is recovered in the bleach and stimulated emission signal when the off-diagonal anharmonicity becomes very small as is the case in the D_{2d} oxalate dianion. The component states of the degenerate level are strongly coupled by the solvent fluctuations and energy flows rapidly between them. The relaxation of one component state can be thought of as vibrational energy relaxation into the other component state. The relaxation is described as one state responding to the solvent forces exerted on the other. These relaxation times are routinely subpicosecond and in the case of oxalate dianions, they compete effectively with the carboxylate relaxation time of 300 fs. The canonical value for the limiting anisotropy is 0.1.

We showed that coherence transfer is a factor that is worthy of consideration in treating nonlinear IR experiments. The coherence transfer rates contain density matrix elements in common with the energy relaxation and so are also expected to be in the subpicosecond range, so we have shown how they will influence the measurements of anisotropy. In particular, the anisotropy during the detection time regime is expected to drop if the coherence transfer becomes fast enough and its effect on pump-probe signal is not then negligible.

Finally, we used oxalate dianions as an example of where inclusion of large transition dipole magnitude fluctuations can influence the anisotropy. The solution of the dynamics equations for the density matrix describing the exciton dynamics that ensues as the carboxylates of oxalate twist from D_{2d} toward the coplanar configuration was presented in detail. We showed that the fluctuations in the frequency of the carboxylate modes are likely to be the main contribution to the degenerate vibrational relaxation and the concomitant loss of polarization anisotropy.

Rotational diffusion was not given a serious place in this chapter, but overall motion is a factor that must be considered. It turns out that the interdegenerate state vibrational relaxation is much faster than the overall motion. Furthermore, the attainment of the limiting anisotropies of 0.1 is finally achieved in a time regime where rotational diffusion must also be considered in treating the experimental data.

ACKNOWLEDGMENTS

We would like to thank Dr. Prabhat Singh, Dr. Matthew Tucker, and Dr. Dmitriy Vorobyev for acquiring some of the data presented here. We are indebted to Dr. Ayanjeet Ghosh for his careful contributions to some of the formulas and figures presented here. The research was supported by NIH through RO1GM12592, P41RR001348, and 9P41GM104605-31 and by NSF CHEM.

REFERENCES

1. Lebrero MCG, Bikiel DE, Elola MD, Estrin DA, and Roitberg AE. 2002. Solvent-induced symmetry breaking of nitrate ion in aqueous clusters: A quantum-classical simulation study. *J Chem Phys* 117(6):2718–2725.
2. Ramesh SG, Re S, Boisson J, and Hynes JT. 2010. Vibrational symmetry breaking of NO_3^- in aqueous solution: NO asymmetric stretch frequency distribution and mean splitting. *J Phys Chem A* 114(3):1255–1269.
3. Vchirawongkwin V, Kritayakornupong C, Tongraar A, and Rode BM. 2011. Symmetry breaking and hydration structure of carbonate and nitrate in aqueous solutions: A study by *ab initio* quantum mechanical charge field molecular dynamics. *J Phys Chem B* 115(43):12527–12536.
4. England AH et al. 2011. On the hydration and hydrolysis of carbon dioxide. *Chem Phys Lett* 514(4–6):187–195.

5. Megyes T et al. 2009. Solution structure of $NaNO_3$ in water: Diffraction and molecular dynamics simulation study. *J Phys Chem B* 113(13):4054–4064.
6. Waterland MR and Kelley AM. 2000. Far-ultraviolet resonance Raman spectroscopy of nitrate ion in solution. *J Chem Phys* 113(16):6760–6773.
7. Vorobyev DY et al. 2009. Ultrafast vibrational spectroscopy of a degenerate mode of guanidinium chloride. *J Phys Chem B* 113(46):15382–15391.
8. Vorobyev DY et al. 2010. Water-induced relaxation of a degenerate vibration of guanidium using 2D IR echo spectroscopy. *J Phys Chem B* 114(8):2944–2953.
9. Ghosh A, Tucker MJ, and Hochstrasser RM. 2011. Identification of arginine residues in peptides by 2D IR echo spectroscopy. *J Phys Chem A* 115(34):9731–9738.
10. Kuroda DG and Hochstrasser RM. 2011. Two-dimensional infrared spectral signature and hydration of the oxalate dianion. *J Chem Phys* 135(20): 044501.
11. Kuroda DG, Singh PK, and Hochstrasser RM. 2012. Differential hydration of tricyanomethanide observed by time resolved vibrational spectroscopy. *J Chem Phys*. DOI: 10.1021/JP3069333.
12. Hamm P, Lim M, and Hochstrasser RM. 1998. Non-Markovian dynamics of the vibrations of ions in water from femtosecond infrared three-pulse photon echoes. *Phys Rev Lett* 81(24):5326–5329.
13. Kuo CH, Vorobyev DY, Chen JX, and Hochstrasser RM. 2007. Correlation of the vibrations of the aqueous azide ion with the O-H modes of bound water molecules. *J Phys Chem B* 111(50):14028–14033.
14. Li SZ, Schmidt JR, and Skinner JL. 2006. Vibrational energy relaxation of azide in water. *J Chem Phys* 125(24): 244507.
15. Li SZ, Schmidt JR, Piryatinski A, Lawrence CP, and Skinner JL. 2006. Vibrational spectral diffusion of azide in water. *J Phys Chem B* 110(38):18933–18938.
16. Zanni MT, Ge NH, Kim YS, and Hochstrasser RM. 2001. Two-dimensional IR spectroscopy can be designed to eliminate the diagonal peaks and expose only the crosspeaks needed for structure determination. *Proc Natl Acad Sci USA* 98(20):11265–11270.
17. Golonzka O, Khalil M, Demirdoven N, and Tokmakoff A. 2001. Coupling and orientation between anharmonic vibrations characterized with two-dimensional infrared vibrational echo spectroscopy. *J Chem Phys* 115(23):10814–10828.
18. Khalil M and Tokmakoff A. 2001. Signatures of vibrational interactions in coherent two-dimensional infrared spectroscopy. *Chem Phys* 266(2–3):213–230.
19. Wilson EB, Decius JC, and Cross PC. 1980. *Molecular Vibrations: The Theory of Infrared and Raman Vibrational Spectra* (Dover Publications, New York), pp xi, 388.
20. Hochstrasser RM. 2001. Two-dimensional IR-spectroscopy: Polarization anisotropy effects. *Chem Phys* 266(2–3):273–284.
21. Hamm P and Zanni MT. 2011. *Concepts and Methods of 2D Infrared Spectroscopy* (Cambridge University Press, Cambridge, New York), pp ix, 286.
22. Hamm P, Lim MH, and Hochstrasser RM. 1998. Structure of the amide I band of peptides measured by femtosecond nonlinear-infrared spectroscopy. *J Phys Chem B* 102(31):6123–6138.
23. Kim YS and Hochstrasser RM. 2005. Chemical exchange 2D IR of hydrogen-bond making and breaking. *Proc Natl Acad Sci USA* 102(32):11185–11190.
24. Galli C, Wynne K, Lecours SM, Therien MJ, and Hochstrasser RM. 1993. Direct measurement of electronic dephasing using anisotropy. *Chem Phys Lett* 206(5–6):493–499.
25. Ferro AA and Jonas DM. 2001. Pump-probe polarization anisotropy study of doubly degenerate electronic reorientation in silicon naphthalocyanine. *J Chem Phys* 115(14):6281–6284.
26. Smith ER and Jonas DM. 2011. Alignment, vibronic level splitting, and coherent coupling effects on the pump-probe polarization anisotropy. *J Phys Chem A* 115(16):4101–4113.
27. Sando GM, Zhong Q, and Owrutsky JC. 2004. Vibrational and rotational dynamics of cyanoferrates in solution. *J Chem Phys* 121(5):2158–2168.
28. Banno M, Iwata K, and Hamaguchi H. 2007. Intra- and intermolecular vibrational energy transfer in tungsten carbonyl complexes W(CO)(5)(X) (X = CO, CS, CH3CN, and CD3CN). *J Chem Phys* 126(20): 204501.
29. Banno M, Sato S, Iwata K, and Hamaguchi H. 2005. Solvent-dependent intra- and intermolecular vibrational energy transfer of W(CO)(6) probed with sub-picosecond time-resolved infrared spectroscopy. *Chem Phys Lett* 412(4–6):464–469.
30. Tokmakoff A and Fayer MD. 1995. Homogeneous vibrational dynamics and inhomogeneous broadening in glass-forming liquids—Infrared photon-echo experiments from room-temperature to 10 K. *J Chem Phys* 103(8):2810–2826.
31. Smith PG et al. 1994. Electronic coupling and conformational barrier crossing of 9,9′-bifluorenyl studied in a supersonic jet. *J Chem Phys* 100(5):3384–3393.

32. Wertheimer R and Silbey R. 1980. On excitation transfer and relaxation models in low-temperature systems. *Chem Phys Lett* 75(2):243–248.
33. Oxtoby DW. 1979. Hydrodynamic theory for vibrational dephasing in liquids. *J Chem Phys* 70(6):2605–2610.
34. Mukamel S. 1995. *Principles of Nonlinear Optical Spectroscopy* (Oxford University Press, New York), pp xviii, 543.
35. Cho M. 2009. *Two-Dimensional Optical Spectroscopy* (CRC Press, Boca Raton), p 378.
36. Khalil M, Demirdöven N, and Tokmakoff A. 2004. Vibrational coherence transfer characterized with Fourier-transform 2DIR. *J Chem Phys* 121(1): 362.
37. Hamm P, Lim M, DeGrado WF, and Hochstrasser RM. 1999. The two-dimensional IR nonlinear spectroscopy of a cyclic penta-peptide in relation to its three-dimensional structure. *Proc Natl Acad Sci USA* 96(5):2036–2041.
38. Ham S, Kim JH, Lee H, and Cho MH. 2003. Correlation between electronic and molecular structure distortions and vibrational properties. II. Amide I modes of NMA-nD(2)O complexes. *J Chem Phys* 118(8):3491–3498.
39. Corcelli SA, Lawrence CP, and Skinner JL. 2004. Combined electronic structure/molecular dynamics approach for ultrafast infrared spectroscopy of dilute HOD in liquid H_2O and D_2O. *J Chem Phys* 120(17):8107–8117.
40. Oh KI et al. 2008. Nitrile and thiocyanate IR probes: Molecular dynamics simulation studies. *J Chem Phys* 128(15): 154504.
41. Waegele MM and Gai F. 2010. Computational modeling of the nitrile stretching vibration of 5-cyanoindole in water. *J Phys Chem Lett* 1(4):781–786.
42. Lindquist BA, Haws RT, and Corcelli SA. 2008. Optimized quantum mechanics/molecular mechanics strategies for nitrile vibrational probes: Acetonitrile and para-tolunitrile in water and tetrahydrofuran. *J Phys Chem B* 112(44):13991–14001.
43. Bagchi S, Falvo C, Mukamel S, and Hochstrasser RM. 2009. 2D IR experiments and simulations of the coupling between amide-I and ionizable side chains in proteins: Application to the villin headpiece. *J Phys Chem B* 113(32):11260–11273.
44. Li SZ, Schmidt JR, Corcelli SA, Lawrence CP, and Skinner JL. 2006. Approaches for the calculation of vibrational frequencies in liquids: Comparison to benchmarks for azide/water clusters. *J Chem Phys* 124(20): 204110.
45. Kwac K and Cho MH. 2003. Molecular dynamics simulation study of *N*-methylacetamide in water. I. Amide I mode frequency fluctuation. *J Chem Phys* 119(4):2247–2255.
46. Schmidt JR, Corcelli SA, and Skinner JL. 2004. Ultrafast vibrational spectroscopy of water and aqueous N-methylacetamide: Comparison of different electronic structure/molecular dynamics approaches. *J Chem Phys* 121(18):8887–8896.
47. Lin YS, Shorb JM, Mukherjee P, Zanni MT, and Skinner JL. 2009. Empirical amide I vibrational frequency map: Application to 2D IR line shapes for isotope-edited membrane peptide bundles. *J Phys Chem B* 113(3):592–602.
48. Hayashi T, Zhuang W, and Mukamel S. 2005. Electrostatic DFT map for the complete vibrational amide band of NMA. *J Phys Chem A* 109(43):9747–9759.
49. Jansen TL and Knoester J. 2006. A transferable electrostatic map for solvation effects on amide I vibrations and its application to linear and two-dimensional spectroscopy. *J Chem Phys* 124(4): 044502.
50. Bloem R, Dijkstra AG, Jansen TLC, and Knoester J. 2008. Simulation of vibrational energy transfer in two-dimensional infrared spectroscopy of amide I and amide II modes in solution. *J Chem Phys* 129(5): 055101.
51. Watson TM and Hirst JD. 2005. Theoretical studies of the amide I vibrational frequencies of [Leu]-enkephalin. *Mol Phys* 103(11–12):1531–1546.
52. Maekawa H and Ge NH. 2010. Comparative study of electrostatic models for the amide-I and -II modes: Linear and two-dimensional infrared spectra. *J Phys Chem B* 114(3):1434–1446.
53. Kuroda DG, Vorobyev DY, and Hochstrasser RM. 2010. Ultrafast relaxation and 2D IR of the aqueous trifluorocarboxylate ion. *J Chem Phys* 132(4): 044501.
54. Kobus M, Nguyen PH, and Stock G. 2011. Coherent vibrational energy transfer along a peptide helix. *J Chem Phys* 134(12): 124518.

7 Polarization-Controlled Chiroptical and 2D Optical Spectroscopy

Kyung-Won Kwak, Kwang-Hee Park, and Minhaeng Cho

CONTENTS

7.1 INTRODUCTION

On a microscopic level, many molecular processes occur on femtosecond (10^{-15} s) and picosecond (10^{-12} s) timescales. For example, electron and proton transfers take place in a few femtoseconds and picoseconds, respectively. Water molecules reorient in a few picoseconds in liquids and exchange their hydrogen-bond partners in several picoseconds in aqueous solutions [1,2]. Carbon–carbon bond rotates in tens of picoseconds [3]. Solvent molecules around a solute can reorganize on subpicosecond and picosecond timescales upon an electronic perturbation of the solute [4,5].

Ultrafast spectroscopy has thus proven to be a powerful experimental tool to understand such molecular processes taking place on picosecond timescales in chemistry, physics, and biology. Over the last two decades, ultrafast nonlinear spectroscopy has been developed in concert with rapid advances in ultrashort and high-power laser technology. Nowadays, Ti:sapphire oscillator and amplifier systems producing ~40 fs in pulse duration with a few mJ of pulse energy are commercially available. Femtosecond pulses in the range from visible to mid-IR region can be readily generated by using various nonlinear optical processes. A variety of ultrafast nonlinear spectroscopic techniques have been employed to investigate the structure and dynamics of molecular systems.

Recently, an interesting advancement in chiroptical spectroscopy, which is an exceptional tool for characterizing the molecular chirality, has been made [6,7]. Here, a molecule is *chiral* if its mirror image does not superimpose onto the original [8,9]. In fact, almost all natural products, biomolecules, and synthetic drugs are chiral and their handedness plays an essential role in biological function, asymmetric catalytic reaction, drug binding, and so on. Such chiral molecules exhibit distinctive optical properties known as *optical activity* (OA). Thus, a variety of OA measurement methods such as circular dichroism (CD), optical rotatory dispersion (ORD), circular luminescence, Raman optical activity, and so on have been extensively used to elucidate the conformations and dynamics of biologically important molecules (protein, DNA, RNA, etc.) and to determine the absolute configurations of asymmetric compounds (drug, catalyst, etc.) [9].

Our recent OA measurement method has been referred to as femtosecond OA free-induction-decay (OA-FID) technique because the chiral FID field is detected using either active or self-heterodyning scheme [6,7,10–14]. Here, typical electric dipole FID field refers to the electric field radiated by a collection of dipoles. Essentially, much like the pulse-based NMR spectroscopy [15,16], the time-domain OA-FID created and detected by a femtosecond optical pulse is directly Fourier-transformed to obtain the frequency-domain CD and ORD spectra. For the success of this measurement method, precise controls of the polarization states of not only the incident field but also the transmitted chiral signal field are prerequisite. It turns out that this novel approach based on a combination of polarization-controlling and heterodyne-detection techniques has the following advantages over the conventional methods utilizing both left- and right-handed radiations: (i) background-free measurement, (ii) nondifferential detection, (iii) single-shot measurability, (iv) simultaneous CD/ORD (imaginary and real parts of the complex chiral susceptibility) measurement, and (v) ultrafast time-resolving capability. A few proof-of-principle experiments have been carried out in both vibrational and electronic transition frequency ranges to demonstrate its experimental feasibility and enhanced sensitivity. In this chapter, we shall present detailed discussions on how important it is to control polarization states of electromagnetic fields in linear chiroptical spectroscopy.

In addition to the polarization-controlled chiroptical spectroscopy, the time-resolved pump-probe (PP) and two-dimensional (2D) optical spectroscopy have also utilized various polarization-controlling methods [17,18]. In the pump-probe spectroscopy, a strong pump pulse coherently excites the molecular system of interest and subsequently a time-delayed probe pulse is used to monitor the relaxation of the molecular system as a function of time, which thus provides dynamic information on the molecular system of interest. More recently, ultrafast multidimensional spectroscopy involving multiple visible and/or infrared pulses has been developed and used to study the molecular dynamics of complex systems that could not be studied by using conventional one-dimensional spectroscopic methods [19]. Congested dynamic information of the molecular system hidden in the one-dimensional spectrum is disentangled in a multidimensional frequency domain. In the present chapter, we shall not provide detailed theoretical and experimental results on multidimensional spectroscopy based on either pump-probe or photon echo measurement scheme, since there already exist thorough reviews and books on this topic published over the years [17–27]. Instead, we shall focus on how controlling polarization states of incident beams enable us to extract critical information on molecular structures and possibly intra- or intermolecular dynamics of coupled multichromophore systems.

A conventional approach to measuring rotational dynamics of chromophores in solutions is to obtain the anisotropic signal that is defined as the ratio of the difference between parallel and perpendicular polarization PP signals to the isotropic signal, that is,

$$R(t) = \frac{S_{\parallel}(t) - S_{\perp}(t)}{S_{\parallel}(t) + 2S_{\perp}(t)}. \tag{7.1}$$

Here, the parallel (perpendicular) PP signal is obtained by controlling the polarization direction of the probe beam to be parallel (perpendicular) to that of the pump beam. Despite the fact that the parallel and perpendicular PP signals have been used to monitor the time evolution of transition dipoles for varying pump-probe delay time, it should be noted that the more general PP signal can be a function of the relative angle φ between the pump and probe polarization directions as $S_{PP} = S_{PP}(t;\varphi)$. Bearing this aspect in mind, let us briefly discuss the experimental geometry used to measure the 2D spectroscopic signals. In general, three incident laser pulses are used to create the third-order polarization in the material and the coherent signal electric field is often characterized by employing an interferometric detection with yet another field called local oscillator (LO). Therefore, totally four different electric fields whose polarization states can be experimentally controlled are used to carry out 2D optical measurements in general. The 2D optical signal is therefore a function of three independent angles of the polarization directions, when one of the four is fixed, that is, $S_{2D} = S_{2D}(t_1, t_2, t_3; \varphi_1, \varphi_2, \varphi_3)$, where t_i are the delay times between pulses. Recently, we have shown that the polarization-angle-scanning method in the 2D optical measurement geometry is quite useful to determine the relative angles between two different transition dipole moments, which are essentially related to the molecular structure [17,28–30].

Here, in this chapter, we shall first present a discussion on the polarization-controlled chiroptical spectroscopy with an emphasis on the coherent optical activity measurement technique (Section 7.2). Particularly, detailed descriptions on the *active*- and *self*-heterodyned-detection methods specifically designed for the phase-and-amplitude characterization of the chiral FID field are presented. We shall show that the active- and self-heterodyne-detection schemes for the linear chiroptical measurements are analogous to the pump-probe and photon echo techniques used in the 2D optical spectroscopy. In Section 7.3, the polarization-controlled pump-probe and 2D optical spectroscopy is discussed in detail. In Section 7.4, novel chiroptical 2D spectroscopic methods are theoretically proposed. Finally, the summary and a few concluding remarks are given in Section 7.5.

7.2 LINEAR CHIROPTICAL SPECTROSCOPY

The property of light propagating through a medium differs from that through a vacuum because its speed and intensity are affected by the frequency-dependent refractive index $n(\omega)$ and absorption coefficient $\kappa(\omega)$ of the medium, respectively, which reflect the intrinsic optical properties of the medium. Consequently, a given electromagnetic wave is subject to frequency-dependent phase retardation and attenuation simultaneously. In the case that the optical medium is spatially isotropic and contains no chiral molecules, these quantities at a given frequency remain the same irrespective of radiation polarization state, meaning that the transmitted light polarization state remains the same. However, for a solution containing chiral molecules, this is not the case for chiral fields because the material properties, $n(\omega)$ and $\kappa(\omega)$, leading to the dispersion and absorption processes depend on its (left or right) handedness, that is, $n_L(\omega) \neq n_R(\omega)$ and $\kappa_L(\omega) \neq \kappa_R(\omega)$. The circular birefringence (CB) and CD that are commonly referred to as *optical activity* are directly related to the frequency-dependent differential absorption coefficient, $\Delta\kappa(\omega) = \kappa_L(\omega) - \kappa_R(\omega)$, and the differential refractive index, $\Delta n(\omega) = n_L(\omega) - n_R(\omega)$, respectively. In principle, $\kappa(\omega)$ and $n(\omega)$ are related to each other via the *Kramers–Kronig* relations, but in practice, the two observables, CD and ORD spectra, should be measured independently due to the limitation in the experimentally tunable frequency range. The CD spectroscopy has been more widely used than the ORD measurement in studying structural details of chiral molecules. That is not only because the measurement is comparatively easy, but also because direct comparisons between experimental and quantum mechanically calculated results are possible. Despite the success of the conventional electronic and vibrational CD (ECD and VCD) measurement methods, they still pose certain limitations that prevented further methodological advancements for a wide range of applications including time-resolved ECD and VCD studies of biomolecules.

7.2.1 Polarization States of Radiation

The electromagnetic radiation used is a transverse oscillating wave of electric and magnetic fields, where the two are in phase and mutually perpendicular. The spatial dependence of an oscillating electric field is given by

$$\tilde{E}(\boldsymbol{r},t) = \tilde{\boldsymbol{e}}E_0 \exp[i(\boldsymbol{k} \cdot \boldsymbol{r} - \omega t)], \tag{7.2}$$

where $\tilde{E}(\boldsymbol{r},t)$ is a complex vector representing the electric field and its real part is the physical observable. E_0 is a scalar representing the maximum magnitude of the field, and $\tilde{\boldsymbol{e}}$ is the complex polarization vector of unit magnitude representing the polarization state of the radiation. Note that this unit vector $\tilde{\boldsymbol{e}}$ is perpendicular to the propagation direction of the transverse electromagnetic wave. Throughout this chapter, we assume that the radiation is traveling in the laboratory Z-direction so that $\boldsymbol{k} = (\omega/c)\tilde{\boldsymbol{e}}_Z$ and the state of polarization is confined to the XY-plane as the wave is transverse.

Then, the electric field with an arbitrary polarization state can be generally written as

$$\tilde{E}(\boldsymbol{r},t) = \frac{1}{\sqrt{|a_X|^2 + |a_Y|^2}}(a_X\tilde{\boldsymbol{e}}_X + a_Y\tilde{\boldsymbol{e}}_Y)E_0 \exp[i(\boldsymbol{k} \cdot \boldsymbol{r} - \omega t)], \tag{7.3}$$

where the factors a_X and a_Y can be complex numbers. In the case of linearly polarized (LP) radiation, we have

$$\tilde{E}_{\text{LP}}(\boldsymbol{r},t) = \frac{1}{\sqrt{a_X^2 + a_Y^2}}(a_X\tilde{\boldsymbol{e}}_X + a_Y\tilde{\boldsymbol{e}}_Y)E_0 \exp[i(\boldsymbol{k} \cdot \boldsymbol{r} - \omega t)], \tag{7.4}$$

where a_X and a_Y are purely real numbers. On the other hand, the electric fields of the left- and right-circularly polarized (LCP and RCP) radiations are given as

$$\begin{aligned}\tilde{E}_{\text{LCP}}(\boldsymbol{r},t) &= \frac{1}{\sqrt{2}}(\tilde{\boldsymbol{e}}_X + i\tilde{\boldsymbol{e}}_Y)E_0 \exp[i(\boldsymbol{k} \cdot \boldsymbol{r} - \omega t)] \\ \tilde{E}_{\text{RCP}}(\boldsymbol{r},t) &= \frac{1}{\sqrt{2}}(\tilde{\boldsymbol{e}}_X - i\tilde{\boldsymbol{e}}_Y)E_0 \exp[i(\boldsymbol{k} \cdot \boldsymbol{r} - \omega t)],\end{aligned} \tag{7.5}$$

where $a_X = 1$ and $a_Y = \pm i$ depending on the handedness of the chiral electric field. Elliptically polarized radiation can then be written as a linear combination of linearly polarized radiation and circularly polarized radiation with proper weighting factors.

7.2.2 Difference Intensity Measurement: Conventional Approaches

The conventional differential intensity measurement of the CD signal utilizes both LCP and RCP radiations. Often, an equal amount of LCP or RCP beam (I_0) is alternately generated by a phase-retarder (PR) or polarization modulation device and then injected into the chiral sample solution (see Figure 7.1a). Each intensity ($I_{L,R}$) of the transmitted field is attenuated by the optical sample due to the absorption that is related to the imaginary part of the susceptibility. The two intensities are separately recorded at the spectrometer, which are then converted into the corresponding absorption spectra by taking their logarithmic values as

$$\Delta A = A_L - A_R = -\log\left(\frac{I_L}{I_0}\right) + \log\left(\frac{I_R}{I_0}\right) = \log\left(\frac{I_R}{I_L}\right). \qquad (7.6)$$

Owing to the frequency dependence of I_L and I_R, the difference absorption spectrum ΔA exhibits variations in amplitude and sign. With $\Delta I = I_L - I_R$ and $I = (I_L + I_R)/2$, ΔA is approximately given by $-\Delta I/(2.303 \times I)$. In the case of the electronic CD, this intensity ratio is about 10^{-4}–10^{-3}, whereas the ratio for the vibrational CD is even one or two orders of magnitude smaller than the electronic CD. When the absorption intensity of LCP or RCP beam is measured, most of the absorbed photons contribute to the achiral signal, which is considered to be a noise, originating from the electric dipole FID field. When we measure the difference intensity ΔI, these extra photons act as a fluctuating noise in addition to the light source fluctuation, increasing shot noise. They combine to deteriorate the signal-to-noise ratio significantly. Since ΔA value is typically about 10^{-3}~10^{-5} at absorbance A~1, even in the case that the incident radiation is fairly stable and its intensity fluctuation amplitude level is just about 0.1% of its average, it is still very difficult to discriminate such a weak chiral signal (ΔI) from such

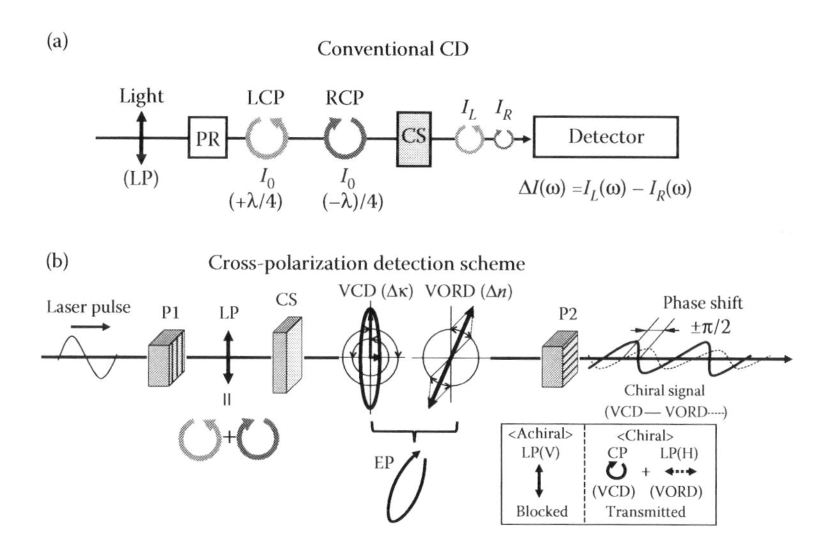

FIGURE 7.1 (a) Conventional differential CD intensity measurement scheme. PR, phase retarder; CS, chiral sample. The PR converts an incident linearly polarized radiation into LCP and RCP radiations by alternately controlling the phase retardation (±λ/4). The attenuated intensity spectra are separately measured and their difference corresponds to the CD spectrum. (b) Cross-polarization detection (CPD) scheme, where P1 and P2 are linear polarizers. The LP radiation after the P1 (vertical) can be viewed as a linear combination of 50% LCP and 50% RCP radiations. After passing through the CS, unequal absorption (Δκ, CD) and phase shift (Δn, ORD) of the two oppositely handed field components lead to a circular-to-elliptical polarization change and at the same time its optical rotation, respectively. The P2 (horizontal) is used to block the vertical LP (achiral) component and allows only the CP (CD) and the horizontal LP (ORD) components to be transmitted through it. The CD and ORD signal fields are in quadrature (±π/2) relation to each other.

a large fluctuating background noise. This is therefore one of the most notorious and fundamental problems of the differential measurement method. We have recently demonstrated that an alternative time-domain approach based on a spectral interferometry is capable of overcoming these difficulties.

7.2.3 Phase-and-Amplitude Measurements of Optical Activity FID Field

The electric field approach we have developed over the years is a nondifferential amplitude-level detection technique, where the OA-FID field is separately detected without undesired achiral background field contribution. The transmitted electric field through the cross-polarization analyzer, where the polarization direction of the detected field is orthogonal to that of the incident radiation, accounts for the response of chiral solution sample. Then, the phase-and-amplitude measurement of the OA-FID field is achieved by employing a Fourier transform spectral interferometry (FTSI). The FTSI [31] is a useful method for characterizing an unknown weak electric field in terms of its spectral phase and amplitude with respect to a reference field called local oscillator and has been widely used in heterodyne-detected 2D optical spectroscopy [18,19,32,33]. Such a heterodyne-detection of the OA-FID using a modified Mach–Zehnder interferometer has allowed us to obtain information on both CD and ORD spectra simultaneously [6]. Furthermore, the cross-polarization detection method is especially useful to remove the huge achiral background noise as well as to reduce additional achiral noise originating from incident light intensity fluctuation [14].

In Figure 7.1b, the basic concept of the cross-polarization detection technique is schematically drawn. A linear polarizer (P1) ensures that an incident field becomes a vertical LP beam (=50% LCP + 50% RCP). While traveling through the chiral sample, each opposite handed field component experiences different chiroptical responses so that one of the two components is more attenuated (intensity change) and delayed (phase change) with respect to the other depending on the molecular chirality. As a result, the incident LP beam is transformed into an elliptically polarized (EP) one whose major axis is slightly rotated left or right from the vertical axis. With a close inspection of the transmitted EP field at amplitude level, it becomes possible to decompose it into three different polarization components: vertical LP(V), horizontal LP(H), and circular polarization (CP). The LP(V) represents the achiral FID electric field generated by a collection of oscillating electric dipoles in the sample. However, this LP(V) component can be selectively removed by the second linear polarizer (P2) placed after the sample cell—note that the optic axis of P2 is perpendicular to that of P1. Thus, the transmitted field after the P2 contains both the LP(H) and CP components and they are, respectively, related to the molecular chiral responses, Δn (ORD) and $\Delta\kappa$ (CD).

Here, it is the phase relationship between the polarization components that should be characterized in a subsequent measurement step. The LP(H) component related to the ORD has a phase angle 0° or 180° with respect to the electric dipole FID field, denoted as LP(V), where the precise phase angle depends on the rotation direction of the generated EP field. On the other hand, the horizontal component of the CP is phase-shifted by either +90° or −90° depending on molecular chirality. Therefore, the CD and ORD components in the transmitted field satisfy a well-defined in-quadrature phase relationship between the two, since the optical rotation is often negligibly small. Then, by using a proper electric field measurement technique that is phase-sensitive, the handedness and intensity of CD and ORD signals can be simultaneously obtained by directly measuring the phase (handedness) and amplitude (intensity) of the chiral electric field. This is essentially the basis of the enhanced chiral selectivity achievable by the present electric field approach.

In the cross-polarization detection (CPD) scheme, the unit vector of the incident radiation is given as $\tilde{e} = \tilde{e}_Y$ (vertical direction) and the (horizontal) X-component of the linear polarization $P_X^{CPD}(t)$ is given as [10]

$$P_X^{CPD}(t) = -\frac{i}{2}\int_0^\infty d\tau \Delta\chi(\tau) E(t-\tau), \tag{7.7}$$

where $\Delta\chi(\tau)$ is the chiral response function. In practice, the experimentally measured quantity is however not the polarization itself but the electric field $\mathbf{E}(t)$. For the X- and Y-components of the transmitted signal electric field at position z inside the sample, $E_X(z,t)$ and $E_Y(z,t)$, we found that the two are coupled to each other in the corresponding Maxwell equation, that is,

$$\nabla^2 E_X(z,t) - \frac{1}{c^2}\frac{\partial^2}{\partial t^2}E_X(z,t) = \frac{4\pi}{c^2}\frac{\partial^2}{\partial t^2}P_X^{CPD}(z,t), \tag{7.8}$$

where

$$P_X^{CPD}(z,t) = -\frac{i}{2}\int_0^\infty d\tau \Delta\chi(\tau)E_Y(z,t-\tau) + \int_0^\infty d\tau \chi_{\mu\mu}(\tau)E_X(z,t-\tau). \tag{7.9}$$

Note that the $P_X^{CPD}(z,t)$, determined by both $E_X(z,t)$ and $E_Y(z,t)$, acts as the source generating $E_X(z,t)$. Solving Equation 7.8 in the frequency domain, one can find that the X-component of the signal electric field, that is, OA-FID, after the sample length L is given as [10]

$$E_X(\omega) = \left(\frac{\pi\omega L}{cn(\omega)}\right)\Delta\chi(\omega)E_Y(\omega), \tag{7.10}$$

where $n(\omega)$ and c are the refractive index and the speed of light, respectively. Here, $E_X(\omega)$ contains information on the chiroptical properties of the molecules and the complex function $\Delta\chi(\omega)$ is the linear chiral susceptibility. Equation 7.10 shows that the phase-and-amplitude relationship between $E_X(\omega)$ and $E_Y(\omega)$ provides direct information on $\Delta\chi(\omega)$ whose imaginary and real parts correspond to the CD and ORD spectra, respectively.

One of the most efficient methods to characterize the spectral phase-and-amplitude of an unknown electric field is the Fourier transform spectral interferometry (FTSI), which involves (i) a heterodyned interferometric detection and (ii) an inverse Fourier transform–Fourier transformation of the measured spectral interferogram (see Figure 7.2). The signal field is allowed to interfere with a reference field called *local oscillator* (Figure 7.2a). The experimentally measured spectrum called spectral interferogram exhibits highly oscillating features due to the temporal delay between the signal and local oscillator fields [7]. Often, the phase and amplitude of the local oscillator field should be predetermined by other interferometric method, or a proper phase-correction scheme should be used. However, the present chiroptical spectroscopic method relying on a self-referencing technique does not require *a priori* determination of the phase and amplitude of the local oscillator field at all. Here, the spectral interferogram between the pulsed signal (E_s) and local oscillator (E_{LO}) fields is given as

$$S^{het}(\omega) = 2\,\mathrm{Re}\left[E_s(\omega)E_{LO}^*(\omega)\exp(i\omega\tau_d)\right], \tag{7.11}$$

where τ_d represents the delay time between the signal and LO fields. Since the measured spectral interferogram $S^{het}(\omega)$ itself is a real function, it does not directly provide spectral phase information of $E_s(\omega)$. The standard inverse-FT (F^{-1}) and FT (F) procedure enables one to convert such real function into its complex form. The stepwise procedure is as follows: (i) inverse-FT of $S^{het}(\omega) \rightarrow F^{-1}\{S^{het}(\omega)\}$, (ii) multiplying the time-domain signal $F^{-1}\{S^{het}(\omega)\}$ by a heavyside step function $\theta(t) \rightarrow \theta(t)F^{-1}\{S^{het}(\omega)\}$, and (iii) FT of the positive time domain signal $\rightarrow F[\theta(t)F^{-1}\{S^{het}(\omega)\}]$. Therefore, the complex electric field $E_s(\omega)$ can be retrieved as by using the following formula:

$$E_s(\omega) = \frac{F\left[\theta(t)F^{-1}\{S^{het}(\omega)\}\right]}{2E_{LO}^*(\omega)\exp(i\omega\tau_d)}. \tag{7.12}$$

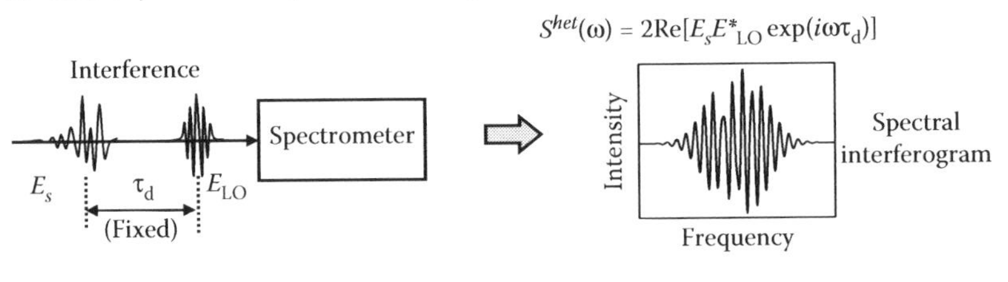

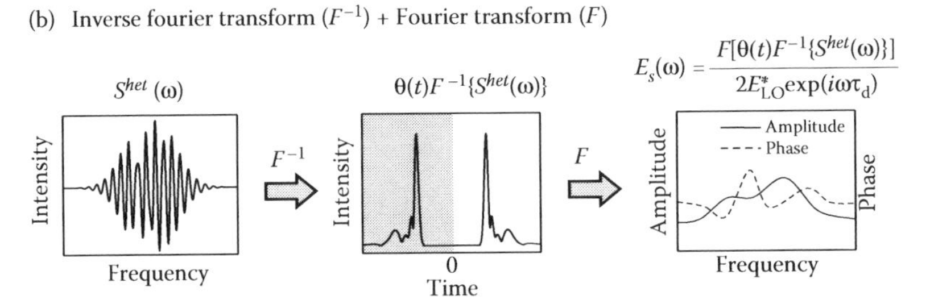

FIGURE 7.2 Standard Fourier transform spectral interferometry (FTSI) procedure for a phase-and-amplitude measurement of an unknown signal electric field (E_s). (a) Frequency-domain-heterodyned interferometric detection. The signal field is time-separated from the local oscillator field, but the two fields interfere with each other in the frequency domain. The interference spectral called spectral interferograms is detected with a spectrograph. (b) Stepwise inverse-FT-and-FT transformation procedure. The spectral interferograms is Fourier-transformed to the corresponding time-domain signal. Inverse Fourier transformation of the positive time-domain signal gives the complex signal spectrum.

In principle, a complete characterization of the local oscillator field $E_{LO}(\omega)$ is possible by using a few well-known nonlinear optical techniques like FROG [34], SPIDER [35], and so on. However, such field characterization requires additional complicated measurement method and is often a difficult task. Furthermore, a precise determination of τ_d within an optical period (less than a few femtoseconds) can be a challenging issue. Fortunately, it was shown that such difficulties can be overcome by using the linear relationship (Equation 7.10) between the complex chiral susceptibility $\Delta\chi(\omega)$ and the ratio of the chiral field ($E_X(\omega)$) to the achiral field ($E_Y(\omega)$), that is, $\Delta\chi(\omega) \propto E_X(\omega)/E_Y(\omega)$—note that the measurements of $E_X(\omega)$ and $E_Y(\omega)$ can be easily achieved by controlling the second linear polarizer P2 in Figure 7.1b. Thus, we have

$$E_{X,Y}(\omega) = \frac{F\left[\theta(t)F^{-1}\{S^{het}_{X,Y}(\omega)\}\right]}{2E^*_{LO}(\omega)\exp(i\omega\tau_d)}, \tag{7.13}$$

where $S^{het}_X(\omega)$ and $S^{het}_Y(\omega)$ are the detected spectral interferograms of the signal fields that are perpendicular and parallel to the polarization direction of the incident LP radiation. It should be emphasized that the complex spectra, $E_X(\omega)$ and $E_Y(\omega)$, have the common factor in the denominator of Equation 7.13. Consequently, the ratio $E_X(\omega)/E_Y(\omega)$ does not depend on the detailed local oscillator field spectrum $E_{LO}(\omega)$ or τ_d. As long as the phase stability of the spectrometer is maintained during the measurements of $S^{het}_X(\omega)$ and $S^{het}_Y(\omega)$, the ratio spectrum $E_X(\omega)/E_Y(\omega)$ gives direct information on $\Delta\chi(\omega)$. Combining these results, we find that the chiral susceptibility is experimentally measured as

$$\Delta\chi(\omega) \propto \frac{F\left[\theta(t)F^{-1}\{S^{het}_X(\omega)\}\right]}{F\left[\theta(t)F^{-1}\{S^{het}_Y(\omega)\}\right]}. \tag{7.14}$$

This formula connecting the linear chiral susceptibility to the heterodyne-detected spectral interferograms shows that the electric field approach is capable of characterizing the complex $\Delta\chi(\omega)$ without precise characterizations of $E_{LO}(\omega)$ and τ_d so that the extremely weak chiroptical signal could be measured with just a single laser pulse [14].

7.2.4 Active- and Self-Heterodyne-Detection Methods

The electric field approach discussed above can be considered to be an active heterodyne-detection technique because the signal field itself is in interference with an additional reference field added to detector. The cross-polarization geometry for the purpose of selectively removing the achiral background field is one of the important elements for the success of this measurement method. In fact, there is a related but different experimental method utilizing such an interference between chiral signal field and incident field, which has been referred to as ellipsometric technique employing a quasi-null geometry with two linear polarizers (Figure 7.3a) [36]. The elliptometric chiroptical spectroscopy was pioneered by Kliger and coworkers [36,37]. Much like the cross-polarization scheme, two crossed linear polarizers were used. However, instead of a linearly polarized radiation, an elliptically polarized beam with vertical major axis (Y-axis), which can be produced by a phase-retarder, was used to generate an electronic OA-FID field in visible frequency domain. Here, it should be noted that an elliptically polarized radiation is a linear combination of a major Y-axis-polarized LP radiation and a minor X-polarized LP radiation phase-shifted by $\pm\pi/2$, depending on the handedness of the elliptically polarized radiation. The major axis (Y-polarized) component of the electric field is used to create the OA-FID field whose polarization direction is parallel to the X-axis. Then, this OA-FID field interferes with the minor X-polarized but phase-shifted component of the incident elliptically polarized beam. Then, the intensity of the X-polarized interference

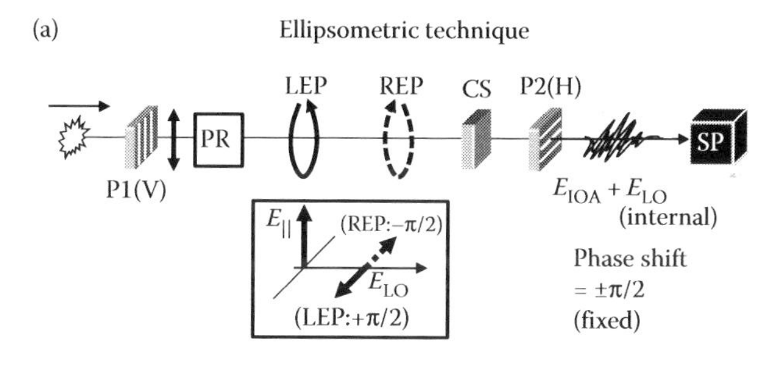

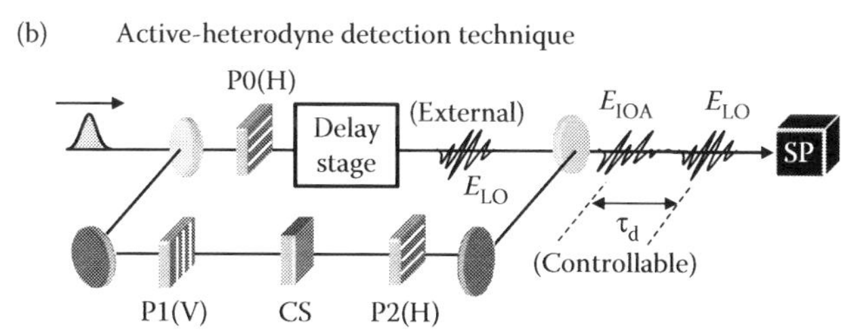

FIGURE 7.3 Comparison between self-heterodyne-detection method (a) using an elliptically polarized radiation and active-heterodyne-detection method (b). P0-2, linear polarizers; V, vertical; H, horizontal; PR, phase retarder; SP, spectrometer, CS, chiral sample. In the ellipsometric self-heterodyne-detection scheme, the major axis component of the incident elliptically polarized radiation is used to create linear polarization in the sample. Then, the optical activity signal field whose polarization direction is perpendicular to the major axis of the incident radiation interferes with the minor axis component of the incident radiation. On the other hand, in the active-heterodyning scheme, the local oscillator (E_{LO}) is externally controlled so that the time delay between E_{LO} and the signal is controllable.

field is selectively measured. Although this technique is still an intensity (not phase-and-amplitude) measurement method, it can be considered to be a self-heterodyne-detection scheme because the in-quadrature phase-different X-polarized component essentially acts like a local oscillator interfering with the generated chiral signal field whose polarization direction is also parallel to the X-axis. Recently, Helbing and coworkers and we experimentally demonstrated that such an ellipsometric technique can be used to detect vibrational and electronic CD spectra, respectively, with a significantly enhanced detection sensitivity [38,39].

However, there is a major difference between the ellipsometric method and our active heterodyne-detection method, which is related to the way of controlling the relative phase between signal and reference (local oscillator) fields during the heterodyning process [7]. In the ellipsometric detection geometry, the chiral signal field interferes with the incident X-polarized electric field component itself. Consequently, the phase shift between the chiral signal and intrinsically present local oscillator fields is not experimentally controllable. As a result, the imaginary (CD) and real (ORD) parts of the OA response should be measured separately. On the other hand, the cross-polarization interferometric technique utilizing a modified Mach–Zehnder interferometer shown in Figure 7.3b uses an external local oscillator (for active-heterodyning) so that both the imaginary and real parts of $\Delta\chi(\omega)$ can be simultaneously obtained via the FTSI procedure discussed above.

Despite the fact that the active- and self-heterodyne-detection techniques have certain advantages and disadvantages, it is important to note that controlling polarization states of incident radiation as well as of the transmitted signal electric field are prerequisite for the success of signal enhancement and precise characterization of the chiroptical properties of chiral molecules in solutions.

7.2.5 Comparisons with Coherent 2D Optical Spectroscopy

We have discussed two different interference (heterodyned) detection methods for characterizing the chiral signal field. Note that the two differ from each other by how to control the relative phase between the signal and LO fields for each interference measurement. Interestingly, the relationship between the two methods is quite similar to that between *active-heterodyne-detected* stimulated photon echo (PE) with four pulses and *self-heterodyne-detected* pump-probe (PP) with two pulses, which have been extensively used in coherent 2D optical spectroscopy. In the active-heterodyned 2D PE spectroscopy, three incident optical pulses with wavevectors, $\mathbf{k}_1$, $\mathbf{k}_2$, and $\mathbf{k}_3$, are used to create the third-order PE polarization and the resulting signal electric field (E_{PE}) in the specific phase-matching direction is detected by making it interfere with another local oscillator pulse. Consequently, the spectral interferogram recorded at a spectrometer is given as

$$S_{PE}(\omega) = 2\,\mathrm{Re}\left[E_{PE}(\omega)E^{*}_{LO}(\omega)\exp(i\omega\tau_{d})\right]. \tag{7.15}$$

The same Fourier–inverse Fourier transformation procedure is used to convert it to the complex photon echo spectrum $E_{PE}(\omega)$.

On the other hand, the self-heterodyned pump-probe spectroscopy basically utilizes two pulses, where the two pump field–matter interactions and one probe field–matter interaction create the third-order PP signal field. This in turn interferes with the probe field itself. The resulting PP spectrum is thus given as

$$S_{PP}(\omega) = 2\,\mathrm{Re}\left[E_{PP}(\omega)\cdot E^{*}_{pr}(\omega)\right], \tag{7.16}$$

where $E_{PP}(\omega)$ and $E_{pr}(\omega)$ are the complex pump-probe spectrum and the incident probe field spectrum, respectively. Note that the $E_{PP}(t)$ field is always in- or out-of-phase relation to the probe field $E_{pr}(t)$. Therefore, the underlying interference phenomenon in the 2D PP spectroscopy is quite similar to the self-heterodyned chiroptical measurement method discussed above.

A notable difference between the active- and self-heterodyned methods should however be noted. The active-heterodyne techniques for the photon echo measurements have the so-called *phasing* problem originating from experimental difficulties in maintaining the absolute phase difference between the signal field and the LO field throughout the whole measurements. In 2D spectroscopy, this problem was overcome by directly comparing the projected 2D spectrum with the dispersed pump-probe spectrum. However, in the present case of the active-heterodyne-detection of the chiroptical signal field, such phase-fluctuation noise is canceled out by taking the ratio of the chiral spectrum to the achiral spectrum (see Equations 7.10 and 7.14), which allows one to perform an accurate characterization of the complex chiral signal field. Second, similar to the fact that the 2D PE provides both real and imaginary parts of the complex 2D spectrum, the active-heterodyne-detection method is of use to measure both CD and ORD spectra, which are related to the imaginary and real parts of the chiral susceptibility. Now, we shall change the topic and present detailed descriptions on the polarization angle dependences of the 2D PE and PP spectroscopy.

7.3 POLARIZATION-CONTROLLED 2D OPTICAL SPECTROSCOPY

In nonlinear optical, particularly two-dimensional (2D) optical spectroscopy, the molecular system is irradiated with three incident light pulses and the emitted signal electric field is measured and analyzed [17–19]. Therefore, it belongs to the general class known as the four-wave-mixing spectroscopy [40]. The interaction of these four electric fields with a given molecular system can be theoretically analyzed in terms of the nonlinear response function of the system.

7.3.1 A Brief Theoretical Background

In general, any spectroscopic signals are generated by the polarization (dipole density) $\mathbf{P}(\mathbf{r},t)$ of the molecular system that is induced by the radiation–matter interactions. The radiation–matter interaction can be written in a general form as a product of external field and its conjugate molecular operator [17]. In 2D optical spectroscopy, the system interacts with external electric field $\mathbf{E}(\mathbf{r},t)$ and, in the electric dipole approximation, the conjugate molecular property is the system dipole moment $\hat{\boldsymbol{\mu}}$. Then, the interaction energy is given as

$$H_{\text{int}}(t) = -\hat{\boldsymbol{\mu}} \cdot \mathbf{E}(\mathbf{r},t), \tag{7.17}$$

where $\mathbf{E}(\mathbf{r},t)$ is the superposition of the three incident pulses $\mathbf{E}_1$, $\mathbf{E}_2$, and $\mathbf{E}_3$. The total Hamiltonian H of the system is given as $H_0 + H_{\text{int}}$ with H_0 the system Hamiltonian in the absence of radiation. The time evolution of the system is governed by the quantum Liouville equation for the density operator $\rho(t)$

$$\frac{\partial \rho(t)}{\partial t} = -\frac{i}{\hbar}\left[H_0,\rho\right] - \frac{i}{\hbar}\left[H_{\text{int}},\rho\right]. \tag{7.18}$$

Once this equation is solved for $\rho(t)$, any physical observable of the system, say $A(t)$, can be obtained by taking the average $\text{Tr}[\hat{A}\rho(t)]$, where $\hat{A}$ is the quantum mechanical operator for observable A and Tr denotes the trace of a matrix. A diagonal element of the density matrix, ρ_{aa}, is called population because it represents the probability that the system is in state a, whereas the off-diagonal element ρ_{ab} is called coherence. The time evolution of a system in coherence ρ_{ab} exhibits an oscillation with frequency $\omega \approx \omega_{ab} \equiv (E_a - E_b)/\hbar$ determined by the energy difference of the two states involved.

Equation 7.18 can be solved by using the time-dependent perturbation theory, where H_{int} is treated as the perturbation to the reference system characterized by H_0. This provides a power series expansion of $\rho(t)$ with the n-th order term $\rho^{(n)}(t)$ containing n factors of H_{int}. The zeroth-order term is the equilibrium density operator for the unperturbed system, $\rho^{(0)}(t) = \rho_{eq}$, and the higher-order terms can be expressed as [40]

$$\rho^{(n)}(t) = \left(-\frac{i}{\hbar}\right)^n \int_{t_0}^{t} d\tau_n \int_{t_0}^{\tau_n} d\tau_{n-1} \cdots \int_{t_0}^{\tau_2} d\tau_1 G_0(t-\tau_n) L_{int}(\tau_n) G_0(\tau_n - \tau_{n-1}) L_{int}(\tau_{n-1}) \cdots G_0(\tau_2 - \tau_1) L_{int}(\tau_1) G_0(\tau_1 - t_0) \rho(t_0), \tag{7.19}$$

where $G_0(t) = \exp(-iL_0 t/\hbar)$ is the time evolution operator in the absence of the external radiation and the Liouville operators are defined as $L_a A = [H_a, A]$ for $a = 0$, int. Equation 7.19 provides us a clear interpretation of the density matrix evolution: the initial state $\rho(t_0)$ at time t_0 propagates freely without perturbation for $\tau_1 - t_0$, that is, $G_0(\tau_1 - t_0)$, and at $t = \tau_1$, the first radiation–matter interaction takes place, which corresponds to the action of $L_{int}(\tau_1)$. This is repeated n times until the final interaction at $t = \tau_n$, which is represented by $L_{int}(\tau_n)$. Finally, the system evolves freely up to the observation time t, and its time evolution is determined by $G_0(t - \tau_n)$. All the possible interaction times are allowed for by the multiple integrals over $\tau_1, \ldots, \tau_n$ under the time ordering condition $t_0 \leq \tau_1 \leq \cdots \leq \tau_n \leq t$.

Each of the density matrices obtained in Equation 7.19 provides the corresponding n-th order nonlinear polarization $\mathbf{P}^{(n)}(\mathbf{r},t)$ as

$$\mathbf{P}^{(n)}(\mathbf{r},t) = \mathrm{Tr}\left[\hat{\mu}\rho^{(n)}(t)\right] = \int_0^{\infty} dt_n \cdots \int_0^{\infty} dt_1 \mathbf{R}^{(n)}(t_n,\ldots,t_1) \vdots \mathbf{E}(\mathbf{r},t-t_n) \cdots \mathbf{E}(\mathbf{r},t-t_n \cdots - t_1) \tag{7.20}$$

with the nonlinear response function $\mathbf{R}^{(n)}(t_n, \ldots, t_1)$ defined as

$$\mathbf{R}^{(n)}(t_n,\ldots,t_1) = \left(\frac{i}{\hbar}\right)^n \theta(t_n)\cdots\theta(t_1)\left\langle \mu(t_n + \cdots + t_1)[\mu(t_{n-1} + \cdots + t_1),[\cdots[\mu(t_1),[\mu(0),\rho_{eq}]]\cdots]]\right\rangle, \tag{7.21}$$

where $\mu(t) = \exp(iH_0 t/\hbar)\mu \exp(-iH_0 t/\hbar)$ is the dipole operator in the interaction picture and the angular bracket denotes the trace of a matrix. The time variables $\tau_1, \ldots, \tau_n$ in Equation 7.19 represent the times at which the radiation–matter interactions take place, whereas $t_1, \ldots, t_n$ in Equations 7.20 and 7.21 are time intervals between them such that $t_m = \tau_{m+1} - \tau_m$ $(1 \leq m \leq n-1)$ and $t_n = t - \tau_n$. According to the time ordering in Equation 7.20, $t_1, \ldots, t_n$ are all positive and the response function must vanish if any of its time arguments is negative. This reflects the causality principle and it is imposed in the response function in Equation 7.21 by the heavyside step function $\theta(t)$. In addition, according to Equation 7.20, the response function connects two real quantities, $\mathbf{P}^{(n)}(t)$ and $\mathbf{E}(\mathbf{r},t)$, and therefore must also be real.

In the third-order optical spectroscopy such as 2D PE and PP, there are a number of experimentally controllable variables of incident radiations, such as (i) center frequencies, (ii) pulse-to-pulse delay times, (iii) relative optical phase of each pulse, (iv) propagation directions, and (v) polarization states. Recently, Cho and coworkers showed that scanning polarization directions of pulses involved in 2D photon echo measurement enables one to increase the sensitivity in determining the molecular structures of coupled multioscillator systems in condensed phase [17,28–30]. Particularly, it was experimentally demonstrated that the angle between two transition dipoles of coupled oscillators can be measured by examining the set of incident beam polarization directions resulting in the suppression of the corresponding cross peak in a given 2D spectrum. Thereby, this extended version of 2D spectroscopy was referred to as polarization-angle-scanning two-dimensional spectroscopy or shortly PAS 2D spectroscopy. Numerical simulation studies combined with quantum chemistry calculation results showed that the PAS 2D spectroscopy can provide intricate structural details of polypeptides such as extended β-sheets [30].

Here, it should however be mentioned that discrete controls of the polarization directions of incident radiations have been widely used to extract information on rotational dynamics and angles between a pair of transition dipole vectors. In this case, the linear combination or the ratio of measured amplitudes in two orthogonal polarization directions has been used. For example, the parallel and perpendicular polarization 2D spectra were used to measure the so-called cross-peak anisotropy, which in turn provide information on the angle between transition dipoles associated with the cross peak [41]. Furthermore, it was shown by Hochstrasser and coworkers that all the diagonal peaks can be simultaneously eliminated by taking a proper difference between the parallel and perpendicular 2D spectra, which allow one to enhance frequency resolutions of the cross peaks [42]. In the pump-probe spectroscopy, the isotropic and anisotropic signals were used to determine the lifetime of an excited state and rotational relaxation rate, respectively.

Instead of measuring just two independent signals, for example, parallel and perpendicular 2D spectra, one can scan one of the polarization directions with all the other three fixed. Such PAS 2D spectrum is in principle written as a linear combination of the parallel (XXXX) and perpendicular (XYYX) 2D spectra, where the weighting factors are determined by the relative angles of the three beam polarization directions. However, in practice, it is not easy to measure the absolute intensities of these two third-order spectra due to the intrinsic laser intensity fluctuation. Consequently, the interpolation with only two (XXXX and XYYX) 2D spectra to describe arbitrary PAS 2D spectra should be of limited success. In the present section, the polarization angle dependence of the 2D PE and PP spectra of a coupled two-oscillator system is discussed. In addition, we shall show that the amplitude and phase of the oscillatory quantum beat signal arising from vibrational coupling are of critical use for estimating the relative angle between a pair of transition dipoles. Even one can selectively remove such oscillatory quantum beat contribution to the signal by properly adjusting the electric field polarization directions.

Now, to describe the orientational contribution to the nonlinear response function separately, we assume that the coupling between the vibronic and rotational degrees of freedom is negligible, that is, Born–Oppenheimer approximation [43]. Then, the molecular Hamiltonian can be partitioned as $H_{\mathrm{mol}} = H_{\mathrm{vib}} + H_{\mathrm{rot}}$ and the transition dipole moment is given as

$$\hat{\mu} = \mu\hat{\mu}, \tag{7.22}$$

where μ is the electric dipole operator associated with vibronic transitions and $\hat{\mu}$ is the unit vector representing the transition dipole direction in the laboratory-fixed frame. Then, each nonlinear response function component can then be factorized into the rotational and vibrational terms, that is [43]

$$\mathbf{R}_\alpha(t_3,t_2,t_1) = \mathbf{Y}_\alpha(t_3,t_2,t_1)R_\alpha(t_3,t_2,t_1). \tag{7.23}$$

Here, the fourth-rank tensorial function $\mathbf{Y}_\alpha(t_3, t_2, t_1)$ represents the orientational contribution to the nonlinear response function, which is treated as a classical function.

The reorientational motion of molecules in solution has been successfully described as a random walk over small angular orientations so that the Fokker–Plank equation for the conditional probability density in the rotational phase space (angular coordinate and velocity) was found to be useful. Since the initial distribution of angular velocities is given by the Maxwell–Boltzmann expression, after taking the ensemble average over the angular velocities, it is possible to recast the orientational function [41,43,44], $\mathbf{Y}_\alpha(t_3, t_2, t_1)$, in the following form:

$$\begin{aligned}\mathbf{Y}_\alpha(t_3,t_2,t_1) = &\int d\nu_3\int d\nu_2\int d\nu_1\int d\nu_0\hat{\mu}_3(\nu_3)W\left(\nu_3,t_3|\nu_2\right)\hat{\mu}_2(\nu_2)\\ &\times W\left(\nu_2,t_2|\nu_1\right)\hat{\mu}_1(\nu_1)W\left(\nu_1,t_1|\nu_0\right)\hat{\mu}_0(\nu_0)P_0(\nu_0),\end{aligned} \tag{7.24}$$

where the molecular orientation is specified by the Euler angles $\nu \equiv (\phi, \theta, \chi)$ and the conditional probability function in an angular configurational space is denoted as $W(\nu_{j+1}, t_{j+1} \mid \nu_j)$. The initial probability distribution of the molecular orientation is denoted as $P_0(\nu_0)$, and it is simply equal to $1/8\pi^2$ for an isotropic system containing randomly oriented molecules, that is, solute molecules in solutions.

In the third-order nonlinear spectroscopy of multilevel systems, each nonlinear response function might involve four different electric dipole transitions, that is,

$$\hat{\mu}_3 = \hat{\mu}_d \; \hat{\mu}_2 = \hat{\mu}_c \; \hat{\mu}_1 = \hat{\mu}_b \; \hat{\mu}_0 = \hat{\mu}_a. \tag{7.25}$$

Therefore, $Y_{ijkl}^{dcba}(t_3,t_2,t_1)$ will specifically denote the $[i,j,k,l]$'th element of the orientational part of the nonlinear response function, where the indices i, j, k, and l represent the Cartesian coordinates. The $[i,j,k,l]$'th tensor element of the corresponding nonlinear response function component can then be written as a product form, $Y_{ijkl}^{dcba}(t_3,t_2,t_1)R^{dcba}(t_3,t_2,t_1)$. Throughout this section, we assume that the four beams including the signal field propagate along the Z-axis and the X-component of the signal field is selectively detected. The measurement of the X-component of the signal field can be achieved by controlling the polarization direction of the local oscillator field to be in the X-direction in the case of the heterodyne-detected 2D PE or by controlling the probe field polarization direction to be parallel to the X-axis in the case of the PP spectroscopy. In general, the unit vectors of the three incident radiations can always be written as

$$\begin{aligned} \tilde{e}_1 &= \mathrm{Y}\sin\varphi_1 + \mathrm{X}\cos\varphi_1 & \tilde{e}_2 &= \mathrm{Y}\sin\varphi_2 + \mathrm{X}\cos\varphi_2 \\ \tilde{e}_3 &= \mathrm{Y}\sin\varphi_3 + \mathrm{X}\cos\varphi_3 & \tilde{e}_4 &= \mathrm{X} \end{aligned}. \tag{7.26}$$

Consequently, the 2D PE and PP signals become functions of φ_i (for $i = 1$–3).

7.3.2 Polarization Angle Scanning Pump-Probe Spectroscopy

The incident electric field for a PP measurement is composed of two pulses separated by T, that is,

$$\begin{aligned} \mathbf{E}(\mathbf{r},t) = {} & \tilde{e}_{pu}E_{pu}(t+T)\exp(i\mathbf{k}_{pu}\cdot\mathbf{r} - i\omega_{pu}t) \\ & + \tilde{e}_{pr}E_{pr}(t)\exp(i\mathbf{k}_{pr}\cdot\mathbf{r} - i\omega_{pr}t) + c.c. \end{aligned} \tag{7.27}$$

The first two field–matter interactions occur with the pump pulse and the signal field emitted in the direction $\mathbf{k}_s = -\mathbf{k}_{pu} + \mathbf{k}_{pu} + \mathbf{k}_{pr} = \mathbf{k}_{pr}$ is self-heterodyne-detected by using the probe pulse. The resulting third-order polarization for the PP signal is [17]

$$\begin{aligned} \mathbf{P}_{PP}^{(3)}(\mathbf{r},t) = {} & e^{i\mathbf{k}_{pr}\cdot\mathbf{r}-i\omega t}\int_0^\infty dt_3\int_0^\infty dt_2\int_0^\infty dt_1 \mathbf{R}^{(3)}(t_3,t_2,t_1)\tilde{e}_{pr}\tilde{e}_{pu}\tilde{e}_{pu}^{*} \times \mathbf{E}_{pr}(t-t_3) \\ & \times \mathbf{E}_{pu}(t+T-t_3-t_2)\mathbf{E}_{pu}(t+T-t_3-t_2-t_1)\exp(i\omega t_{pr} - i\omega t_{pu}). \end{aligned} \tag{7.28}$$

The frequency-resolved PP signal corresponds to a Fourier–Laplace transform of the time-domain signal as

$$E_{\mathrm{sig}}^{PP}(T,\omega_t) = \int_0^\infty dt\, E_{\mathrm{sig}}^{PP}(T,t)\exp(i\omega_t t). \tag{7.29}$$

In the pump-probe measurements, since only two pulses are used, there is a single controllable polarization direction angle φ_{pu}, which is the angle between the polarization directions of the pump and probe fields. In the PAS PP spectroscopy, one can measure the PP signals for varying φ_{pu}, that is, $E_{\mathrm{sig}}^{PP} = E_{\mathrm{sig}}^{PP}(T, \omega_{pr}; \varphi_{pu})$.

Traditionally, only two cases that are $\varphi_{pu} = 0$ (parallel PP) and $\varphi_{pu} = \pi/2$ (perpendicular PP) have been considered. The isotropic and anisotropic signals are defined as, respectively

$$PP_{\mathrm{iso}} = PP_{XXXX}(\varphi_{pu} = 0) + 2PP_{XXYY}\left(\varphi_{pu} = \frac{\pi}{2}\right)$$
$$PP_{\mathrm{aniso}} = \frac{\left(PP_{XXXX} - PP_{XXYY}\right)}{PP_{\mathrm{iso}}}. \tag{7.30}$$

For a single oscillator system, the isotropic (anisotropic) signal does not depend on the rotational dynamics (population relaxation) so that they have been used to extract information on lifetime (rotational relaxation time). However, if the target molecule is a coupled multioscillator system, due to the additional Liouville pathways contributing to the PP signal, the physical interpretation of the isotropic and anisotropic signals can be quite complicated. For the sake of simplicity, let us consider a vibrationally coupled two-oscillator system, which has symmetric and antisymmetric normal modes with frequencies of ω_s and ω_a, respectively. The angle between the two vibrational transition dipoles is now denoted as θ. Without loss of generality, one can assume that the direction of the transition dipole moment $\mu_s (= \langle v_s = 1 | \mu | v_s = 0 \rangle)$ of the symmetric mode is parallel to the Z-axis in a molecule-fixed frame, that is, $\mu_s = \hat{z}$, and that μ_a $(= \langle v_a = 1 | \mu | v_a = 0 \rangle)$ is on the z–x plane as

$$\mu_a = \hat{z}\cos\theta + \hat{x}\sin\theta. \tag{7.31}$$

In Figure 7.4, the eight Liouville pathways contributing to the pump-probe signal at the probe frequency of $\omega_{pr} = \omega_s$ are depicted. Hereafter, we will consider the pump-probe signal at a positive waiting

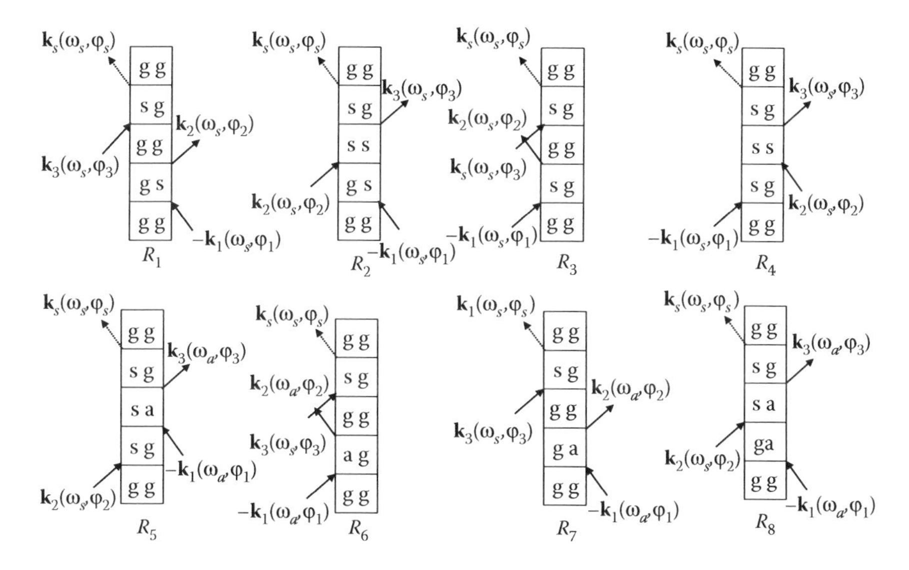

FIGURE 7.4 Double-sided Feynman diagrams associated with the third-order optical signal from a coupled two-oscillator system. In the polarization-angle-scanning PP and 2D PE spectroscopy, the polarization direction angles, φ_1, φ_2, and φ_3, of the three incident radiations can be controlled in a continuous manner. For two-pulse pump-probe measurements, it is assumed that the polarization direction of the probe beam is parallel to the X-axis whereas that of the pump beam is rotated. The pump-probe signal is determined by all the eight Liouville pathways, $R_1 - R_8$. For 2D photon echo experiments, the five response function components, $R_1 - R_5$, are needed to describe the diagonal peak at $\omega_\tau = \omega_s$ and $\omega_t = \omega_a$. The cross peak at $\omega_\tau = \omega_a$ and $\omega_t = \omega_s$ is associated with the $R_6 - R_8$ pathways. Among them, the eighth component R_8 makes the cross oscillate in the waiting time.

time, T. Although the PP signal at zero waiting time does not depend on vibrational and rotational relaxations, the third-order PP signal at $T = 0$ is often contaminated by the coherent artifact as well as additional Liouville pathway contributions originating from the pulse overlapping effects [41,45].

In the impulsive pulse limit, the eight response functions corresponding to each Liouville pathway in Figure 7.4 can be written as

$$\begin{aligned} R_{1\sim4}(0,T,t_3) &= \mu_s^4 Y_{ZZ\varphi\varphi}^{ssss}(0,T,t_3)R(t_3)e^{-T/T_{1,s}} \\ R_5(0,T,t_3) &= \mu_s^2\mu_a^2 Y_{ZZ\varphi\varphi}^{saas}(0,T,t_3)R_{QB}(t_3)e^{i\omega_{sa}T} \\ R_{6\sim7}(0,T,t_3) &= \mu_s^2\mu_a^2 Y_{ZZ\varphi\varphi}^{ssaa}(0,T,t_3)R(t_3)e^{-T/T_{1,a}} \\ R_8(0,T,t_3) &= \mu_s^2\mu_a^2 Y_{ZZ\varphi\varphi}^{sasa}(0,T,t_3)R_{QB}(t_3)e^{i\omega_{sa}T}. \end{aligned} \tag{7.32}$$

The first four response function components are expressed in terms of the same response function determined by the transition dipole, μ_s, and its vibrational lifetime denoted as $T_{1,s}$. The vibrational part $R(t_3)$ of the response function depends on the final coherence evolution time, t_3, in the impulsive pulse limit. Both the rephasing and nonrephasing pathways contribute equally to the PP signal. The fifth response function component describes the quantum beat signal and it is observed only when the pump pulse is spectrally broad enough to excite the symmetric and antisymmetric modes simultaneously. Similarly, the eighth response function component represents the other quantum beat contribution. Note that R_5 $(0,T,t_3)$ differs from R_8 $(0,T,t_3)$ by the orientational part. The amplitudes of the sixth and seventh response function components are determined by the dipole strengths of the symmetric and antisymmetric modes and they also depend on the relative angle θ between the two transition dipoles. In general, the orientational part of the nonlinear response function component in Equation 7.32 can be written as a product of three functions as, for a spherical rotor, [41]

$$Y_\alpha(t_1,T,t_3) = C_1(t_1)y_\alpha(T)C_1(t_3), \tag{7.33}$$

where C_l $(t) = \exp\,[-l(l+1)D_{or}\,t]$ and D_{or} is the orientational diffusion coefficient. For the pump-probe geometry, since $t_1 = 0$, we have $Y_\alpha(0,T,t_3) = y_\alpha(T)C_1(t_3)$. The orientational response functions, $y_\alpha(T)$, appearing in Equation 7.32, are found to be

$$\begin{aligned} y_{XX\varphi\varphi}^{ssss}(T) &= \frac{1}{9}\left[1+\frac{4}{5}C_2(T)\right]\cos^2\varphi + \frac{1}{9}\left[1-\frac{2}{5}C_2(T)\right]\sin^2\varphi \\ y_{XX\varphi\varphi}^{ssaa}(T) &= \frac{1}{9}\left\{1+\frac{4}{5}C_2(T)\left[\cos^2\theta-\frac{1}{2}\sin^2\theta\right]\right\}\cos^2\varphi \\ &\quad + \frac{1}{9}\left\{1-\frac{2}{5}C_2(T)\left[\cos^2\theta-\frac{1}{2}\sin^2\theta\right]\right\}\sin^2\varphi \\ y_{XX\varphi\varphi}^{sasa}(T) &= \frac{1}{9}\left\{\cos^2\theta+\frac{1}{5}C_2(T)\left[4\cos^2\theta+3\sin^2\theta\right]\right\}\cos^2\varphi \\ &\quad + \frac{1}{9}\left\{\cos^2\theta-\frac{1}{10}C_2(T)\left[4\cos^2\theta+3\sin^2\theta\right]\right\}\sin^2\varphi. \end{aligned} \tag{7.34}$$

Then, the frequency-resolved broadband PP signal at $\omega_{pr} = \omega_s$ is given as a function of φ, that is

$$\begin{aligned} PP_{XX\varphi\varphi}(\omega_s,T) &= \left[4\mu_s^4 y_{XX\varphi\varphi}^{ssss}(T)e^{-T/T_{1,s}} + 2\mu_a^2\mu_s^2 y_{XX\varphi\varphi}^{ssaa}(T)e^{-T/T_{1,a}}\right]R^{PP}(\omega_s) \\ &\quad + 2\mu_a^2\mu_s^2 y_{XX\varphi\varphi}^{sasa}(T)e^{-T/\tau_q}\cos\left(\omega_{sa}T\right)R_{QB}^{PP}(\omega_s), \end{aligned} \tag{7.35}$$

where the auxiliary line shape functions are defined as

$$R^{PP}(\omega_s) = \mathrm{Im}\left[\int_0^\infty R(t_3)C_1(t_3)e^{i\omega_s t_3}dt_3\right] \quad \text{and} \quad R_{QB}^{PP}(\omega_s) = \mathrm{Im}\left[\int_0^\infty R_{QB}(t_3)C_1(t_3)e^{i\omega_s t_3}dt_3\right] \tag{7.36}$$

In Equation 7.35, τ_q is the decoherence time for the coherence, $|v_s = 1\rangle\langle v_a = 1|$. Using the general expression for the φ-dependent PP signal, we shall consider some limiting cases.

7.3.3 Isotropic and Anisotropic PP Signals of a Coupled Two-Oscillator System

For a single oscillator system, only the first term in Equation 7.35 is important and the isotropic and anisotropic PP signals are simply given as $PP_{iso} \cong 4/3\mu_s^4 e^{-T/T_1}$ and $PP_{aniso} = 0.4C_2(T)$. However, in the case of the coupled two-oscillator system, two additional terms arise from the other Liouville pathways so that the isotropic PP signal is given as

$$\begin{aligned} PP_{iso}(\omega_s, T) &= \left[\frac{4}{3}\mu_s^4 e^{-T/T_{1,s}} + \frac{2}{3}\mu_a^2\mu_s^2 e^{-T/T_{1,a}}\right] R^{PP}(\omega_s) \\ &+ \frac{2}{3}\cos^2\theta\mu_a^2\mu_s^2 e^{-T/\tau_q}\cos(\omega_{sa}T) R_{QB}^{PP}(\omega_s). \end{aligned} \tag{7.37}$$

Note that the last term describing the oscillatory component of the measured isotropic PP signal decays in time τ_q and its magnitude is proportional to $\cos^2\theta$. Therefore, if the isotropic PP signal shows an oscillatory behavior with frequency of ω_{sa}, it indicates that the two transition dipoles are not orthogonal to each other.

To verify the above theoretical results, we shall consider the experimentally measured dispersed PP spectra of [dicarbonylacetylacetonato]rhodium(I) ($Rh(CO)_2C_5H_7O_2$:RDC) dissolved in chloroform. Figure 7.5 depicts the temporal profile of the PP signal when the probe frequency is equal to the frequency of the symmetric CO stretch normal mode. In the figure, only the first 5 ps time profile is shown, even though we measured the PP signal over 100 ps. One can immediately find

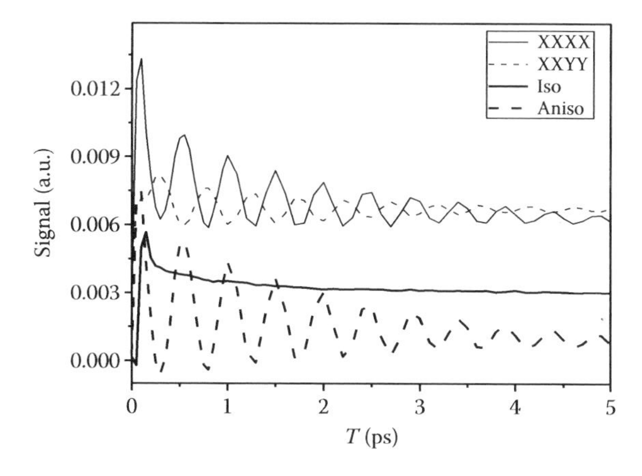

FIGURE 7.5 Parallel (XXXX) and perpendicular (XXYY) pump-probe signals of RDC are plotted with thus calculated isotropic and anisotropic signals at $\omega_{pr} = \omega_s$. The oscillating (quantum beat) components in the parallel and perpendicular PP signals are in out-of-phase relation to each other. Isotropic signal shows no oscillatory behavior, because the out-of-phase quantum beat signals cancel with each other when the isotropic signal is obtained.

that, in the case of the isotropic signal (thick solid line in Figure 7.5), the oscillating component of the parallel signal cancels out with that of the perpendicular signal, since they are in out-of-phase relation to each other (compare the thin solid line (XXXX) with the thin dashed line (XXYY) in Figure 7.5). On the other hand, the anisotropic signal clearly exhibits an oscillating feature. These observations result from the fact that the two transition dipoles are orthogonal to each other, that is, $\cos^2 \theta = 0$ in this specific case. Inversely, an observation of oscillatory features in the isotropic signal indicates that the angle between the two transition dipoles deviates from $\pi/2$. In that case, to measure the excited state lifetime from the isotropic signal, one should eliminate the oscillatory component first.

Here, Fourier transform filtering method for this purpose is discussed. The method called Fourier transform filtering has been widely used (see Figure 7.6) [46,47]. First, the oscillatory signal (Figure 7.6a) is Fourier-transformed to the corresponding spectrum (Figure 7.6b). The high-frequency peak can be identified in the resulting spectrum. After removing the peak (zero-filling the corresponding data), the inverse Fourier transformation will give us the isotropic signal without the oscillating component, which should correspond to the first two terms in Equation 7.37. Note that the isotropic signal at ω_s may decay bi-exponentially if the lifetime of the symmetric mode significantly differs from that of the antisymmetric mode. In that case, the preexponential factors in Equation 7.37 can be predetermined by using the dipole strengths of the two modes and the biexponential decay time constants can be assigned to the lifetimes of each mode. Nonetheless, the isotropic signal is not affected by the rotational dynamics even in this case of a coupled two-oscillator

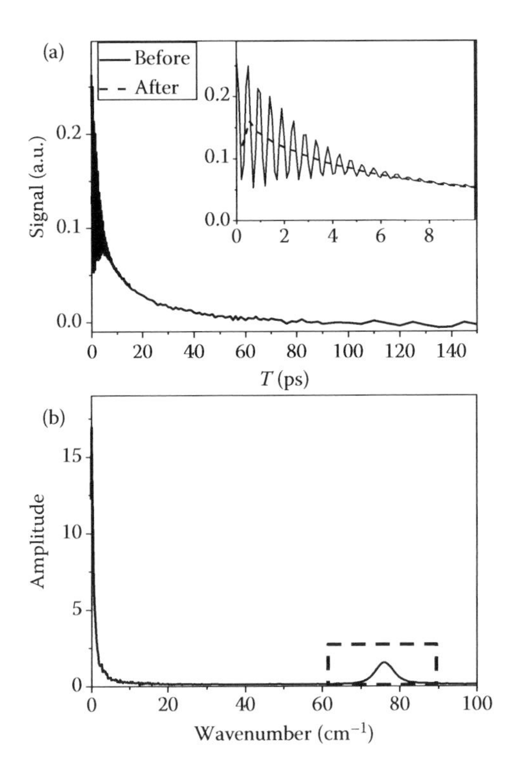

FIGURE 7.6 Fourier transform filtering procedure. (a) The parallel pump-probe signal of RDC in chloroform is shown. The oscillating component originates from the quantum beat (or vibrational coherence between the symmetric and antisymmetric CO stretching vibrations). (b) Direct Fourier-transformed spectrum of the parallel PP signal in (a) is shown. The peak at around 75 cm^{-1} is removed from the FT spectrum and then it is inverse-Fourier-transformed to retrieve the time-domain signal (see dashed line in (a)) without the oscillating component. The first 10 ps time profiles are shown in the inset of (a).

system. For the isotropic signal of RDC in chloroform (Figure 7.5), we found that a single exponential function fits the experimental data almost perfectly, indicating that the vibrational excited state lifetimes of the symmetric and antisymmetric modes are the same.

Unlike the isotropic PP signal, the anisotropic signal depends on not only the transition dipole angle but also transition dipole ratio $\mu_{as} = \mu_a/\mu_s$, and rotational dynamics in a complicated manner. Furthermore, the anisotropic signal does depend on the lifetimes of the two modes if they differ from each other. More specifically, using Equation 7.35, we find that

$$PP_{\text{aniso}}(\omega_s, T) = \frac{PP_{XXXX}(\omega_s, T) - PP_{XXYY}(\omega_s, T)}{PP_{\text{iso}}(\omega_s, T)} = \frac{2}{5}C_2(T)\frac{2e^{-T/T_{1,s}} + \mu_{as}^2 e^{-T/T_{1,a}}\left(3\cos^2\theta - 1\right)/2}{2e^{-T/T_{1,s}} + \mu_{as}^2 e^{-T/T_{1,a}}}. \tag{7.38}$$

Note that, only if $T_{1,s} = T_{1,a}$ and $\theta = 0$ or π, the anisotropic signal is purely determined by the second-order rotational function, $C_2(T)$. Otherwise, the decaying pattern of the anisotropic signal is complicated. In general, the initial anisotropy value, which can be obtained by extrapolating the anisotropic signal to $T = 0$, deviates from the familiar value of 0.4. More specifically, the initial anisotropy is given as

$$PP_{\text{aniso}}(\omega_s, 0) = 0.4\left\{\frac{2 + \mu_{as}^2\left(3\cos^2\theta - 1\right)/2}{2 + \mu_{as}^2}\right\}. \tag{7.39}$$

If the two transition dipoles are orthogonal to each other, that is, $\theta = \pi/2$, and also the transition dipole ratio μ_{as} is unity, the initial anisotropy value becomes 0.2. However, if the two transition dipole moments are different, that is, $\mu_{as} \neq 1$, the anisotropy value becomes a probe frequency-dependent function. To explain this, note that the transition dipole ratio required to describe the initial anisotropy value at $\omega_{pr} = \omega_s$ is $\mu_{as} = \mu_a/\mu_s$, whereas that at $\omega_{pr} = \omega_a$ is $\mu_{sa} = \mu_s/\mu_a$. As an example, we estimated the transition dipoles of the symmetric and asymmetric CO stretch normal modes of RDC, analyzing the FT-IR spectrum. With transition dipole ratio and angle, the initial anisotropy value was found to be 0.18 at $\omega_{pr} = \omega_s$ and 0.22 at $\omega_{pr} = \omega_a$. Thus, a caution is needed in quantitatively analyzing the initial anisotropy value. In fact, a similar equation for the anisotropy value was reported by Tokmakoff and coworkers and they suggested that the transition dipole angle can be estimated by using the initial anisotropy value [41]. However, it should be noted that the transition dipole ratio and the frequency dependence of the initial anisotropy value are to be correctly taken into account to precisely determine the transition dipole angle θ using the polarization-controlled PP method.

Quite often, any deviation of the initial anisotropy value from 0.4 has been explained in terms of ultrafast inertial rotational motion [48,49]. However, for a general coupled multioscillator system, the anisotropic PP signal becomes a complicated function of not only the usual rotational relaxation but also transition dipole angles and dipole strength ratios. Although the above argument and that presented by Tokmakoff and coworkers suggest that the initial anisotropy value can be used to estimate the transition dipole angle, it may not be so easy because the ultrafast inertial rotational motion would also make the initial anisotropy value deviate from 0.4.

7.3.4 Polarization-Angle-Scanning Pump-Probe Spectroscopy

For a single oscillator system, the isotropic and anisotropic PP signals provide information on rotational dynamics-free lifetime and population relaxation-free rotational time. In the case of a coupled two-oscillator system, despite the fact that the initial anisotropy value can be in principle related to the transition dipole angle, it is still difficult to accurately estimate the transition dipole angle for reasons mentioned above. Now, let us consider the quantum beat contribution to the PP signal more in depth. It is noted that, from Equation 7.35, the amplitude of the quantum beat contribution is given as, for an arbitrary polarization angle φ of the probe pulse with respect to that of the pump

$$PP_{XX\varphi\varphi}^{QB}(\omega_s,T) = 2\mu_a^2\mu_s^2 y_{XX\varphi\varphi}^{sasa}(T)e^{-T/\tau_q}R_{QB}^{PP}(\omega_s)\cos(\omega_{sa}T). \tag{7.40}$$

The φ-dependence of the quantum beam amplitude is determined by the orientation function $y_{XX\varphi\varphi}^{sasa}(T)$, which is given in Equation 7.34. Depending on the angle φ, the amplitude of the quantum beat signal changes. Therefore, it should be possible to find out the precise polarization direction angle φ^* at which the quantum beam amplitude vanishes. Note that the quantum beat term decays more rapidly than the population relaxation, since τ_q is likely to be smaller than the lifetimes [17]. As a consequence, at a short time, the rotational relaxation can be ignored. In this case, we find that the transition dipole angle is related to the polarization direction angle φ^* at which the quantum beat amplitude vanishes as

$$\cos^2\theta = \frac{1-3\cos^2\varphi^*}{3+\cos^2\varphi^*}. \tag{7.41}$$

Owing to the fact that $\cos^2\theta$ varies from 0 to 1, it is not necessary to scan φ from 0 to $\pi/2$. Instead, the polarization angle φ should be scanned from the magic angle to 90°, that is, $54.7° \le \varphi \le 90°$. For the sake of quantitative analysis of the quantum beat amplitudes, one can use the Fourier transform method. First, the temporal profile of the PP signal at a given φ is Fourier-transformed. One can identify the peak associated with the corresponding vibrational coherence. The peak height or volume would be a function of φ. Therefore, using a nonlinear curve fitting method, one can find the φ^* value. Then, inserting thus determined φ^* to the right-hand side of Equation 7.41 would allow us to determine the transition dipole angle θ.

7.3.5 Polarization-Angle-Scanning 2D PE Spectroscopy

In the above, we presented a detailed discussion about the polarization-controlled PP spectroscopy. Cho and coworkers have shown that the polarization-angle-scanning 2D spectroscopy can be of use to determine molecular structures [28–30]. In a photon echo-type 2D spectroscopy, the incident electric field is composed of three pulses temporarily separated by τ and T. The coherent photon echo signal field emitted in the direction of $\mathbf{k}_s = -\mathbf{k}_1 + \mathbf{k}_2 + \mathbf{k}_3$ is measured at amplitude level with a heterodyne-detection method. The resulting third-order polarization within the rotating wave approximation is given as

$$\begin{aligned}\mathrm{P}^{(3)}(r,t) = e^{i\mathbf{k}_s\cdot\mathbf{r}-i\omega t}\int_0^\infty dt_3\int_0^\infty dt_2\int_0^\infty dt_1 R^{(3)}(t_3,t_2,t_1)\tilde{e}_3\tilde{e}_2\tilde{e}_1^* E_3(t-t_3)\\ \times E_2(t+T-t_3-t_2)E_1(t+T+\tau-t_3-t_2-t_1)\exp(i\omega t_3 - i\omega t_1).\end{aligned} \tag{7.42}$$

The 2D PE spectrum is then obtained by performing the 2D Fourier–Laplace transformation of $E_{PE}(t, T, \tau)$ with respect to τ and t as

$$S_{ijkl}^{2D}(\omega_t,T,\omega_\tau) = \int_0^\infty dt\int_0^\infty d\tau E_{PE}(\tau,T,t)\exp(i\omega_t t + i\omega_\tau\tau). \tag{7.43}$$

Here, the subscript *ijkl* specifies the polarization directions of the four radiations. As emphasized in this chapter, the 2D spectrum is a function of the three polarization direction angles of the three incident beams, when the polarization direction of the detected signal field is fixed. One can measure the anisotropy values of the diagonal and cross peaks, properly controlling polarization directions of the second and third pulsed radiations to be either parallel or perpendicular to that of the first pulse. The parallel and perpendicular 2D PE signals have often been measured to determine transition dipole angles. Hereafter, we discuss about the polarization angle dependence of the diagonal and cross peaks in a given 2D spectrum. Particularly, the anisotropy measurement method is compared with the more general PAS 2D spectroscopy.

7.3.6 Polarization Angle Dependence of Diagonal and Cross Peaks

In this section, we shall consider the 2D spectra at nonzero waiting time to avoid any possible complications due to coherent artifact contributions. For the sake of simplicity, it is assumed that the anharmonic frequency shift is sufficiently larger than the vibrational coupling constant. The orientational contribution to the diagonal peak amplitude is then determined by $Y_{X\varphi_3\varphi_2X}(t_3,t_2,t_1)$ for the rephasing signal, whereas it is $Y_{X\varphi_3X\varphi_2}(t_3,t_2,t_1)$ for the nonrephasing signal in the impulsive limit. Here, it should be noted that the spectral diffusion dynamics makes the relative magnitudes of the rephasing and nonrephasing signals change in time. Moreover, the absorptive peak can be distorted due to the imbalance between the rephasing and nonrephasing signals, which may in turn cause an inaccuracy in evaluating the peak amplitude. The two rephasing pathways, $R_1(t_1, T, t_3)$ and $R_2(t_1, T, t_3)$, that are associated with a given diagonal peak are the ground-state bleaching and stimulated emission contributions. The corresponding contributions to the nonrephasing signal are $R_3(t_1, T, t_3)$ and $R_4(t_1, T, t_3)$, respectively (see Figure 7.4). The other component contributing to the diagonal peak, which is related to the quantum beat term, is $R_5(t_1, T, t_3)$. They are given as

$$\begin{aligned}
R_1(t_1,T,t_3) &= \mu_s^4 Y^{ssss}_{X\varphi_3\varphi_2X}(t_1,T,t_3)R^R_{GB}(t_1,T,t_3) \\
R_2(t_1,T,t_3) &= \mu_s^4 Y^{ssss}_{X\varphi_3\varphi_2X}(t_1,T,t_3)R^R_{SE}(t_1,T,t_3) \\
R_3(t_1,T,t_3) &= \mu_s^4 Y^{ssss}_{X\varphi_3X\varphi_2}(t_1,T,t_3)R^{NR}_{GB}(t_1,T,t_3) \\
R_4(t_1,T,t_3) &= \mu_s^4 Y^{ssss}_{X\varphi_3X\varphi_2}(t_1,T,t_3)R^{NR}_{SE}(t_1,T,t_3) \\
R_5(t_1,T,t_3) &= \mu_s^2\mu_a^2 Y^{saas}_{X\varphi_3X\varphi_2}(t_1,T,t_3)R^{NR}_5(t_1,T,t_3)e^{i\omega_{sa}T}.
\end{aligned} \tag{7.44}$$

The orientational part of the four response function components, $R_1 - R_4$, are the same and it is given as

$$\begin{aligned}
Y^{ssss}_{X\varphi_3\varphi_2X}(t_1,T,t_3) &= Y^{ssss}_{X\varphi_3X\varphi_2}(t_1,T,t_3) \\
&= C_1(t_1)C_1(t_3)\left\{\frac{1}{9}\left[1+\frac{4}{5}C_2(T)\right]\cos\varphi_3\cos\varphi_2 + \frac{1}{15}C_2(T)\sin\varphi_3\sin\varphi_2\right\}.
\end{aligned} \tag{7.45}$$

The fifth response function component $R_5(t_1, T, t_3)$ oscillates in frequency of ω_{sa}. It differs from the other four terms by not only the transition dipole factor, but also the orientational part. The orientational function $Y^{saas}_{X\varphi_3X\varphi_2}(t_1,T,t_3)$ for the quantum beam term depends on the transition dipole angle θ and the polarization direction angles φ_2 and φ_3 as

$$Y^{saas}_{X\varphi_3X\varphi_2}\left(t_1,T,t_3\right) = C_1(t_1)C_1(t_3)\left\{\frac{1}{9}\left[1+\frac{4}{5}C_2(T)\right]\cos^2\theta\cos\varphi_3\cos\varphi_2 + \frac{1}{15}C_2(T)\sin^2\theta\cos\varphi_3\cos\varphi_2 + \frac{1}{15}C_2(T)\cos^2\theta\sin\varphi_3\sin\varphi_2 + \left[\frac{1}{20}C_2(T)-\frac{1}{12}C_1(T)\right]\sin^2\theta\sin\varphi_3\sin\varphi_2\right\}. \quad (7.46)$$

Now, the absorptive 2D spectrum is given as the sum of rephasing and nonrephasing terms, that is

$$S^{C}_{2D}\left(\omega_\tau,\omega_t,T\right) \propto \mathrm{Re}\left[S^R\left(\omega_\tau,\omega_t,T\right)+S^{NR}\left(\omega_\tau,\omega_t,T\right)\right], \quad (7.47)$$

where the two contributions are

$$\begin{aligned} S^R\left(\omega_\tau,\omega_t,T\right) &= y_\alpha(T)\int_0^\infty dt_1\int_0^\infty dt_3 \exp(i\omega_t t_3 - i\omega_\tau t_1)C_1(t_1)C_1(t_3)R^R(t_1,T,t_3) \\ &= y_\alpha(T)R^R(\omega_\tau,T,\omega_t) \\ S^{NR}\left(\omega_\tau,\omega_t,T\right) &= y_\alpha(T)\int_0^\infty dt_1\int_0^\infty dt_3 \exp(i\omega_t t_3 - i\omega_\tau t_1)C_1(t_1)C_1(t_3)R^{NR}(t_1,T,t_3) \\ &= y_\alpha(T)R^{NR}(\omega_\tau,T,\omega_t). \end{aligned} \quad (7.48)$$

Then, the amplitude of the diagonal peak at $\omega_\tau = \omega_s$ and $\omega_m = \omega_s$ is found to be

$$\begin{aligned} S_{2D}\left(\omega_\tau = \omega_s,\omega_t = \omega_s,T\right) &\propto \mathrm{Re}\left[\sum_{i=1}^{5} R_i\left(\omega_\tau,\omega_t,T\right)\right] \\ &= \mu_s^4 y^{ssss}_{X\varphi_3\varphi_2X}(T)S^C_{2D}\left(\omega_\tau,\omega_t,T\right)e^{-T/T_{1,s}} \\ &\quad + \mu_s^2\mu_a^2 y^{saas}_{X\varphi_3X\varphi_2}(T)\,\mathrm{Re}\left[R_5^{NR}\left(\omega_\tau,\omega_t,T\right)e^{-T/\tau_q}e^{i\omega_{sa}T}\right]. \end{aligned} \quad (7.49)$$

Often, the last term associated with the quantum beat contribution has been ignored because it decays rapidly. However, if the 2D spectrum at a short waiting time is used to extract structural information, the quantum beat contribution should be properly taken into account.

We next consider the amplitudes of the cross peaks that are related to the ground-state bleach rephasing (R_6) and nonrephasing (R_7) terms as well as the quantum beat term (R_8) that are given as

$$\begin{aligned} R_6\left(t_1,T,t_3\right) &= \mu_s^2\mu_a^2 Y^{aass}_{X\varphi_3\varphi_2X}\left(t_1,T,t_3\right)R^{\mathrm{R}}_{GB}\left(t_1,T,t_3\right) \\ R_7\left(t_1,T,t_3\right) &= \mu_s^2\mu_a^2 Y^{aass}_{X\varphi_3X\varphi_2}\left(t_1,T,t_3\right)R^{\mathrm{NR}}_{GB}\left(t_1,T,t_3\right) \\ R_8\left(t_1,T,t_3\right) &= \mu_s^2\mu_a^2 Y^{asas}_{X\varphi_3\varphi_2X}\left(t_1,T,t_3\right)R^{\mathrm{R}}\left(t_1,T,t_3\right)e^{i\omega_{as}T}. \end{aligned} \quad (7.50)$$

Then, the amplitudes of the two cross peaks are

$$S_{2D}(\omega_\tau = \omega_a,\omega_t = \omega_s,T) \propto \mathrm{Re}\left[\sum_{i=6}^{8} R_i\left(\omega_\tau,\omega_t,T\right)\right]$$

$$
\begin{aligned}
&= \mu_s^2\mu_a^2 y_{X\varphi_3\varphi_2X}^{ssaa}(T)S_{GB}^{C}\left(\omega_\tau,\omega_t,T\right)e^{-T/T_{1,a}} \\
&\quad + \mu_s^2\mu_a^2 y_{X\varphi_3\varphi_2X}^{sasa}(T)\,\mathrm{Re}\left[R_8^{NR}\left(\omega_\tau,\omega_t,T\right)e^{-T/\tau_q}e^{i\omega_{sa}T}\right] \\
&= \mu_s^2\mu_a^2 y_{X\varphi_3\varphi_2X}^{ssaa}(T)S_{GB}^{C}\left(\omega_\tau,\omega_t,T\right)e^{-T/T_{1,a}} \\
&\quad + \mu_s^2\mu_a^2 y_{X\varphi_3\varphi_2X}^{sasa}(T)e^{-T/\tau_q}\begin{bmatrix}\mathrm{Re}\left\{R_8^{NR}\left(\omega_\tau,\omega_t,T\right)\right\}\cos\left(\omega_{sa}T\right)\\ +\,\mathrm{Im}\left\{R_8^{NR}\left(\omega_\tau,\omega_t,T\right)\right\}\sin\left(\omega_{sa}T\right)\end{bmatrix}
\end{aligned}
\tag{7.51}
$$

$$
\begin{aligned}
S_{2D}&\left(\omega_\tau=\omega_s,\omega_t=\omega_a,T\right) \\
&= \mu_s^2\mu_a^2 y_{X\varphi_3\varphi_2X}^{aass}(T)S_{GB}^{C}\left(\omega_\tau,\omega_t,T\right)e^{-T/T_{1,s}} \\
&\quad + \mu_s^2\mu_a^2 y_{X\varphi_3\varphi_2X}^{asas}(T)\,\mathrm{Re}\left[R_8^{NR}\left(\omega_\tau,\omega_t,T\right)e^{-T/\tau_q}e^{-i\omega_{sa}T}\right] \\
&= \mu_s^2\mu_a^2 y_{X\varphi_3\varphi_2X}^{ssaa}(T)S_{GB}^{C}\left(\omega_\tau,\omega_t,T\right)e^{-T/T_{1,a}} \\
&\quad + \mu_s^2\mu_a^2 y_{X\varphi_3\varphi_2X}^{sasa}(T)e^{-T/\tau_q}\begin{bmatrix}\mathrm{Re}\left\{R_8^{NR}\left(\omega_\tau,\omega_t,T\right)\right\}\cos\left(\omega_{sa}T\right)\\ -\,\mathrm{Im}\left\{R_8^{NR}\left(\omega_\tau,\omega_t,T\right)\right\}\sin\left(\omega_{sa}T\right)\end{bmatrix}.
\end{aligned}
$$

These two equations with Equation 7.49 are the general expressions for the amplitudes of the cross and diagonal peaks. Here, it should be noted that the oscillatory quantum beat contributions to the two cross peaks make the amplitude of two cross peaks unequal and the difference between the two depends on the waiting time and oscillation frequency.

7.3.7 Parallel and Perpendicular 2D Measurements

Similar to the measurements of the isotropic and anisotropic PP signals, one can perform polarization-controlled 2D measurements to obtain the anisotropic cross and diagonal peaks, where the parallel and perpendicular 2D signals will be denoted as S_{XXXX}^{2D} and S_{XYYX}^{2D}. Note that, for a coupled two-oscillator system, the quantum beat contribution can be effectively removed by examining the signals at $T^* = (2n+1)\pi/2\omega_{sa}$ (for an integer n) after adding two cross-peak amplitude. The ratio of the sum of the two cross peaks in the perpendicular 2D spectrum to that in the parallel 2D spectrum is found to be

$$
\frac{S_{XYYX}^{2D}\left(\omega_a,\omega_s,T^*\right)+S_{XYYX}^{2D}\left(\omega_s,\omega_a,T^*\right)}{S_{XXXX}^{2D}\left(\omega_a,\omega_s,T^*\right)+S_{XYYX}^{2D}\left(\omega_s,\omega_a,T^*\right)} = \frac{10}{3}\frac{C_2(T^*)\left\{2\cos^2\theta-\sin^2\theta\right\}}{1+0.4C_2(T^*)\left\{2\cos^2\theta-\sin^2\theta\right\}}. \tag{7.52}
$$

Tokmakoff and coworkers considered the amplitude ratio $S_{XYYX}^{2D}/S_{XXXX}^{2D}$ at $T = 0$ in the limiting case that the rotational dynamics is slow [41]. However, as shown in Equations 7.49 and 7.51, the quantum beat contribution cannot be ignored even at $T = 0$. Thus, we suggest that the ratio in Equation 7.52 would be more useful than the anisotropic signal $S_{XYYX}^{2D}/S_{XXXX}^{2D}$.

7.3.8 Selective Elimination of Diagonal Peaks

Zanni and Hochstrasser showed that one can selectively eliminate all the diagonal peaks by properly controlling the polarization directions of the incident pulses [42]. More recently, we found that there is a general expression for the relationship between the polarization directions of the pulses at which all the diagonal peaks vanish [17]. Such relationship is given as, for $\varphi_1 = \varphi_4 = 0$

$$
3\cos\varphi_3^*\cos\varphi_2^* + \sin\varphi_3^*\sin\varphi_2^* = 0. \tag{7.53}
$$

However, this is valid only at a short waiting time. Furthermore, there is a nonzero contribution from the quantum beat to the diagonal peak amplitude (see Equation 7.49), that is

$$S_{2D}\left(\omega_\tau = \omega_s, \omega_t = \omega_s, T, \varphi_2^*, \varphi_3^*\right)$$
$$= \mu_s^2 \mu_a^2 \frac{\sin^2\theta}{6} \cos\varphi_2^* \cos\varphi_3^* \, \mathrm{Re}\left[R_5^{NR}\left(\omega_\tau, \omega_t, T\right) e^{-T/T_{1,s}} e^{i\omega_{sa}T}\right]. \quad (7.54)$$

In Figure 7.7, the 2D IR spectra of RDC in chloroform at specific angles φ_2^* and φ_3^* that satisfy Equation 7.53 are plotted and compared with the parallel polarization 2D IR spectrum (see the upper left one in Figure 7.7). The diagonal peaks in the 2D IR spectra at φ_2^* and φ_3^* are indeed significantly smaller than not only those of the parallel 2D IR spectrum, but also the cross peaks in the spectra at φ_2^* and φ_3^*. However, still there appear notable intensities on the diagonal. It is believed that such

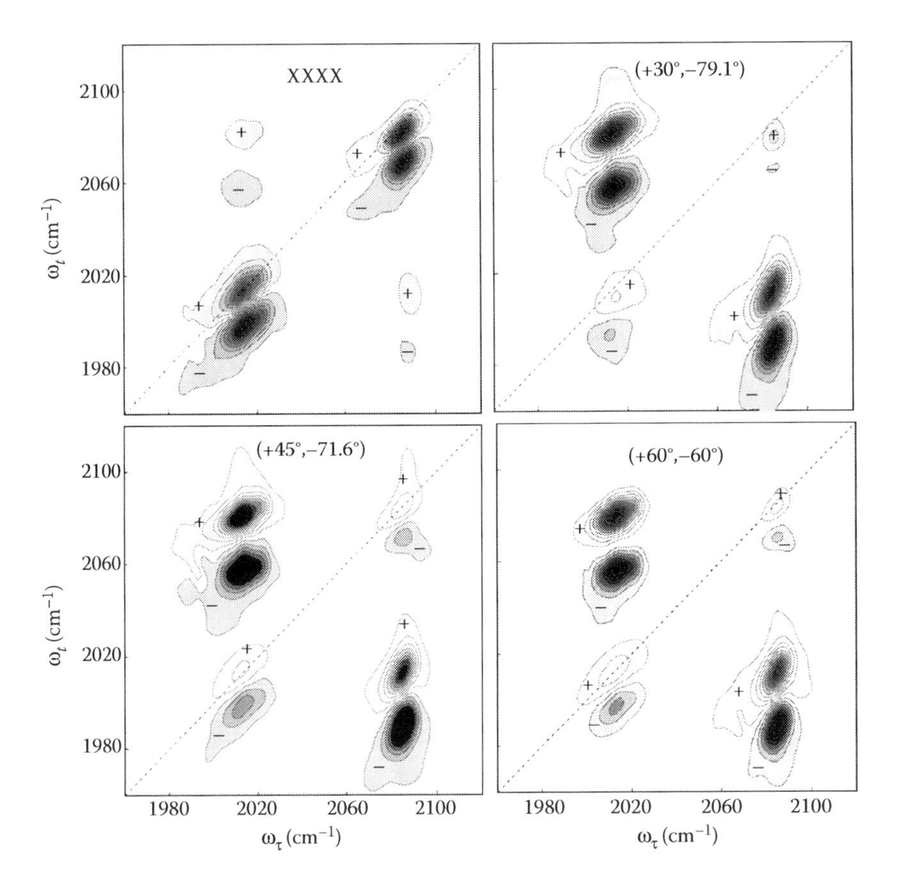

FIGURE 7.7 Polarization-controlled 2D IR spectra at the waiting time of $T = 0.2$ ps. The polarization directions of the first pulse and the detected signal (local oscillator) field are parallel to the X-axis in the laboratory-fixed frame. The two polarization direction angles φ_2 and φ_3 are varied to make all the diagonal peaks vanish. The φ_2 and φ_3 angles are shown in the legend and they are (0°, 0°), (+30°, –79.1°), (+45°, –71.6°), and (+60°, –60°). Among them, (0°, 0°) corresponds to the all-parallel-polarization 2D IR spectrum, which is denoted as XXXX. Note that Z-axis scale of the parallel polarization 2D IR spectrum differs from those of other three spectra. More specifically, the diagonal peak amplitudes in the other 2D IR spectra are <5% of those in the parallel polarization 2D IR spectra. Despite the fact that the diagonal peaks are largely eliminated in the polarization-controlled 2D IR spectra, there are still finite diagonal features that originate from the quantum beat signals.

residual diagonal peak intensities in the 2D IR spectra even at φ_2^* and φ_3^* originate from the quantum beat contributions. When the 2D IR spectra shown in Figure 7.7 were measured, the waiting time was fixed at 0.2 ps [28]. At this waiting time, in general, the quantum beat contributions do not vanish (see Figure 7.5). Interestingly, we find that the residual diagonal peak originating from the quantum beat contribution (Equation 7.54) is a function of the transition dipole angle θ as $\sin^2\theta$. Thus, it is suggested that, to completely remove all the diagonal peaks in a given 2D spectrum, one should control not only the two polarization direction angles φ_2 and φ_3 to be φ_2^* and φ_3^*, but also the waiting time.

7.3.9 Selective Elimination of Cross Peaks

Observation of cross peaks in a given 2D spectrum is the direct evidence of vibrational coupling between two oscillators. Recently, examining the polarization direction angle dependence of the cross-peak amplitudes, we showed that the PAS 2D method can be used to selectively eliminate particular cross peaks, which in turn provides direct information on the transition dipole angles. As a proof-of-principle experiment, we performed the PAS 2D IR experiments for a coupled two-oscillator system, RDC in chloroform [28]. The relationship between the transition dipole angle and the polarization direction angles at which the corresponding cross peak vanishes was found to be

$$\cos^2\theta = \frac{\tan\varphi_2^*\tan\varphi_3^* - 2}{4 + 3\tan\varphi_2^*\tan\varphi_3^*}. \qquad (7.55)$$

The experimentally measured 2D IR spectra are shown in Figure 7.8, where the waiting time was zero. Often, the cross-peak amplitudes in the zero waiting time 2D spectra have been used to estimate the transition dipole angles. At $T = 0$, due to the finite pulse width, it is necessary to include the nonrephasing contribution to the cross peaks, in addition to the rephrasing terms. For arbitrary polarization direction angles, we found that the amplitude of the cross peak at $\omega_\tau = \omega_a$ and $\omega_t = \omega_s$ is given as

$$\begin{aligned} &S_{2D}\left(\omega_\tau = \omega_a, \omega_t = \omega_s, T = 0\right) \\ &= \mu_a^2\mu_s^2 y_{X\varphi_3\varphi_2X}^{ssaa}(0)\left\{S_{GB}^{C}\left(\omega_\tau,\omega_t,0\right) + \mathrm{Re}\left[R^{\mathrm{R}}\left(\omega_\tau,\omega_t,0\right)\right]\right\} \\ &\quad + 2\mu_a^2\mu_s^2 y_{X\varphi_3\varphi_2X}^{sasa}(0)\,\mathrm{Re}\left[R^{\mathrm{R}}\left(\omega_\tau,\omega_t,0\right)\right] + y_{XX\varphi_3\varphi_2}^{asas}(0)\,\mathrm{Re}\left[R^{\mathrm{NR}}\left(\omega_\tau,\omega_t,0\right)\right]. \end{aligned} \qquad (7.56)$$

The three orientational functions appearing on the right-hand side of Equation 7.56 appear to be different from one another, but they are the same at $T = 0$ as

$$\begin{aligned} y_{Z\varphi_3\varphi_2Z}^{aass}(0) &= y_{Z\varphi_3\varphi_2Z}^{asas}(0) = y_{ZZ\varphi_3\varphi_2}^{saas}(0) \\ &= \frac{1}{15}\left\{2\cos^2\theta + 1\right\}\cos\varphi_3\cos\varphi_2 + \frac{1}{30}\left\{3\cos^2\theta - 1\right\}\sin\varphi_3\sin\varphi_2. \end{aligned} \qquad (7.57)$$

Owing to the fact that different terms in Equation 7.56 depend on the same orientational function, the polarization direction angle dependences of all the cross peaks are commonly described by Equation 7.57. Consequently, one can determine the transition dipole angle by following the procedure: (i) scan the polarization direction angles, (ii) examine the cross-peak amplitudes in the 2D spectra at $T = 0$, (iii) determine the φ_2^* and φ_3^* angles at which the cross peak vanishes, and (iv) use Equation 7.55 to determine θ.

Although the two polarization angles can be independently scanned, without loss of generality one can fix one of the two to carry out the PAS 2D IR measurements. In our recent experiments,

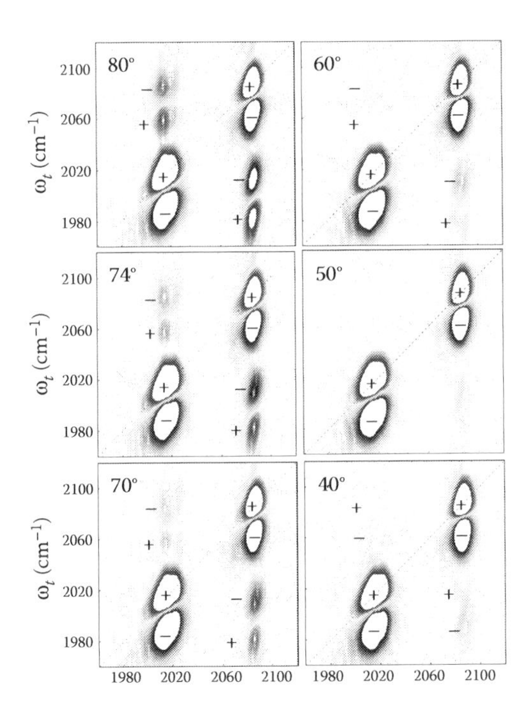

FIGURE 7.8 PAS 2D IR spectra ($T = 0$ ps) of RDC in chloroform for varying φ_3 angle. The polarization directions of the first pulse and signal (local oscillator) field are parallel to the X-axis ($\varphi_1 = \varphi_s = 0°$). For the sake of simplicity, the polarization angle φ_2 of the second pulse is fixed to be 60° with respect to the X-axis. Then, as φ_3 decreases from 80° to 40°, the cross-peak amplitude (on the upper-left region of each 2D IR spectrum) becomes smaller and vanishes at around $\varphi_3 = 50°$. As φ_3 is decreased further below 49°, the cross-peak amplitude with opposite sign increases again.

the polarization direction angle of the second pulse was set to be 60° with respect to that of the first pulse and the angle φ_3 was varied. Since the value of $\cos^2\theta$ is in the range from 0 to 1, the scanning range of the angle φ_3 is from 49.1° to 120°. In our PAS 2D IR experiments, the φ_3 angle was varied from 80° to 40° with an interval of 2°. In Figure 7.8, the real part of the PAS 2D IR spectrum is plotted. As φ_3 decreases from 80° to 40°, the amplitude of the cross peak at $\omega_\tau = 2011$ cm^{-1} and $\omega_t = 2087$ cm^{-1} decreases and approaches zero at about 50°. Then, as φ_3 decreases further below 50°, the sign of the cross peak changes as expected from Equation 7.57. In Figure 7.9, the amplitude of the cross peak is plotted with respect to the scanning angle φ_3, where the solid line represents Equation 7.57. From the interpolated line, we found that $\varphi_2^* = 60°$ and $\varphi_3^* = 49°$. Inserting these values to Equation 7.55, we found that the relative angle between the two transition dipole vectors of the symmetric and antisymmetric CO stretching normal modes is indeed equal to 90°. This demonstrates that the PAS 2D spectroscopy is of use to estimate the relative angle between two different transition dipoles.

Despite the success of the PAS 2D measurement method, at a nonzero waiting time, the cross-peak amplitudes can have oscillatory components originating from vibrational coherences, for example, $|v_s = 1\rangle\langle v_a = 1|$, and furthermore they are strongly affected by vibrational and rotational relaxations. However, note that the oscillation of a given cross peak can be removed by adding two cross peaks in the off-diagonal region of the 2D spectrum at the waiting time, T^*. That is to say, the sum of the conjugate pair of cross peaks at $T = T^*$ is found to be

$$S_{2D}\left(\omega_\tau = \omega_a, \omega_m = \omega_s, T^*\right) + S_{2D}\left(\omega_\tau = \omega_s, \omega_m = \omega_a, T^*\right)$$
$$= \mu_s^2\mu_a^2 y_{Z\varphi_3\varphi_2 Z}^{ssaa}(T^*)(e^{-T^*/T_{1,a}} + e^{-T^*/T_{1,s}}) S_{GB}^{C}\left(\omega_\tau, \omega_m, T^*\right), \quad (7.58)$$

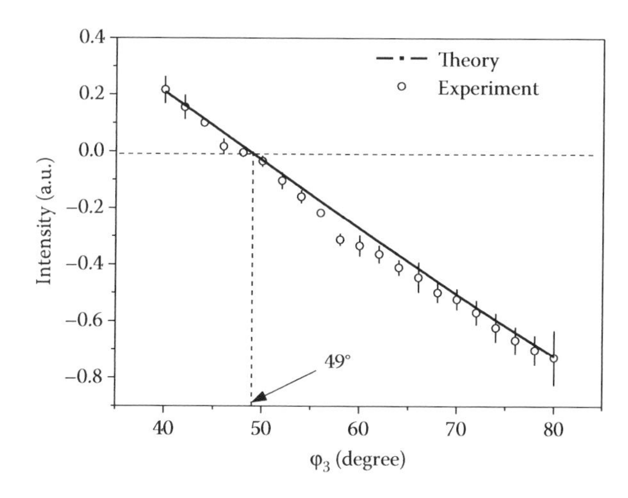

FIGURE 7.9 Amplitude of the cross peak at $\omega_\tau = 2011$ cm^{-1} and $\omega_t = 2087$ cm^{-1} in real part 2D IR spectra of RDC is plotted with respect to the polarization direction angle φ_3 of the third pulse. Here, we have $\varphi_1 = \varphi_s = 0$ and $\varphi_2 = 60$, and the cross-peak intensity is given in an arbitrary unit. The line in this figure is the theoretical line predicted with Equation 7.55.

where the orientational function is

$$y^{ssaa}_{Z\varphi_3\varphi_2 Z}(T) = \frac{1}{9}\left\{1 + \frac{4}{5}C_2(T)\left[\cos^2\theta - \frac{1}{2}\sin^2\theta\right]\right\}\cos\varphi_3\cos\varphi_2$$
$$+\frac{1}{15}C_2(T)\left[\cos^2\theta - \frac{1}{2}\sin^2\theta\right]\sin\varphi_3\sin\varphi_2. \qquad (7.59)$$

At a short waiting time, one can ignore the rotational relaxation contribution to the cross-peak amplitude, that is, $C_2(T) = 1$, and in this limit Equation 7.59 reduces to Equation 7.57.

In summary, controlling beam polarization directions of multiple pulses in third-order nonlinear optical spectroscopic measurements, one can extract vital information on molecular structure and dynamics. In particular, we showed that the selective elimination of the cross peak of interest is possible by properly adjusting the beam polarization configuration. This in turn allows us to extract information on the relative angles between coupled transition dipoles from the polarization-angle-scanning nonlinear spectroscopic experiments.

7.4 CHIROPTICAL 2D SPECTROSCOPY

In Section 7.2, we discussed about the linear optical activity spectroscopy for CD and ORD measurements. The dichroic or refractive index difference is essentially determined by the so-called rotatory strength defined as Im $[\boldsymbol{\mu}_{ge} \cdot \mathbf{m}_{eg}]$. Not only the magnitude of the rotatory strength, but its sign also provides information on the absolute configuration of a given chiral molecule. This renders the CD or any other optical activity spectroscopies very useful for studying molecular chirality. Nevertheless, these linear optical activity measurement methods, even though they are superior in frequency resolution to the other linear spectroscopic methods, are still one-dimensional spectroscopic method. In contrast, the 2D spectrum with significantly increased information density provides far more detailed information on weak features like vibrational anharmonicities and couplings that are highly sensitive to molecular structures. However, the 2D spectroscopy based on four-wave-mixing techniques cannot be of use to distinguish two different optical isomers because the 2D optical transition amplitude is determined by products of achiral transition electric dipole moments. Thus, it is highly desirable to develop potentially useful 2D optical activity measurement techniques in the near future.

7.4.1 Nonlinear Optical Activity Measurement Methods

Note that the optical activity is, by definition, related to the differential interaction of a chiral molecule with left- and right-handed chiral radiations. Therefore, the optical activity signal, denoted as ΔS, can always be defined as the difference between two signals obtained with left- and right-handed radiations as

$$\Delta S = S_L - S_R. \tag{7.60}$$

Instead of using LCP and RCP lights, one can use left- and right-elliptically polarized lights or even purely linearly polarized light, depending on the optical activity property of interest and the specific measurement method used. Recently, Cho and coworkers theoretically proposed both circularly polarized PP and PE spectroscopic methods, where the polarization state of one of the incident beams is modulated between left- and right-handed radiations and the difference PP or PE signal should be measured [17,50–54].

Instead of using a differential measurement scheme utilizing left- and right-handed radiations, one can carry out nonlinear optical activity measurements by selectively detecting the spectroscopic response functions that are not rotationally invariant within the electric dipole approximation, for example, S_{XXXY}—note that the number of "X" or "Y" index is odd [17,55–57]. In the case of the all-electric-dipole-allowed four-wave-mixing spectroscopy, the elements of the fourth-rank tensorial response function, which contain rotationally invariant isomers like $\delta_{l_1l_2}\delta_{l_3l_4}$, $\delta_{l_1l_3}\delta_{l_2l_4}$, and $\delta_{l_1l_4}\delta_{l_2l_3}$, can be experimentally measured. For example, the parallel- and perpendicular-polarization photon echo signals, denoted as E^{PE}_{XXXX} and E^{PE}_{XYYX}, are associated with the XXXX- and $XYYX$-tensor elements of the photon echo response function, respectively, when the incident beams and emitted photon echo field propagate along the Z-axis in a space-fixed frame. In contrast, the $XXXY$-tensor element of the all-electric-dipole-allowed third-order response function vanishes. However, the same $XXXY$-tensor element of the generalized response function beyond the electric dipole approximation (or long wavelength limit) does not vanish for chiral molecules because the magnetic dipole and electric quadrupole contributions are finite in such cases. Here, it should be noted that the magnetic field vector is orthogonal to the electric field vector in a given electromagnetic field. If the electric field component of the first pulse is X-polarized, its magnetic field vector is parallel to the Y-axis. Then, even when the $YYYX$-tensor element is measured, there is a third-order response function component $[\boldsymbol{\mu\mu\mu m}]_{YYYY}$ that is rotationally invariant, where $\mathbf{m}$ represents the transition magnetic dipole. Thus, for the measurement of the $YYYX$-signal, there is no need to use chiral fields. In this regard, the cross-polarization detection method developed for linear chiroptical spectroscopy would be of critical use.

7.4.2 Chiroptical 2D Pump-Probe

Despite the fact that the broadband 2D pump-probe spectroscopy has been widely used, again it is not a nonlinear chiroptical measurement method. It can be however extended by considering additional field–matter interaction terms beyond the electric dipole approximation. Experimentally, such measurement is feasible by taking the difference between the PP signals obtained with LCP and RCP radiations. The all-electric-dipole-allowed achiral PP signal is removed by the subtraction.

The lowest-order field–matter interaction Hamiltonian that should be taken into consideration for a theoretical description of the nonlinear chiroptical spectroscopy is given as [17]

$$H_{rad-mat} = -\{\boldsymbol{\mu} + (\mathbf{m} \times \hat{\mathbf{k}}) + (i/2)\mathbf{k} \cdot \mathbf{Q}\} \cdot eE(t)e^{i\mathbf{k}\cdot\mathbf{r}-i\omega t} - \{\boldsymbol{\mu} + (\mathbf{m} \times \mathbf{k}) - (i/2)\mathbf{k} \cdot \mathbf{Q}\} \cdot e^* E^*(t)e^{-i\mathbf{k}\cdot\mathbf{r}+i\omega t}, \tag{7.61}$$

where $\mathbf{Q}$ represents the electric quadrupole operator. Including the magnetic dipole–magnetic field and electric quadrupole–electric field interactions, one can expand the pump-probe polarization in a power series of $\mathbf{m}$ and $\mathbf{Q}$ as

$$\mathbf{P}_{PP}(t) = \mathbf{P}_{PP}^{(0)}(t) + \mathbf{P}_{PP}^{(1)}(t;m) + \mathbf{P}_{PP}^{(1)}(t;Q) + \cdots. \tag{7.62}$$

The first zero-order term on the right-hand side of Equation 7.62, which is the all-electric-dipole PP polarization, is about two to three orders of magnitude larger than the second and third terms. Here, the second term is linearly proportional to the transition magnetic dipole moment and the third is to the transition electric quadrupole moment. The higher-order terms can be safely ignored because of the following inequalities:

$$|\mathbf{P}_{PP}^{(0)}(t)| \gg |\mathbf{P}_{PP}^{(1)}(t;m)| \approx |\mathbf{P}_{PP}^{(1)}(t;Q)| \gg |\mathbf{P}_{PP}^{(2)}| \gg |\mathbf{P}_{PP}^{(3)}|. \tag{7.63}$$

Although a variety of different ways to measure the above pump-probe polarization in Equation 7.62, for example, transient grating, transient dichroism (TD), and transient birefringence, we shall specifically consider the self-heterodyne-detected TD signal defined as

$$S_{TD}(\omega_{pu},\omega_{pr};T) = \mathrm{Im}\left[\int_{-\infty}^{\infty} dt\, \mathrm{E}_{pr}^*(t) \cdot \mathbf{P}_{PP}(t)\right]. \tag{7.64}$$

By inserting Equation 7.62 into 7.64, the experimentally measured TD spectrum can also be written in a power series form, that is

$$S_{TD}(\omega_{pu},\omega_{pr};T) \cong S_{TD}^{(0)}(\omega_{pu},\omega_{pr};T) + S_{TD}^{(1)}(\omega_{pu},\omega_{pr};T;m) + S_{TD}^{(1)}(\omega_{pu},\omega_{pr};T;Q). \tag{7.65}$$

The first term in Equation 7.65 is the usual electric-dipole-allowed TD, and for a simple two-level system, it is given as

$$S_{TD}^{(0)}(\omega_{pu},\omega_{pr};T) \propto [\boldsymbol{\mu}_{ge}\boldsymbol{\mu}_{eg}\boldsymbol{\mu}_{ge}\boldsymbol{\mu}_{eg}] \otimes \tilde{\boldsymbol{e}}_{pr}^*\tilde{\boldsymbol{e}}_{pr}\tilde{\boldsymbol{e}}_{pu}^*\tilde{\boldsymbol{e}}_{pu}\Gamma_{TD}^{2LS}(\omega_{pu},\omega_{pr};T), \tag{7.66}$$

where the normalized 2D peak shape function is denoted as $\Gamma_{TD}^{2LS}(\omega_{pu},\omega_{pr};T)$ and its functional form is not important here (see Ref. [17] for detailed descriptions on the peak shape functions). Once the chiral contributions to the TD signal are included, we found that the 2D TD spectrum up to the first order in m or Q is given as

$$\begin{aligned} S_{TD}(\omega_{pu},\omega_{pr};T) \propto\ & ([\boldsymbol{\mu}_{ge}\boldsymbol{\mu}_{eg}\boldsymbol{\mu}_{ge}\boldsymbol{\mu}_{eg}] + \{[\boldsymbol{\mu}_{ge}\boldsymbol{\mu}_{eg}\boldsymbol{\mu}_{ge}(\mathbf{m}_{eg} \times \hat{\mathbf{k}}_{pu})] + [\boldsymbol{\mu}_{ge}\boldsymbol{\mu}_{eg}(\mathbf{m}_{ge} \times \hat{\mathbf{k}}_{pu})\boldsymbol{\mu}_{eg}] \\ & + [\boldsymbol{\mu}_{ge}(\mathbf{m}_{eg} \times \hat{\mathbf{k}}_{pr})\boldsymbol{\mu}_{ge}\boldsymbol{\mu}_{eg}] + [(\mathbf{m}_{ge} \times \hat{\mathbf{k}}_{pr})\boldsymbol{\mu}_{eg}\boldsymbol{\mu}_{ge}\boldsymbol{\mu}_{eg}]\} + (i/2)\{[\boldsymbol{\mu}_{ge}\boldsymbol{\mu}_{eg}\boldsymbol{\mu}_{ge}(\mathbf{k}_{pu} \cdot \mathbf{Q}_{eg})] \\ & - [\boldsymbol{\mu}_{ge}\boldsymbol{\mu}_{eg}(\mathbf{k}_{pu} \cdot \mathbf{Q}_{ge})\boldsymbol{\mu}_{eg}] + [\boldsymbol{\mu}_{ge}(\mathbf{k}_{pr} \cdot \mathbf{Q}_{eg})\boldsymbol{\mu}_{ge}\boldsymbol{\mu}_{eg}] - [(\mathbf{k}_{pr} \cdot \mathbf{Q}_{ge})\boldsymbol{\mu}_{eg}\boldsymbol{\mu}_{ge}\boldsymbol{\mu}_{eg}]\}) \\ & \otimes \tilde{\boldsymbol{e}}_{pr}^*\tilde{\boldsymbol{e}}_{pr}\tilde{\boldsymbol{e}}_{pu}^*\tilde{\boldsymbol{e}}_{pu}\Gamma_{TD}^{2LS}(\omega_{pu},\omega_{pr};T). \end{aligned} \tag{7.67}$$

In the above equation, it should be emphasized that the unit vectors specifying the polarization states of the fields can be in general complex. This result in Equation 7.67 is quite general for any arbitrary beam configuration.

Hereafter, we shall specifically consider the case that the pump and probe beam propagation directions are parallel to the Z-axis in a space-fixed frame, that is, $\hat{\mathbf{k}}_{pu} = \hat{\mathbf{k}}_{pr} = \hat{Z}$. The polarization direction of the linearly polarized probe (and local oscillator) beam is set to be parallel to the X-axis, that is, $\tilde{\boldsymbol{e}}_{pr} = \hat{X}$. Then, if we use a polarization modulation technique to generate LCP and RCP pump pulses, one can measure the chiroptical 2D TD signal $\Delta S_{TD}(\omega_{pu}, \omega_{pr}; T)$ defined as

$$\Delta S_{TD}(\omega_{pu},\omega_{pr};T) = S_{TD}^{LCP-pump}(\omega_{pu},\omega_{pr};T) - S_{TD}^{RCP-pump}(\omega_{pu},\omega_{pr};T). \tag{7.68}$$

Examining the rotational averages of the fourth-rank tensorial functions, we showed that the difference TD spectrum can be related to the $XXXY$- and $XXYX$-components of the TD signal, that is

$$\Delta S_{TD}(\omega_{pu},\omega_{pr};T) = \sqrt{2}i\{S_{TD}^{XXXY}(\omega_{pu},\omega_{pr};T) - S_{TD}^{XXYX}(\omega_{pu},\omega_{pr};T)\}. \tag{7.69}$$

The above results indicate that one can use either conventional approaches utilizing both left- and right-handed fields or cross-polarization detection approach using linearly polarized lights to measure the nonlinear chiroptical signal. After carrying out tedious algebraic calculations, we have found that the electric quadrupole contribution to the chiroptical TD signal vanishes in the case of a two-level system. The final result for $\Delta S_{TD}(\omega_{pu}, \omega_{pr}; T)$ is then given as, for a simple two-level system

$$\Delta S_{TD}(\omega_{pu},\omega_{pr};T) \propto D_{ee}^{g} R_{ee}^{g} \Gamma_{TD}^{2LS}(\omega_{pu},\omega_{pr};T), \tag{7.70}$$

where the dipole and rotatory strengths of the transition between e and g are defined as

$$D_{ee}^{g} \equiv [\boldsymbol{\mu}_{ge} \cdot \boldsymbol{\mu}_{eg}]^{M} \quad \text{and} \quad R_{ee}^{g} \equiv \mathrm{Im}[\boldsymbol{\mu}_{ge} \cdot \mathbf{m}_{eg}]^{M}. \tag{7.71}$$

Here, $\mathbf{m}_{eg}$ and $\mathbf{m}_{ge}$ are purely imaginary. The transition electric and magnetic dipole matrix elements inside the square bracket $[\cdots]^{M}$ are those in a molecule-fixed frame. It is interesting to note that the entire transition strength of the TD signal is determined by the product of the molecular dipole and rotatory strengths. Therefore, the sign of the chiroptical 2D TD peak is determined by the rotatory strength, which can be either positive or negative depending on the chirality of the corresponding quantum transition. Here, the dipole and rotatory strengths in Equation 7.71 are those of ground-state properties and they determine the all-electric-dipole-induced transition probability and electric dipole–magnetic dipole-induced transition probability from g (ground) to e (excited) states, respectively. In the case of the general nonlinear optical spectroscopy, not only the transitions between g and e but also those between e and other higher-lying (doubly) excited states should be taken into account. Then, the dipole and rotatory strengths of excited states become important in quantitatively describing other nonlinear optical activity properties.

In the present subsection, we focused on the chiroptical 2D PP spectroscopy of a two-level system. The rotational averages of the fourth- and fifth-rank tensors involving magnetic dipole and electric quadrupole transition matrix elements were discussed. We next consider the chiroptical 2D PE spectroscopy, which is the direct analog of 2D COSY (correlation spectroscopy) NMR.

7.4.3 Chiroptical 2D Photon Echo

To carry out such chiroptical 2D PE measurements, the polarization state of one of the three incident fields should be controlled or modulated between two oppositely handed fields. Here, we shall specifically consider the case that the first pulsed field is circularly polarized to obtain the LCP- and

RCP-PE signals. In this case, again by following the same arguments presented above for the chiroptical 2D PP spectroscopy, the 2D PE spectrum can be written in a power series form

$$\tilde{E}_{PE}(\omega_t,T,\omega_\tau) = \tilde{E}_{PE}^{(0)}(\omega_t,T,\omega_\tau) + \tilde{E}_{PE}^{(1)}(\omega_t,T,\omega_\tau;m) + \tilde{E}_{PE}^{(1)}(\omega_t,T,\omega_\tau;Q) + \cdots. \tag{7.72}$$

For a two-level system, the all-electric-dipole-allowed PE spectrum, $\tilde{E}_{PE}^{(0)}(\omega_t,T,\omega_\tau)$, is given as

$$\tilde{E}_{PE}^{(0)}(\omega_t,T,\omega_\tau) = 2[\boldsymbol{\mu}_{ge}\boldsymbol{\mu}_{eg}\boldsymbol{\mu}_{eg}\boldsymbol{\mu}_{ge}] \otimes \tilde{\boldsymbol{e}}_s^*\tilde{\boldsymbol{e}}_3\tilde{\boldsymbol{e}}_2\tilde{\boldsymbol{e}}_1^*\Gamma(\omega_t = \bar{\omega}_{eg},\omega_\tau = \bar{\omega}_{eg}), \tag{7.73}$$

where only the two resphasing terms were taken into consideration and the 2D peak shape function at $(\omega_t = \bar{\omega}_{eg},\omega_\tau = \bar{\omega}_{eg})$ is denoted as $\Gamma(\omega_t = \bar{\omega}_{eg},\omega_\tau = \bar{\omega}_{eg})$.

In the expanded form of the PE signal in Equation 7.72, we found that the magnetic dipole and electric quadrupole terms that are the second and third terms on the right-hand side of Equation 7.72 are found to be

$$\begin{aligned}\tilde{E}_{PE}^{(1)}(\omega_t,T,\omega_\tau;m) = 2\{&[\boldsymbol{\mu}_{ge}\boldsymbol{\mu}_{eg}\boldsymbol{\mu}_{eg}(\mathbf{m}_{ge}\times\hat{\mathbf{k}}_1)] + [\boldsymbol{\mu}_{ge}\boldsymbol{\mu}_{eg}(\mathbf{m}_{eg}\times\hat{\mathbf{k}}_2)\boldsymbol{\mu}_{ge}]\\ &+[\boldsymbol{\mu}_{ge}(\mathbf{m}_{eg}\times\hat{\mathbf{k}}_3)\,\boldsymbol{\mu}_{eg}\boldsymbol{\mu}_{ge}] + [(\mathbf{m}_{ge}\times\hat{\mathbf{k}}_s)\boldsymbol{\mu}_{eg}\boldsymbol{\mu}_{eg}\boldsymbol{\mu}_{ge}]\}\\ &\otimes \tilde{\boldsymbol{e}}_s^*\tilde{\boldsymbol{e}}_3\tilde{\boldsymbol{e}}_2\tilde{\boldsymbol{e}}_1^*\Gamma(\omega_t = \bar{\omega}_{eg},\omega_\tau = \bar{\omega}_{eg})\end{aligned} \tag{7.74}$$

$$\begin{aligned}\tilde{E}_{PE}^{(1)}(\omega_t,T,\omega_\tau;Q) = i\{&-[\boldsymbol{\mu}_{ge}\boldsymbol{\mu}_{eg}\boldsymbol{\mu}_{eg}(\mathbf{k}_1\cdot\mathbf{Q}_{ge})] + [\boldsymbol{\mu}_{ge}\boldsymbol{\mu}_{eg}(\mathbf{k}_2\cdot\mathbf{Q}_{eg})\boldsymbol{\mu}_{ge}]\\ &+[\boldsymbol{\mu}_{ge}(\mathbf{k}_3\cdot\mathbf{Q}_{eg})\boldsymbol{\mu}_{eg}\boldsymbol{\mu}_{ge}] - [(\mathbf{k}_s\cdot\mathbf{Q}_{ge})\boldsymbol{\mu}_{eg}\boldsymbol{\mu}_{eg}\boldsymbol{\mu}_{ge}]\}\\ &\otimes \tilde{\boldsymbol{e}}_s^*\tilde{\boldsymbol{e}}_3\tilde{\boldsymbol{e}}_2\tilde{\boldsymbol{e}}_1^*\Gamma(\omega_t = \bar{\omega}_{eg},\omega_\tau = \bar{\omega}_{eg}).\end{aligned} \tag{7.75}$$

The above expressions are valid for any arbitrary beam polarization configurations.

Now, for the sake of simplicity, it is assumed that the three incident beam propagation directions are almost collinear so that $\mathbf{k}_j = \hat{Z}$ for all j and, except for the first pulse, the polarization directions of the second, third, and echo fields are all parallel to the X-axis in the laboratory-fixed frame. Then, the difference 2D PE spectrum that carries information on molecular chirality should be considered:

$$\Delta\tilde{E}_{PE}(\omega_t,T,\omega_\tau) = \tilde{E}_{PE}^{LCP}(\omega_t,T,\omega_\tau) - \tilde{E}_{PE}^{RCP}(\omega_t,T,\omega_\tau). \tag{7.76}$$

From the general arguments on the rotationally invariant tensor properties, we found that $\Delta\tilde{E}_{PE}(\omega_t,T,\omega_\tau)$ is given as

$$\Delta\tilde{E}_{PE}(\omega_t,T,\omega_\tau) = -\sqrt{2}i\tilde{E}_{PE}^{XXXY}(\omega_t,T,\omega_\tau). \tag{7.77}$$

Here, $\tilde{E}_{PE}^{XXXY}(\omega_t,T,\omega_\tau)$ is the 2D photon echo spectrum obtained by using Y-polarized beam 1, and X-polarized beams 2 and 3, and by detecting the X-component of the coherent PE signal field. The resulting expression for the chiroptical 2D PE spectrum of a two-level-system is

$$\Delta\tilde{E}_{PE}(\omega_t,T,\omega_\tau) = \frac{8\sqrt{2}}{15}D_{ee}^{g}R_{ee}^{g}\Gamma(\omega_t = \bar{\omega}_{eg},\omega_\tau = \bar{\omega}_{eg}). \tag{7.78}$$

In the case of a weakly anharmonic oscillator system, the negative peak originating from the excited state ($v = 1 \rightarrow v = 2$) absorption is determined by the product of the dipole and rotatory strengths of the excited state instead of those of the ground state.

7.4.4 Chiroptical 2D PE of a Coupled Two-Level System Dimer

One of the interesting model systems that can be studied by using the chiroptical 2D PE spectroscopy would be a coupled two-level system dimer, where each monomer is modeled as a two-level system. In this case, there are ground state (g), two singly excited states (e_1 and e_2), and one doubly excited state (f). Even though this system consists of just four eigenstates, there are many dynamic timescales determining the entire time-resolved 2D spectra. At a very short time $T < \tau_{decoh}$, where τ_{decoh} is the decoherence time, the quantum beat contributions influence the diagonal and cross-peak amplitudes. At a long waiting time, the diagonal and cross-peak amplitudes change in time due to the population (excitation) transfer between the two singly excited states. Of course, the overall population decay (τ_{pop}) and rotational relaxation make the 2D spectral features depends on the waiting time in a complicated manner. However, for the sake of simplicity, let us focus on the intermediate time regime, that is, $\tau_{decoh} < T < \tau_{pop}$. In this case, the two diagonal peaks in the 2D chiroptical spectrum are found to be

$$\begin{aligned} \Delta\tilde{E}_{D1}(\omega_t,T,\omega_\tau) &= \frac{8\sqrt{2}}{15} D^g_{e_1e_1} R^g_{e_1e_1} \Gamma(\omega_t = \bar{\omega}_{e_1g}, \omega_\tau = \bar{\omega}_{e_1g}) \\ \Delta\tilde{E}_{D2}(\omega_t,T,\omega_\tau) &= \frac{8\sqrt{2}}{15} D^g_{e_2e_2} R^g_{e_2e_2} \Gamma(\omega_t = \bar{\omega}_{e_2g}, \omega_\tau = \bar{\omega}_{e_2g}). \end{aligned} \quad (7.79)$$

The cross peaks solely originating from the first-order magnetic dipole terms are

$$\begin{aligned} \Delta\tilde{E}_{C12}(\omega_t,T,\omega_\tau;m) &= \frac{\sqrt{2}}{30}\{6D^g_{e_2e_2}R^g_{e_1e_1} + 2D^g_{e_2e_1}R^g_{e_2e_1}\}\Gamma(\omega_t = \bar{\omega}_{e_2g}, \omega_\tau = \bar{\omega}_{e_1g}) \\ &- \frac{\sqrt{2}}{30}\{6D^{e_1}_{ff}R^g_{e_1e_1} + 2[\boldsymbol{\mu}_{fe_1}\cdot\boldsymbol{\mu}_{e_1g}]^M \,\mathrm{Im}[\boldsymbol{\mu}_{fe_1}\cdot\mathbf{m}_{e_1g}]^M\}\Gamma(\omega_t = \bar{\omega}_{fe_1}, \omega_\tau = \bar{\omega}_{e_1g}) \\ \Delta\tilde{E}_{C21}(\omega_t,T,\omega_\tau;m) &= \frac{\sqrt{2}}{30}\{6D^g_{e_1e_1}R^g_{e_2e_2} + 2D^g_{e_1e_2}R^g_{e_1e_2}\}\Gamma(\omega_t = \bar{\omega}_{e_1g}, \omega_\tau = \bar{\omega}_{e_2g}) \\ &- \frac{\sqrt{2}}{30}\{6D^{e_2}_{ff}R^g_{e_2e_2} + 2[\boldsymbol{\mu}_{fe_2}\cdot\boldsymbol{\mu}_{e_2g}]^M \,\mathrm{Im}[\boldsymbol{\mu}_{fe_2}\cdot\mathbf{m}_{e_2g}]^M\}\Gamma(\omega_t = \bar{\omega}_{fe_2}, \omega_\tau = \bar{\omega}_{e_2g}). \end{aligned} \quad (7.80)$$

On the other hand, the cross peaks originating from the first-order electric quadrupole terms are

$$\begin{aligned} \Delta\tilde{E}_{C12}(\omega_t,T,\omega_\tau;Q) &= \frac{\sqrt{2}k}{30}[\boldsymbol{\mu}_{ge_2}\cdot\{\boldsymbol{\mu}_{ge_1}\times(\mathbf{Q}_{e_1g}\cdot\boldsymbol{\mu}_{e_2g})\}]^M\,\Gamma(\omega_t = \bar{\omega}_{e_2g}, \omega_\tau = \bar{\omega}_{e_1g}) \\ &- \frac{\sqrt{2}k}{30}[\boldsymbol{\mu}_{e_1f}\cdot\{\boldsymbol{\mu}_{ge_1}\times(\mathbf{Q}_{e_1g}\cdot\boldsymbol{\mu}_{fe_1})\}]^M\,\Gamma(\omega_t = \bar{\omega}_{fe_1}, \omega_\tau = \bar{\omega}_{e_1g}) \\ \Delta\tilde{E}_{C21}(\omega_t,T,\omega_\tau;Q) &= \frac{\sqrt{2}k}{30}[\boldsymbol{\mu}_{ge_1}\cdot\{\boldsymbol{\mu}_{ge_2}\times(\mathbf{Q}_{e_2g}\cdot\boldsymbol{\mu}_{e_1g})\}]^M\,\Gamma(\omega_t = \bar{\omega}_{e_1g}, \omega_\tau = \bar{\omega}_{e_2g}) \\ &- \frac{\sqrt{2}k}{30}[\boldsymbol{\mu}_{e_2f}\cdot\{\boldsymbol{\mu}_{ge_2}\times(\mathbf{Q}_{e_2g}\cdot\boldsymbol{\mu}_{fe_2})\}]^M\,\Gamma(\omega_t = \bar{\omega}_{fe_2}, \omega_\tau = \bar{\omega}_{e_2g}), \end{aligned} \quad (7.81)$$

where $k = |\mathbf{k}|$. In the above expressions, the dipole and rotatory strengths are defined as

$$\begin{aligned} D^g_{e_je_k} &\equiv [\boldsymbol{\mu}_{e_jg}\cdot\boldsymbol{\mu}_{e_kg}]^M \\ R^g_{e_je_k} &\equiv \mathrm{Im}[\boldsymbol{\mu}_{e_jg}\cdot\mathbf{m}_{e_kg}]^M \\ D^{e_j}_{ff} &\equiv [\boldsymbol{\mu}_{fe_j}\cdot\boldsymbol{\mu}_{fe_j}]^M. \end{aligned} \quad (7.82)$$

The dipole strength $D^{g}_{e_je_k}$ for $j \neq k$ represents the transition dipole strength associated with the creation of excited state coherence $\rho^{(2)}_{e_je_k}$ via two electric dipole–electric field interactions, whereas $R^{g}_{e_je_k}$ does the same strength via both electric dipole–electric field and magnetic dipole–magnetic field interactions. Note that they should not be considered as transition probability because the product state is not a population but a coherence. The dipole strength $D^{e_j}_{ff}$ is however the transition probability of finding the population on the fth doubly excited state when the system was initially on the jth singly excited state.

Now, combining all the magnetic dipole and electric quadrupole contributions to the diagonal and cross peaks, one can obtain the general expression for the chiroptical 2D PE spectroscopy of coupled multichromophore systems:

$$\begin{aligned}\Delta\tilde{E}(\omega_t,T,\omega_\tau) = \frac{\sqrt{2}}{30}\Bigg[&\sum_{j,k}\{6D^{g}_{e_ke_k}R^{g}_{e_je_j} + 2D^{g}_{e_ke_j}R^{g}_{e_ke_j}\}\Gamma(\omega_t = \bar{\omega}_{e_kg},\omega_\tau = \bar{\omega}_{e_jg}) \\ &-\sum_{j,k}\{6D^{e_j}_{f_kf_k}R^{g}_{e_je_j} + 2[\boldsymbol{\mu}_{f_ke_j}\cdot\boldsymbol{\mu}_{e_jg}]^M\,\mathrm{Im}[\boldsymbol{\mu}_{f_ke_j}\cdot\mathbf{m}_{e_jg}]^M\}\,\Gamma(\omega_t = \bar{\omega}_{f_ke_j},\omega_\tau = \bar{\omega}_{e_jg}) \\ &+k\sum_{j,k\neq j}[\boldsymbol{\mu}_{ge_k}\cdot\{\boldsymbol{\mu}_{ge_j}\times(\mathbf{Q}_{e_jg}\cdot\boldsymbol{\mu}_{e_kg})\}]^M\,\Gamma(\omega_t = \bar{\omega}_{e_kg},\omega_\tau = \bar{\omega}_{e_jg}) \\ &-k\sum_{j,k}[\boldsymbol{\mu}_{e_jf_k}\cdot\{\boldsymbol{\mu}_{ge_j}\times(\mathbf{Q}_{e_jg}\cdot\boldsymbol{\mu}_{f_ke_j})\}]^M\,\Gamma(\omega_t = \bar{\omega}_{f_ke_j},\omega_\tau = \bar{\omega}_{e_jg})\Bigg].\end{aligned} \tag{7.83}$$

The above result is valid within a few approximations used. The short-time quantum beat contributions, which originate from coherences created on the singly excited state manifold, and the slow population relaxations were ignored. However, one can easily incorporate such contributions by considering the general theory of 2D PE spectroscopy and by properly taking into consideration rotational averages for magnetic dipole and electric quadrupole terms in the expanded nonlinear response functions with respect to m and Q.

7.5 SUMMARY AND A FEW CONCLUDING REMARKS

In this chapter, we presented theoretical descriptions on the linear and nonlinear chiroptical spectroscopies utilizing a variety of polarization-controlled and polarization-selective detection methods. The heterodyne-detected OA-FID techniques for CD and ORD measurements of chiral molecules in isotropic media were described in detail and their distinctive characteristics and advantages over the conventional intensity differential measurement methods were discussed. Analogies between the present heterodyned OA-FID techniques and the 2D spectroscopic techniques based on a four-wave mixing scheme were clarified. To demonstrate experimental feasibilities of these electric field approaches to the measurement of chiral signal field, we carried out both vibrational CD/ORD measurements for small organic optical isomers and electronic CD/ORD measurements for chiral organometallic compounds and DNA–dye complexes in mid-IR, near-IR, and even visible frequency regions. Although we considered the steady-state optical activity measurements under equilibrium conditions, it is believed that the present OA-FID method can be extended to time-resolved applications if it is combined with a proper triggering method instantaneously initiating a certain nonequilibrium dynamical process such as temperature-jump, pH-jump, photo-cleavage, and so on.

Over the last decade, a variety of coherent multidimensional vibrational or electronic spectroscopic methods, which are analogous to pulsed multidimensional NMR method detecting radio-frequency domain FID field produced by the relaxation of magnetization, have been developed and applied to a wide range of chemical and biological systems. As discussed in this chapter, one can control the polarization states of incident radiations to determine transition dipole angles, which are in turn closely related to 3D molecular structure. The polarization-angle-scanning 2D spectroscopy would be of particular use because, unlike the conventional anisotropic measurement methods, its capability of selectively eliminating either all the diagonal peaks or specific cross peaks enables one to accurately determine the transition dipole angles and rotational dynamics.

Despite the fact that the 2D spectroscopy has provided invaluable information on molecular structure and dynamics, none of them is sensitive to the handedness of a given chiral molecule. In this regard, we believe that the pulse-based OA-FID detection technique developed for the linear chiroptical measurement will be of critical use in further developing even chiroptical 2D spectroscopy. Then, this novel technique can provide information on correlation between achiral (electric-dipole-allowed) quantum transitions and chiral transitions in 2D frequency space. Cho theoretically proposed a two-dimensional circularly polarized pump-probe spectroscopy about a decade ago. Later, a few theoretical studies on other types of nonlinear OA spectroscopic techniques have been reported. However, no relevant experiment has been successfully performed yet due to the weakness of the corresponding signal, that is, $\Delta A_{CD}/A \sim 10^{-4}$, $S_{2D}{-}_{PE}/A \sim 10^{-4}$, and thus $\Delta S_{2D-PE}/A \sim 10^{-8}$, where A, ΔA_{CD}, S_{2D-PE}, and ΔS_{2D-PE} denote the absorbance, CD, 2D photon echo signal, and chiral 2D photon echo signal, respectively. A key factor for the success of such experiments is therefore to precisely control polarization state of incident, transmitted, and scattered radiations as well as to effectively eliminate linear and nonlinear achiral background noises. Furthermore, a single-pulse measurement technique that is free from phase and power fluctuations of a train of laser pulses will be essential to the realization of chiroptical 2D vibrational or electronic coupling measurements. In this regard, we anticipate that the nondifferential and background-free chiroptical measurement methods discussed in this chapter would play a critical role in further developing novel nonlinear chiroptical tools for studying structure and dynamics of chiral systems.

ACKNOWLEDGMENT

This work was supported by the National Research Foundation of Korea (NRF) grants (No. 20090078897 and 20110020033) to MC, which was funded by the Korean government (MEST).

REFERENCES

1. Laage D and Hynes JT. 2006. A molecular jump mechanism of water reorientation. *Science* 311:832–835.
2. Cowan ML et al. 2005. Ultrafast memory loss and energy redistribution in the hydrogen bond network of liquid H_2O. *Nature* 434:199–202.
3. Zheng J, Kwak K, Xie J, and Fayer MD. 2006. Ultrafast carbon-carbon single-bond rotational isomerization in room-temperature solution. *Science* 313:1951–1955.
4. Fleming GR and Cho M. 1996. Chromophore-solvent dynamics. *Ann. Rev. Phys. Chem.* 47:109–134.
5. Jimenez R, Fleming GR, Kumar PV, and Maroncelli M. 1994. Femtosecond solvation dynamics of water. *Nature (London)* 369:471–473.
6. Rhee HJ et al. 2009. Femtosecond characterization of vibrational optical activity of chiral molecules. *Nature* 458:310–313.
7. Rhee H, Choi JH, and Cho M. 2010. Infrared optical activity: Electric field approaches in time domain. *Acc. Chem. Res.* 43:1527–1536.
8. Barron LD. 2004. *Molecular Light Scattering and Optical Activity* (Cambridge University Press, New York).
9. Berova N, Nakanishi K, and Woody RW. 2000. *Circular Dichroism: Principles and Applications* (Wiley-VCH, New York).
10. Rhee H, Ha JH, Jeon SJ, and Cho M. 2008. Femtosecond spectral interferometry of optical activity: Theory. *J. Chem. Phys.* 129:094507.
11. Rhee H, June YG, Kim ZH, Jeon SJ, and Cho M. 2009. Phase sensitive detection of vibrational optical activity free-induction-decay: Vibrational CD and ORD. *J. Opt. Soc. Am. B* 26:1008–1017.
12. Rhee H, Kim SS, Jeon SJ, and Cho M. 2009. Femtosecond measurements of vibrational circular dichroism and optical rotatory dispersion spectra. *Chem. Phys. Chem.* 10:2209–2211.
13. Eom I, Ahn SH, Rhee H, and Cho M. 2011. Broadband near UV to visible optical activity measurement using self-heterodyned method. *Opt. Express* 19:10017–10028.

14. Eom I, Ahn SH, Rhee H, and Cho M. 2012. Single-shot electronic optical activity interferometry: Power and phase fluctuation-free measurement. *Phys. Rev. Lett.* 108:103901.
15. Ernst RR, Bodenhausen G, and Wokaun A. 1987. *Nuclear Magnetic Resonance in One and Two Dimensions* (Oxford University Press, Oxford).
16. Wuthrich K. 1986. *NMR of Proteins and Nucleic Acids* (John Wiley & Sons, New York).
17. Cho M. 2009. *Two-Dimensional Optical Spectroscopy* (CRC Press, Boca Raton).
18. Hamm P and Zanni M. 2011. *Concepts and Methods of 2D Infrared Spectroscopy* (Cambridge University Press, UK).
19. Cho M. 2008. Coherent two-dimensional optical spectroscopy. *Chem. Rev.* 108:1331–1418.
20. Cho M. 1999. Two-dimensional vibrational spectroscopy. In: *Advances in Multi-Photon Processes and Spectroscopy*, ed Lin SH, Villaeys AA, Fujimura Y (World Scientific Publishing Co., Singapore), Vol 12, pp. 229–300.
21. Mukamel S. 2000. Multidimensional femtosecond correlation spectroscopies of electronic and vibrational excitations. *Ann. Rev. Phys. Chem.* 51:691–729.
22. Ganim Z et al. 2008. Amide I two-dimensional infrared spectroscopy of proteins. *Acc. Chem. Res.* 41:432–441.
23. Khalil M, Demirdoven N, and Tokmakoff A. 2003. Coherent 2D IR spectroscopy: Molecular structure and dynamics in solution. *J. Phys. Chem. A* 107:5258.
24. Zanni MT and Hochstrasser RM. 2001. Two-dimensional infrared spectroscopy: A promising new method for the time resolution of structures. *Curr. Opin. Chem. Biol.* 11:516–522.
25. Cho M, Brixner T, Stiopkin I, Vaswani H, and Fleming GR. 2006. Two dimensional electronic spectroscopy of molecular complexes. *J. Chin. Chem. Soc.* 53:15–24.
26. Cho M, Vaswani HM, Brixner T, Stenger J, and Fleming GR. 2005. Exciton analysis in 2D electronic spectroscopy. *J. Phys. Chem. B* 109:10542–10556.
27. Zhuang W, Hayashi T, and Mukamel S. 2009. Coherent multidimensional vibrational spectroscopy of biomolecules: Concepts, simulations, and challenges. *Angew. Chem. Int. Ed.* 48:3750–3781.
28. Lee KK, Park KH, Park S, Jeon SJ, and Cho M. 2011. Polarization-angle-scanning 2D IR spectroscopy of coupled anharmonic oscillators: A polarization null angle method. *J. Phys. Chem. B* 115:5456–5464.
29. Choi JH and Cho M. 2011. Polarization-angle-scanning two-dimensional spectroscopy: Application to dipeptide structure determination. *J. Phys. Chem. A* 115:3766–3777.
30. Choi JH and Cho M. 2010. Polarization-angle-scanning two-dimensional infrared spectroscopy of antiparallel beta-sheet polypeptide: Additional dimensions in two-dimensional optical spectroscopy. *J. Chem. Phys.* 133:241102.
31. Lepetit L, Cheriaux G, and Joffre M. 1995. Linear techniques of phase measurement by femtosecond spectral interferometry for applications in spectroscopy. *J. Opt. Soc. Am. B* 12:2467–2474.
32. Jonas DM. 2003. Two-dimensional femtosecond spectroscopy. *Annu. Rev. Phys. Chem.* 54:425–463.
33. Brixner T, Stiopkin IV, and Fleming GR. 2004. Tunable two-dimensional femtosecond spectroscopy. *Opt. Lett.* 29:884–886.
34. Kane DJ and Trebino R. 1993. Characterization of arbitrary femtosecond pulses using frequency-resolved optical gating. *IEEE J. Quantum Electron.* 29:571–579.
35. Iaconis C and Walmsley IA. 1998. Spectral phase interferometry for direct electric-field reconstruction of ultrashort optical pulses. *Opt. Lett.* 23:792–794.
36. Goldbeck RA, Kim-Shapiro DB, and Kliger DS. 1997. Fast natural and magnetic circular dichroism spectroscopy. *Ann. Rev. Phys. Chem.* 48:453–479.
37. Lewis JW et al. 1985. New technique for measuring circular-dichroism changes on a nanosecond time scale—application to (carbonmonoxy)myoglobin and (carbonmonoxy)hemoglobin. *J. Phys. Chem.* 89:289–294.
38. Helbing J and Bonmarin M. 2009. Vibrational circular dichroism signal enhancement using self-heterodyning with elliptically polarized laser pulses. *J. Chem. Phys.* 131:174507.
39. Bonmarin M and Helbing J. 2008. A picosecond time-resolved vibrational circular dichroism spectrometer. *Opt. Lett.* 33:2086–2088.
40. Mukamel S. 1995. *Principles of Nonlinear Optical Spectroscopy* (Oxford University Press, Oxford).
41. Golonzka O and Tokmakoff A. 2001. Polarization-selective third-order spectroscopy of coupled vibronic states. *J. Chem. Phys.* 115:297–309.
42. Zanni MT, Ge NH, Kim YS, and Hochstrasser RM. 2001. Two-dimensional IR spectroscopy can be designed to eliminate the diagonal peaks and expose only the crosspeaks needed for structure determination. *Proc. Natl. Acad. Sci. U. S. A.* 98:11265–11270.

43. Cho MH, Fleming GR, and Mukamel S. 1993. Nonlinear response functions for birefringence and dichroism measurements in condensed phases. *J. Chem. Phys.* 98:5314–5326.
44. Sung JY and Silbey RJ. 2001. Four wave mixing spectroscopy for a multilevel system. *J. Chem. Phys.* 115:9266–9287.
45. Ferwerda HA, Terpstra J, and Wiersma DA. 1989. Discussion of a coherent artifact in 4-wave mixing experiments. *J. Chem. Phys.* 91:3296–3305.
46. Baiz CR, McRobbie PL, Anna JM, Geva E, and Kubarych KJ. 2009. Two-dimensional infrared spectroscopy of metal carbonyls. *Acc. Chem. Res.* 42:1395–1404.
47. Bracewell RN. 1965. *The Fourier Transform and Its Applications* (McGraw-Hill Book Company, New York).
48. Rezus YLA and Bakker HJ. 2005. On the orientational relaxation of HDO in liquid water. *J. Chem. Phys.* 123:114502.
49. Steinel T, Asbury JB, Zheng JR, and Fayer MD. 2004. Watching hydrogen bonds break: A transient absorption study of water. *J. Phys. Chem. A* 108:10957–10964.
50. Cho M. 2003. Two-dimensional circularly polarized pump-probe spectroscopy. *J. Chem. Phys.* 119:7003–7016.
51. Cheon S and Cho M. 2005. Circularly polarized infrared and visible sum-frequency-generation spectroscopy: Vibrational optical activity measurement. *Phys. Rev. A* 71:013808.
52. Choi JH and Cho M. 2007. Two-dimensional circularly polarized IR photon echo spectroscopy of polypeptides: Four-wave-mixing optical activity measurement. *J. Phys. Chem. A* 111:5176–5184.
53. Choi JH and Cho M. 2007. Nonlinear optical activity measurement spectroscopy of coupled multi-chromophore systems. *Chem. Phys.* 341:57–70.
54. Choi JH, Cheon S, Lee H, and Cho M. 2008. Two-dimensional nonlinear optical activity spectroscopy of coupled multi-chromophore system. *Phys. Chem. Chem. Phys.* 10:3839–3856.
55. Choi JH and Cho M. 2007. Quadrupole contribution to the third-order optical activity spectroscopy. *J. Chem. Phys.* 127:024507.
56. Abramavicius D and Mukamel S. 2005. Coherent third-order spectroscopic probes of molecular chirality. *J. Chem. Phys.* 122:134305.
57. Abramavicius D and Mukamel S. 2006. Chirality-induced signals in coherent multidimensional spectroscopy of excitons. *J. Chem. Phys.* 124:034113.

8 Ultrafast Infrared Probes of Electronic Processes in Materials

John B. Asbury

CONTENTS

8.1 INTRODUCTION

Ultrafast visible and near-infrared spectroscopies have been used to examine electronic processes in emerging electronic materials for many years [1–27] with particular focus on the formation and evolution of primary photoexcitations such as excitons and polarons. Many emerging materials targeting applications in flexible electronics and inexpensive photovoltaics are molecular in nature. As a consequence, the interactions of excitations and charge carriers with molecular species figure prominently in the photophysics and photochemistry of these materials [28–32]. Electronic transitions in these materials tend to be broad and inhomogeneously broadened due to conformational flexibility and thermally induced disorder. As a consequence, limited molecular structural information can be extracted from spectroscopic studies focusing solely on electronic transitions in these materials.

Ultrafast infrared (IR) spectroscopy is uniquely positioned as a probe of electronic processes in emerging electronic materials because it combines ultrafast time resolution with measurement of transient vibrational spectra [33–35]. Many emerging electronic materials are nanocrystalline or glassy solids with vibrational features exhibiting static inhomogeneity arising from morphological variations in molecular order and composition [36]. While this vibrational inhomogeneity would seem to complicate efforts to extract compositional and morphological information about electronic processes in materials, quite the opposite is true. The vibrational inhomogeneity provides a spectroscopic handle to identify unique compositional and morphological environments, thus permitting them to be examined with structural specificity on length scales that are inaccessible by other techniques [28]. In liquids, the interconversion of local molecular interactions and compositions gives rise to spectral diffusion within inhomogeneously broadened vibrational line shapes. However, because spectral diffusion occurs on much longer timescales in glassy solids, the inhomogeneity of vibrational features in emerging electronic materials is essentially static [37]—providing a local probe of molecular structure and morphology.

In addition to providing a probe of transient vibrational features that yield compositional and morphological information about electronic processes, ultrafast IR spectroscopy is also positioned to directly probe the electronic structure and trap state distribution of emerging electronic materials [31]. The mid-IR spectral region corresponds to energies that are typical of charge trapping energies in disordered electronic materials (806.6–8066 cm^{-1} corresponds to 0.1–1.0 eV). For example, disordered materials with charge trap depths >0.5 eV have very low charge carrier mobility and are not considered electrically active (they are insulators). Charge trap depths significantly <0.1 eV are less than the energetic disorder in many disordered semiconductors such that charge trapping does not limit transport. Therefore, ultrafast IR spectroscopy can provide a direct probe of charge traps relevant to transport in emerging electronic materials [38] while simultaneously uncovering molecular information about species involved in formation of the traps.

The purpose of this chapter is to illustrate some of the unique capabilities that ultrafast IR spectroscopy brings to the electronic materials community using two examples from recent studies of organic and colloidal quantum dot (CQD) photovoltaic materials. The chapter begins by outlining the experimental methods used to extract vibrational features from broad electronic transitions resulting from transient electronic species. Vibrational solvatochromism [39] and static inhomogeneity found [36,37] in many emerging electronic materials are described with particular emphasis on organic photovoltaic materials. Then, vibrational spectroscopy of ligand-exchanged CQD materials is discussed to highlight the use of IR spectroscopy to examine nanocrystal surface chemistry [31,38]. Finally, two case studies are presented to illustrate the unique combination of molecular and electronic structural information provided by ultrafast IR spectroscopy. In the first case, electronic processes in organic photovoltaics are described with emphasis on the influence of molecular structure on charge separation mechanisms [32]. In the second case, ultrafast IR spectroscopy is used to study the influence of nanocrystal surface chemistry and electronic structure on charge transport and recombination [31].

8.2 EXPERIMENTAL METHODS

Excited or charged electronic states in materials exhibit broad electronic transitions in the IR spectral region. The nature of these electronic transitions depends on the properties of the materials. For example, electrons or holes in crystalline semiconductors absorb strongly in the IR region via free carrier absorptions that increase approximately as λ^3 with increasing wavelength, λ [40,41]. The free carrier absorption results from coupling of free carriers to phonons that permit them to change momentum states by interaction with photons. The precise variation of free carrier absorption intensity with wavelength depends on which types of phonons dominate the coupling to electronic degrees of freedom [41]. In disordered semiconductors, carriers that are initially free may become trapped. Such trapped carriers still exhibit broad electronic transitions in the mid-IR region because they can be optically excited from trap states back into band states. The high density of band states causes trap-to-band transitions to have large extinction coefficients in the mid-IR [31]. Organic semiconductors typically do not support free carriers. Strong electron–phonon coupling in these materials localizes carrier wavefunctions into polaronic states [42]. However, charge carriers in the organic semiconductor still absorb strongly and broadly in the IR spectra region via a variety of electronic transitions. In the mid-IR region, these transitions are known as polaron absorptions in which localized charge carriers occupying mid-gap states are optically excited back into states in the valence or conduction levels of the semiconductors [5,6].

To examine electronic processes in materials through the vibrational modes of molecules, it is necessary to accurately measure molecular vibrational line shapes that are superimposed onto broad electronic transitions in the mid-IR region [30,31]. Ultrafast IR methods having high spectral sensitivity are needed to extract vibrational line shapes from transient IR spectra because the oscillator strength of mid-IR electronic transitions is typically much greater than the oscillator strengths

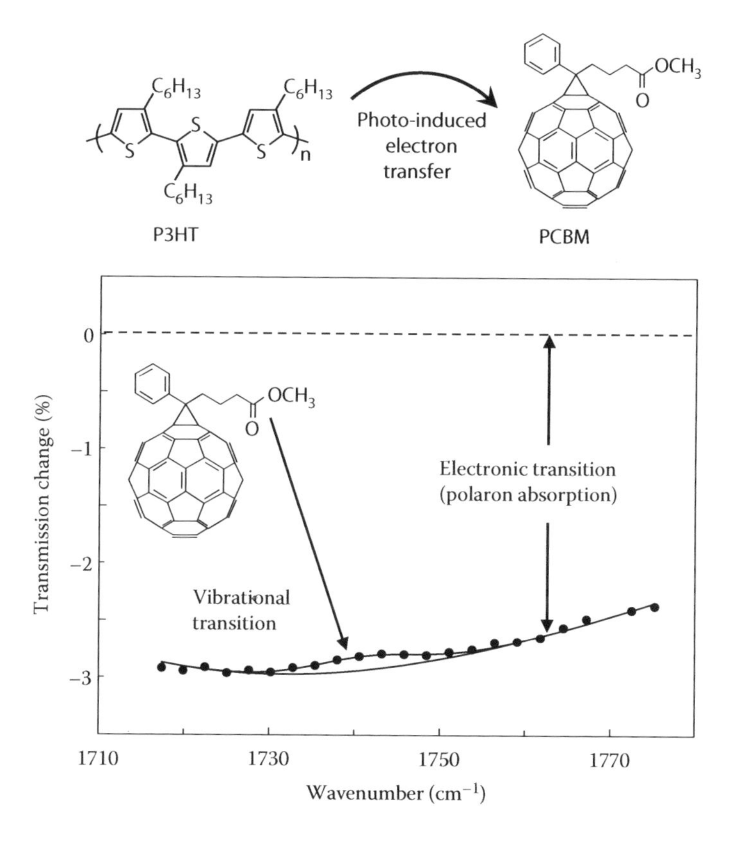

FIGURE 8.1 Photo-induced electron transfer induces a broad electronic absorption feature (negative offset) in the mid-infrared resulting from the formation of polarons in the conjugated polymer, P3HT, and in the fullerene, PCBM. Superimposed on the broad electronic transition is a small vibrational feature (positive feature) of the carbonyl stretch mode of the fullerene. The dots represent experimental data while smooth lines represent fits to the data. High spectral sensitivity is required to accurately measure the vibrational feature because the oscillator strength of the transition is much smaller than the oscillator strength of the electronic transition.

even of strong vibrational modes. Figure 8.1 illustrates the mismatch of transient vibrational and electronic transition intensities in a transient IR spectrum of the conjugated polymer, poly(3-hexylthiophene) (P3HT), with the electron-accepting functionalized fullerene, [6,6]-phenyl-C_{61}-butyric acid methyl ester (PCBM). The vibrational feature corresponds to the carbonyl stretch mode of PCBM following optical excitation of the polymer. Modern ultrafast mid-IR laser sources combined with multichannel detection and normalization techniques are sufficient to obtain the needed spectral sensitivity [28,30,31,43]. The instrumentation at the Pennsylvania State University consists of an ultrafast Ti:sapphire laser that pumps two optical parametric amplifiers (OPAs) to conduct three different experiments. One OPA is used to generate mid-IR pulses at 5.8 μm with 6 μJ pulse energy and 100 fs duration for two-dimensional infrared (2D IR) and polarization-resolved IR pump-probe experiments and as the probe for visible pump-infrared (Vis-IR) probe experiments. A second OPA generates visible pump pulses for the aforementioned experiments. In all cases, a 64-element mercury cadmium telluride dual array detector (Infrared Systems/Infrared Associates) is utilized to obtain the spectral sensitivity needed to separate vibrational line shapes from broad electronic transitions. The dual array permits 32 probe frequencies to be measured simultaneously through a spectrograph (JY Horiba) while facilitating single shot normalization. These measurements also permit low excitation densities of 100 μJ/cm^2 to be used at the sample. Such excitation densities are near the limit where nonlinear relaxation processes (such as biexciton annihilation) cease to dominate the transient excited-state dynamics [32].

To obtain information about electronic states in materials through the vibrational dynamics of the molecules, it is essential to measure the vibrational dynamics of the molecules in their ground electronic states. Ultrafast 2D IR and polarization-resolved broadband IR pump-probe spectroscopies provide the needed measurements of vibrational dynamics. The self-heterodyned pump-probe beam geometry is utilized for both types of measurements [36,44]. IR pulses from one of the OPAs of the ultrafast system are split into two pulses with a 30:1 intensity ratio. The less intense probe pulse is focused at the sample with a 200 μm diameter spot size. The more intense pump pulse passes through a Fabry–Perot interferometer to create a continuously adjustable pump spectrum with a full width at half maximum of ~7 cm^{-1} that is stabilized to within ±1 cm^{-1}. The spot size at the sample is 250 μm. Polarization-resolved IR pump-probe studies are carried out by passing the probe beam through a polarizer that is held by a computer-controlled optical rotator (Newport Corp.). The optical rotator toggles between parallel and perpendicular polarizations relative to the pump beam during data collection. Polarizers are located before and after the sample and in as close proximity to the sample as possible to minimize the depolarizing interactions with optical elements such as mirrors and lenses [45]. Polarization-resolved measurements of the probe beam are made either parallel or perpendicular to the pump beam polarization to avoid polarization rotation problems associated with polarization selective optical elements such as gratings and mirrors.

8.3 MOLECULAR MORPHOLOGY AND COMPOSITION CORRELATED WITH VIBRATIONAL FREQUENCIES

8.3.1 Vibrational Solvatochromism

The sensitivity of vibrational frequencies of molecules to their local molecular or solvent environments provides a unique pathway to examine electronic processes at interfaces in nanoscale-structured materials [39]. In this section, a specific example is described in which this sensitivity, termed vibrational solvatochromism, is used to identify the unique spectroscopic signatures of molecules at organic donor–acceptor interfaces. These findings when coupled with ultrafast IR spectroscopy enable detailed investigation of the primary events leading to electron transfer and charge separation in organic photovoltaic materials (Section 8.4.1) [32].

Solvatochromism results from the influence that surrounding molecular or solvent environments have on the absorption spectra of embedded or dissolved molecular species [46]. Early studies of solvatochromism focused on the influence of solvents on the ultraviolet, visible, and near-IR absorption spectra of solutes. It was recognized that solvatochromism arises from intermolecular solute–solvent interactions of both specific (such as hydrogen bonding) and nonspecific types that in aggregate define the polarity of a solvent [47,48]. The magnitude of a solvatochromic shift is determined by the difference in solvation energy of the equilibrium ground state versus the Franck–Condon excited state of the solute in the solvent. In general, the influence of solvent–solute intermolecular interactions on the absorption spectra of molecules cannot be defined by a single physical property of the solvent such as the dielectric constant or dipole moment [46]. As a consequence, a variety of multiparameter empirical scales have been developed to correlate solvatochromic shifts with solvent properties such as linear free energy relationships [49] or the acceptor number [50,51]. Initially developed to predict frequency shifts of visible absorption bands of molecular species in liquids, the empirical scales have been extended to describe frequency shifts of vibrational transitions in a variety of solvent environments [52,53].

The origin of vibrational solvatochromism may be traced at a molecular level to variation of the local electrostatic potential of inhomogeneously distributed molecules (or solvent) surrounding a vibrational mode of interest [54–56]. Two theoretical frameworks have dominated much of the thinking in the field that has focused largely on the amide-I vibration of proteins or their model systems [57–65] and on the hydroxyl stretch vibration of water [66–72]. In one approach, the transition frequency, $\omega(\varphi_s)$, of a molecule dissolved in a solvent is given by $\omega(\varphi_s) = \omega(\varphi = 0) - \Delta\omega(\varphi_s)$ [57–63].

The transition frequency, $\omega(\varphi = 0)$, describes the isolated molecule in the gas phase, and the perturbation of the transition frequency resulting from the local electrostatic potential, φ_s, is given by $\Delta\omega(\varphi_s)$. The second approach is similar with the exception that the local electrostatic potential is parameterized by the solvent electric field, E_s, evaluated at the sites of the atoms undergoing the vibrational motion [62–65,73–78]. A linear or quadratic correlation between the local solvent electric field and the instantaneous vibrational frequency is assumed. In the case of a linear correlation, the perturbation of the transition frequency is given by $\Delta\omega(E_s) = \alpha E_s$ where the parameter, α, is the proportionality constant, which is not necessarily the Stark tuning rate of the vibrational transition. It was found that other interactions such as hydrogen bonding significantly influence the vibrational frequencies of a number of systems.

The evidence for vibrational solvatochromism at electron donor–acceptor interfaces in organic photovoltaic materials is found in a comparison of IR absorption spectra of polymer blends of an electron-donating conjugated polymer, poly[2-methoxy-5-(2′-ethylhexyloxy)-1,4-(1-cyanovinylene) phenylene (CN–PPV), with an electron-accepting functionalized fullerene, PCBM (Figure 8.2) [39]. The IR spectra exhibit a gradual shift to higher frequency of the carbonyl stretch of the methyl ester group of PCBM with increasing polymer content. Of particular note is the increase in width and growth of the vibrational line shape on the higher frequency side of the transition. As the density of PCBM molecules in contact with CN–PPV increases with greater polymer loading in the film, the molecules in contact with the polymer contribute a larger fraction to the total vibrational line shape. The gradual shift of the vibrational transition to higher frequency indicates that PCBM molecules in contact with the polymer exhibit higher-frequency carbonyl stretch vibrations in comparison to PCBM molecules embedded in the interior of fullerene-rich clusters. In this case, the difference in molecular environment comparing polymer-rich versus fullerene-rich phases gives rise to solvatochromism in the carbonyl stretch vibration of PCBM.

Similar solvatochromic shifts of carbonyl vibrational modes have been reported in other electron donor–acceptor blend systems. For example, Panels A and B of Figure 8.3 display IR absorption spectra of polymer blend films of two electron acceptors, perylene diimide (PDI), and PCBM, with

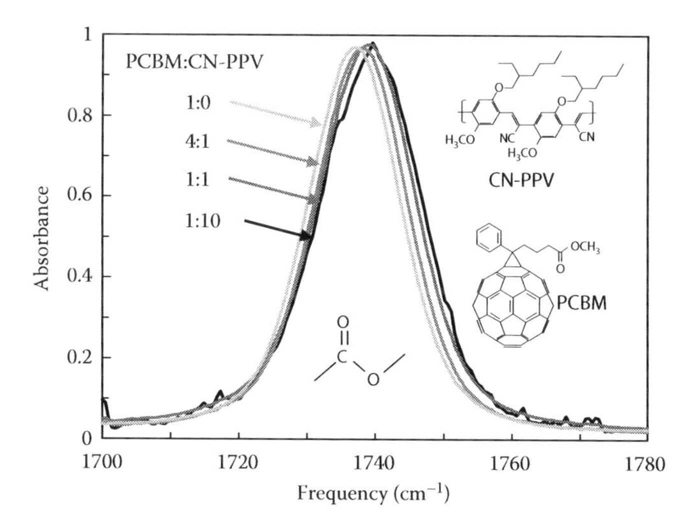

FIGURE 8.2 Comparison of infrared absorption spectra of a series of organic films composed of various mixtures of the conjugated polymer, CN–PPV, and the fullerene derivative, PCBM, focusing on the carbonyl stretch absorption of PCBM. The spectra exhibit a gradual shift to higher frequency with increasing polymer content. The frequency shift arises from solvatochromism and indicates that the carbonyl stretch frequency of PCBM molecules is higher when they are in close proximity to the polymer. (Adapted from Pensack, R. D., Banyas, K. M., and Asbury, J. B. 2010. *Phys. Chem. Chem. Phys.* 12:14144–14152.)

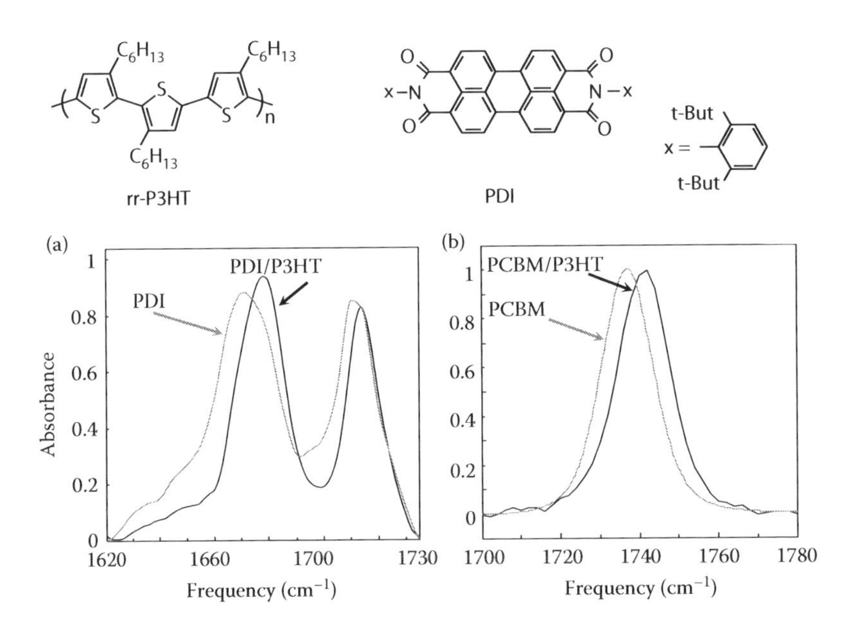

FIGURE 8.3 Infrared absorption spectra of organic thin films of two types of electron acceptors, PDI (panel a), a perylene derivative, and PCBM (panel b), blended with regioregular poly(3-hexylthiophene), P3HT. Infrared absorption spectra of thin films of the neat acceptors focusing on the carbonyl stretch absorptions are represented as well. The comparison reveals that the carbonyl stretch modes of both acceptors exhibit solvatochromic shifts to higher frequency when in contact with the conjugated polymer. The frequency shift provides a means to distinguish acceptor molecules at polymer–acceptor interfaces from acceptor molecules embedded in acceptor clusters. (Adapted from Pensack, R. D., Banyas, K. M., and Asbury, J. B. 2010. *Phys. Chem. Chem. Phys.* 12:14144–14152.)

the conjugated polymer P3HT versus the IR absorption spectra of films of the pure electron acceptors [39]. In both cases, the carbonyl vibrational frequencies shift to higher values in the blends with P3HT—an indication of solvatochromism in these systems.

The vibrational solvatochromism observed in the polymer blend spectra represented in Figures 8.2 and 8.3 may arise from variations in the electrostatic potentials across electron donor–acceptor interfaces in a manner similar to the amide-I and hydroxyl stretch vibrational modes. In addition, variations in the molecular density and polarizability due to disruption of intermolecular order particularly at heterojunctions with the conjugated polymers may be additional influences on the frequencies of vibrational modes at these interfaces.

8.3.2 Static Vibrational Inhomogeneity

Correlation of the vibrational dynamics measured in ultrafast IR spectroscopy experiments with the underlying photophysics and photochemistry of electronic materials requires the ability to discriminate between vibrational dynamics that occur at thermal equilibrium versus dynamics that result from transient excited-state or photoproduct species. This requirement places significant restrictions on the vibrational modes used to examine the electronic processes in materials. For example, the absorption spectrum of the vibrational modes must not exhibit strong temperature dependence such that thermal redistribution of electronic excitation energy does not cause large frequency changes in the vibrational modes of interest [37]. In addition, the vibrational modes should not undergo complete spectral diffusion on the timescale over which electronic processes are to be examined. Indeed, vibrational solvatochromic effects (described above) permit the observation of transient electronic species in various morphologies and molecular compositions within the materials—but only if exchange of the vibrational frequencies unique to these molecular environments occurs more slowly than the electronic species evolve within and among these environments.

The restrictions described above are generally not met by vibrational modes of molecules in liquid environments in which facile interchange of inhomogeneously distributed molecular environments and interactions is the norm. For example, the hydroxyl stretch of water and alcohols displays significant temperature dependence because the molecules are coupled to their surroundings via hydrogen-bonding interactions [79–81]. The amide-I vibration of proteins exhibits pronounced temperature sensitivity for similar reasons—temperature-induced changes in secondary and tertiary structure influence vibrational frequencies due to coupling of amide groups to their molecular environments [82–86]. In both cases, the vibrational chromophores of interest are coupled to their environments via weak intermolecular interactions with dissociation energies on the order of 20 kJ/mol [87–89]. The vibrational occupation of the weak interactions is strongly temperature dependent. Because they are coupled to the high-frequency hydroxyl or amide-I modes, these vibrational modes exhibit pronounced temperature dependence.

The strong coupling of the hydroxyl and amide-I vibrational modes to the surrounding environments coupled with facile interchange of molecular conformations in liquids also gives rise to fast spectral diffusion within the inhomogeneously broadened vibrational absorption bands. For example, fast hydrogen bond network reorganization in water gives rise to complete spectral diffusion in the hydroxyl stretch vibrational line shape on the few picosecond timescale [73,74,76,78,90–92]. Vibrational probes of proteins exhibit some slower timescale motions that are influenced by local fluctuations and by evolution of secondary and tertiary structure [86,93–98]. However, embedding proteins in glasses such as trehalose has been shown to prevent complete randomization of the spectral inhomogeneity of the amide-I vibration largely by arresting secondary and tertiary structural evolution of the proteins [99,100].

Organic photovoltaic polymer blends are examples of electronic materials that satisfy the requirements to examine photophysical and photochemical processes through their vibrational features [37]. Figure 8.4 displays 2D IR spectra of such a material. The spectra focus on the carbonyl vibration of the methyl ester group of the electron acceptor, PCBM, that is blended within the conjugated polymer, CN–PPV. 2D IR spectra were measured using the self-heterodyned pump-probe approach such that a narrow band IR pump pulse (7 cm^{-1} full width at half maximum) was used to excite the sample followed by a broadband probe pulse covering the full carbonyl absorption spectrum. 2D IR spectra measured at time delays of 1, 3, and 10 ps between the narrow band pump pulse and the broadband probe pulse are represented at two temperatures, 350 and 200 K. The two-dimensional line shapes of the ground to first excited state (0–1) transitions appear on the diagonals of the 2D IR spectra while the first to second excited state (1–2) transitions appear off-diagonal and with a negative sign. The data reveal that no spectral diffusion occurs in the vibrational transitions on the 1–10 ps timescale. At lower temperature, the antidiagonal widths are smaller, and the center line slopes (indicated by the black lines) are larger. These observations indicate that the dynamically broadened line shapes of the carbonyl stretch mode are narrower at lower temperature.

The lack of spectral diffusion in the polymer blend was quantified by measuring the center line slopes [101] of the 0–1 transitions in the 2D IR spectra versus time delay and temperature. Figure 8.5 displays the center line slopes of the 0–1 transitions for all time delays and temperatures revealing negligible change in slope with increasing time delay. The inhomogeneous broadening mechanism is dominated by vibrational solvatochromism in which different morphologies and molecular compositions exhibit unique carbonyl vibrational frequencies [39]. The lack of spectral diffusion confirms that these environments do not interconvert on ultrafast timescales at either temperature. Because the polymer blends are glassy solids, their inhomogeneous molecular environments indeed interconvert on slow timescales—timescales longer than the ultrafast time regime in which electronic processes in the materials are examined [37].

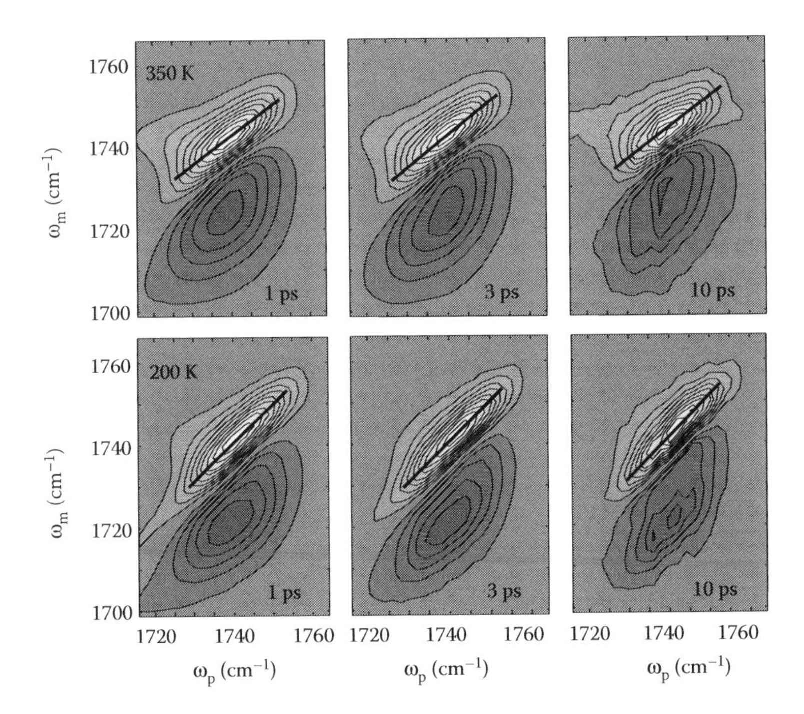

FIGURE 8.4 2D IR spectra of the carbonyl (C=O) stretch of the methyl ester group of PCBM in a blend with CN–PPV measured at different time delays at two temperatures. The elongation of the peaks on the diagonal (0–1 transition, lighter shades) and off of the diagonal (1–2 transition, darker shades) indicate that the carbonyl stretch mode is inhomogeneously broadened. The peaks do not grow in antidiagonal width on the 10 ps timescale, indicating that negligible spectral diffusion occurs within this time range. At lower temperature, the antidiagonal widths are smaller, and the center line slopes (indicated by the black lines) are larger. These observations indicate that the dynamically broadened line shapes are narrower at lower temperature. (Adapted from Pensack, R. D., Banyas, K. M., and Asbury, J. B. 2010. *J. Phys. Chem. B* 114:12242–12251.)

The carbonyl stretch mode of the methyl ester group of PCBM in the polymer blend also exhibits the weak temperature dependence needed to accurately measure the dynamics of transient electronic species without interference from thermal redistribution processes. Figure 8.6 represents a collection of eight IR absorption spectra of polymer blends of PCBM with CN–PPV measured at temperatures ranging from 390 to 184 K. The spectra exhibit a shift of the peak frequency of 1.5 cm^{-1} over the temperature range with only a 0.5 cm^{-1} shift occurring from 300 to 390 K. Ultrafast polarization-resolved broadband IR pump-probe measurements of the carbonyl stretch mode permit separate measurement of the vibrational excited-state lifetime and the orientational diffusion time of the methyl ester group [37]. Kinetic decay traces measured with parallel, $S_{\parallel}(t)$, and perpendicular, $S_{\perp}(t)$, pump and probe pulse polarizations near the peak of the 0–1 transition of the carbonyl group at 1740 cm^{-1} are displayed in Figure 8.7. The inset indicates the frequency at which the kinetics were measured relative to the transient IR pump-probe spectrum of the sample. Combinations of the parallel and perpendicular polarization kinetics traces are used to construct the excited-state population relaxation kinetics, $P(t)$, according to $P(t) = 1/3(S_{\parallel}(t) + 2S_{\perp}(t))$. The excited-state dynamics reveal biphasic population relaxation occurring on 300 fs and 1.7 ps timescales at room temperature. Temperature-dependent measurements of the excited-state population dynamics indicate that the vibrational lifetime has little variation with temperature down to 200 K [37]. However, calculation of the orientational diffusion dynamics from the polarization-resolved kinetics traces according to $r(t) = 0.4C_2(t) = (S_{\parallel}(t) - S_{\perp}(t))/(S_{\parallel}(t) + 2S_{\perp}(t))$ at several temperatures demonstrates that the orientational motion changes with temperature. Figure 8.8 displays anisotropy decay traces of the

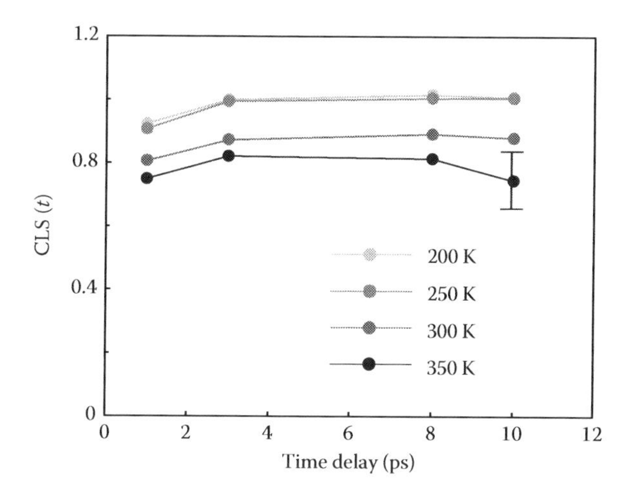

FIGURE 8.5 Plot of the center line slopes of two-dimensional peak shapes of the 0–1 transitions in 2D IR spectra of the carbonyl stretch mode of PCBM in a blend with CN–PPV measured at several temperatures. The absence of a decay of the center line slopes toward zero confirms that negligible spectral diffusion occurs in the carbonyl stretch mode on the 10 ps timescale. The lack of spectral diffusion arises from the solvatochromic inhomogeneous broadening mechanism. PCBM molecules must diffuse within the solid matrix to randomize the frequency variation due to solvatochromism. Such diffusional processes are very slow for large molecules in glasses. (Adapted from Pensack, R. D., Banyas, K. M., and Asbury, J. B. 2010. *J. Phys. Chem. B* 114:12242–12251.)

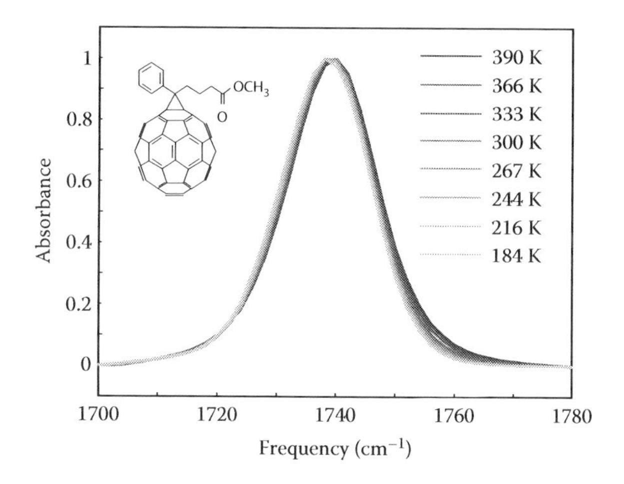

FIGURE 8.6 Collection of infrared absorption spectra of the carbonyl stretch mode of PCBM in a blend with CN–PPV measured at a range of temperatures. The spectra exhibit weak temperature dependence with a 1.5 cm^{-1} shift of the peak to higher frequency within the temperature range indicated. Most of the shift to higher frequency occurs between 184 and 300 K (1 cm^{-1}) with the remaining 0.5 cm^{-1} shift occurring between 300 and 390 K. (Adapted from Pensack, R. D., Banyas, K. M., and Asbury, J. B. 2010. *J. Phys. Chem. B* 114:12242–12251.)

carbonyl group of PCBM in the polymer blend measured at several temperatures. The data reveal fast wobbling in cone motion over half angles that decrease from 34° at 350 K (larger subpicosecond decay component) to 24° at 200 K (smaller subpicosecond decay component). The change in wobbling in the cone angle results from loss of free volume in the blend associated with an increase in the density of the film at lower temperature [102].

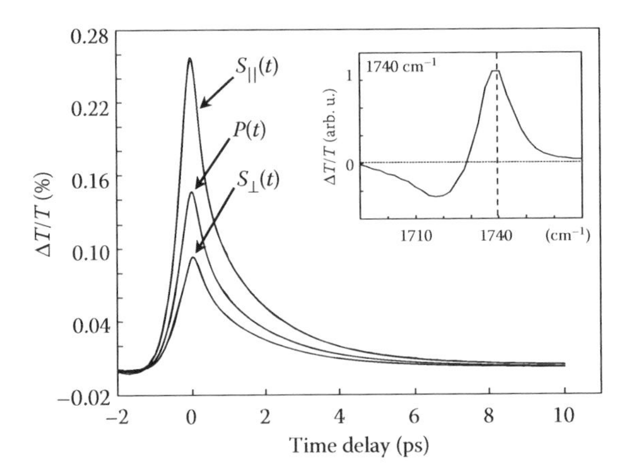

FIGURE 8.7 Polarization-resolved ultrafast infrared pump-probe kinetics traces measured near the peak of the 0–1 transition of the carbonyl stretch mode of PCBM in a blend with CN–PPV. The population kinetics trace measured at 1740 cm^{-1}, $P(t)$, exhibits a biexponential decay. The inset indicates the frequency at which the kinetics traces were measured relative to the peak of the 0–1 transition in the infrared pump-probe spectrum measured at 100 fs time delay. (Adapted from Pensack, R. D., Banyas, K. M., and Asbury, J. B. 2010. *J. Phys. Chem. B* 114:12242–12251.)

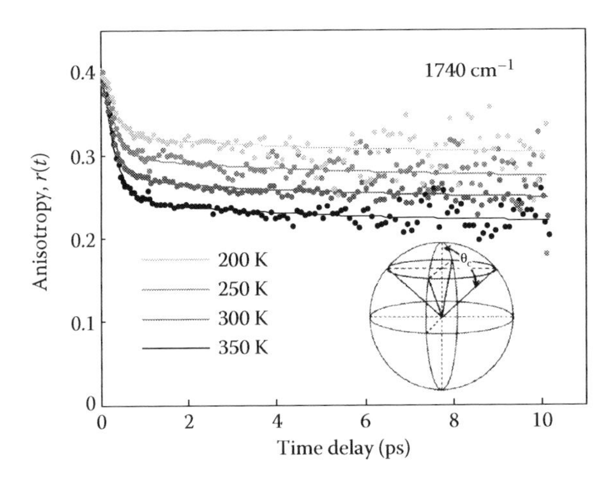

FIGURE 8.8 Anisotropy decay kinetics of the carbonyl stretch of PCBM in a blend with CN–PPV measured at 1740 cm^{-1} at several temperatures. The data reveal biphasic anisotropy decay arising from fast wobbling in cone orientational motion (see inset) on the subpicosecond timescale followed by slow diffusional reorientation on timescales much greater than 10 ps. The wobbling in the cone angle increases with increasing temperature due to thermal expansion of the polymer blend. (Adapted from Pensack, R. D., Banyas, K. M., and Asbury, J. B. 2010. *J. Phys. Chem. B* 114:12242–12251.)

8.3.3 Vibrational Spectroscopy of Ligand Exchange

CQD or nanocrystal materials offer the promise of combining favorable properties of inorganic semiconductors such as delocalized wavefunctions and high dielectric permittivity with the benefits of low-temperature solution processability [103]. To enable nanocrystalline materials to be competent electronic materials, it is necessary to passivate dangling bonds at their surfaces while closely packing the nanocrystals into dense (ideally) ordered lattices [69,104–107]. Closely packing nanocrystals into dense films maximizes the overlap of electronic wavefunctions in neighboring nanocrystals for facile charge transport among the nanocrystals. Passivation of nanocrystal surfaces minimizes the density and energetic depth of charge traps to enable high charge carrier mobility [31].

Molecules or inorganic species are typically used as ligands to passivate dangling bonds [66–72,105,108–117]; therefore, the shortest possible molecular or inorganic ligands are desired. These small ligands do not correspond to those used during synthesis of nanocrystals because the original ligands must contain long aliphatic chains to maintain colloidal stability. Therefore, it is necessary to exchange the original ligands of as-synthesized nanocrystals with smaller ligands suitable for closely packing nanocrystals into dense solid films. The extent to which the ligand exchange reaction occurs can be challenging to characterize [118]. Therefore, the development of methods to quantitatively analyze the extent of ligand exchange is of considerable importance to understanding and controlling the surface chemistry and electronic properties of colloidal nanocrystalline electronic materials.

IR spectroscopy provides a means to probe the identity of ligands bound to nanocrystal surfaces through their unique vibrational features [31]. Not only do common surface active functional groups have unique vibrational frequencies enabling chemical identification of the molecules on the surfaces of quantum dots, but also the frequencies of those vibrational modes change when the functional groups are bonded to the surfaces versus when they are not bonded. Figure 8.9 illustrates this capability in the case of four ligands that were used in the fabrication of PbS CQD photovoltaics [38,68,108]. The original oleate (OA) ligands used during synthesis of the quantum dots exhibit vibrational features in the 1400 and 1500 cm^{-1} region of their IR absorption spectrum

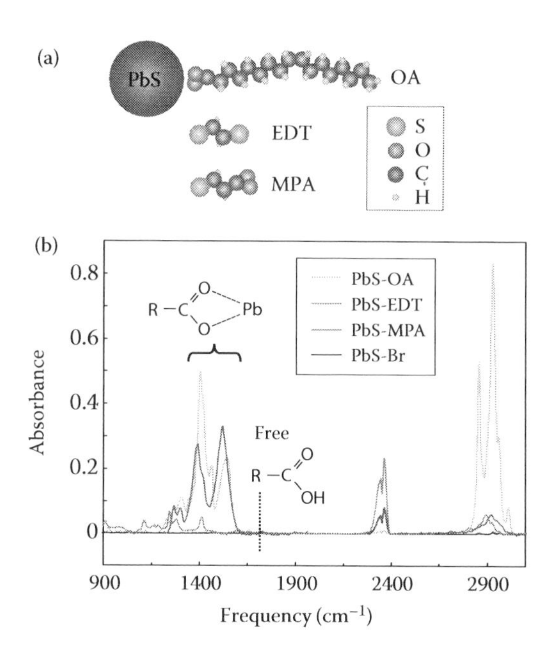

FIGURE 8.9 (a) Structures of short ligands used to replace original oleate ligands (OA) used during synthesis of PbS colloidal quantum dots. (b) Infrared absorption spectra of colloidal quantum dot films treated with several ligands, including OA, EDT, MPA, and Br. The vibrational features of the ligands provide a means to examine the surface chemistry of the quantum dots. (Adapted from Jeong, K. S. et al. 2012. *ACS Nano* 6:89–99.)

that are indicative of a carboxylate group bound to lead surface sites [119]. Strong absorption is observed in the C–H stretch region around 2900 cm^{-1} because of the long aliphatic chains of OA ligands. When the quantum dots are treated with 3-mercaptopropionic acid (MPA) that also contains a carboxylate group, the same vibrational features around 1400 and 1500 cm^{-1} are observed. However, the absorption amplitude in the 2900 cm^{-1} region is strongly reduced due to the smaller number of C–H groups of MPA. Ligand exchange with ethane dithiol (EDT) instead of MPA results in loss of the carboxylate group stretch modes at 1400 and 1500 cm^{-1}—indicating complete ligand exchange in this case. The evidence for complete ligand exchange in the MPA-treated film is found by noting that treatment of PbS films with EDT and MPA, which each have the same number of methylene groups, results in the same absorption strength at 2900 cm^{-1}. Because ligand exchange with EDT is complete as evidenced by the loss of the carboxylate groups, it is concluded that ligand exchange with MPA is also complete [31]. Any residual OA ligands remaining in the MPA-treated film would result in a larger C–H stretch absorption band at 2900 cm^{-1} in comparison to the band in the EDT-treated film. An all-inorganic passivation strategy utilizing halide ions results in complete removal of organic species from the PbS surfaces [38]. The corresponding IR absorption spectrum of the CQD film (PbS–Br) reveals the absence of any vibrational modes of molecules in the film.

IR spectroscopy can also be used to examine the surface chemistry of nanocrystalline materials during the process of ligand exchange. To illustrate, Figure 8.10 displays IR absorption spectra of a PbS CQD film at various stages of the ligand exchange process. A spectrum of the film prior to ligand exchange (PbS–OA) displays the characteristic vibrational features of OA ligands bound to the as-synthesized PbS nanocrystals. Immediately after the chemical treatment with a solution of MPA in methanol, the IR spectrum (PbS–MPA before drying) demonstrates that OA ligands are removed as evidenced by the loss of the intense C–H stretching absorption bands at 2900 cm^{-1}. However, the carboxylate groups of MPA have not fully attached to the nanocrystal surfaces as indicated by the high-frequency carboxylate stretch mode at 1700 cm^{-1}. A vibrational feature at this frequency indicates the presence of a carboxylate group that is protonated and thus has not yet formed a bond with the lead atoms on the surfaces of the nanocrystals. Alternatively,

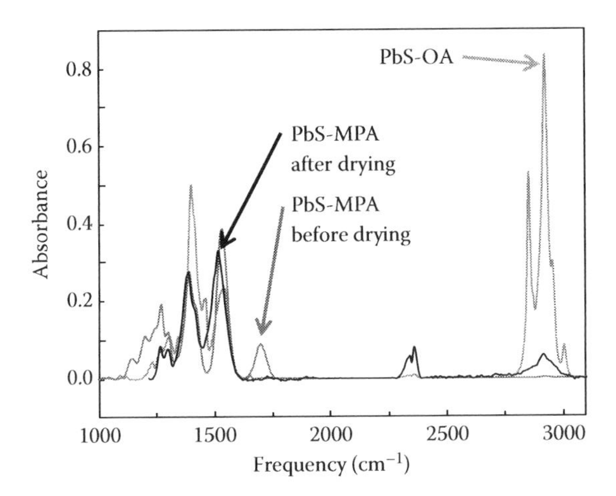

FIGURE 8.10 Infrared absorption spectra of an MPA-treated PbS CQD film at various stages of ligand exchange. Before ligand exchange, the film exhibits vibrational features of oleate ligands (PbS–OA). Immediately after ligand exchange, the oleate ligands are removed, but the carboxylate groups of MPA have not fully bonded to the PbS surfaces resulting in the observation of free carboxylate absorption at 1700 cm^{-1} (PbS–MPA before drying). After a period of drying in vacuum, the carboxylate groups of MPA are fully bonded to PbS surfaces as evidenced by the loss of the free carboxylate peak (PbS–MPA after drying). (Adapted from Jeong, K. S. et al. 2012. *ACS Nano* 6:89–99.)

the 1700 cm^{-1} feature could arise from a carboxylic group that is bonded in a unidentate geometry (through the single-bonded C–O group) to a surface lead atom [119]. This spectrum was recorded prior to incubation in vacuum in which residual methanol solvent molecules bound to the nanocrystal surfaces are driven off. Following this vacuum incubation step, all the carboxylate groups form bidentate bonds to the nanocrystal surfaces in either bridging or chelating geometries as indicated by the loss of the high-frequency carbonyl stretch mode of the carboxylate groups (PbS–MPA after drying).

8.4 ELECTRONIC PROCESSES IN EMERGING PHOTOVOLTAIC MATERIALS

8.4.1 Charge Separation in Organic Photovoltaics

Ultrafast IR spectroscopy is particularly well suited to examine the primary events leading to charge separation and photocurrent generation in organic photovoltaic materials because they consist of nanoscale phase-separated blends of electron-donating and electron-accepting species that exhibit vibrational solvatochromism. Figure 8.11a represents an energy-filtered transmission electron micrograph of a polymer blend of the conjugated polymer P3HT with PCBM in which the light regions correspond to the sulfur-rich (polymer) phase [120]. The image reveals the nanoscale interpenetration of electron donor and acceptor phases that characterize organic photovoltaic materials. Understanding the mechanisms of charge photogeneration in these materials requires a detailed knowledge of electronic processes at electron donor–acceptor interfaces that can be difficult to study directly because they are buried in the interiors of the polymer blend films.

Figure 8.11b illustrates some of the photophysical processes that are involved in charge carrier generation at interfaces in organic photovoltaic materials [121,122]. These processes include exciton

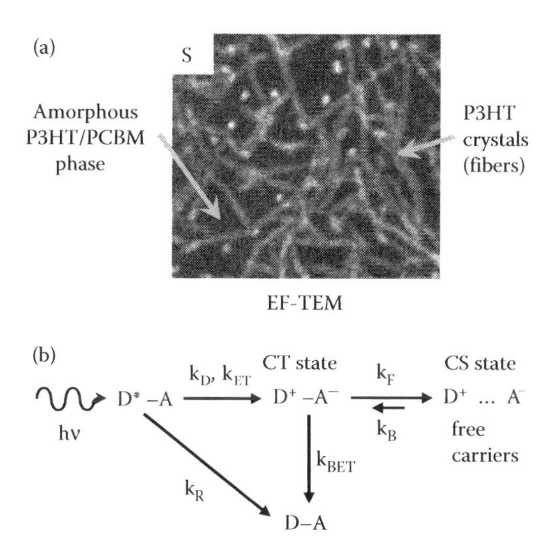

FIGURE 8.11 (a) Energy-filtered transmission electron microscopy (TEM) image showing the sulfur-rich regions of a polymer blend of regioregular P3HT with PCBM. P3HT crystallizes into high aspect ratio fibers that are embedded in an amorphous phase consisting of a mixture of P3HT with PCBM. The image represents an example of the nanoscale phase separation needed to efficiently split excitons at electron donor–acceptor interfaces. (b) Schematic diagram of photophysical processes at electron donor (D) and acceptor (A) interfaces in organic photovoltaic materials. Following exciton diffusion and electron transfer, charge transfer (CT) states form at D–A interfaces. Subsequently, CT states dissociate to form separated charge carriers (CS states). The latter CT state dissociation process is the focus of the current section. (Adapted from Pensack, R. D. et al. 2012. *J. Phys. Chem. C* 116:4824–4831.)

formation by absorption of light, charge transfer (CT) state formation by dissociation of excitons via electron transfer at electron donor–acceptor interfaces, and charge-separated (CS) state formation by dissociation of CT states. Unambiguous observation of the latter process, CT state dissociation to form CS states, has been challenging because of the similarity in electronic properties of both states. Fortunately, the sensitivity of vibrational modes to their local molecular environments (solvatochromism, see Section 8.3.1) causes molecules at electron donor–acceptor interfaces that are involved in the formation of CT states to be distinguishable from molecules involved in CS states through their vibrational frequencies [39]. A method utilizing this sensitivity, ultrafast solvatochromism-assisted vibrational spectroscopy (SAVS), was recently developed to examine the mechanism of charge separation in organic photovoltaic polymer blends [30]. The key elements of the technique and principal conclusions about the charge separation mechanism in such materials are described here.

The ultrafast SAVS approach is a combination of several established ultrafast IR spectroscopy techniques that collectively provide unique insights into fundamental photophysical processes occurring in nanostructured materials. The approach combines 2D IR [36,37] and other IR third-order techniques [28,29,43,123,124] to characterize the dynamics of the vibrational mode(s) of interest on their ground-state electronic potentials. The photophysics of the nanostructured materials are examined using Vis-IR probe spectroscopy [28,29,43,123,124]. Applied specifically to organic photovoltaic materials, the vibrational dynamics associated with CT state formation and subsequent dissociation are measured by an ultrafast IR probe pulse tuned to the vibrational mode(s) of interest [123]. The experimental procedures for the use of these techniques have been described in Section 8.2 and elsewhere [28,37,39,43].

The ultrafast SAVS approach was recently used to elucidate the influence that electron acceptor structure has on the timescale and energetics of charge separation in organic photovoltaic materials [32]. Two classes of electron acceptors were examined: a PDI derivative that represents the class of conjugated molecules having planar conjugated frameworks (Figure 8.3), and a functionalized fullerene (PCBM) representing the class of molecules whose conjugated frameworks have three-dimensional topology (Figure 8.2). Each acceptor was blended with P3HT and cast into a solid film of 300–500 nm thickness. Visible excitation wavelengths were selected for each polymer blend to selectively excite P3HT with minimal direct excitation of the acceptor such that no perturbations of the vibrational modes of the acceptors were observed prior to electron transfer from excitons in the polymer [32].

Typical ultrafast transient absorption spectra focused on the carbonyl absorption features of PDI and PCBM molecules in their respective polymer blends measured around 300 K are represented in Figure 8.12. The spectra are displayed at several time delays between the visible pump and IR probe pulses showing the time evolution of the carbonyl bleach features that are superimposed onto broad electronic transitions. The electronic transitions give rise to the time-dependent offsets in the spectra. The carbonyl bleach features are visible by comparing the best fits of the transient spectra (smooth curves through the transient spectra) with the best fits of the broad electronic transitions (smooth curves underneath the bleach features). Because the excitation wavelengths were chosen to selectively excite P3HT in both cases, the appearance of carbonyl bleach features indicates that electrons have transferred to the acceptors on ultrafast timescales. In both cases, the initial carbonyl bleach features appear at frequencies higher than the equilibrium spectra of the polymer blends and shift toward and reach the equilibrium spectra with increasing time delay. The dashed curves in both sets of transient vibrational spectra highlight the time-dependent shift of the carbonyl bleach features toward the equilibrium center frequencies. It has been demonstrated that the shifts of the carbonyl bleach features toward the equilibrium spectra result from dissociation of CT states at electron donor–acceptor interfaces to form CS states [123]. A fitting routine was developed to extract the carbonyl bleach spectra from the transient absorption spectra to quantify the charge separation dynamics. A two-dimensional frequency–time plot of the carbonyl bleach spectra of the P3HT polymer blend with PCBM obtained from the fitting routine is represented in Figure 8.13.

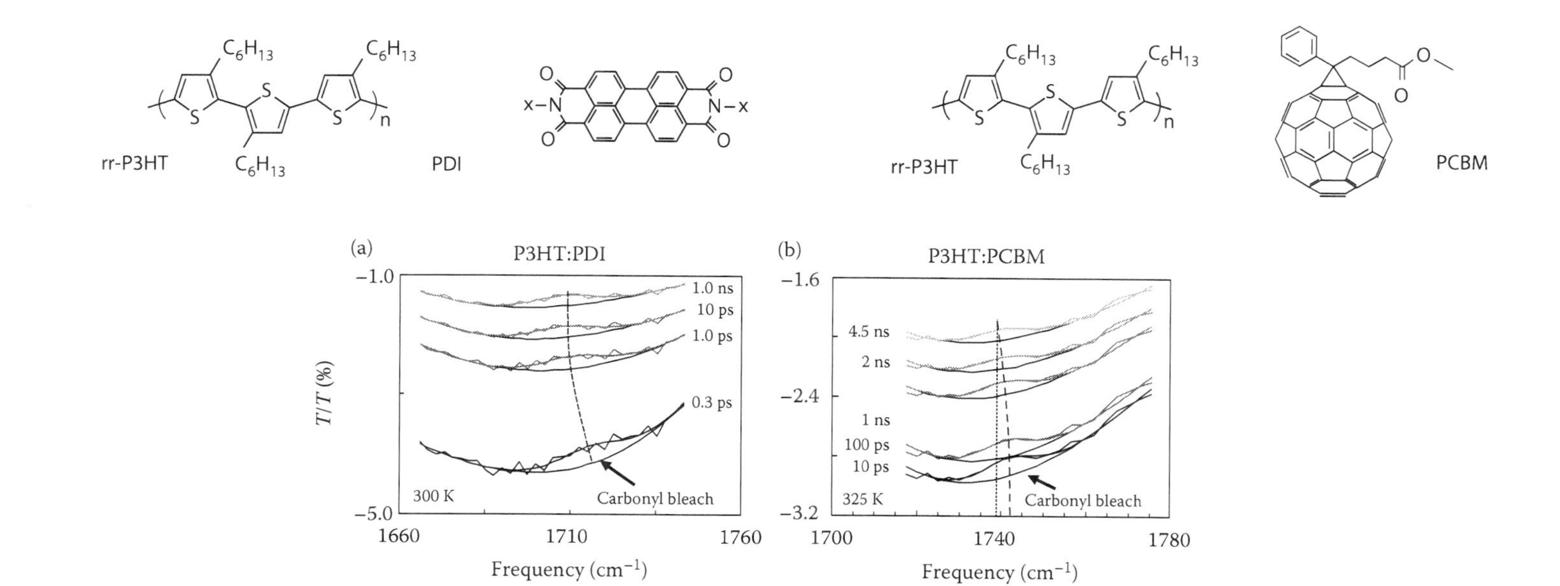

FIGURE 8.12 Transient infrared spectra following photoexcitation of polymer blends of (a) P3HT with PDI measured at 300 K and (b) P3HT with PCBM measured at 325 K. Carbonyl bleach features of each acceptor are superimposed onto broad polaron absorptions. Both features arise from the transfer of electrons from photoexcited P3HT to the acceptors. The carbonyl bleach features initially appear at higher frequencies due to solvatochromic shifts of the acceptor vibrational modes when in contact with P3HT. The time-dependent shifts of the carbonyl bleach center frequencies to lower values are highlighted by the dashed curves. (Adapted from Pensack, R. D. et al. 2012. *J. Phys. Chem. C* 116:4824–4831.)

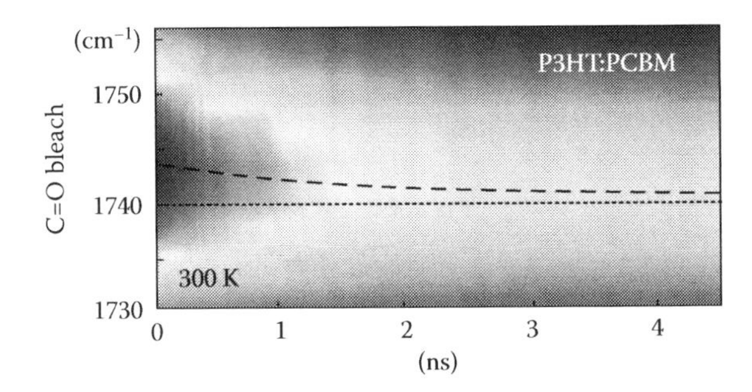

FIGURE 8.13 Two-dimensional frequency–time plot of carbonyl absorption features of PCBM in a blend with P3HT versus corresponding time delay following photoexcitation of P3HT. The carbonyl bleach feature appears initially with large amplitude above 1740 cm^{-1}. With increasing time delay, the feature broadens, decreases in maximum amplitude, and shifts toward the equilibrium center frequency (dashed line) on the nanosecond timescale. The shift of the carbonyl bleach feature is highlighted by the dashed line. (Adapted from Pensack, R. D. and Asbury, J. B. 2011. *Chem. Phys. Lett.* 515:197–205.)

As seen in the transient spectra, the carbonyl bleach features initially appear at higher frequencies than the equilibrium band center (indicated by the dotted horizontal line at 1740 cm^{-1}) and shift toward the equilibrium center frequency on the nanosecond timescale.

To examine the influence that acceptor structure has on the energetic barriers to charge separation, charge separation dynamics were examined at several temperatures in both systems. The time-dependent center frequencies of carbonyl bleach spectra measured in both polymer blends at various temperatures are represented in Figures 8.14a and 8.14b. The center frequencies of the carbonyl bleach spectra are plotted versus the corresponding time delays at which the transient spectra were measured. In both cases, the time-dependent shift of the center frequencies toward lower values is indicative of charge separation because electrons initially occupy CT states involving molecules with higher frequency carbonyl stretch vibrations. Dissociation of the CT states to form CS states gives rise to the shift of the carbonyl bleach features to lower frequency. A logarithmic time axis is used to represent the frequency shift kinetics measured in the polymer blend having PDI as the electron acceptor, while a linear scale is used in the case of the blend containing PCBM. The difference in time axis format was chosen to most clearly represent the data. The inset in Figure 8.14b indicates that negligible charge separation occurs on the 100 ps timescale in the P3HT–PCBM polymer blend.

The temperature dependences of the average rates of charge separation in the P3HT–PDI and P3HT–PCBM polymer blends are represented in Figures 8.14c and 8.14d, respectively. In both cases, the logarithm of the average rates is plotted versus inverse temperature such that Arrhenius behavior is represented as a straight line whose slope indicates the activation barrier for the reaction. The average rates were calculated from the average time constants obtained from the frequency shift kinetics, $G(t)$, according to $\langle\tau\rangle = \int t\left(G(t) - g(\infty)\right)\mathrm{d}t/\int G(t) - g(\infty)\mathrm{d}t$, where $g(\infty)$ represents the asymptomatic frequency to which the carbonyl bleach spectra approach at long time delays. A confidence limit of ±30% of the average rate is represented by the error bar displayed in Panel D. The error bars are smaller than the symbols in Panel C.

The rate of charge separation in the P3HT–PDI polymer blend depends sensitively on temperature, indicating that the process is activated in this polymer blend. Using the slope defined by the data measured at the two lowest temperatures, an activation energy of approximately 0.1 eV is obtained. Although the kinetics decays from which this activation energy is obtained are clearly different (Figure 8.14a), this value should be taken only as an estimate because it is derived from a limited data set. The transition from stronger to weaker temperature dependence occurs when the timescale of charge separation decreases from the >10 ps timescale at lower temperature to the 1 ps timescale at higher temperature. This seemingly non-Arrhenius behavior is a result of the

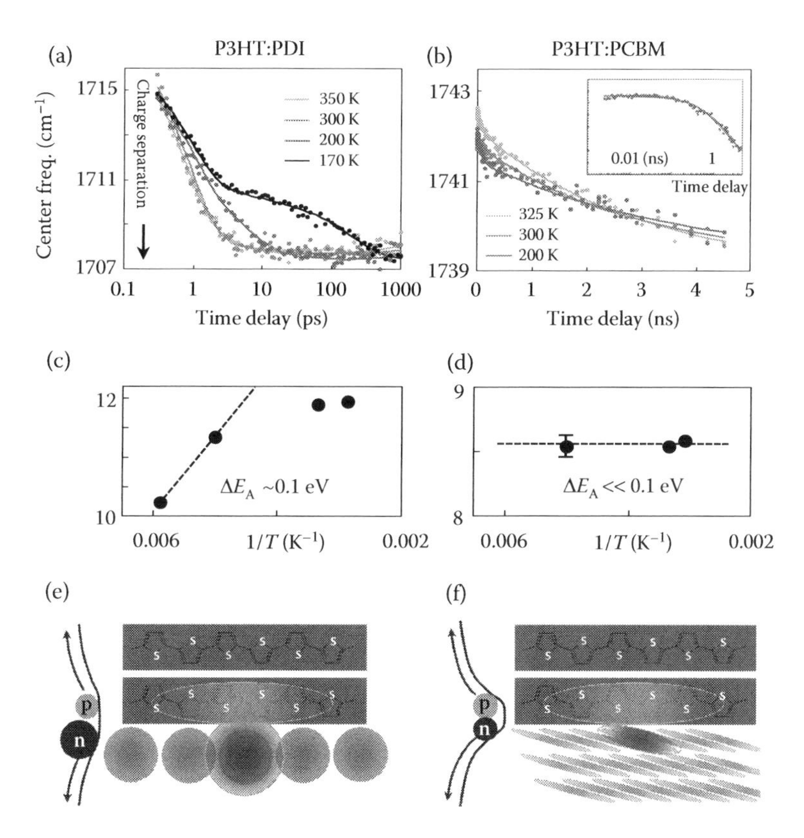

FIGURE 8.14 Charge separation kinetics traces measured from carbonyl bleach frequency shifts in polymer blends of P3HT with PDI (a) and PCBM (b) following photoexcitation of the polymer. Charge separation dynamics are measured at different temperatures to assess the effective barriers to the process. Plots of the average rates of charge separation versus inverse temperature of PDI (c) and PCBM (d) polymer blends. The PDI blend exhibits activated charge separation while the PCBM blend does not. Schematic diagram of the influence of molecular structure on the free energy barriers to charge separation in the PDI (e) and PCBM (f) blends. The larger and three-dimensional conjugated framework of PCBM enables greater electron delocalization in comparison to PDI molecules. The comparison suggests that acceptor molecules with two-dimensional topology exhibit stronger Coulomb forces and larger reorganization energies to charge separation in comparison to molecules with three-dimensional topologies such as fullerenes. (Adapted from Pensack, R. D. et al. 2012. *J. Phys. Chem. C* 116:4824–4831.)

finite bandwidth of the carbonyl vibrational mode that is used to measure the fast charge separation dynamics [32]. Charge separation occurring faster than 1 ps is obscured by the ~1 ps free-induction decay of the carbonyl bleach feature (full width at half maximum is 15 cm^{-1}). Probing the dynamics with a vibrational mode having a greater bandwidth will likely reveal that the dynamics follow the expected Arrhenius behavior at higher temperatures.

In contrast to the P3HT–PDI polymer blend, the rate of charge separation in the P3HT–PCBM blend does not change appreciably with temperature, indicating that this polymer blend undergoes barrierless charge separation. Barrierless charge separation has been observed in a variety of systems, including dye-sensitized solar cells [125–128], dye-sensitized silver halide crystals [129,130], and photosynthetic reaction centers [131–133]. These studies have identified a number of mechanisms giving rise to weak temperature dependence [125–130,132–137]. The most applicable mechanism has its origin in the influence of electron delocalization on the Coulomb potentials and on the reorganization energy associated with charge separation in the polymer blend. Because CT states consist of oppositely charged electrons and holes in close proximity, separation of these charges requires that their mutual Coulombic attraction must be overcome. Increased delocalization of electronic wavefunctions decreases this Coulombic attraction because the distributed charge densities average over the Coulomb potential, thus avoiding the strongest attraction that occurs at close proximity. The temperature dependence of the rates of electron transfer processes giving rise to CT state dissociation also depends on the reorganization energy associated with moving electrons from one molecule to another. Here again, increased delocalization of electronic wavefunctions decreases the reorganization energy by spreading the charge distribution over a larger volume. Lesser changes in charge distribution result in smaller changes in nuclear positions and lower energies associated with their reorganization. The different influences of electron delocalization on barriers to charge separation in the polymer blends are represented schematically in Figures 8.14e and 8.14f where the electron charge distribution at the P3HT–PDI interface is represented as smaller in comparison to the P3HT–PCBM interface.

The distinctly different barriers to charge separation exhibited by the P3HT–PDI and P3HT–PCBM polymer blends can be understood by considering the molecular structures of the acceptors and their blend morphologies. For example, an electron transferred to a PCBM molecule is capable of delocalizing over the entire 60-atom conjugated framework of the fullerene cage. Thus, electronic wavefunctions are quite diffuse even if localized on individual fullerene molecules. If PCBM molecules are in close proximity (x-ray diffraction studies indicate that some of them are), then electronic wavefunctions can delocalize over neighboring molecules as well. Thus, electron delocalization onto several molecules enables charge densities that are already close to the Coulomb capture radius in organic semiconductors (~10 nm). At this separation, Coulomb forces become negligible. In addition, because the molecules are isotropic, they can experience substantial deviations from crystalline packing and still support electron delocalization as long as fullerenes are densely packed.

Unlike PCBM molecules with their large three-dimensional topology, the conjugated frameworks of PDI molecules are smaller, planar, and anisotropic. Electron density must be delocalized over many more PDI molecules to achieve the same degree of spatial delocalization as can be achieved with fullerenes. However, x-ray diffraction studies indicate that the PDI variant examined in this study does not form ordered crystals. The planar, anisotropic topology of the conjugated frameworks of PDI molecules causes the intermolecular coupling between them to vary strongly with deviations from crystalline packing. These properties result in more localized wavefunctions with corresponding larger reorganization energies and stronger Coulomb forces. The faster rate of charge separation observed in the PDI system in comparison to the PCBM system suggests that the electronic coupling between molecules is higher. But a high degree of delocalization may be difficult to support without the formation of an ordered phase of pure PDI.

8.4.2 Influence of Surface Chemistry on Charge Transport and Recombination in CQD Photovoltaics

Ultrafast IR spectroscopy is uniquely positioned to address some of the most pressing issues related to the development of CQD photovoltaics because it provides a unique combination of ultrafast time resolution, molecular structural sensitivity, and the ability to directly probe the electronic structure of charge traps. This combination of capabilities is particularly important for the development of CQD photovoltaics because the materials have a high density of interfaces. For example, nearly half of the Pb and S atoms in a densely packed film of PbS CQDs are associated with the surfaces of the nanocrystals used in optimized CQD solar cells [103]. These surface atoms are bonded to small ligands that passivate the dangling bonds and enable neighboring quantum dots to be in close electrical communication. Consequently, understanding the electronic structure of nanocrystal surfaces and how ligand interactions determine that electronic structure is an important focus area in the development of high-efficiency CQD photovoltaic materials. Ultrafast IR spectroscopy provides a pathway to probe all these properties simultaneously [31].

There is a pressing need to understand and control the interaction of ligands with nanocrystal surfaces in CQD photovoltaics because the original ligands used to render nanocrystals colloidally stable during synthesis do not support facile electrical communication between nanocrystals. These ligands must be replaced by smaller ligands that allow nanocrystals to pack densely and in close proximity such that electrical charges are able to tunnel readily between neighboring nanocrystals. Several ligand exchange strategies have been explored, including replacement of the original ligands using ethanedithiol [66–72], benzenedithiol [113–115], hydrazine [105,116], and pyridine [117], among others [66–72]. These strategies have led to significant enhancements of carrier mobilities in dense CQD films as determined by thin-film transistor measurements. However, more relevant to CQD photovoltaics is the combined mobility and charge carrier lifetime that determines the minority carrier diffusion and drift lengths [103]. These properties, which are influenced both by the mobility of charge carriers and by their recombination and trapping timescales, have received comparatively little attention by the CQD photovoltaic community.

Recently, time-resolved IR spectroscopy was combined with electrical measurements of PbS CQD photovoltaic materials to examine the charge carrier mobility and recombination lifetime in the materials and to identify the molecular interactions of ligands that determine those properties [31,38]. Transient IR spectra resulting from 532 nm excitation of PbS CQD films deposited under conditions identical to the state of the devices are displayed in Figure 8.15. Three spectra are displayed corresponding to films treated with different ligands, EDT (PbS–EDT), 3-MPA (PbS–MPA), and bromide ions (PbS–Br). The spectra, recorded 500 ns following a nanosecond duration excitation pulse, have been scaled according to the number of photons absorbed by the films such that the signal amplitudes are quantitatively comparable. The spectra contain two principal components, a broad electronic transition covering the entire mid-IR region and narrow vibrational features corresponding to functional groups of the various ligands.

The electronic and vibronic components of the transient spectra are intimately linked and reveal one of the most useful (and surprising) applications of ultrafast IR spectroscopy to the study of CQD photovoltaics. Recalling the photophysical sequence, excitons are initially created by bandgap excitation of CQDs by the 532 nm excitation pulse. The carriers subsequently become trapped (Figure 8.15, cartoon, dashed horizontal arrow) at localized surface states of the nanocrystals. The broad electronic transitions arise from excitation of these trapped carriers (electrons in the case of p-doped PbS films) back into the core or band states of the nanocrystals (cartoon, diagonal solid arrow). These transitions are termed trap-to-band transitions [31]. The spectrum of trap-to-band transitions provides information about the average energetic depth of the charge traps. Localizing charge carriers in surface trap states changes the charge distribution experienced by surface-bound ligands associated with the states. As a result, the IR probe pulse simultaneously excites trap-to-band transitions in the nanocrystals and probes the changes of the associated vibrational frequencies of the ligands. Consequently, the vibrational spectra of ligands measured before charges are trapped and the spectra of ligands located at charged traps both appear in the transient vibrational spectra [31]. It is the information about the electronic structure of charge traps combined with vibrational information about the interactions of ligands with nanocrystal surfaces giving rise to the traps that uniquely positions ultrafast vibrational spectroscopy as an important tool in the development of CQD photovoltaic technology.

An example of the use of ultrafast IR spectroscopy combined with electrical measurements to examine charge transport in PbS CQD photovoltaic films follows. The time-dependent amplitudes of trap-to-band transitions such as those appearing in Figure 8.15 provide a measure of the charge recombination kinetics in a CQD film [38]. The kinetics traces appearing in Figure 8.16a represent the dynamics of charge recombination in EDT-, hexanedithiol (HDT)-, MPA-, and Br^--treated PbS CQD films following excitation at 532 nm. From these kinetics traces, the average charge recombination lifetimes were calculated and are represented on the vertical axis in Figure 8.16b. The horizontal axis indicates the average charge trap depth determined from the trap-to-band transitions of the corresponding PbS CQD films. The correlation reveals that the

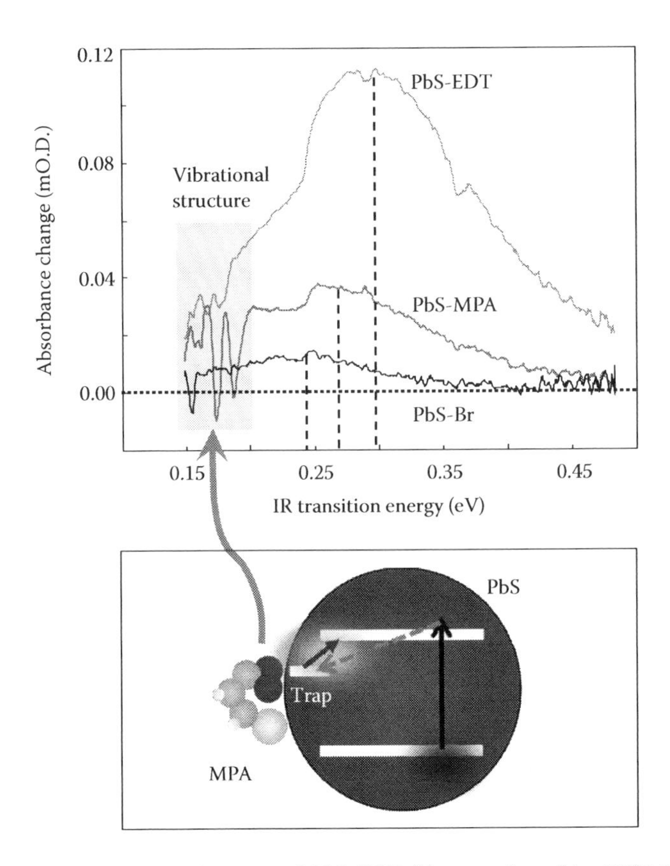

FIGURE 8.15 Time-resolved infrared spectra of PbS CQD films passivated by EDT, MPA, and Br^-. The spectra were measured at 500 ns following optical excitation of the bandgap of the films. Narrow vibrational features are superimposed onto broad electronic transitions. As illustrated in the cartoon, the broad electronic transitions correspond to excitation of carriers out of trap states near the quantum dot surfaces back into band states. The shape of the broad absorption features provides information about the average trap energy. Excitation of the carriers changes the charge distribution experienced by the surface-bound ligands resulting in the appearance of vibrational features in the transient spectra. The frequencies of the vibrational features provide information that can be correlated to the nanocrystal–ligand interactions determining the trap energy. (Adapted from Tang, J. et al. 2011. *Nat. Mater.* 10:765–771.)

average charge recombination lifetime of a PbS CQD film is related to the average charge trap depth of the nanocrystals.

The IR spectroscopy measurements were also coupled to thin-film transistor measurements to assess the minority carrier mobility (Figure 8.17) [31]. These measurements reveal that the electron mobility is inversely correlated to the charge recombination lifetime measured in the PbS CQD films as expected from diffusion-controlled charge recombination theories. The minority carrier (electron) mobility of Br^--passivated CQD films is 200-fold larger than the mobility of EDT-treated films. Importantly, the 200-fold increase in mobility moving from EDT- to Br^--treated films is much larger than the 10-fold decrease in charge recombination lifetime among the same films. The data reveal that the product of the electron mobility and recombination lifetime increased by a factor of 20 within the series of ligand-treated films. This increase corresponds to a more than fourfold enhancement of the diffusion length of electrons in the Br-treated PbS CQD films. It is instructive to point out that the power conversion efficiency (PCE) of the corresponding PbS CQD photovoltaic devices increases threefold within the series: EDT-treated devices (2% PCE) [108], MPA-treated

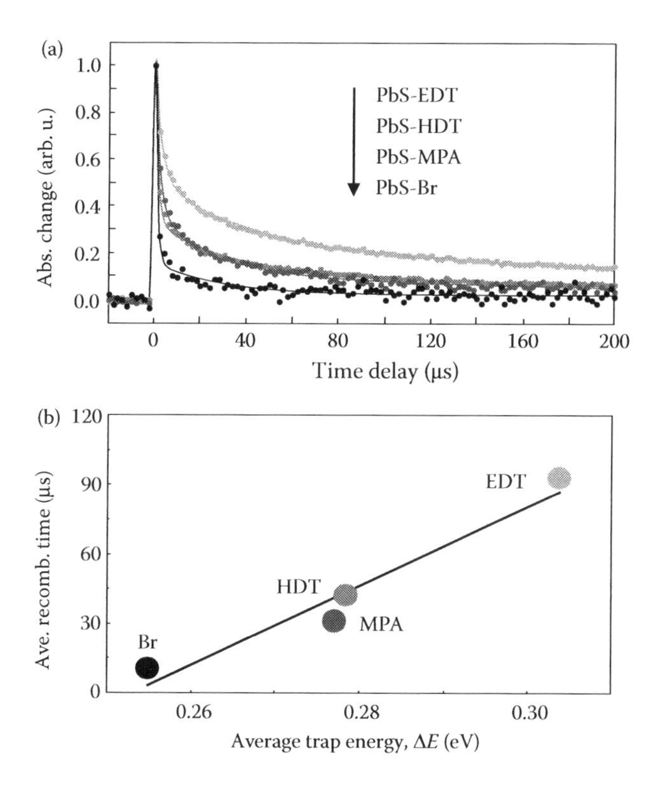

FIGURE 8.16 (a) Time-resolved infrared kinetics traces measured at the peaks of trap-to-band transitions of PbS CQD films passivated by various ligands following optical excitation of the bandgap of the films. The amplitude decays provide a measure of charge recombination dynamics in the films. The choice of ligand passivation strategy strongly influences the charge recombination lifetime of the photovoltaic films. (b) Plot of the average charge recombination lifetime from the kinetics traces in Panel A versus the average trap energy measured from the transient infrared spectra. The recombination lifetime of Br^--passivated films is reduced by factor of 10 relative to EDT-treated films. (Adapted from Tang, J. et al. 2011. *Nat. Mater.* 10:765–771.)

devices (5% PCE) [68], and Br^--treated devices (6% PCE) [38]. Reduction of the charge trap density and enhancement of electron diffusion length in the PbS CQD photovoltaic materials results in significant gains in PCE [31,38].

One of the strengths of ultrafast IR spectroscopy used to examine charge carrier transport in CQD photovoltaic materials is the capacity to probe molecular species associated with defects. This capacity is particularly important for guiding continued development of higher efficiency devices in the future on the basis of molecular information from present devices. An illustration of this capacity is represented in Figure 8.18 that depicts in the upper panel IR absorption spectra of two PbS CQD films treated with MPA and Br^- ligands. The spectrum of the MPA-treated film exhibits the characteristic absorption bands of the carboxylic stretch modes around 1400 and 1500 cm^{-1} and the C–H stretch modes around 2900 cm^{-1}. The spectrum of the Br^--treated film contains no significant vibrational bands in the 1100–3100 cm^{-1} range because all organic molecules are removed by the ligand exchange process. The lower panel represents time-resolved IR spectra of the MPA- and Br^--treated films measured at 500 ns following 532 nm excitation of the films. The transient spectrum of the MPA-treated film exhibits narrow vibrational features of the carboxylic group superimposed on a broad trap-to-band transition feature as described above. The transient spectrum of the Br^--treated film exhibits the same type of trap-to-band transition although lower in amplitude. Similar to the MPA-treated film, the transient spectrum of the Br^--treated film exhibits a narrow vibrational feature around 1300 cm^{-1} that is indicative

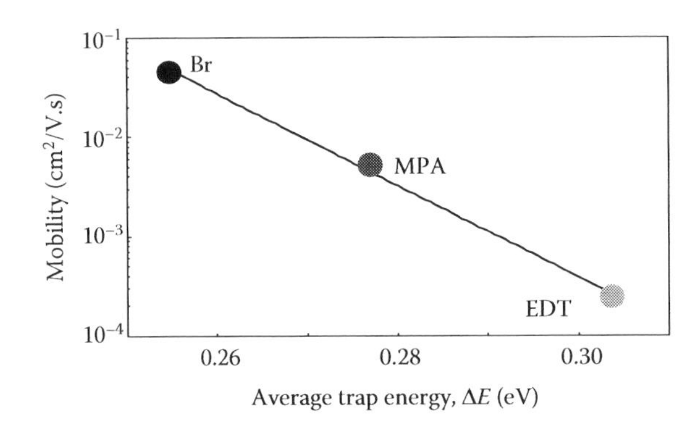

FIGURE 8.17 Plot of the minority carrier (electron) mobility of PbS CQD films passivated with various ligands versus the average charge trap energy of the films. The correlation reveals that the mobility of the Br^--passivated film is 200-fold larger than the mobility of the EDT-treated film. (Adapted from Tang, J. et al. 2011. *Nat. Mater.* 10:765–771.)

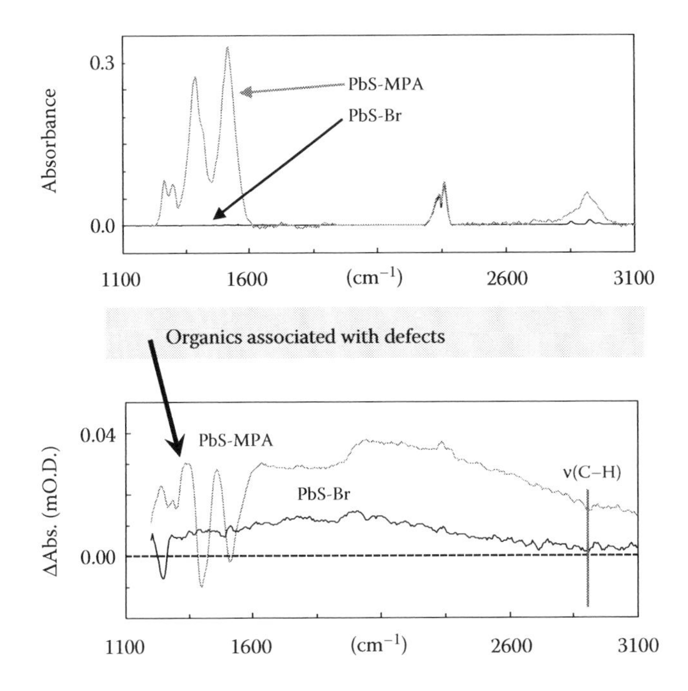

FIGURE 8.18 The upper panel displays infrared absorption spectra of PbS CQD films passivated by MPA and Br^- ligands. The characteristic carboxylate stretch vibrations of MPA are evident around 1400 and 1500 cm^{-1}. No significant vibrational features are observed in the Br^--passivated film because the ligand exchange process removes all organic material from the film. The lower panel displays transient infrared spectra of the corresponding PbS CQD films measured at 500 ns following bandgap excitation. Trap-to-band transitions appear in the spectra throughout the mid-infrared along with transient vibrational features. In the MPA-treated film, these features correspond to the carboxylate groups of MPA molecules as expected. Interestingly, the Br^--treated film also exhibits a vibrational feature around 1300 cm^{-1} that is suggestive of a C–H bending or out-of-plane wagging mode. The data reveal that organic impurities remain in the all-inorganic Br-passivated PbS CQD film and that the impurities are associated with defects in the film. This finding suggests that a more complete extraction of organic intermediates used during the synthesis of the PbS CQD film may result in decreased defect density for improved photovoltaic efficiency.

of a C–H bending or out-of-plane wagging mode. This observation is surprising because the Br^--treated film should contain no organic species that could give rise to a vibration at 1300 cm^{-1}. The transient spectra reveal that [1] residual organic species exist in the Br^--treated film and [2] these residual species are closely associated with charge traps that remain in the film. Therefore, the transient IR spectra reveal a pathway to further improve the PCE of PbS CQD devices. A more complete removal of residual organic species will both decrease the trap density and increase the PCE of the devices.

8.5 CONCLUSIONS AND FUTURE DIRECTIONS

The need to understand and ultimately control the interactions of excitations and charge carriers with molecular species is being driven by the emergence of new materials for flexible electronics and inexpensive photovoltaics applications. Although many of these materials possess nanostructured morphologies and compositions, few experimental techniques have been devised to unambiguously examine the influence that molecular structure and morphology have on their excited-state and electrical properties. Ultrafast IR spectroscopy provides a unique probe of electronic processes in materials since it probes the nature and dynamics of transient electronic species through their vibrational features. Because the vibrational features are sensitive to the local composition and morphology of molecules, the technique also provides information about the nanostructure of the materials in which transient electronic species evolve.

Two examples from recent work on organic and CQD photovoltaic materials are used to illustrate some of the unique capabilities that ultrafast IR spectroscopy brings to the study of electronic materials. In one example, ultrafast IR methods are used to examine the influence of molecular structure on charge separation mechanisms in organic photovoltaics. The sensitivity of the vibrational frequencies of organic semiconductor molecules to their local molecular and morphological environment (vibrational solvatochromism) provides a means to spectroscopically distinguish molecules at electron donor–acceptor interfaces from molecules embedded in the interiors of either donor or acceptor phase. This sensitivity enabled the first direct measure of the energetic barriers to charge separation and provided insights about how the energy barriers depend on molecular structure.

In a second example, the influence of nanocrystal surface chemistry and electronic structure on charge transport and recombination is examined using ultrafast IR spectroscopy. Energies typical of charge trap depths in disordered electronic materials (0.1–1.0 eV) correspond to the mid-IR spectral range (806.6–8066 cm^{-1}). Consequently, ultrafast IR spectroscopy provides a means to directly probe the electronic structure and trap state distribution of emerging electronic materials while simultaneously examining transient electronic species through the vibrational modes of the molecules. The charge trap energy distributions of a variety of CQD photovoltaic materials were examined and correlated with the corresponding charge trap density and molecular interactions of ligands with nanocrystal surfaces. These studies led to unique insights about charge transport and, importantly, about how to further reduce the density and energetic distribution of defects in future CQD photovoltaic materials.

As these examples demonstrate, ultrafast IR spectroscopy provides unique opportunities to explore fundamental charge carrier dynamics in emerging electronic materials that are of direct relevance to ongoing efforts to understand and ultimately control their properties. There is considerable room for expansion in the nascent ultrafast IR spectroscopy of materials field because of diversity of materials and variety of outstanding issues that can be addressed. In addition to expansion of the materials basis, higher-order ultrafast spectroscopy methods such as variations of transient 2D IR spectroscopy, sum-frequency and transient sum-frequency spectroscopies, and time-resolved Raman spectroscopies will open new opportunity space for examination of electronic processes in emerging materials.

ACKNOWLEDGMENTS

The spectroscopy discussed in this chapter is the result of the efforts of many exceptional people with whom the author has been blessed to work. Special thanks go to Dr. Ryan Pensack, Kwang Jeong, Jihye Kim, and Larry Barbour from the author's group and to Professor Enrique Gomez, Changhe Guo, and Kiarash Vakhshouri from the Pennsylvania State University. The author thanks Professor Edward Sargent, Dr. Jiang Tang, Kyle Kemp, and Dr. Huan Liu at the University of Toronto for supplying PbS CQDs, film deposition procedures, and electrical characterization of PbS CQD films. This research was supported by the National Science Foundation (CHE-0846241 and DMR-0820404), the Office of Naval Research (N00014-11-1-0239), and the Petroleum Research Fund (49639-ND6).

REFERENCES

1. Sariciftci, N. S., Smilowitz, L., Heeger, A. J., and Wudl, F. 1992. Photoinduced electron-transfer from a conducting polymer to buckminsterfullerene. *Science* 258:1474–1476.
2. Hwang, I. W., Moses, D., and Heeger, A. J. 2008. Photoinduced carrier generation in P3HT/PCBM bulk heterojunction materials. *J. Phys. Chem. C* 112:4350–4354.
3. Beljonne, D., Pourtois, G., Silva, C., Hennebicq, E., Herz, L. M., Friend, R. H., Scholes, G. D., Setayesh, S., Mullen, K., and Bredas, J. L. 2002. Interchain vs. intrachain energy transfer in acceptor-capped conjugated polymers. *Proc. Natl. Acad. Sci. USA* 99:10982–10987.
4. Silva, C. D., Russel, A. S., Stevens, D. M., Arias, M. A., MacKenzie, A. C., Greenham, N. C., Friend, R. H., Setayesh, S., and Mullen, K. 2001. Efficient exciton dissociation via two-step photoexcitation in polymeric semiconductors. *Phys. Rev. B* 64:125211.
5. Jiang, X.M., Osterbacka, R., Korovyanko, O., An, C. P., Horovitz, B., Janssen, R. A. J., and Vardeny, Z. V. 2002. Spectroscopic studies of photoexcitations in regioregular and regiorandom polythiophene films. *Adv. Funct. Mater.* 12:587–597.
6. Sheng, C. X., Tong, M., Singh, S., and Vardeny, Z. V. 2007. Experimental determination of the charge/neutral braning ration in the photoexcitation of pi-conjugated polymers by broadband ultrafast spectroscopy. *Phys. Rev. B* 75:085206(7).
7. Muller, J. G., Lupton, J. M., Feldmann, J., Lemmer, U., Scharber, M. C., Sariciftci, N. S., Brabec, C. J. and Scherf, U. 2005. Ultrafast dynamics of charge carrier photogeneration and geminate recombination in conjugated polymer: Fullerene solar cells. *Phys. Rev. B* 72:195208(10).
8. Kersting, R., Lemmer, U., Mahrt, R. F., Leo, K., Kurz, H., Bassler, H., and Gobel, E. O. 1993. Femtosecond energy relaxation in pi-conjugated polymers. *Phys. Rev. Lett.* 70:3820–3823.
9. Ma, Y. Z., Stenger, J., Zimmermann, J., Bachilo, S. M., Smalley, R. E., Weisman, R. B., and Fleming, G. R. 2004. Ultrafast carrier dynamics in single-walled carbon nanotubes probed by femtosecond spectroscopy. *J. Chem. Phys.* 120:3368–3373.
10. Graham, M. W., Ma, Y. Z., and Fleming, G. R. 2008. Femtosecond photon echo spectroscopy of semiconducting single-walled carbon nanotubes. *Nano Lett.* 8:3936–3941.
11. Scholes, G. D., and Rumbles, G. 2006. Excitons in nanoscale systems. *Nat. Mater.* 5:683–696.
12. Collini, E., and Scholes, G. D. 2009. Coherent intrachain energy migration in a conjugated polymer at room temperature. *Science* 323:369–373.
13. Wehrenberg, B. L., Wang, C. J., and Guyot-Sionnest, P. 2002. Interband and intraband optical studies of PbSe colloidal quantum dots. *J. Phys. Chem. B* 106:10634–10640.
14. Yu, D., Wang, C., Wehrenberg, B. L., and Guyot-Sionnest, P. 2004. Variable range hopping conduction in semiconductor nanocrystal solids. *Phys. Rev. Lett.* 92:216802(4).
15. Pandey, A., and Guyot-Sionnest, P. 2010. Hot electron extraction from colloidal quantum dots. *J. Phys. Chem. Lett.* 1:45–47.
16. Trinh, M. T., Houtepen, A. J., Schins, J. M., Hanrath, T., Piris, J., Knulst, W., Goosens, A. P. L. M., and Siebbeles, L. D. A. 2008. In spite of recent doubts carrier multiplication does occur in PbSe nanocrystals. *Nano Lett.* 8:1713–1718.
17. Piris, J., Dykstra, T. E., Bakulin, A. A., van Loosdrecht, P. H. M., Knulst, W., Trinh, M. T., Schins, J. M., and Siebbeles, L. D. A. 2009. Photogeneration and ultrafast dynamics of excitons and charges in P3HT/PCBM blends. *J. Phys. Chem. C* 113:14500–14506.
18. Grzegorczyk, W. J., Savenije, T. J., Dykstra, T. E., Piris, J., Schins, J. M., and Siebbeles, L. D. A. 2010. Temperature-independent charge carrier photogeneration in P3HT-PCBM blends with different morphology. *J. Phys. Chem. C* 114:5182–5186.

19. Ai, X., Beard, M. C., Knutsen, K. P., Shaheen, S. E., Rumbles, G., and Ellingson, R. J. 2006. Photoinduced charge carrier generation in a poly(3-hexylthiophene) and methanofullerene bulk heterojunction investigated by time-resolved terahertz spectroscopy. *J. Phys. Chem. B* 110:25462–25471.
20. Coffey, D. C., Ferguson, A. J., Kopidakis, N., and Rumbles, G. 2010. Photovoltaic charge generation in organic semiconductors based on long-range energy transfer. *ACS Nano* 4:5437–5445.
21. Klimov, V. I. 2000. Optical nonlinearities and ultrafast carrier dynamics in semiconductor nanocrystals. *J. Phys. Chem. B* 104:6112–6123.
22. Schaller, R. D., and Klimov, V. I. 2004. High efficiency carrier multiplication in PbSe nanocrystals: Implications for solar energy conversion. *Phys. Rev. Lett.* 92:186601.
23. Asbury, J. B., Ellingson, R. J., Ghosh, H. N., Ferrere, S., Nozik, A. J., and Lian, T. 1999. Femtosecond IR study of excited-state relaxation and electron-injection dynamics of $Ru(dcbpy)_2(NCS)_2$ in solution and on nanocrystalline TiO_2 and Al_2O_3 thin films. *J. Phys. Chem. B* 103:3110–3119.
24. Asbury, J. B., Hao, E., Wang, Y., Ghosh, H. N., and Lian, T. 2001. Ultrafast electron transfer dynamics from molecular adsorbates to semiconductor nanocrystalline thin films. *J. Phys. Chem. B* 105:4545–4557.
25. Hsu, J. W. P., Yan, M., Jedju, T. M., and Rothberg, L. J. 1994. Assignment of the picosecond photoinduced absorption in phenylene vinylene polymers. *Phys. Rev. B* 49:712–715.
26. Rothberg, L. J., Yan, M., Papadimitrakopoulos, F., Galvin, M. E., Kwock, E. W., and Miller, T. M. 1996. Photophysics of phenylenevinylene polymers. *Synth. Met.* 80:41–58.
27. Cuppoletti, C. M., and Rothberg, L. J. 2003. Persistent photoluminescence in conjugated polymers. *Synth. Met.* 139:867–871.
28. Barbour, L. W., Hegadorn, M., and Asbury, J. B. 2007. Watching electrons move in real time: Ultrafast infrared spectroscopy of a polymer blend photovoltaic material. *J. Am. Chem. Soc.* 129:15884–15894.
29. Pensack, R. D., Banyas, K. M., and Asbury, J. B. 2010. Charge trapping in organic photovoltaic materials examined with time resolved vibrational spectroscopy. *J. Phys. Chem. C* 114:5344–5350.
30. Pensack, R. D., and Asbury, J. B. 2011. Ultrafast probes of charge transfer states in organic photovoltaic materials. *Chem. Phys. Lett.* 515:197–205.
31. Jeong, K. S., Tang, J., Liu, H., Kim, J., Schaefer, A. W., Kemp, K., Levina, L. et al. 2012. Enhanced mobility-lifetime products in colloidal quantum dot photovoltaics. *ACS Nano* 6:89–99.
32. Pensack, R. D., Guo, C., Vakhshouri, K., Gomez, E. D., and Asbury, J. B. 2012. Influence of acceptor structure on barriers to charge separation in organic photovoltaic materials. *J. Phys. Chem. C* 116: 4824–4831.
33. Anglin, T. C., Sohrabpour, Z., and Massari, A. M. 2011. Nonlinear spectroscopic markers of structural change during charge accumulation in organic field-effect transistors. *J. Phys. Chem. C* 115:20258–20266.
34. Anglin, T. C., Speros, J. C., and Massari, A. M. 2011. Interfacial ring orientation in polythiophene field-effect transistors on functionalized dielectrics. *J. Phys. Chem. C* 115:16027–16036.
35. Anglin, T. C., O'Brien, D. B., and Massari, A. M. 2010. Monitoring the charge accumulation process in polymeric field-effect transistors via *in situ* sum frequency generation. *J. Phys. Chem. C* 114:17629–17637.
36. Barbour, L. W., Hegadorn, M., and Asbury, J. B. 2006. Microscopic inhomogeneity and ultrafast orientational motion in an organic photovoltaic bulk heterojunction thin film studied with 2D IR vibrational spectroscopy. *J. Phys. Chem. B* 110:24281–24286.
37. Pensack, R. D., Banyas, K. M., and Asbury, J. B. 2010. Temperature independent vibrational dynamics in an organic photovoltaic material. *J. Phys. Chem. B* 114:12242–12251.
38. Tang, J., Kemp, K., Hoogland, S., Jeong, K. S., Liu, H., Levina, L., Furukawa, M. et al. 2011. Colloidal quantum dot photovoltaics using atomic ligand passivation. *Nat. Mater.* 10:765–771.
39. Pensack, R. D., Banyas, K. M., and Asbury, J. B. 2010. Vibrational solvatochromism in organic photovoltaic materials: Method to distinguish molecules at donor/acceptor interfaces. *Phys. Chem. Chem. Phys.* 12:14144–14152.
40. Baer, W. S. 1966. Free-carrier absorption in reduced $SrTiO_3$. *Phys. Rev.* 144:734–738.
41. Walukiewicz, W., Lagowski, L., Jastrzebski, L., Lichtensteiger, M., and Gatos, H. C. 1979. Electron mobility and free carrier absorption in GaAs: Determination of the compensation ratio. *J. Appl. Phys.* 50:899–908.
42. Pope, M., and Swenberg, C. E. 1999. *Electronic Processes in Organic Crystals and Polymers* (Oxford University Press, New York).
43. Barbour, L. W., Pensack, R. D., Hegadorn, M., Arzhantsev, S., and Asbury, J. B. 2008. Excitation transport and charge separation in an organic photovoltaic material: Watching excitations diffuse to interfaces. *J. Phys. Chem. C* 112:3926–3934.

44. Hamm, P., Lim, M., DeGrado, W. F., and Hochstrasser, R. M. 2000. Pump/probe self heterodyned 2D spectroscopy of vibrational transitions of a small globular peptide. *J. Chem. Phys.* 112:1907–1916.
45. Tan, H. S., Piletic, I. R., and Fayer, M. D. 2005. Polarization selective spectroscopy experiments: Methodology and pitfalls. *J. Opt. Soc. Am. B* 22:2009–2017.
46. Reichardt, C. 1994. Solvatochromic dyes as solvent polarity indicators. *Chem. Rev.* 94:2319–2358.
47. Reichardt, C. 1988. *Solvents and Solvent Effects in Organic Chemistry* (VCH Publishers, Weinheim).
48. Reichardt, C. 1965. Empirical parameters of the polarity of solvents. *Angew. Chem. Int. Ed.* 4:29–40.
49. Kamlet, M. J., Abboud, J. L. M., and Taft, R. W. 1981. An examination of linear solvation energy relationships, In *Progress in Physical Organic Chemistry*, ed. Taft, R. W. (John Wiley and Sons, New York), Vol. 13, pp. 485–630.
50. Mayer, U., Gutmann, V., and Gerger, W. 1975. Acceptor number—Quantitative empirical parameter for the electrophilic properties of solvents. *Monatschefte Chem.* 106:1235–1257.
51. Gutmann, V. 1978. *The Donor-Acceptor Approach to Molecular Interactions* (Plenum Press, New York).
52. Engberts, J. B. F. N., Famini, G. R., Perjessy, A., and Wilson, L. Y. 1998. Solvent effects on C=O stretching frequencies of some 1-substituted 2-pyrrolidinones. *J. Phys. Org. Chem.* 11:261–272.
53. Vdovenko, S. I., Gerus, I. I., and Kuhmar, V. P. 2009. Solvent influence on the infrared spectra of β-alkoxyvinyl methyl ketones. II. Stretching vibrations and integrated intensities of carbonyl and vinyl bands of (3*Z*,*E*)-4-ethoxy-1,1,1-trifluoro-5,5-dimethylhex-3-en-2-one. *Spectrochim. Acta, Part A: Molec. Biomolec. Spec.* 72:229–235.
54. Lee, H., Lee, G., Jeon, J., and Cho, M. 2012. Vibrational spectroscopic determination of local solvent electric field, solute-solvent electrostatic interaction energy, and their fluctuation amplitudes. *J. Phys. Chem. A* 116:347–357.
55. Choi, J. H., and Cho, M. 2011. Vibrational solvatochromism and electrochromism of infrared probe molecules containing CO, CN, C=O, or C–F vibrational chromophore. *J. Chem. Phys.* 134:154513(12).
56. Cho, M. 2009. Vibrational solvatochromism and electrochromism: Coarse-grained models and their relationships. *J. Chem. Phys.* 130:094505(15).
57. Ham, S., Kim, J. H., Lee, H., and Cho, M. 2003. Correlation between electronic and molecular structure distortions and vibrational properties. II. Amide I modes of NMA-ND_2O complexes. *J. Chem. Phys.* 118: 3491–3498.
58. Kwac, K., and Cho, M. 2003. Molecular dynamics simulation study of *N*-methylacetamide in water. I. Amide I mode frequency fluctuation. *J. Chem. Phys.* 119:2247–2255.
59. Choi, J. H., Hahn, S., and Cho, M. 2005. Amide I IR, VCD, and 2D IR spectra of isotope-labeled alpha-helix in liquid water: Numerical simulation studies. *Int. J. Quantum Chem.* 104:616–634.
60. Choi, J. H., Ham, S., and Cho, M. 2003. Local amide I mode frequencies and coupling constants in polypeptides. *J. Phys. Chem. B* 107:9132–9138.
61. DeCamp, M. F., Deflores, L. P., McCracken, J. M., Tokmakoff, A., Kwac, K., and Cho, M. 2005. Amide I vibrational dynamics of *N*-methylacetamide in polar solvents: The role of electrostatic interactions. *J. Phys. Chem. B* 109:11016–11026.
62. la Cour Jansen, T., Dijkstra, A., Watson, T. M., Hirst, J. D., and Knoester, J. 2006. Modeling the amide I bands of small peptides. *J. Chem. Phys.* 125:044312(9).
63. Schmidt, J. R., Corcelli, S. A., and Skinner, J. L. 2004. Ultrafast vibrational spectroscopy of water and aqueous *N*-methylacetamide: Comparison of different electronic structure/molecular dynamics approaches. *J. Chem. Phys.* 121:8887–8896.
64. la Cour Jansen, T., and Knoester, J. 2006. A transferable electrostatic map for solvation effects on amide I vibrations and its application to linear and two-dimensional spectroscopy. *J. Chem. Phys.* 124: 044502(11).
65. Hayashi, T., Zhuang, W., and Mukamel, S. 2005. Electrostatic DFT map for the complete vibrational amide band of NMA. *J. Phys. Chem. A* 109:9747–9759.
66. Johnston, K. W., Pattantyus-Abraham, A. G., Clifford, J. P., Myrskog, S. H., Hoogland, S., Shukla, H., Klem, E. J. D., Levina, L., and Sargent, E. H. 2008. Efficient Schottky-quantum-dot photovoltaics: The roles of depletion, drift, and diffusion. *Appl. Phys. Lett.* 92:122111(3).
67. Debnath, R., Tang, J., Barkhouse, D. A. R., Wang, X., Pattantyus-Abraham, A. G., Brzozowski, L., Levina, L., and Sargent, E. H. 2010. Ambient-processed colloidal quantum dot solar cells via individual pre-encapsulation of nanoparticles. *J. Am. Chem. Soc.* 132:5952–5953.
68. Pattantyus-Abraham, A. G., Kramer, I. J., Barkhouse, A. R., Wang, X., Konstantatos, G., Debnath, R., Levina, L., Nazeeruddin, M. K., Gratzel, M., and Sargent, E. H. 2010. Depleted-heterojunction colloidal quantum dot solar cells. *ACS Nano* 4:3374–3380.
69. Kovalenko, M. V., Scheele, M., and Talapin, D. V. 2009. Colloidal nanocrystals with molecular metal chalcogenide surface ligands. *Science* 324:1417–1420.

70. Kovalenko, M. V., Bodnarchuk, M. I., Zaumseil, J., Lee, J. S., and Talapin, D. V. 2010. Expanding the chemical versatility of colloidal nanocrystals capped with molecular metal chalcogenide ligands. *J. Am. Chem. Soc.* 132:10085–10092.
71. Porter, V. J., Geyer, S., Halpert, J. E., Kastner, M. A., and Bawendi, M. G. 2008. Photoconductivity in annealed and chemically treated CdSe/ZnS inorganic nanocrystal films. *J. Phys. Chem. C* 112:2308–2316.
72. Geyer, S., Porter, V. J., Halpert, J. E., Mentzel, T. S., Kastner, M. A., and Bawendi, M. G. 2010. Charge transport in mixed CdSe and CdTe colloidal nanocrystal films. *Phys. Rev. B* 82:155201(8).
73. Asbury, J. B., Steinel, T., Stromberg, C., Corcelli, S. A., Lawrence, C. P., Skinner, J. L., and Fayer, M. D. 2004. Dynamics of water probed with vibrational echo correlation spectroscopy. *J. Chem. Phys.* 121:12431–12446.
74. Asbury, J. B., Steinel, T., Stromberg, C., Corcelli, S. A., Lawrence, C. P., Skinner, J. L., and Fayer, M. D. 2004. Water dynamics: Vibrational echo correlation spectroscopy and comparison to molecular dynamics simulations. *J. Phys. Chem. A* 108:1107–1119.
75. Corcelli, S. A., Lawrence, C. P., and Skinner, J. L. 2004. Combined electronic structure/molecular dynamics approach for ultrafast infrared spectroscopy of dilute HOD in liquid H_2O and D_2O. *J. Chem. Phys.* 120:8107–8117.
76. Fecko, C. J., Eaves, J. D., Loparo, J. J., Tokmakoff, A., and Geissler, P. L. 2003. Ultrafast hydrogen-bond dynamics in the infrared spectroscopy of water. *Science* 301:1698–1702.
77. Hayashi, T., Jansen, T. l. C., Zhuang, W., and Mukamel, S. 2005. Collective solvent coordinates for the infrared spectrum of HOD in D_2O based on an ab initio electrostatic map. *J. Phys. Chem. A* 109:64–82.
78. Auer, B., Kumar, R., Schmidt, J. R., and Skinner, J. L. 2007. Hydrogen bonding and Raman, IR, and 2D IR spectroscopy of dilute HOD in liquid D_2O. *Proc. Natl. Acad. Sci. USA* 104:14215–14220.
79. Fishman, E., and Saumagne, P. 1965. Near-infrared spectrum of liquid water. *J. Phys. Chem.* 69:3671.
80. Falk, M., and Ford, T. A. 1966. Infrared spectrum and structure of liquid water. *Can. J. Chem.* 44:1699–1707.
81. Libnau, F. O., Toft, J., Christy, A. A., and Kvalheim, O. M. 1994. Structure of liquid water determined from infrared temperature profiling and evolutionary curve resolution. *J. Am. Chem. Soc.* 116: 8311–8316.
82. Wang, J., and El-Sayed, M. A. 1999. Temperature jump-induced secondary structural change of the membrane protein bacteriorhodopsin in the premelting temperature region: A nanosecond time-resolved Fourier transform infrared study. *Biophys. J.* 76:2777–2783.
83. Fabian, H., Schultz, C., Naimann, D., Landt, O., Hahn, U., and Saenger, W. 1993. Secondary structure and temperature-induced unfolding and refolding of ribonuclease T_1 in aqueous solution: A Fourier transform infrared spectroscopic study. *J. Mol. Biol.* 232:967–981.
84. Reinstadler, D., Fabian, H., Backmann, J., and Naumann, D. 1996. Refolding of thermally and urea-denatured ribonuclease A monitored by time-resolved FTIR spectroscopy. *Biochemistry* 35:15822–15830.
85. Chung, H. S., Khalil, M., Smith, A. W., Ganim, Z., and Tokmakoff, A. 2005. Conformational changes during the nanosecond-to-millisecond unfolding of ubiquitin. *Proc. Natl. Acad. Sci. USA* 102:612–617.
86. Ganim, Z., Chung, H. S., Smith, A. W., Deflores, L. P., Jones, K. C., and Tokmakoff, A. 2008. Amide I two-dimensional infrared spectroscopy of proteins. *Acc. Chem. Res.* 41:432–441.
87. Solomonov, B. N., Novikov, V. B., Varfolomeev, M. A., and Klimovitskii, A. E. 2005. Colorimetric determination of hydrogen-bonding enthalpy for near aliphatic alcohols. *J. Phys. Org. Chem.* 18:1132–1137.
88. Khan, A. 2000. A liquid water model: Density variation from supercooled to superheated states, Prediction of H-bonds, and temperature limits. *J. Phys. Chem. B* 104:11268–11274.
89. Suresh, S. J., and Naik, V. M. 2000. Hydrogen bond thermodynamic properties of water from dielectric constant data. *J. Chem. Phys.* 113:9727–9732.
90. Steinel, T., Asbury, J. B., Zheng, J., and Fayer, M. D. 2004. Watching hydrogen bonds break: A transient absorption study of water. *J. Phys. Chem. A* 108:10957–10964.
91. Eaves, J. D., Loparo, J. J., Fecko, C. J., Roberts, S. T., Tokmakoff, A., and Geissler, P. L. 2005. Hydrogen bonds in liquid water are broken only fleetingly. *Proc. Natl. Acad. Sci. USA* 102:13019–13022.
92. Loparo, J. J., Roberts, S. T., and Tokmakoff, A. 2006. Multidimensional infrared spectroscopy of water. I. vibrational dynamics in two-dimensional IR line shapes. *J. Chem. Phys.* 125:194521(13).
93. Fayer, M. D. 2001. Fast protein dynamics probed with infrared vibrational echo experiments. *Annu. Rev. Phys. Chem.* 52:315–356.
94. Merchant, K. A., Noid, W. G., Akiyama, R., Finkelstein, I. J., Goun, A., McClain, B. L., Loring, R. F. and Fayer, M. D. 2003. Myoglobin-CO substate structures and dynamics: Multidimensional vibrational echoes and molecular dynamics simulations. *J. Am. Chem. Soc.* 125:13804–13818.
95. Chung, H. S., and Tokmakoff, A. 2006. Visualization and characterization of the infrared active amide I vibrations of proteins. *J. Phys. Chem. B* 110:2888–2898.

96. Lim, M. H., Hamm, P., and Hochstrasser, R. M. 1998. Protein fluctuations are sensed by stimulated infrared echoes of the vibrations of carbon monoxide and azide probes. *Proc. Natl. Acad. Sci. USA* 95:15315–15320.
97. Dutta, S., Li, Y. L., Rock, W., Houtman, J. C. D., Kohen, A., and Cheatum, C. M. 2012. 3-Picolyl azide adenine dinucleotide as a probe of femtosecond to picosecond enzyme dynamics. *J. Phys. Chem. B* 116: 542–548.
98. Bandaria, J. N., Dutta, S., Nydegger, M. W., Rock, W., Kohen, A., and Cheatum, C. M. 2010. Characterizing the dynamics of functionally relevant complexes of formate dehydrogenase. *Proc. Natl. Acad. Sci. USA* 107:17974–17979.
99. Massari, A. M., Finkelstein, I. J., McClain, B. L., Goj, A., Wen, X., Bren, K. L., Loring, R. F., and Fayer, M. D. 2005. The influence of aqueous versus glassy solvents on protein dynamics: Vibrational echo experiments and molecular dynamics simulations. *J. Am. Chem. Soc.* 127:14279–14289.
100. Londergan, C. H., Kim, Y. S., and Hochstrasser, R. M. 2005. Two-dimensional infrared spectroscopy of dipeptides in trehalose glass. *Mol. Phys.* 103:1547–1553.
101. Kwak, K., Park, S., Finkelstein, I. J., and Fayer, M. D. 2007. Frequency-frequency correlation functions and apodization in two-dimensional infrared vibrational echo spectroscopy: A new approach. *J. Chem. Phys.* 127:124503(17).
102. Brandrup, J., Immergut, E. H., and Grulke, E. A. 1999. *Polymer Handbook* (John Wiley and Sons, New York).
103. Tang, J., and Sargent, E. H. 2011. Infrared colloidal quantum dots for photovoltaics: Fundamentals and recent progress. *Adv. Mater.* 23:12–29.
104. Fafarman, A. T., Koh, W. K., Diroll, B. T., Kim, D. K., Ko, D. K., Oh, S. J., Ye, X. et al. 2011. Thiocyanate-capped nanocrystal colloids: Vibrational reporter of surface chemistry and solution-based route to enhanced coupling in nanocrystal solids. *J. Am. Chem. Soc.* 133:15753–15761.
105. Talapin, D. V., and Murray, C. B. 2005. PbSe nanocrystal solids for n- and p-channel thin film field-effect transistors. *Science* 310:86–89.
106. Choi, J. J., Bealing, C. R., Bian, K., Hughes, K. J., Zhang, W., Smilgies, D. M., Hennig, R. G., Engstrom, J. R., and Hanrath, T. 2011. Controlling nanocrystal superlattice symmetry and shape-anisotropic interactions through variable ligand surface coverage. *J. Am. Chem. Soc.* 133:3131–3138.
107. Hanrath, T., Choi, J. J., and Smilgies, D. M. 2009. Structure/property relationships of highly ordered lead salt nanocrystal superlattices. *ACS Nano* 3:2975–2988.
108. Tang, J., Brzozowski, L., Barkhouse, D. A. R., Wang, X., Debnath, R., Wolowiec, R., Palmiano, E. et al. 2010. Quantum dot photovoltaics in the extreme quantum confinement regime: The surface-chemical origins of exceptional air- and light-stability. *ACS Nano* 4:869–878.
109. Luther, J. M., Law, M., Beard, M. C., Song, Q., Reese, M. O., Ellingson, R. J., and Nozik, A. J. 2008. Schottky solar cells based on colloidal nanocrystal films. *Nano Lett.* 8:3488–3492.
110. Luther, J. M., Gao, J., Lloyd, M. T., Semonin, O. E., Beard, M. C., and Nozik, A. J. 2010. Stability assessment on a 3% bilayer PbS/ZnO quantum dot heterojunction solar cell. *Adv. Mater.* 22:3704–3707.
111. Choi, J. J., Lim, Y.F., Santiago-Berrios, M. B., Oh, M., Hyun, B.R., Sun, L., Bartnik, A. C. et al. 2009. PbSe nanocrystal excitonic solar cells. *Nano Lett.* 9:3749–3755.
112. Leschkies, K. S., Beatty, T. J., Kang, M. S., Norris, D. J., and Aydil, E. S. 2009. Solar cells based on junctions between colloidal PbSe nanocrystals and thin ZnO films. *ACS Nano* 3:3638–3648.
113. Koleilat, G. I., Levina, L., Shukla, H., Myrskog, S. H., Hinds, S., Pattantyus-Abraham, A. G., and Sargent, E. H. 2008. Efficient, stable infrared photovoltaics based on solution-cast colloidal quantum dots. *ACS Nano* 2:833–840.
114. Ma, W., Luther, J. M., Zheng, H., Wu, Y., and Alivisatos, A. P. 2009. Photovoltaic devices employing ternary PbS_xSe_{x-1} nanocrystals. *Nano Lett.* 9:1699–1703.
115. Tsang, S. W., Fu, H., Wang, R., Lu, J., Yu, K., and Tao, Y. 2009. Highly efficient cross-linked PbS nanocrystal/C_{60} hybrid heterojunction photovoltaic cells. *Appl. Phys. Lett.* 95:183505(3).
116. Urban, J. J., Talapin, D. V., Shevchenko, E. V., and Murray, C. B. 2006. Self-assembly of PbTe quantum dots into nanocrystal superlattices and glassy films. *J. Am. Chem. Soc.* 128:3248–3255.
117. Sun, B., Findikoglu, A. T., Sykora, M., Werder, D. J., and Klimov, V. I. 2009. Hybrid photovoltaics based on semiconductor nanocrystals and amorphous silicon. *Nano Lett.* 9:1235–1241.
118. Owen, J. S., Park, J., Trudeau, P.E., and Alivisatos, A. P. 2008. Reaction chemistry and ligand exchange at cadmium-selenide nanocrystal surfaces. *J. Am. Chem. Soc.* 130:12279–12281.
119. Deacon, G. B., and Phillips, R. J. 1980. Relationships between the carbon-oxygen stretching frequencies of carboxylato complexes and the type of carboxylate coordination. *Coord. Chem. Rev.* 33:227–250.

120. Kozub, D. R., Vakhshouri, K., Orme, L. M., Wang, C., Hexemer, A., and Gomez, E. D. 2011. Polymer crystallization of partially miscible polythiophene/fullerene mixtures controls morphology. *Macromolecules* 44:5722–5726.
121. Bredas, J. L., Norton, J. E., Cornil, J., and Coropceanu, V. 2009. Molecular understanding of organic solar cells: The challenges. *Acc. Chem. Res.* 42:1691–1699.
122. Clarke, T. M., and Durrant, J. R. 2010. Charge photogeneration in organic solar cells. *Chem. Rev.* 110: 6736–6767.
123. Pensack, R. D., Banyas, K. M., Barbour, L. W., Hegadorn, M., and Asbury, J. B. 2009. Ultrafast vibrational spectroscopy of charge carrier dynamics in organic photovoltaic materials. *Phys. Chem. Chem. Phys.* 11:2575–2591.
124. Pensack, R. D., and Asbury, J. B. 2009. Barrierless free carrier formation in an organic photovoltaic material measured with ultrafast vibrational spectroscopy. *J. Am. Chem. Soc.* 131:15986–15987.
125. Hashimoto, K., Hiramoto, M., and Sakata, T. 1988. Temperature-independent electron transfer: Rhodamine B/oxide semiconductor dye-sensitization system. *J. Phys. Chem.* 92:4272–4274.
126. Burfeindt, B., Hannappel, T., Storck, W., and Willig, F. 1996. Measurement of temperature-independent femtosecond interfacial electron transfer from an anchored molecular electron donor to a semiconductor as acceptor. *J. Phys. Chem.* 100:16463–16465.
127. Ramakrishnan, S., and Willig, F. 2000. Pump-probe spectroscopy of ultrafast electron injection from the excited state of an anchored chromophore to a semiconductor surface in UHV: A theoretical model. *J. Phys. Chem. B* 104:68–77.
128. Duncan, W. R., and Prezhdo, O. V. 2008. Temperature independence of the photoinduced electron injection in dye-sensitized TiO_2 rationalized by ab initio time-domain density functional theory. *J. Am. Chem. Soc.* 130:9756–9762.
129. Trosken, B., Willig, F., Schwarzburg, K., Ehret, A., and Spitler, M. 1995. The primary steps in photography: Excited J-aggregates on AgBr microcystals. *Adv. Mater.* 7:448–450.
130. Trosken, B., Willig, F., Schwarzburg, K., Ehert, A., and Spitler, M. 1995. Electron transfer quenching of excited J-aggregate dyes on AgBr microcrystals between 300 and 5 K. *J. Phys. Chem.* 99:5152–5160.
131. Bixon, M., and Jortner, J. 1989. Activationless and pseudoactivationless primary electron transfer in photosynthetic bacterial reaction centers. *Chem. Phys. Lett.* 159:17–20.
132. Haffa, A. L. M., Lin, S., Katilius, E., Williams, J. C., Taguchi, A. K. W., Allen, J. P., and Woodbury, N. W. 2002. The dependence of the initial electron-transfer rate on driving force in *Rhodobacter sphaeroides* reaction centers. *J. Phys. Chem. B* 106:7376–7384.
133. Chuang, J. I., Boxer, S. G., Holten, D., and Kirmaier, C. 2008. Temperature dependence of electron transfer to the M-side bacteriopheophytin in rhodobacter capsulatus reaction centers. *J. Phys. Chem. B* 112:5487–5499.
134. Bixon, M., and Jortner, J. 1991. Non-Arrhenius temperature dependence of electron transfer rates. *J. Phys. Chem.* 95:1941–1944.
135. Bixon, M., and Jortner, J. 1993. Charge separation and recombination in isolated supermolecules. *J. Phys. Chem.* 97:13061–13066.
136. Ramakrishnan, S., Willig, F., and May, V. 2001. Theory of ultrafast photoinduced heterogeneous electron transfer: Decay of vibrational coherence into a finite electronic-vibrational quasicontinuum. *J. Chem. Phys.* 115:2743–2756.
137. Khundkar, L. R., Perry, J. W., Hanson, J. E., and Dervan, P. B. 1994. Weak temperature dependence of electron transfer rates in fixed-distance porphyrin-quinone model systems. *J. Am. Chem. Soc.* 116: 9700–9709.

9 Vibrational Energy and Molecular Thermometers in Liquids

Ultrafast IR-Raman Spectroscopy

Brandt C. Pein and Dana D. Dlott

CONTENTS

9.1 INTRODUCTION

In this update to our 2001 article [1,2], we describe advances in instrumentation for IR-Raman spectroscopy and measurements on vibrational energy in molecular liquids. We will emphasize quantitative determination of the efficiency of energy transfer pathways and the use of molecular thermometers [3–9], which, as vibrationally excited molecules lose their excess energy to the bath, can sense the growing levels of bath excitation. The systems that will be discussed in detail are water under extreme conditions, biologically relevant molecules in aqueous solution, where the water itself serves as the thermometer, and benzene and d_6-benzene, where dissolved CCl_4 is the molecular thermometer.

Throughout this chapter, we will use the following consistent notation. Vibrational relaxation (VR) refers to vibrational energy relaxation, as opposed to phase relaxation. Intramolecular vibrational relaxation (IVR) refers to a VR process that dissipates little or no energy to the bath. The lifetime of an individual state due to VR or IVR is denoted T_1. Vibrational cooling (VC) refers to a process [10–12] involving multiple VR steps and possibly some IVR steps, in which a vibrationally excited molecule attains thermal equilibrium with the bath. The term "thermalization" is also sometimes used to describe VC. Here, we use VC when details of state-to-state transitions are resolved and thermalization when an aggregate process such as the growth of a temperature increase is observed.

In IR-Raman measurements, an ultrashort infrared (IR) pulses are used to deliver energy to a selected molecular transition (the "parent" excitation) and time-delayed visible probe pulses are used to generate a time series of Raman spectra from the parent and all its Raman-active daughters. A daughter state is one that becomes excited, or temporarily excited, as a result of the VC process initiated by parent excitation. For molecules lacking a center of symmetry, in principle, the parent and all daughters can be probed. With our current laser apparatus, we simultaneously detect both Stokes and anti-Stokes Raman spectra, and from the intensities of the various Raman transitions, we can quantitatively determine the instantaneous occupation numbers of the states corresponding to each Raman transition. The power of the IR-Raman method is its ability to track the energy as it leaves the parent and as it flows through the vibrations of an excited molecule into the bath [13]. The weakness of the IR-Raman method stems from the small magnitudes of Raman cross-sections, which necessitate the detection of quite weak scattering signals that may be obscured by a variety of interfering factors such as even the weakest fluorescence, or light generated by nonlinear processes involving the intense pump and probe pulses and the sample or windows in a sample cell [6,14,15]. For these reasons, IR-Raman experiments are usually—but not always—limited to liquids that can be flowed in a windowless jet, where the liquid consists of small molecules (large molecular number densities) that have only high-lying electronic transitions.

In the early days of picosecond lasers, which produced pulses at 1 and 0.5 μm wavelengths, stimulated Raman pumping and coherent or incoherent scattering were used to study vibrations in liquids [13]. However, a coherent Raman probe is sensitive only to molecular dephasing processes that are also responsible for the Raman lineshape, so those experiments mostly provided information already available from Raman scattering [16]. Later, stimulated Raman pumping combined with incoherent anti-Stokes Raman probing was used [13], but the weak incoherent signal could easily be contaminated by probe pulse scattering or signals generated by nonlinear interactions of the pulses and the medium. It would ultimately prove much better to use vibrational IR pulses for excitation. In 1974, Laubereau and coworkers generated intense picosecond pulses in the vibrational IR near 3 μm where the fundamental transitions of CH-stretch and OH-stretch are located [17], and in 1978, Spanner, Laubereau, and Kaiser made the first IR-Raman measurements on ethanol and chloroform [18]. Unfortunately, the laser technologies available in 1976, which generated ~3 ps pulses at a rate no greater than one every minute, made these experiments extremely difficult; only a few molecular liquids were studied as summarized in two excellent reviews [13,19]. In retrospect, it appears that the results were often contaminated with an artifact that appears near $t = 0$ due to nonlinear light scattering (NLS) [20]. The NLS effect resulted in the generation of a signal at the pump + probe frequency [21], which also happens to be the frequency of the anti-Stokes signal from the parent excitation.

Since 1978, improvements in laser technologies have enabled a few groups other than ours to perform accurate IR-Raman measurements on a few systems [22–26], most notably the work by Graener and coworkers [27–32] who used a 50 Hz Nd:YLF laser system producing 4 ps pulses. One of the most important improvements in ultrafast laser technology has been the development of chirp-pulse-amplified Ti:sapphire lasers, which generate millijoule ultrashort pulses at 1 kHz that, when combined with optical parametric amplifiers, can generate synchronized visible and tunable vibrational IR pulses. In fact, the challenge with Ti:sapphire is increasing the pulse duration from the usual <100 fs, where the pulse optical bandwidth is >140 cm^{-1}, to create pulses about 1 ps in duration with widths of 10–30 cm^{-1} suitable to resolve Raman linewidths in molecular liquids. In 1997, our group built such a 1 kHz laser system that featured tunable IR pump pulses and fixed 532 nm probe pulses with ~25 cm^{-1} spectral bandwidths and ~1 ps pulse durations, with energies of ~50 μJ per pulse [33]. This laser system, with subsequent improvements, became the workhorse of our IR-Raman efforts and allowed us to apply this versatile technique to a number of chemically interesting systems, to greatly improve our ability to discern and remove artifacts such as NLS, and to generate high-quality, quantitative measurements of vibrational energy flow in liquids.

Today, the majority of condensed-phase vibrational energy studies are performed using IR pump-probe methods. To a great extent, this is due to much larger IR cross-sections. A typical IR cross-section is 10^{-20} cm^2 whereas a typical Raman cross-section is 10^{-30} cm^2 Sr^{-1}. One-color IR pump-probe methods can measure the loss of vibrational energy from the parent state but cannot *directly* monitor the arrival of this energy at daughter states as we do with IR-Raman. Both IR and Raman methods can indirectly observe vibrational energy after it leaves the parent state by the effects it has on the parent state spectrum. For instance, an IR pump pulse will, after thermalization is complete, create a bulk temperature jump (T-jump) ΔT in the sample, which will broaden the parent spectrum. But apart from this caveat, it takes two-color IR methods to probe energy flow from a parent to a daughter state. It is not practical to review all time-resolved IR studies in this chapter, and other contributors to this volume discuss these methods in detail, so for brevity's sake we will mention just a few recent and notable two-color IR studies that have contributed greatly to our understanding of VR. The T. Elsaesser group used two-color IR pump-probe method to study energy flow in liquid water, from a parent OH-stretch to daughter bend and libration states [34–37], and energy flow from water to the phosphate groups of hydrated DNA [38] and from confined water in phospholipid micelles [39]. The H. Bakker group studied VR of amide I and amide II groups in *N*-methylacetamide clusters [40] and vibrational Förster transfer between OH-stretching modes of H_2O and HOD in ice Ih [41]. Because of the relatively large IR cross-sections, a variety of IR optical coherence and coherent multidimensional techniques are possible, for instance, 2D IR [42], as described in other chapters of this volume. Two-color multidimensional IR techniques are well suited to measure vibrational energy flow from parent to daughter. For instance the J. R. Zheng group has measured vibrational energy transfer from deuterated chloroform to phenyl selenocyanate [43], nonresonant and resonant mode-specific vibrational energy exchange in aqueous electrolyte solutions [44], and intramolecular energy flow and conformational relaxation of 1-cyanovinyl acetate in CCl_4 [45]. The I. Rubtsov group has introduced an ingenious relaxation-assisted 2D IR method that can track the effects of vibrational energy bursts in molecules over long distances [46–50], up to 6 nm in polyethylene glycol polymer chains [47].

9.2 INSTRUMENTATION FOR IR-RAMAN

The instrumentation used presently in our lab is a second-generation setup [51] that has many improvements over our previous version. Both versions are based on a picosecond chirp-pulse amplified Ti:sapphire laser, in order to get narrow enough IR and Raman pulse spectra, ~25 cm^{-1}, for selective pumping and probing of vibrational transitions in molecular liquids. In a picosecond laser of this type, the Ti:sapphire oscillator produces femtosecond pulses, but in the pulse stretcher the spectrum is narrowed using a mask. Computer calculations were used to determine the mask shape to give the desired Gaussian spectrum. This shape was then fit to a polynomial that was fed into a computer-controlled milling machine. With this hard mask, the spectrum is Gaussian but the laser beam has a spatial chirp; however, this spatial chirp is inconsequential after several passes through the regenerative amplifier. In our system, the Ti:sapphire amplifier (Quantronix Titan) operates at 1 kHz repetition rate; it is pumped by an Nd:YLF laser (Quantronix Darwin) generating 15 mJ pulses at 527 nm and the output 800 nm pulses after compression are 3 mJ. The 800 nm autocorrelation full-width at half maximum (FWHM) was 1.2 ps, corresponding to an approximate pulse width of 0.9 ps, and the bandwidth was about 20 cm^{-1}.

In the older experimental setup [33], the 800 nm Ti:sapphire laser output was mixed in KTA crystals with 20 ns pulses signal from a Q-switched single longitudinal mode Nd:YAG laser at 1.064 μm. This mixing produced a picosecond idler mid-IR pump pulse and a picosecond signal 1.064 μm pulse that was frequency doubled to 532 nm for the Raman probe. The mid-IR pulses were tuned in the 2800–3650 cm^{-1} range by tuning the Ti:sapphire laser in the 766–820 nm range, while the probe pulse stayed at 532 nm. It is useful to have a fixed-frequency probe pulse due to the need for expensive optical notch filters to eliminate elastically scattered light. Unfortunately, tuning the chirp-pulse amplifier and reoptimizing the stretcher and compressor were laborious. Our newer version has higher powers, shorter pulse duration, a greater IR tuning range, and—most importantly—facile IR tuning.

As shown in Figure 9.1, in the current setup, the 800 nm pulses were split into two parts. One part with ~1.5 mJ energy was frequency doubled to produce 0.84 mJ pulses at 400 nm, which were then

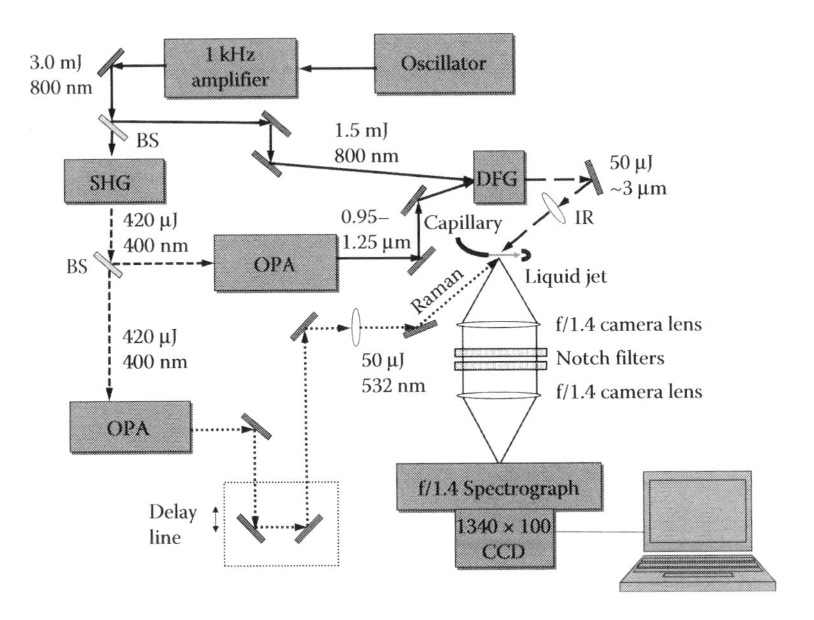

FIGURE 9.1 Schematic of apparatus for IR-Raman spectroscopy. Key: BS = beam splitter; SHG = second-harmonic generation crystal; OPA = optical parametric amplifier; DFG = difference-frequency generation crystal; CCD = charge-coupled detector array. (Adapted with permission from Fang Y et al. Vibrational energy dynamics of glycine, *N*-methyl acetamide and benzoate anion in aqueous (D_2O) solution. *J. Phys. Chem. A* 113:75–84, Copyright 2009, American Chemical Society.)

split into two parts to pump two optical parametric amplifiers (OPAs; Light Conversion TOPAS 400-ps). The probe pulse generator was fixed at 532 nm, where it generated 0.7 ps duration ~50 μJ pulses with a bandwidth of 30 cm^{-1}. The Raman probe pulses were transmitted through a 0.8 nm bandpass filter (Omega Optics) to clean up any stray light not at 532 nm that was generated in the OPA optics or steering optics. After the filter, the probe pulses had a 28 cm^{-1} bandwidth. The second part of the 800 nm pulses was mixed in a KTA crystal with the idler of the second OPA, which was tuned in the 0.95–1.15 μm range, to generate tunable IR pulses in the 2000–3800 cm^{-1} range. The pulse energy in the IR was ~50 μJ at 3000 cm^{-1}. The TOPAS OPA has a system of computer-controlled stepper motors that can be used to tune the wavelength. We built an additional system that lets us continuously scan the mid-IR using one additional computer-controlled stepper to tune the KTA crystal angle and a second stepper to rotate a mirror right after the KTA crystal to compensate for beam wandering. The apparatus time response function was measured by cross-correlating IR and visible pulses in water, where the NLS signal [21] was detected as a function of time delay. The apparatus time response function had 1.0 ps FWHM.

The Raman detection system consisted of a unique f/1.4 large-aperture spectrograph that was designed to simultaneously collect both Stokes and anti-Stokes spectra over the approximate range −3800 to +3800 cm^{-1} with spectral resolution of ~20 cm^{-1}, so the spectrograph would not appreciably broaden Raman spectra generated with 30 cm^{-1} bandwidth probe pulses. The spectrograph, manufactured by Kaiser Optical (Ann Arbor, MI), had an f/1.4 aperture and a holographic volume diffraction grating that imaged the spectral region 443–671 nm onto a Princeton Instruments CCD detector with 1340 × 100 elements having a pixel pitch of 20 μm, which gives a dispersion of 5.7 cm^{-1}/pixel. The detection system was corrected for its wavelength-dependent response using a calibrated blackbody source (Ocean Optics). The Raman light was collected from a 60 μm liquid jet and it was imaged 1:1 onto the 50 μm spectrograph slit using two Nikon f/1.4 camera lenses. Figure 9.2 shows a representative ambient temperature Raman spectrum of chlorobenzene obtained using the picosecond probe pulses and a pair of Raman notch filters.

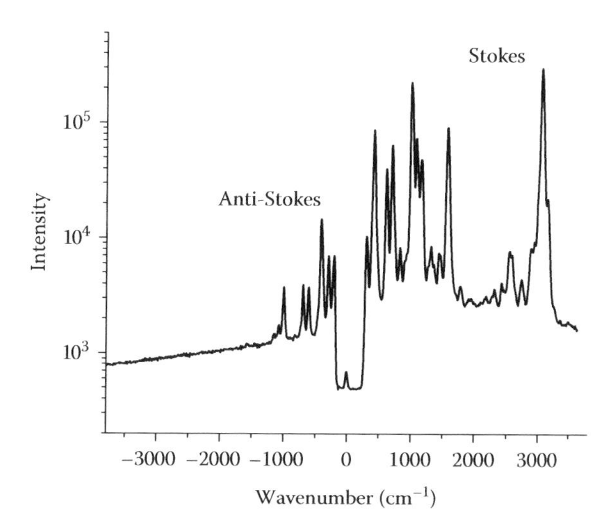

FIGURE 9.2 Raman spectrum of a flowing 60-μm-diameter jet of liquid chlorobenzene obtained using a 1 s integration time, 1 ps duration 532 nm pulses, a pair of Raman notch filters, and an f/1.4 spectrograph that detects Stokes and anti-Stokes signals simultaneously.

Over several years, many different sample configurations were used. The best results were obtained with flowing jets having no optical windows. We also wanted to minimize the flow rates to reduce sample volumes. We tried a number of different jet nozzles, usually with rectangular cross-sections, which were difficult to obtain or fabricate. Ultimately, we found that excellent results were obtained using simple 60 μm diameter stainless-steel capillaries for the jet nozzle. With this simple jet, the biggest remaining problem was pump-induced variations in the liquid flow rate that caused the liquid stream diameter and the Raman intensity to vary. We tried a high-pressure liquid chromatograph syringe pump, but the pullback cycle when the syringe reset was a problem. Ultimately, to minimize variations, we obtained a high-quality liquid chromatography system (HP 1090) with a dual-syringe pump. The dual-syringe practically eliminates pump pressure variations when the syringe reaches the end of its stroke.

Raman spectra were obtained in pairs, a signal at the indicated positive time delay and a background obtained at negative time delay (probe precedes IR pump). The Stokes spectrum was monitored during data acquisition to verify that the jet did not wander or the sample did not or, if a solution is used, change concentration due to evaporation. Stokes and anti-Stokes signals were combined to determine quantitative occupation numbers as follows. When the occupation number of a vibration of frequency ω is n_ω and the laser frequency is ω_L, the Stokes intensity is

$$I_{\mathrm{ST}} \propto \omega_L(\omega_L - \omega)^3\,[n_\omega + 1]\sigma_R, \tag{9.1}$$

and the anti-Stokes intensity is

$$I_{\mathrm{AS}} \propto \omega_L(\omega_L + \omega)^3 n_\omega \sigma_R, \tag{9.2}$$

where intensity is taken to mean the integrated area of the transition, and σ_R is the Raman cross-section at ω_L. The integrated areas were obtained by fitting each vibrational transition to a Voigt lineshape function using Microcal Origin software [51]. The Voigt function was used because it provided an excellent fit to all transitions. The use of a Voigt is not intended to convey a specific opinion about the nature of vibrational dephasing [51]. Equations 9.1 and 9.2 can be combined to eliminate the proportionality constant, which is a function of the detection system and is the same for both Stokes and anti-Stokes measurements, and when $n_\omega \ll 1$, the fraction of molecules in the excited state is given by the fraction I_{AS}/I_{ST}. For multiply degenerate modes, Equations 9.1 and 9.2 give the combined occupation number of the degenerate modes.

9.3 PRESSURE AND TEMPERATURE JUMP

Since rather intense IR pump pulses are used to produce strong Raman transient signals, there is a process where a nonequilibrium state is produced that decays within picoseconds into a state with a temporary pressure jump (P-jump) ΔP and T-jump ΔT [52]. These jumps occur because on the timescale of a few picoseconds, the heating process is approximately adiabatic and isochoric. The jumps have quite complicated spatial distributions due to the approximately Gaussian profile of the pump pulses and the exponential nature of the Beer's law absorption. In general, we will refer to the peak values of ΔT and ΔP at the center of the pump beam since the probe pulses interrogate the center of the pump beam near-Gaussian profile. The P-jumps decay within a few hundred picoseconds as a result of volume expansion and the T-jumps decay within a few hundred microseconds as a result of thermal diffusion. Both T-jump and P-jump have thoroughly decayed away and the sample has been refreshed before the next laser shot, 1 ms later.

It is possible to estimate the peak values of the P-jump and T-jump with good accuracy [52]. The peak values occur on the sample surface facing the IR pulses at the center of the Gaussian beam profile. Later in this chapter, we will illustrate these estimates using water as an example. With a Gaussian profile IR pulse of energy E_p, the *average* fluence $J_{avg} = E_p/(\pi r_0^2)$, where r_0 is the $(1/e^2)$ Gaussian beam radius. The *peak* fluence at the beam center is *twice this value*, $J_c = 2E_p/(\pi r_0^2)$. The peak energy density E_v in the water near the irradiated surface is $E_v = J_c\alpha$, where α is the absorption coefficient. Values of α for pure water are tabulated in Ref. [53]. The peak temperature and pressure jumps are [54]

$$\Delta T = \frac{J_c\alpha}{\rho C_v} = \frac{2E_p\alpha}{\pi r_0^2 \rho C_v}, \tag{9.3}$$

and

$$\Delta P \approx \left(\frac{\partial P}{\partial T}\right)_V \Delta T = \left(\frac{\beta}{\kappa_T}\right)\Delta T, \tag{9.4}$$

where β is the coefficient of thermal expansion and κ_T the isothermal compressibility.

For ambient water, $\rho C_v = 4.2$ MJ K^{-1} m^{-3} and $(\partial P/\partial T)_V = 1.8$ MPa K^{-1}, isochoric heating of water by 1 K results in a pressure increase of 18 bar. With IR pulses having $E_p = 9.5$ μJ and $r_0 = 150$ μm, tuned to 3310 cm^{-1} where the absorption coefficient $\alpha = 4660$ cm^{-1} [53], the peak value of $\Delta T = 30$ K and the corresponding ΔP would be 54 MPa (~0.5 kilobars).

9.4 IR-RAMAN AND IR-PUMP PROBE

There are quite a few differences between probing molecular vibrations with Raman or IR [55,56]. The fundamental reason for this is that in Raman, both ground and excited states contribute to the Stokes (downshifted) part of the spectrum, but only excited states contribute to the anti-Stokes part (upshifted). In IR, there are absorption and stimulated emission processes, and both ground and excited states contribute to each process.

With a Raman probe, as depicted in Figure 9.3a, a laser pulse at ω_L is incident on the sample and the inelastically scattered light is collected, spectrally resolved, and detected with a multichannel array [57]. With no pump pulses, the Stokes signals of higher-frequency vibrational transitions ($h\nu/k_BT \gg 1$) are much more intense than the anti-Stokes signals. In our earlier studies [21,58–62], we acquired only anti-Stokes spectra, but as described in the Section 9.2, we now simultaneously

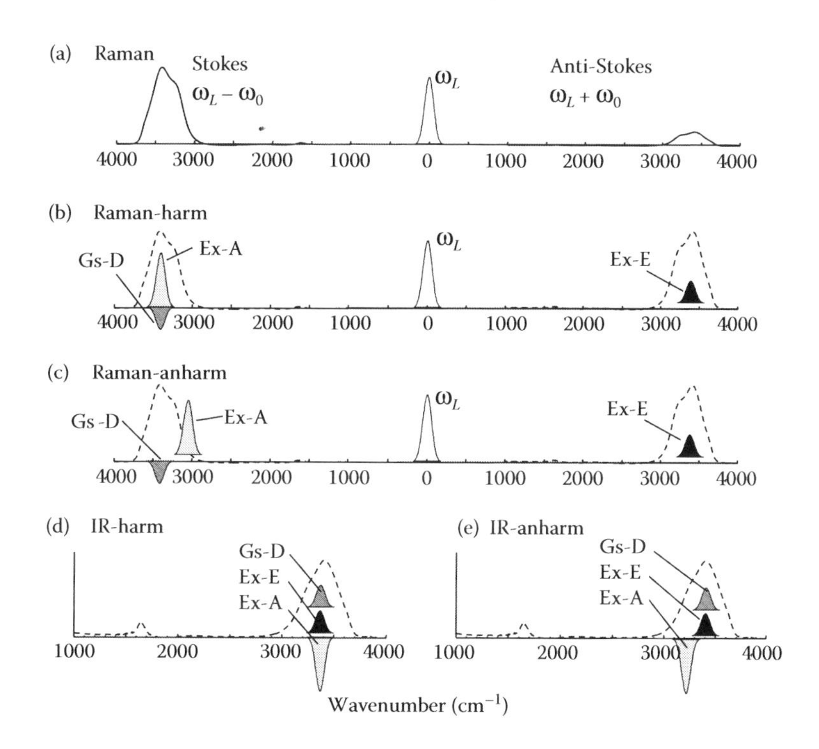

FIGURE 9.3 An IR pump pulse creates an excitation in ν_{OH} that can be monitored by a Raman or IR probe. Both probe methods see ground-state absorption depletion (Gs-D), excited-state emission (Ex-E), and excited-state absorption (Ex-A). (a) Raman spectra have a Stokes and an anti-Stokes branch. (b,c) Ex-E appears in the anti-Stokes region and Ex-A and Gs-D in the Stokes branch. In harmonic approximation (b), the Ex-A is partially offset by Gs-D, but for anharmonic vibrations (c), Ex-A is redshifted from Gs-D. (d) In IR with harmonic vibrations, Ex-E, Ex-A, and Gs-D cancel each other, so there is no IR signal. (e) With anharmonicity, Ex-A is redshifted and a bipolar IR signal will be observed. (Reproduced with permission from Wang Z, Pang Y, Dlott DD. Hydrogen-bond disruption by vibrational excitations in water. *J. Phys. Chem. A* 111:3196–3208, Copyright 2007, American Chemical Society.)

acquire both Stokes and anti-Stokes spectra [63], as in Figure 9.2. A *Raman transient* is defined to be the pump-induced change in the Raman intensity relative to a signal obtained with a probe pulse that precedes the pump pulse. Anti-Stokes transients arise from a single source, excited-state emission, and with higher-frequency vibrations, the excited states are detected against a nearly dark background [13,19]. Stokes transients are more difficult to interpret [32,63]. Stokes transients arise from two sources, ground-state depletion and excited-state Stokes scattering, and both effects are observed as small changes against a much larger background arising from the ground-state Stokes scattering [63].

In a two-color IR pump-probe experiment, the signal is the pump-induced change in the transmitted probe pulse intensity [64]. Pump-induced changes in the probe intensity are small effects against a large background, but excellent signal-to-noise ratios are achievable provided the laser pulses have good intensity stability. IR probe pulse signals can be more difficult to interpret than either Stokes or anti-Stokes Raman signals, since the probe pulse transmission is simultaneously sensitive to three processes, excited-state emission, ground-state absorption depletion, and excited-state absorption [63,64].

It is useful to consider what Raman and IR experiments would see in a baseline approximation where all vibrations were harmonic, and also in the more germane framework of somewhat anharmonic excitations. In many molecules, the diagonal anharmonicity (the frequency difference between the $0 \rightarrow 1$ and $1 \rightarrow 2$ transitions) is at most a few percent. For instance, in water, the diagonal anharmonicity is ~40 cm^{-1} for δ_{H_2O} (~2.4% of the transition frequency) [65], but ν_{OH} has an exceptionally large anharmonicity of ~250 cm^{-1} (~7% of the transition frequency).

With the Raman experiment in the harmonic approximation (Figure 9.3b), excited states created by the pump would create a sudden intensity jump in the anti-Stokes region [13,63]. In the Stokes region, the ground-state depletion leads to an intensity dip but the excited-state absorption leads to an intensity increase twice as large as the dip, so the net result is a sudden Stokes intensity jump of the same amplitude as the anti-Stokes jump [63]. This happens because in the harmonic approximation, the cross-section for excited-state absorption is twice that of ground-state absorption. When anharmonicity is introduced, as depicted in Figure 9.3c, the anti-Stokes transients are not affected but in the Stokes spectrum, the excited-state absorption becomes redshifted away from the ground-state absorption. The Stokes transient in this case becomes bipolar, with a negative-going part from the ground-state depletion and a positive-going part of approximately twice the amplitude due to the excited-state absorption [63]. When the anharmonic shift is smaller than the vibrational linewidth, clearly separating these two contributions to the Stokes transient can become a problem.

With IR probing, an interesting and well-known result [19,64] is obtained in the harmonic approximation (Figure 9.3d): there is no signal in an IR probe transmission measurement. The ground-state depletion increases the transmitted probe. The excited-state emission further increases the probe by the same amount. But the excited-state absorption decreases the probe enough to exactly offset the other two effects. In the anharmonic IR case (Figure 9.3e), the IR signal becomes bipolar, with a negative-going part due to excited-state absorption and a positive-going part due to the combination of ground-state depletion and excited-state emission [66].

9.5 ROADMAP FOR VIBRATIONAL ENERGY IN POLYATOMIC (LIQUID OR SOLID) MOLECULES

After studying many different molecular liquids and solutions, where the molecules or solutes are of "intermediate size," a roadmap has been developed by us to describe the VC process triggered by exciting a higher-energy vibrational fundamental, usually a CH-stretch or OH-stretch excitation near 3000 cm^{-1} or a CD-stretch near 2000 cm^{-1}. "Intermediate size" means bigger than a small molecule and smaller than a large molecule [67]. In a small molecule (usually two or three or perhaps four atoms), the vibrational levels are widely separated compared to the fundamental frequencies of bath excitations, say 50–100 cm^{-1}. In a large molecule, such as a polymer chain, different modes can be localized on distant parts of the molecule so excitation transfer can be limited by distance and propagation effects.

The VC process in polyatomic liquids has traditionally been viewed as a vibrational cascade, a multistep descent down a vibrational ladder [68]. Vibrational cascade is a good description for high-lying vibrational states of diatomic molecules in condensed phases, where a molecule such as XeF in solid Ar might be prepared in $v = 20$ and then cascade down the ladder of lower v-states until cooled [69]. But with larger polyatomic molecules, IR-Raman and hot fluorescence measurements did not observe this type of successive-step ladder descent [68]. Energy from the parent excitation tended to randomize rather than descend a ladder. Our group has introduced a three-stage model [51,70,71] for VC of condensed-phase molecules, which is used to describe situations where the initial excitation has enough energy for the primary decay to occur via IVR [51,70], that is, the initial excitation is above the energy threshold for IVR. This model was discussed extensively in publications from our group [51,71]. As depicted in Figure 9.4, we divide the molecular vibrational excitations into three tiers: the parent P, the midrange levels M, and the lower energy levels L, along with a bath consisting of a continuum of lower-energy collective excitations. This division is based on the 1979 paper by Nitzan and Jortner [67].

The first stage is relaxation of the excitation P, a bright state produced by laser excitation. The pumping rate is $\alpha J(t)$, where α is the absorption coefficient and $J(t)$ the time-dependent fluence

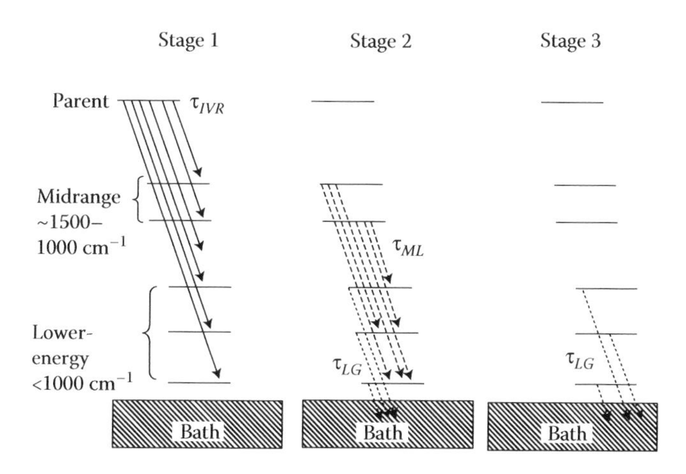

FIGURE 9.4 Schematic of the three-stage roadmap to describe vibrational cooling (VC) after excitation of a higher-frequency vibration above the IVR threshold. Stage 1: A parent vibration *P* such as a CH-stretch is excited. *P* undergoes a fast intramolecular redistribution consisting of a coherent transfer part on the timescale of T_2 and an incoherent intramolecular vibrational relaxation (IVR) with lifetime τ_{IVR}. These processes populate most or all lower-energy vibrations. Stage 2: The midrange tier *M* undergoes vibrational relaxation (VR) with lifetime τ_{ML} by exciting lower-energy vibrations *L* plus the bath, while *L* undergoes VR with time constant τ_{LG} (*G* denotes ground vibrational state) by exciting only bath excitations. Stage 3: Second-generation lower-energy vibrations created in stage 2 undergo relaxation with time constant τ_{LG}. (Reproduced with permission from Seong N-H, Fang Y, Dlott DD. Vibrational energy dynamics of normal and deuterated liquid benzene. *J. Phys. Chem. A* 113:1445–1452, Copyright 2009, American Chemical Society.)

(*photons* m^{-2}) of the IR pump pulse. Other vibrations that contribute to the character of the bright state *P* will also be pumped directly by the laser, and these states are described [7,8,57,72] as "coherently coupled" to *P*.

We use the term "coherently coupled" to describe vibrational excitations that are coupled to the parent bright state by an interaction, which is about equal to or greater than $(T_2)^{-1}$, where T_2 is the time constant for dephasing [7,57]. In order to understand the effects of such coupling, consider the familiar case of a 2:1 Fermi resonance of CH-stretch and bend. The laser initially excites a state with mainly stretch character, but this state evolves into a mixed stretch–bend state. This is a kind of IVR process. When the time resolution is high, this time evolution usually takes the form of damped quantum beats, which are well known in both vibronic spectroscopy [73], and ultrafast IR spectroscopy [74]. As discussed previously [75], in our anti-Stokes experiments, IR excitation is semi-impulsive, that is, the pulse duration is long compared to the vibrational period, comparable to T_2, and shorter than T_1, so we are probing vibrational populations in the parent and coupled states averaged over a time period ~T_2 and usually not resolving quantum beats. Ordinarily, in quantum beat measurements, the probe pulse is sensitive to either the parent or the coupled states, so as population oscillates among them, the signal intensity also oscillates. In anti-Stokes Raman, the probe situation is a bit different because in anti-Stokes Raman where the probe bandwidth is greater than the anharmonic coupling, an overtone looks like a fundamental excitation with twice the amplitude and a combination band appears as excitation of both fundamentals [30]. For instance, with the 2:1 Fermi resonance, we would see a CH-stretch at ~3000 cm^{-1} via the $v = 1 \rightarrow 0$ transition and a CH-bend at ~1500 cm^{-1} via the $v = 2 \rightarrow 1$ transition. Thus, in our anti-Stokes Raman experiments, "coherent coupling" is observed as instantaneous excitation of lower-energy vibrations that can combine with other vibrations to produce states near the parent energy. The parameters α' and α'', *which most often are zero*, characterize the rates at which *M* or *L* vibrations respectively become excited by coherent coupling with IR pumping of *P*.

After the initial excitation is prepared, the population in P decays via IVR with time constant τ_{IVR}. This IVR process results in little or no energy dissipated to the bath. But it is not necessarily true that this liquid-state IVR process has to be the same IVR as in the isolated molecule, since the bath can modulate the molecule's energy levels without dissipating energy. Parent IVR excites many M and L vibrations. However, the M and L vibrations become excited to different extents, which depend on the detailed intermolecular couplings. Let the subscript i denote vibrations in the M tier and the subscript j vibrations in the L tier. Then, ϕ_{PMi} is the quantum efficiency for IVR from P to mode i of the M tier, and ϕ_{PLj} is the quantum efficiency from P to mode j of the L tier.

In the second stage, the excited M and L vibrations decay, but with different mechanisms. The M vibrations decay with time constant τ_{ML} by exciting L vibrations plus the bath. The lower-energy vibrations decay with time constant τ_{LG} (G denotes vibrational ground state) by exciting the bath only. The quantum efficiency for transfer from mode i of the midrange tier to mode j of the lower-energy tier is ϕ_{MiLj}. However, we do not experimentally determine which specific M vibration excites a specific L vibration, so we can measure only the net quantum efficiency of transfer *from all M modes* to mode L_i, $\phi_{MLj} \sum M_i(t) = \sum \phi_{MiLj} M_i(t)$.

In the third stage, there is no excitation left in either P or M. The remaining L excitations decay into the bath with time constant τ_{LG}. To simplify the model to eliminate vast numbers of fitting parameters, we assume all vibrations in the M tier have the same lifetime τ_{ML} and all vibrations in the L tier the same lifetime τ_{LG}. We are saying lifetime variations within the M and L tiers are small, and if they are not, the model fails and additional parameters must be introduced. Thus, the VR process is described by three global rate constants $k_{IVR} = (\tau_{IVR})^{-1}$, $k_{ML} = (\tau_{ML})^{-1}$, and $k_{LG} = (\tau_{LG})^{-1}$.

The three-stage model is summarized by the following set of Equations 9.5:

$$\begin{aligned}
\frac{dP(t)}{dt} &= -k_{IVR}\left[P(t) - P^{eq}\right] + \alpha J(t) \\
\frac{dM_i(t)}{dt} &= k_{IVR}\phi_{PMi}\left[P(t) - P^{eq}\right] - k_{ML}\left[M_i(t) - M_i^{eq}\right] + \alpha_i' J(t) \\
\frac{dL_j(t)}{dt} &= k_{IVR}\phi_{PLj}\left[P(t) - P^{eq}\right] + k_{ML}\phi_{MLj}\sum_i\left[M_i(t) - M_i^{eq}\right] - k_{LG}\left[L_j(t) - L_j^{eq}\right] + \alpha_j'' J(t)
\end{aligned} \tag{9.5}$$

In Equations 9.5, the superscript *eq* denotes thermal equilibrium population at the final temperature T_f, and these terms account for the T-jump. Although we determine the quantum efficiencies ϕ_{PMi}, ϕ_{PLj}, and ϕ_{MLj} directly, since we do not detect many of the molecular vibrations, we have found it convenient to normalize the *observed vibrational energy* using the conditions $\Sigma\phi_{PMi} + \Sigma\phi_{PLj} = 1$ and $\Sigma\phi_{MLj} = 1$. By way of example, if we observe two M vibrations and one L vibration having the same quantum efficiency for transfer, then $\phi = 0.33$ for each vibration, even though a great deal of energy might also be present in unobserved vibrations.

The three stages participate differently in the VC process. The first stage is purely an IVR process, so little or no energy is dissipated to the bath. The second $M \rightarrow L$ stage is responsible for the smaller part of VC because the energies being dissipated are the difference between the M and L energies. For instance, an $M \rightarrow L$ relaxation might involve a 1500 cm^{-1} M vibration, which decays by exciting a 1000 cm^{-1} L vibration plus 500 cm^{-1} of bath excitation. The greater part of VC is the $L \rightarrow G$ relaxation process, since all the energy of the L excitation is converted into bath excitations.

9.6 HYDROGEN-BOND DISRUPTION BY VIBRATIONAL EXCITATIONS IN WATER

Water is an exceedingly complicated liquid, so it should come as no surprise that vibrational excitations of water exhibit complex and unusual properties. The studies previously made of the response of hydrogen bonds in water to ultrashort pulses that excite ν_{OH} are too numerous to be mentioned here, but a recent review is Ref. [76], and many references to individual studies can be found in the "A Brief History of Water Vibrational Relaxation Measurements" section of Ref. [55]. The focus in this section is the discovery that an excitation of the OH-stretch ν_{OH} of water has unique disruptive effects on the local hydrogen bonding [36,55]. The disruption is not an immediate vibrational predissociation, as originally proposed by Staib and Hynes [77], and which is frequently the case with hydrogen-bonded clusters, but instead is a delayed disruption caused by a burst of energy from a vibrationally excited water molecule. The disruptive effects are the result of a fragile hydrogen-bonding network subjected to a large amount of vibrational energy released in a small volume in a short time. The effects are especially large because in water the energy of the relatively high frequency (~3400 cm^{-1}) OH-stretch excitations is spread among just three atoms, and the lifetimes of the ν_{OH} stretching and δ_{H_2O} bending excitations are extremely short, each about 0.2 ps, so the energy is released very rapidly since the burst is created on the roughly 0.4 ps timescale [55].

The energy burst process is described schematically in Figure 9.5. A few water molecules with OH-stretch excitations (Figure 9.5a) emit a burst of energy that creates a transient nonequilibrium state with locally weakened hydrogen bonding denoted H_2O^* (Figure 9.5b). H_2O^* does not have a single configuration; instead, experiments show that it represents a broad distribution of configurations. The first direct observation of a delayed nonequilibrium cleavage of hydrogen bonds should be attributed; we believe to the Fayer group, who studied OD stretches of alcohol oligomers in CCl_4 [78–80], and OD-stretch excitations of HOD in H_2O [81]. Eventually, this nonequilibrium state relaxes into a thermalized state where the region surrounding each previously excited water molecule is in thermal equilibrium at an elevated temperature (Figure 9.5c), and ultimately thermal diffusion leads to a locally uniform T-jump. The magnitude of the T-jump depends on the original concentration of ν_{OH} excitations.

Generally speaking, as T is increased, water's hydrogen bonds weaken and become less abundant, so the states depicted in Figures 9.5c and 9.5d represent states of hydrogen-bond weakening, but we emphasize here that H_2O^* is a transient and extreme state that cannot be achieved by simply varying T and/or P in equilibrium.

The discovery of H_2O^* was unexpected and serendipitous. It occurred [82] because the laser pulses we used for IR-Raman, 0.8 ps, were longer in duration than the 0.2 ps lifetimes of either the OH-stretch of water bend, so that when we excited ν_{OH} of water, during the pulses there was time

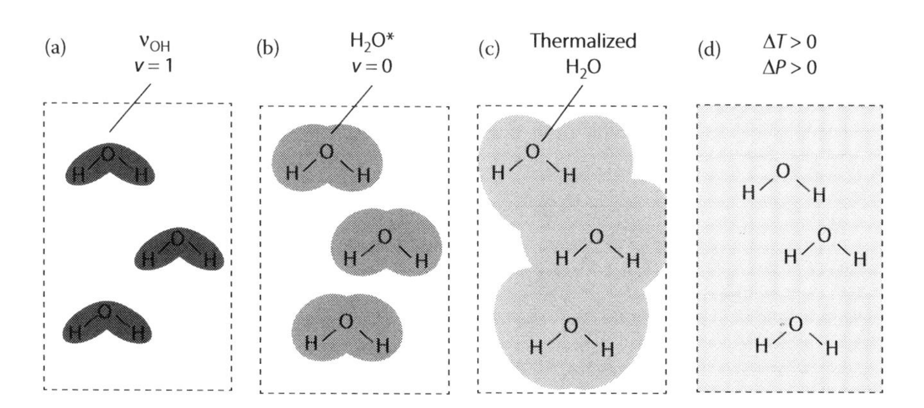

FIGURE 9.5 Transient disruption of hydrogen bonding by vibrationally excited water. (a) Water molecules with excited OH-stretch, ν_{OH}. (b) Vibrational relaxation of ν_{OH} generates an energy burst from each excited molecule that disrupts the local hydrogen bonding to create H_2O^* in the vibrational ground state. (c),(d) The H_2O^* thermalizes and the excess energy becomes uniformly distributed in time and space to produce an equilibrium state with a temperature jump ΔT and a pressure jump ΔP. (Reproduced with permission from Wang Z, Pang Y, Dlott DD. Hydrogen-bond disruption by vibrational excitations in water. *J. Phys. Chem. A* 111:3196–3208, Copyright 2007, American Chemical Society.)

for the vibrational energy burst to create H_2O^*, which then could become excited, and therefore detectable by anti-Stokes Raman, by the trailing edge of our IR pump pulses.

The energy bursts are intense. To put some numbers into this discussion, we use a crude but effective example. If a CH-stretch with 3400 cm^{-1} of energy were added to a classical nonlinear triatomic molecule ($C_v = 6\ k_B$), the local temperature would be 1100 K (Figure 9.5a). If this energy were spread out over two or three adjacent molecules (Figures 9.5b and 9.5c), the local temperatures would be 700 or 570 K, respectively [55]. Obviously, we could get different answers using other models for the local heat capacity, for instance, to include the hydrogen bonding, but the picture of high local temperatures would be unchanged. Keep in mind that with 3400 cm^{-1} ν_{OH} excitations, the eventual equilibrium T-jump depends not on the excitation frequency but on the initial density of excitations, according to Equation 9.3, which can be controlled by varying the IR pulse energy, but the properties of the energy burst depend solely on water dynamics and the IR pump wavenumber regardless of whether ΔT is large or small. *Even the weakest IR sources such as the glow bars used in FTIR spectrometers will create these energy bursts.*

Figure 9.6 depicts, for reference, the effects of temperature on the water Raman spectra in the ν_{OH} region. As T is increased, the spectral intensity decreases on the lower-wavenumber side. This is in accordance with a general decrease in hydrogen-bond strength with increasing T and the well-known correlation between the hydrogen-bond strength and the vibrational redshift, that is, the most redshifted OH-stretch oscillators are those with the strongest hydrogen bonds. Figure 9.6b shows Raman difference spectra resulting from heating water. The Raman difference spectrum has a characteristic bipolar shape and it makes an excellent thermometer for measuring the T-jumps.

In the IR-Raman experiments described here, the IR excitation pulses were adjusted to produce $\Delta T = 30$ K. Figure 9.7 shows a typical pair of anti-Stokes and Stokes transients obtained at the time delay $t = 0$. A coherent artifact due to NLS that appears near $t = 0$ has been subtracted away [20,21,83], as described previously [61,75]. The narrow-band pump pulse was centered at 3140 cm^{-1}, on the red edge of the water ν_{OH} transition (*dashed line* in Figure 9.7). The transient Stokes signal is bipolar. It has two peaks and one dip. The dip is caused by ground-state depletion. The redshifted peak is caused by excited-state absorption. The blueshifted peak is caused by H_2O^* absorption ($v = 0 \rightarrow 1$). Keep in mind that the laser pump pulse duration of 0.8 ps is longer

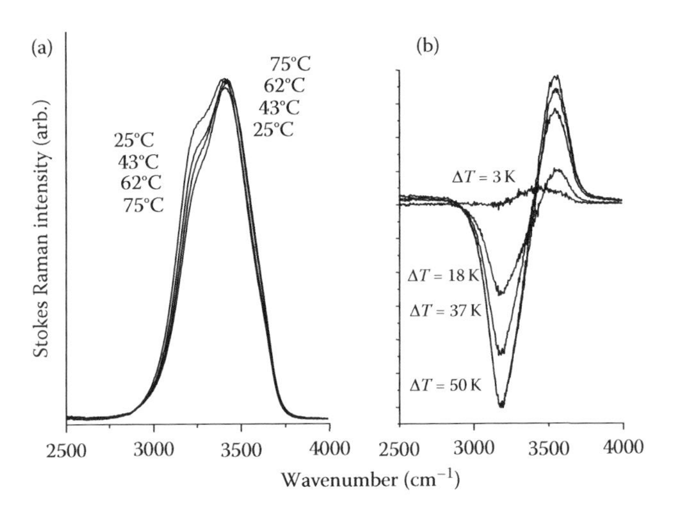

FIGURE 9.6 (a) Raman spectra of water at the indicated temperatures, all normalized to the same peak amplitudes. (b) Raman difference spectra at the indicated values of T-jump ΔT, showing how the bipolar pattern can be used as a molecular thermometer.

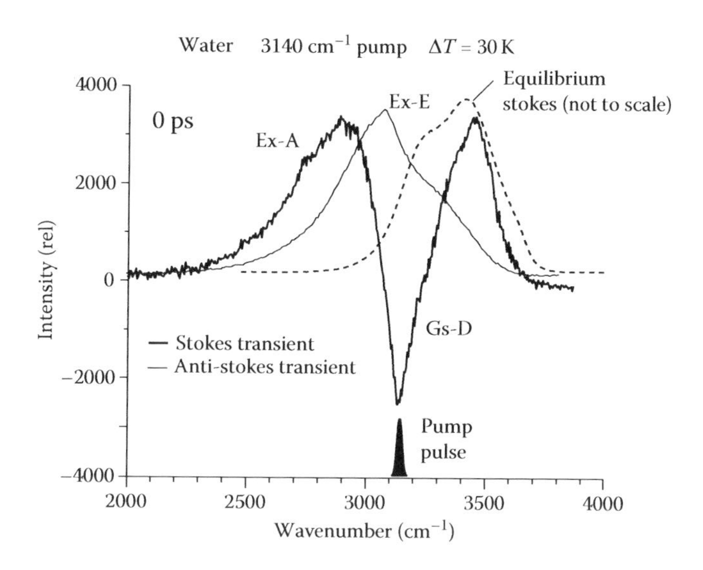

FIGURE 9.7 The Raman spectrum of water (dashed curve) compared to Stokes and anti-Stokes transients obtained with 3140 cm^{-1} pump pulses, at a delay $t = 0$, with a final temperature increase $\Delta T = 30$ K. The anti-Stokes transient arises from excited-state emission (Ex-E) from ν_{OH} and a narrower blueshifted band attributed to ($v = 1$) states of H_2O^*. Excited states of H_2O^* are created by a two-step process: excitation of ν_{OH} followed by relaxation to H_2O^* ($v = 0$) and subsequent IR pumping of H_2O^* to the ($v = 1$) state. (Reproduced with permission from Wang Z, Pang Y, Dlott DD. Hydrogen-bond disruption by vibrational excitations in water. *J. Phys. Chem. A* 111:3196–3208, Copyright 2007, American Chemical Society.)

than the ~0.4 ps time to generate H_2O^* from ν_{OH}, so with this laser it is possible to create H_2O^* in the OH-stretch ground state and then subsequently excite its OH-stretch during the same pump pulse. The anti-Stokes transient is entirely due to excited-state emission. The anti-Stokes spectrum has an unresolved double-peak structure. The redshifted peak is due to excited-state anti-Stokes scattering of water, and the blueshifted peak to excited-state anti-Stokes scattering of H_2O^* in the ($v = 1$) state.

Figure 9.8 provides information about vibrationally excited H_2O and H_2O^* and their decay pathways. In Figure 9.8, the sharper feature-labeled NLS is a coherent artifact due to NLS, which was subtracted away from the data in Figure 9.7. NLS is an IR + visible sum-frequency signal that arises from the bulk of the water sample [21]. Ordinarily sum-frequency generation (SFG) is thought to be forbidden in bulk centrosymmetric liquids such as water, but this is true only within the dipole approximation. The NLS signal is present only when the pump and probe pulses overlap temporally. Figure 9.8 shows anti-Stokes spectra of water with red-edge 3300 cm^{-1} pumping (Figure 9.8a) and with blue-edge 3600 cm^{-1} pumping. Recalling that the anti-Stokes spectra show only vibrationally excited water, with blue-edge pumping, much more excited H_2O^* is produced, since the weakened hydrogen bonding associated with H_2O^* results in its having a blueshifted absorption. The appearance of excited δ_{OH} bending vibrations near 1640 cm^{-1} in Figure 9.8a but not in Figure 9.8b indicates that ν_{OH} of water decays by generating δ_{OH} (which is known independently from two-color pump-probe measurements), but ν_{OH} of H_2O^* does not.

Using the IR-Raman technique with 0.8 ps pulses, we see two kinds of water, ordinary water and H_2O^*. H_2O^* anti-Stokes scattering from excited ν_{OH}, with a Gaussian lineshape peaked at ~3500 cm^{-1} having FWHM of ~200 cm^{-1} is blueshifted and narrowed compared to water, and its lifetime is 0.8 ps versus 0.2 ps for water. Its relaxation occurs by a characteristically different mechanism than excited ν_{OH} of water, which appears to produce much less δ_{OH} excitation.

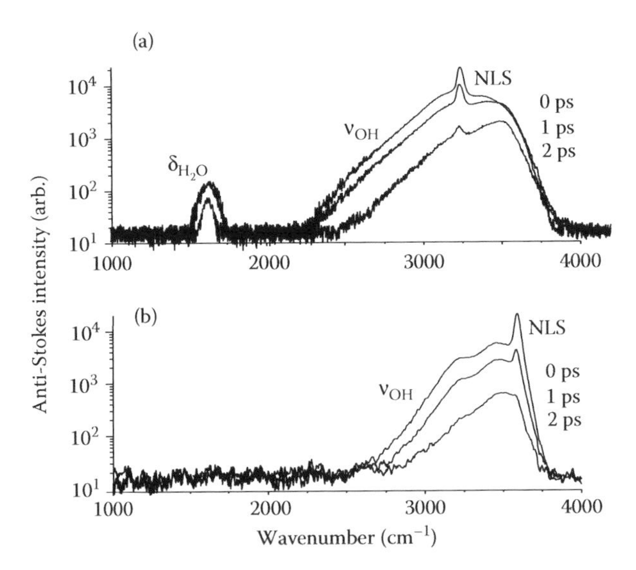

FIGURE 9.8 (a) Water red-edge pump; (b) water blue-edge pump. Transient anti-Stokes spectra of water with red-edge pumping at 3200 cm^{-1} and blue-edge pumping at 3600 cm^{-1} and $\Delta T = 30$ K. The narrower peak at the pump wavenumber is an artifact due to nonlinear light scattering (NLS). δ_{H_2O} is observed with red-edge pumping but not with blue-edge pumping where mainly H_2O^* is produced. (Reproduced with permission from Wang Z et al. Vibrational substructure in the OH stretching transition of water and HOD. *J. Phys. Chem. A* 108:9054–9063, Copyright 2004, American Chemical Society.)

We can understand a great deal about H_2O^* by looking at equilibrium water spectra as a function of temperature and pressure. Even a brief inspection of water Raman spectra under extreme conditions [84,85] makes a strong case that H_2O^* looks a great deal like supercritical water. For instance, the Raman spectrum [84] at $P = 40$ MPa, $T = 300$°C quite closely resembles our H_2O^* spectra, with a peak at 3550 cm^{-1} and an FWHM of 200 cm^{-1}.

Now, some caveats are in order, since the H_2O^* state is metastable and cannot be replicated in an equilibrium experiment. The local conditions due to the energy burst from the ν_{OH} relaxation approximate a state of high T and P, but the density is fixed by inertial confinement at 1000 kg/m^3. A 1998 study on Raman spectroscopy of supercritical water [84] concluded that the strength of hydrogen bonding does not vary much with temperature up to 500°C *at constant density*. In other words, the usual constant-pressure blueshifting and narrowing of the water equilibrium ν_{OH} spectrum with increasing temperature is attributed mainly to the density decrease. Very close to the critical point ($P_c = 22.1$ MPa, $T_c = 374$°C, $\rho_c = 322$ kg/m^3), the ν_{OH} spectrum is peaked at 3620 cm^{-1} with an FWHM of just 60 cm^{-1}. Conditions where a spectrum very much like H_2O^* is observed [84,85] ($P = 40$ MPa, $T = 300$°C) correspond to a greater density $\rho \approx 0.6$ g cm^{-3}. So we think it is fair to conclude that H_2O^* is characterized by a local environment with the strength of hydrogen bonding similar to water at a density of 0.6 g cm^{-3}.

The spectrum we associate with H_2O^* clearly represents an inhomogeneous distribution of structures but we have to wonder why all our results seem to be consistent with mainly two, rather than a multitude of water species [61]. The most satisfying resolution of this question is to think of H_2O^* as being water with a single broken donor hydrogen bond, as suggested by the Fayer group [81], along with a broad distribution of other structural parameters. For us, it is difficult to understand why there would be a sharp distinction between "intact" and "broken" hydrogen bonds in the dense liquid state. However, it should be noted that models that posit a sharp cut-off for hydrogen bonding, when used to calculate the spectrum of water with a single broken donor hydrogen bond [86,87], do output results that closely resemble our anti-Stokes spectra [61].

9.7 LONG-LIVED INTERFACIAL VIBRATIONS OF WATER PRODUCED BY LASER ABLATION

In the previous section, we described vibrational excitations of H_2O^* created by a vibrational energy burst, which had blueshifted OH-stretch spectra and a lifetime 0.8 ps that was about four times longer than in water. Here, we describe the vibrational excitations of water created by an even more violent process, ultrashort-laser ablation, which creates a massive disruption of the hydrogen-bonding network. In the ablation process, bulk water undergoes a phase explosion that causes a rapid decrease in mean density as it transitions to an ablation plume of tiny water droplets. Based on laser ablation simulations, Zhigilei et al. [88] describe the state present at ~200 ps as a "foamy transient structure of interconnected liquid clusters" [47].

Water ν_{OH} vibrations at liquid–vapor interfaces have been studied extensively by vibrational SFG spectroscopy [89–91] and water vibrations at cluster–vacuum interfaces have been studied by IR cluster predissociation spectroscopy [92]. In both types of measurements, a sharper ~20 cm^{-1} width feature is observed near 3700 cm^{-1}, which is attributed to free surface OH groups. Broader features are seen at nearby lower wavenumbers that are attributed to various ice-like and liquid-like structures and structures with broken hydrogen bonds [89,90,92]. The vibrational lifetime of the free OH group at the water/vapor interface is 0.8 ps [93], four times longer than the OH-stretch lifetime of bulk water.

In our experiments, ablation of a flowing jet of water occurred when 3310 cm^{-1} IR pulses focused to a ($1/e^2$) beam radius $r_0 = 150$ μm had an energy of 30 μJ. At threshold, a sudden increase in probe (green) light scattering, accompanied by large fluctuations, was observed along with a fine spray of droplets. The ablation measurements reported here [52] were made a bit above threshold, at $E_p = 43$ μJ, where the ablation process was more stable. The tip of a pipette connected to a vacuum was placed near the water surface so that the droplets could be sucked away during the 1 ms interval between laser shots.

At 3310 cm^{-1}, the absorption coefficient of water is quite large, $\alpha = 4660$ cm^{-1} [53], so the pump pulse strongly heated a slab of water ~1 μm deep. The peak energy density E_v in the pumped volume of water is given by $E_v = J_c\alpha$. The vibrational excitations become thermalized within a few picoseconds [34,94–96], and on this timescale, heating is both adiabatic and isochoric. The peak energy density at threshold is $E_v \approx 400$ J/cm^3, and in the ablation measurements presented here, $E_v = 570$ J/cm^3. As a point of reference, water at 25°C would be heated to 100°C with $E_v = 315$ J/cm^3, and water at 25°C would be totally vaporized with $E_v = 2575$ J/cm^3. Thus, the ablation regime corresponds to weakly superheated water, where the IR pulses have enough energy to convert up to ~10% of the heated liquid to vapor. The pressure jump in the water is ~0.2 GPa.

Figure 9.9 shows anti-Stokes data in the ν_{OH} region over a wide range of intensities. The sharper feature at 3310 cm^{-1} is again due to NLS [21]. Figures 9.9a through 9.9c compare results obtained with progressively higher-intensity pump pulses. Each time the pulse intensity is doubled, the signal level also doubles. With $\Delta T = 35$ K and $\Delta T = 70$ K (Figures 9.9a and 9.9b), the results are similar, and similar to what has been reported previously by us [58,60,61]. After 3–4 ps, the ν_{OH} excitation level has dropped below our detection limit. With data obtained above ablation threshold, as in Figure 9.9c, a new feature is observed. The anti-Stokes Raman spectrum of ν_{OH} excitations continues to narrow and blueshift until about 10 ps. Subsequently, the spectrum, peaked near 3600 cm^{-1}, remains unchanged out to at least 200 ps. Judging from relative intensities, this long-lived excited-state ν_{OH} population comprises several percent of the initial water excited states. In Figure 9.10, we compare the spectrum of this long-lived excited state with the spectrum of water.

There can be no doubt that the 200 ps anti-Stokes signals in Figure 9.9c are associated with long-lived ν_{OH} ($v = 1$) excitations of water. If this spectrum were an artifact due to a nonlinear interaction between the pump and probe pulses, it would be centered at 3310 cm^{-1} and would disappear at time delays past 2 ps. The spectrum also has the wrong shape and wrong intensity to result from anti-Stokes scattering from equilibrium water heated by the pump pulse. The peak energy density $E_v = 570$ J/cm^3 corresponds to ~25% of the water molecules at the surface being excited to $v = 1$, so

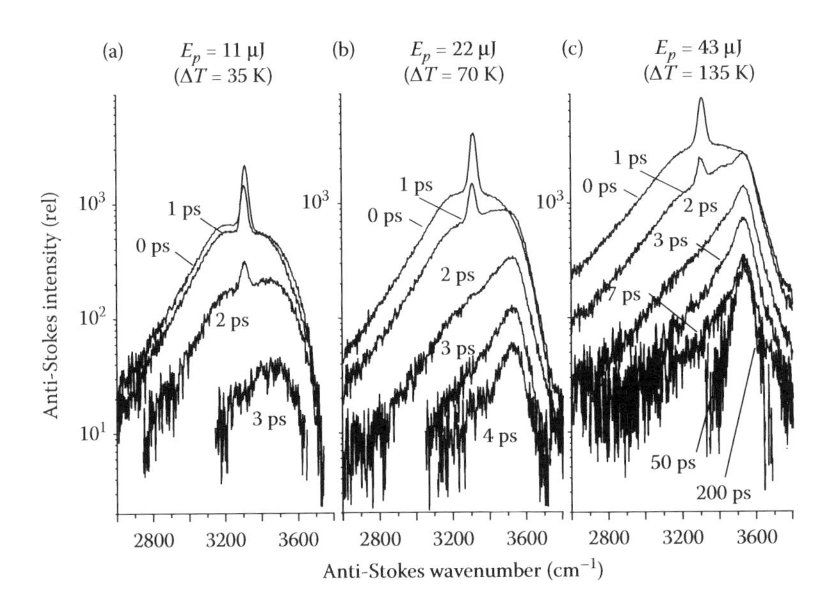

FIGURE 9.9 Transient anti-Stokes spectra with 3310 cm^{-1} pump pulses at the indicated energies. Water ν_{OH} excitations are created and detected. (a,b) Below the ~30 μJ ablation threshold, redshifted excitations decay more quickly, leading to a time-dependent spectral blueshift. After ~5 ps, all ν_{OH} excitations have decayed. (c) Above ablation threshold, a long-lived blueshifted spectrum is observed that persists beyond 200 ps. (Reproduced with permission from Wang Z, Pang Y, Dlott DD. Long-lived interfacial vibrations of water. *J. Phys. Chem. B* 110:20115–20117, Copyright 2006, American Chemical Society.)

the long-lived spectrum in Figure 9.9c, which is less intense than the $t = 0$ signal by a factor of 50, represents an occupation number $n \approx 5 \times 10^{-3}$. If this were due to ordinary heating processes, the temperature would have to be ~900 K, which is inconceivable.

We attribute the long-lived spectrum in Figure 9.9c to ν_{OH} excitations associated with liquid–vapor interfaces created by the rapid expansion. These interfaces have a complicated three-dimensional structure in contrast to the sedate two-dimensional liquid water–vapor interface. The long-lived

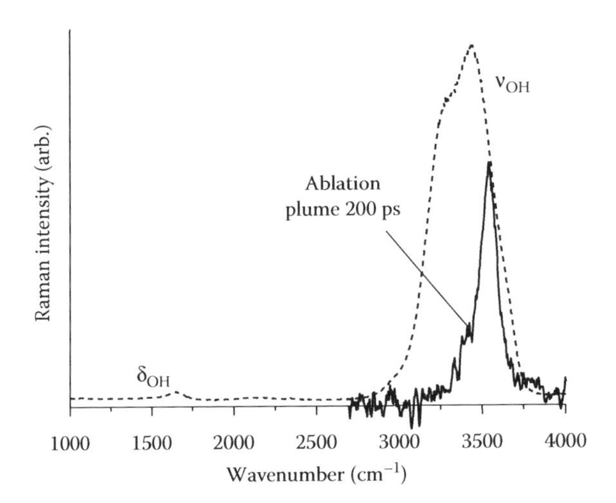

FIGURE 9.10 Comparison of the transient Anti-Stokes Raman spectrum of the long-lived water ν_{OH} excitations observed during laser ablation with the equilibrium Stokes Raman spectrum of water. (Reproduced with permission from Wang Z, Pang Y, Dlott DD. Long-lived interfacial vibrations of water. *J. Phys. Chem. B* 110:20115–20117, Copyright 2006, American Chemical Society.)

spectrum is too broad and too redshifted to result from water vapor or isolated water molecules. The association of this spectrum with interfacial water follows from its appearance during ablation, from the blueshift in Figure 9.10 that indicates broken hydrogen bonds, and from the weak coupling to the bath of bulk states implied by the enormous (>1000-fold) increase in the OH-stretch lifetime.

Features similar to the 100 cm^{-1} width transition at 3600 cm^{-1} shown in Figure 9.10 are frequently seen in SFG and water cluster studies. In the SFG literature of water–air interfaces [89,90], these are described as resulting from the bonded OH stretch of surface water molecules with one bonded OH and one free surface OH. In the water simulation literature [86,87,97], spectra have been computed for subensembles having different hydrogen bonding. Our spectra quite closely resemble water molecules having a nonhydrogen-bonded H atom plus a nonhydrogen-bonded O atom [86]. In other words, the transient spectrum in Figure 9.10 is best described as representing liquid-state water molecules with multiple broken hydrogen bonds. These bonds remain broken for much longer than in water, since within the ablation plume the rapid volume expansion impedes hydrogen bond reformation.

9.8 VIBRATIONAL ENERGY IN BIOLOGICAL BUILDING BLOCKS

In this section, we discuss a "bottom-up" approach to the problem of vibrational energy in proteins via a detailed study of simpler building blocks. The emphasis is on the power of IR-Raman to detect the entire VC process that results from pumping a higher-energy vibrational fundamental, a CH-stretch transition, and the use of the aqueous medium [55,63] as a molecular thermometer [3,5,98]. We investigated vibrational energy flow of three biologically relevant molecules in aqueous (D_2O) solution, glycine (GLY)—the simplest amino acid in the form d_3-glycine zwitterion, d_1-*N*-methyl acetamide (NMA)—one of the simplest molecules having a peptide bond, and benzoate (BZ)—a model for peptide aromatic side chains. The results are understood in the context of the vibrational roadmap presented in Section 9.5.

Most prior studies of vibrational energy in proteins have utilized a "top-down" approach to energy dissipation mechanisms, which might be relevant to biological function such as enzyme catalysis. Well-known examples include myoglobin, hemoglobin, or cytochrome *c* where heme was electronically excited [98–105]. Owing to ultrarapid internal conversion, the electronic excitations were converted to heme vibrational energy in ~5 ps, and the hot heme subsequently cooled by 20–40 ps energy transfer through the protein into the aqueous medium [98–102,106]. Heme cooling has been monitored using transient gratings [107], resonance anti-Stokes Raman measurements of the heme [100–103,108,109], ultraviolet resonance Raman measurements of the globin [110–112], and IR absorption of the aqueous medium used as a molecular thermometer [98]. Experimental [111] and theoretical [106,113–115] studies have suggested that energy can be funneled from a hot porphyrin, through its side chains, into specific parts of the protein. "Bottom-up" approaches to protein vibrational energy have used IR pump-probe techniques to study the VR of small ligands such as CO bound to the active sites of heme proteins [116–119], or the peptide backbone [120] itself via the amide I mode [120–123], which is predominantly a CO stretching excitation.

The molecules studied here, GLY, NMA, and BZ, have 10, 12, and 14 atoms, giving 24, 30, and 36 normal modes of vibration, but in the most favorable case (NMA), we have the sensitivity to probe only nine of them. The aqueous molecular thermometer here can be used to determine whether the observed vibrational energy is representative or nonrepresentative of the total molecular energy. A representative molecule would be a more useful probe of protein vibrational energy.

The Stokes spectrum of D_2O was used as the molecular thermometer. As water temperature is increased from T_i to T_f, the Raman difference spectrum in the OH-stretch or OD-stretch region evidences a characteristic bipolar shape, as in H_2O (cf. Figure 9.6b). In D_2O, the most striking feature in the difference spectrum is the dip near 2330 cm^{-1} whose amplitude increases with increasing temperature. This response originates from the blueshift due to weakened hydrogen bonding at higher temperature, as in H_2O (cf. Figure 9.6a). The molecular thermometer employs

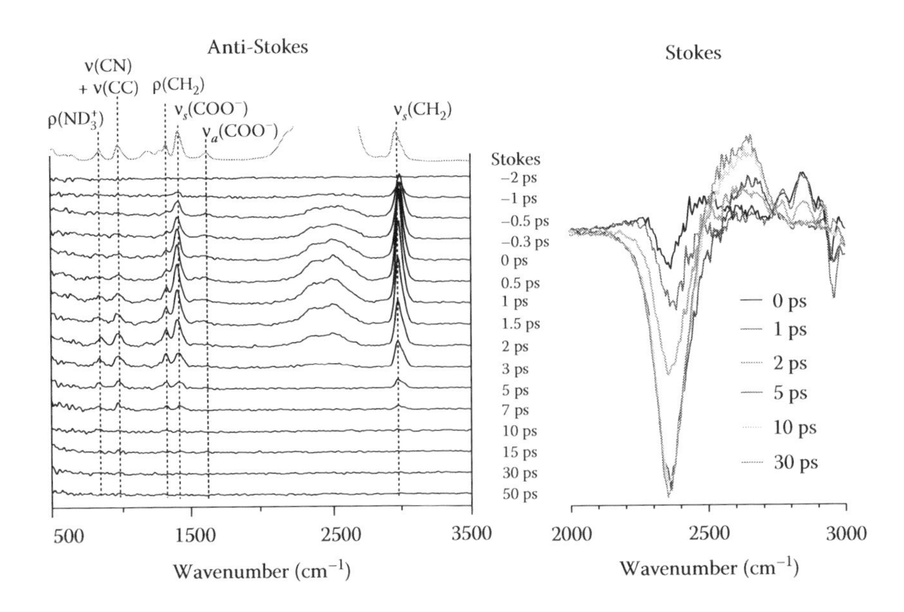

FIGURE 9.11 Transient Raman data for d_3-glycine zwitterion in D_2O (GLY) with $\nu_s(CH_2)$ excitation. The Stokes spectrum with vibrational assignments (dotted curve) is plotted at the top for reference. The parent and five daughter vibrations are seen in the anti-Stokes region (left) after CH-stretch excitation. The response of the D_2O molecular thermometer is shown at right. (Adapted with permission from Fang Y et al. Vibrational energy dynamics of glycine, *N*-methyl acetamide and benzoate anion in aqueous (D_2O) solution. *J. Phys. Chem. A* 113:75–84, Copyright 2009, American Chemical Society.)

temperature shifting of ground vibrational state transitions. At shorter times, there may also be vibrationally excited states created by the IR pump pulses, so that ground-state depletion and excited-state absorption effects may also be present in the Stokes spectra [63]. In that case, the aqueous Stokes spectrum will not look exactly like thermalized reference spectra. Previously [70], we showed how a singular-value decomposition (SVD) analysis could be used to separate the transient Stokes spectra into an excited-state part that we do not use, and a thermalized ground-state (molecular thermometer) part. We did not need to use this SVD method here. There was little overlap between the parent CH-stretch pumped by the IR pulses and the D_2O stretching spectrum, so the pump pulses produced negligible amounts of vibrational excitation in the D_2O molecular thermometer.

Figure 9.11 summarizes the GLY response following IR pumping of the methylene CH-stretch. At the top of the figure is a reference Stokes spectrum with the D_2O stretch transition truncated. In the anti-Stokes transient spectra in Figure 9.11, we see the parent $\nu_s(CH_2)$ and five of the daughter vibrations (in order of descending wavenumber), $\nu_a(COO^-)$, $\nu_s(COO^-)$, $\rho(CH_2)$, $\nu(CN) + \nu(CC)$, and $\rho(ND_3^+)$. Figure 9.11 also shows Stokes transients in the ν_{OD} region of the D_2O medium. By combining transient Stokes and anti-Stokes spectra (not shown), we found that the pump pulse excited 2.1% of the GLY solute and 0.2% of the D_2O solvent.

The d_1-NMA data are summarized in Figures 9.12 and 9.13. In the anti-Stokes data in Figure 9.12, we see the parent $\nu_s(CH_3)$ along with eight daughter vibrations, in order of descending wavenumber, amide I′, amide II′, CCH_3 *ab*, NCH_3 *ab*, $\nu_s(NC)$, amide III′, skeletal deformation, and amide IV′. Prime designates amide deuteration. The vibrational assignments [122,124,125] are given in Table 2 of Ref. [71]. Also, in Figure 9.12 are the D_2O molecular thermometer data. The anti-Stokes transients are shown in Figure 9.13. In Figures 9.13a through 9.13d, the transients are plotted with their absolute populations and the smooth curves are fits to the three-stage model with parameters listed in Ref. [71]. In Figures 9.13e and 9.13f, the transients were all normalized to the same maximum intensity to facility comparisons of the rise times. Notice how the rise of the midrange *M* vibrations

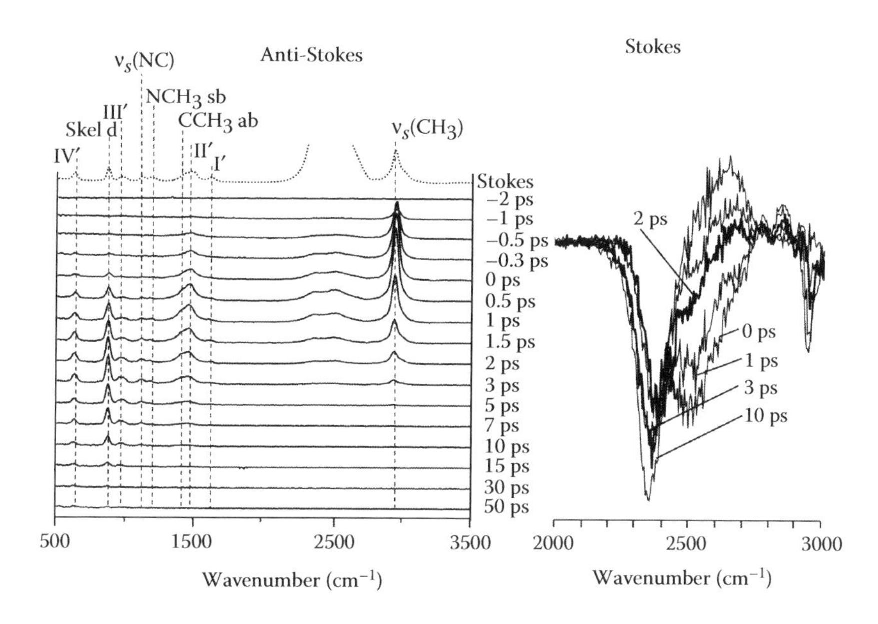

FIGURE 9.12 Transient Raman data for d_1-*N*-methylacetamide (NMA) in D_2O with $\nu_s(CH_3)$ excitation. The Stokes spectrum with vibrational assignments (dotted curve) is plotted at the top for reference. The parent and eight daughter vibrations are seen in the anti-Stokes region (left). The response of the D_2O molecular thermometer is shown at right. (Adapted with permission from Fang Y et al. Vibrational energy dynamics of glycine, *N*-methyl acetamide and benzoate anion in aqueous (D_2O) solution. *J. Phys. Chem. A* 113:75–84, Copyright 2009, American Chemical Society.)

tracks the decay of the parent excitation in Figure 9.13e and, by comparing Figures 9.13e and 9.13f, how the rise of the *M* vibrations precedes the rise of the lower-energy *L* vibrations. The parent decay in Figure 9.13a gives $\tau_{IVR} = 1.2$ (±0.2) ps. The observed midrange daughter transients in Figure 9.13e gave $\tau_{ML} = 1.7$ (±0.2) ps. The amide II′ vibration evidences a faster rise than the other *M* vibrations indicating a degree of coherent coupling with the parent. The lower-energy transients in Figure 9.13f yielded $\tau_{LG} = 2.8$ (±0.3) ps. The smooth curves used to fit the data were calculated using the model described in Section 9.5 with parameters listed in Ref. [71].

The BZ data are summarized in Figure 9.14. In the anti-Stokes data, we observe the parent ν_s(CH) along with six daughter vibrations, in descending order ν_s(CC), $\nu_s(COO^-)$, ρ(CH), ν_s(phenyl), δ_{oop}(CH), and ρ(CCC). The parent decay gave $\tau_{IVR} = 1.0$ (±0.2) ps. The midrange vibrations gave $\tau_{ML} = 2.5$(±0.2) ps, and the lower-energy vibrations gave $\tau_{LG} = 2.7$ (±0.3) ps. The most dramatic observation from the BZ data is the large difference in timescales for relaxation of phenyl vibrational excitations in an aqueous medium versus neat benzene [126], whose VR is described in detail in the next section. The BZ lifetimes were all in the 1–3 ps time range, whereas in liquid benzene observed vibrational lifetimes were in the 8–100 ps time range [126].

Figure 9.15 shows the molecular thermometer time dependences. If the thermometer response resulted from the exponential decay of a single level, then the thermometer rise would be a rising exponential. If the response resulted from the accumulation of many individual level decay processes, each an exponential in time, then the thermometer rise should approach an error function [11]. We fit the thermometer rise with single-exponential functions, since the quality of the data does not justify a more sophisticated treatment. Figure 9.15a is the result for pure D_2O. The intrinsic response of the D_2O molecular thermometer, measured by pumping ν(OD) of pure D_2O at 2950 cm^{-1} with an excited-state concentration of 0.7%, is characterized by 1.8 ps time constant. This is just about what would be expected based on a ν_{OD} lifetime [81] of 1.4 ps and a D_2O thermalization time constant [35,36,55,95,127] of ~0.5 ps. When solutes were added, the fraction of ν(OD) excited by the

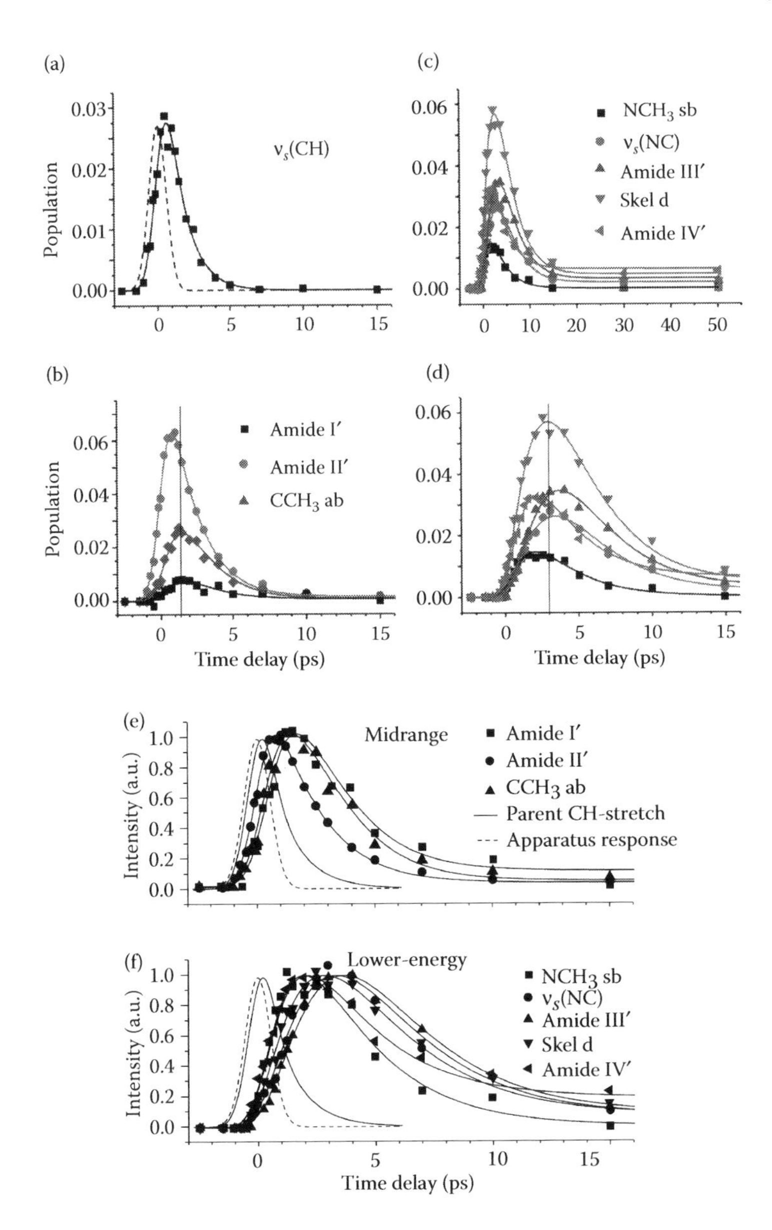

FIGURE 9.13 Anti-Stokes transients for d_1-*N*-methylacetamide (NMA) in D_2O with $\nu_s(CH_2)$ excitation. The smooth curves are the fit to the three-stage model described in Section 9.5. The dashed curve in (a) is the apparatus time response. The vertical lines in (b) and (d) are visual guides. (d) shows the data in (c) on an expanded timescale. (e,f) The midrange *M* transients and lower-energy *L* transients with intensities normalized to facilitate comparisons. (Adapted with permission from Fang Y et al. Vibrational energy dynamics of glycine, *N*-methyl acetamide and benzoate anion in aqueous (D_2O) solution. *J. Phys. Chem. A* 113:75–84, Copyright 2009, American Chemical Society.)

IR pulse decreased due to the competition with solute absorption. The ν(OD) excitation fraction and the 1.8 ps time dependence for pure D_2O were used to subtract the part of the thermometer response due to direct ν(OD) pumping. The resulting molecular thermometer responses for GLY, NMA, and BZ due to solute pumping alone are shown in Figures 9.15b through 9.15d. The fastest VC denoted by the thermometer rise, 4.9 ps, was observed with NMA. With GLY and BZ, the time constants for VC were 7.2 and 8.0 ps, respectively.

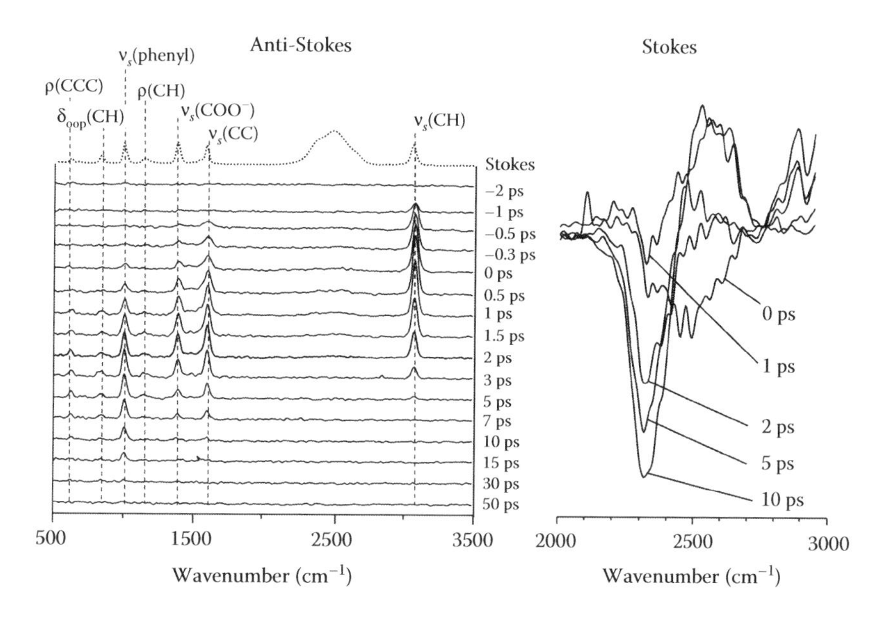

FIGURE 9.14 Transient Raman data for benzoate anion (BZ) in D_2O with ν_s(CH) excitation. The Stokes spectrum with vibrational assignments (dotted curve) is plotted at the top for reference. The parent and five daughter vibrations are seen in the anti-Stokes region (left). The response of the D_2O molecular thermometer is shown at right. (Adapted with permission from Fang Y et al. Vibrational energy dynamics of glycine, *N*-methyl acetamide and benzoate anion in aqueous (D_2O) solution. *J. Phys. Chem. A* 113:75–84, Copyright 2009, American Chemical Society.)

Using the anti-Stokes data in Figures 9.11, 9.12, and 9.14, we determined the time dependence of the observed vibrational energy using the formula

$$E_{\text{vib}}^{\text{obs}}(t) = \sum_{i=1}^{\#\text{obs}} h\nu_i n_i(t). \tag{9.6}$$

The results were decaying curves for each solute that could be reasonably fit to exponential decays. The exponential fits to E_{obs} are shown in Figures 9.15b through 9.15d. The time constants for energy loss in the observed vibrations were 0.8 ps for D_2O, 2.6 ps for glycine, 5.1 ps for NMA, and 3.6 ps for sodium BZ.

The overall time constant for thermalization of the parent vibration energy is ~5 ps for NMA and ~8 ps for GLY and BZ. Thus, informed speculation would suggest that energy deposited in the backbone or in flexible sidechains of polypeptides would be thermalized in ~5 ps, and even energy deposited in rigid sidechain structures would thermalize within <10 ps.

Dissipation of the observed energy in the strongly Raman-active vibrations of NMA is representative of the overall thermalization process monitored by the molecular thermometer (cf. Table 1 of Ref. [71]), in the sense that the E_{obs} and the molecular thermometer have the same exponential time constant. However, the GLY and BZ observed vibrational energy loss is about three time faster than the thermalization. Thus, the strongly Raman-active NMA vibrations do an excellent job of representing energy flow through the solute, whereas the GLY and BZ vibrations do not. The GLY and BZ molecular thermometer response is slower than the observed vibrational energy. This means there must be at least one unobserved state that is releasing energy to the surroundings more slowly than the Raman-active states we observe. The representative versus nonrepresentative nature of the observed NMA vibrations versus GLY and BZ might result from the nature of specific vibrational pathways, but there might be a simpler statistical explanation. NMA is the molecule where

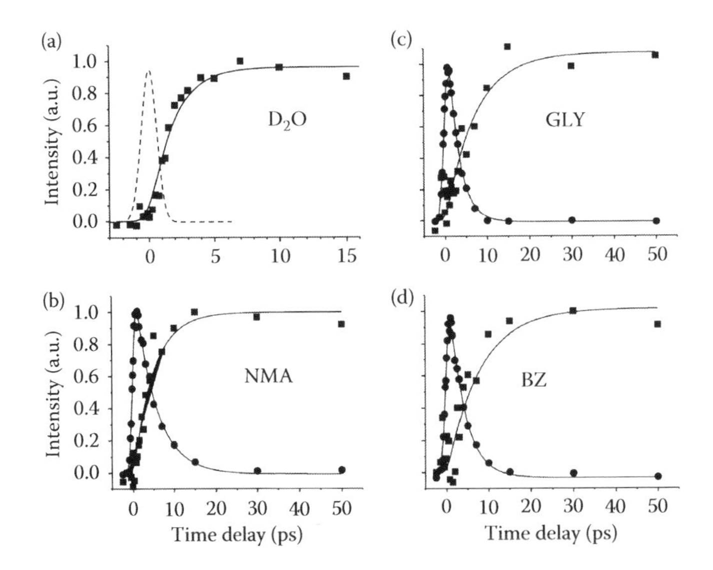

FIGURE 9.15 Molecular thermometer response. The molecular thermometer uses Stokes spectroscopy of the OD-stretch region of D_2O. Hydrogen bond weakening causes the transition to blueshift. (a) Response of pure D_2O pumped at 2950 cm^{-1} can be fit to an exponential with a 1.8 ps time constant. The dashed curve is the laser apparatus time response. (b–d) Thermometer data for NMA, GLY, and BZ after CH-stretch pumping (squares). The circles represent the total amount of vibrational energy observed via anti-Stokes probing of the solutes' vibrational transitions. In NMA, the observed energy is representative of the solute cooling process detected by the thermometer response. In GLY and BZ, the observed energy decays faster than the thermometer heats up, indicating energy is stored in a reservoir of unobserved vibrations. The observed energy is not representative of the solute cooling process for GLY and BZ. (Reproduced with permission from Fang Y et al. Vibrational energy dynamics of glycine, *N*-methyl acetamide and benzoate anion in aqueous (D_2O) solution. *J. Phys. Chem. A* 113:75–84, Copyright 2009, American Chemical Society.)

we observe the largest number of vibrations, nine, compared to five for GLY and six for BZ. The larger the number of observed vibrations, the more likely the observed vibrational energy will be representative of the entire molecule.

9.9 VIBRATIONAL ENERGY CALORIMETRY OF BENZENE AND PERDEUTEROBENZENE

In these studies, we probed vibrational energy in either benzene or d_6-benzene [2]. We also spiked the benzene with a CCl_4 molecular thermometer. The CCl_4 was present at a volume fraction of ~15%, but we showed by direct comparison to neat benzene that even this quantity of CCl_4 has negligible effect on the VR of benzene. The CCl_4 is apparently a nearly inert spectator, and it has only a small effect on the fluctuating forces acting on vibrationally excited benzene.

Benzene has a high-molecular-symmetry D_{6h} and a skeletal framework more rigid than the molecules discussed above. Raman intensity is concentrated in a rather small number of transitions (six or seven). Unlike previous studies on [30] $CHCl_3$ or [51] CH_3NO_2 where most or even all vibrations could be probed, in benzene, much of the vibrational energy will be in Raman-inactive modes we cannot observe. As in the biological building block studies in the preceding section, one issue we will address is to what extent is the observable vibrational energy representative of the total energy. In order to do this, we have made an advance in the molecular thermometer method to create an "ultrafast Raman calorimetry" method. We have a known energy input from the laser pulse, and we can total up all the observed vibrational energy and bath energy from anti-Stokes measurements of benzene and CCl_4. Based on energy conservation, what remains represents the time-dependent aggregate invisible vibrational energy.

Benzene vibrational energy has previously been studied in isolated molecules, liquids, and low-temperature crystals. In isolated molecules, the CH-stretch fundamentals near 3050 cm^{-1} have VR that is so slow that the states decay by IR emission [128]. Some of the higher CH-stretch overtones [129–133] of isolated molecules do have facile IVR because the density of states is so much greater than for the fundamental. Prior studies of low-temperature crystals [130,134,135] involved indirect methods that probed the IR or Raman linewidths in the frequency or time domains. Vibrational linewidths in ambient liquids are ordinarily dominated by pure dephasing processes [136,137] and sometimes inhomogeneous broadening [68], so T_1 cannot be confirmed from linewidth measurements. However, in low-temperature isotopically pure crystals [138], T_1 processes are believed to be dominant [139] so that T_1 can be determined from the Raman linewidth. In the liquid state, Fendt and coworkers [140] studied benzene with time-resolved anti-Stokes Raman. However, those studies probed only the parent relaxation process, and in fact the parent CH-stretch lifetime was incorrectly determined, as the signals detected appear to have originated from a coherence artifact due to NLS [126]. Iwaki and coworkers [126] measured $T_1 = 8$ ps for the parent CH-stretch and also observed signals originating from daughter excitations at 1584 cm^{-1}, 991 cm^{-1}, and 606 cm^{-1}. CCl_4 molecular thermometer data were noisy but suggested an overall time constant for VC of ~80 ps.

In our experiments, the IR pulses were tuned to either a CH-stretch absorption near 3050 cm^{-1} or a CD-stretch absorption near 2280 cm^{-1}. The most intense transition near the pump pulse has been assigned to ν_{12}. In the benzene Raman spectrum, six transitions dominate: ν_1, ν_2, ν_{11}, ν_{16}, ν_{17}, and ν_{18}. ν_1 and ν_2 are singly degenerate and the other four are doubly degenerate, so a total of 10 modes were observed.

Figure 9.16 shows a time series of anti-Stokes spectra following IR pumping at 3063 cm^{-1} of benzene with CCl_4, and Figure 9.17 shows the time dependence of the vibrational populations

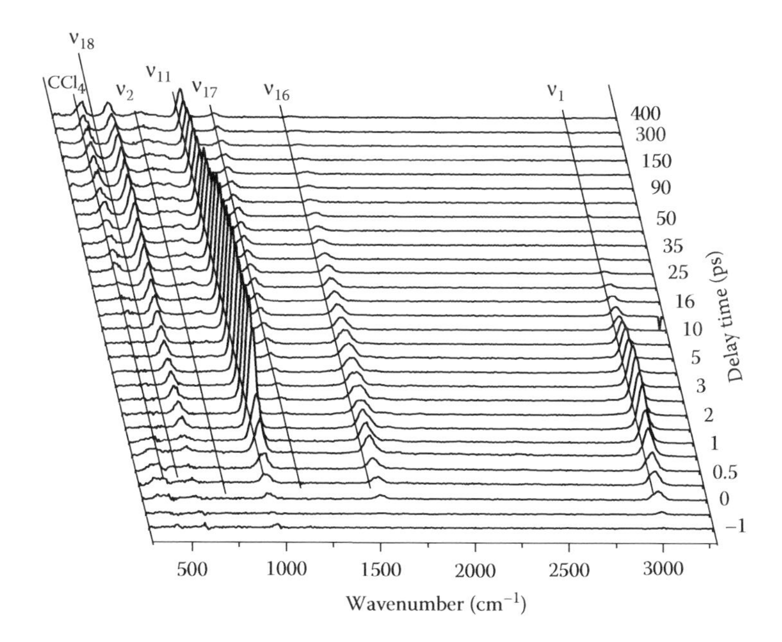

FIGURE 9.16 Time-series of anti-Stokes spectra after IR pumping at 3053 cm^{-1} from a mixture of benzene + 17% CCl_4 (v/v). Note the delay time axis is not a linear scale. (Reproduced with permission from Seong N-H, Fang Y, Dlott DD. Vibrational energy dynamics of normal and deuterated liquid benzene. *J. Phys. Chem. A* 113:1445–1452, Copyright 2009, American Chemical Society.)

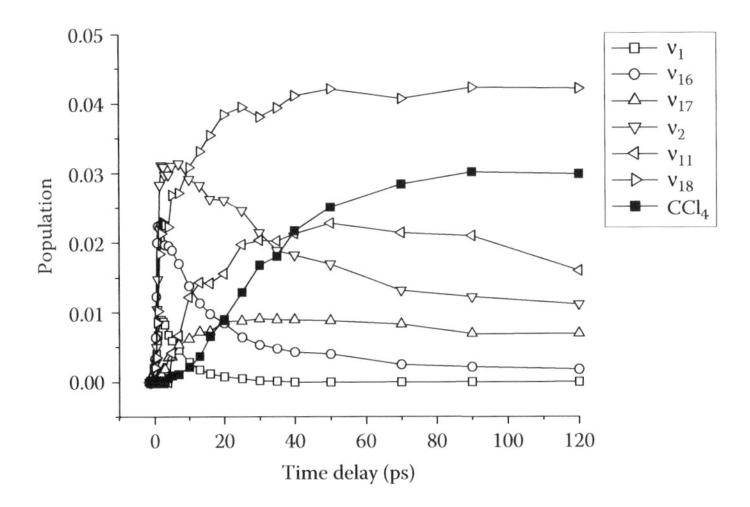

FIGURE 9.17 Population transients for benzene with CCl_4 determined using time-dependent Stokes and anti-Stokes spectra. (Reproduced with permission from Seong N-H, Fang Y, Dlott DD. Vibrational energy dynamics of normal and deuterated liquid benzene. *J. Phys. Chem. A* 113:1445–1452, Copyright 2009, American Chemical Society.)

(occupation numbers) of the six observed benzene transitions and the CCl_4 extracted from the data in Figure 9.16. CCl_4 has a trio of lower-frequency Raman-active vibrations and all three act equally well as molecular thermometers. We used the 459 cm^{-1} mode for our thermometer. For doubly degenerate states, the occupation number is the aggregate population. All the vibrations except ν_{11} and ν_{17} evidence an instantaneous component on the rising edge. This instantaneous component indicates that the parent "bright" state, which is excited by tuning the IR pulse into ν_{12} consists of an admixture of ν_{12} with ν_1, ν_2, ν_{16}, ν_{18}, and possibly other states we could not observe. We were able to fit the benzene data quite well using the three-stage model described in Section 9.5. The parameters used to fit the data are given in Table 9.1, taken from Ref. [2].

After the VC process was complete, the sample came to equilibrium at a final temperature T_f, where T_f represents an average over a spatially inhomogeneous region of the liquid pumped by the IR pulses [51]. Figure 9.18 shows how we determine T_f for our calorimetry measurements. The figure compares benzene + CCl_4 anti-Stokes spectra obtained at negative delay (probe precedes pump) where the sample is at ambient temperature, and a spectrum obtained at a longer delay time after the parent excitation has thermalized. The accuracy of determining T_f for an anti-Stokes transitions is greatest for higher-frequency transitions with larger Raman cross-sections [14,141]. We were able to determine T_f for ν_{18}, ν_2, and ν_{17} with an estimated error of 2–4 K, as shown in Figure 9.18, and when the results for each vibration were averaged, we obtained $\Delta T = 40$ K.

With d_6-benzene, we observed the same transitions, ν_1, ν_2, ν_{11}, ν_{16}, ν_{17}, ν_{18}, as in benzene and we saw an additional CD-stretch transition ν_{15}. Since five of these seven modes are doubly degenerate, we observed a total of 12 vibrations. Figure 9.19 shows the time-resolved anti-Stokes data in d_6-benzene, and Figure 9.20 shows the time-dependent population transients. The parameters used to fit the data are given in Table 9.1, taken from Ref. [2]. When we performed the temperature jump analysis for d_6-benzene using the method described in Figure 9.18, we obtained $\Delta T = 10$ K. The T-jump was smaller than with benzene because the CD-stretch absorption coefficient and the IR laser energy were both smaller.

In the benzene-CCl_4 experiments, we observed the part of the benzene vibrational energy $E_{obs}(t)$ in the strongly Raman-active vibrations. We also knew the total energy input to the system, which can be computed either from the IR pulse properties and sample absorption coefficient, or more conveniently from ΔT and the solution heat capacity. The CCl_4 molecular thermometer measures

TABLE 9.1
Parameters for Benzene and d_6-Benzene Vibrational Cooling

Benzene

Wavenumber (cm^{-1})	Assignment	Lifetime T_1 (ps)	Source of Excitation: IR Pump Pulse[a]	Parent	*M* to *L*
3063	ν_1: ν_s (CH)	6.2	17%		
1589	ν_{16}: ν (CC)/(992 + 606)	20	28%		
1176	ν_{17}: δ_{ip} (CH)	146	0	42%	8%
992	ν_2: ring breathing	55	34%	39%	0
850	ν_{11}: δ_{oop} (CH)	125	0	0	44%
606	ν_{18}: δ_{ip} (CCC)	300	21%	19%	48%

d_6-Benzene

Wavenumber (cm^{-1})	Assignment	Lifetime T_1 (ps)	Source of Excitation: IR Pump Pulse[b]	Parent	X-State[c]
2282	ν_1: ν_s (CD)	6.4	27%		
2254	ν_{15}: ν_{as} (CD)	4.5	24%		
1551	ν_{16}: ν (CC)	25	0	22%	0
937	ν_2: ring breathing	26	48%	0	0
868	ν_{17}: δ_{ip} (CD)	53	0	13%	19%
653	ν_{11}: δ_{oop} (CD)	137	0	36%	26%
578	ν_{18}: δ_{ip} (CCC)	91	0	29%	55%

Source: Reprinted with permission form Seong N-H, Fang Y, Dlott DD. Vibrational energy dynamics of normal and deuterated liquid benzene. *J. Phys. Chem.* A 113:1445–1452. Copyright 2009, American Chemical Society.

[a] The absolute fraction of CH-stretch excitations generated by the IR pump pulse is 1.3% in benzene.

[b] The absolute fraction of CD-stretch excitations generated by the IR pump pulse is 0.5% in d_6-benzene.

[c] The lifetime of state X is 80 ps.

the rate that energy was dissipated into the bath [5]. The lower-frequency continuum of collective bath states of the solution is strongly coupled to the lowest-frequency E-symmetry CCl_4 vibrations [5], so that CCl_4 becomes excited within a few picoseconds, allowing the molecular thermometer to respond quickly to a build-up of bath excitation. By properly normalizing the total energy, the observed energy, and the bath energy, we can determine by energy conservation how much energy resides in the invisible benzene vibrations. This ultrafast Raman calorimetry determination of the invisible vibrational energy $E_{invis}(t)$ is an obvious idea, but until now we did not have good enough data to implement it.

When energy is input by an IR pulse to a solution of benzene and CCl_4 at ambient temperature, after thermalization a temperature jump ΔT is created, which here was 40 K for benzene and or 10 K for d_6-benzene, respectively. For a δ-function excitation, the total energy increase corresponding to a particular value of ΔT would be [2]

$$\Delta E_{tot}(t) = 0 \quad t < 0 \tag{9.7}$$

$$\Delta E_{tot}(t) = C_{tot}\Delta T \quad t \geq 0,$$

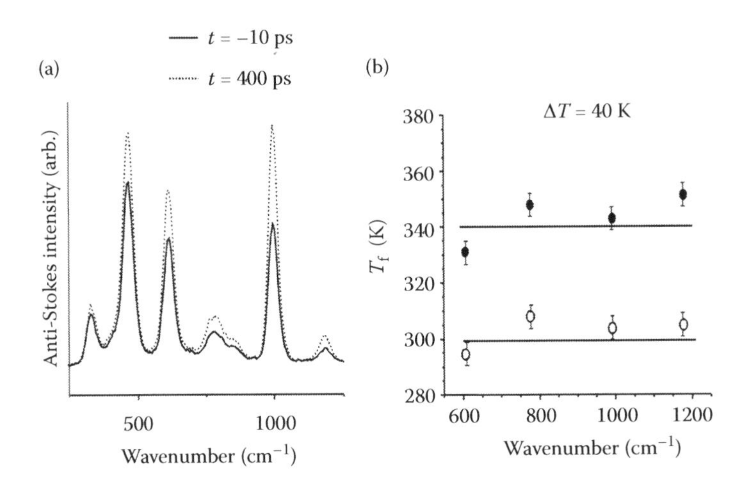

FIGURE 9.18 (a) Anti-Stokes spectra of benzene + 17% CCl_4 (v/v) at −10 ps prior to the IR pump pulse where the solution is in equilibrium at ambient temperature, and at 400 ps after the IR pump pulse when the solution has thermalized. (b) Determination of ΔT. Solid circle is calculated temperature using 400 ps data, and open circle is calculated temperature using −10 ps data as reference. (Reproduced with permission from Seong N-H, Fang Y, Dlott DD. Vibrational energy dynamics of normal and deuterated liquid benzene. *J. Phys. Chem. A* 113:1445–1452, Copyright 2009, American Chemical Society.)

where C_{tot} is the heat capacity of the solution. We used tabulated heat capacity data [142] for benzene and CCl_4, and given the quality of anti-Stokes data, we felt it was adequate to assume that $C_{tot} = C_{benzene} + C_{CCl4}$, and that C_{tot} could be treated as constant despite the temperature jump and the associated pressure jump with small volume expansion. Let $P(t)$ be the laser apparatus function, a Gaussian with FWHM of 1.4 ps in this case normalized so $\int P(t)\,dt = 1$. Then, the time-dependent total energy increase with this finite-duration pulse would be

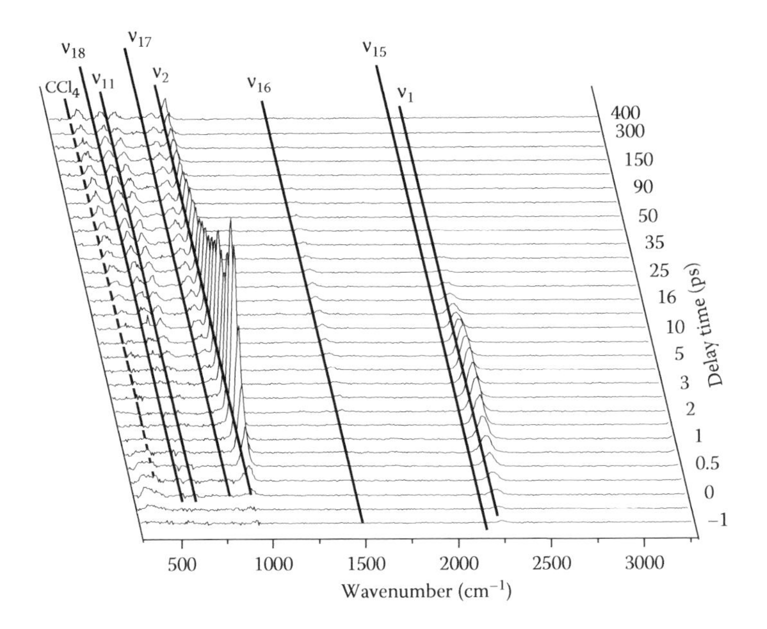

FIGURE 9.19 (Left) Anti-Stokes spectra of d_6-benzene + 17% CCl_4 (v/v) after pumping a CD-stretch transition near 2280 cm^{-1}. Note the delay time axis is not a linear scale.

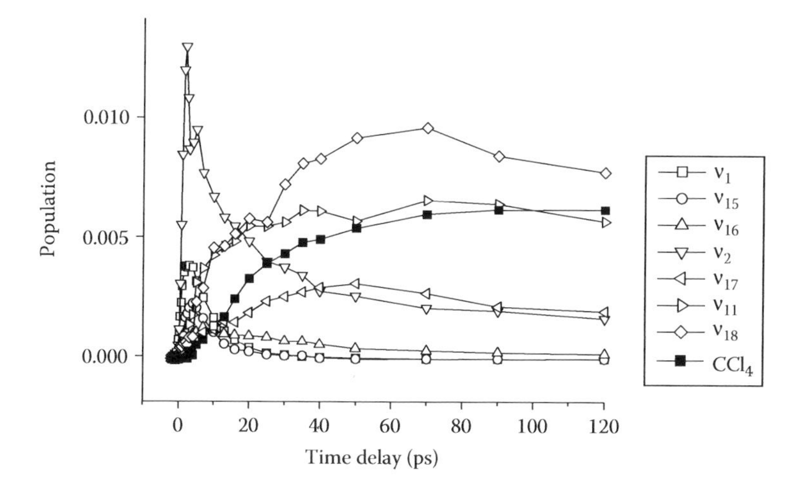

FIGURE 9.20 Population transients for d_6-benzene with CCl_4 determined using time-dependent Stokes and anti-Stokes spectra.

$$\Delta E_{\text{tot}}(t) = \int_{-\infty}^{t} P(t')\Delta E_{\text{tot}}(t')\,\text{d}t'. \tag{9.8}$$

The solution is now taken to consist of two parts, a "system" of benzene vibrations and a "bath" of everything else: the lower-energy continuum of collective states of the two-component solution plus CCl_4 vibrations. Then the heat capacity of the bath C_{bath} can be written as [2]

$$C_{\text{bath}} = C_{\text{tot}} - \sum_{i=1}^{n}\left(\frac{h\nu_i}{k_B T}\right)\frac{\exp(-h\nu_i/k_B T)}{\left[1-\exp(-h\nu_i/k_B T)\right]^2}, \tag{9.9}$$

where the sum is over all benzene vibrations ($n = 30$). To determine C_{bath}, once again avoiding the complications of a temperature-dependent heat capacity, Equation 9.9 was evaluated at the average temperature $T_i + \Delta T/2$. The benzene vibrational frequencies ν_i were taken from the literature [143].

The normalized time response of the CCl_4 molecular thermometer is denoted $T_{\text{th}}(t)$, where $T_{\text{th}}(t)$ is zero before the IR pulse and unity at long time. Then, the time-dependent energy in the bath is given by [2]

$$E_{\text{bath}}(t) = T_{\text{th}}(t)C_{\text{bath}}\Delta T, \tag{9.10}$$

and the invisible vibrational energy is given by

$$E_{\text{invis}}(t) = E_{\text{tot}}(t) - E_{\text{obs}}(t) - E_{\text{bath}}(t). \tag{9.11}$$

In Figures 9.21 and 9.22, we have plotted $E_{\text{tot}}(t)$, $E_{\text{obs}}(t)$, $E_{\text{bath}}(t)$, and $E_{\text{invis}}(t)$ for benzene and d_6-benzene. In Figure 9.21, the build-up of bath energy monitored by CCl_4 does not commence until a time delay of ~5 ps, which is consistent with the 6 ps parent decay of benzene resulting from an IVR process that does not involve significant dissipation into the bath. We do not believe this time delay is caused by sluggish response of the molecular thermometer, since a significantly faster CCl_4 response was observed previously in a study of acetonitrile [7]. The half-life for the bath build-up

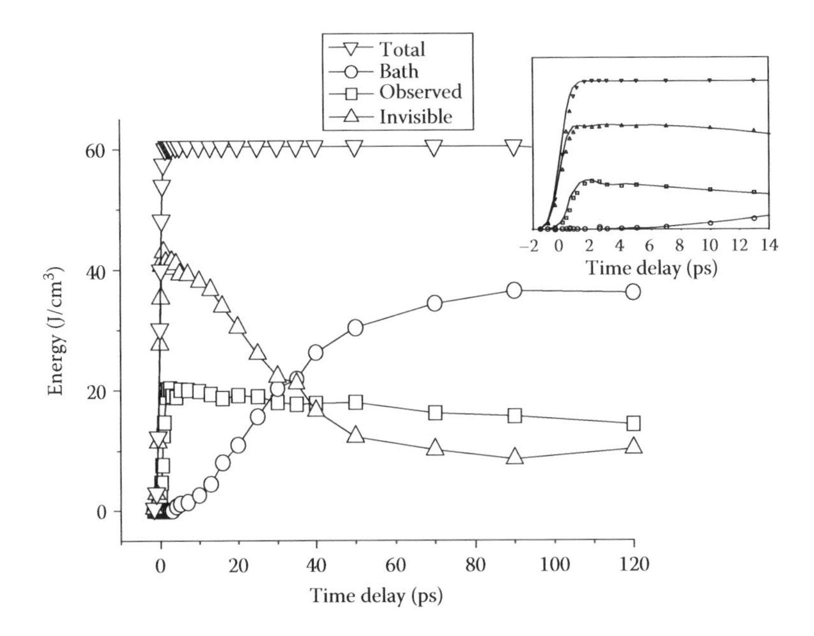

FIGURE 9.21 Ultrafast Raman calorimetry of benzene + 17% (v/v) CCl_4. The total energy input rises with the apparatus time response. The observed vibrational energy is the sum of the energy in 10 Raman-active vibrations probed by Raman. The bath, consisting of all excitations except the benzene vibrations, is probed using the CCl_4 molecular thermometer. The bath energy is determined using the long-time tail and the solution heat capacity. The invisible energy is the unobserved energy in benzene vibrations, determined using energy conservation. The *inset* shows the different rise times for E_{obs}, E_{invis}, and E_{bath}. (Reproduced with permission from Seong N-H, Fang Y, Dlott DD. Vibrational energy dynamics of normal and deuterated liquid benzene. *J. Phys. Chem. A* 113:1445–1452, Copyright 2009, American Chemical Society.)

is 30 ps and thermalization is essentially complete by 100 ps. Although the ν_{11} and ν_{18} vibrations have lifetimes in excess of 100 ps, they are lower-energy excitations, and after 100 ps their aggregate contributions to the total dissipated energy are small. These measurements of the benzene VC process should be considered more accurate than our previous study [126] due to improved instrumentation and methodology.

In Figure 9.21 (*inset*), the rising edge of $E_{invis}(t)$ is clearly faster than the rise of $E_{obs}(t)$, and at a shorter delay time of 2–3 ps immediately after the IR pulse stops pumping the benzene, the ratio of energies $E_{invis}/E_{obs} = 2.1$. The IR-active vibrations excited by the laser are invisible vibrations to a Raman probe, so at shorter times the IR pulse is directly pumping more energy into the invisible vibrations. By 3 ps, about twice as much energy has been pumped into the invisible vibrations as into the observed vibrations. This might be expected since the IR pulse is pumping IR-active vibrations near 3050 cm^{-1} such as ν_{12}, while we are probing IR-inactive vibrations such as ν_1 and ν_2. In other words, the coherent admixture of normal modes that constitutes the bright state has about twice as much of the IR-active mode character than IR-inactive mode character. The decay rate of $E_{invis}(t)$ seen in Figure 9.21 is perhaps 50% faster than $E_{obs}(t)$.

In the benzene experiments, the number of observed vibrations was 10 and the total number of vibrations was 30. We observed 33% of the vibrations, and the maximum value of E_{obs}/E_{tot} (from Figure 9.12) was also 33%. So, in benzene, the amount of vibrational energy per mode is the same in the strongly Raman-active "observed" vibrations as in the "invisible" vibrations.

Figure 9.22 shows d_6-benzene vibrational energy after CD-stretch pumping. In Figure 9.22, the rising edge of $E_{invis}(t)$ is again clearly faster than the rise of $E_{obs}(t)$, and at a shorter delay time of 2 ps, the ratio of energies $E_{invis}/E_{obs} = 2.1$. That indicates that, as in benzene, the IR pulse is directly pumping twice as much energy into the invisible vibrations than the observed vibrations. There is

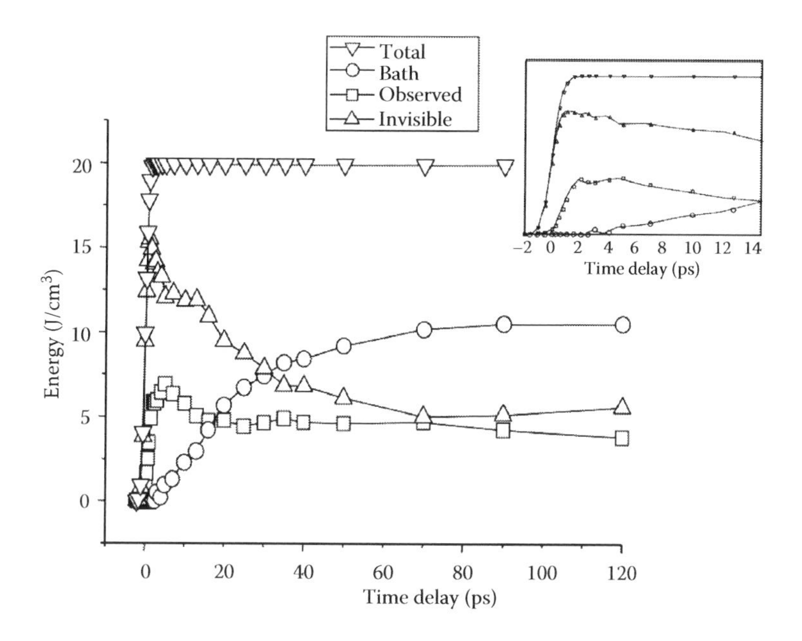

FIGURE 9.22 Anti-Stokes Raman calorimetry of d_6-benzene + 17% (v/v) CCl_4 showing the total energy input by the IR pulses, the energy in the 12 observed d_6-benzene vibrations, energy in the bath monitored with the CCl_4 molecular thermometer, and the invisible energy in d_6-benzene vibrations not observed by Raman. The *inset* shows the different rise times for E_{obs}, E_{invis}, and E_{bath}. (Reproduced with permission from Seong N-H, Fang Y, Dlott DD. Vibrational energy dynamics of normal and deuterated liquid benzene. *J. Phys. Chem. A* 113:1445–1452, Copyright 2009, American Chemical Society.)

an ~5 ps time delay before the molecular thermometer begins to rise, indicating that the decay of the parent CD-stretch is, as in benzene, primarily intramolecular. The half-life of the bath build-up is noticeably shorter than in benzene, 20 ps compared to 30 ps. Thermalization is essentially complete by 100 ps. The decay rates of $E_{invis}(t)$ and $E_{obs}(t)$ are similar enough to be indistinguishable. With d_6-benzene, the number of observed vibrations was 12 and the total number of vibrations was 30, so we observed 40% of the vibrations, and the maximum value of E_{obs}/E_{tot} (from Figure 9.13) was 36%. Thus, the amount of energy per vibration in d_6-benzene is 10% less for the "observed" strongly Raman-active vibrations than for the "invisible" vibrations.

Here, we have presented the most detailed investigation to date of vibrational energy dynamics in liquid benzene, and the first investigation of vibrational energy in liquid d_6-benzene. Since these molecules have inversion symmetry, the IR pump must excite states that are Raman-inactive. The Raman intensity is concentrated into a small number of transitions of which several are doubly degenerate. Nevertheless, we observe only a fraction of the 30 vibrations, 33% in benzene and 40% in d_6-benzene. Using a CCl_4 molecular thermometer, we can monitor the total energy dissipated from vibrationally excited benzene to the bath and thereby infer the time dependence of vibrational energy in the aggregate unobserved vibrations. The detailed vibration-to-vibration relaxation pathways were analyzed in the context of the three-state model described previously, keeping in mind that the rigid molecular framework might give rise to relaxation pathways that are more specific than what was previously observed in other species such as the aqueous glycine zwitterion [70,71] discussed above.

Our measurements show that the parent excitations in benzene and d_6-benzene, nominally a CH-stretch (3050 cm^{-1} pumping with ~40 cm^{-1} bandwidth) or CD-stretch excitation (2280 cm^{-1} pumping with ~40 cm^{-1} bandwidth), evidence a great deal of coherent coupling with other states, where coherent coupling as described in the introduction means the redistribution is faster than T_2. IR-Raman measurements have frequently observed this type of coupling with CH-bend and

CD-bend vibrations [56], and occasionally with other vibrations at about one-half the parent frequency such as NO_2 stretching of CH_3NO_2 [8,51], but the number of strongly coupled modes we observe by anti-Stokes Raman is much greater in benzene than in any other molecule yet studied. Additionally, the ultrafast calorimetry data indicate that even more of the pump pulse energy is coherently coupled to the invisible vibrations.

We observed only 33% or 40% of the benzene or d_6-benzene vibrations, but we compensated in part with ultrafast calorimetry, which provides some new insights to the question of whether observing just the Raman-active vibrations provides an accurate picture of benzene VC. In d_6-benzene, this is clearly true. The average energy (per mode) is within 10% in observed and invisible vibrations and the decay rates are quite similar. In benzene, we concluded that it is almost true that the Raman-active vibrations are representative of the entirety of vibrational energy. The average energy per mode is the same in both observed and invisible vibrations but the invisible vibrational energy rises a bit faster and decays a bit faster. The slower decay of energy from Raman-active vibrations in benzene but not in d_6-benzene is consistent with many low-temperature coherent Raman studies of crystalline naphthalene [10,139,144–146], anthracene [144], and pentacene [147,148], which showed that A_g vibrations, also those with the largest Raman cross-sections, had significantly longer lifetimes in the proto but not in the deutero species [149].

One point worth emphasizing is the very different timescales for decay of the initial state, ~6 ps, and the overall VC process [11]. The cooling process is characterized by a half-life of 30 [20] ps in benzene (d_6-benzene) while there is clear evidence for vibrationally hot molecules even at 80 ps.

9.10 SUMMARY AND CONCLUSIONS

In this update to our 2001 chapter, we have focused on advanced applications of IR-Raman spectroscopy. Water under extreme conditions was studied in two ways. The first involved the energy bursts that are created whenever water stretching vibrations are excited by IR photons. For a brief time, the water has weakened hydrogen bonding that looks like equilibrium water at a density 60% of the ambient density. The second way involved laser ablation of water, where the high-speed expansion produces a foamy medium with many broken hydrogen bonds. In fact, there are so many broken bonds that the decay lifetime of water increases by a factor of at least 1000.

Molecules that represent the building blocks of proteins, namely an amino acid, a molecule with a peptide bond, and a molecule with an aromatic side chain, were also studied. A molecular thermometer using the aqueous medium showed that the overall rate of VC was quite a bit longer than the VR lifetime of the parent states. This thermometer also showed that the vibrations of NMA that can be seen in the Raman spectrum are representative of the totality of vibrational energy that was not true for GLY or BZ. The three-stage model for VC introduced in Section 9.5 was shown to describe VR well.

Our most detailed study to date involved liquid benzene and its deuterated analog. We should mention that a similar study with equivalent detail has also been made for nitromethane and its deuterated analog. In these studies, we showed that we can determine the absolute occupation number of all observable vibrations, we can fit the transients using the three-stage model, and by lumping together the lifetimes and quantum efficiencies of the midrange M and lower-energy L vibrations we could quantitatively determine the rate constants and quantum efficiencies of all the processes depicted in Figure 9.4. Ultrafast vibrational calorimetry was introduced to quantitatively determine how much energy was present at any instant in the observed, invisible and bath vibrations. In benzene, the thermalization process takes about 10 times as long as the decay of the parent state, so studies of individual-level VR tell us little about the full VC process. When benzene is pumped by an IR pulse in the CH-stretch region, it takes a few picoseconds for the energy to be transferred from the "invisible" IR-active vibrations to the "observed" Raman-active vibrations. The attribution of an IVR mechanism to the parent state decay is confirmed by seeing the lack of energy build-up in the bath.

With the sensitivity and precision now available to us with IR-Raman spectroscopy, it is possible to take this research into many new directions. One direction that is proving particularly fruitful is the study of energy transfer in monosubstituted benzenes [150], where it is possible to watch energy flow from the substituent to the phenyl group and vice versa.

ACKNOWLEDGMENTS

The research described in this study was based on work supported by the National Science Foundation under award DMR-09-55259 and the US Air Force Office of Scientific Research under award number FA9550-09-1-0163.

REFERENCES

1. Iwaki LK, Deàk JC, Rhea ST, Dlott DD. 2000. Vibrational energy redistribution in polyatomic liquids: Ultrafast IR-Raman spectroscopy. *Ultrafast Infrared and Raman Spectroscopy*, ed Fayer MD (Marcel Dekker, New York), pp 541–592.
2. Seong N-H, Fang Y, Dlott DD. 2009. Vibrational energy dynamics of normal and deuterated liquid benzene. *J. Phys. Chem. A* 113:1445–1452.
3. Seilmeier A, Scherer POJ, Kaiser W. 1984. Ultrafast energy dissipation in solutions measured by a molecular thermometer. *Chem. Phys. Lett.* 105(2):140–146.
4. Lee I-YS, Wen X, Tolbert WA, Dlott DD. 1992. Direct measurement of polymer temperature during laser ablation using a molecular thermometer. *J. Appl. Phys.* 72:2440–2448.
5. Graham PB, Matus KJM, Stratt RM. 2004. The workings of a molecular thermometer: The vibrational excitation of carbon tetrachloride by a solvent. *J. Chem. Phys.* 121:5348–5354.
6. Chen S, Lee I-YS, Tolbert W, Wen X, Dlott DD. 1992. Applications of ultrafast temperature jump spectroscopy to condensed phase molecular dynamics. *J. Phys. Chem.* 96:7178–7186.
7. Deàk JC, Iwaki LK, Dlott DD. 1998. Vibrational energy relaxation of polyatomic molecules in liquids: Acetonitrile. *J. Phys. Chem.* 102:8193–8201.
8. Deàk JC, Iwaki LK, Dlott DD. 1999. Vibrational energy redistribution in polyatomic liquids: Ultrafast IR-Raman spectroscopy of nitromethane. *J. Phys. Chem. A* 103:971–979.
9. Dlott DD. 2001. Vibrational energy redistribution in polyatomic liquids: 3D infrared-Raman spectroscopy. *Chem. Phys.* 266:149–166.
10. Hill JR, Chronister EL, Chang T-C, Kim H, Postlewaite JC, Dlott DD. 1988. Vibrational relaxation and vibrational cooling in low temperature molecular crystals. *J. Chem. Phys.* 88:949–967.
11. Hill JR, Dlott DD. 1988. A model for ultrafast vibrational cooling in molecular crystals. *J. Chem. Phys.* 89(2):830–841.
12. Hill JR, Dlott DD. 1988. Theory of vibrational cooling in molecular crystals: Application to crystalline naphthalene. *J. Chem. Phys.* 89:842–858.
13. Laubereau A, Kaiser W. 1978. Vibrational dynamics of liquids and solids investigated by picosecond light pulses. *Rev. Mod. Phys.* 50(3):607–665.
14. Chen S, Hong X, Hill JR, Dlott DD. 1995. Ultrafast energy transfer in high explosives: Vibrational cooling. *J. Phys. Chem.* 99:4525–4530.
15. Hong X, Chen S, Dlott DD. 1995. Ultrafast mode-specific intermolecular vibrational energy transfer to liquid nitromethane. *J. Phys. Chem.* 99:9102–9109.
16. Mukamel S. 1995. *Principles of Nonlinear Optical Spectroscopy* (Oxford University Press, New York).
17. Laubereau A, Greiter L, Kaiser W. 1974. Intense tunable picosecond pulses in the infrared. *Appl. Phys. Lett.* 25(1):87–89.
18. Spanner K, Laubereau A, Kaiser W. 1976. Vibrational energy redistribution of polyatomic molecules in liquids after ultrashort infrared excitation. *Chem. Phys. Lett.* 44(1):88–92.
19. Seilmeier A, Kaiser W. 1988. Ultrashort intramolecular and intermolecular vibrational energy transfer of polyatomic molecules in liquids. *Ultrashort Laser Pulses and Applications, Topics in Applied Physics*, ed Kaiser W (Springer Verlag, Berlin), Vol 60, pp 279–315.
20. Terhune RW, Maker PD, Savage CM. 1965. Measurements of nonlinear light scattering. *Phys. Rev. Lett.* 14:681–684.
21. Deàk JC, Rhea ST, Iwaki LK, Dlott DD. 2000. Vibrational energy relaxation and vibrational spectral diffusion in liquid water and deuterated water. *J. Phys. Chem. A* 104:4866–4875.

22. Tokmakoff A, Sauter B, Kwok AS, Fayer MD. 1994. Phonon-induced scattering between vibrations and multiphoton vibrational up-pumping in liquid solution. *Chem. Phys. Lett.* 221:412–418.
23. Ambroseo JR, Hochstrasser RM. 1988. Pathways of relaxation of the N-H stretching vibration of pyrrole in liquids. *J. Chem. Phys.* 89(9):5956–5957.
24. Ambroseo JR, Hochstrasser RM. 1988. Vibrational relaxation pathways of the N-H stretch of pyrrole in liquids. *Ultrafast Phenomena VI,* Springer Series in Chemical Physics, eds Yajima T, Yoshihara K, Harris CB, Shionoya S (Springer-Verlag, Berlin Heidelberg), Vol 48, pp 450–451.
25. Kozich V, Dreyer J, Werncke W. 2009. Mode-selective vibrational redistribution after spectrally selective N–H stretching mode excitation in intermolecular hydrogen bonds. *J. Chem. Phys.* 130(3):034505.
26. Kozich V, Szyc Ł, Nibbering ETJ, Werncke W, Elsaesser T. 2009. Ultrafast redistribution of vibrational energy after excitation of NH stretching modes in DNA oligomers. *Chem. Phys. Lett.* 473(1–3):171–175.
27. Graener H. 1990. The equilibration of vibrational excess energy. *Chem. Phys. Lett.* 165(1):110–114.
28. Graener H, Laubereau A. 1982. New results on vibrational population decay in simple liquids. *Appl. Phys. B* 29:213–218.
29. Graener H, Laubereau A. 1983. Ultrafast overtone excitation for the study of vibrational population decay in liquids. *Chem. Phys. Lett.* 102:100–104.
30. Graener H, Zürl R, Hofmann M. 1997. Vibrational relaxation of liquid chloroform. *J. Phys. Chem.* 101:1745–1749.
31. Hofmann M, Graener H. 1995. Time resolved incoherent anti-Stokes Raman spectroscopy of dichloromethane. *Chem. Phys.* 206:129–137.
32. Seifert G, Zürl R, Graener H. 1999. Novel information about vibrational relaxation in liquids using time-resolved Stokes probing after picosecond IR excitation. *J. Phys. Chem. A* 103(50):10749–10754.
33. Deàk JC, Iwaki LK, Dlott DD. 1997. High power picosecond mid-infrared optical parametric amplifier for infrared-Raman spectroscopy. *Opt. Lett.* 22:1796–1798.
34. Huse N, Ashihara S, Nibbering ETJ, Elsaesser T. 2005. Ultrafast vibrational relaxation of O-H bending and librational excitations in liquid H_2O. *Chem. Phys. Lett.* 404:389–393.
35. Ashihara S, Huse N, Espagne A, Nibbering ETJ, Elsaesser T. 2006. Vibrational couplings and ultrafast relaxation of the O-H bending mode in liquid H_2O. *Chem. Phys. Lett.* 424:66–70.
36. Elsaesser T, Ashihara S, Huse N, Espagne A, Nibbering E. 2007. Ultrafast structural dynamics of water induced by dissipation of vibrational energy. *J. Phys. Chem. A* 111:743–746.
37. Rey R, Ingrosso F, Elsaesser T, Hynes JT. 2009. Pathways for H_2O bend vibrational relaxation in liquid water. *J. Phys. Chem. A* 113(31):8949–8962.
38. Szyc L, Yang M, Elsaesser T. 2010. Ultrafast energy exchange via water-phosphate interactions in hydrated DNA. *J. Phys. Chem. B* 114(23):7951–7957.
39. Levinger NE, Costard R, Nibbering ETJ, Elsaesser T. 2011. Ultrafast energy migration pathways in self-assembled phospholipids Interacting with confined water. *J. Phys. Chem. A* 115(43):11952–11959.
40. Piatkowski L, Bakker HJ. 2010. Vibrational relaxation pathways of AI and AII modes in *N*-methylacetamide clusters. *J. Phys. Chem. A* 114(43):11462–11470.
41. Timmer RLA, Bakker HJ. 2010. Vibrational Förster transfer in ice Ih. *J. Phys. Chem. A* 114(12):4148–4155.
42. Hamm P, Zanni MT. 2011. *Concepts and Methods of 2D Infrared Spectroscopy* (Cambridge University Press, Cambridge).
43. Bian HT, Li JB, Wen XW, Zheng JR. 2010. Mode-specific intermolecular vibrational energy transfer. I. Phenyl selenocyanate and deuterated chloroform mixture. *J. Chem. Phys.* 132(18):184505.
44. Bian HT, Chen HL, Li JB, Wen XW, Zheng JR. 2011. Nonresonant and resonant mode-specific intermolecular vibrational energy transfers in electrolyte aqueous solutions. *J. Phys. Chem. A* 115(42):11657–11664.
45. Bian HT, Li JB, Wen XW, Sun ZG, Song JA, Zhuang W, Zheng JR. 2011. Mapping molecular conformations with multiple-mode two-dimensional infrared spectroscopy. *J. Phys. Chem. A* 115(15):3357–3365.
46. Naraharisetty SRG, Kasyanenko VM, Rubtsov IV. 2008. Bond connectivity measured via relaxation-assisted two-dimensional infrared spectroscopy *J. Chem. Phys.* 128:104502.
47. Lin ZW, Rubtsov IV. 2012. Constant-speed vibrational signaling along polyethyleneglycol chain up to 60-angstrom distance. *Proc. Natl. Acad. Sci. USA* 109(5):1413–1418.
48. Kasyanenko VM, Tesar SL, Rubtsov GI, Burin AL, Rubtsov IV. 2011. Structure dependent energy transport: Relaxation-assisted 2D IR measurements and theoretical studies. *J. Phys. Chem. B* 115(38):11063–11073.
49. Rubtsov IV. 2009. Relaxation-assisted two-dimensional infrared (RA 2D IR) method: Accessing distances over 10 angstrom and measuring bond connectivity patterns. *Acct. Chem. Res.* 42(9):1385–1394.
50. Kasyanenko VM, Lin ZW, Rubtsov GI, Donahue JP, Rubtsov IV. 2009. Energy transport via coordination bonds. *J. Chem. Phys.* 131(15):154508.

51. Shigeto S, Pang Y, Fang Y, Dlott DD. 2008. Vibrational relaxation of normal and deuterated liquid nitromethane. *J. Phys. Chem. B* 112:232–241.
52. Wang Z, Pang Y, Dlott DD. 2006. Long-lived interfacial vibrations of water. *J. Phys. Chem. B* 110: 20115–20117.
53. Bertie JE, Lan Z. 1996. Infrared intensities of liquids XX: The intensity of the OH stretching band of liquid water revisited, and the best current values of the optical constants of $H_2O(l)$ at 25°C between 15,000 and 1 cm^{-1}. *Appl. Spectrosc.* 50:1047–1057.
54. Hare DE, Franken J, Dlott DD. 1995. Coherent Raman measurements of polymer thin film pressure and temperature during picosecond laser ablation. *J. Appl. Phys.* 77:5950–5960.
55. Wang Z, Pang Y, Dlott DD. 2007. Hydrogen-bond disruption by vibrational excitations in water. *J. Phys. Chem. A* 111:3196–3208.
56. Iwaki LK, Deàk JC, Rhea ST, Dlott DD. 2001. Vibrational energy redistribution in polyatomic liquids: Ultrafast IR-Raman spectroscopy. *Ultrafast Infrared and Raman Spectroscopy*, ed Fayer MD (Marcel Dekker, New York), pp 541–592.
57. Deàk JC, Iwaki LK, Rhea ST, Dlott DD. 2000. Ultrafast infrared-Raman studies of vibrational energy redistribution in polyatomic liquids. *J. Raman Spectrosc.* 31:263–274.
58. Pakoulev A, Wang Z, Dlott DD. 2003. Vibrational relaxation and spectral evolution following ultrafast OH stretch excitation of water. *Chem. Phys. Lett.* 371:594–600.
59. Pakoulev A, Wang Z, Pang Y, Dlott DD. 2003. Vibrational energy relaxation pathways of water. *Chem. Phys. Lett.* 380:404–410.
60. Wang Z, Pakoulev A, Pang Y, Dlott DD. 2003. Vibrational substructure in the OH stretching band of water. *Chem. Phys. Lett.* 378:281–288.
61. Wang Z, Pakoulev A, Pang Y, Dlott DD. 2004. Vibrational substructure in the OH stretching transition of water and HOD. *J. Phys. Chem. A* 108:9054–9063.
62. Wang Z, Pang Y, Dlott DD. 2004. The vibrational Stokes shift of water (HOD in D_2O). *J. Chem. Phys.* 120:8345–8348.
63. Wang Z, Pang Y, Dlott DD. 2004. Vibrational energy dynamics of water studied with ultrafast Stokes and anti-Stokes Raman spectroscopy. *Chem. Phys. Lett.* 397:40–45.
64. Fayer MD. 2001. *Ultrafast Infrared and Raman Spectroscopy* (Marcel Dekker, Inc., New York).
65. Herzberg G. 1945. *Molecular Spectra and Molecular Structure II. Infrared and Raman Spectra of Polyatomic Molecules* (Van Nostrand Reinhold, New York).
66. Graener H, Seifert G, Laubereau A. 1991. New spectroscopy of water using tunable picosecond pulses in the infrared. *Phys. Rev. Lett.* 66(16):2092–2095.
67. Nitzan A, Jortner J. 1973. Vibrational relaxation of a molecule in a dense medium. *Molec. Phys.* 25(3): 713–734.
68. Iwaki L, Dlott DD. 2001. Vibrational energy transfer in condensed phases. *Encyclopedia of Chemical Physics and Physical Chemistry*, eds Moore JH, Spencer ND (IOP Publishing Ltd., London), pp. 2717–2736.
69. Hoffman GJ, Imre DG, Zadoyan R, Schwentner N, Apkarian VA. 1993. Relaxation dynamics in the B(1/2) and C(3/2) charge transfer states of XeF in solid Ar. *J. Chem. Phys.* 98(12):9233–9240.
70. Shigeto S, Dlott DD. 2007. Vibrational relaxation of an amino acid in aqueous solution. *Chem. Phys. Lett.* 447:134–139.
71. Fang Y, Shigeto S, Seong N-H, Dlott DD. 2009. Vibrational energy dynamics of glycine, *N*-methyl acetamide and benzoate anion in aqueous (D_2O) solution. *J. Phys. Chem. A* 113:75–84.
72. Deàk JC, Iwaki LK, Dlott DD. 1998. When vibrations interact: Ultrafast energy relaxation of vibrational pairs in polyatomic liquids. *Chem. Phys. Lett.* 293:405–411.
73. Felker PM, Zewail AH. 1984. Direct observation of nonchaotic multilevel vibrational energy flow in isolated polyatomic molecules. *Phys. Rev. Lett.* 53:501–504.
74. Tokmakoff A, Kowk AS, Urdahl RS, Francis RS, Fayer MD. 1995. Multilevel vibrational dephasing and vibrational anharmonicity from infrared photon echo beats. *Chem. Phys. Lett.* 234:289–295.
75. Iwaki LK, Dlott DD. 2000. Three-dimensional spectroscopy of vibrational energy relaxation in liquid methanol. *J. Phys. Chem. A* 104:9101–9112.
76. Skinner JL, Auer BM, Lin YS. 2009. Vibrational line shapes, spectral diffusion, and hydrogen bonding in liquid water. *Advances in Chemical Physics, Vol 142,* Advances in Chemical Physics, ed Rice SA (John Wiley & Sons Inc, New York), Vol 142, pp. 59–103.
77. Staib A, Hynes JT. 1993. Vibrational predissociation in hydrogen-bonded OH...O complexes via OH stretch-OO stretch energy transfer. *Chem. Phys. Lett.* 204:197–205.
78. Asbury JB, Steinel T, Fayer MD. 2004. Hydrogen bond networks: Structure and evolution after hydrogen bond breaking. *J. Phys. Chem. B* 108:6544–6554.

79. Asbury JB, Steinel T, Stromberg C, Gaffney KJ, Piletic IR, Fayer MD. 2003. Hydrogen bond breaking probed with multidimensional stimulated vibrational echo correlation spectroscopy. *J. Chem. Phys.* 119:12981–12997.
80. Gaffney JJ, Piletic IR, Fayer MD. 2002. Hydrogen bond breaking and reformation in alcohol oligomers following vibrational relaxation of a non-hydrogen-bond donating hydroxyl stretch. *J. Phys. Chem. A* 106:9428–9435.
81. Steinel T, Asbury JB, Zheng JR, Fayer MD. 2004. Watching hydrogen bonds break: A transient absorption study of water. *J. Phys. Chem. A* 108:10957–10964.
82. Bakker HJ, Lock AJ, Madsen D. 2004. Strong feedback effect in the vibrational relaxation of liquid water. *Chem. Phys. Lett.* 384:236–241.
83. Kauranen M, Persoons P. 1996. Theory of polarization measurements of second-order nonlinear light scattering. *J. Chem. Phys.* 104:3445–3456.
84. Ikushima Y, Hatakeda K, Saito N. 1998. An *in situ* Raman spectroscopy study of subcritical and supercritical water: The peculiarity of hydrogen bonding near the critical point. *J. Chem. Phys.* 198:5855–5860.
85. Lin J-F, Militzer B, Struzhkin VV, Gregoryanz E, Hemley RJ, Mao H. 2004. High pressure-temperature Raman measurements of H_2O melting to 22 GPa and 900 K. *J. Chem. Phys.* 121:8423–8427.
86. Lawrence CP, Skinner JL. 2003. Ultrafast infrared spectroscopy probes hydrogen bonding dynamics in liquid water. *Chem. Phys. Lett.* 369:472–477.
87. Rey R, Møller KB, Hynes JT. 2002. Hydrogen bond dynamics in water and ultrafast infrared spectroscopy. *J. Phys. Chem. A* 106:11993–11996.
88. Zhigilei LV, Leveugle E, Garrison BJ. 2003. Computer simulations of laser ablation of molecular substrates. *Chem. Rev.* 103:321–347.
89. Richmond GL. 2002. Molecular bonding and interactions at aqueous surfaces as probed by vibrational sum frequency spectroscopy. *Chem. Rev.* 102:2693–2724.
90. Shen YR, Ostroverkhov V. 2006. Sum-frequency vibrational spectroscopy on water interfaces: Polar orientation of water molecules at interfaces. *Chem. Rev.* 106:1140–1154.
91. Skinner JL, Pieniazek PA, Gruenbaum SM. 2012. Vibrational spectroscopy of water at interfaces. *Acct. Chem. Res.* 45(1):93–100.
92. Steinbach C, Andersson P, Kazimirski JK, Buck U, Buch V, Beu TA. 2004. Infrared predissociation spectroscopy of large water clusters: A unique probe of cluster surfaces. *J. Phys. Chem. A* 108:6165–6174.
93. Hsieh C-S, Campen RK, Vila Verde AC, Bolhuis P, Nienhuys H-K, Bonn M. 2011. Ultrafast reorientation of dangling OH groups at the air-water interface using femtosecond vibrational Spectroscopy. *Phys. Rev. Lett.* 107(11):116102.
94. Lock AJ, Bakker HJ. 2002. Temperature dependence of vibrational relaxation in liquid H_2O. *J. Chem. Phys.* 117:1708–1713.
95. Lock AJ, Woutersen S, Bakker HJ. 2001. Ultrafast energy equilibration in hydrogen-bonded liquids. *J. Phys. Chem. A* 105:1238–1243.
96. Cringus D, Lindner J, Milder MTW, Pshenichnikov MS, Vöhringer P, Wiersma DA. 2005. Femtosecond water dynamics in reverse-micellar nanodroplets. *Chem. Phys. Lett.* 408:162–168.
97. Lawrence CP, Skinner JL. 2002. Vibrational spectroscopy of HOD in liquid D_2O. II. Infrared line shapes and vibrational Stokes shift. *J. Chem. Phys.* 117:8847–8854.
98. Lian T, Locke B, Kholodenko Y, Hochstrasser RM. 1994. Energy flow from solute to solvent probed by femtosecond IR spectroscopy: Malachite green and heme protein solutions. *J. Phys. Chem.* 98:11648–11656.
99. Henry ER, Eaton WA, Hochstrasser RM. 1986. Molecular dynamics simulations of cooling in laser-excited heme proteins. *Proc. Natl. Acad. Sci. USA* 83:8982–8986.
100. Uchida T, Kitagawa T. 2005. Mechanism for transduction of the ligand-binding signal in heme-based gas sensory proteins revealed by resonance Raman spectroscopy. *Acct. Chem. Res.* 2005:662–670.
101. Lingle RJ, Xu X, Zhu H, Yu S-C, Hopkins JB. 1991. Picosecond Raman study of energy flow in a photoexcited heme protein. *J. Phys. Chem.* 95:9320–9331.
102. Lingle RJ, Xu XB, Zhu HP, Yu S-C, Hopkins JB. 1991. Direct observation of hot vibrations in photoexcited deoxyhemoglobin using picosecond Raman spectroscopy. *J. Am. Chem. Soc.* 113:3992–3994.
103. Li P, Sage JT, Champion PM. 1992. Probing picosecond processes with nanosecond lasers: Electronic and vibrational relaxation dynamics of heme proteins. *J. Chem. Phys.* 97:3214–3227.
104. Li P, Champion PM. 1994. Investigations of the thermal response of laser-excited biomolecules. *Biophys. J.* 66:430–436.
105. Ye X, Demidov A, Rosca F, Wang W, Kumar A, Ionascu D, Zhu L, Barrick D, Wharton D, Champion PM. 2003. Investigation of heme protein absorption lineshapes, vibrational relaxation and resonance Raman scattering on ultrafast time scales. *J. Phys. Chem. A* 107:8156–8165.

106. Fujisaki H, Straub JE. 2005. Vibrational energy relaxation in proteins. *Proc. Natl. Acad. Sci. USA* 102: 6726–6731.
107. Miller RJD. 1991. Vibrational-energy relaxation and structural dynamics of heme proteins. *Annu. Rev. Phys. Chem.* 42:581–614.
108. Simpson MC, Peterson ES, Shannon CF, Eads DD, Friedman JM, Cheatum CM, Ondrias MR. 1997. Transient Raman observations of heme electronic and vibrational photodynamics in deoxyhemoglobin *J. Am. Chem. Soc.* 119:5110–5117.
109. Challa JR, Gunaratne TC, Simpson MC. 2006. State preparation and excited electronic and vibrational behavior in hemes *J. Phys. Chem. B* 110:19956–19965.
110. Sato A, Gao Y, Kitagawa T. 2007. Primary protein resonse after ligand photodissociation in carbonmonoxy myoglobin. *Proc. Natl. Acad. Sci. USA* 104:9627–9632.
111. Gao Y, Koyama M, El-Mashtoly SF, Hayashi T, Harada K, Mizutani Y, Kitagawa T. 2006. Time-resolved Raman evidence for energy "funneling" through propionate side chains in heme "cooling" upon photolysis of carbonmonoxy myoglobin. *Chem. Phys. Lett.* 429:239–243.
112. Gao Y, El-Mashtoly SF, Pal B, Hayashi T, Harada K, Kitagawa T. 2006. Pathway of information transmission from heme to protein upon ligand binding/dissociation in myoglobin revealed by UV resonance Raman spectroscopy. *J. Biol. Chem.* 281 24637–24646.
113. Bu L, Straub JE. 2003. Vibrational energy relaxation of "tailored" hemes in myoglobin followed ligand photolysis supports energy funneling mechanism of heme "cooling". *J. Phys. Chem. B* 107:10634–10639.
114. Zhang Y, Fujisaki H, Straub JE. 2007. Molecular dynamics study on the solvent dependent heme cooling following ligand photolysis in carbonmonoxy myoglobin *J. Phys. Chem. B* 111:3243–3250.
115. Sagnella DE, Straub JE. 2001. Directed energy "funneling" mechanism for heme cooling following ligand photolysis or direct excitation in solvated carbonmonoxy myoglobin. *J. Phys. Chem. B* 105:7057–7063.
116. Hill JR, Tokmakoff A, Peterson KA, Sauter B, Zimdars D, Dlott DD, Fayer MD. 1994. Vibrational dynamics of carbon monoxide at the active site of myoglobin: Picosecond infrared free-electron laser pump-probe experiments. *J. Phys. Chem.* 98:11213–11219.
117. Peterson KA, Hill JR, Tokmakoff A, Sauter B, Zimdars D, Dlott DD, Fayer MD. 1994. Vibrational dynamics at the active site of myoglobin: Picosecond infrared free-electron-laser experiments. *Ultrafast Phenomena IX,* Springer Series in Chemical Physics, ed Barbara PF (Springer-Verlag, Berlin, Heidelberg, New York), Vol 60, pp 445–447.
118. Peterson KA, Boxer SG, Decatur SM, Dlott DD, Fayer MD, Hill JR, Rella CW, Rosenblatt MM, Suslick KS, Ziegler CJ. 1996. Vibrational relaxation of carbonmonoxide in myoglobin mutants and model heme compounds. *Time-Resolved Vibrational Spectroscopy VII*, (Los Alamos National Laboratory Technical Report LA-13290-C, Los Alamos, NM), pp 173–177.
119. Owrutsky JC, Li M, Locke B, Hochstrasser RM. 1995. Vibrational relaxation of the CO stretch vibration in hemoglobin-CO, myoglobin-CO, and protoheme-CO. *J. Phys. Chem.* 99:4842–4846.
120. Fujisaki H, Straub JE. 2007. Vibrational energy relaxation of isotopically labeled amide I modes in cytochrome c: Theoretical investigation of vibrational energy relaxation rates and pathways *J. Phys. Chem. B* 111:12017–12023.
121. Hamm P, Lim M, Hochstrasser RM. 1998. Ultrafast dynamics of amide-I vibrations. *Biophys. J.* 74:A332–A332.
122. DeFlores LP, Ganim Z, Ackley SF, Chung HS, Tokmakoff A. 2006. The anharmonic vibrational potential and relaxation pathways of the Amide I and II modes of *N*-methylacetamide. *J. Phys. Chem. B* 110:18973–18980.
123. Peterson KA, Rella CW, Engholm JR, Schwettman HA. 1999. Ultrafast vibrational dynamics of the myoglobin amide I band. *J. Phys. Chem. B* 103:557–561.
124. Chen XG, Schweitzer-Stenner R, Asher SA, Mirkin NG, Krimm S. 1995. Vibrational assignments of *trans-N*-methylacetamide and some of its deuterated isotopomers from band decomposition of IR, visible, and resonance Raman spectra. *J. Phys. Chem.* 99:3074–3083.
125. Kubelka J, Keiderling TA. 2001. Ab initio calculation of amide carbonyl stretch vibrational frequencies in solution with modified basis sets. 1. *N*-Methyl acetamide *J. Phys. Chem. A* 105:10922–10928.
126. Iwaki LK, Deàk JC, Rhea ST, Dlott DD. 1999. Vibrational energy redistribution in liquid benzene. *Chem. Phys. Lett.* 303:176–182.
127. Kropman MF, Nienhuys H-K, Woutersen S, Bakker HJ. 2001. Vibrational relaxation and hydrogen-bond dynamics of HDO:H_2O. *J. Phys. Chem. A* 105:4622–4626.
128. Stewart GM, McDonald JD. 1983. Intramolecular vibrational relaxation from C-H stretch fundamentals. *J. Chem. Phys.* 78:3907–3915.

129. Callegari A, Srivastava HK, Merker U, Lehmann KK, Scoles G, Davis MJ. 1997. Eigenstate resolved infrared-infrared double-resonance study of intramolecular vibrational relaxation in benzene: First overtone of the CH stretch. *J. Chem. Phys.* 106:432–435.
130. Reddy KV, Heller DF, Berry MJ. 1982. Highly vibrationally excited benzene: Overtone spectroscopy and intramolecular dynamics of C_6H_6, C_6D_6 and partially deuterated or substituted benzenes. *J. Chem. Phys.* 76:2814–2837.
131. Sibert III EL, Hynes JT, WReinhardt WP. 1984. Classical dynamics of highly excited CH and CD overtones in benzene and perdeuterobenzene. *J. Chem. Phys.* 81:1135–1144.
132. Sibert III EL, Reinhardt WP, Hynes JT. 1982. Intramolecular vibrational-relaxation of CH overtones in benzene. *Chem. Phys. Lett.* 92:455–458.
133. Sibert III EL, Reinhardt WP, Hynes JT. 1984. Intramolecular vibrational relaxation and spectra of CH and CD overtones in benzene and perdeuterobenzene. *J. Chem. Phys.* 81(3):1115–1134.
134. Ho F, Tsay W-S, Trout J, Velsko S, Hochstrasser RM. 1983. Picosecond time-resolved CARS in isotopically mixed crystals of benzene. *Chem. Phys. Lett.* 97:141–146.
135. Velsko S, Hochstrasser RM. 1985. Studies of vibrational relaxation in low-temperature molecular crystals using coherent Raman spectroscopy. *J. Phys. Chem.* 89:2240–2253.
136. Neuman MN, Tabisz GC. 1976. On a Raman linewidth study of molecular motion in liquid benzene. *Chem. Phys.* 15:195–200.
137. Tanabe K, Jonas J. 1977. Raman study of vibrational relaxation of benzene in solution. *Chem. Phys. Lett.* 53:278–281.
138. Trout TJ, Velsko S, Bozio R, Decola PL, Hochstrasser RM. 1984. Nonlinear Raman study of line shapes and relaxation of vibrational states of isotopically pure and mixed crystals of benzene. *J. Chem. Phys.* 81(11):4746–4759.
139. Decola PL, Hochstrasser RM, Trommsdorff HP. 1980. Vibrational relaxation in molecular crystals by four-wave mixing: Naphthalene. *Chem. Phys. Lett.* 72:1–4.
140. Fendt A, Fischer SF, Kaiser W. 1981. Vibrational lifetime and Fermi resonance in polyatomic molecules. *Chem. Phys.* 57:55–64.
141. Chen S, Tolbert WA, Dlott DD. 1994. Direct measurement of ultrafast multiphonon up pumping in high explosives. *J. Phys. Chem.* 98:7759–7766.
142. Watanabe H, Kato H. 2004. Thermal conductivity and thermal diffusivity of twenty-nine liquids: Alkenes, cyclic (alkanes, alkenes, alkadienes, aromatics) and deuterated hydrocarbons. *J. Chem. Eng. Ref. Data* 49:809–825.
143. Shimanouchi T. 1972. *Tables of Molecular Vibrational Frequencies. Consolidated Volume I.* (US Government Printing Office, Washington, D. C.).
144. Schosser CL, Dlott DD. 1984. A picosecond CARS study of vibron dynamics in molecular crystals: Temperature dependence of homogeneous and inhomogeneous linewidths. *J. Chem. Phys.* 80:1394–1406.
145. Bellows JC, Prasad PN. 1979. Dephasing times and linewidths of optical transitions in molecular crystals: Temperature dependence of line shapes, linewidths, and frequencies of raman active phonons in naphthalene. *J. Chem. Phys.* 70(4):1864–1871.
146. Hesp BH, Wiersma DA. 1980. Vibrational relaxation in neat crystals of naphthalene by picosecond cars. *Chem. Phys. Lett.* 75:423–426.
147. Hill JR, Chronister EL, Chang T-C, Kim H, Postlewaite JC, Dlott DD. 1988. Vibrational relaxation of guest and host in mixed molecular crystals. *J. Chem. Phys.* 88:2361–2371.
148. Hesselink WH, Wiersma DA. 1980. Optical dephasing and vibronic relaxation in molecular mixed crystals: A picosecond photon echo and optical study of pentacene in naphthalene and *p*-terphenyl. *J. Chem. Phys.* 73(2):648–663.
149. Dlott DD. 1988. Dynamics of molecular crystal vibrations. *Laser Spectroscopy of Solids II*, ed Yen W (Springer Verlag, Berlin), pp 167–200.
150. Pein BC, Seong N-H, Dlott DD. 2010. Vibrational energy relaxation of liquid aryl-halides X-C_6H_5 (X = F, Cl, Br, I). *J. Phys. Chem. A* 114(39):10500–10507.

10 Ultrafast Processes at Liquid Interfaces Investigated with Time-Resolved Sum Frequency Generation

Yi Rao, Benjamin Doughty, Nicholas J. Turro, and Kenneth B. Eisenthal

CONTENTS

10.1 INTRODUCTION

The molecules at the interface between two phases are subject to inherently anisotropic forces that result in its unique properties, structure, and dynamics [1–7]. The asymmetric environment is manifested in the equilibrium characteristics of the interface such as molecular orientation [8–10], chemical composition [11–13], and polarity [14–20]. When subject to outside stimuli, such as photoexcitation or an external electric field [1–7,21–23], the response of the interfacial molecules often differs from the dynamics observed in bulk media; for instance, transport kinetics [21–25], the dynamics of molecular rotation [26–33], solvation [29,34–40], energy relaxation [41,42], electron transfer [43], and chemical reactions [23,44,45] can be different at interfaces as compared to the bulk material. Since numerous practical devices rely on junctions between two materials for charge separation and biological systems make use of interfaces for specificity in transport and chemical reactions, it is of fundamental and pragmatic importance that equilibrium and dynamical aspects of interfaces be explored [1–3,46,47].

Since the interface represents only a very small subset of the total number of molecules in solution, traditional spectroscopic techniques, such as ultraciolet-visible (UV), infrared (IR), Raman, nuclear magnetic resonance (NMR), and electron paramagnetic resonance (EPR) spectroscopies [48], are overwhelmed with background signals from the bulk [49–53]. To circumvent the problems associated with traditional spectroscopic methods in measuring surface properties, second-order nonlinear spectroscopies [49–51], such as second harmonic generation (SHG) [10,17,21,50,54–70] and sum frequency generation (SFG) [9,45,71–99], are used to selectively probe anisotropic distributions of molecules at interfaces. SHG is primarily sensitive to the electronic structure of the interfacial molecules, whereas SFG is sensitive to the electronic and vibrational states of the molecules. These spectroscopies can detect the spectral signatures of molecules in the molecularly thin layer separating the two bulk phases [57,72,78,84,87,100]. When combined with optical excitation pulses, SHG and SFG are capable of tracking ultrafast photo-dynamics, including rotational dynamics, solvation, and electron transfer [26–28,30–40,43,101], as will be discussed individually in the sections below.

Static SHG and SFG spectroscopies provide interfacial information about the molecular species, orientational ordering, acid–base equilibrium, and so on [9,10,17,21,45,50,54–99]. Time-resolved measurements are advantageous in situations where one wants to directly determine the timescales for multiple dynamical processes induced by a single pump pulse, which cannot be extracted from frequency-resolved measurements [26–29,31–39,41,43,101]. To this end, time-resolved SHG (TR-SHG) has been used to study ultrafast interfacial processes such as solvation and rotational dynamics, electron, energy and proton transfer, and excited-state lifetimes [26–29,31–39,41,43,101]. Less work has been reported using time-resolved SFG (TR-SFG) with much of the published results focusing on the dynamics occurring after IR-excitation of vibrational modes in ground electronic states [42,102–104]. The work reviewed here focuses on the use of TR-SFG to track the time evolution of electronically excited states by selectively probing vibrational modes associated with distinct functional groups in the interfacial molecules.

In the experiments described below, coumarin 314 (C314, structure in Scheme 10.1) at the air/water interface was excited to the first singlet excited state with an optical pump pulse and subsequently probed with SFG to track the time evolution of the solvation and orientational dynamics of C314. The electron transfer reaction between photoexcited C314 and *N,N*-dimethylaniline (DMA, structure in Scheme 10.1) at a water/DMA monolayer interface was also monitored by TR-SFG. In general, the results show that some aspects of the bulk are preserved at interfaces, which might

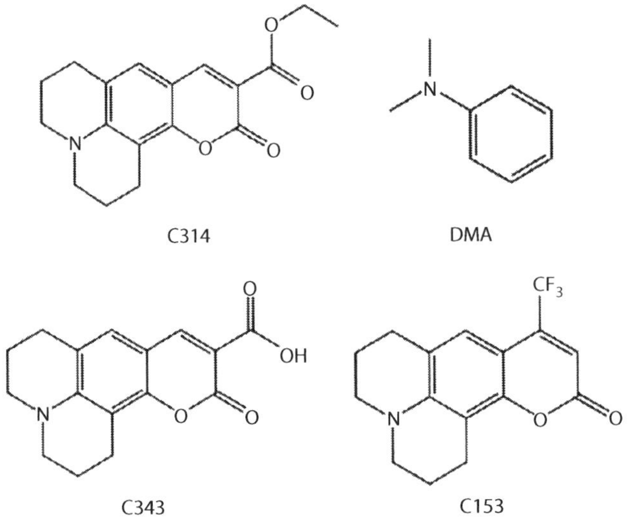

C314 DMA

C343 C153

SCHEME 10.1 Chemical structures of coumarin 314 (C314), *N,N*-dimethylaniline (DMA), coumarin 343 (C343), and coumarin 153 (C153).

be surprising, but that other dynamics, such as electron transfer, are significantly different from those observed in the bulk. The results demonstrate that detailed studies of molecular excited-state dynamics at the air/water interface benefit from previous work studying the relaxation processes in bulk, but cannot simply be explained, nor the results predicted, with the same models and mechanisms, illustrating the need for an in-depth understanding of interfacial properties. Before addressing specific experiments and their findings, a brief theoretical framework of TR-SFG is given.

10.2 THEORETICAL CONSIDERATIONS

TR-SFG is a form of pump-probe spectroscopy where interfacial molecules are resonantly excited by an optical pump pulse and the subsequent time evolution is tracked by time-delayed SFG probe pulses that are tuned to monitor specific chemical groups on the molecules [31,40]. Although the pump pulse may excite solute molecules beyond the interface, the technique selectively monitors the interfacial molecules due to the surface selectivity of SFG.

We consider two incident laser pulses, one visible pulse and an IR pulse that is resonant with a vibrational transition of an interfacial molecule. The SFG intensity, $I(\omega_{SF})$, is proportional to the square of the sum of all sum frequency susceptibility terms separated into a resonant part, $\chi_R^{(2)}$, where the IR frequency is resonant with the vibration of interest, and a nonresonant part, $\chi_{NR}^{(2)}$. The SFG intensity is given by [9,54,55]

$$I(\omega_{SF}) \propto \left|\chi_{NR}^{(2)} + \chi_R^{(2)}\right|^2 I(\omega_{vis})I(\omega_{IR}) = \left|\chi_{NR}^{(2)} + \sum_q \frac{A_q}{\omega_{IR} - \omega_q + i\Gamma}\right|^2 I(\omega_{vis})I(\omega_{IR}) \quad (10.1)$$

where A_q contains the product of the Raman and IR matrix elements of the qth normal mode, ω_q denotes the resonant frequency of the qth vibrational normal mode, ω_{IR} is the frequency of the IR light, and Γ_q is the spectral width of the vibrational state.

The resonant second-order susceptibility is related to molecular hyperpolarizability, $\alpha_{ijk}^{(2)}$, by a spatial transformation from a molecular coordinate frame to the laboratory coordinate frame [10,100].

$$\chi_{R,IJK}^{(2)} = N\sum_{ijk} < R_{Ii}R_{Jj}R_{Kk} > \alpha_{ijk}^{(2)} \quad (10.2)$$

where R_{Ii}, R_{Jj}, and R_{Kk} are the direction cosine matrix elements that transform the laboratory coordinate frame ($I, J, K = X, Y, Z$) into the molecular coordinate frame ($i, j, k = x, y, z$), N is the surface density of ground-state molecules at equilibrium, and $< >$ denotes the ensemble-averaged orientational distribution. The hyperpolarizability can be assumed to be uniaxial with only one element, $\alpha_{zzz}^{(2)}$, in the case when the SHG wavelength is strongly resonant with a transition along the molecular z-axis. However, in the case of vibrationally resonant SFG experiments, the hyperpolarizability of a vibrational dipole may not be simply considered as a single element, depending upon the symmetry of chemical groups of interest. This arises because the electronic polarizabilities of a chemical bond, which are a measure of the response of the electrons to an external field, are usually anisotropic [73,74,105–108]. Of special interest are $C_{\infty v}$, C_{2v}, and C_{3v} symmetries, which represent a majority of chemical functional groups. There are 11 nonzero microscopic hyperpolarizability elements for the C_{3v} symmetry group, which include $-CH_3$ and $-CF_3$ groups; of these 11 elements, three correspond to symmetric stretching modes and eight to asymmetric stretching modes [9,55,106–111]. For C_{2v} symmetry groups, such as the $-CH_2$ group, there are seven nonvanishing molecular hyperpolarizability tensor elements, three contributing to symmetric stretching modes and four to asymmetric stretching modes [12,86,106–108]. The nonzero hyperpolarizability terms for a $C_{\infty v}$ symmetry group, such as –C=O, –CN, and –CH functional groups, include $\alpha^{(2)}_{xxz}$, $\alpha^{(2)}_{yyz}$, and $\alpha^{(2)}_{zzz}$, where r is equal to the hyperpolarizability ratio $r = \alpha^{(2)}_{xxz}/\alpha^{(2)}_{ZZZ}$ [73,74,110,111]. The ratio r can be obtained from measurements of the Raman depolarization ratio and the relative magnitude of the SFG intensities when both the IR and visible light are polarized parallel and then perpendicular to each other [73,74,110,111].

10.2.1 Ground-State Susceptibility and Orientational Order Parameter

At a liquid surface, only the seven elements $\chi^{(2)}_{XZX} = \chi^{(2)}_{YZY}$, $\chi^{(2)}_{ZXX} = \chi^{(2)}_{ZYY}$, $\chi^{(2)}_{XXZ} = \chi^{(2)}_{YYZ}$, and $\chi^{(2)}_{ZZZ}$ are nonzero (the X and Y axes are equivalent for an isotropic surface) [10,50,54,100]. The second-order susceptibilities of the symmetric stretching modes for a $C_{\infty v}$ functional group, such as –C=O, before photoexcitation are given by [86,110,111]

$$\chi^{(2)}_{XXZ}(-\infty) = \chi^{(2)}_{YYZ}(-\infty) = \frac{1}{2}N\alpha^{(2)}_{zzz}\left[(1+r) < \cos\theta(-\infty) > -(1-r) < \cos^3\theta\,(-\infty) >\right]$$

$$\chi^{(2)}_{XZX}(-\infty) = \chi^{(2)}_{YZY}(-\infty) = \chi^{(2)}_{ZYY}(-\infty) = \chi^{(2)}_{ZXX}(-\infty) = \frac{1}{2}N\alpha^{(2)}_{zzz}(1-r)\left[< \cos\theta\,(-\infty) > - < \cos^3\theta\,(-\infty) >\right]$$

$$\chi^{(2)}_{ZZZ}(-\infty) = N\alpha^{(2)}_{zzz}\left[r < \cos\theta\,(-\infty) > +(1-r) < \cos^3\theta\,(-\infty) >\right] \tag{10.3}$$

Typically, four polarization configurations are measured in SFG experiments using different polarization combinations for the incident and the generated light, that is, SSP, PPP, SPS, and PSS (S and P are mutually orthogonal and perpendicular to the light propagation direction, S is perpendicular to the incident plane, and P is in the incident plane); here, the first letter denotes the polarization of the sum frequency light, the second represents the polarization of the visible light, and the last one indicates the polarization of the IR light. The four polarization combinations are related to the seven second-order susceptibilities by local field factors L_{ii} as [10,86,110–114]

$$\begin{aligned}
\chi^{(2)}_{SSP}(-\infty) &= L_{XX}L_{XX}L_{ZZ}\sin\beta_3\chi^{(2)}_{XXZ}(-\infty)\\
\chi^{(2)}_{SPS}(-\infty) &= L_{XX}L_{ZZ}L_{XX}\sin\beta_2\chi^{(2)}_{XZX}(-\infty)\\
\chi^{(2)}_{PSS}(-\infty) &= L_{ZZ}L_{XX}L_{XX}\sin\beta_1\chi^{(2)}_{ZXX}(-\infty)\\
\chi^{(2)}_{PPP}(-\infty) &= L_{ZZ}L_{ZZ}L_{ZZ}\sin\beta_1\sin\beta_2\sin\beta_3\chi^{(2)}_{ZZZ}(-\infty) + L_{ZZ}L_{YY}L_{YY}\sin\beta_1\cos\beta_2\cos\beta_3\chi^{(2)}_{ZYY}(-\infty)\\
&\quad - L_{YY}L_{ZZ}L_{YY}\cos\beta_1\sin\beta_2\cos\beta_3\chi^{(2)}_{YZY}(-\infty) - L_{YY}L_{YY}L_{ZZ}\cos\beta_1\cos\beta_2\sin\beta_3\chi^{(2)}_{YYZ}(-\infty)
\end{aligned} \tag{10.4}$$

By combining Equations 10.1, 10.3, and 10.4, the SFG intensity of any polarization combination before photoexcitation, $I^{\delta\eta\xi}_{SFG}(-\infty)(\delta,\eta,\xi = \text{S or P})$, can be formulated to give the generalized expression [31,33,40]

$$I^{\delta\eta\xi}_{SFG}(-\infty) \propto \left|\chi^{(2)}_{\delta\eta\xi}(-\infty)\right|^2 I^{\eta}_{Vis}I^{\xi}_{IR} \tag{10.5}$$

Accordingly, one can measure an interfacial order parameter, $D(-\infty)$, at equilibrium, before photoexcitation, from the ratio of any two of the four polarization SFG intensities or null-angle measurements [10,33,54,62,64,75,86,100,110,111,115–117].

$$D(-\infty) = \frac{< \cos\theta(-\infty) >}{< \cos^3\theta(-\infty) >} \tag{10.6}$$

10.2.2 Time-Dependent Sum Frequency Susceptibility

After photoexcitation, the total SFG susceptibility consists of contributions from both ground- and excited-state molecules. The time-dependent SFG intensity is given by [28,31,33,40]

$$I_{SF}(t) \propto \left|\chi_g^{(2)}(t) + \chi_e^{(2)}(t)\right|^2 = \left|\left\langle (N - n(t))\alpha_g^{(2)} \right\rangle_{\rho_g(\Omega,t)} + \left\langle n(t)\alpha_e^{(2)} \right\rangle_{\rho_e(\Omega,t)}\right|^2 \tag{10.7}$$

where $n(t)$ is the density of excited interfacial molecules that decay with a lifetime on the order of a few nanoseconds; the orientational averages, $\langle\rangle_{\rho g(\Omega,t)}$ and $\langle\rangle_{\rho e(\Omega,t)}$, are time dependent because orientational probability functions, $\rho_g(\Omega,t)$ and $\rho_e(\Omega,t)$, of the ground and excited state, respectively, evolve in time. The timescale for the evolution of the orientational probability functions is on the order of hundreds of picoseconds [28,30,32]. The time-dependent susceptibility of the excited state, $\chi_e^{(2)}(t)$, evolves as the solvent reorganizes about the excited interfacial molecules at early time delays (on the order of picoseconds). Because the SFG is a coherent optical process, the contributions from ground and excited states interfere constructively or destructively, depending on their relative phases.

10.3 EXPERIMENTAL TECHNIQUES

10.3.1 Optical System

In broadband SFG experiments, the spectral bandwidth of the incident IR pulse is often much larger than that of the vibrational resonance being studied [75,88,118–122]. Thus, vibrations that lie within the several hundred wave-number IR pulse spectral bandwidth can be detected with each laser pulse, making it possible to rapidly obtain vibrational spectra of interfacial molecules. Good spectral resolution is maintained by implementing a spectrally narrowband visible laser pulse that can be generated directly by a separate picosecond laser system [79,123] or by manipulating the output of a commercially available femtosecond Ti:sapphire regenerative or multipass amplifier system [75,88,118–122]. Broadband SFG, when coupled with an optical pump, allows for the simultaneous acquisition of time- and spectrally resolved SFG signals [31,40]. The SFG wavelengths obtained from an 800 nm picosecond pulse and an IR pulse with center wavelengths ranging from 2.8 to 10 μm fall into the wavelength region of 622–740 nm. Visible-pump SFG-probe experiments that aim to investigate structural dynamics of organic chromophores often implement excitation frequencies (i.e., ~400 nm) that induce large fluorescence background signals that overwhelm the weak SFG signals. The broad fluorescence generated from the samples investigated is often centered around 450 nm, and extends to longer wavelength regions, such as 750 nm, where the SFG is generated. The combination of a 400 nm picosecond laser pulse with the broadband IR pulse leads to SFG wavelengths ranging from 340 to 380 nm, thus avoiding the fluorescence generated by the visible pump beam. To generate a 400 nm narrowband picosecond pulse from a femtosecond 800 nm laser, a phase conjugation technique can be used, which will be described below [31,40].

Figure 10.1 illustrates typical experimental configurations used to obtain time-resolved SFG data [31,40]. A regeneratively amplified Ti:sapphire laser (Spitfire, SpectraPhysics) operating at a repetition rate of 1 kHz with a center wavelength of 800 nm was seeded with a MaiTai Ti:sapphire femtosecond oscillator operating at 80 MHz. For the SFG measurements using 800 nm and broadband IR pulses, a homemade pulse shaper was used to produce a 2.5 ps 800 nm pulse with a 10 cm^{-1} bandwidth. To generate picosecond 400 nm laser pulses, a portion of the femtosecond amplifier output was separated and subsequently split into two parts: one positively chirped to 10 ps and another being negatively chirped to 10 ps. The two chirped beams were spatially collimated and temporally overlapped in a 1 mm BBO (type I) to produce 400 nm 10 ps pulses with a typical frequency

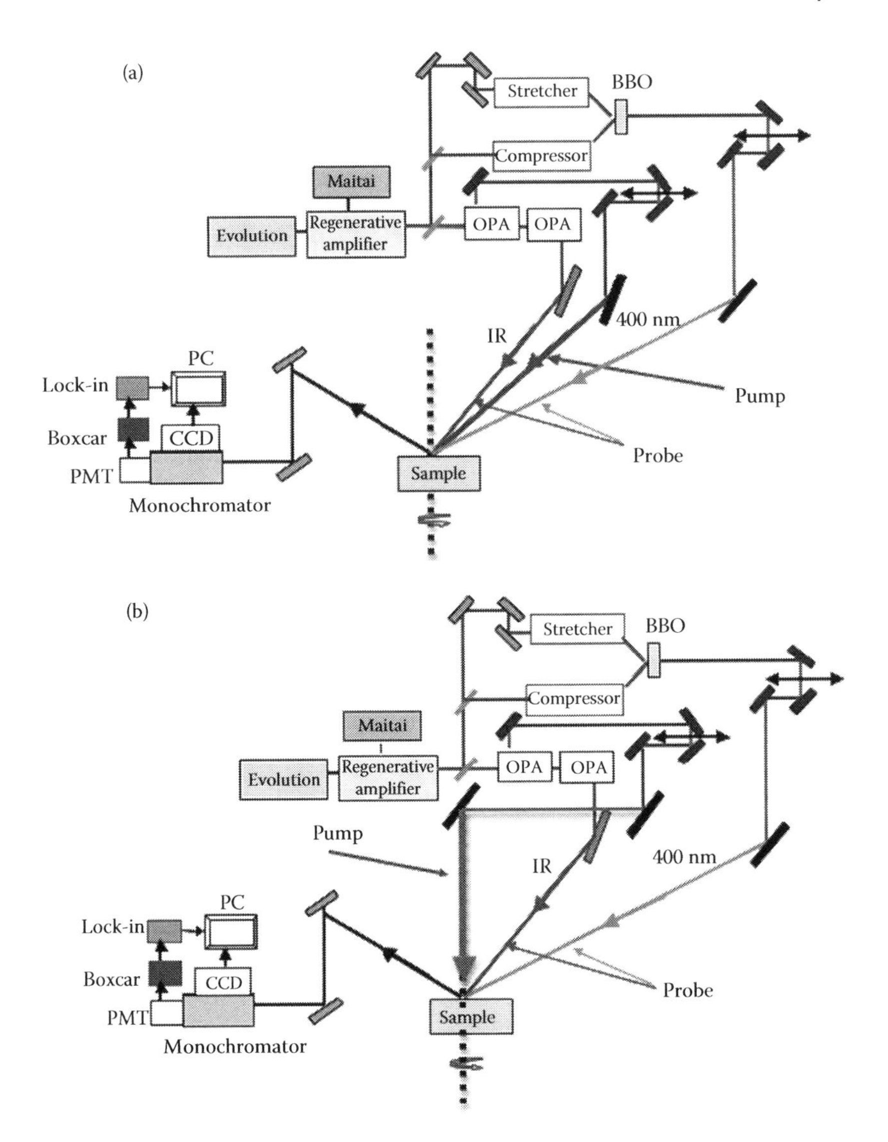

FIGURE 10.1 A schematic of the experimental setups used in the time-resolved SFG experiments. Details of the apparatus can be found in the text. The setup in (a) was used for investigations of solvation dynamics and electron transfer. The setup in (b) is suited for studies of orientational dynamics, where the pump is incident along the surface normal so that a circularly polarized pump can be used to isolate the out-of-plane motions from in-plane motion. Two linearly polarized beams in this configuration can be used for tracking in-plane orientational motions.

bandwidth of 12 cm^{-1} and pulse energy of 7 μJ. The IR beam, with a typical power of 1.5 μJ per pulse at 5.7 μm, was focused onto the sample at an angle of 67° relative to the surface normal using a BaF_2 lens. In both experiments, the 800 nm and the 400 nm picosecond pulses were focused on the sample at an angle of 76° relative to the surface normal. The incident IR and visible pulses were temporally and spatially overlapped at the interface.

Femtosecond pump pulses centered at 423 nm were produced by fourth-harmonic generation of the idler output from an optical parametric amplifier (OPA). There are two geometries for visible-pump SFG probe experiments shown in Figure 10.1. One geometry is shown in Figure 10.1a where the pump beam lies in between the visible picosecond pulse and the broadband femtosecond IR pulse, which is suitable for experiments requiring higher time-resolution such as those studying ultrafast solvation dynamics and interfacial electron transfer [40]. The configuration shown in Figure 10.1b shows the pump light incident along the surface normal [31], which is a configuration appropriate for isolating out-of-plane rotational motion from in-plane rotation motion in experiments using circularly polarized excitation pulses; experiments using two linearly polarized pump beams can be used to extract in-plane rotational dynamics.

The instrumental time-resolution is determined by the cross-correlation of the IR pulse and the visible pump pulse. The cross-correlation of the pump and probe beams was measured independently using difference frequency generation at a BBO or GaAs interface. The instrument response was found to be 185 fs at full width at half maximum. The sample beaker was mounted to a stage rotating at 2.5 rpm to minimize the heating and degradation effects.

10.3.2 Detection System

A 300 mm spectrograph with one entry and two exits (Acton Research, three gratings, including 1200 grooves/mm with 450 nm blazed, 1200 grooves/mm with 500 nm blazed, and 600 grooves/mm with 4 μm blazed) was used to disperse the SFG radiation onto a liquid-nitrogen-cooled, back-thinned charged coupled device (CCD) camera (Roper Scientific, 1340 × 400 pixels) operating at −120°C. The 450 nm blazed grating was used to disperse SFG signals. For some time-resolved experiments, the SFG signal was focused into the monochromator and detected by a photomultiplier tube (PMT) (Hamamatsu). The signal from the PMT was sent into a BOXCAR integrator and then into a lock-in amplifier referenced to a 500 Hz chopper frequency in the pump arm. A translation stage and the signal from the lock-in amplifier were controlled and read out by a computer running LabView software.

10.4 INTERFACIAL SOLVATION DYNAMICS

Solvation is a complicated process that depends on nonspecific interactions, arising from electrostatic and polarization forces, as well as specific interactions, such as hydrogen bonding, which are manifested in both equilibrium and time-dependent processes [29,35,36,40,101,124–142]. In general, solvation properties describe the rearrangement of the solvent molecules to accommodate the solute, thus minimizing the total free energy [29,35,36,40,101,124–145]. Water, being the most abundant liquid on Earth, has been extensively studied using various theoretical and experimental approaches [3,146]. Numerous investigations demonstrate the complexity of dynamical processes that take place in water, in that they involve the motions of water molecules acting as a part in a hydrogen-bonding network of water [127,137,147–151]. It is the dynamics of the formation and breaking of hydrogen bonds that underlies the rotation and translational motions of water molecules and solutes.

A commonly used experimental approach to investigate bulk solvation dynamics makes use of the ultrafast photoexcitation of bulk solute molecules to electronic states that have a charge distribution differing from the ground electronic states, for example, different dipole moments [125,127,152]. The reorganization of the water molecules surrounding the excited-state solute molecules can be tracked in time by a time-delayed probe pulse, which is sensitive to the change in the energy of the excited-state molecules as they evolve to their lowest solvation energy. These experiments have shown that solvation dynamics of organic chromophores, chiefly coumarins, in bulk phase aqueous solutions, can be characterized by three time constants [125,127,152–155]; a sub-50 fs component that is attributed to librational water motions, and two slower timescales, one of which is roughly several hundred femtoseconds and the other on the order of a few picoseconds, that are referred to as diffusive [125,127,152–155]. The diffusive timescales are generally attributed to the breaking and the formation of hydrogen bonds and water-network reorganization, although experimental observations of the water molecular motions have been unable to directly reveal the origin of the two diffusive solvation timescales [149,151,156,157].

Similar to bulk processes [29,35,36,40,43,101,124–142], the energetics and dynamics of interfacial solvation have a profound effect on adsorption, chemical equilibria, polarity, as well as on reaction dynamics at liquid interfaces [13–15,17,18,20,40,43,105,158–161]. Surfaces of biomembranes host many fundamental biological processes, including electron transfer, energy transduction, biosynthesis, and molecular recognition [1–3,162]. Solvent properties of water near these interfaces profoundly affect both the energetics and the dynamics of chemical reactions. For example, the dynamics of solvation can be a rate-limiting step in interfacial electron transfer. Owing to the interactions with the multitude of interfacial moieties present at a biological interface, properties of water near the interface may be drastically different from bulk water. In fact, the term "biological water" has been coined to describe a thin layer of water adjacent to biological structures such as cell membranes, organelles, and proteins [3,143–145]. Intrinsic to the interface is the asymmetrical environment experienced by the chemical species at the interface, whether they are in the form of a solid, liquid, or gas, or are interface charges made up of electrons or ions [133,143–145,163–167]. The solvation dynamics of species within an interfacial environment, although an important aspect of adsorbate–solvent interactions, has been relatively unexplored.

Prior to the research presented here, there have been studies using the pump-time-resolved SHG method to investigate solvation dynamics of C314 at the neat air/water, neutral, positively charged, and negatively charged surfactant/water interfaces [34–39]. An interesting result was the finding that the chemical composition of negatively charged surfactant head groups did not affect the solvation dynamics [35,36,38,39]. The surfactants used were the negatively charged dodecyl sulfate $CH_3(CH_2)_{11}OSO^-_3$ (SDS) and the negatively charged state, $CH_3(CH_2)_{16}COO^-$. It was also found that at the same coverage the charged form, $CH_3(CH_2)_{16}COO^-$, is four times slower than the neutral form, $CH_3(CH_2)_{16}COOH$, indicating the strong effect of surface charge on interfacial water network structure. A comparison of the effects of a positively charged surfactant with negatively charged surfactant molecules indicates a significant difference in their static and dynamic interfacial properties, for example, adsorbate orientation and solvation dynamics. The electrostatic interactions due to the oppositely charged surfactants would tend to align the water molecules in opposite directions and could result in the C314 molecules being at different interfacial locations, one below the layer of surfactant head group for the positively charged dodecyl trimethylammonium bromide (DTAB) and the other between the surfactant layer head groups of the negatively charged SDS.

The investigation of the solvation process is initiated by pumping C314 molecules at the air/water interface from their ground electronic state, S_0, which has a permanent dipole moment of approximately 8 Debye, to the lowest singlet excited state, S_1, which has a permanent dipole moment of approximately 12 Debye [28,34]. The ground- and excited-state energies of the C314 molecules are dependent on the organization of the solvent surrounding the solute molecules. On photoexcitation of interfacial molecules, a nonequilibrium interfacial solvent configuration is induced; that is, the solvent structure corresponding to the ground-state solute now surrounds the excited-state molecule. The subsequent relaxation of the nonequilibrium solvent configuration to one that corresponds to the charge distribution in the S_1 state lowers the energy of the S_1 state.

In previous studies, SFG has been used to probe vibrational relaxation in molecules that have been IR-pumped to excited vibrational states in the ground electronic state [42,102,168], whereas in the work discussed here it is excited electronic states that are generated by the pump light at 423 nm. SFG has also been used with great success in measuring the dynamics of vibrational energy transfer along a hydrocarbon chain [169]. In the SFG work reported here, the interfacial solvation dynamics of C314 at the air/water interface was probed by tuning the frequency of the incident femtosecond IR light into resonance with the ring carbonyl symmetric stretch mode.

Near a vibrational resonance, the SFG hyperpolarizability is given by [170–172]

$$\alpha^{(2)}_{IJK}(\omega_{SF}) = \alpha^{(2)}_{NR}(\omega_{SF}) + \alpha^{(2)}_{R}(\omega_{SF})$$

$$\alpha^{(2)}_{R}(\omega_{SF}) \propto \frac{\mu_{gv',gv}(I)}{\omega_{gv',gv} - \omega_{IR} + i\Gamma_{gv',gv}} \sum_{eu}\left[\frac{\mu_{gv,eu}(J)\mu_{eu,gv'}(K)}{\omega_{eu,gv} - \omega_{IR} - \omega_{Vis}} + \frac{\mu_{gv,eu}(K)\mu_{eu,gv'}(J)}{\omega_{eu,gv'} + \omega_{IR} + \omega_{Vis}}\right] \tag{10.8}$$

where gv, gv', and eu denote the initial, the IR excited vibrational state, and the intermediate vibronic electronic states, respectively. $\mu_{gv',gv}(I)$, $\mu_{gv,eu}(J)$, and $\mu_{eu,gv}(K)$ are the ground-state vibrational transition moment and the electronic transition moments, $\omega_{gv',gv}$ is the vibrational transition frequency of a ground state, and $\Gamma_{gv',gv}$ is the line width for the vibrational transition. It is seen that the Raman part of the SFG hyperpolarizability in Equation 10.8 is dependent on the energies of the electronic states, which is similar to the SHG hyperpolarizability as given by [170–172]

$$\alpha^{(2)}_{IJK}(\omega_{SH}) = \alpha^{(2)}_{NR}(\omega_{SH}) + \alpha^{(2)}_{R}(\omega_{SH})$$

$$\alpha^{(2)}_{R}(\omega_{SH}) \propto \frac{\mu_{gv,eu}(I)}{\omega_{eu,gv} - 2\omega_{Vis} + i\Gamma_{ge}} \sum_{mw}\left[\frac{\mu_{mw,gv}(J)\mu_{eu,mw}(K)}{\omega_{mw,gv} - \omega_{Vis}} + \frac{\mu_{mw,gv}(K)\mu_{eu,mw}(J)}{\omega_{mw,gv} - \omega_{Vis}}\right] \quad (10.9)$$

where gv, mw, and eu denote the initial, intermediate, and final vibronic electronic states. $\mu_{gv,eu}(I)$, $\mu_{mw,gv}(J)$, and $\mu_{eu,mw}(K)$ are electronic transition moments, $\omega_{eu,gv}$ is the transition frequency between the ground and excited state for a molecule, and Γ_{ge} is the line width for the electronic transition.

Figure 10.2 shows the SFG spectra of the –C=O group of C314 at the air/water interface employing the SSP polarization combination for two experiments, one mixing 800 nm with IR and the other mixing 400 nm with IR. The IR light was centered at 5.7 μm with a bandwidth of 150 cm^{-1}. The SFG spectra shown in Figure 10.2 exhibit two peaks: one at 1738 cm^{-1} and the other, smaller peak, at 1680 cm^{-1}. To assign the main peak, a control experiment was performed on coumarin 153 (C153), the structure of which is similar to C314 except that it contains only one carbonyl group as shown in Scheme 10.1. The SFG spectrum of interfacial C153 showed only one peak at 1723 cm^{-1}; accordingly, the larger peak at 1738 cm^{-1} was assigned to the symmetric stretch of the carbonyl in the ring of C314 and the peak at 1680 cm^{-1} is assigned to the symmetric stretch mode of the carbonyl group of the ester group of C314. In Figure 10.2, it is seen that the SFG intensity is much larger when using 400 nm picosecond pulses to generate SFG as compared to experiments using 800 nm picosecond pulses even though the peak laser intensity of the 800 nm laser is a factor of three larger than that of the 400 nm visible pulse. The larger SFG signal obtained with the 400 nm pulse is attributed to the near resonance of the 400 nm light and the SFG wavelength, which enhances the generated SFG light.

The time-dependent excited-state contribution to the SFG signal, $\chi^{(2)}_{e}(t)$, evolves as the solvent reorganizes around the photoexcited molecule (Figure 10.3a). The effect of the excitation pulse is

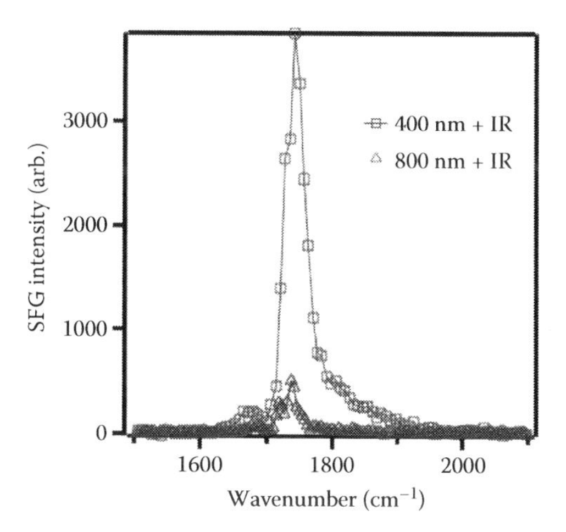

FIGURE 10.2 SFG spectra of the –C=O group of C314 at the air/water interface for the SSP polarized combination. The square data points represent the SFG spectrum of C314 using 400 nm and IR light. The triangle data points denote the SFG spectrum of C314 using 800 nm and IR light.

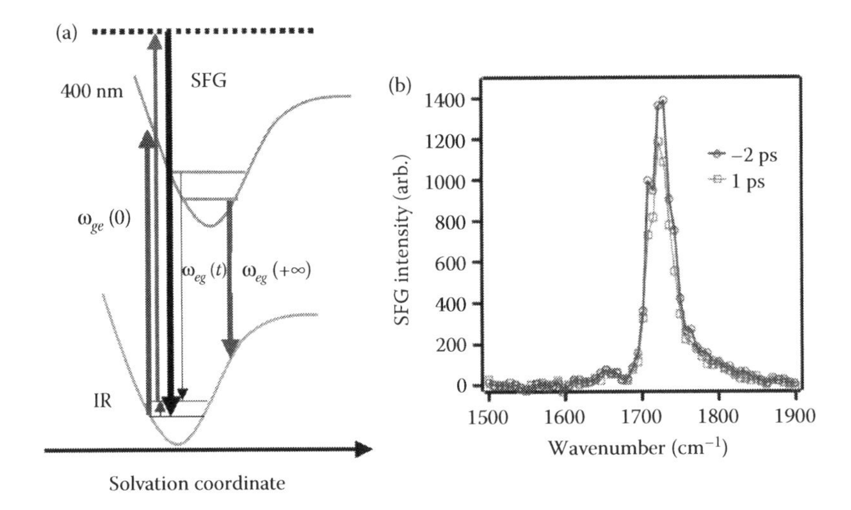

FIGURE 10.3 (a) A schematic of the energy levels of C314 versus solvation coordinate. (b) The SSP SFG spectra of the –C=O group of C314 at two different time delays.

observed in the experiments as a decrease in the SFG intensity at positive time delays (+1 ps) after the pump pulse, as seen in Figure 10.3b. This is to be compared with the SFG signal measured at –2 ps, that is, when the pump pulse arrived after the probe SFG pulses. With respect to the carbonyl chromophore, there was no observable frequency shift in the time-resolved SFG experiments, which indicates that the carbonyl vibrational frequency in the excited singlet state is the same as that in the ground state within the experimental accuracy of 6 cm^{-1}. In addition, no detectable shift was found in experiments probing the carbonyl stretch frequency in the excited singlet state of the structurally similar coumarin 337 in bulk dichloromethane and in bulk DMA [173].

Figure 10.4 shows the time-resolved SFG electric fields observed after excitation of C314 at the air/water interface and the subsequent solvation dynamics. The initial drop of the SFG signal is due to the decrease in the C314 ground-state population and possible interference effects due to

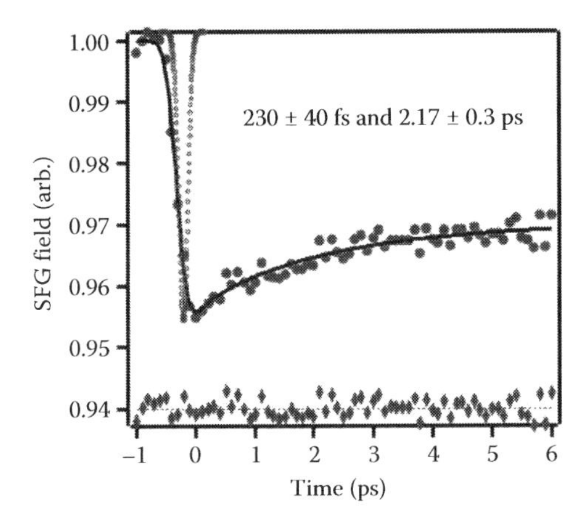

FIGURE 10.4 Electronic excited-state solvation dynamics of the –C=O group in the ring of coumarin 314 at the air/water interface. The cross-correlation time (185 fs, open circles) was measured by difference frequency mixing of the pump and the IR pulses. The data (filled circles) were fit to a sum of two exponential decays (solid line) giving time constants of $\tau_1 = 230 \pm 40$ fs and $\tau_2 = 2.17 \pm 0.3$ ps. Residuals are shown as diamonds to illustrate the quality of the fit.

the phase difference between ground- and excited-state susceptibilities. Using TR-SFG to track the solvation of C314 at the air/water interface, it was found that the time constants obtained from the SFG and SHG experiments are the same, that is, 230 ± 40 fs and 2.17 ± 0.3 ps from TR-SFG, and 250 ± 50 fs and 2.0 ± 0.4 ps from TR-SHG. Similarly, the relative amplitudes of the individual components were found to be the same, that is, SFG: fast component = 0.46, slow component = 0.3 and for SHG: fast component = 0.5, and the slow component = 0.24 [34–36,38,39]. The amplitude-averaged solvation time for the two measurements is 1.0 and 0.88 ps, for SFG and SHG, respectively [34–36,38,39]. An explanation of the SHG and SFG results is that the solvation dynamics express the time-dependent shifts in electronic state energies, which are contained in a similar form for both the SHG hyperpolarizability and the Raman part of the SFG hyperpolarizability (see Equations 10.8 and 10.9). The measured time constants are in good agreement with molecular dynamics simulations of C314 solvation at the air/water interface [174,175].

10.5 INTERFACIAL ELECTRON TRANSFER

Electron transfer is a fundamentally important process intimately tied to the fields of light harvesting in synthetic and biological systems, photo-electrochemistry, liquid/semiconductor junctions, and aerosol interfaces [176–179]. The forward and back electron transfer reactions have been thoroughly studied since the discovery of excited-state charge transfer complexes [163,180–183]. The electron transfer reaction represents a channel for excited molecules to dissipate their energy. This can result in the formation of radical ions or excited-state charge transfer complexes that can relax through radiative or nonradiative decay channels [163,180–183]. The richness of the dynamics and the broad range of potential reaction pathways and intermediates make electron transfer reactions of fundamental scientific interest [163,176–183]. Furthermore, a molecular-level description of interfacial electron transfer dynamics is essential to the continued development of applications in technological fields that make use of electron transfer processes [177,178].

To date, numerous measurements have been made in an effort to discern the reaction mechanisms and ultrafast dynamics involved in excited-state electron transfer reactions in bulk media [163–165,184–188], including fluorescence up-conversion and transient absorption measurements; likewise, transient vibrational spectroscopies, such as IR and Raman, have been used to correlate structural changes in the charge transfer species to the electron transfer reaction [163,164,173,177,178,183,186,188–191]. Theoretical models have also greatly improved the understanding and interpretation of the fundamental physical processes involved in electron transfer reactions [176–178,192,193]. Electron transfer dynamics at liquid/liquid interfaces have been carried out using spectroscopic and electrochemical techniques [46]. However, in the electrochemical technique, both liquids require supporting electrolytes, whose diffusion must be explicitly accounted for in the data analysis.

In the work reviewed here, the dynamics of excited-state electron transfer at the air/water interface are investigated using TR-SFG. The reaction makes use of photoexcited C314 molecules, which act as electron acceptors, and DMA molecules in the ground electronic state that serve as electron donors. DMA was chosen as the electron donor because it is present as a monolayer and therefore negates issues of translational diffusion to encounter and react with interfacial acceptor molecules. A similar approach was used in earlier picosecond studies on photoexcited anthracene reacting with solvent diethylaniline molecules [164,194].

The electronic spectrum of C314 at the water/DMA monolayer interface was measured using SHG and shown in Figure 10.5. The SHG spectrum of interfacial C314 is blue-shifted relative to the absorption maximum in bulk DMA and water, indicating that the interfacial C314 molecules experience a lower polarity than in the bulk liquids. Figure 10.6 shows the vibrational SFG spectrum collected with the SSP polarization combination of C314 at the water/DMA monolayer interface; as in the previous section, the feature at 1738 cm^{-1} was attributed to the symmetric stretching mode of the –C=O in the ring of C314 while the weaker feature at 1680 cm^{-1} was attributed to the stretching

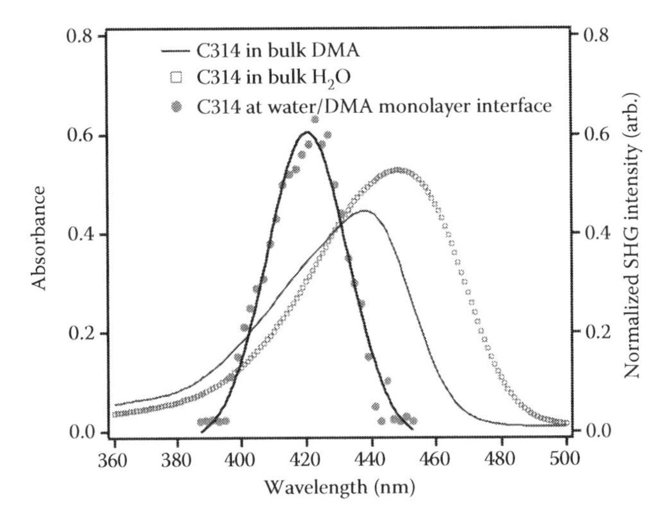

FIGURE 10.5 The SHG spectrum of C314 molecules at the water/DMA monolayer interface (solid circles) and optical absorbance spectra of C314 in bulk DMA (solid line) and in water (open squares).

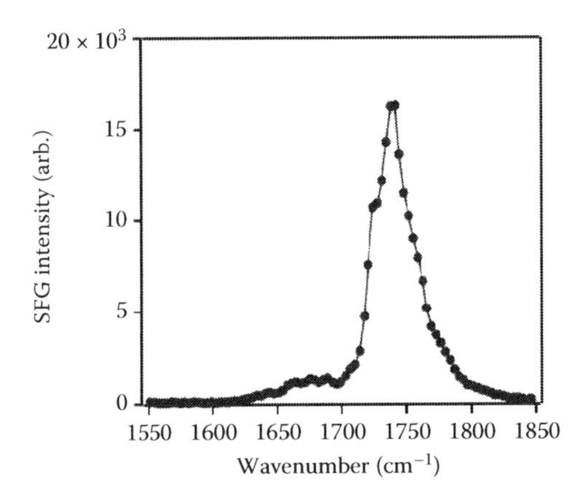

FIGURE 10.6 SFG spectrum of the –C=O group of C314 at the water/DMA monolayer interface for the SSP polarized combination.

mode of the carbonyl group of the ethyl ester [31]. Using the ratio of SSP to PPP SFG intensities and assuming a narrow orientational distribution [75,86], the carbonyl group in the ring was found to be oriented approximately 73° relative to the surface normal and points toward the bulk water in the presence of DMA.

As in the solvation dynamics experiments, an optical pump laser pulse was employed to photoexcite C314. The time-dependent evolution of the interfacial species was obtained by monitoring the SFG signal as a function of time delay between the pump and probe beams and is given by [28,31,33]

$$I_{SF}(t) \propto \left|\chi_g^{(2)}(t) + \chi_e^{(2)}(t)\right|^2 I(\omega_{Vis})I(\omega_{IR}) \tag{10.10}$$

As was the case in the solvation dynamics section, the time-dependent excited-state susceptibility, $\chi_e^{(2)}(t)$, evolves as the solvent reorganizes about the newly excited C314 molecules,

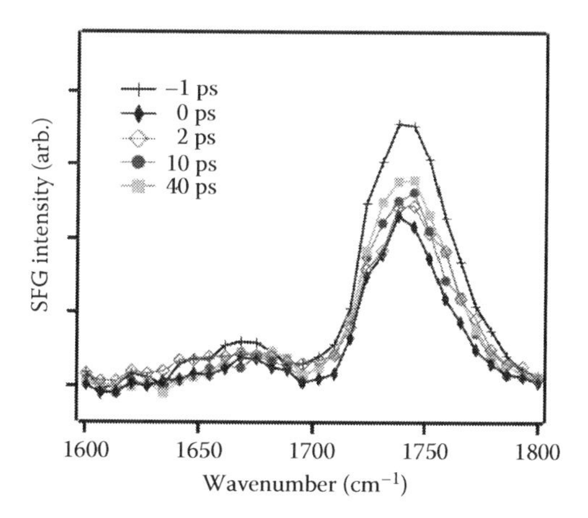

FIGURE 10.7 SFG spectra of the –C=O group of C314 at the water/DMA monolayer interface for the SSP polarized combination at different time delays.

and includes contributions from the orientational motion of C314 and the electron transfer reaction. Figure 10.7 illustrates the change in the SFG intensity as a function of time delay between the pump and probe SFG pulses. As the delay between the pump and SFG probe pulses was varied, the SFG intensity decreased. The decrease in intensity is due to photo-bleaching of the ground-state C314 molecules and the possible interference of the SFG electric fields originating from ground- and excited-state molecules. The time traces of SFG electric fields for C314 at the air/water and water/DMA monolayers interfaces are given in Figure 10.8. In both experiments, the instrument response was found to be 185 fs. After photoexcitation of C314, the following processes occur:

i. Solvation dynamics of C314
ii. Electron transfer from DMA to an excited-state C314 molecule
iii. Rotation of the ground and photoexcited molecules to their equilibrium orientations
iv. Back electron transfer from the C314 radical anion, $C314^{-\bullet}$, to the DMA radical cation, $DMA^{+\bullet}$ as shown in Scheme 10.2

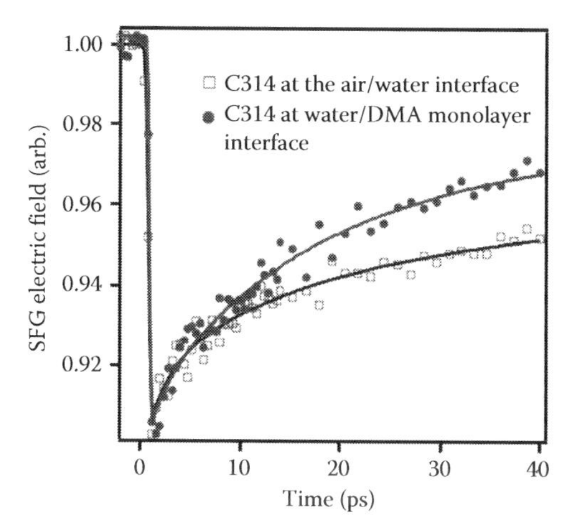

FIGURE 10.8 Time profiles of the SFG electric fields of the –C=O group in the ring of C314 at the water/DMA monolayer (solid circles) and air/water (open squares) interfaces.

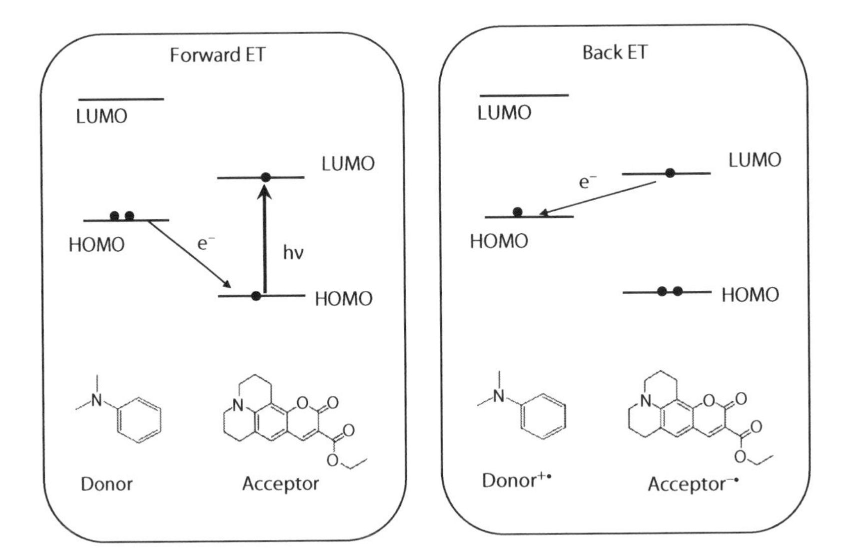

SCHEME 10.2 A schematic representation of interfacial forward electron transfer and back electron transfer between photoexcited C314 molecules and ground-state DMA molecules is given here.

As seen in Figure 10.8, the recovery of the SFG field was found to be faster at the water/DMA monolayer interface than at the air/water interface; however, careful inspection of the transient signals at early time delays, that is, less than ~5 ps, shows that the transient signals are indistinguishable within the uncertainty of the measurements. The reason for this is that the early time reaction dynamics in both experiments are dominated by solvation of excited-state C314 molecules. The differences between the two signals become more apparent at longer time delays where forward electron transfer can take place and solvation has ended. At longer times, both the time constants for the back electron transfer and the rotational motions of the interfacial molecules are similar, hence the observed dynamics contain contributions from both. To separate the back electron transfer timescale from the rotational dynamics of the C314 molecules, experiments were carried out in which only rotational dynamics occur. To do this, the DMA monolayer was replaced with a benzonitrile monolayer, which does not undergo an electron transfer reaction with photoexcited C314 molecules. In addition, benzonitrile has a room temperature viscosity of 1.24 cp, which is close to that of DMA of 1.30 cp. The rotational dynamics in the benzonitrile experiments were used as a substitute for the C314 rotational dynamics in the experiments with DMA. In this way, we sought to separate the back electron transfer dynamics from the orientational relaxation dynamics in the C314-DMA reaction. The rotational relaxation time of C314 at the water/benzonitrile monolayer interface was found to be 247 ± 20 ps.

The back electron transfer dynamics at longer time delays, seen in Figure 10.9, show that at ~600 ps the ground- and excited-state populations have almost completely evolved to the equilibrium configuration that was observed before excitation. The data were fit to a triple exponential function with time constants and amplitudes that describe distinct mechanistic processes in the ground-state recovery. The fastest timescale was found to be 16 ± 2 ps (amplitude of 0.7), which was attributed to forward electron transfer. The second time constant was found to be 174 ± 21 ps (amplitude of 0.2), and was attributed to the back electron transfer dynamics. The third component was the orientational relaxation obtained in the benzonitrile experiment, 247 ± 20 ps (amplitude 0.1). The resulting timescales and amplitudes are in excellent agreement with previous SHG measurements of the electron transfer dynamics of the same system [43].

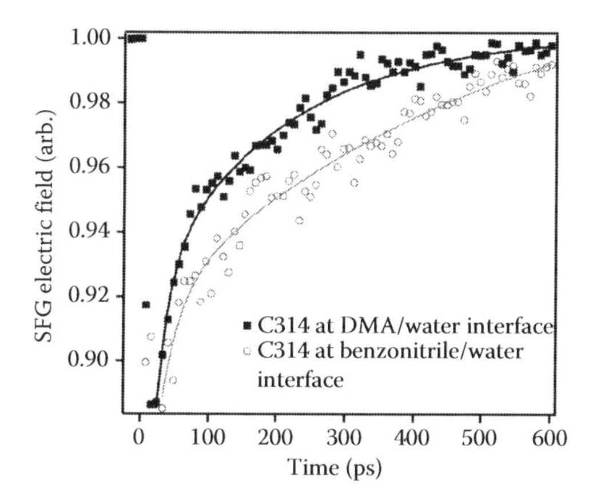

FIGURE 10.9 Time profiles of the SFG electric fields of the –C=O group in the ring of C314 at the water/DMA monolayer (filled squares) and water/benzonitrile monolayer (open circles) interfaces.

Previously published TR-SHG results probed the forward electron transfer rate in two different ways [43,195]; one experiment was performed in which the SHG radiation was resonant with the $S_0 \rightarrow S_1$ transition in C314, essentially tracking the dynamics of the C314 molecule, and the second measurement monitored the dynamics of the formation of the $DMA^{+\bullet}$ radical cation, which directly yields the electron transfer reaction dynamics. This was achieved by tuning the SHG wavelength to match the optical transition in the DMA radical cation, and thereby avoiding contributions from C314 to the SHG signal. The SHG experiments probing the C314 resonance found the forward electron transfer lifetime to be 14 ± 2 ps [43], in agreement with the SFG findings of 16 ± 2 ps. The SHG and SFG experiments confirm that the observed dynamics were that of the electron transfer and not some other dissipative process. The SHG and SFG timescales are in excellent agreement with one another, but there are no known reports of the electron transfer timescales in bulk media with which the interfacial timescales can be directly compared.

To qualitatively compare the forward electron transfer time constant at the interface with the bulk, the structurally similar C153 (its reduction potential is only 0.01 eV larger than C314) can be considered for which both interfacial and bulk electron transfer timescales are known [43,195,196]. In separate experiments, SHG results for C153 showed that the forward electron transfer dynamics are faster at the water/DMA monolayer interface than electron transfer in bulk DMA. Based on the findings that the electron transfer time constants for C314 and C153 at the water/DMA monolayer interface are the same, it was inferred that the electron transfer rate of C314 at the water/DMA monolayer interface is faster than would be observed in bulk DMA.

The dynamics of electron transfer are dependent on both the reorganization energy and the intermolecular interactions of donor and acceptor molecules, which are a function of relative molecular orientation and distance between them. The lower polarity at the interface, as inferred from the separate SHG experiments (see Figure 10.5), suggests that the reorganization free energy is smaller at the interface than in bulk DMA, thus resulting in a faster electron transfer reaction. On the other hand, the higher density of DMA in the bulk solution relative to the interface means that among the larger number of DMA molecules surrounding a C314 molecule there would be more that are oriented such that they can serve as electron donors, than is the case at the water/DMA monolayer interface. Thus, the greater density in the bulk solution would favor a faster electron transfer in bulk DMA than at the interface. The conclusion that the lower reorganization energy at the interface is responsible for the faster electron transfer at the interface should be qualified, noting that both the C314 and DMA molecules are oriented with respect to the surface normal and some of the donor/acceptor pairs might already have favorable reaction geometries before photoexcitation.

10.6 INTERFACIAL ORIENTATIONAL MOTIONS

Dissolved solute molecules in bulk water can freely rotate and diffuse in all directions, in contrast to the degrees of freedom available to a solute molecule at the air/water interface, which are restricted by the asymmetric potential at the interface [26–28,30–33]. Some parts of the molecule may prefer to project down into the water due to hydrophilic interactions, while other parts may preferentially project up into the air due to hydrophobicity. Thus, the asymmetric surface potential induces anisotropic orientational motions at the interface. In other words, the in-plane rotational motions will differ from the out-of-plane motions [3,28,30–32].

It was mentioned previously that SHG pump-probe measurements monitored the rotations of the $S_0 \rightarrow S_1$ transition dipole moment with respect to the interface normal; the transition dipole moment is parallel to the C314 permanent dipole moment axis [26–28,30,33]. In the work discussed here, we use the carbonyl stretching mode in the ring of C314 at the air/water interface to observe the rotational dynamics of the carbonyl group.

Figure 10.10a depicts the coordinate system used in the analysis; the out-of-plane angle, θ, subtends the molecular z-axis and the Z-axis normal to the surface plane; the in-plane angle, φ, subtends the projection of the molecular z-axis in the surface plane [33]. It should be noted that the in-plane angular distribution of ground-state species is isotropic at equilibrium because the time-averaged intermolecular forces are isotropic. This is in contrast to the out-of-plane forces, which are anisotropic and are responsible for the interfacial alignment. To separate the in-plane and out-of-plane orientational evolution, a circularly polarized pump pulse is directed normal to the interfacial plane as will be discussed below.

10.6.1 Absolute Orientation of C314 at the Air/Water Interface

The SFG spectra of C314 at the air/water interface were recorded with SSP and PPP polarization combinations (shown in Figure 10.11), which determine the orientation of the ring carbonyl group at the air/water interface. It was found to be ~110° relative to the surface normal, assuming a narrow orientational distribution. Previous SHG results indicated that the transition dipole axis of the $S_0 \rightarrow S_1$ resonance in C314 is oriented at an angle of 70° with respect to the surface normal [28,38,39,195]. Using a previously described procedure, the absolute orientation of the C314 molecule at the air/water interface was calculated [75]; the results show that the angle between the molecular plane normal and the interfacial normal is 20°, as depicted in Figure 10.12.

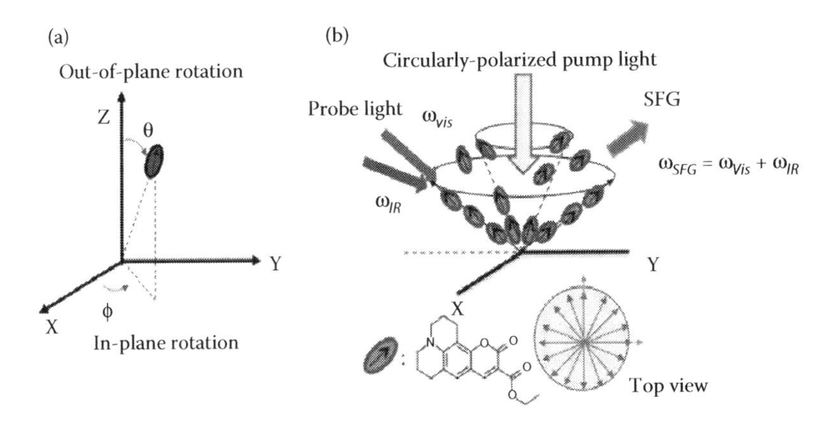

FIGURE 10.10 (a) A schematic showing the out-of-plane orientation angle, θ, and in-plane orientation angles, ϕ. (b) A schematic of the orientational distribution after photoexcitation and a top–down view of the surface.

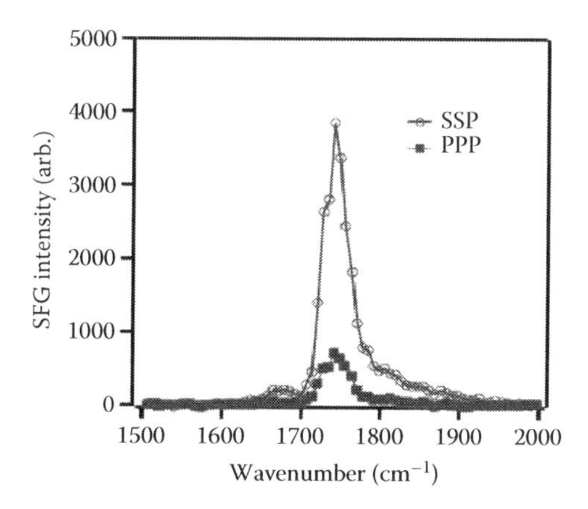

FIGURE 10.11 SFG spectra of the –C=O group of C314 at the air/water interface for the SSP and PPP polarized combination.

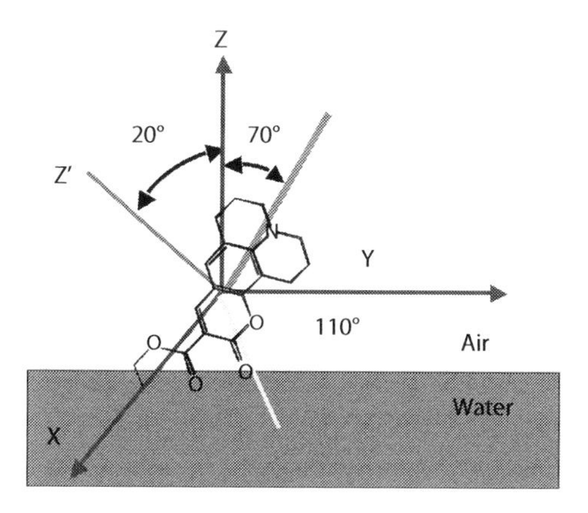

FIGURE 10.12 The absolute orientation of C314 at the air/water interface. The angle between the molecular plane normal and the surface normal is 20°. The electronic dipole direction with respect to the surface normal is 70°, and the carbonyl group is 110° to the surface normal.

10.6.2 Pump-Induced Nonequilibrium Orientational Distribution

The excitation pulse photo-bleaches the ground-state population, generating a nonequilibrium orientational distribution of excited- and the remaining ground-state molecules. Because the circularly polarized pump pulse is incident normal to the interface, it has equal intensity in all directions in the interface plane [28,30–33]. As a consequence, ground- and excited-state molecules are isotropically oriented in the interfacial plane, whereas the out-of-plane orientations are not isotropically distributed and will change in time. Thus, the dynamics represent the time evolution of the out-of-place rotational motions. The excitation probability for a ground-state molecule whose transition dipole moment, μ, is oriented at an angle, θ, relative to the interface normal is given by [28]

$$\left|\mu \vec{E}_c\right|^2 = \left|\mu\right|^2 \left|E\right|^2 \sin^2\theta \tag{10.11}$$

where $\vec{E}_c$ is the incident electric field of the circularly polarized pump. The perturbed ground-state orientational distribution, $\rho_g(\theta_0,0)$, at $t = 0$, induced by photoexcitation, results in a new ground-state orientational distribution given by $\rho_g(\theta_0,t<0)[1-\zeta|\mu|^2|E|^2\sin^2\theta_0]$, where ζ is a collection of constants associated with photoexcitation [33]. It is noted that due to the isotropic excitation in the plane, only the polar angle θ_0 is considered here. Similarly, the orientational distribution of the excited-state molecules at the time of excitation is given by $\rho_g(\theta_0,t<0)\zeta|\mu|^2|E|^2\sin^2\theta_0$. The orientational probability functions for the ground- and excited-state distributions evolve in time as the system relaxes to the equilibrium orientational distributions. The time evolution of $\rho_g(\theta,t)$ and $\rho_e(\theta,t)$ can be written as [28,31,33]

$$\rho_g(\theta,t)=\int G(\theta,t;\theta_0,0)\rho_g(\theta_0,0)\sin\theta_0\,d\theta_0=\int G(\theta,t;\theta_0,0)\rho_g(\theta_0,t<0)[1-\zeta|\mu|^2|E|^2\sin^2\theta_0]\sin\theta_0\,d\theta_0$$
$$\rho_e(\theta,t)=\int G(\theta,t;\theta_0,0)\rho_e(\theta_0,0)\sin\theta_0\,d\theta_0=\int G(\theta,t;\theta_0,0)\rho_g(\theta_0,t<0)\zeta|\mu|^2|E|^2\sin^2\theta_0\sin\theta_0\,d\theta_0 \tag{10.12}$$

where the time evolution function $G(\theta,t;\theta_0,0)$ describes the rotation of a molecule from an orientation angle, θ_0, at $t=0$ to the orientational angle, θ, at a later time t.

The time-dependent ground- and excited-state SFG susceptibilities for any polarization combinations ($\delta\eta\xi$) can be written as [31,33]

$$\chi^{\delta\eta\xi}_{g,\,SFG}(t)=\alpha^{(2)}_{g,\,zzz}\int A(\cos\theta-c\cos^3\theta)\rho_g(\theta,t)\sin\theta\,d\theta$$
$$\chi^{\delta\eta\xi}_{e,\,SFG}(t)=\alpha^{(2)}_{e,\,zzz}\int A(\cos\theta-c\cos^3\theta)\rho_e(\theta,t)\sin\theta\,d\theta \tag{10.13}$$

where the parameters A and c are functions of the incident laser and outgoing SFG angles relative to the surface normal, the polarization angles relative to the incident plane, and the wavelength-dependent dielectric constants of the bulk media and molecular monolayer.

Although the rotational diffusion equation can be solved in bulk systems, where the forces are isotropic, to yield the orientational evolution function [197–200], such an equation does not exist for the anisotropic rotational motions at an interface [199,201]. Currently, molecular dynamics simulations offer a promising approach to the description of orientational motions at interfaces [160,202].

10.6.3 Out-of-Plane Orientational Motion of the –C=O Group

The TR-SFG signal acquired in measurements of the out-of-plane rotational dynamics is given in Figure 10.13. The time evolution is best fit to a single exponential to yield a time constant of 220 ± 20 ps; a multiexponential fit did not improve the fit quality, which could imply that the ground- and excited-state rotational dynamics are different. Since the data are best described by a single decay, it is likely that the rotational dynamics of the ground- and excited-state C314 molecules are similar and that the larger dipole moment of the excited state does not sufficiently reorganize the surrounding water molecules to alter the rotational friction. Previous TR-SHG measurements found that the rotational timescale for the C314 dipole moment at the air/water interface was 343 ± 13 ps [28,43], which is significantly slower than the SFG-measured time constant. It should be recalled that the SHG and SFG measurements make use of two different aspects of the molecular structure, namely, SHG tracks the rotation of the $S_0 \rightarrow S_1$ electronic transition dipole while vibrational SFG tracks the rotation of the carbonyl vibrational bond axis, which in C314 represent two different molecular axes. Thus, it is not surprising that the reorientational dynamics differ significantly between the SHG and SFG measurements, considering that the respective susceptibility elements are different and the rotational motions of different axes will have different energetic barriers. As an example, consider the carbonyl group in the ring of C314 that is pointed toward the bulk

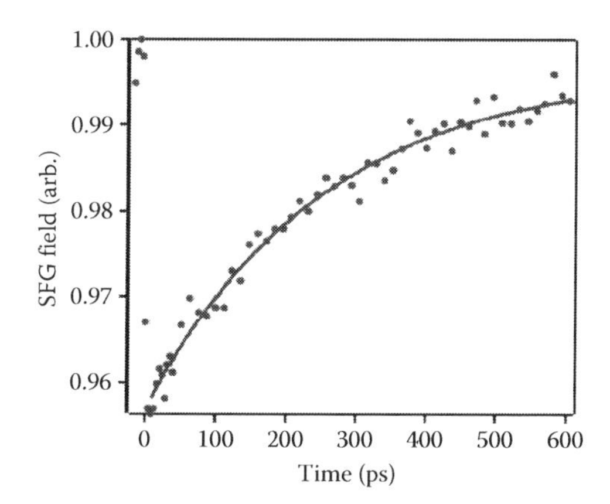

FIGURE 10.13 The SFG electric field of the C=O group in the ring of C314 as a function of pump-probe time delay after excitation of C314 with femtosecond 423 nm circularly polarized light incident along the surface normal. The solid line is a fit to single exponential, yielding a recovery time of 220 ± 20 ps. The data from time delays earlier than 12 ps are not included to eliminate contributions from the solvation dynamics.

solution and is hydrogen bonded. Rotations of the bond axis toward the interface are likely to be energetically unfavorable. In other words, the frictional landscape for rotation about various molecular axes is different and therefore will have different dynamics.

A rigorous treatment of interfacial molecular rotation has not, to date, been reported in the literature; however, the diffusion in a cone model provides qualitative insight to the orientational dynamics. As the name suggests, the model confines molecular rotation to a conical volume [199,200], and predicts a multiexponential decay of the rotational dynamics. The predicted multiexponential decay was not observed in the TR-SFG or TR-SHG data, indicating the limitations of this model to quantitatively predict rotational dynamics at an interface. However, the model does predict that rotational motions for molecules confined to smaller cones will go faster than rotations in a broader cone, fitting in with physical intuition, that is, the smaller the range of polar angles, the faster the interfacial motion [200].

10.7 CONCLUDING REMARKS AND OUTLOOK

TR-SFG has been used to observe the ultrafast dynamics of molecular solvation, rotational motion, and intermolecular electron transfer at aqueous interfaces. In the solvation experiments, two time constants were obtained, 230 ± 40 fs and 2.17 ± 0.3 ps. Within the experimental error, the time constants obtained from SFG experiments are the same as those obtained from TR-SHG experiments. This is attributed to the similar frequency dependence of the SHG hyperpolarizability and the Raman part of the SFG hyperpolarizability.

The ultrafast dynamics of interfacial electron transfer from ground-state DMA to photoexcited coumarin 314 (C314) at a water/DMA monolayer interface was obtained using TR-SFG. The forward electron transfer time constant was found to be 16 ± 2 ps, which is approximately twice as fast as the structurally similar coumarin 153 in bulk DMA. The key factors that determine the rate of an electron transfer are the reorganization free energy and the electronic coupling between the donor and the acceptor molecules. The SFG results showed faster dynamics at the interface relative to that of the bulk, indicating that the reorganization free energy, which reflects the lower polarity of the interface relative to the bulk water, is the dominant factor in the electron transfer rate. The time constant of the back electron transfer process was found to be 174 ± 21 ps and was separated from rotational timescales through comparison with experiments that replaced DMA with benzonitrile, which does not donate an electron to excited-state C314.

The out-of-plane rotation dynamics of photoexcited C314 were probed using the carbonyl vibrational mode to measure the time-dependent changes in its polar angle at the air/water interface. The orientational relaxation time was found to be 220 ± 20 ps, which is much faster than the orientational relaxation time of 343 ± 13 ps of the permanent dipole moment axis of C314 at the same interface, as obtained from pump-second harmonic probe experiments. Possible effects on the rotation of the –C=O bond axis due to the carbonyl group hydrogen bonding with interfacial water are discussed. From the measured equilibrium orientation of the permanent dipole moment axis and the carbonyl axis, and knowledge of their relative orientation in the molecule, the absolute orientation of C314 at the air/water interface was obtained.

The time-resolved pump-probe SFG techniques have the potential to open up new avenues to investigate interfacial structural dynamics of material, biological, and environmental relevance. The structural and dynamical information extracted would advance the understanding of the subtle yet crucial roles that interfaces play.

ACKNOWLEDGMENTS

K.B.E. thanks the National Science Foundation through grant CHE-1057483, and Office of Basic Energy Sciences, Office of Science, US Department of Energy, and DTRA (W911NF-07-1-0116). N.J.T. thanks National Science Foundation through grants CHE-11-11398 and DRM 02-13774. The authors thank Professor Ilan Benjamin, Professor Tony F. Heinz, Professor Hongfei Wang, Professor Feng Wang, Dr. Daohua Song, Dr. Hugen Yan, Dr. Salvo Mamone, Dr. Mahamud Subir, Dr. Xiaoming Shang, Dr. Jian Liu, Dr. Eric A. McArthur, Dr. Soohwan Sul, Dr. Man Xu, Dr. Steffen Jockusch, and Sung-Young Hong for their discussion.

REFERENCES

1. Adam NK. 1991. *The Physics and Chemistry of Surfaces* (Oxford University Press, London).
2. Adamson SW and Cast AP ed. 1997. *Physical Chemistry of Surfaces* (Wiley, New York), 6th Ed.
3. Volkov AG, Deamer DW, Tanelian DL, and Markin VSS. 1998. *Liquid Interfaces in Chemistry and Biology* (Wiley, New York) pp. x, 551 ill. 524 cm.
4. Birdi DK ed. 1984. *Adsorption and the Gibbs Surface Excess* (Plenum Press, New York).
5. Israelachvili JN. 1991. *Intermolecular and Surface Forces* (Academic Press, London) 2nd Ed, pp. xxi, 450 ill. 424 cm.
6. Lyklema J. 1991. *Fundamentals of Interface and Colloid Science* (Academic Press, London). pp. v., ill. 26 cm.
7. Rosen MJ ed. 2004. *Surfactants and Interfacial Phenomena* (Wiley, Chichester, UK), 3rd. Ed, pp. 65–80.
8. Goh MC et al. 1988. Absolute orientation of water-molecules at the neat water-surface. *J. Phys. Chem.* 92(18):5074–5075.
9. Superfine R, Huang JY, and Shen YR. 1991. Nonlinear optical studies of the pure liquid vapor interface—Vibrational-spectra and polar ordering. *Phys. Rev. Lett.* 66(8):1066–1069.
10. Heinz TF, Tom HWK, and Shen YR. 1983. Determination of molecular-orientation of monolayer adsorbates by optical 2nd-harmonic generation. *Phys. Rev. A* 28(3):1883–1885.
11. Vogel V and Shen YR. 1991. Air liquid interfaces and adsorbed molecular monolayers studied with nonlinear optical techniques. *Annu. Rev. Mater. Sci.* 21:515–534.
12. Vogel V, Mullin CS, Shen YR, and Kim MW. 1991. Surface-density of soluble surfactants at the air-water-interface—Adsorption equilibrium studied by 2nd harmonic-generation. *J. Chem. Phys.* 95(6): 4620–4625.
13. Castro A, Bhattacharyya K, and Eisenthal KB. 1991. Energetics of adsorption of neutral and charged molecules at the air-water-interface by 2nd harmonic-generation—Hydrophobic and solvation effects. *J. Chem. Phys.* 95(2):1310–1315.
14. Wang HF, Borguet E, and Eisenthal KB. 1997. Polarity of liquid interfaces by second harmonic generation spectroscopy. *J. Phys. Chem. A* 101(4):713–718.
15. Wang HF, Borguet E, and Eisenthal KB. 1998. Generalized interface polarity scale based on second harmonic spectroscopy. *J. Phys. Chem. B* 102(25):4927–4932.

16. Steel WH, Damkaci F, Nolan R, and Walker RA. 2002. Molecular rulers: New families of molecules for measuring interfacial widths. *J. Am. Chem. Soc.* 124(17):4824–4831.
17. Steel WH and Walker RA. 2003. Measuring dipolar width across liquid-liquid interfaces with 'molecular rulers'. *Nature* 424(6946):296–299.
18. Zhang X, Steel WH, and Walker RA. 2003. Probing solvent polarity across strongly associating solid/liquid interfaces using molecular rulers. *J. Phys. Chem. B* 107(16):3829–3836.
19. Zhang XY, Cunningham MM, and Walker RA. 2003. Solvent polarity at polar solid surfaces: The role of solvent structure. *J. Phys. Chem. B* 107(14):3183–3195.
20. Steel WH, Beildeck CL, and Walker RA. 2004. Solvent polarity across strongly associating interfaces. *J. Phys. Chem. B* 108(41):16107–16116.
21. Eisenthal KB. 1992. Equilibrium and dynamic processes at interfaces by 2nd harmonic and sum frequency generation. *Annu. Rev. Phys. Chem.* 43:627–661.
22. Eisenthal KB. 1996. Photochemistry and photophysics of liquid interfaces by second harmonic spectroscopy. *J. Phys. Chem.* 100(31):12997–13006.
23. Corn RM and Higgins DA. 1994. Optical 2nd-harmonic generation as S probe of surface-chemistry. *Chem. Rev. (Washington, DC, U.S.A.)* 94(1):107–125.
24. Garrett BC, Schenter GK, and Morita A. 2006. Molecular simulations of the transport of molecules across the liquid/vapor interface of water. *Chem. Rev. (Washington, DC, U.S.A.)* 106(4):1355–1374.
25. Davidovits P, Kolb CE, Williams LR, Jayne JT, and Worsnop DR. 2006. Mass accommodation and chemical reactions at gas-liquid interfaces. *Chem. Rev. (Washington, DC, U.S.A.)* 106(4):1323–1354.
26. Castro A, Sitzmann EV, Zhang D, and Eisenthal KB. 1991. Rotational relaxation at the air-water-interface by time-resolved 2nd harmonic-generation. *J. Phys. Chem.* 95(18):6752–6753.
27. Shi X, Borguet E, Tarnovsky AN, and Eisenthal KB. 1996. Ultrafast dynamics and structure at aqueous interfaces by second harmonic generation. *Chem. Phys.* 205(1–2):167–178.
28. Zimdars D, Dadap JI, Eisenthal KB, and Heinz TF. 1999. Anisotropic orientational motion of molecular adsorbates at the air-water interface. *J. Phys. Chem. B* 103(17):3425–3433.
29. Zimdars D and Eisenthal KB. 1999. Effect of solute orientation on solvation dynamics at the air/water interface. *J. Phys. Chem. A* 103(49):10567–10570.
30. Nguyen KT, Shang XM, and Eisenthal KB. 2006. Molecular rotation at negatively charged surfactant/aqueous interfaces. *J. Phys. Chem. B* 110(40):19788–19792.
31. Rao Y, Song DH, Turro NJ, and Eisenthal KB. 2008. Orientational motions of vibrational chromophores in molecules at the air/water interface with time-resolved sum frequency generation. *J. Phys. Chem. B* 112(43):13572–13576.
32. Shang XM, Nguyen K, Rao Y, and Eisenthal KB. 2008. In-plane molecular rotational dynamics at a negatively charged surfactant/aqueous interface. *J. Phys. Chem. C* 112(51):20375–20381.
33. Rao Y, Hong SY, Turro NJ, and Eisenthal KB. 2011. Molecular orientational distribution at interfaces using second harmonic generation. *J. Phys. Chem. C* 115(23):11678–11683.
34. Zimdars D, Dadap JI, Eisenthal KB, and Heinz TF. 1999. Femtosecond dynamics of solvation at the air/water interface. *Chem. Phys. Lett.* 301(1–2):112–120.
35. Benderskii AV and Eisenthal KB. 2000. Effect of organic surfactant on femtosecond solvation dynamics at the air-water interface. *J. Phys. Chem. B* 104(49):11723–11728.
36. Benderskii AV and Eisenthal KB. 2001. Aqueous solvation dynamics at the anionic surfactant air/water interface. *J. Phys. Chem. B* 105(28):6698–6703.
37. Zimdars D and Eisenthal KB. 2001. Static and dynamic solvation at the air/water interface. *J. Phys. Chem. B* 105(28):3993–4002.
38. Benderskii AV and Eisenthal KB. 2002. Dynamical time scales of aqueous solvation at negatively charged lipid/water interfaces. *J. Phys. Chem. A* 106(33):7482–7490.
39. Benderskii AV, Henzie J, Basu S, Shang XM, and Eisenthal KB. 2004. Femtosecond aqueous solvation at a positively charged surfactant/water interface. *J. Phys. Chem. B* 108(37):14017–14024.
40. Rao Y, Turro NJ, and Eisenthal KB. 2010. Solvation dynamics at the air/water interface with time-resolved sum-frequency generation. *J. Phys. Chem. C* 114(41):17703–17708.
41. Sitzmann EV and Eisenthal KB. 1989. Dynamics of intermolecular electronic-energy transfer at an air liquid interface. *J. Chem. Phys.* 90(5):2831–2832.
42. McGuire JA and Shen YR. 2006. Ultrafast vibrational dynamics at water interfaces. *Science* 313(5795): 1945–1948.
43. McArthur EA and Eisenthal KB. 2006. Ultrafast excited-state electron transfer at an organic liquid/aqueous interface. *J. Am. Chem. Soc.* 128(4):1068–1069.

44. Zhao XL, Ong SW, Wang HF, and Eisenthal KB. 1993. New method for determination of surface pk(a) using 2nd-harmonic generation. *Chem. Phys. Lett.* 214(2):203–207.
45. Rao Y, Subir M, McArthur EA, Turro NJ, and Eisenthal KB. 2009. Organic ions at the air/water interface. *Chem. Phys. Lett.* 477(4–6):241–244.
46. Bard AJ and Faulkner LR ed. 2000. *Electrochemical Methods: Fundamentals and Applications* (Wiley, New York), 2nd Ed.
47. Lyklema J. 1991. *Fundamentals of Interface and Colloid Science—Fundamentals* (Academic Press, New York).
48. Levine IN. 1975. *Molecular Spectroscopy* (Wiley, New York) pp. x, 491 ill. 423 cm.
49. Boyd RW. 1992. *Nonlinear optics* (Academic Press, Boston) pp. xiii, 439 ill. 424 cm.
50. Shen YR. 1984. *The Principles of Nonlinear Optics* (Wiley, New York) pp. xii, 563 ill. 525 cm.
51. Mills DL. 1991. *Nonlinear Optics: Basic Concepts* (Springer-Verlag, Berlin) pp. viii, 184 ill. 124 cm.
52. Nelson DR. 2004. Statistical mechanics of membranes and surfaces [Electronic resource]. (World Scientific Publishing, Singapore), pp. xvi, 426 ill.
53. Rosen MJ. 1989. *Surfactants and Interfacial Phenomena* (Wiley, New York), 2nd Ed, pp. xv, 431 ill. 423 cm.
54. Shen YR. 1989. Surface-properties probed by 2nd-harmonic and sum-frequency generation. *Nature* 337(6207):519–525.
55. Vogel V, Mullin CS, and Shen YR. 1991. Probing the structure of the adsorption layer of soluble amphiphilic molecules at the air-water-interface. *Langmuir* 7(6):1222–1224.
56. Kemnitz K et al. 1986. The phase of 2nd-harmonic light generated at an interface and its relation to absolute molecular-orientation. *Chem. Phys. Lett.* 131(4–5):285–290.
57. Eisenthal KB. 1996. Liquid interfaces probed by second-harmonic and sum-frequency spectroscopy. *Chem. Rev. (Washington, DC, U.S.A.)* 96(4):1343–1360.
58. Georgiadis R and Richmond GL. 1991. Wavelength-dependent 2nd harmonic-generation from ag(111) in solution. *J. Phys. Chem.* 95(7):2895–2899.
59. Li JW, He G, and Xu Z. 1997. Determination of the ratio of nonlinear optical tensor components at solid-liquid interfaces using transmission second-harmonic generation (TSHG). *J. Phys. Chem. B* 101(18): 3523–3529.
60. Wark A et al. 1997. In-situ ellipsometry and SHG measurements of the growth of CdS layers on CdxHg1-xTe. *J. Electroanal. Chem.* 435(1–2):173–178.
61. Yamada S and Lee IYS. 1998. Recent progress in analytical SHG spectroscopy. *Anal. Sci.* 14(6):1045–1051.
62. Simpson GJ and Rowlen KL. 1999. An SHG magic angle: Dependence of second harmonic generation orientation measurements on the width of the orientation distribution. *J. Am. Chem. Soc.* 121(11):2635–2636.
63. Petersen PB, Saykally RJ, Mucha M, and Jungwirth P. 2005. Enhanced concentration of polarizable anions at the liquid water surface: SHG spectroscopy and MD simulations of sodium thiocyanide. *J. Phys. Chem. B* 109(21):10915–10921.
64. Rao Y, Tao YS, and Wang HF. 2003. Quantitative analysis of orientational order in the molecular monolayer by surface second harmonic generation. *J. Chem. Phys.* 119(10):5226–5236.
65. Xu YY et al. 2009. Inhomogeneous and spontaneous formation of chirality in the langmuir monolayer of achiral molecules at the air/water interface probed by *in situ* surface second harmonic generation linear dichroism. *J. Phys. Chem. C* 113(10):4088–4098.
66. Wang H, Yan ECY, Borguet E, and Eisenthal KB. 1996. Second harmonic generation from the surface of centrosymmetric particles in bulk solution. *Chem. Phys. Lett.* 259(1–2):15–20.
67. Kriech MA and Conboy JC. 2005. Using the intrinsic chirality of a molecule as a label-free probe to detect molecular adsorption to a surface by second harmonic generation. *Appl. Spectrosc.* 59(6):746–753.
68. Mifflin AL, Konek CT, and Geiger FM. 2006. Tracking oxytetracyline mobility across environmental interfaces by second harmonic generation. *J. Phys. Chem. B* 110(45):22577–22585.
69. Jen SH, Gonella G, and Dai HL. 2009. The effect of particle size in second harmonic generation from the surface of spherical colloidal particles. I: Experimental observations. *J. Phys. Chem. A* 113(16):4758–4762.
70. Fomenko V, Gusev EP, and Borguet E. 2005. Optical second harmonic generation studies of ultrathin high-k dielectric stacks. *J. Appl. Phys.* 97(8):083711.
71. Du Q, Superfine R, Freysz E, and Shen YR. 1993. Vibrational spectroscopy of water at the vapor water interface. *Phys. Rev. Lett.* 70(15):2313–2316.
72. Miranda PB and Shen YR. 1999. Liquid interfaces: A study by sum-frequency vibrational spectroscopy. *J. Phys. Chem. B* 103(17):3292–3307.

73. Zhang D, Gutow J, and Eisenthal KB. 1994. Vibrational-spectra, orientations, and phase-transitions in long-chain amphiphiles at the air-water-interface—Probing the head and tail groups by sum-frequency generation. *J. Phys. Chem.* 98(51):13729–13734.
74. Zhang D, Gutow JH, and Eisenthal KB. 1996. Structural phase transitions of small molecules at air/water interfaces. *J. Chem. Soc.-Faraday Trans.* 92(4):539–543.
75. Rao Y, Comstock M, and Eisenthal KB. 2006. Absolute orientation of molecules at interfaces. *J. Phys. Chem. B* 110(4):1727–1732.
76. Rao Y, Turro NJ, and Eisenthal KB. 2009. Water structure at air/acetonitrile aqueous solution interfaces. *J. Phys. Chem. C* 113(32):14384–14389.
77. Messmer MC, Conboy JC, and Richmond GL. 1995. A resonant sum-frequency generation study of surfactant conformation at the liquid-liquid interface as a function of alkyl chain-length. *Abstr. Pap. Am. Chem. Soc.* 210:116.
78. Richmond GL. 2001. Structure and bonding of molecules at aqueous surfaces. *Annu. Rev. Phys. Chem.* 52:357–389.
79. Hommel EL, Ma G, and Allen HC. 2001. Broadband vibrational sum frequency generation spectroscopy of a liquid surface. *Anal. Sci.* 17(11):1325–1329.
80. Gopalakrishnan S, Liu DF, Allen HC, Kuo M, and Shultz MJ. 2006. Vibrational spectroscopic studies of aqueous interfaces: Salts, acids, bases, and nanodrops. *Chem. Rev. (Washington, DC, U.S.A.)* 106(4):1155–1175.
81. Baldelli S, Schnitzer C, Campbell DJ, and Shultz MJ. 1999. Effect of H_2SO_4 and alkali metal $SO4_2$-/ HSO_4—Salt solutions on surface water molecules using sum frequency generation. *J. Phys. Chem. B* 103(14):2789–2795.
82. Wang CY, Groenzin H, and Shultz MJ. 2005. Comparative study of acetic acid, methanol, and water adsorbed on anatase TiO_2 probed by sum frequency generation spectroscopy. *J. Am. Chem. Soc.* 127(27): 9736–9744.
83. Baldelli S. 2005. Probing electric fields at the ionic liquid-electrode interface using sum frequency generation spectroscopy and electrochemistry. *J. Phys. Chem. B* 109(27):13049–13051.
84. Chen Z, Shen YR, and Somorjai GA. 2002. Studies of polymer surfaces by sum frequency generation vibrational spectroscopy. *Annu. Rev. Phys. Chem.* 53:437–465.
85. Lu R, Gan W, Wu BH, Chen H, and Wang HF. 2004. Vibrational polarization spectroscopy of CH stretching modes of the methylene goup at the vapor/liquid interfaces with sum frequency generation. *J. Phys. Chem. B* 108(22):7297–7306.
86. Wang HF, Gan W, Lu R, Rao Y, and Wu BH. 2005. Quantitative spectral and orientational analysis in surface sum frequency generation vibrational spectroscopy (SFG-VS). *Int. Rev. Phys. Chem.* 24(2):191–256.
87. Geiger FM. 2009. Second harmonic generation, sum frequency generation, and chi((3)): Dissecting environmental interfaces with a nonlinear optical Swiss army knife. *Annu. Rev. Phys. Chem.* 60:61–83.
88. Richter LJ, Petralli-Mallow TP, and Stephenson JC. 1998. Vibrationally resolved sum-frequency generation with broad-bandwidth infrared pulses. *Opt. Lett.* 23(20):1594–1596.
89. Can SZ, Mago DD, Esenturk O, and Walker RA. 2007. Balancing hydrophobic and hydrophilic forces at the water/vapor interface: Surface structure of soluble alcohol monolayers. *J. Phys. Chem. C* 111(25): 8739–8748.
90. Fourkas JT, Walker RA, Can SZ, and Gershgoren E. 2007. Effects of reorientation in vibrational sum-frequency spectroscopy. *J. Phys. Chem. C* 111(25):8902–8915.
91. Ding F et al. 2010. Interfacial organization of acetonitrile: Simulation and experiment. *J. Phys. Chem. C* 114(41):17651–17659.
92. Roke S, Kleyn AW, and Bonn M. 2005. Femtosecond sum frequency generation at the metal-liquid interface. *Surf. Sci.* 593(1–3):79–88.
93. Wurpel GWH, Sovago M, and Bonn M. 2007. Sensitive probing of DNA binding to a cationic lipid monolayer. *J. Am. Chem. Soc.* 129(27):8420.
94. Yamaguchi S and Taharaa T. 2008. Heterodyne-detected electronic sum frequency generation: "Up" versus "down" alignment of interfacial molecules. *J. Chem. Phys.* 129(10):101102.
95. Ye S and Osawa M. 2009. Molecular structures on solid substrates probed by sum frequency generation (SFG) vibration spectroscopy. *Chem. Lett.* 38(5):386–391.
96. Liu J and Conboy JC. 2004. Phase transition of a single lipid bilayer measured by sum-frequency vibrational spectroscopy. *J. Am. Chem. Soc.* 126(29):8894–8895.
97. Fan YB, Chen X, Yang LJ, Cremer PS, and Gao YQ. 2009. On the structure of water at the aqueous/air interface. *J. Phys. Chem. B* 113(34):11672–11679.

98. Bordenyuk AN and Benderskii AV. 2005. Spectrally- and time-resolved vibrational surface spectroscopy: Ultrafast hydrogen-bonding dynamics at D_2O/CaF_2 interface. *J. Chem. Phys.* 122(13):134713.
99. Fu L, Liu J, and Yan ECY. 2011. Chiral sum frequency generation spectroscopy for characterizing protein secondary structures at interfaces. *J. Am. Chem. Soc.* 133(21):8094–8097.
100. Shen YR. 1989. Optical 2nd harmonic-generation at interfaces. *Annu. Rev. Phys. Chem.* 40:327–350.
101. Shang XM, Benderskii AV, and Eisenthal KB. 2001. Ultrafast solvation dynamics at silica/liquid interfaces probed by time-resolved second harmonic generation. *J. Phys. Chem. B* 105(47):11578–11585.
102. Harris AL and Rothberg L. 1991. Surface vibrational-energy relaxation by sum frequency generation—5-wave mixing and coherent transients. *J. Chem. Phys.* 94(4):2449–2457.
103. Ghosh A et al. 2008. Ultrafast vibrational dynamics of interfacial water. *Chem. Phys.* 350(1–3):23–30.
104. Eftekhari-Bafrooei A and Borguet E. 2010. Effect of hydrogen-bond strength on the vibrational relaxation of interfacial water. *J. Am. Chem. Soc.* 132(11):3756–3761.
105. Zhang D, Gutow JH, Eisenthal KB, and Heinz TF. 1993. Sudden structural-change at an air binary-liquid interface—Sum frequency study of the air acetonitrile-water interface. *J. Chem. Phys.* 98(6):5099–5101.
106. Hirose C, Akamatsu N, and Domen K. 1992. Formulas for the analysis of the surface sfg spectrum and transformation coefficients of cartesian sfg tensor components. *Appl. Spectrosc.* 46(6):1051–1072.
107. Hirose C, Akamatsu N, and Domen K. 1992. Formulas for the analysis of surface sum-frequency generation spectrum by ch stretching modes of methyl and methylene groups. *J. Chem. Phys.* 96(2):997–1004.
108. Hirose C, Yamamoto H, Akamatsu N, and Domen K. 1993. Orientation analysis by simulation of vibrational sum-frequency generation spectrum—ch stretching bands of the methyl-group. *J. Phys. Chem.* 97(39):10064–10069.
109. Stanners CD et al. 1995. Polar ordering at the liquid-vapor interface of N-alcohols (C-1-C-8). *Chem. Phys. Lett.* 232(4):407–413.
110. Zhuang X, Miranda PB, Kim D, and Shen YR. 1999. Mapping molecular orientation and conformation at interfaces by surface nonlinear optics. *Phys. Rev. B* 59(19):12632–12640.
111. Wei X, Hong SC, Zhuang XW, Goto T, and Shen YR. 2000. Nonlinear optical studies of liquid crystal alignment on a rubbed polyvinyl alcohol surface. *Phys. Rev. E* 62(4):5160–5172.
112. Luca AAT, Hebert P, Brevet PF, and Girault HH. 1995. Surface 2nd-harmonic generation at air/solvent and solvent/solvent interfaces. *J.Chem. Soc.-Faraday Trans.* 91(12):1763–1768.
113. Brevet PF. 1996. Phenomenological three-layer model for surface second-harmonic generation at the interface between two centrosymmetric media. *J. Chem. Soc.-Faraday Trans.* 92(22):4547–4554.
114. Vidal F and Tadjeddine A. 2005. Sum-firequency generation spectroscopy of interfaces. *Rep. Prog. Phys.* 68(5):1095–1127.
115. Simpson GJ and Rowlen KL. 2000. Orientation-insensitive methodology for second harmonic generation. 1. Theory. *Anal. Chem.* 72(15):3399–3406.
116. Simpson GJ, Westerbuhr SG, and Rowlen KL. 2000. Molecular orientation and angular distribution probed by angle-resolved absorbance and second harmonic generation. *Anal. Chem.* 72(5):887–898.
117. Simpson GJ. 2001. New tools for surface second-harmonic generation. *Appl. Spectrosc.* 55(1):16A–32A.
118. Esenturk O and Walker RA. 2006. Surface vibrational structure at alkane liquid/vapor interfaces. *J. Chem. Phys.* 125(17):174701.
119. Voges AB et al. 2004. Carboxylic acid- and ester-functionalized siloxane scaffolds on glass studied by broadband sum frequency generation. *J. Phys. Chem. B* 108(48):18675–18682.
120. Smits M et al. 2007. Polarization-resolved broad-bandwidth sum-frequency generation spectroscopy of monolayer relaxation. *J. Phys. Chem. C* 111(25):8878–8883.
121. Wang ZH et al. 2008. Ultrafast dynamics of heat flow across molecules. *Chem. Phys.* 350(1–3):31–44.
122. Jayathilake HD et al. 2009. Molecular order in langmuir-blodgett monolayers of metal-ligand surfactants probed by sum frequency generation. *Langmuir* 25(12):6880–6886.
123. Velarde L et al. 2011. Communication: Spectroscopic phase and lineshapes in high-resolution broadband sum frequency vibrational spectroscopy: Resolving interfacial inhomogeneities of “identical” molecular groups. *J. Chem. Phys.* 135(24):241102.
124. Maroncelli M and Fleming GR. 1987. Picosecond solvation dynamics of coumarin-153—The importance of molecular aspects of solvation. *J. Chem. Phys.* 86(11):6221–6239.
125. Kahlow MA, Jarzeba W, Kang TJ, and Barbara PF. 1989. Femtosecond resolved solvation dynamics in polar-solvents. *J. Chem. Phys.* 90(1):151–158.
126. Kang TJ, Jarzeba W, Barbara PF, and Fonseca T. 1990. A photodynamical model for the excited-state electron-transfer of bianthryl and related molecules. *Chem. Phys.* 149(1–2):81–95.
127. Jimenez R, Fleming GR, Kumar PV, and Maroncelli M. 1994. Femtosecond solvation dynamics of water. *Nature* 369(6480):471–473.

128. Rosenthal SJ, Jimenez R, Fleming GR, Kumar PV, and Maroncelli M. 1994. Solvation dynamics in methanol—Experimental and molecular-dynamics simulation studies. *J. Mol. Liq.* 60(1–3):25–56.
129. Gardecki J, Horng ML, Papazyan A, and Maroncelli M. 1995. Ultrafast measurements of the dynamics of solvation in polar and nondipolar solvents. *J. Mol. Liq.* 65–66, 49–57.
130. Horng ML, Gardecki JA, Papazyan A, and Maroncelli M. 1995. Subpicosecond measurements of polar solvation dynamics—Coumarin-153 revisited. *J. Phys. Chem.* 99(48):17311–17337.
131. Kumar PV and Maroncelli M. 1995. Polar solvation dynamics of polyatomic solutes—Simulation studies in acetonitrile and methanol. *J. Chem. Phys.* 103(8):3038–3060.
132. Reid PJ and Barbara PF. 1995. Dynamic solvent effect on betaine-30 electron-transfer kinetics in alcohols. *J. Phys. Chem.* 99(11):3554–3565.
133. Shi XL, Long FH, and Eisenthal KB. 1995. Electron solvation in neat alcohols. *J. Phys. Chem.* 99(18): 6917–6922.
134. Sarkar N, Datta A, Das S, and Bhattacharyya K. 1996. Solvation dynamics of coumarin 480 in micelles. *J. Phys. Chem.* 100(38):15483–15486.
135. Passino SA, Nagasawa Y, Joo T, and Fleming GR. 1997. Three-pulse echo peak shift studies of polar solvation dynamics. *J. Phys. Chem. A* 101(4):725–731.
136. de Boeij WP, Pshenichnikov MS, and Wiersma DA. 1998. Ultrafast solvation dynamics explored by femtosecond photon echo spectroscopies. *Annu. Rev. Phys. Chem.* 49:99–123.
137. Lang MJ, Jordanides XJ, Song X, and Fleming GR. 1999. Aqueous solvation dynamics studied by photon echo spectroscopy. *J. Chem. Phys.* 110(12):5884–5892.
138. Levinger NE. 2000. Ultrafast dynamics in reverse micelles, microemulsions, and vesicles. *Curr. Opin. Colloid Interface Sci.* 5(1–2):118–124.
139. Pant D and Levinger NE. 2000. Polar solvation dynamics in nonionic reverse micelles and model polymer solutions. *Langmuir* 16(26):10123–10130.
140. Faeder J and Ladanyi BM. 2001. Solvation dynamics in aqueous reverse micelles: A computer simulation study. *J. Phys. Chem. B* 105(45):11148–11158.
141. Hazra P and Sarkar N. 2002. Solvation dynamics of Coumarin 490 in methanol and acetonitrile reverse micelles. *Phys. Chem. Chem. Phys.* 4(6):1040–1045.
142. Matyushov DV. 2005. On the microscopic theory of polar solvation dynamics. *J. Chem. Phys.* 122(4):044502.
143. Gauduel Y and Rossky PJ ed. 1994. *Ultrafast Reaction Dynamics and Solvent Effects: Royaumont, France 1993* (American Institute of Physics, New York) pp. x, 564 ill. 524 cm.
144. Simon JD ed. 1994. *Ultrafast Dynamics of Chemical Systems* (Kluwer Academic Publishers, Dordrecht) pp. vi, 385 ill. 325 cm.
145. Dogonadze RR. 1985. *The Chemical Physics of Solvation* (Elsevier, Amsterdam) pp. 3, v. ill. 25 cm.
146. Marechal Y. 2007. *The Hydrogen Bond and the Water Molecule: The Physics and Chemistry of Water, Aqueous and Bio Media* (Elsevier, Amsterdam) 1st Ed, pp. xiii, 318 ill. 325 cm.
147. Auer BM and Skinner JL. 2009. Water: Hydrogen bonding and vibrational spectroscopy, in the bulk liquid and at the liquid/vapor interface. *Chem. Phys. Lett.* 470(1–3):13–20.
148. Gale GM, Gallot G, Hache F, and Lascoux N. 1999. Femtosecond dynamics of hydrogen bonds in liquid water: A real time study. *Phys. Rev. Lett.* 82:1–4.
149. Koffas TS, Kim J, Lawrence CC, and Somorjai GA. 2003. Detection of immobilized protein on latex microspheres by IR-visible sum frequency generation and scanning force microscopy. *Langmuir* 19(9):3563–3566.
150. Tokmakoff A. 2003. Coherent 2D IR spectroscopy: Molecular structure and dynamics in solution. *J. Phys. Chem. A* 107(27):5258–5279.
151. Yeremenko S, Pshenichnikov MS, and Wiersma DA. 2003. Hydrogen-bond dynamics in water explored by heterodyne-detected photon echo. *Chem. Phys. Lett.* 369:107–113.
152. Nagarajan V, Brearley AM, Kang TJ, and Barbara PF. 1987. Time-resolved spectroscopic measurements on microscopic solvation dynamics. *J. Chem. Phys.* 86(6):3183–3196.
153. Chandler D et al. 1988. Solvation–General discussion. *Faraday Discuss.* 85:77–106.
154. Jarzeba W, Walker GC, Johnson AE, Kahlow MA, and Barbara PF. 1988. Femtosecond microscopic solvation dynamics of aqueous-solutions. *J. Phys. Chem.* 92(25):7039–7041.
155. Kahlow MA, Kang TJ, and Barbara PF. 1988. Transient solvation of polar dye molecules in polar aprotic-solvents. *J. Chem. Phys.* 88(4):2372–2378.
156. Asbury JB, Steinel T, and Fayer MD. 2004. Hydrogen bond networks: Structure and evolution after hydrogen bond breaking. *J. Phys. Chem. B* 108(21):6544–6554.
157. Woutersen S and Bakker HJ. 1999. Hydrogen bond in liquid water as a Brownian oscillator. *Phys. Rev. Lett.* 83(10):2077–2080.

158. Benjamin I. 1991. Theoretical-study of ion solvation at the water liquid-vapor interface. *J. Chem. Phys.* 95(5):3698–3709.
159. Benjamin I. 2002. Chemical reaction dynamics at liquid interfaces: A computational approach. *Prog. React. Kinet. Mech.* 27(2):87–126.
160. Benjamin I. 2009. Solute dynamics at aqueous interfaces. *Chem. Phys. Lett.* 469(4–6):229–241.
161. Esenturk O and Walker RA. 2004. Surface structure at hexadecane and halo-hexadecane liquid/vapor interfaces. *J. Phys. Chem. B* 108(30):10631–10635.
162. Volkov AG ed. 2001. Liquid interfaces in chemical, biological, and pharmaceutical applications [Electronic resource]. In *Surfactant Science Series; v. 95.* (Marcel Dekker, New York).
163. Chuang TJ and Eisenthal KB. 1975. Studies of excited-state charge-transfer interactions with picosecond laser pulses. *J. Chem. Phys.* 62(6):2213–2222.
164. Gnadig K and Eisenthal KB. 1977. Picosecond kinetics of excited charge-transfer interactions. *Chem. Phys. Lett.* 46(2):339–342.
165. Wang Y, Crawford MK, McAuliffe MJ, and Eisenthal KB. 1980. Picosecond laser studies of electron solvation in alcohols. *Chem. Phys. Lett.* 74(1):160–165.
166. Long FH, Shi XL, Lu H, and Eisenthal KB. 1994. Electron photodetachment from halide-ions in solution—Excited-state dynamics in the polarization well. *J. Phys. Chem.* 98(30):7252–7255.
167. Long FH, Lu H, and Eisenthal KB. 1995. Femtosecond transient absorption studies of electrons in liquid alkanes. *J. Phys. Chem.* 99(19):7436–7438.
168. Sovago M et al. 2008. Vibrational response of hydrogen-bonded interfacial water is dominated by intramolecular coupling. *Phys. Rev. Lett.* 100(17):173901.
169. Wang ZH et al. 2008. Ultrafast dynamics of heat flow across molecules. *Chem. Phys. Lett.* 350(1–3): 31–44.
170. Lin SH and Villaeys AA. 1994. Theoretical description of steady-state sum-frequency generation in molecular adsorbates. *Phys. Rev. A* 50(6):5134–5144.
171. Villaeys AA, Pflumio V, and Lin SH. 1994. Theory of 2nd-harmonic generation of molecular-systems—The case of coincident pulses. *Phys. Rev. A* 49(6):4996–5014.
172. Lin SH et al. 1996. Molecular theory of second-order sum-frequency generation. *Physica B* 222(1–3): 191–208.
173. Wang CF, Akhremitchev B, and Walker GC. 1997. Femtosecond infrared and visible spectroscopy of photoinduced intermolecular electron transfer dynamics and solvent-solute reaction geometries: Coumarin 337 in dimethylaniline. *J. Phys. Chem. A* 101(15):2735–2738.
174. Pantano DA and Laria D. 2003. Molecular dynamics study of solvation of coumarin 314 at the water/air interface. *J. Phys. Chem. B* 107:2971–2977.
175. Pantano DA, Sonoda MT, Skaf MS, and Laria D. 2005. Solvation of coumarin 314 at water/air interfaces containing anionic surfactants. I. Low coverage. *J. Phys. Chem. B* 109(15):7365–7372.
176. Marcus RA. 1956. On the theory of oxidation-reduction reactions involving electron transfer. 1. *J. Chem. Phys.* 24(5):966–978.
177. Bolton JR, Mataga N, McLendon G ed. 1991. *Electron Transfer in Inorganic, Organic, and Biological Systems* (American Chemical Society, Washington, DC) pp. viii, 295 ill. 224 cm.
178. Bixon M and Jortner J ed. 1999. *Electron Transfer—From Isolated Molecules to Biomolecules* (Wiley, New York) pp. 2, v. ill. 24 cm.
179. Gratzel M. 1989. *Heterogeneous Photochemical Electron Transfer* (CRC Press, Boca Raton, FL) pp. 159, ill. 127 cm.
180. Leonhardt H and Weller A. 1963. Elektronenubertragungsreaktionen des angeregten perylens [Electron transfer reaction of the excited perylene]. *Berichte Der Bunsen-Gesellschaft Fur Physikalische Chemie* 67(8):791–795.
181. Syage JA, Felker PM, and Zewail AH. 1984. Picosecond excitation and selective intramolecular rates in supersonic molecular-beams.3. Photochemistry and rates of a charge-transfer reaction. *J. Chem. Phys.* 81(5):2233–2256.
182. Chuang TJ and Eisentha KB. 1973. Measurements of rate of excited charge-transfer complex-formation using picosecond laser pulses. *J. Chem. Phys.* 59(4):2140–2141.
183. Eisenthal KB. 1975. Studies of chemical and physical processes with picosecond lasers. *Acc. Chem. Res.* 8(4):118–123.
184. Wang Y, Crawford MK, and Eisenthal KB. 1980. Intramolecular excited-state charge-transfer interactions and the role of ground-state conformations. *J. Phys. Chem.* 84(21):2696–2698.
185. Wang Y, Crawford MC, and Eisenthal KB. 1982. Picosecond laser studies of intramolecular excited-state charge-transfer dynamics and small-chain relaxation. *J. Am. Chem. Soc.* 104(22):5874–5878.

186. Weidemaier K, Tavernier HL, Swallen SF, and Fayer MD. 1997. Photoinduced electron transfer and geminate recombination in liquids. *J. Phys. Chem. A* 101(10):1887–1902.
187. Castner EW, Kennedy D, and Cave RJ. 2000. Solvent as electron donor: Donor/acceptor electronic coupling is a dynamical variable. *J. Phys. Chem. A* 104(13):2869–2885.
188. Fox MA. 1988. *Photoinduced Electron Transfer* (Elsevier, New York) pp. 4, v. ill. 25 cm.
189. Frontiera RR, Dasgupta J, and Mathies RA. 2009. Probing interfacial electron transfer in coumarin 343 sensitized TiO_2 nanoparticles with femtosecond stimulated Raman. *J. Am. Chem. Soc.* 131(43):15630–15632.
190. Huber R, Moser JE, Gratzel M, and Wachtveitl J. 2002. Observation of photoinduced electron transfer in dye/semiconductor colloidal systems with different coupling strengths. *Chem. Phys.* 285(1): 39–45.
191. Johnson AE, Tominaga K, Walker GC, Jarzeba W, and Barbara PF. 1993. Femtosecond electron-transfer—Experiment and theory. *Pure Appl. Chem.* 65(8): 1677–1680.
192. Benjamin I and Pollak E. 1996. Variational transition state theory for electron transfer reactions in solution. *J. Chem. Phys.* 105(20):9093–9103.
193. Vieceli J and Benjamin I. 2004. Electron transfer at the interface between water and self-assembled monolayers. *Chem. Phys. Lett.* 385(1–2):79–84.
194. Chuang TJ, Cox RJ, and Eisentha KB. 1974. Picosecond studies of excited charge-transfer interactions in anthracene-$(CH_2)3$-*N,N*-dimethylaniline systems. *J. Am. Chem. Soc.* 96(22):6828–6831.
195. McArthur EA. 2008. Time-resolved second harmonic generation (TRSHG) studies at aqueous liquid interfaces. PhD thesis.
196. Shirota H, Pal H, Tominaga K, and Yoshihara K. 1998. Substituent effect and deuterium isotope effect of ultrafast intermolecular electron transfer: Coumarin in electron-donating solvent. *J. Phys. Chem. A* 102(18):3089–3102.
197. Chuang TJ and Eisentha KB. 1972. Theory of fluorescence depolarization by anisotropic rotational diffusion. *J. Chem. Phys.* 57(12):5094.
198. Tao T. 1969. Time-dependent fluorescence depolarization and brownian rotational diffusion coefficients of macromolecules. *Biopolymers* 8(5):609–632.
199. Szabo A. 1984. Theory of fluorescence depolarization in macromolecules and membranes. *J. Chem. Phys.* 81(1):150–167.
200. Wang CC and Pecora R. 1980. Time-correlation functions for restricted rotational diffusion. *J. Chem. Phys.* 72(10):5333–5340.
201. Gengeliczki Z, Rosenfeld DE, and Fayer MD. 2010. Theory of interfacial orientational relaxation spectroscopic observables. *J. Chem. Phys.* 132(24):244703.
202. Johnson ML, Rodriguez C, and Benjamin I. 2009. Rotational dynamics of strongly adsorbed solute at the water surface. *J. Phys. Chem. A* 113(10):2086–2091.

11 Energy Transport in Molecules Studied by Relaxation-Assisted 2D IR Spectroscopy

Igor V. Rubtsov

CONTENTS

11.1 INTRODUCTION

The two-dimensional infrared (2D IR) method allows correlating the frequency distributions of different vibrators in a molecular system, thus providing structural constraints as well as important insight into molecular dynamics [1–6]. The applicability of the 2D IR method as an analytical technique will be decided by several factors, such as its sensitivity, availability of convenient spatially localized vibrational labels, and ability of collecting efficiently a large number of structural constraints for any given molecular system. Dual-frequency 2D IR measurements using a three-pulse noncollinear geometry with heterodyned detection provide the highest sensitivity and allow collecting a large number of cross peaks, while permitting a suppression of the diagonal peaks [7–9]. The details of such measurements are discussed in the next section focusing on sensitivity and phase stability issues. A recently proposed relaxation-assisted 2D IR (RA 2D IR) approach [10] permits enhancing the cross-peak amplitudes by over an order of magnitude as well as delivers additional structural constraints that can be linked to bond connectivity patterns and vibrational mode delocalization extents. The principles and examples illustrating the advantages of RA 2D IR spectroscopy are given in Section 11.3. RA 2D IR permits measuring energy transport on a molecular level; such transport molecular systems are discussed in Section 11.4. 2D IR measurements involving delocalized modes in the fingerprint region are discussed in Section 11.5.

11.2 DUAL-FREQUENCY 2D IR MEASUREMENTS

The schematic of the dual-frequency 2D IR setup is shown in Figure 11.1. Two mid-IR beams (mIR1 and mIR2) are generated using two optical parametric amplifiers (OPAs) and two different-frequency generation (DFGs) units. Both mid-IR beams are split into two parts, forming k_1, k_2, k_3, and local oscillator (LO) beams. These beams then pass through computer-controlled delay stages, which allow setting the delays between the IR pulses with high precision. After the delay stages, the polarizations of all four beams are set using pairs of waveplates and wire-grid polarizers. The k_1, k_2, and k_3 beams are focused into the sample (Figure 11.1) so that the phase-matching conditions are satisfied. The phase-matching geometry depends on the frequency of the third-order response. Examples of phase-matching geometries for the third-order response at $\omega_{EF} = \omega_3$ are shown in Figure 11.2 for the case of $\omega_3 < (\omega_1 = \omega_2)$ (Figure 11.2a) and $\omega_3 > (\omega_1 = \omega_2)$ (Figure 11.2b); here, $\hbar\omega_{1-3}$ are the energies of the states of the molecular system excited by the respective k_{1-3} beams. The beam directions should satisfy the energy and momentum conservation relations: $\omega_{EF} = -\omega_1 + \omega_2 + \omega_3$, $k_{EF} = -k_1 + k_2 + k_3$. Here, each beam is labeled in the laboratory frame using its wave vector. Note that the labeling of a beam as k_1 does not necessarily mean that the laser pulse of this beam interacts first with the sample. The actual pulse ordering is determined by the time delays introduced for each beam, which is traditionally represented by double-sided Feynman diagrams [4]. Figure 11.3 shows Feynman diagrams describing a cross-peak pair for the so-called rephasing experiments (R_1 and R_2) where the k_1 pulse comes first and nonrephasing experiments (N_1 and N_2) where the k_2 pulse comes first. Notice that the geometry of the three beams before the focusing parabolic reflector (P1) needs to be changed if a different cross peak is to be targeted (Figure 11.2). This requirement complicates the scanning of the 2D IR spectrum over a large spectral region. The third-order signal emitted in the phase-matching direction is recollimated by the parabolic reflector (P2), mixed with the LO using a 50/50 beam splitter (BS), and measured either by a pair of single-channel detectors in a balanced detection scheme [11] or with an array detector (D3) attached to a monochromator. With the pair of single-channel detectors, the interferograms of the third-order field and the LO are measured as a function of two delay times ($M(\tau, t)$), the dephasing time τ, which is the delay between the first two pulses, k_1 and k_2, and the detection time, t, the delay between the k_3 pulse and the LO. The waiting time, T, is kept constant in these measurements. 2D IR spectrum is obtained by a double Fourier transformation of the $M(\tau, t)$ data sets and plotted as a contour plot with ω_t as abscissa and ω_τ as

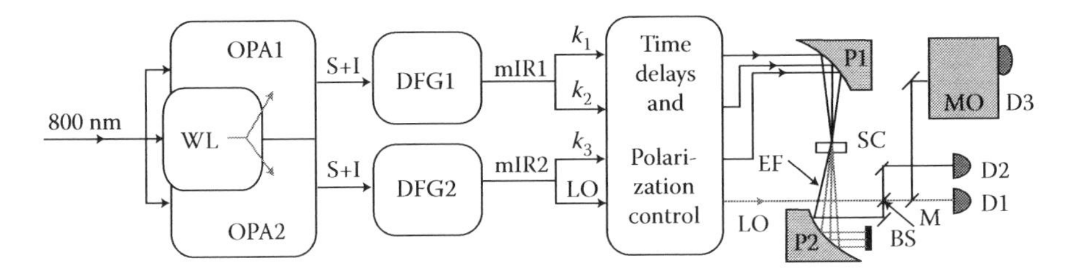

FIGURE 11.1 Scheme of the dual-frequency 2D IR setup with heterodyned detection. Two OPAs pumped by the output of a common Ti:S regenerative amplifier (ω_F) and seeded by a common white light (WL) source. Two pairs of near-IR pulses, signal (ω_S) and idler (ω_I), are generated in a parametric generation process: $\omega_I = \omega_F - \omega_S$. A difference frequency between the signal and idler pulses is generated in each pair: $\omega_{mIR} = \omega_S - \omega_I$. The resulting mid-IR pulses (mIR1 and mIR2) are tunable from 2.5 to 20 μm and have pulse durations approximately matching that of the pulses at 800 nm. Parabolic reflectors, P1 and P2, are used to focus three m-IR beams onto the sample and recollimating the third-order signal, respectively. The third-order signal, denoted here as emitted field (EF), is directed at the pair of single-channel detectors (D1 and D2) using a 50/50% beam splitter (BS). The LO is also directed to the detectors, collinear with EF. Note that mirror M should be removed for the single-channel balanced detection. Alternatively, an array detector (D3) attached to a monochromator (MO) (and mirror M) can be used for detecting 2D IR spectra via spectral interferometry.

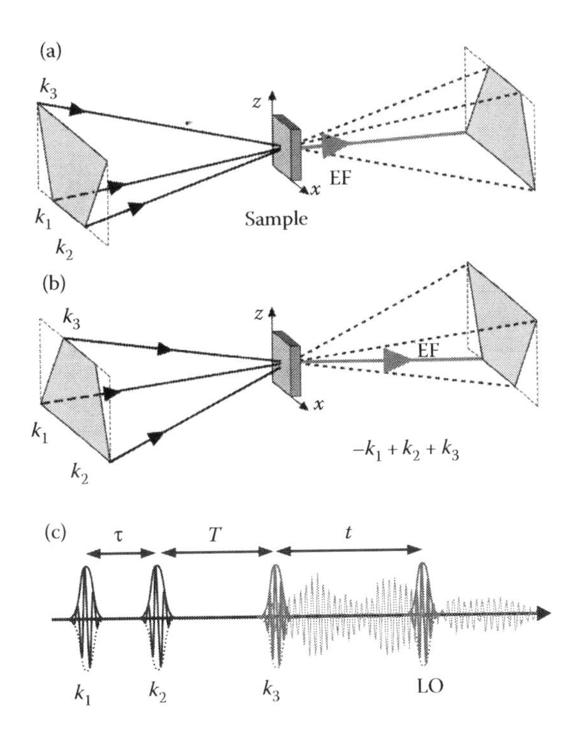

FIGURE 11.2 Beam geometry satisfying the phase-matching conditions ($\vec{k}_{EF} = -\vec{k}_1 + \vec{k}_2 + \vec{k}_3$ and $\omega_{EF} = \omega_1 - \omega_2 + \omega_3$) for the case when $\omega_3 < \omega_1$ (a) and when $\omega_3 > \omega_1$ (b). The EF vector indicates the direction of the dual-frequency third-order signal. (c) An example of pulse sequence used in the experiments where k_1 pulse comes first, k_2 comes second, and k_3 comes third. While the order of the pulses could be different in a particular experiment, τ is always used to denote the delay between the pulse arriving first and the pulse arriving second to the sample. Likewise, T denotes the delay between the pulse arriving second and the pulse arriving third.

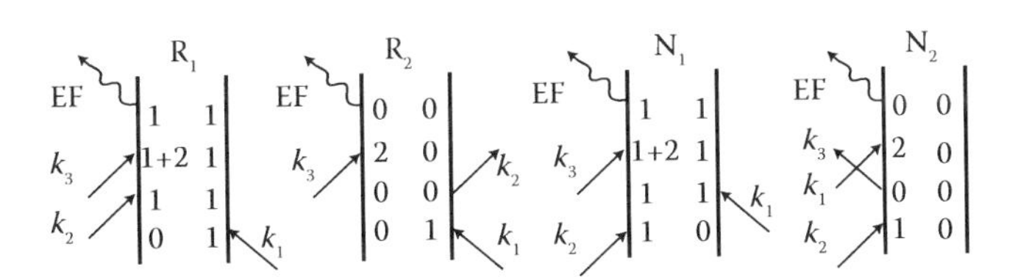

FIGURE 11.3 Liouville pathways describing cross peaks between mode 1 and mode 2 in a dual-frequency 2D IR spectrum measured with the pulse sequence where the k_1 and k_2 pulses access mode 1 and k_3 and LO pulses access mode 2. The diagrams describing rephasing (R_1 and R_2) and nonrephasing (N_1 and N_2) experiments are shown.

ordinate. With the use of a multichannel detector, the ω_t axis is obtained directly from the detector, with a single time axis (τ) scanned. 2D IR spectra at different waiting times can be measured.

The 2D IR method is a phase-sensitive technique and as such it requires that the phases of the pulses during the experiment are known with high precision. The phase differences in the pulse pairs (k_1, k_2) and (k_3, LO) are determined by the differences in the pathlengths they travel from their splitting point to the sample cell. The delay stages, which are often mechanical, introduce the largest uncertainty for the relative phases in pulse pairs originated from a common mid-IR beam. A superior alternative to very expensive translation stages is an external phase measurement system, which is typically based on interferometric position measurement using a continuous wave (CW) HeNe laser. Two types of schemes were demonstrated. In one, the beam of the HeNe laser was introduced into the mid-IR track before the beam is split into two parts, say k_1 and k_2. When the mid-IR beam is split into the k_1 and k_2 beams, the HeNe beam is split into two parts as well. The two HeNe beams were reflected off the k_1 and k_2 beams just before the parabolic reflector (P1) for interferometric measurements with fringe counting [12]. In such approach, fluctuations of all elements copropagated by the HeNe beam are included in evaluation of the relative phase for the two mid-IR pulses.

An alternative approach is to use an external HeNe interferometer for each translation stage [13]. In this case, the absolute position of each translation stage is measured accurately. It has been shown that such external position measuring system provides a sufficient accuracy for the measurements in mid-IR. The performance of the position control system was tested by introducing another HeNe beam into the IR path and measuring its spectrum via recording its interference with itself while the delays of the delay stage were measured using the external interferometric position measurement system. The line width ca. 1 cm^{-1} (*fwhm*) was recorded for the HeNe laser having a coherent length of ca. 10 cm (Figure 11.4). The accuracy of the measured delays is found to be better than 7 nm (<50 as). The results obtained using the external position measurement system demonstrate that regular mirror mounts can provide sufficient stability for the phase-sensitive measurements in the mid-IR region and confirm that the largest instability originates from mechanical translation stages.

The dual-frequency 2D IR measurements use two pulse pairs that originate from two different OPAs. How stable the relative phase of such pulses is? The phase stability of the pulses can be tested by observing interference between the pulses generated from two OPAs when they are tuned to the same frequency. Alternatively, the two mid-IR beams can be tuned so that their central frequencies differ by a factor of two. The generation of a second harmonic for one of the mid-IR beams brings them to the same frequency so that their interference can be observed. The instability of the relative phase is then converted into the amplitude instability of the interference pattern. The mode quality of the fundamental laser pulses was found to be important for having stable phase relations of the output of two OPAs. The use of a common white light for both OPAs (Figure 11.1) is not only convenient, but also helps achieving stable phase relations for the pulses of the two OPAs. Although the phase stability among all four pulses is important in general, there are several types of experiments (pulse sequences) that require only a pairwise phase stability among the pulses of the same central frequencies (Figure 11.3)—such pulse sequences were predominantly used in the reported dual-frequency 2D IR experiments [7,14].

Several inherently phase-stable designs were implemented for "single-color" 2D IR measurements where all mid-IR beams originate from a single OPA; these include approaches using diffraction optics [15,16] and pulse shaping [17]. An elegant 2D IR spectrometer has been developed by Zanni

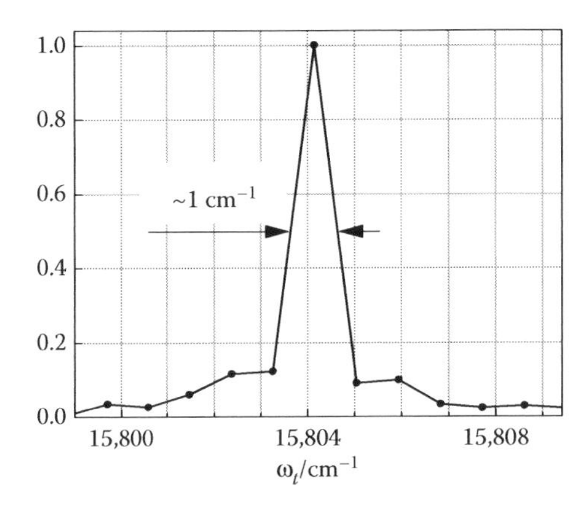

FIGURE 11.4 The spectrum of the HeNe laser recovered from a 1-cm-long interferogram using the external position measurement system. (Adapted from Kasyanenko VM et al. 2009. *J. Chem. Phys.* 131:154508/154501–154508/154512.)

group [17]. A mid-IR pulse shaper was used to generate copropagating pulse pairs with the delay between them controlled electronically in the pulse shaper [17]. These two pulses were used as k_1 and k_2 pulses. A third (probe) pulse intersects the first two pulses in the sample at some angle and its intensity is detected using spectral interferometry. In addition to phase stability, the scheme features very rapid data acquisition and allows suppressing the noise efficiently. It would be interesting to see the developments of these approaches toward the dual-frequency 2D IR measurements.

11.2.1 Sensitivity

An important requirement for any analytical technique is its sensitivity. The 2D IR approach with three noncollinear mid-IR pulses has a superior sensitivity as it is formally a background-free technique and thus permits controlling independently the light intensity of the third pulse and the local oscillator. Noncollinearity of the first two mid-IR pulses results in a transient grating formed in the sample by them (Figure 11.2). The period of the grating is characterized by a wavevector, which is a vector difference of the wavevectors of the first two pulses. A population or phase gratings can dominate depending on the mutual polarization of the first two pulses. The third beam diffracts off the grating into a direction not coinciding with the directions of the three laser pulses, allowing background-free measurements. The emitted field of the third-order signal (EF) is directed to the detector where it overlaps collinearly with the LO beam. Since the LO pulse is different from the third pulse, as opposed to the schematics with collinear k_1 and k_2 pulses, the k_3 pulse can have a power similar to that of the first two pulses, increasing the amplitude of the diffracted third-order signal. The LO can be independently tuned to avoid the detector saturation. As a result, the measured signal is not a fraction of the third (probe) pulse intensity ($\sim E_3^2$), as in the case of the homodyne detection and pump-probe schemes, but depends on $E_3 E_{LO}$, where E_3 and E_{LO} are the electric field amplitudes of the third pulse and LO, respectively. Owing to this separation of functions of the third pulse and LO, the sensitivity of the measurements in the noncollinear three-pulse geometry with heterodyned detection is superior. An ability to measure off-diagonal anharmonicities smaller than 0.005 cm^{-1} has been reported using such approach [18].

The waiting-time dependence of the 2D IR spectrum can provide valuable data about the structure and dynamics of the molecular system. For example, structural changes in a molecular system can be followed by 2D IR spectroscopy if they occur on the timescale comparable to the lifetime of the excited vibrational mode, using the so-called chemical exchange approach [19–21]. Vibrational and electronic hole-burning and fluorescence peak-shift measurements have the same general principle: The frequency of the excited oscillator is followed spectroscopically as a function of the waiting time reporting on the changes associated with the oscillator. The frequency of a selected oscillator is typically observed in such measurements: the frequency changes when conformational or chemical changes occur in the system.

The cross peaks accessible via 2D IR spectroscopy also change their shapes and amplitudes as a function of the waiting time. Important information about the system can be obtained from their waiting-time dependences, including the mode connectivity patterns, mode spatial locations, energy transport pathways, efficiencies, and so on, via the approach named relaxation-assisted 2D IR (RA 2D IR) spectroscopy [10].

11.3 RELAXATION-ASSISTED 2D IR: NEW STRUCTURAL REPORTERS

11.3.1 Principles

How the 2D IR cross peaks depend on the waiting time? The cross-peak amplitude and shape at $T = 0$ is determined by the coupling of the modes involved, which results in a shift of the combination band level (Figure 11.5a). In the experiment where the first two IR pulses access the mode at ω_1 and the third pulse probes the transition at ca. ω_2, a cross-peak pair is observed in the region

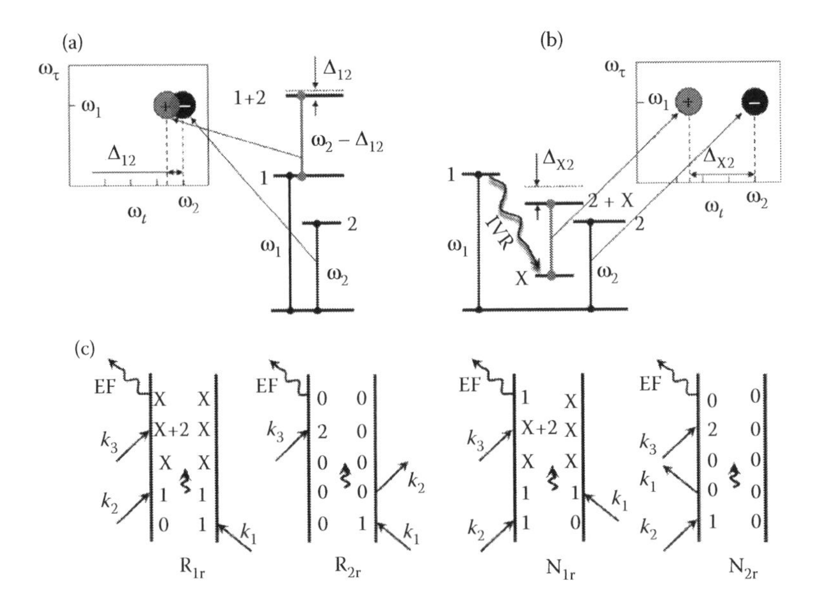

FIGURE 11.5 Energy diagram involving three coupled vibrational oscillators (ω_1, ω_2, ω_X) and two-dimensional spectrum for the mid-IR pulse sequence where the k_1 and k_2 pulses excite ω_1 mode and k_3 pulse excites transitions at ω_2, $\omega_2 - \Delta_{12}$, and $\omega_2 - \Delta_{X2}$, for traditional 2D IR (a) and RA 2D IR (b). The transitions resulting in the peaks in the cross-peak pairs are indicated with thin arrows. (c) Liouville pathways dominating the RA 2D IR cross peaks. The diagrams describing rephasing experiments (R_{1r} and R_{2r}) and nonrephasing experiments (N_{1r} and N_{2r}) are shown.

$(\omega_\tau, \omega_t) = (\omega_1, \omega_2)$ of the 2D IR spectrum (Figures 11.5a and 11.3). The transitions contributing along ω_t (probe) direction involve that at ω_2 (often called the ground-state bleach, diagrams R_1 or N_1) and that at $\omega_2 - \Delta_{12}$ (excited state absorption; diagram R_2 or N_2). Note that the stimulated emission signal will not appear in the cross-peak region as the ω_2 frequencies cannot stimulate transitions at the ω_1 frequency. The phases of the two contributions (pathways) differ by π, which results in different signs for the two peaks in the 2D IR spectrum. While the overall sign in the 2D IR spectrum can be selected arbitrarily, we follow the convention of the pump-probe spectroscopy, where the 2D IR peak amplitude has units of absorbance and, as such, is positive when less of the probe light reaches the detector (excited state absorption). The mode excited by the first two pulses is called throughout the chapter as a *tag* mode, and the mode accessed by the third pulse is called a *reporter* mode.

Notice that if the two modes are well separated spatially, their coupling (anharmonic shift, Δ_{12}) is small (Figure 11.5a). A small separation of the two peaks in the 2D IR spectrum (Figure 11.5a) results in a cancellation of the negative (at $\omega_t = \omega_2$) and positive (at $\omega_t = \omega_2$) cross peaks of the cross-peak pair described by either R_1 and R_2 or N_1 and N_2 diagrams in Figure 11.3. This cancellation results in a substantial decrease of the amplitudes of the two peaks in the cross-peak pair. Interestingly, the separation between the resulting cross peaks (Figure 11.6a) under the condition $\Delta_{12} \ll \delta\omega$ is independent of Δ_{12} but is a simple function of $\delta\omega$, where $\delta\omega$ is the *fwhm* width of the transition [22]. For example, if the transition can be approximated with a Lorentz profile, the frequency separation between the minimum and maximum in the cross-peak pair equals ca. $1.58*\delta\omega$ when $\Delta_{12} \ll \delta\omega$ (Figure 11.6a). While for small Δ_{12}, the peak separation does not characterize Δ_{12}, the amplitude of the cross peak does. The amplitude of the cross peak depends linearly on Δ_{12} if $\Delta_{12}/\delta\omega \ll 1$ (Figure 11.6b).

The cross-peak amplitude changes with the waiting time due to the relaxation of the tag mode and due to the excitation of other modes in the molecular system. The signals originated from the Liouville pathways shown in Figure 11.3 diminish and new signals associated with pathways shown

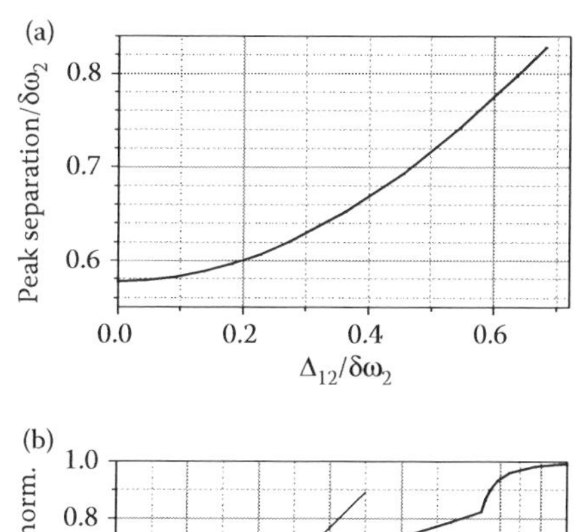

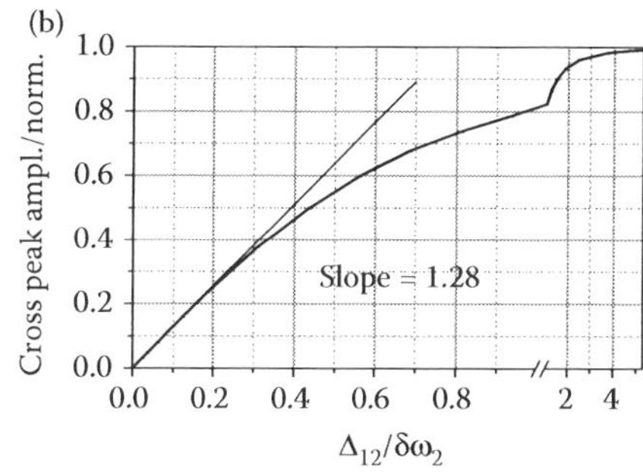

FIGURE 11.6 Apparent separation between the extrema in a cross-peak pair (a) along the ω_t axis and resulting cross-peak amplitude as a function of the off-diagonal anharmonicity (Δ_{12}) normalized to the width of the reporter transition ($\delta\omega_2$).

in Figure 11.5c appear. The dynamics can be understood as a frequency change of the reporter mode caused by excited modes (X) in the molecule. If a mode (X) coupled strongly to the reporter mode is populated via vibrational relaxation, the shift of the positive peak of the cross-peak pair can be easily observed (large $\Delta_{X/report.}$). Figure 11.7 shows the CN/CO cross-peak pairs in 3-cyanocoumarin at three waiting times. The ω_t of the zero contour line between the positive and negative peaks in the peak pair (ω_0^i) at $\omega_\tau = \omega_{tag}$ can serve as an indicator of the changes in the origin of the signal. Notice that the spectral envelope of the $<01| \rightarrow <11|$ transition tends to that of the $<00| \rightarrow <10|$ transition when the mode coupling (Δ_{12}) tends to zero as the two transition became indistinguishable. At the waiting time close to zero (Figure 11.7a), the $\omega_0^{cr.p.}$ is close to the ω_{CO} frequency; the amount of shift from it is small and difficult to evaluate precisely. The diagonal CO peak pair serves as a convenient reference in each graph; a large diagonal anharmonicity of the C=O mode (~15 cm^{-1}) results in apparent peak separation in the peak pair of ca. 15 cm^{-1} and appearance of the negative (ground state bleach) peak at $\omega_t \sim \omega_{CO}$. At larger time delays, the cross-peak pair shifts to smaller frequencies (Figures 11.7b and 11.7c), which indicates that a mode(s) with larger anharmonicity with the CO reporter is(are) populated via vibrational relaxation (VR) and intramolecular vibrational energy redistribution (IVR) processes. The VR and IVR processes have a spatial component, which determines the waiting-time dependence of the cross peak. In this example, the CN and CO modes were in sufficient proximity, so that their direct coupling was strong enough and no cross-peak growth was found as a function of the waiting time (the cross peak showed essentially a plateau at time delays <5 ps) [10]. Another example where the tag mode (CN) and three reporters, amide I, amide II, and C=O stretching mode, reside on the same molecule is shown in Figure 11.8. Here, the reporters are separated from the tag (C≡N) by substantial distances of 5.8 Å (Am-I), 7.1 Å (Am-II), and 11.4 Å (C=O). The given distances are based on DFT calculated structure and were measured between the carbon atom of the cyano group and the carbon atom of the amide for the Am-I mode, the nitrogen atom of the amide for the Am-II mode, and the carbon atom of the carbonyl group for the C=O mode. Note that the distances across phenyl and piperidine rings were taken as

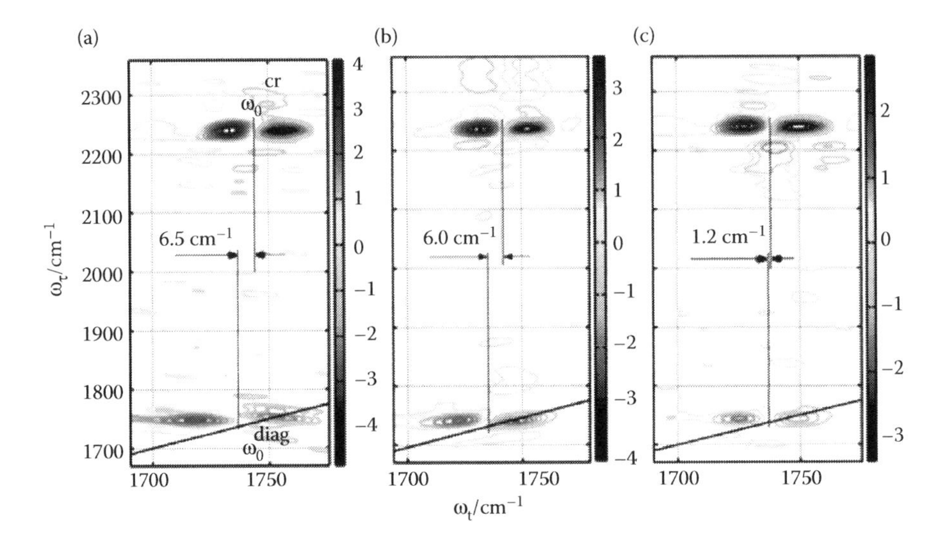

FIGURE 11.7 Absorptive 2D IR spectra of 3-cyanocoumarin in dichloromethane measured at waiting times of 0.670 (a), 2 (b), and 4 ps (c). Note that k_1, k_2 pulses were centered at ca. 2200 cm^{-1} and had very small intensity at 1750 cm^{-1}, resulting in a strongly suppressed C=O diagonal peaks. The vertical lines indicate the positions of zero contour line between the peaks in the peak pairs for both the diagonal (ω_0^{diag}) and cross (ω_0^{cr}) peaks. (Adapted from Kurochkin DV, Naraharisetty SG, and Rubtsov IV 2007. *Proc. Natl. Acad. Sci. U.S.A.* 104:14209–14214.)

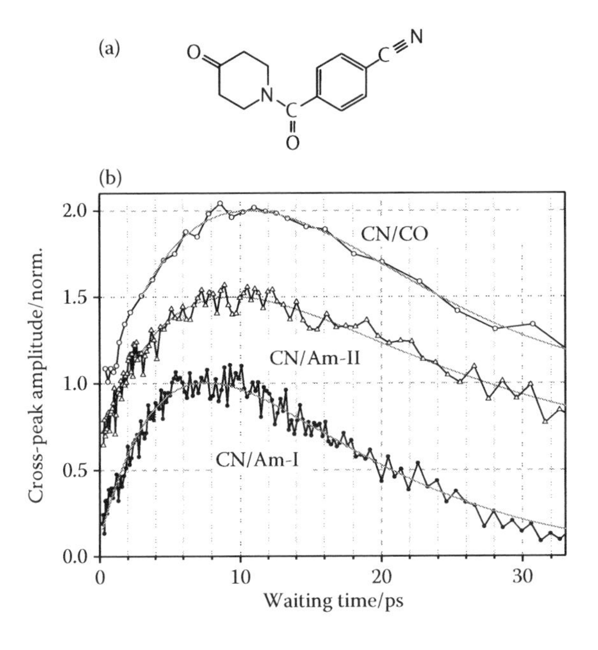

FIGURE 11.8 (a) Structure of 4-(4-oxo-piperidine-1-carbonyl)-benzonitrile (PBN). (b) Normalized absolute-value amplitudes of the CN/Am-I, CN/Am-II, and CN/CO cross peaks for PBN as a function of the waiting time. The fits with two-exponential functions are shown with gray lines. Note that the baselines for the CN/Am-II and CN/CO cross peaks were shifted for clarity by 0.5 and 1.0, respectively. (Adapted from Naraharisetty SG, Kasyanenko VM, and Rubtsov IV 2008. *J. Chem. Phys.* 128:104502/104501–104502/104507.)

the through-space distances between the ring atoms in the para positions. Because the direct tag–reporter coupling is weak enough for all these reporters, the vibrational energy transport causes large enhancement for all three cross peaks (Figure 11.8b). A spectacular 18-fold amplification of the cross peak was observed for the CN/CO cross peak. Vibrational relaxation of a high-frequency mode in a molecule with more than 10 atoms is often efficient due to a large mechanical coupling between different modes of the molecule and a substantial density of states at the site where the high-frequency mode resides [23,24]. This intramolecular coupling results in a typical lifetime of high-frequency modes of ca. 1–5 ps. After relaxation, the excess energy is trapped in the molecule; with time it propagates in the molecule and dissipates into the solvent. The rate of the IVR process depends on the square of the interaction matrix element, which, in turn, depends on the spatial overlap of the modes involved. As a result, the energy propagation is not instantaneous, but occurs as hopping from site to site, until complete thermalization in the molecule. Notice that despite partially delocalized character of many molecular vibrations, the IVR process is still a hopping process due to a requirement of spatial overlap of the modes involved; the energy is transferred between the modes overlapping at least partially. Importantly, the tag and reporter modes typically used in RA 2D IR experiments are localized in a sense that their delocalization size is much smaller than the intermode distance. Since the reporter mode is localized, the strongest coupling is expected with the other modes localized at the same site. Excitation of such modes via energy transport is required to cause the largest frequency shift of the reporter mode resulting in the largest cross peak. Therefore, the IVR process is expected to occur in the diffusion-like or Brownian-like fashion where IVR steps in the forward and backward directions occur with almost the same probability. Note that the energy excess per molecule in these experiments is small (~2000 cm^{-1}).

11.3.2 Accessing Angles between Transition Dipoles

Although an RA 2D IR cross peak originates not from the coupling of the tag and reporter modes, the RA 2D IR cross-peak anisotropy still reports on the angle between the transition dipoles of the tag and the reporter. That is because the cross-peak anisotropy, measured as $r = (S_{ZZZZ} - S_{ZZXX})/(S_{ZZZZ} + 2S_{ZZXX})$, is sensitive only to the mutual orientation of the modes accessed by the polarized IR pulses but insensitive to the origin of the frequency shift of the reporter; here, S_{ZZZZ} and S_{ZZXX} are the cross-peak amplitudes measured using the indicated polarizations of all four pulses (k_1, k_2, k_3, and LO), which are parallel to either the z or x axis (Figure 11.2a). In the pump-probe measurements (where $\tau = 0$), the cross-peak anisotropy (r_{12}) can easily be related to the angle between the transition moments (θ_{12})

$$r_{12} = \frac{2}{5}\langle P_2(\cos\theta_{12})\rangle, \tag{11.1}$$

where $P_2(x) = (3\cos^2 x - 1)/2$ is the second Legendre polynomial and averaging is taken over the distribution of molecular structures. In the three pulse measurements, the validity of Equation 11.1 requires negligible depolarization during the dephasing time, τ [25]. This depolarization can include rotation of the group carrying the mode or rotation of the whole molecule as well as pseudo-rotation where transitions between several resonant states with different orientations occur [26,27]. Note, however, that the fast depolarization of the tags will cause fast anisotropy decay during the waiting time as well, resulting in less useful data. Therefore, the anisotropy measurements at large waiting time are practically useful only if the depolarization is much slower than the waiting time, permitting the use of Equation 11.1. If the depolarization is slow, as, for example, in large molecules such as proteins, the angle between transition moments of the tag and the reporter can be determined from the cross-peak measurements at large waiting time, taking advantage of the cross-peak amplification and therefore accessing much larger distances between the tag and the reporter for the angular measurements.

11.3.3 Connectivity Patterns

The spatial component of the IVR process results in a correlation of the energy transport time with the distance between the groups bearing the tag and reporter modes. Such correlation can be clearly seen in Figure 11.8b, where a substantial difference is found between the T_{max} values for the CN/Am-I and CN/CO cross peaks [9]. The T_{max} value, which is the waiting time at which the cross-peak maximum is reached, is found to be a convenient parameter for characterizing the energy transport time. Alternatively, one could use a rise time in the waiting-time dependence to characterize the transport. The difficulty though is in a complex shape of the waiting-time dynamics, which often cannot fit to an exponentially growing function. For example, a rise dynamics with the induction period can be seen for the CN/CO cross peak in Figure 11.8b. It is also possible to use an infliction point for the signal growth [28], which is very clear, for example, in the CN/CO dynamics (Figure 11.8). However, for the modes at shorter distances, such as, for example, CN/Am-I and CN/Am-II (Figure 11.8), the infliction point cannot be accurately determined. A fit of the T-dependence to a theoretical model would be the most appropriate approach, although very involved. The T_{max} value is easily accessible for accurate experimental evaluation and therefore is currently used as a characteristic of the energy transport. The reason the T_{max}-value approach works is based on similarity of the energy dissipation dynamics as seen by different reporters. Indeed, the decay tails for different cross peaks observed at larger waiting times are similar (although not the same), which results in their similar influence on the T_{max} values.

The T_{max} value for the CN/CO cross peaks (10.6 ps) is substantially larger than that for the CN/Am-I peak (7.5 ps) in correlation with the through-bond distances between the respective groups (ca. 11 and 6.5 Å). Comparison of the T_{max} values for the CN/Am-I (7.5 ps) and CN/Am-II (9.0 ps) indicates that the energy accepting modes for the Am-II reporter are located further away from the C≡N tag. Notice that both Am-I and Am-II reside on the same amide and one could consider their affective locations to be similar. More detailed consideration shows that on the molecular backbone the Am-I mode involves motion of the carbon atom of the amide, while the Am-II mode involves motions of carbon and nitrogen atoms. In addition, the CH_2 bending motion at the two carbons adjacent to the nitrogen atom is contributing to the Am-II mode, which shifts the effective mode location toward the nitrogen atom of the amide. This conclusion is also supported by a smaller T_{max} value for the C=O/Am-II cross peak than that for the C=O/Am-I cross peak (*vide infra*, Figure 11.18b). Thus, a monotonic correlation of T_{max} with distance is found for all three reporters. The monotonic correlation of the energy transport time with distance permits assessing connectivity patterns in molecules, very much similar to that for the methods of multidimensional NMR spectroscopy, TOCSY and HMBC.

Importantly, the amplification factor also correlates with the distance for these three modes with the values of 4.3-, 5.4-, and 18-fold for the CN/Am-I, CN/Am-II, and CN/CO cross peaks, respectively. The origin of this correlation is in a sharp decrease of the direct coupling with distance, which ensures a weak coupling at $T = 0$ for remote reporters. At the same time, the intramolecular energy transport appears to be efficient enough even to larger distances. Assuming that the intramolecular energy transport can be approximated by a macroscopic Fourier law, the maximal temperature that is reached at certain distance from the heat source is reversely proportional to the cube, square, or the first power of the distance for the heat conduction in three-, two-, and one-dimensional space, respectively. Even in the three-dimensional space, this distance dependence is much softer than that for the direct mode coupling for localized modes (Δ_{12}). The latter can have two contributions: the through-space transition dipole interaction, which depends on distance (R) as R^{-6}, and through-bond mechanical coupling, which, at large distances, decays exponentially. The difference in the distance dependences for the direct mode coupling and the energy transport efficiency explains the correlation of the amplification factor with distance. Note, however, that the through-space interaction depends on orientation of the transition dipoles of interacting modes; a combination of through-bond and through-space interactions can result in variations of Δ_{12} anharmonicity so that it does not correlate with the distance. Therefore, the correlation of the amplification factor with distance is not as general as that for the energy transport time.

11.4 ATOMIC-SCALE ENERGY TRANSPORT STUDIED BY RA 2D IR

Energy dissipation from molecules to the solvent has been studied extensively over the past several decades [29–31] but only recently, with the development of IR-pump/anti-Stokes Raman probe [32–34] and RA 2D IR methods [10], it became possible to study specific transport pathways. Using the RA 2IR approach, a correlation of the energy transport time with distance has been observed for many molecular systems, including medium-size model compounds [9,14,35,36], peptides [37,38], transition metal complexes [13,39], and polymers [40,41]. Several characteristic examples will be considered in this section.

11.4.1 Structure-Dependent Energy Transport

To understand how the vibrational energy transport depends on the molecular structure, the RA 2D IR experiments were performed for ortho-, meta-, and para-isomers of acetylbenzonitrile (AcPhCN), focusing at the cross peaks between the cyano group and carbonyl group stretching modes. The waiting-time behavior is found to be very different for these three compounds (Figure 11.9) [36]. Moreover, the CN mode lifetimes are found different as well, measured at 3.4, 7.1, and 7.2 ps for *o*-, *m*-, and *p*-AcPhCN, respectively. To eliminate the influence of the lifetime of the tag mode onto the T_{max} value, a deconvolution can be performed with the function representing the CN excited state dynamics (Figure 11.9). It is more practical to implement a fit of the experimental data by the function that is a convolution of the CN excited state dynamics and a function representing pure energy transport properties, taken as a two-exponential (rise–decay) function. The functions obtained represent a true energy transport corresponding to the case of instantaneous release of the tag energy (Figure 11.9, thin black lines). The true transport dynamics for *o*-AcPhCN revealed that the rising time is essentially zero (<0.1 ps), indicating that the very first step of the CN mode relaxation populates the modes that are contributing the most to the cross peak. Thus, relatively high-frequency

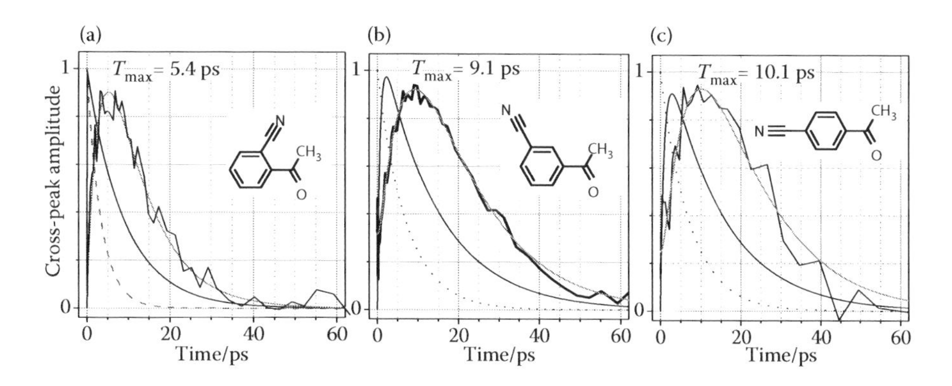

FIGURE 11.9 Experimental waiting-time dependences (thick black line) for ortho- (a), meta- (b), and para- (d) AcPhCN. The fit of the data was performed by a function that is a convolution (thick gray lines) of the two-exponential (rise–decay) function (thin black lines) with an exponential decay function (dashed lines) corresponding to the lifetime of the CN mode in each compound, measured at 3.4 ps (a), 7.1 ps (b), and 7.2 ps (c). The obtained deconvoluted functions (thin black lines) have characteristic times of <0.1 and 9 ± 1 ps (a), 0.7 ± 0.4 and 14 ± 1 ps (b), and 1.0 ± 0.3 and 14 ± 1 ps (c). They reach their maxima at (a) <0.2 ps, (b) 2.2 ps, and (c) 2.8 ps, respectively. (Adapted from Kasyanenko VM et al. 2011. *J. Phys. Chem. B* 115:11063–11073.)

modes are contributing to the cross peak at time delays comparable to T_{max}. The rise time for the *m*- and *p*-AcPhCN are found to be 0.7 and 1.0 ps, respectively, indicating that more relaxation steps are needed for para-isomer compared to the meta-isomer. The characteristic lifetime of the modes in the fingerprint region is about 0.5–1 ps, except for very localized modes. Therefore, the rise-time values report that it takes only one or two IVR steps for energy to reach the reporter mode region in both meta- and para-isomers. The difference in the transport times for the three isomers indicates that the phenyl ring is substantially perturbed by the acetyl and cyano group substituents; the modes of the ring cannot be considered as uniformly delocalized over all six carbon sites of the ring. The energy transport time taken as the rise time of the deconvoluted function is found to correlate with distance, even at such small distances and such delocalized spacer as the phenyl ring.

The time-dependent frequency shifts of the CO reporter in three AcPhCN isomers were recently modeled using a new approach proposed by A. Burin. The approach uses a Marcus-type equation for evaluating the transition probabilities between harmonic vibrational states (normal modes), computed for an isolated molecule using the DFT anharmonic calculations [42]. The low-frequency modes of the molecule itself serve as a bath facilitating IVR transitions between the vibrational states. The overall cooling of the molecule was introduced phenomenologically, which was the only free parameter of the theory. The CN mode lifetimes for the ortho-, meta-, and para-isomers were computed as 1.2, 3.8, and 3.4 ps, respectively, which matches the trend observed experimentally (3.4, 7.1, and 7.2 ps) [36]. The computed T_{max} values of 2.9, 6.7, and 6.9 ps for *o*-, *m*-, and *p*-AcPhCN, respectively, reproduce well the experimental trend of 5.4, 9.1, and 10.1 ps. It seems that the difference between the theory and experiment is largely due to the deviations in the computed CN lifetimes. The theory was also used to compute energy transport dynamics in PBN (Figure 11.8). The T_{max} values computed for the CN/Am-I and CN/CO cross peaks as 6.9 and 11.5 ps, respectively, matched closely the experimental values of 7.5 and 10.5 ps [43].

In general, the efficiency of the energy transport via different bridging motifs is expected to be different. The IVR process, where the quantum numbers of three or more oscillators change, is governed by anharmonic interactions, which scale with the bond strength. Typical coordination bonds are much weaker than typical covalent bonds; it is interesting to investigate the efficiency of the energy transport via coordination bonds.

11.4.2 Energy Transport via Coordination Bonds

Vibrational energy transport between ligands in transition metal complexes involves stages where energy has to cross relatively weak coordination bonds between the coordinating metal atom and the ligands. Tetraethylammonium bis(maleonitriledithiolate)iron(III)nitrosyl (FNS) complex (Figure 11.10), synthesized by Dr. J. Donahue's group, was studied [13]. In one set of RA 2D IR measurements, the N=O stretching mode of the nitrosyl ligand served as a tag and the C≡N and C–C stretching modes served as the reporters (Figure 11.10c–e). In another set of measurements, the roles were switched and the CN mode served as the tag and NO served as the reporter (Figure 11.10b) [13]. Note that the energy transfer between ligands in FNS does not have a resonant pathway involving only fundamental transitions at high frequencies, as in the case of energy transfer between carbonyl ligands in metal carbonyls [44].

All three cross peaks show a substantial amplification due to vibrational energy transport from the tag toward the reporter (Figure 11.10b–d), including a record amplification of 27-fold observed for the C≡N/N=O cross peak (Figure 11.10b). As expected, the finite lifetime of the tag modes affects the waiting-time dynamics. For the tag lifetimes, much shorter than the characteristic time of the energy transport, the energy release from the tag can be considered instantaneous and the measured *T*-dynamics represents essentially a pure energy transport dynamics. When the CN mode serves as the tag, the energy release is not instantaneous, as the CN lifetime (2.9 ± 0.1 ps) is only ca. three times smaller than T_{max} (9.8 ps). If the tag decay dynamics is slower than the transport dynamics, the *T*-dependence is strongly affected by the tag lifetime; the T_{max} value then does not

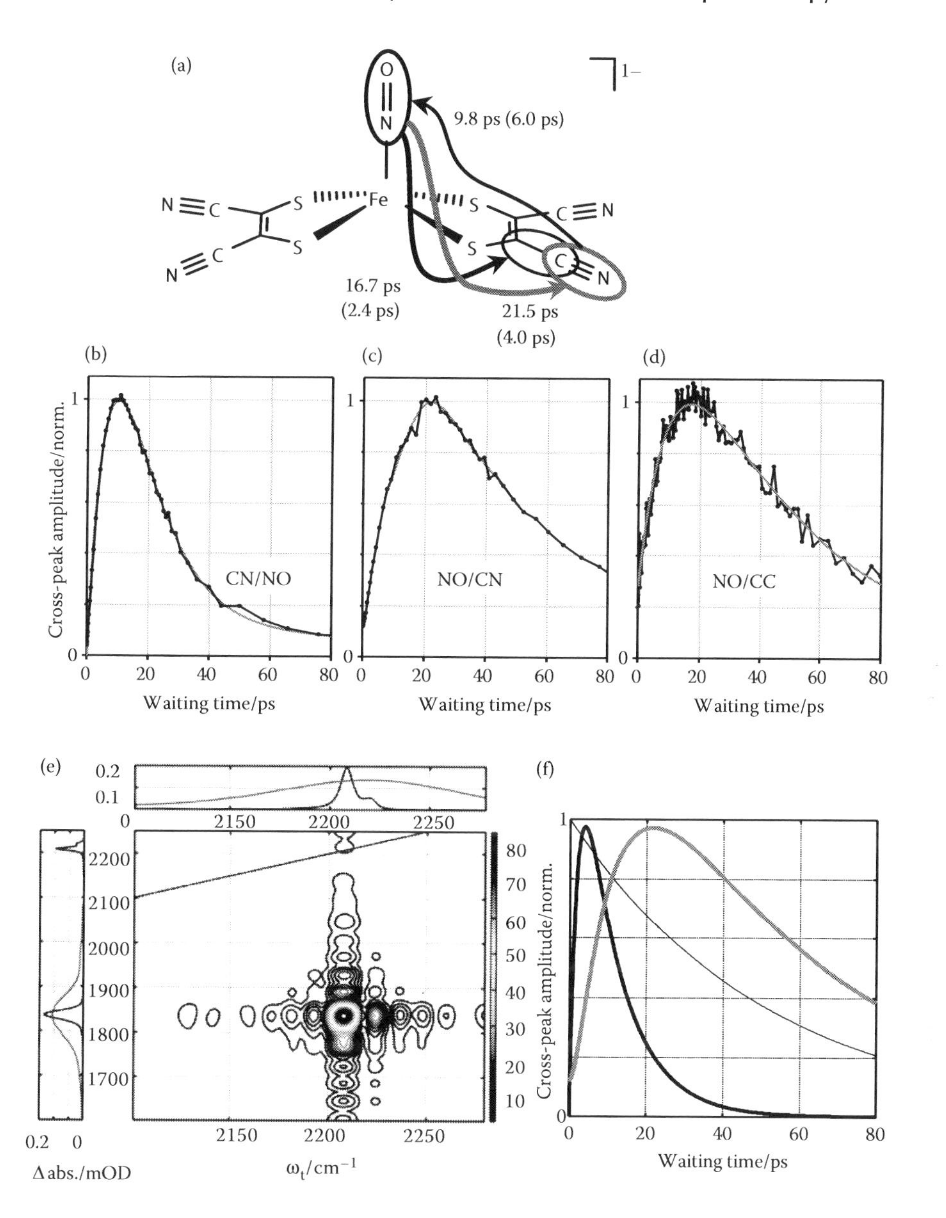

FIGURE 11.10 (a) Structure of FNS with indicated experimental T_{max} values for three cross peaks. The numbers in parentheses show pure energy transport times (T_{max}) obtained by deconvoluting the T-dependences with the respective tag lifetimes. (b,c,d) Waiting-time dependence of the amplitude of three cross peaks indicated in the graphs. The gray lines show the fit with two-exponential functions performed in the vicinity of the maximum, which were used for determining the T_{max} values. (e) Dual-frequency 2D IR spectrum of the FNS complex focusing on the NO/CN cross peak. The linear spectrum and the spectra of the IR pulses used in the measurements are shown in the attached panels. (f) Results of deconvolution (fit with convolution) of the data shown in panel C. The thick gray curve is the convolution of the two-exponential (rise-decay) function (thick black line) with an exponential decay function (thin black line) of 51 ps. The deconvoluted function obtained has characteristic times of 2 and 10 ps and reaches maximum at 4.0 ps. (Adapted from Kasyanenko VM et al. 2009. *J. Chem. Phys.* 131:154508/154501–154508/154512.)

represent the energy transport time. This is the case when NO mode serves as a tag (Figures 11.10c and 11.10d). The lifetime of the excited NO mode in FNS (51.2 ± 0.3 ps) is not only slower than the characteristic energy transport times in FNS but even slower than the characteristic cooling time of the maleonitriledithiolate ligands of the complex, which is ca. 20 ps (Figure 11.10b). As a result, the decay time for the NO/CN and NO/CC cross peaks matches well the lifetime of the NO mode, as the NO relaxation represents the slowest step in the overall energy dissipation process. A deconvolution performed to eliminate the influence of the tag lifetime and extract the true energy transport dynamics from the T-dependences results in the characteristic rise and decay times of 3 and 15 ps (CN/NO), 2 and 10 ps (NO/CN, Figure 11.10f), and 1 and 8 ps (NO/CC). The error bars are substantial for the parameters of the deconvoluted function as the two values are partially linked, but T_{max} for the deconvoluted function is more accurate; the T_{max} values for the pure transport are found to be ca. 6.0 ps (CN/NO), 4.0 ps (NO/CN), and 2.4 ps (NO/CC), as indicated in Figure 11.10a in parentheses. Both the rise-time values and the deconvoluted T_{max} values correlate with the transport distance: it takes longer to get from the NO group to the CN groups than to the CC groups.

Notice that the N=O/C≡N cross peak, where the reporter mode frequency is larger than that of the tag, shows a ninefold amplification. This proves unequivocally that the excitation of the reporter itself via energy transport from the tag is not required for observing the cross-peak enhancement and that the lower-frequency modes serve as the energy-accepting modes. The transitions located at the counterion (tetraethylammonium cation), at 1400–1490 cm^{-1} and around 1180 cm^{-1}, were tested as reporters in the RA 2D IR experiments: no sizable cross peaks with either NO or CN tags were observed. This result emphasizes the importance of the coordination bonds with the metal and covalent bonds between the tag and the reporter for efficient energy transport.

These experiments demonstrate that the energy transport through weak coordination bonds is efficient enough to compete with the overall cooling of the molecule to the solvent, which makes the use of the RA 2D IR method advantageous for structural interrogation of transition metal complexes. Interestingly, the energy transfer time from NO to CN is not the same as that from CN to NO. The asymmetry can be understood considering a difference in the fraction of pathways leading to the desired processes: most of the relaxation pathways from NO lead to the energy transfer to the other ligands, while only a few pathways from dithiolate ligand lead to the NO ligand. Such asymmetric energy transport is fundamentally interesting and can lead to developing practically important devices.

11.4.3 Energy Dissipation to the Solvent

In addition to the through-bond energy transport pathways, energy dissipation to the solvent is always present. While the direct relaxation of high-frequency modes of the solute into high-frequency modes of the solvent has a very low probability, the solvent can facilitate the IVR process by offering or absorbing small quanta of energy [45,46]. The energy dissipation to the solvent becomes more efficient with time as the region excited excessively in the solute increases and the excess energy relaxes into modes with smaller energies. The excess energy dissipates into the first solvation shell of the solute and then propagates further into the solvent. A diffusive energy dissipation mechanism into the bulk solvent is expected [47,48]. At larger time delays, the excess energy introduced by excitation of the tag is thermalized within the exited region. The mean volume to which each excited tag is thermalized equals $V_T = (f{*}C{*}N_A)^{-1}$, where f is the fraction of the excited tags, C is the tag concentration, and N_A is the Avogadro's number. At concentrations of 50 mM and excitation fraction of 0.2, the resulting volume equals to a volume of a sphere with a radius of 34 Å. Thermalization into such volume occurs on a timescale of several tens of ps, as apparent from large number of experimental results (see, e.g., Figures 11.8 through 11.10). The final temperature increase (ΔT) when thermalization is completed can be calculated as $\Delta T = Q/(c_V{*}V_T)$, where Q is the photon energy absorbed by a single tag and c_V is the specific heat capacity per unit volume (equals, e.g., 1.42 J cm^{-3} K^{-1} for chloroform). A typical temperature increase in the sample encountered in 2D IR measurements is of the order of 0.1 K (18). Such temperature increase is easily measurable

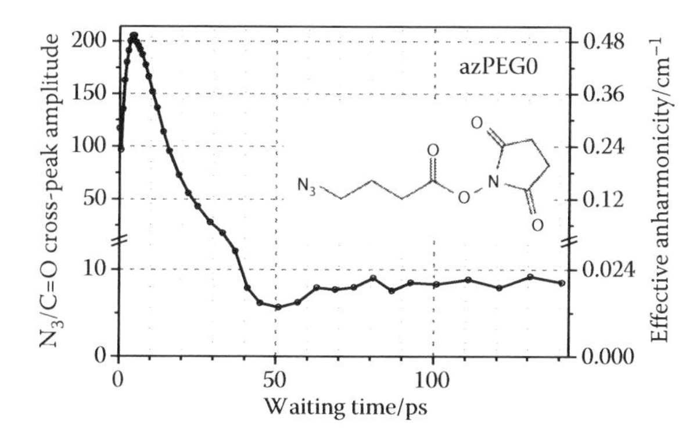

FIGURE 11.11 Waiting-time dependence for the N≡N/1742 cm^{-1} cross peak for azPEG0 (see inset) in chloroform. (From Lin Z and Rubtsov IV. 2012. *Proc. Natl. Acad. Sci. U.S.A.* 109:1413–1418.)

if the mode frequency is sufficiently sensitive to temperature. A typical sensitivity of a vibrational mode central frequency to temperature ranges from 0.01 to 0.06 cm^{-1}/K [18,49], which results in a typical reporter mode frequency shift of ca. 0.001–0.006 cm^{-1} in RA 2D IR measurements at large waiting times. Notice, however, that the frequencies of all reporters in the excited volume are shifted after thermalization, not only of those paired with the excited tags; this effect increases the cross peak at the plateau by a factor of f^{-1}, compared to that at $T \sim 0$.

Figure 11.11 shows the waiting-time dependence for the N≡N/1742 cm^{-1} cross peak for azPEG0 (Figure 11.11 inset). The cross peak at $T = 0$, associated with the direct N_3–CO coupling, is changed to RA 2D IR contribution associated with the through-bond energy transport reaching maximum at ca. 4.5 ps and then decaying due to energy dissipation to the solvent, which occurs with a characteristic time of ca. 15–20 ps. At $T > 50$ ps, a plateau is reached, which is due to complete thermalization of the excess energy in the excitation region. The sensitivity of the mode at 1742 cm^{-1}, originating from the asymmetric C=O stretching motion at succinimide moiety (Figure 11.11 inset), to temperature is substantial ($\eta_{1742} = 0.058$ cm^{-1}/K), resulting in an easily measurable plateau caused by the temperature increase after thermalization. The plateau can also be seen in the diagonal peak for the N≡N stretching mode (Figure 11.12b), which also in temperature sensitive but shifts to lower frequencies upon heating ($\eta_{N3} = -0.035$ cm^{-1}/K, Figure 11.13). Notice that both the cross and diagonal peaks at the plateau are due to resonant signals (Figures 11.12c and 11.12d).

An interesting dip is observed in the waiting-time dependence for the N_3/1742 cm^{-1} cross peak at ca. 50 ps (Figure 11.11). This dip has been assigned to a cancellation effect of the intramolecular anharmonic coupling (IAC) contribution shifting the CO frequency to lower values, which dominates the signal for the time delays smaller than 40 ps and the thermal contribution shifting the CO frequency to larger values, which dominates at $T > 50$ ps [18]. Such switch has been observed in multiple systems [18,36,50]. It is expected that such cancellation will be stronger if the frequency shifts involved (positive and negative) are smaller in the absolute values [36]. The dip provides a clear indicator of the sign change for the involved cross-peak contributions. Notice that there is no such dip in the diagonal N_3 signal (Figure 11.12b), indicating that the intramolecular anharmonic coupling contribution and the thermal contributions have the same signs. Indeed, the N_3 diagonal anharmonicity and the N_3 thermal sensitivity ($\eta_{N3} = -0.035$ cm^{-1}/K, Figure 11.13a) are both negative.

The plateau associated with the excess-energy thermalization can help measuring accurately small off-diagonal anharmonicities. As we discussed earlier, the cross-peak shape is a poor indicator of the anharmonicity value for small anharmonicities; anharmonicities as small as 0.005 cm^{-1} are currently accessible [40]. While the cross-peak amplitude is proportional to the anharmonicity, the reference point for the cross-peak amplitude corresponding to a known anharmonicity is difficult to obtain. One of the natural references is provided by the diagonal peak, which is governed by a known, or easily measurable, diagonal anharmonicity. The switch in the 2D IR measurements from the (ω_1, ω_2)

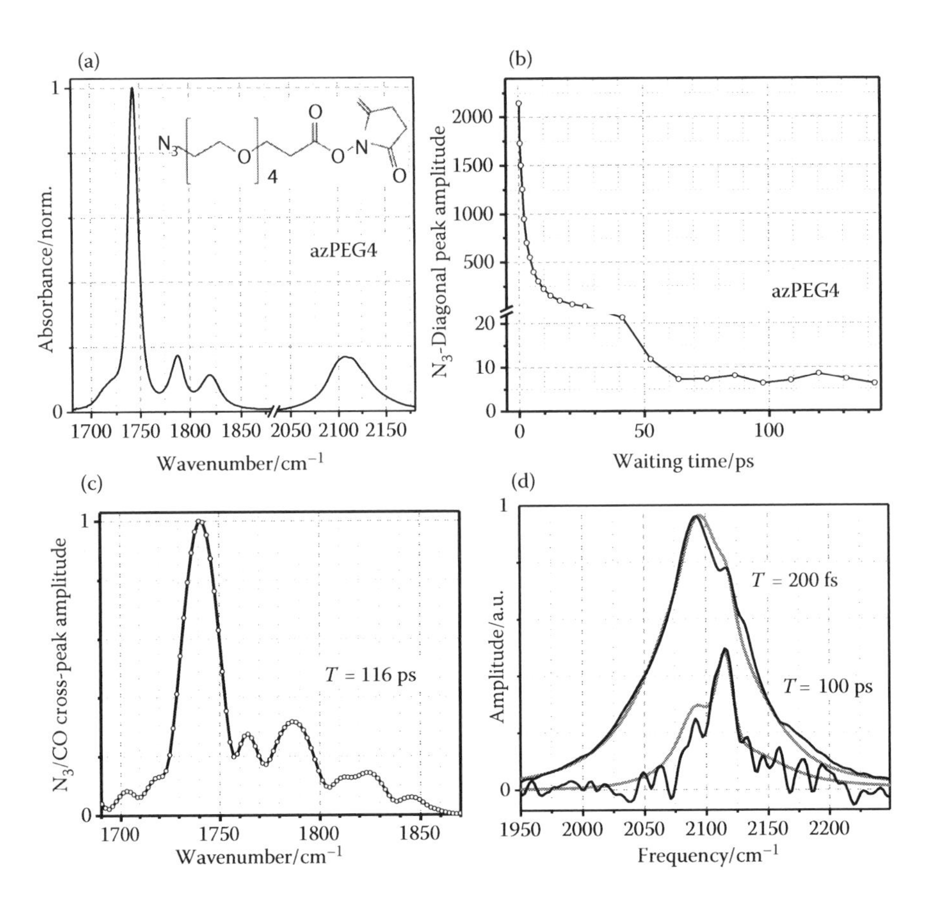

FIGURE 11.12 (a) Linear absorption spectrum of azPEG4 in chloroform. (b) The magnitude of the N_3 diagonal peak as a function of the waiting time, measured at the magic angle conditions ($\tau = 0$). (c) N_3/CO magnitude spectrum measured at $\tau = 0$ and $T = 116$ ps. The k_1 and k_3 spectra were centered at 2110 and 1770 cm^{-1}, respectively. (d) N_3 diagonal magnitude spectra measured at the waiting times of 200 fs (larger spectrum) and at 100 ps and $\tau = 0$. The gray lines show the results of the modeling. (Adapted from Lin Z, Keiffer P, and Rubtsov IV 2011. *J. Phys. Chem. B* 115:5347–5353.)

cross peak to the (ω_1, ω_1) diagonal peak requires significant changes in the experimental conditions that include the frequency tuning of one of the OPAs and changing the geometry of the three mid-IR pulses to satisfy the phase-matching conditions. These changes are sufficiently elaborate to limit the accuracy in comparing the absolute amplitudes of the cross peak and the diagonal peak.

Notice that the plateaus in these diagonal and cross peaks are caused by the same temperature increase in the sample as the same k_1, k_2 pulses excite the same transition (ω_1) in both measurements. Therefore, the plateaus can be used for relating the amplitudes of the 2D IR features in the two experiments [18]. Since the known diagonal anharmonicity governs the diagonal peak amplitude at $T = 0$, one can determine the off-diagonal anharmonicity that is responsible for the cross peak at $T = 0$. Under the assumption that each oscillator, ω_1 and ω_2, consists of a single transition and that both anharmonicities are smaller than the respective width of the transition, the off-diagonal anharmonicity can be expressed as follows:

$$\Delta_{12} = \Delta_{11} \frac{\eta_2}{\eta_1} \frac{A_{pl}^{\text{diag}}}{A_0^{\text{diag}}} \frac{A_0^{\text{cross}}}{A_{pl}^{\text{cross}}}$$

Here, A_0^{diag}, A_0^{cross}, A_{pl}^{diag}, and A_{pl}^{cross} are the diagonal and cross-peak amplitudes at $T = 0$ and at the plateau, respectively, Δ_{11} is the diagonal anharmonicity of the mode ω_1, and η_i is the slope

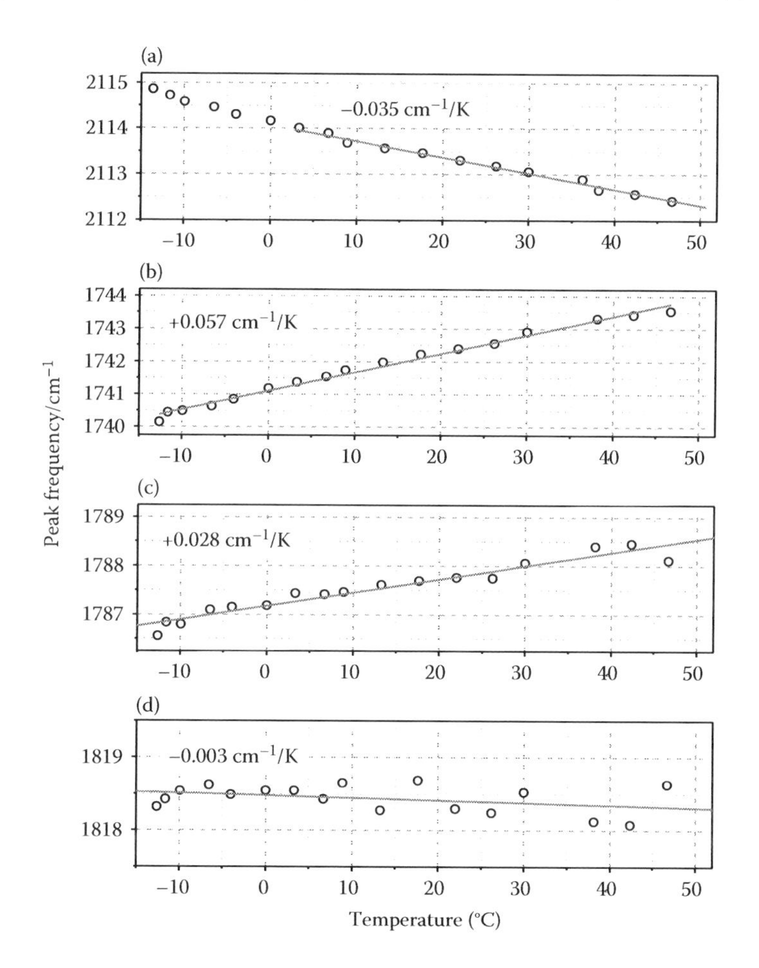

FIGURE 11.13 Temperature dependences of the peak frequency for N≡N peak at ca. 2113 cm^{-1} (a) and for three CO transitions at 1742 cm^{-1} (b), at 1788 cm^{-1} (c), and at 1819 cm^{-1} (d) in azPEG4. The temperature sensitivities evaluated at room temperature are shown in the insets. (Adapted from Lin Z, Keiffer P, and Rubtsov IV 2011. *J. Phys. Chem. B* 115:5347–5353.)

in the temperature dependence for the mode i. A straightforward modeling in the time domain can be used if the above assumptions break [18]. This approach permits calibrating the cross-peak waiting-time dependences in terms of efficient anharmonicity and replacing the cross-peak amplitude axis by that of the effective anharmonicity. An example of such replacement is shown in Figure 11.11 with a right-side axis.

The described approach permits evaluating small anharmonicity values using only relative-amplitude cross- and diagonal-peak measurements. It also requires knowing the sensitivities to temperature for both modes involved. While the sensitivity of the approach depends on various specific parameters of the molecular system, such as transition dipoles and temperature sensitivity of the high-frequency modes involved, the anharmonicities as small as 0.005 cm^{-1} are accessible with the N_3 and C=O mode pair.

11.4.4 Evaluating Distance Distributions

The thermal response in 2D IR cross peaks can be seen for mixtures of compounds where the tag and reporter reside on different compounds. Figure 11.14 shows an example of such dynamics for

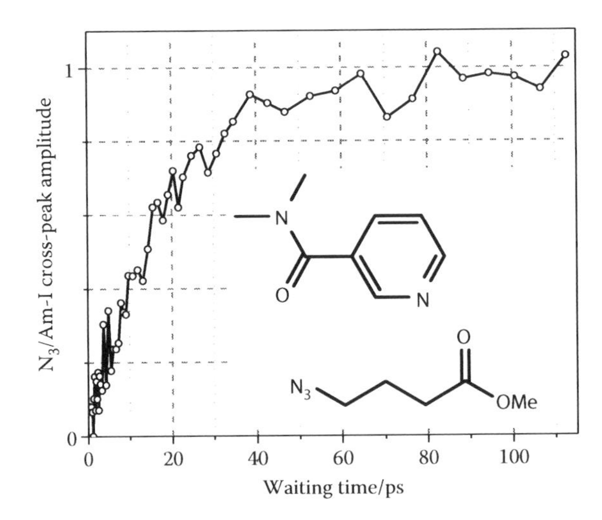

FIGURE 11.14 Waiting-time dependence for the N≡N/amide-I cross peak in the *N,N*-dimethyl nicotinamide and methyl 4-azidobutanoate mixture. (Adapted from Lin Z et al. 2012. *Phys. Chem. Chem. Phys.* 30(14): 10445–10454. DOI: 10.1039/c2cp40187h.)

the mixture of methyl-4-azidobutanoate featuring N≡N tag and *N,N*-dimethyl nicotinamide featuring amide-I reporter [41]. The data shows a very small cross peak at $T = 0$, suggesting the absence of aggregation of the two compounds. The N≡N/Am-I cross peak grows essentially exponentially until the plateau associated with complete thermalization is reached at $T > 50$ ps. The cross-peak amplitude at $T = 0$ reflects the distribution of distances between the IR labels involved—again, the initial mean anharmonicity can be measured by measuring the cross peak and diagonal peak waiting-time dependences and calibrating the cross-peak amplitudes in wavenumbers of effective anharmonicity using the plateau values from the two data sets. The mean anharmonicity observed at $T = 0$ can then be related to the equilibrium constant of the two compounds. The approach is expected to be particularly useful for evaluating small association constants.

11.5 2D IR AND RA 2D IR MEASUREMENTS OVER LARGER SPECTRAL REGIONS

The dual-frequency 2D IR and RA 2D IR methods permit measuring a large number of cross peaks, which, with appropriate modeling, can result in a large number of structural constraints. The fingerprint spectral region brings ample opportunities for such measurements. There are well-known complications of using the fingerprint region for structural evaluation, which stem from the complexity of the peaks there. First, the region is congested; the spectrum consists of numerous largely overlapping peaks. Moreover, the modes involved are often delocalized over several groups connected covalently, which are all involved in the motion. This results in a substantial size for a typical mode in the fingerprint region.

The congestion difficulty can be overcome by performing 2D IR spectroscopy using a single label, which is spectrally resolved and localized. The 2D IR spectrum focusing on the interaction of such tag and modes in the congested region will automatically select the modes that are spatially close to the spectrally isolated tag (interacting strongly with it), resulting in a simplified 2D IR spectrum [51,52]. The mode delocalization is harder to overcome. Quantum chemistry calculations provide a good estimation for the mode content, but typically do not provide it with sufficient accuracy. The computed absorption spectra in the fingerprint region can provide a good guidance of the main absorption contributors, but it is challenging to reproduce accurately the absorption spectrum in the fingerprint region. Two ideas can help the assignment. First, if there is a strong

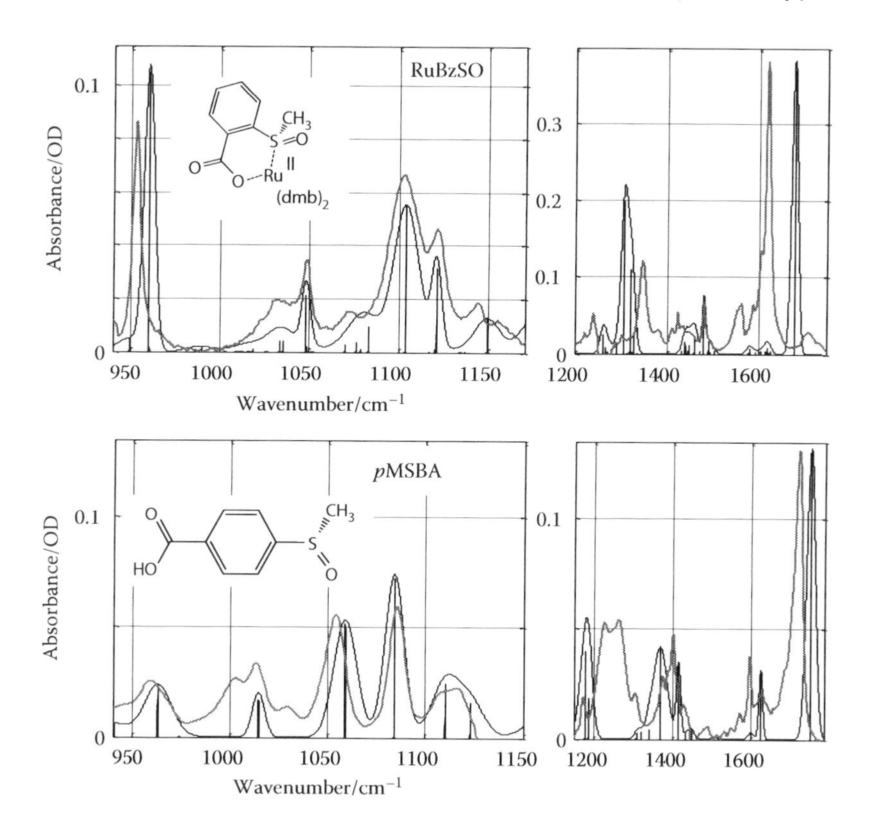

FIGURE 11.15 Linear absorption spectra (thick gray line) of (a) the bis(4,4′-dimethyl-2,2′-bipyridyl) (*o*-methylsulfinylbenzoate) ruthenium II complex (RuBzSO) in CH_2Cl_2 and (b), *p*-methylsulfinyl-benzoic acid (*p*MSBA) in CH_3CN/D_2O (v/v = 95/5). The DFT computed frequencies are shown with bars. The theoretical spectrum (thin black line) was generated by broadening of the bar spectrum using Gaussian line shape functions. (Adapted from Keating CS, et al. 2010. *J. Chem. Phys.* 133:144513.)

local mode with dominating transition moment in the FP region, the absorption spectrum in the vicinity of its frequency will be dominated by its contributions to various delocalized modes. This can simplify the absorption spectrum significantly, and lead to a better quantitative description using computational methods of quantum chemistry. For example, the presence of a strong S=O stretching mode in a set of sulfoxides results in good quality modeling of the absorption spectrum around 1100 cm^{-1} (Figure 11.15) [39]. The use of a single localized mode as a tag, such as the C=O stretching mode of *o*-methylsulfinylbenzoate ligand of RuBzSO (Figure 11.15a, inset), leads to the 2D IR spectrum reporting essentially on the structure of this ligand (Figure 11.16). It has been shown that the 2D IR spectrum changes dramatically when the complex structure changes from the S-bonded to O-bonded conformations [35].

The extent of delocalization can be measured experimentally using the RA 2D IR method [9,39]. For example, the waiting-time dependences of the amplitudes of the cross peaks among C=O stretching mode and four modes in the fingerprint region for RuBzSO are shown in Figure 11.17a. The T_{max} values obtained correlate with the effective distance between the modes involved. The effective distance between a point in space, selected as the center of the C=O group, and a mode (k) can be computed using the following equation:

$$\chi_k^{C=O} = \left(\sum_{i,\text{atoms}} \frac{dx_{ik}^2 + dy_{ik}^2 + dz_{ik}^2}{l_i} \right)^{-1}. \tag{11.2}$$

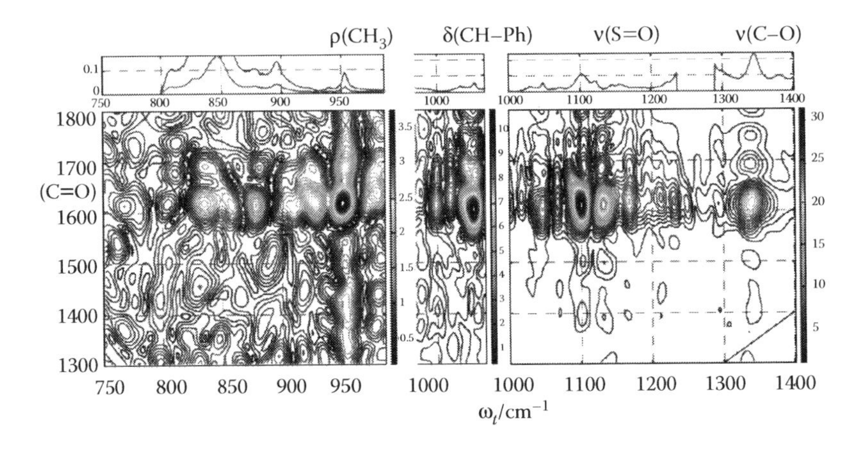

FIGURE 11.16 Nonrephasing absolute-value 2D IR spectrum of RuBzSO at $T = 267$ fs. The k_1 and k_3 spectra were centered at 1620 and 1050 cm^{-1}, respectively. (Adapted from Keating CS et al. 2010. *J. Chem. Phys.* 133:144513.)

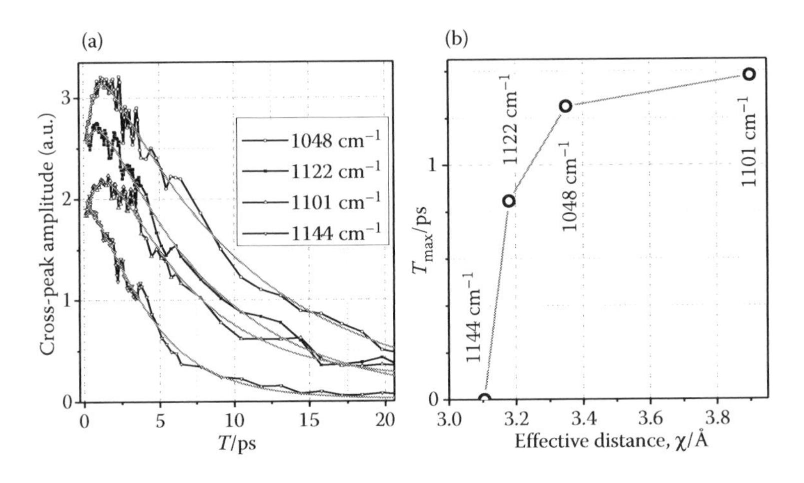

FIGURE 11.17 (a) The waiting-time dependences of the amplitudes of the cross peaks among C=O stretching mode and the modes indicated in the inset for RuBzSO. The fits with two-exponential functions are shown with gray lines. (b) The T_{max} values are plotted as a function of the effective intermode distance. (Adapted from Keating CS et al. 2010. *J. Chem. Phys.* 133:144513.)

Here, dx_{ik}, dy_{ik}, and dz_{ik} are the displacements of the ith atom in the normal mode k, and l_i is the distance between an ith atom and the center of the C=O group. The $\chi_k^{C=O}$ parameter has a dimension of length because the numerator in Equation 11.2 is dimensionless, as $\sum_{i,atoms} dx_{ik}^2 + dy_{ik}^2 + dz_{ik}^2 = 1$. Note that the effective distance equals the actual distance for the remote localized modes (Equation 11.2). The effective distance of 3.9 Å obtained between C=O and the mode at 1101 cm^{-1}, which has the dominant SO contribution, is similar but smaller than the distance between the centers of the C=O and SO bonds, which is ca. 4.4 Å, indicating a delocalized character of the latter. The T_{max} values for several modes in the fingerprint region were plotted as a function of the effective distance from the C=O group (Figure 11.17b). A clear correlation of T_{max} with the effective distance permits assessing the special location of a mode in the compound. Not surprisingly, the mode with dominant SO stretching contribution (1101 cm^{-1})

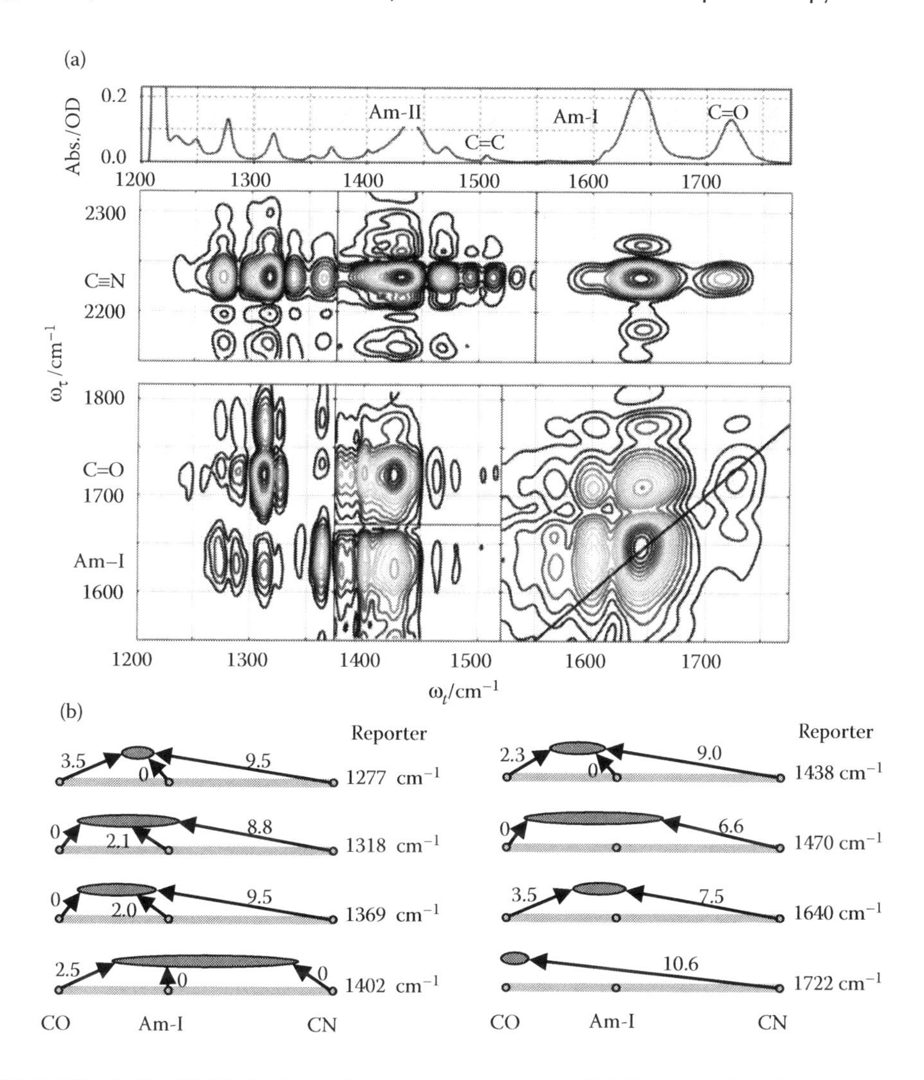

FIGURE 11.18 (a) RA 2D IR absolute-value rephasing spectrum of PBN measured at 10 ps. (b) Summary of the energy transport times, given in ps above the arrows, measured for PBN (see text). The arrows start at three initially excited modes, CN, Am-I, and CO, and indicate the energy transport times to the reporter modes indicated at the right side of each diagram. (Adapted from Naraharisetty SG, Kasyanenko VM, and Rubtsov IV. 2008. *J. Chem. Phys.* 128:104502/104501–104502/104507.)

appears as the most remote mode from the C=O group and dictates the range of possible T_{max} values for the modes at the same ligand.

The correlation of T_{max} with distance allows evaluating the spatial location for modes in the fingerprint region for which the assignment is less certain. Figure 11.18a shows the absolute-value rephasing RA 2D IR spectrum of PBN. The diagram in Figure 11.18 combines the RA 2D IR data for three tag modes (CN, Am-I, and CO) and a range of reporters in PBN (indicated at the right). For example, the top diagram in Figure 11.18b shows that the T_{max} values for the CN/1277 cm^{-1}, CO/1277 cm^{-1}, and Am-I/1277 cm^{-1} cross peaks are 9.5, 3.5, and 0 ps, respectively, where zero indicates that no peak is observed in the waiting-time dependence. The diagram gives a qualitative assessment of the region of delocalization for a number of modes in the fingerprint region. More quantitative analysis, for example, using Equation 11.2, is possible but requires a good mode assignment. The experiment predicts a large delocalization across the molecule for the mode giving rise to

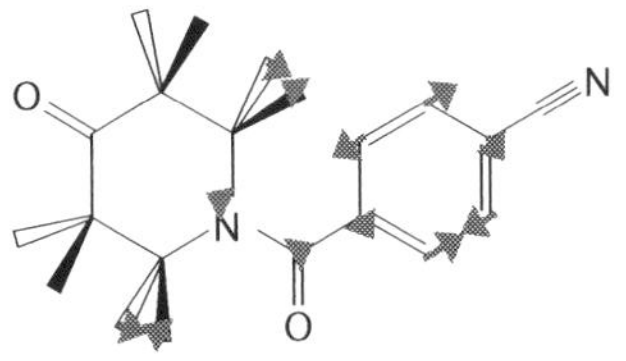

FIGURE 11.19 Atom displacements of the normal mode in PBN assigned to the peak at 1402 cm^{-1}. (Adapted from Naraharisetty SG, Kasyanenko VM, and Rubtsov IV. 2008. *J. Chem. Phys.* 128:104502/104501–104502/104507.)

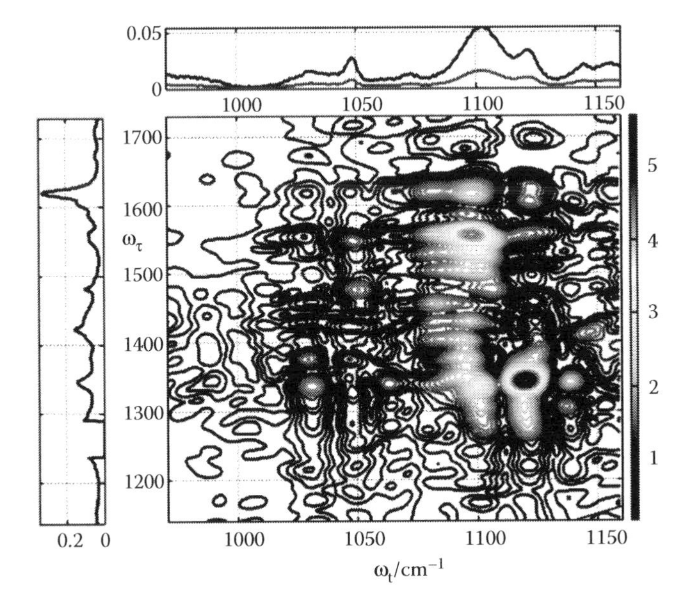

FIGURE 11.20 Absolute-value rephasing 2D IR spectrum of RuBzSO at T = 267 fs. The IR spectra of the k_1 and k_3 pulses were centered at 1420 and 1100 cm^{-1}, respectively.

the absorption peak at 1402 cm^{-1}. The DFT calculations confirm its large delocalization; the mode at 1402 cm^{-1} is assigned to a combination of CC stretching in the phenyl ring and CH_2 bending motions, as indicated in Figure 11.19. This example illustrates how RA 2D IR spectroscopy can help the mode assignment that is based on DFT calculations.

The 2D IR and RA 2D IR spectra focusing at the coupling between the modes in the fingerprint region provide rich peak patterns that are characteristic for the molecule (Figure 11.20). It is expected that such patterns, combined with quantum-chemistry-based modeling can provide valuable information about the molecular structure.

2D IR correlation spectroscopy has recently been used to identify the characteristic spectral features of two anomers of acetylated 2-azido-2-deoxy-d-glucopyranose, which differ at a single stereochemical center (Figures 11.21 and 11.22) [50]. While the linear absorption spectra for the α and β anomers are distinctive (Figure 11.23a), a substantial difference between them was found only in the spectral region between 1000 and 1200 cm^{-1} (Figures 11.24a). It is not surprising that their 2D correlation spectra involving this spectral region are very different (Figures 11.24b and 11.24c). The advantage of the 2D IR correlation spectroscopy over linear absorption spectroscopy is in insensitivity of the former to background absorption. Given that the tag mode is unique for the anomer, a similar 2D IR spectrum is expected for the anomer under conditions where other compounds absorb substantially in the fingerprint region. Interestingly, the 2D IR correlation spectra are found to be significantly

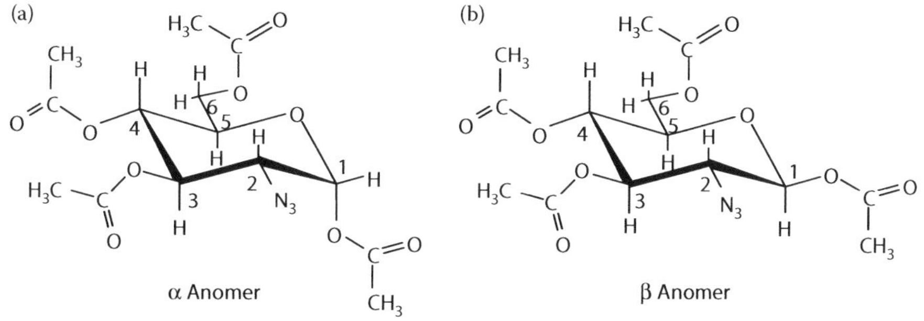

(a) α Anomer (b) β Anomer

FIGURE 11.21 Molecular structures of (a) 1,3,4,6-tetra-*O*-acetyl-2-azido-2-deoxy-α-d-glucopyranose (α anomer) and (b) 1,3,4,6-tetra-*O*-acetyl-2-azido-2-deoxy-β-d-glucopyranose (β anomer).

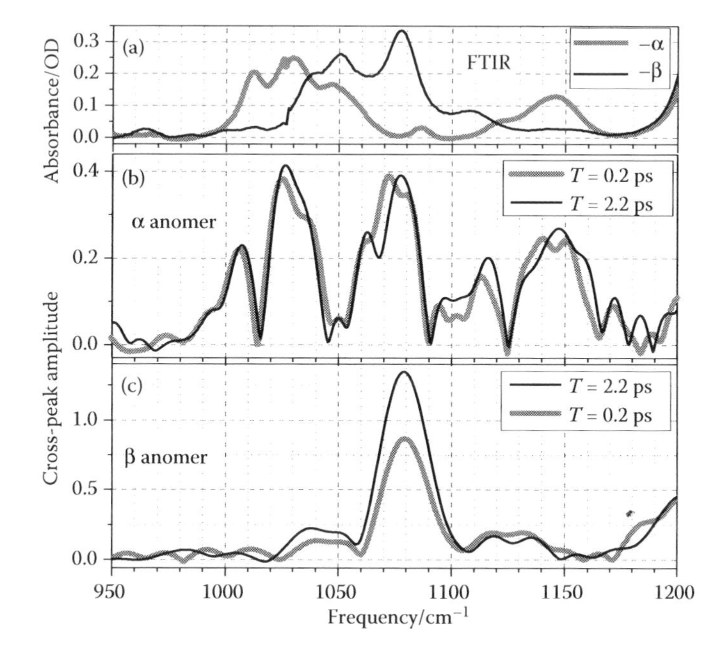

FIGURE 11.22 (a) Linear absorption spectra of the α and β anomers measured at 81 ± 5 mM concentration in deuterated chloroform for both compounds. Correlation spectra of the α (b) and β (c) anomers at the waiting times of 0.2 and 2.2 ps. The IR spectra of the k_1 and k_3 pulses were centered at 2110 and 1060 cm^{-1}, respectively. (Adapted from Lin Z, Bendiak B, and Rubtsov IV 2012. *Phys. Chem. Chem. Phys.* 14:6179–6191.)

different even in the regions where their linear absorption spectra are similar, such as, for example, in the C=O stretching region around 1750 cm^{-1}. The N≡N/C=O cross peaks at $T \sim 0$ are found to be very different (Figure 11.23b). In addition, the energy transport times, originated from the N≡N stretching mode relaxation, were found different for the two anomers by up to 1.8-fold (Figure 11.25). The results demonstrate the capability of 2D IR and RA 2D IR spectroscopies to provide unique spectroscopic data specific to sugar anomers that vary at a single stereochemical center. Unique coupling networks within individual sugar stereochemical units have a potential for identifying these units.

11.6 CONCLUSIONS

Application of the relaxation-assisted 2D IR approach to a variety of molecular systems was demonstrated. Relying on strong cross-peak amplification, the RA 2D IR method offers an

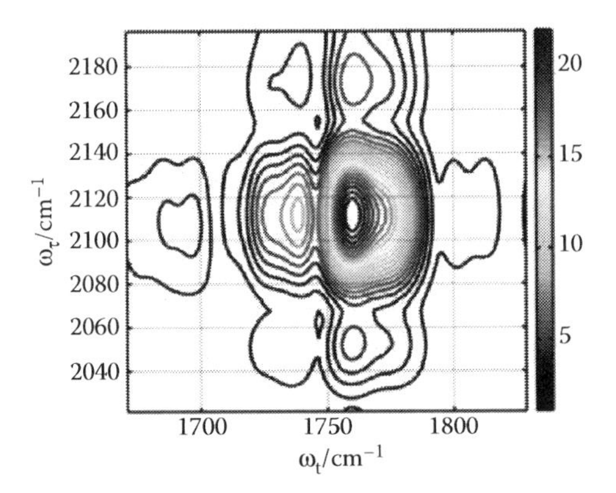

FIGURE 11.23 2D IR correlation spectrum for the β anomer measured at the waiting time of 0.67 ps. The IR spectra of the k_1 and k_3 pulses were centered at 2110 and 1750 cm^{-1}, respectively. (Adapted from Lin Z, Bendiak B, and Rubtsov IV 2012. *Phys. Chem. Chem. Phys.* 14:6179–6191.)

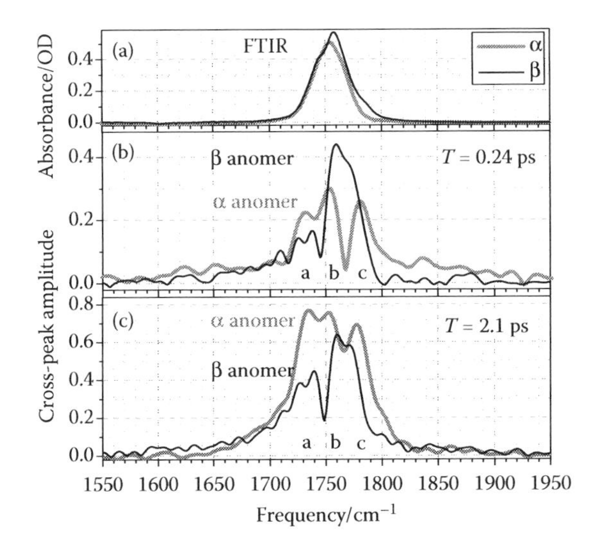

FIGURE 11.24 (a) Linear absorption spectra of the α and β anomers. Correlation spectra of the α (thick gray line) and β (thin black line) anomers at the waiting times of 0.24 ps (b) and 2.1 ps (c). (Adapted from Lin Z, Bendiak B, and Rubtsov IV. 2012. *Phys. Chem. Chem. Phys.* 14:6179–6191.)

increased range of distances accessible for structural measurements. It also permits measuring connectivity patterns in molecules, mode delocalization extents, and distributions of distances between vibrational reporters. To increase the breadth of practical applications of relaxation-assisted 2D IR spectroscopy, a better understanding of the energy transport on a molecular scale is required. Further development of the theoretical methods gaining quantitative predictions for energy transport should include realistic models for energy transport into the solvent and transport via solvent or the surrounding medium. Recent experiments suggested that a part of vibrational excess energy can propagate ballistically in molecules that offer highly delocalized vibrational states [28,40,50,53,54]. Development of the models incorporating simultaneously coherent and incoherent energy transfer pathways suitable for large molecular systems is essential for describing and making use of such regimes of energy transport.

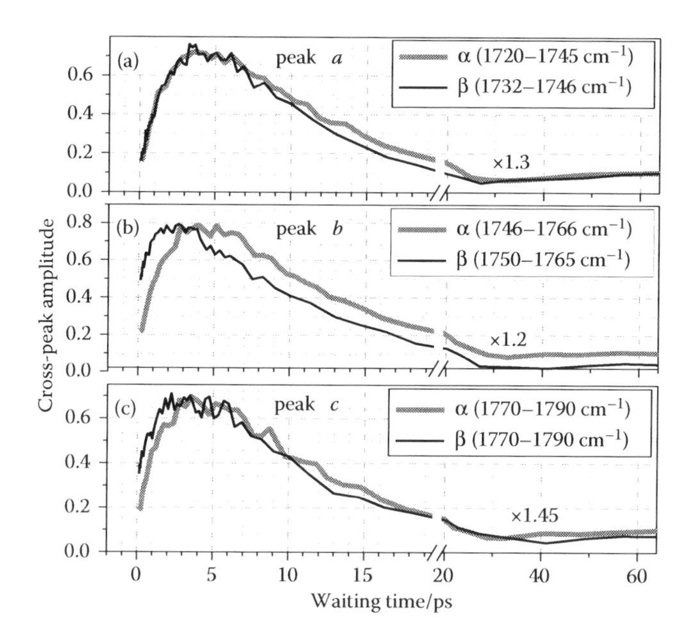

FIGURE 11.25 The waiting-time dynamics for the two anomers at indicated frequencies, which correspond to the peaks *a*, *b*, and *c* in Figure 11.24. The integration ranges and relative normalization factors are shown in the insets. (From Lin Z, Bendiak B, and Rubtsov IV 2012. *Phys. Chem. Chem. Phys.* 14:6179–6191.)

ACKNOWLEDGMENT

Support by the National Science Foundation (CHE-0750415) is gratefully acknowledged.

REFERENCES

1. Hamm P, Lim M, and Hochstrasser RM. 1998. Structure of the amide I band of peptides measured by femtosecond non-linear infrared spectroscopy. *J. Phys. Chem. B* 102:6123–6138.
2. Zimdars D, Tokmakoff A, Chen S, Greenfield SR, and Fayer MD. 1993. Picosecond, infrared, vibrational photon echoes in a liquid and glass using a free-electron laser. *Phys. Rev. Lett.* 70:2718–2721.
3. Asplund MC, Zanni MT, and Hochstrasser RM. 2000. Two-dimensional infrared spectroscopy of peptides by phase-controlled femtosecond vibrational photon echoes. *Proc. Natl. Acad. Sci. U.S.A.* 97:8219–8224.
4. Mukamel S. 1995. *Principles of Nonlinear Spectroscopy* (Oxford University Press, New York).
5. Golonzka O, Khalil M, Demirdoven N, and Tokmakoff A. 2001. Vibrational anharmonicities revealed by coherent two-dimensional infrared spectroscopy. *Phys. Rev. Lett.* 86:2154–2157.
6. Asbury JB et al. 2003. Ultrafast heterodyne detected infrared multidimensional vibrational stimulated echo studies of hydrogen bond dynamics. *Chem. Phys. Lett.* 374:362–371.
7. Rubtsov IV, Wang J, and Hochstrasser RM. 2003. Dual frequency 2D IR heterodyned photon-echo of the peptide bond. *Proc. Natl. Acad. Sci. U.S.A.* 100:5601–5606.
8. Kurochkin DV, Naraharisetty SG, and Rubtsov IV. 2005. Dual-frequency 2D IR on interaction of weak and strong IR modes. *J. Phys. Chem. A* 109:10799–10802.
9. Naraharisetty SG, Kasyanenko VM, and Rubtsov IV. 2008. Bond connectivity measured via relaxation-assisted two-dimensional infrared spectroscopy. *J. Chem. Phys.* 128:104502/104501–104502/104507.
10. Kurochkin DV, Naraharisetty SG, and Rubtsov IV. 2007. Relaxation-assisted 2D IR spectroscopy method. *Proc. Natl. Acad. Sci. U.S.A.* 104:14209–14214.
11. Middleton CT, Strasfeld DB, and Zanni MT. 2009. Polarization shaping in the mid-IR and polarization-based balanced heterodyne detection with application to 2D IR spectroscopy. *Opt. Express* 17:14526–14533.
12. Volkov V, Schanz R, and Hamm P. 2005. Active phase stabilization in Fourier-transform two-dimensional infrared spectroscopy. *Opt. Lett.* 30:2010–2012.

13. Kasyanenko VM, Lin Z, Rubtsov GI, Donahue JP, and Rubtsov IV. 2009. Energy transport via coordination bonds. *J. Chem. Phys.* 131:154508/154501–154508/154512.
14. Rubtsov IV. 2009. Relaxation-assisted 2D IR: Accessing distances over 10 Å and measuring bond connectivity patterns. *Acc. Chem. Res.* 42:1385–1394.
15. Goodno GD, Dadusc G, and Miller RJD. 1998. Ultrafast heterodyne-detected transient-grating spectroscopy using diffractive optics. *J. Opt. Soc. Am.* 15:1791–1794.
16. Turner DB, Stone KW, Gundogdu K, and Nelson KA. 2011. The coherent optical laser beam recombination technique (COLBERT) spectrometer: Coherent multidimensional spectroscopy made easier. *Rev. Sci. Instr.* 82:081301/081301–081301/081322.
17. Shim S-H, Strasfeld DB, Ling YL, and Zanni MT. 2007. Automated two-dimensional IR spectroscopy using a mid-IR pulse shaper and application of this technology to the human islet amyloid polypeptide. *Proc. Natl. Acad. Sci. U.S.A.* 104:14197–14202.
18. Lin Z, Keiffer P, and Rubtsov IV. 2011. A method for determining small anharmonicity values from 2D IR spectra using thermally induced shifts of frequencies of high-frequency modes. *J. Phys. Chem. B* 115:5347–5353.
19. Woutersen S, Mu Y, Stock G, and Hamm P. 2001. Hydrogen-bond lifetime measured by time-resolved 2D IR spectroscopy: *N*-Methylacetamide in methanol. *Chem. Phys.* 266:137–147.
20. Zheng J, et al. 2005. Ultrafast dynamics of solute-solvent complexation observed at thermal equilibrium in real time. *Science* 309:1338–1343.
21. Kim YS and Hochstrasser RM. 2005. Chemical exchange 2D IR of hydrogen-bond making and breaking. *Proc. Natl. Acad. Sci. U.S.A.* 102:11185–11190.
22. Rubtsov IV and Hochstrasser RM. 2002. Vibrational dynamics, mode coupling and structure constraints for acetylproline-NH_2. *J. Phys. Chem. B* 106:9165–9171.
23. Stuchebrukhov AA and Marcus RA. 1993. Theoretical study of intramolecular vibrational relaxation of acetylenic CH vibration for v = 1 and 2 in large polyatomic molecules ethynyltrimethylmethane and ethynyltrimethylsilane ($(CX_3)_3YCCH$, where X = H or D and Y = C or Si). *J. Chem. Phys.* 98:6044–6061.
24. Bigwood R, Gruebele M, Leitner D, and Wolynes P. 1998. The vibrational energy flow transition in organic molecules: Theory meets experiment. *Proc. Natl. Acad. Sci. U.S.A.* 95:5960–5964.
25. Bredenbeck J, Helbing J, and Hamm P. 2004. Transient two-dimensional infrared spectroscopy: Exploring the polarization dependence. *J. Chem. Phys.* 121:5943–5957.
26. Qian W and Jonas DM. 2003. Role of cyclic sets of transition dipoles in the pump-probe polarization anisotropy: Application to square symmetric molecules and perpendicular chromophore pairs. *J. Chem. Phys.* 119:1611–1622.
27. Rubtsov IV, Khudiakov DV, Nadtochenko VA, Lobach AS, and Moravskii AP. 1994. Rotational reorientation dynamics of C-60 in various solvents—picosecond transient grating dynamics. *Chem. Phys. Lett.* 229:517–523.
28. Wang Z et al. 2007. Ultrafast flash thermal conductance of molecular chains. *Science* 317:787–790.
29. Elsaesser T and Kaiser W. 1991. Vibrational and vibronic relaxation of large polyatomic molecules in liquids. *Annu. Rev. Phys. Chem.* 42:83–107.
30. Lian T, Locke B, Kholodenko Y, and Hochstrasser RM. 1994. Energy flow from solute to solvent probed by femtosecond IR spectroscopy: Malachite green and heme protein solutions. *J. Phys. Chem.* 98:11648–11656.
31. Ashihara S, Huse N, Espagne A, Nibbering ETJ, and Elsaesser T. 2007. Ultrafast structural dynamics of water induced by dissipation of vibrational energy. *J. Phys. Chem. A* 111:743–746.
32. Deak JC, Iwaki LK, and Rhea ST. 2000. Ultrafast infrared-Raman studies of vibrational energy redistribution in polyatomic liquids. *J. Raman Spectr.* 31:263–274.
33. Wang Z, Pakoulev A, and Dlott DD. 2002. Watching vibrational energy transfer in liquids with atomic spatial resolution. *Science* 296:2201–2203.
34. Pang Y et al. 2007. Vibrational energy in molecules probed with high time and space resolution. *Int. Rev. Phys. Chem.* 26:223–248.
35. Keating CS, McClure BA, Rack JJ, and Rubtsov IV. 2010. Mode coupling pattern changes drastically upon photoisomerization in Ru II complex. *J. Phys. Chem. C* 114:16740–16745.
36. Kasyanenko VM, Tesar SL, Rubtsov GI, Burin AL, and Rubtsov IV. 2011. Structure dependent energy transport: Relaxation-assisted 2D IR and theoretical studies. *J. Phys. Chem. B* 115:11063–11073.
37. Naraharisetty SRG et al. 2009. C-D modes of deuterated side chain of leucine as structural reporters via dual-frequency two-dimensional infrared spectroscopy. *J. Phys. Chem. B* 113:4940–4946.
38. Backus EHG et al. 2008. Energy transport in peptide helices: A comparison between high- and low-energy excitation. *J. Phys. Chem.* 112:9091–9099.

39. Keating CS, McClure BA, Rack JJ, and Rubtsov IV. 2010. Sulfoxide stretching mode as a structural reporter via dual-frequency two-dimensional infrared spectroscopy. *J. Chem. Phys.* 133:144513.
40. Lin Z and Rubtsov IV. 2012. Constant-speed vibrational signaling along polyethyleneglycol chain up to 60-Å distance. *Proc. Natl. Acad. Sci. U.S.A.* 109:1413–1418.
41. Lin, Z, Zhang, N, Jayawickramarajah, J, Rubtsov, IV. 2012. Ballistic energy transport along PEG chains: Distance dependence of the transport efficiency; (invited) *Phys. Chem. Chem. Phys.* 30(14):10445–10454. DOI: 10.1039/c2cp40187h
42. Burin AL, Tesar SL, Kasyanenko VM, Rubtsov IV, and Rubtsov GI. 2010. Semiclassical model for vibrational dynamics of polyatomic molecules: Investigation of internal vibrational relaxation. *J. Phys. Chem. C* 114:20510–20517.
43. Tesar SL, Kasyanenko VM, Rubtsov IV, Rubtsov GI, and Burin AL. Theoretical study of internal vibrational relaxation and energy transport in polyatomic molecules; submitted to *J. Phys. Chem.*
44. Anna JM, King JT, and Kubarych KJ. 2011. Multiple structures and dynamics of $[CpRu(CO)_2]_2$ and $[CpFe(CO)_2]_2$ in solution revealed with two-dimensional infrared spectroscopy. *Inorg. Chem.* 50:9273–9283.
45. Yu X and Leitner DM. 2003. Vibrational energy transfer and heat conduction in a protein. *J. Phys. Chem. B* 107:1698–1707.
46. Davydov AS. 1985. *Solitons in Molecular Systems* (Kluwer Academic, Dordrecht, Holland).
47. Leitner DM. 2005. Heat transport in molecules and reaction kinetics: The role of quantum energy flow and localization. *Adv. Chem. Phys.* 130 B:205–256.
48. Nitzan A. 2007. Molecules take the heat. *Science* 317:759–760.
49. Kasyanenko VM, Keiffer P, and Rubtsov IV. 2012. Intramolecular contribution to temperature dependence of vibrational modes frequencies. *J. Chem. Phys.* 136:144503/144501–144503/144510.
50. Lin Z, Bendiak B, and Rubtsov IV. 2012. Discrimination between coupling networks of glucopyranosides varying at a single stereocenter using two-dimensional vibrational correlation spectroscopy. *Phys. Chem. Chem. Phys.* 14:6179–6191.
51. Hamm P and Hochstrasser RM. 2000. Structure and dynamics of proteins and peptides: Femtosecond two-dimensional infrared spectroscopy. *Ultrafast Infrared and Raman Spectroscopy*, ed Fayer MD (Marcel Dekker Inc., New York), p 273.
52. Zhuang W, Hayashi T, and Mukamel S. 2009. Coherent multidimensional vibrational spectroscopy of biomolecules: Concepts, simulations, and challenges. *Angewandte Chemie Int. Ed.* 48:3750–3781.
53. Schwarzer D, Hanisch C, Kutne P, and Troe J. 2002. Vibrational energy transfer in highly excited bridged azulene-aryl compounds: Direct observation of energy flow through aliphatic chains and into the solvent. *J. Phys. Chem. A* 106:8019–8028.
54. Backus EHG et al. 2009. Dynamical transition in a small helical peptide and its implication for vibrational energy transport. *J. Phys. Chem. B* 113:13405–13409.

12 An Introduction to Protein 2D IR Spectroscopy

Carlos R. Baiz, Mike Reppert, and Andrei Tokmakoff

CONTENTS

12.1 INTRODUCTION

Proteins are molecules that behave in beautiful and astounding ways in the course of their biological function, and all biological processes involve protein conformational changes. These processes might be enzyme catalysis, transport and signaling, dynamic scaffolding for structures, charge transfer, or mechanical or electrical energy transduction. Our view of such processes is colored by the methods we use to study them, and most of what we know about proteins is based on structural studies and biochemical assays. Biochemists and biologists think of the processes in which proteins engage in terms of directed motion, often illustrating them through movies, but in current experiments, one rarely actually watches the conformational changes that occur directly. The method of two-dimensional infrared (2D IR) spectroscopy [1–3] is providing new approaches that can be used to characterize the features that are masked by traditional methods, particularly for visualizing conformational dynamics on picosecond to millisecond timescales, and characterizing the conformational variation and structural disorder. Furthermore, it is also proving useful for samples that are difficult to study by traditional techniques, such as protein aggregates and amyloid fibers [4,5], intrinsically disordered peptides [6], and membrane proteins [7,8].

Two-dimensional IR spectroscopy was developed as a tool to study transient molecular structure and dynamics in solution. As a vibrational spectroscopy, it directly interrogates the vibrations of chemical bonds and how the many vibrations of a molecule and its environment interact with one another. Inspired by the two-dimensional (2D) techniques first developed in the field of nuclear magnetic resonance (NMR), 2D IR spectroscopy spreads a vibrational spectrum over two frequency axes that report on how excitation of a vibration with a given frequency influences all other vibrations within a detection window, following a waiting time. Spectral features, in the form of frequencies, amplitudes, and lineshapes, and the evolution of these features with time are used to understand structural connectivity in space and time and leads to new avenues for studying the molecular structure, dynamics, and dynamical heterogeneity. Given enough information on the mechanism of vibrational interactions, one can model a spectrum to reveal structural information on the vibrations being observed.

From subpicosecond water fluctuations to hour-long aggregation processes, biophysical processes vary over many orders of magnitude in time, and therefore require methods that can span a wide range of timescales. Figure 12.1 shows a variety of processes along with their corresponding timescales [9,10]. Since the 2D IR measurement is made with a picosecond or faster "shutter speed," it captures information on molecular structure in solution on a fast timescale compared to most dynamics, and is uniquely positioned to probe many of these processes. Correlation 2D IR, in which the waiting time is varied, characterizes dynamics on the picosecond timescale. The nonequilibrium variants of 2D IR, such as temperature-jump 2D IR, expand the dynamic range of the technique from picoseconds to milliseconds and enable the study of transient processes such as protein folding or association [11]. Longer timescales can be reached by rapid-acquisition continuous-scanning 2D IR methods.

Careful consideration of the experimental objectives facilitates the design of 2D IR experiments. For example, short-range structure and dynamics, such as hydrogen bonding environments, are visualized through 2D lineshape analysis [12] or waiting time dynamics experiments [13] on localized vibrations, whereas delocalized amide I spectra give a global view of the protein architecture [14]. Triggered experiments are used to study nonequilibrium conformational dynamics. Finally, interpretation of experimental spectra can be done on multiple levels: empirical rules offer basic structural information, whereas a more sophisticated structural view involves spectral modeling. Analogous to NMR spectroscopy, 2D IR provides a set of structural constraints that can be used with structure-based

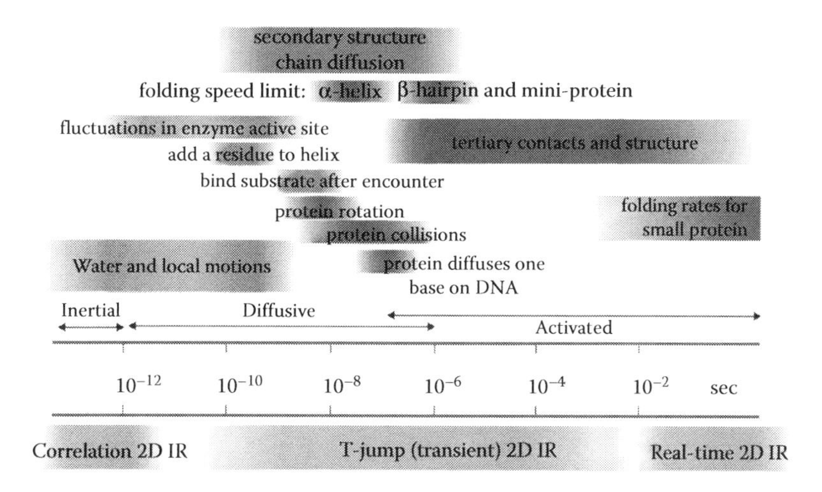

FIGURE 12.1 Timescales of protein dynamics.

modeling methods to provide structures that are consistent with the experimental constraints [12]. One approach describes 2D IR spectra in terms of semiclassical electrostatic maps that describe the structure-dependent vibrational couplings and frequency shifts within the protein [15–18]. These techniques can be interfaced with molecular dynamics (MD) simulations and Markov-state models, or with alternative structure-based computational biophysics simulation models.

This chapter offers an introduction of 2D IR methods aimed at new users who wish to use the technique to shed light on questions related to structure and dynamics of proteins. Sections 12.2 and 12.3 introduce the methods, theory, and applications of backbone 2D IR spectroscopy as a probe of protein structure, protein solvation, folding and binding, and protein–protein interactions. To date, the majority of 2D IR spectroscopy has been focused on amide I vibrations, largely composed by C=O stretching and N–H wag vibrations of the amide moiety. Delocalized over the entire protein backbone, amide I modes are sensitive to the global structure of the protein. However, there are many alternative vibrational probes, which can be tailored to probe localized structure, solvent exposure, and hydrogen bonding. Cross-correlations between different vibrational modes compound the information contained within individual modes to provide new structural insights that are not available from single vibrations. The merits of localized vibrational probes and other backbone modes are discussed in Section 12.3. Vibrational modeling provides the crucial link between structure and spectra. Section 12.4 describes the framework for modeling protein 2D IR spectra and the methods available for protein 2D IR analysis. Section 12.4 is geared toward theoreticians with experience in MD simulations and a basic background in quantum mechanics. Section 12.5 introduces recent examples of 2D IR spectroscopy to illustrate the applications of 2D IR to study protein structure, conformational heterogeneity, and solvent exposure, and transient temperature-jump-induced denaturation.

12.2 BACKGROUND

12.2.1 Amide I Modes

Amide vibrations of the polypeptide backbone are the most commonly used vibrations in infrared (IR) studies of proteins [19–22]. Of these, the amide I vibrations observed between 1600 and 1700 cm^{-1} are of particular interest since they provide distinct spectroscopic signatures of secondary structure and hydrogen bonding contacts in peptides and proteins. As illustrated in Figure 12.2, amide I vibrations are combinations of C=O stretching and in-plane N–H bending vibrations of the backbone amide units that have a strong IR transition dipole moment. The physical interactions or

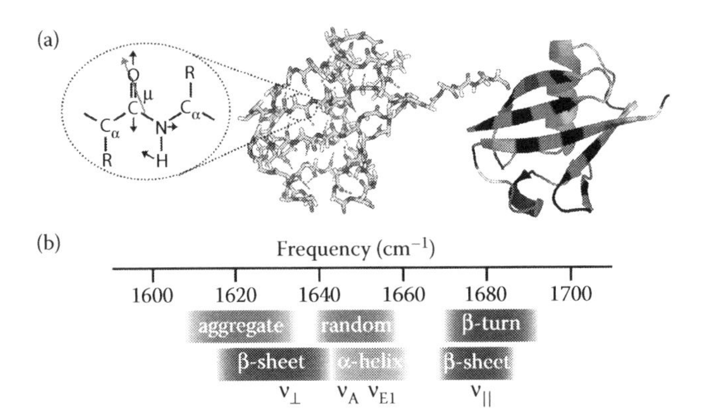

FIGURE 12.2 Amide I spectroscopy of proteins: (a) atom displacements and transition dipole moment associated with amide I vibrations in a single amide unit, along with the contributions of different units to a normal mode in ubiquitin. The amplitude and phase of the vibration is encoded on the grayscale intensity: dark, light are 180 degrees out of phase. (b) Representation of experimentally observed IR bands corresponding to different structural motifs.

couplings between amide I vibrations of the protein's multiple peptide units give rise to delocalized vibrations of the protein backbone. With the exception of proline, the side-chain vibrations do not interact strongly with the amide I vibration. Thus, amide units are chemically identical throughout the entire backbone and have similar vibrational frequencies, allowing for efficient coupling of the local modes on each amide to form delocalized vibrations. We refer to the vibrations of the peptide unit as local amide I modes (or sites) that serve as a basis to describe the delocalized vibrations—called *normal modes* for harmonic systems or more generally *excitons*, a term borrowed from solid-state physics [23]—as linear combinations of the sites [22]. Figure 12.2 shows a typical β-sheet normal mode encoded onto the ribbon diagram of ubiquitin. Similar to vibrations in small molecules, the character of the normal mode is dictated by the local symmetry of the protein backbone: α-helices and β-sheets exhibit modes characteristic of the local arrangement of the residues within the structure [24–28].

As discussed in more detail in Section 12.4, structural models for the coupling interactions between peptide units provide a means of connecting the exciton band information contained in IR spectra to protein structure. Unfortunately, broad absorption lineshapes and the large number of delocalized modes obscure much of the information present in amide I spectra, making the task of extracting structure from absorption spectra particularly difficult. 2D spectroscopy unpacks structural information by spreading the spectral contents onto two frequency axes and measuring frequency correlations between multiple vibrations.

12.2.2 Structural Interpretation of Amide I Infrared Spectra

Figure 12.2 shows an empirical relationship between common secondary structures and spectral features observed in amide I spectra. α-Helices exhibit a single peak centered near 1650 cm^{-1}, while β-sheets exhibit two peaks centered near 1630 and 1680 cm^{-1} whose center frequencies and intensities depend on the length and number of β strands [14,25]. Vibrations associated with unstructured regions also appear as a single broad peak near 1650 cm^{-1}. Figure 12.3 shows Fourier-transform infrared spectroscopy (FTIR) and 2D IR spectra of proteins with different secondary structures: myoglobin (α-helix), ubiquitin (mixed α/β), and concanavalin A (antiparallel β-sheet). Here, we provide a brief interpretation of amide I 2D IR spectra, more detailed descriptions of 2D IR spectroscopy and a step-by-step interpretation of a 2D IR spectrum are provided in the next section.

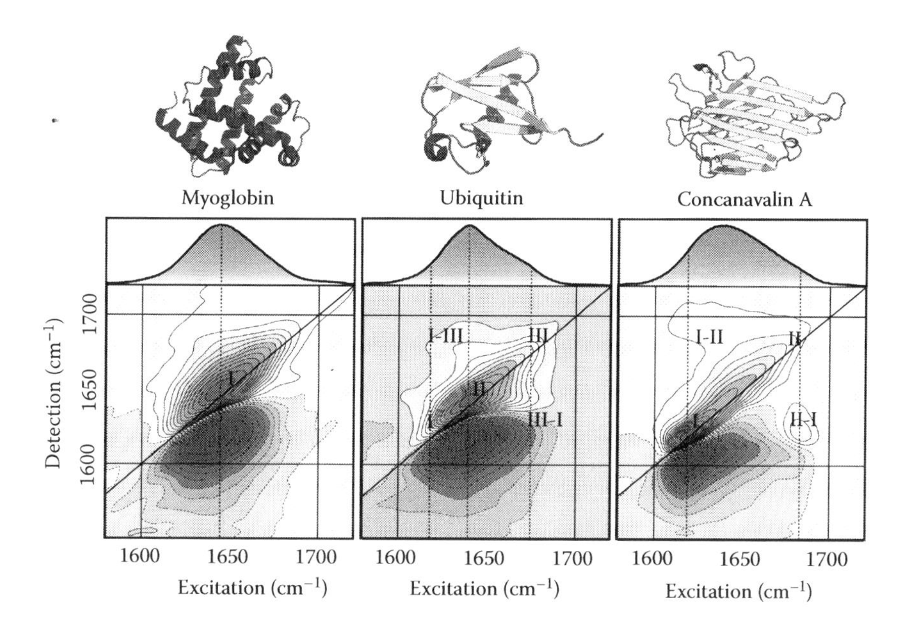

FIGURE 12.3 Amide I absorption and 2D IR spectra of myoglobin, ubiquitin, and concanavalin A. Dashed lines indicate the approximate peak positions. Cartoon structures are shown for reference (PDB codes: 1MBO, 1UBQ, 1JBC). Nonlinear scaling of the contours emphasizes the low-amplitude features in the spectrum.

Two-dimensional spectra can be interpreted as a 2D map that links a set of excitation frequencies (ω_1) to a set of detection frequencies (ω_3). The features observed along the diagonal axis correspond to excitation and detection of the same frequencies, and can be related approximately to the absorption spectrum. Off-diagonal peaks correspond to exciting one particular peak and detecting emission from a different one, which reflects coupling or energy transfer between two modes. As a result of vibrational anharmonicity, each peak in the spectrum appears as a positive–negative doublet. When these peaks are in a congested spectrum, complex lineshapes can arise. Myoglobin exhibits a round diagonal peak centered near 1650 cm^{-1} (denoted as *I*). Concanavalin A shows two peaks centered near 1620 and 1670 cm^{-1}, labeled *I* and *II* in the 2D spectrum. Since the two peaks are broad, the spectrum appears elongated along the diagonal. However, the presence of narrow off-diagonal peaks (*I–II* and *II–I*) indicates the presence of two distinct transitions underneath the broad diagonal lineshape. Beta-sheet amide I spectra provide a clear illustration of the information content gained by 2D spectroscopy over one-dimensional techniques [29]. In the case of ubiquitin, contributions from the α-helix and β-sheet residues show a feature centered near 1640 cm^{-1} (*II*) with a strong diagonal elongation toward the low-frequency (*I*) high-frequency regions (*III*). Cross peaks (*I–III* and *III–I*) are observed near [1640,1680] cm^{-1}, a clear signature of β-sheet contribution. Similarly, two broad peaks centered near 1640 and 1680 cm^{-1} are observed in the corresponding FTIR spectrum. Mixed α/β proteins, such as ubiquitin, tend to exhibit lineshapes resembling the combination of purely α-helix and purely β-sheet spectra. The interpretation of these features has been aided greatly by 2D IR studies [14,29,30] as described below.

12.2.2.1 Spectral Signatures of Secondary Structure

Similar to most spectroscopic techniques, structural assignments are facilitated by theoretical models and simulations [24,31]. The standard modeling approaches are described in Section 12.4. This section provides a qualitative interpretation of the spectral features associated with different secondary structures.

12.2.2.1.1 α-Helices

Simulations for idealized α-helices have revealed the presence of two main IR-active modes with A and E_1 symmetries: The most intense A mode, accounting for approximately 70% of the intensity, involves the in-phase oscillation of all the residues in the helix whereas the E_1 mode has a periodic phase shift of approximately 3.6 residues per cycle [18,26,32]. The A mode red-shifts from ~1660 cm^{-1} for helices of 5–10 residues in length to ~1650 cm^{-1} for helices of greater than 20 residues. In contrast, the E_1 mode shows almost no frequency dependence with respect to the helix length. The A–E_1 frequency splitting and E_1 phase twist are attributed to the large positive coupling observed between adjacent peptide units and the negative coupling between hydrogen-bonded ones. Since the A mode accounts for much of the main band intensity in α-helices, a frequency dependence on the helix length is observed for primarily α-helical proteins. However, the ~10 cm^{-1} shift is relatively small compared to the ~60 cm^{-1} total width of the amide I band. Polarization-controlled 2D IR experiments have been able to resolve the two modes experimentally in short 21-residue helical peptides and the E_1 and A modes are centered at 1638 and 1650 cm^{-1}, respectively [33].

12.2.2.1.2 3_{10} Helices

The spectral character of 3_{10} helices is very similar to those observed in α-helices: low-frequency A modes and higher-frequency E modes are split by 10–15 cm^{-1} depending on the length of the helix [34]. Vibrational circular dichroism and FTIR, as well as computational studies, suggest that the peaks of a 3_{10} helix are red-shifted by 5–10 cm^{-1} compared to α-helices [35,36]. Owing to the broad lineshapes, it can be difficult to distinguish between α and 3–10 helices via IR absorption spectroscopy, but recent studies suggest that amide I 2D IR spectra contain specific spectral signatures for 3_{10} helices that can serve to spectrally separate them from α-helices [37–39].

12.2.2.1.3 β-Sheets

Antiparallel β-sheets are constructed from a repeating rectangular four-peptide unit [25,29]. For idealized infinite sheets, the vibrations of a single unit describe the observed vibrations of the sheet. The unit cell for an antiparallel sheet consists of four oscillators; therefore, four distinct modes are observed, only two of which are IR active. The lowest IR-active mode, which carries most of the intensity, appears as a narrow band near 1630–1640 cm^{-1} (see concanavalin A spectrum in Figure 12.3), and involves the in-phase oscillation of sites on adjacent strands. In the four-peptide unit, this mode involves the in-phase oscillation of units that lie in opposite corners along the diagonal of the rectangle. Vibrations of this character are denoted as $\upsilon_\perp$ since the transition dipole lies perpendicular to the β-strands [40–44]. The high-frequency IR active mode, denoted $\upsilon_{//}$, is centered ~1670 cm^{-1} and has lower overall intensity since it involves the in-phase oscillation of adjacent residues along the β-strands, and is out-of-phase with respect to the hydrogen-bonded neighboring residue on the adjacent strand. Figure 12.4 shows simulated spectra for idealized antiparallel β-sheets. As the size of the sheet increases, $\upsilon_\perp$ red-shifts by ~20 cm^{-1} and gains additional intensity with respect to $\upsilon_{//}$. The high-frequency $\upsilon_{//}$ peak position remains nearly unchanged, and therefore serves as an internal reference marker of β-sheet structure. Thus, the red shift of the $\upsilon_\perp$ band in β-sheets serves as an indicator of the β-sheet size [25].

Parallel β-sheets can be described as a repeating set of two-residue unit cells [28]. Therefore, the vibrational spectrum of an infinite parallel β-sheet will show two main vibrational bands. In a 4 × 4 residue idealized sheet, the high-frequency band appears near 1660 cm^{-1} but carries virtually no intensity. Most of the oscillator strength is shifted to the low-frequency band, centered around 1635 cm^{-1}, described as out-of-phase oscillations of residues within the same strand. Transition dipole modes for vibrations of this character lie perpendicular to the β-strands.

12.2.2.1.4 β-Turns

Contributions from residues in β-turn configurations are difficult to isolate experimentally. For this reason, the spectral assignment of β-turns is largely derived from simulations. In general, β-turn

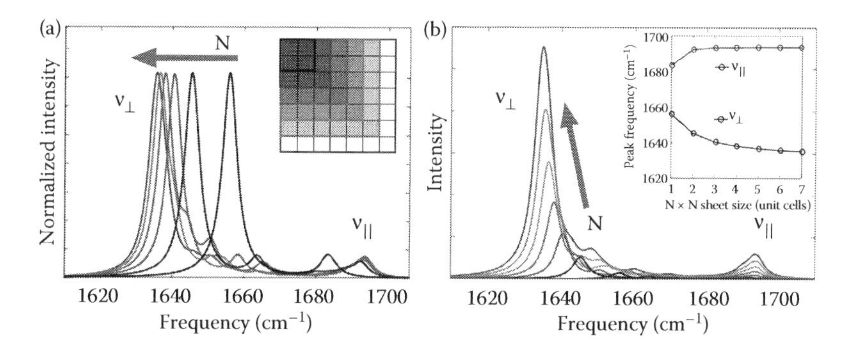

FIGURE 12.4 Simulated spectra of idealized N × N-unit antiparallel β-sheet structures ranging from 2 × 2 to 7 × 7 units. Each unit corresponds contains four residues arranged in a 2 × 2 square configuration. (a) Size-dependence of the $\upsilon_{//}$ and $\upsilon_\perp$ peak positions. (b) Size dependence of the peak intensities. The inset shows the center frequencies of the two peaks as a function of the sheet size.

peaks appear near 1680 cm^{-1} [45]. Since the $\upsilon_{//}$ band of β-sheets also appears in the same frequency range, and both peaks are expected to have low amplitudes, the first due to the relatively few oscillators in β-turn configurations and the second due to the low oscillator strength of $\upsilon_{//}$, peaks in the 1680 cm^{-1} region and above can be easily misassigned. One must therefore be particularly careful when assigning spectral features associated with β-turns.

12.2.2.1.5 Coils

Unstructured coils appear as a broad, featureless peak centered near 1650–1660 cm^{-1}. Lacking long-range order, coils exhibit random coupling patterns that give rise to broadened peaks. In addition, the increased solvent exposure of the backbone contributes to the variation in site frequencies and further broadens the peaks. The difficulty in resolving α-helices from random coils is one of the shortcomings of amide I IR absorption spectroscopy. Since random coils are more solvent exposed, hydrogen/deuterium exchange experiments [46] and waiting-time experiments [47] combined with isotope labeling can aid in assigning the individual residues to secondary structures.

12.2.2.2 Doorway Mode Analysis

Intuitive interpretation of individual amide I modes is difficult for proteins for two main reasons: First, a large number of normal modes is packed into a small spectral region, and modes have partially mixed character, for example, individual modes are often delocalized over α-helices or β-sheets. Second, the character of the normal modes is highly dynamic, and slight changes in structure or solvent environment can remix the modes. Therefore, while the normal mode picture provides a convenient framework for describing the models, individual normal modes have limited relevance for interpreting spectra. Since normal modes within the same frequency region share similar overall characteristics, instead of focusing on individual modes, a more intuitive interpretation is obtained via a doorway mode analysis. In brief, the doorway mode analysis described here relies on singular value decomposition to project out the shared vibrational features within a set of normal modes in a small frequency window [24,27,48]. Intensity-weighted components are referred to as *doorway modes* and provide an intuitive visualization of the normal modes' character within the selected frequency region. The character of the principal component modes is less affected by small changes in structure, allowing for comparison of modes among different conformations of a protein or even different proteins.

Figure 12.5 shows simulated absorption spectra of four proteins with varying α-helix/β-sheet conformation: from 50% β-sheet (concanavalin A) to 75% α-helix (myoglobin). To interpret amide I spectral signatures, each residue is assigned to one of four structures: α-helix, β-sheet, β-turn, and coil. The structural character of the doorway state is calculated by amplitude weighting the

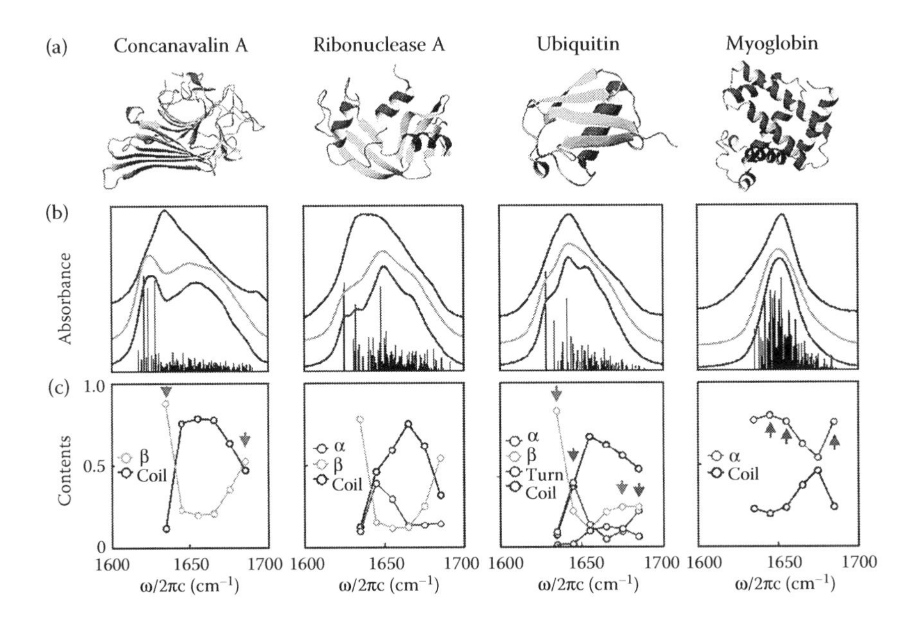

FIGURE 12.5 (a) Crystal structures of the four proteins selected for doorway mode analysis. (b) (Top) experimental absorption spectra of proteins in D_2O; (middle) spectra calculated with site disorder; (bottom) spectra calculated without site disorder. Black lines represent individual mode frequencies and intensities. (c) Amplitude-weighted structure of the doorway modes. (Adapted from Chung HS, Tokmakoff A. 2006. *J Phys Chem B* 110:2888–2898.)

contribution of each oscillator to an individual secondary structure. For example, a β-sheet doorway mode will have a large number of amplitude-weighted residues in a β-sheet conformation contributing to the mode, in comparison to the α-helix. Doorway mode analyses (Figure 12.5c) show that β-sheets contribute mostly in the regions below 1630 cm^{-1} and above 1670 cm^{-1}, whereas α-helices contribute intensity mainly toward the center of the amide I band. Random coils also contribute in the 1650–1670 cm^{-1} region, and overlap some of the helix peaks. These observations are consistent with spectral assignments shown in Figure 12.3.

12.2.2.3 Beyond Secondary Structure

The basic examples described above are not sufficient to explain all the features observed in 2D IR, suggesting that spectra contain more detailed information related to the specific architecture of the protein. For example, 2D IR spectra of proteins with similar percentage of residues in α-helix and β-sheet conformations exhibit very distinct features, suggesting that the three-dimensional architecture of proteins is encoded in the spectrum [14]. To date, the spectral signatures of supersecondary motifs remain largely unexplored. Further investigations will be required to understand questions such as: How does the spectrum of a twisted β-sheet differ from a flat, idealized, β-sheet? How does one distinguish parallel and antiparallel β-sheet contacts in real systems? Do multiple secondary structures in a well-defined registry, for instance, a coiled coil or a protein oligomer, have distinct intermolecular couplings? These are areas of active research.

12.2.2.3.1 Amyloids and Aggregates

Misfolded proteins, such as the extensively studied β-amyloid peptide, aggregate into insoluble fibrils with well-defined stacked parallel cross-β-sheet structures. A high degree of order and limited water penetration into the fibrils contribute to the sharp β-sheet peaks observed in absorption spectra [4,5,49,50]. The kinetics of amyloid aggregation can be easily monitored with IR spectroscopy, and molecular insights into the aggregation mechanism are aided by isotope labeling experiments

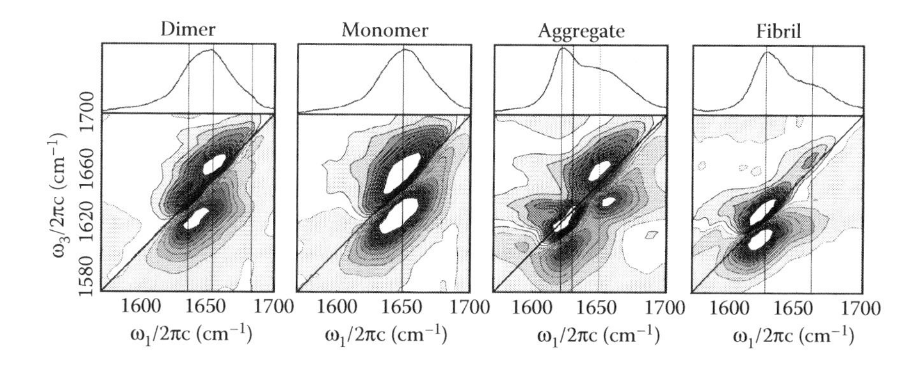

FIGURE 12.6 Two-dimensional spectra of insulin in different forms. Solid and dashed contours represent positive and negative features, respectively.

[4,51]. Other insoluble protein aggregates appear as a single peak centered near 1620 cm^{-1}. Narrow lineshapes and red-shifted frequencies (~1620–1630 cm^{-1}) make IR spectroscopy particularly sensitive to the presence of aggregates in the sample. Nonlinear methods further enhance the spectral features, such that even a small amount of aggregate present in the sample produces intense features that dominate the amide I spectrum.

Spectra of insulin (Figure 12.6) illustrate how multidimensional spectroscopy has the sensitivity to probe the interprotein contacts and interactions responsible for dimerization, oligomerization, and aggregation. In comparison with monomer spectra, dimers show features associated with β-sheets: a two-peak structure and a more pronounced β-ridge, indicative of an interprotein β-sheet present in the dimer and suggesting that the interfacial residues of the monomer remain in a disordered configuration. The aggregate and fibril samples show a two-peak pattern but with different center frequencies and peak intensity ratios, indicating that the aggregate has a significant percentage of residues in α-helix or disordered regions, whereas the fibril exhibits features characteristic of ordered β-sheets. Though qualitative, the simple interpretation of the data helps characterize the structural changes associated with protein association and fibril formation. Recent studies on the amyloid-β peptide illustrate how quantitative modeling, and incorporation of isotope labels, provides detailed insights into the mechanisms of protein aggregation and fibril formation [51].

12.2.3 Spectrally Isolated Sites: Hydrogen Bonding and Isotope Labels

12.2.3.1 Hydrogen Bonding

Hydrogen bonds stabilize protein structure. IR spectroscopy is one of the few experimental techniques that is sensitive to hydrogen bonding: amide I vibrations shift depending on the number of hydrogen bonds accepted by the oxygen atom or donated by the hydrogen atom in the amide unit [52]. A single hydrogen bond between the amide oxygen atom and a water molecule red-shifts the amide I frequency by approximately 16 cm^{-1} [53]. The ability of the oxygen atom to accept multiple hydrogen bonds causes solvent-exposed residues to have strongly red-shifted transitions compared to residues locked into stable protein–protein hydrogen bonds. Understanding the origin of the H-bond-induced shift is straightforward; vibrational frequencies are proportional to the square root of the bond-stretching constant and inversely proportional to the reduced mass of the oscillator. From a chemical-bonding perspective, an H-bond reduces the π-orbital overlap CO imparting partial single bond C–O–H character to the C=O···H double bond. Similarly, the hydrogen atom can be thought of as effectively increasing the mass of the oxygen atom along with the reduced mass of the C=O bond, thus lowering the vibrational frequency. The relationship between H-bond strength and vibrational frequency has been measured for many systems. In general, the observed shift in vibrational frequency is proportional to the strength of the hydrogen bond. The empirical relationship between hydrogen bond distance and C=O frequency shift, $\delta\nu$ in cm^{-1}, for dipeptides in water is given by

$$\delta\nu = 30(r_{OH} - 2.6) \tag{12.1}$$

where r_{OH} is the distance between the carbonyl oxygen and the water hydrogen atom (in Å) [52,54]. This simple relationship illustrates the sub-Angstrom sensitivity of IR spectroscopy.

12.2.3.2 Isotope Labeling

A unique advantage of vibrational spectroscopy is its ability to spectroscopically isolate a single residue or small region of interest with a noninvasive isotope label [22,50,55,56]. Isotope labeling aids in determining the local structure and dynamics of individual residues within a peptide or protein [57]. A single ^{13}C-isotope label red-shifts the local vibration by 35–40 cm^{-1}, largely decoupling the site from other residues. Since amide I peaks are 80–100 cm^{-1} broad, although the spectral shift afforded by a single ^{13}C is not necessarily sufficient to isolate the residue peak from the main band, a ^{18}O label gives the additional 25–35 cm^{-1} needed to resolve single-residue peaks [58].

Figure 12.7 provides a straightforward illustration of how the 2D IR spectrum of a 12-residue β-hairpin peptide TrpZip2 with an isotope label at the K8 position reveals conformational heterogeneity [12]. The K8 label exhibits two peaks separated by the difference of a single hydrogen bond (16 cm^{-1}). The two peaks can be attributed to different β-turn conformations: K8-1 corresponds to a bulged-loop configuration in which the solvent-exposed carbonyl accepts two hydrogen bonds on average, whereas K8-2 is attributed to type-I′ turn with a single K8-W4 internal hydrogen bond. Assignments are based on spectral simulations from MD simulations and Markov-state modeling. The minor structural difference between the two structures illustrates the exquisite structural resolution of vibrational spectroscopy. Conformational dynamics can be extracted from the 2D lineshapes, the diagonal linewidth is related to the conformational disorder, and the antidiagonal describes the timescale in which molecules sample the states (see Section 12.3). K8-2 has a broader diagonal lineshape, indicating that there is an increased amount of disorder in the solvent-exposed conformation compared to the more rigid K8-W4 hydrogen bond. The two β-turn conformations are predicted to interconvert on the microsecond timescale. This example highlights the structural

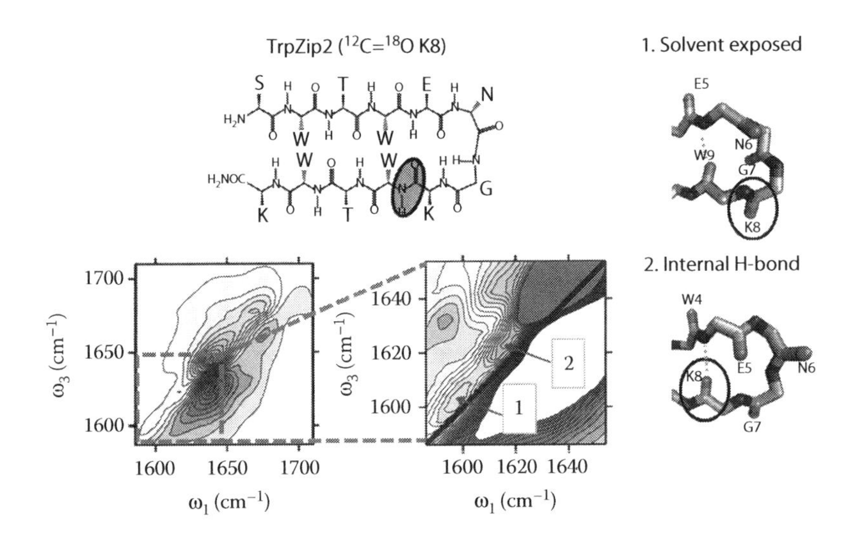

FIGURE 12.7 Structure and 2D IR spectra of $^{12}C{=}^{18}O$ K8 TrpZip2. The two peaks observed in the 1600–1630 cm^{-1} region correspond to two different conformations of the β-turn as shown in the figure.

and inherent ultrafast time resolution of nonlinear IR spectroscopy. In contrast, multidimensional NMR's millisecond time resolution results in an averaged spectrum that makes the two conformations indistinguishable.

12.2.4 Additional Probes of Protein Structure: Amide II and Side-Chain Vibrations

Here, we provide a description of vibrations in the 1400–1800 cm^{-1} region, which contain structural and dynamical information [20,59].

12.2.4.1 Amide II Vibrations

Amide II modes are characterized by large-amplitude C–N bond stretching combined with N–H bending motion. The large motion of the H atom causes the mode to red-shift from ~1550 cm^{-1} (amide II) to ~1450 cm^{-1} (amide II′) upon deuteration. In comparison, amide I, which also includes some N–H bending motion, only red-shifts by ~10 cm^{-1} upon deuteration. Hydrogen/deuterium (H/D) exchange rates are dictated by the solvent exposure and conformational flexibility of the protein. The ratio of amide II to amide II′ peak intensities is proportional to the ratio of protonated to deuterated residues. However, amide II modes are thought to be less sensitive to structure. In addition, a variety of side-chain absorptions overlap the amide II band. Therefore, to date, relatively little experimental and theoretical work has been directed toward exploring the structural sensitivity of amide II modes. Section 12.5 describes a recent 2D IR method whereby combining amide I with amide II spectroscopy enables the direct measurement of the solvent exposure of individual secondary structures within proteins.

12.2.4.2 Other Amide Modes

In addition to amide I and amide II modes, there are a number of other backbone vibrations associated with the amide moiety: Amide III (1200–1400 cm^{-1}) consists of an in-phase combination of the C–C and C–N bond vibrations with contributions from the C=O in-plane bend. Amide IV (~630 cm^{-1}) consists mainly of C=O in-plane bending mode with small contributions from a C–C stretch and a C–C–N deformation. Amide A and amide B modes (~3170 and ~3300 cm^{-1}) are characterized by N–H stretches [59]. For various reasons, these modes have not received much attention as spectroscopic markers of protein structure.

12.2.4.3 Side-Chain Absorptions

Side chains with carboxyl groups, aromatic rings, or the guanidinium moiety have absorptions near the amide I region [60]. These include arginine, asparagine, aspartic acid, glutamic acid, glutamine, tyrosine, and tryptophan. Side-chain vibrations can provide a useful probe of protein structure: carboxylic acids are sensitive to metal coordination and protonation state, and thus are sensitive to pH and solvent exposure. Some modes, particularly those involving large motions of hydrogen atoms, are also sensitive to deuteration, and as such, the frequencies report on the solvent exposure in hydrogen/deuterium exchange experiments. In contrast, aromatic ring modes are characterized by narrow lineshapes and center frequencies that remain insensitive to the environment. Noncovalent interactions, such as electrostatic interactions (i.e., salt bridges), or nonpolar aromatic interactions are crucial to stabilizing the protein structure, as well as providing surface contacts that drive protein recognition and binding. Side chains inside binding pockets determine substrate selectivity and provide the essential interactions needed for catalysis. Side-chain spectra may provide useful insights into structure, heterogeneity, folding, or catalytic mechanisms, especially when combined with amide I or amide II spectroscopy [46]. The broadband pulses generated by ultrafast IR sources facilitate the simultaneous measurement of the backbone amide I and amide II bands as well as side-chain vibrations in the 3–7 μm region (see Figure 12.9). Probing the differences between backbone and side-chain kinetics can be valuable in pinpointing the essential interactions that drive the backbone and side-chain ordering in protein folding [61].

TABLE 12.1
Common Absorption Frequencies Observed Near the Amide I Region for Deuterated Side Chains

AA	Frequency (cm^{-1})	Mode	pK_a
Arg(NH_2^+)	1605	$\nu_{as}CN_3D_5^+$	11.6–12.6
	1586	$\nu_sCN_3D_5^+$	
Asp (COOH)	1713	νC=O	4.0–4.8
Asp (COO^-)	1584	$\nu_{as}COO^-$	
	1404	ν_sCOO^-	
Asn	1648	νC=O	
Cys	1849	νSD	8.0–9.5
Glu (COOH)	1706	νC=O	4.4–4.6
Glu (COO^-)	1567	$\nu_{as}COO^-$	
	1407	ν_sCOO^-	
Gln	1640	νC=O	
	1409	νC–N	
His	1600	νC=C(D_2^+)	6.0–7.0
	1569	νC=C(D)	
	1439	δCD_3, νCN	
Trp	1618	νC=C, νC–C	
	1455	δC–D, νC=C, νC=N	
	1382	δC–D, νC=C, νN–D	
Tyr (OH)	1615	νC=C, νC–D	9.8–10.4
	1590	νC=C	
	1515	νC=C, νC–D	
Tyr (O^-)	1630	νC=C	
	1499	νC=C, δC–H	

Table 12.1 provides a reference of common side-chain absorptions in D_2O, along with the pK_a of the individual side chains. Note that the modes associated with C=O vibrations in carboxylic acids, Asp and Glu, blue-shift by >100 cm^{-1} upon protonation.

12.2.4.4 Vibrational Probes and Unnatural Amino Acids

A relatively recent experiment strategy involves incorporating unnatural side chains as localized probes of structure and dynamics [62]. Vibrational probes provide a minimally perturbative way of obtaining fine control over the spatial and vibrational localization of selected protein sites and take advantage of the quiet spectral region between amide I and the 3 μm C–H, N–H, or O–H stretching regions. Common probes include nitrile (–C≡N), azides (–N=N=N), thiocyanate (–S–C≡N), and C–D bonds [63–65].

In the case of synthetic peptides, vibrational probes can be easily incorporated during solid-phase synthesis. Larger proteins can be produced by recombinant expression. One common method is to substitute an amino acid for a close structural analog in bacterial growing media [66]. However, substitution at multiple sites throughout the protein limits the site selectivity of this technique. New techniques incorporate unnatural amino acids at specific sites by creating a unique tRNA-codon that does not encode for any of the natural amino acids, along with a corresponding aminoacyl-tRNA synthetase [67]. To date, over 30 unnatural amino acids have been incorporated for spectroscopic and reactivity studies. It is also worth mentioning that the first ultrafast IR spectroscopy experiments were carried out on metal-bound C≡O ligands in heme proteins [68]. Narrow lineshapes, long vibrational lifetimes, and large absorption coefficients make metal carbonyls excellent vibrational probes for nonlinear IR spectroscopy. More specifically, vibrational dephasing, and lifetime, and 2D IR measurements of C≡O-bound myoglobin have provided a localized view of the fluctuations at the active site of this heme protein [69]. Recently, ruthenium carbonyls, known as CO-releasing molecules, bound to the aromatic ring of solvent-exposed histidines have provided useful insight on interfacial water dynamics in globular proteins [70].

Small vibrational probes have the advantage of being amenable to high-level theoretical modeling [71]. Spectroscopic observables, such as absorption lineshapes, frequency–frequency time correlation functions, and vibrational relaxation rates, can be extracted from semiclassical simulations [72]. Computational simulations serve a twofold purpose: First, simulated spectra can be compared to experiment. If there is good agreement, the simulations are used to extract an atomistic interpretation of the data. Second, simulations can be used to aid in the strategic development of experiments to maximize the amount of structural information extracted from the data. For example, simulations of label spectra can be particularly informative for selecting unique labeling sites. Electrostatic maps, similar to those described in Section 12.4, have been developed for a number of probes, including nitriles, azide, and thiocyanate probes [63,64]. Within the semiclassical fluctuating frequency approximation, the effects of the environment are captured by the fluctuating frequency of the site, which is calculated through MD simulations combined with spectral maps to translate electrostatic potential trajectories into frequency trajectories. Once this trajectory is obtained, calculating linear and nonlinear spectra is relatively straightforward. Though short vibrational lifetimes limit 2D IR spectroscopy to picosecond timescales, it is becoming clear that fast fluctuations are important for biological function.

12.3 EXPERIMENTAL METHODS

Here we present an overview of the experimental IR spectroscopy from a practical perspective. We introduce the basic background knowledge required to understand the experimental implementations of 2D IR spectroscopy, data processing methods, and interpretation of spectra. The general sample preparation procedures are described, and the main experimental limitations and other practical considerations are outlined.

12.3.1 Infrared Spectroscopy in One and Two Dimensions

IR absorption is a result of the interaction between an electromagnetic field and the oscillating molecular dipole moment. Within a classical picture, the oscillating electric field interacts with the atomic charges in the molecule and amplifies the vibrations that are in resonance with the frequency of the incoming light. In turn, the oscillating charges emit an electromagnetic field out-of-phase with the incoming light, which causes a destructive interference that gives rise to the observed absorption peaks. Large atomic charges give rise to strong oscillating dipole moments, which in turn produce strong absorption peaks.

IR absorption spectroscopy is generally carried out using incoherent light. If short pulses of coherent light are used, the oscillating *polarization* persists after the IR pulse passes through the sample, much like a bell ringing after being knocked by a hammer. One can measure the field radiated by this polarization directly by interfering it with a second pulse that arrives at varying time delays. A Fourier transformation of this time-domain oscillation results in the absorption spectrum. The use of pairs of electric fields generated by an interferometer to measure absorption spectra is Fourier transform spectroscopy, the most commonly used method. This spectrum is the same as what one observes in a frequency-domain experiment, where one measures the change of light intensity through a sample as a function of the frequency of a monochromatic field. These are both one-dimensional experiments since there is one independent time or frequency variable (see Figure 12.8).

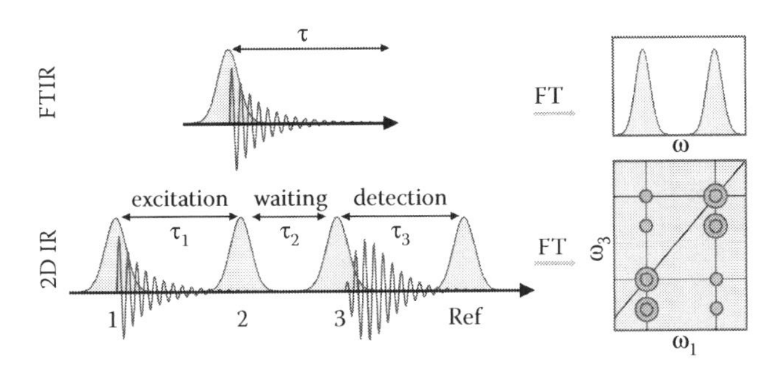

FIGURE 12.8 Pulse sequences for FTIR and 2D IR spectroscopy.

In the case of 2D spectroscopy, one measures the changes to the absorption spectrum induced by excitation at an independent frequency. This can also be performed in the frequency domain as a double-resonance experiment with independent excitation and detection beams; however, here also Fourier transform methods are more commonly applied in one or both dimensions of the 2D spectrum. Following the initial short pulse excitation, subsequent pulses can interact with the sample polarization to generate a *nonlinear* polarization [1,73]. Since multiple pulses can excite/deexcite the sample multiple times, the final nonlinear polarization radiates light-carrying frequencies with new information about the molecular potential not present in linear absorption spectrum.

The pulse sequences used for 2D IR spectroscopy are shown in Figure 12.8. An interferogram is recorded at each τ_1 delay by overlapping the emitted signal with a reference beam, and the amplitude and phase of the emitted electric field are recovered through spectral interferometry. The frequency components of the emitted signal are represented directly in the detection axis of the spectrum, and a Fourier transformation of the signal along the time delay between the first two pulses (τ_1) produces the excitation frequency (ω_1). The delay between the second and third pulses (τ_2) is referred to as the *waiting* or *population* time, and corresponds to the time between the excitation and detection events. The excitation and detection times are often referred to as *coherence* times, since the measured signal involves a coherent superposition of different states during these time periods.

The vibrational modes available for measurement in a 2D IR spectrum are dictated by the carrier frequency, duration, and spectral bandwidth of femtosecond IR pulses used in the experiment. Commercially available laser sources currently generate pulses between 3 and 8 μm (3600–1200 cm^{-1}) with pulse lengths between 50 and 100 fs with a corresponding bandwidth of 300–150 cm^{-1}. To illustrate the capabilities and limitations of these light sources, Figure 12.9 shows the IR absorption spectrum of ubiquitin and *N*-methylacetamide (NMA) along with spectra of typical optical parametric amplifier (OPA) sources. New plasma-based broadband IR (BBIR) generation methods have been recently demonstrated, which generate IR pulses that cover the entire vibrational spectrum from terahertz up to ~4000 cm^{-1} [74,75]. Currently, low pulse energies prevent their use as sources for IR excitation, but provide a new avenue for simultaneously probing multiple transitions, following the interaction with narrowband pump.

Two-dimensional spectra are the most complete representation of the complex third-order IR response of the sample. One-dimensional projections of the response function are measured in the form of *dispersed pump probe* (DPP), *dispersed vibrational echo* (DVE), and *heterodyne-detected vibrational echo* (HDVE) signals. However, these signals contain much of the information represented in the 2D IR spectrum projected onto a single frequency axis. These can be collected significantly faster than full 2D IR spectrum. Therefore, some experiments, such as the T-jump experiments described below, utilize DPP and HDVE as a probe of kinetics, whereas 2D IR spectra are collected at few selected time delays to obtain a structural view of the transient response. Two-dimensional spectra are complex quantities, containing *absorptive* and *dispersive* features; however, it is common to plot only the absorptive component, the spectrum. Dispersed pump-probe, or simply referred to as

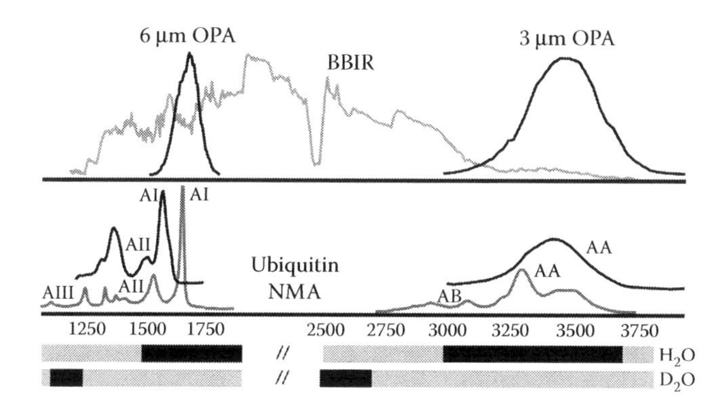

FIGURE 12.9 Absorption spectra of ubiquitin in D_2O and *N*-methylacetamide (NMA) in DMSO. The backbone amide bands (amide I/II/III/A/AB) are indicated in the spectrum. The transparency windows for H_2O and D_2O are shown in gray for reference. The top panel shows typical output spectra for mid-IR optical parametric amplifiers in the 3 μm region (~35 fs pulses) and 6 μm regions (~90 fs pulses) along with a spectrum of a plasma-based broadband IR (BBIR) source.

pump-probe, spectra are absorptive and correspond to the projection of the real part of the 2D IR spectrum onto the ω_3 (detection) axis (Figure 12.10). Complex HDVE spectra correspond to the projection of the complex 2D IR spectrum onto the detection axis, although it is common to plot only the real part of the HDVE spectrum (equivalent to the DPP spectrum). DPP and HDVE signals are measured interferometrically by combining the signal with a reference pulse. DVE spectra represent the power spectrum (absolute-value squared) of the complex HDVE spectrum and is measured without a reference.

In addition to excitation and detection frequencies, an important spectroscopic degree of freedom is the polarization of the four IR pulses—as it refers to the orientation of the pulse electric field vector in space rather than the collective oscillating dipole moment of the sample discussed above. Particular polarization conditions, in which one controls the polarization of the four excitation and detection fields, can be used to suppress or enhance certain spectral features, and can be used to determine the orientation of the different transitions within the molecule, providing additional structural information. The two main polarization conditions are parallel ZZZZ (all four pulses

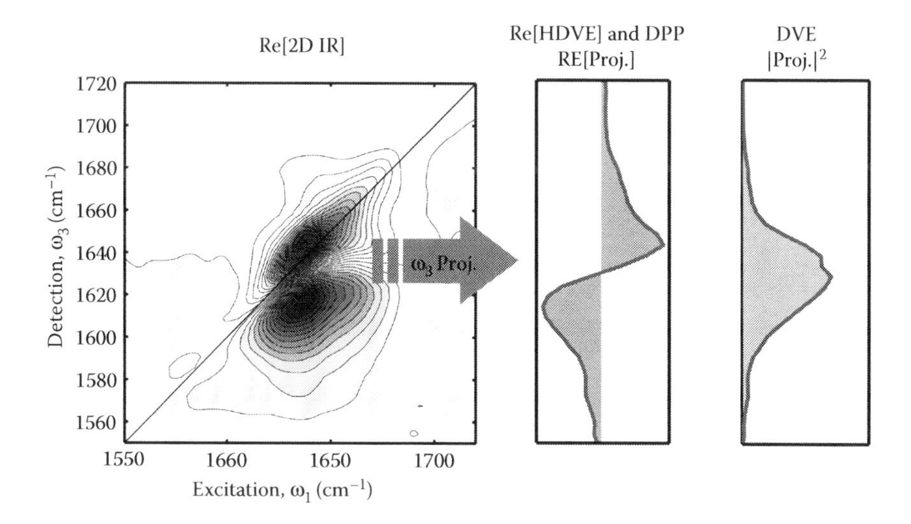

FIGURE 12.10 Absorptive 2D IR, DPP, Real[HDVE], and DVE spectra of ubiquitin in D_2O collected in the ZZYY polarization condition.

polarized parallel with respect to each other) and perpendicular ZZYY (first pair perpendicular to the others) [76]. The parallel and perpendicular polarization conditions enhance cross peaks between vibrations with relative transition dipoles oriented parallel and perpendicular, respectively. The angle between two transition dipoles can be directly derived from the intensities of the peaks under different polarization conditions.

To illustrate the information content in a 2D IR spectrum, Figure 12.11 shows absorption and 2D IR spectra for the carbonyl vibrations of a model compound rhodium acetylacetonato dicarbonyl (RDC) in *n*-hexane [77]. The narrow peaks of the two carbonyl stretches provide a clean illustration of peak patterns typically observed in 2D IR spectra. The absorption spectrum (top) exhibits two narrow peaks at 2014 and 2084 cm^{-1} corresponding to the asymmetric (ω_a) and symmetric (ω_s) stretching vibrations, respectively. In this context, it is important to clarify the distinction between *local mode*, *normal mode*, and *eigenstate*. Local modes are typically associated with a single-bond vibration, and, importantly, they form a basis onto which normal modes are described. In RDC, the two local modes correspond to the stretch of each individual terminal C≡O bond. In proteins, the amide I local mode basis we use corresponds to a combination of C=O stretching and N–H bending vibrations of a peptide group. The local modes are often referred to as *oscillators* or *sites*. Normal modes involve linear combinations of local modes, in-and-out of phase oscillations of individual sites give each normal mode a unique character, and, since the two representations are connected by a linear transformation, the number of normal modes equals the number of local modes. In the case of RDC, the two normal modes correspond to the in-phase and out-of-phase vibration of the C≡O oscillators: symmetric and asymmetric stretches, respectively. Finally, *eigenstates* represent the fully orthogonal set of anharmonic vibrations that are derived from the Hamiltonian for the vibrational system, and are commonly described as perturbative mixtures of the normal modes. For example, the symmetric, asymmetric, and combination state between the asymmetric and symmetric stretches are all eigenstates. Within the normal mode representation, eigenstates are denoted by the number of quanta in each normal mode. For example, the singly and doubly excited symmetric stretches are denoted as $|s\rangle$ and $|2s\rangle$, respectively, whereas the combination state is denoted as $|as\rangle$. Distinguishing

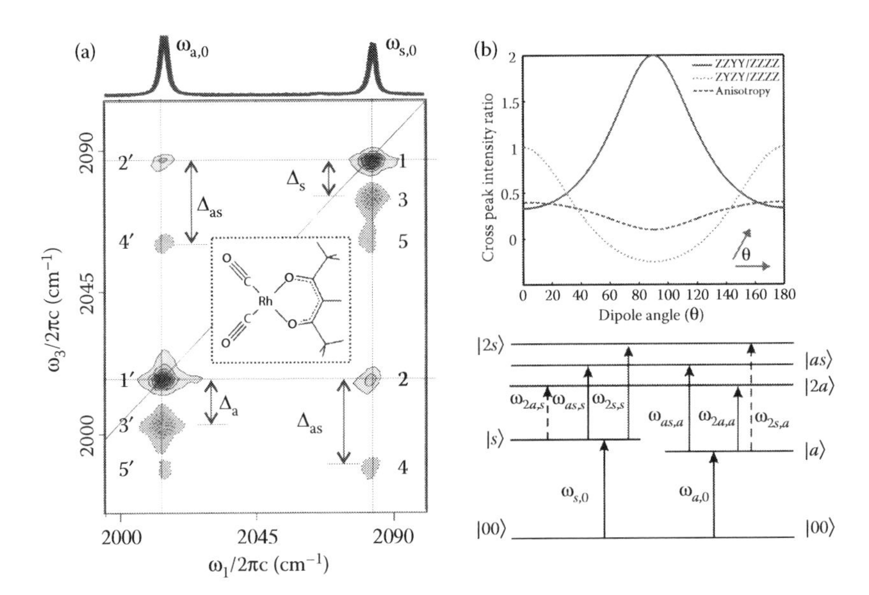

FIGURE 12.11 (a) Infrared absorption and 2D IR spectrum of rhodium acetylacetonato dicarbonyl (RDC, structure shown for reference) in the terminal carbonyl stretching region. Solid and dashed contours represent positive and negative peaks, respectively. (b) Simulated cross-peak intensity ratio between two polarization conditions as a function of the dipole angle between two coupled vibrations.

between normal modes and eigenstates is important in nonlinear spectroscopy and these distinctions arise from the very vibrational couplings that we hope to measure, and as spectra are sensitive to frequency differences and dipole amplitudes for transitions between eigenstates, not normal modes.

Two-dimensional IR spectra can be interpreted as a 2D map with one excitation (ω_1) and one detection axes (ω_3), which are represented on the abscissa and ordinate, respectively. Diagonal peaks (1 and 1′, Figure 12.11) correspond to excitation and detection at the same frequency. Peaks that appear immediately below the diagonal (3 and 3′) correspond to excitation of $|0\rangle \rightarrow |1\rangle$ ($|0\rangle \rightarrow |s\rangle, |0\rangle \rightarrow |a\rangle$) transition and further excitation of the $|1\rangle \rightarrow |2\rangle$ ($|s\rangle \rightarrow |2s\rangle$ or $|a\rangle \rightarrow |2a\rangle$) transition during the detection time. The cross peak (2) represents $|0\rangle \rightarrow |s\rangle$ excitation vibration and stimulated emission from the $|a\rangle \rightarrow |0\rangle$. Analogously, peak 2′ corresponds to excitation of the asymmetric stretch and detection of the symmetric stretch. Peaks 4 and 4′ arise from transitions involving the two quanta combination state. Finally, peaks 5 and 5′ arise from excitation of the symmetric (asymmetric) stretch $|0\rangle \rightarrow |s\rangle$ ($|0\rangle \rightarrow |a\rangle$) and stimulated emission of the asymmetric $|2a\rangle \rightarrow |a\rangle$ (symmetric) stretch ($|2s\rangle \rightarrow |s\rangle$). One and two quanta energy levels of the molecule can be simply read out from the peak positions. Anharmonicities and coupling constants can be determined by spectral modeling. Changes in the intensities of the cross peaks with respect to the polarization of the excitation and detection pulses are determined by the relative angle of the transition dipoles between the two vibrations. To illustrate this example, Figure 12.11 shows the calculated ratio of cross peaks using three different polarization geometries (ZZYY/ZZZZ, ZYZY/ZZZZ) as a function of the transition dipole angle between two coupled oscillators.

Structural information is extracted from the position and intensities of the peaks, whereas dynamics are extracted from the peak shapes, more specifically, the ellipticity of the peaks as well as the change in ellipticity with waiting time [78]. Figure 12.12 shows an example of changes in lineshape along the waiting time: a diagonal elongation at early waiting times, as each conformation within the ensemble has a different transition frequency such that there is a large correlation between the excitation and detection frequencies. Node lines rotate as a function of waiting time, as dynamical effects cause the correlation to decrease. The rate at which frequency correlation is lost can be directly related to the frequency fluctuations of the molecules in the sample, referred to as *spectral diffusion*. Elongation along the diagonal is referred to as *inhomogeneous* broadening

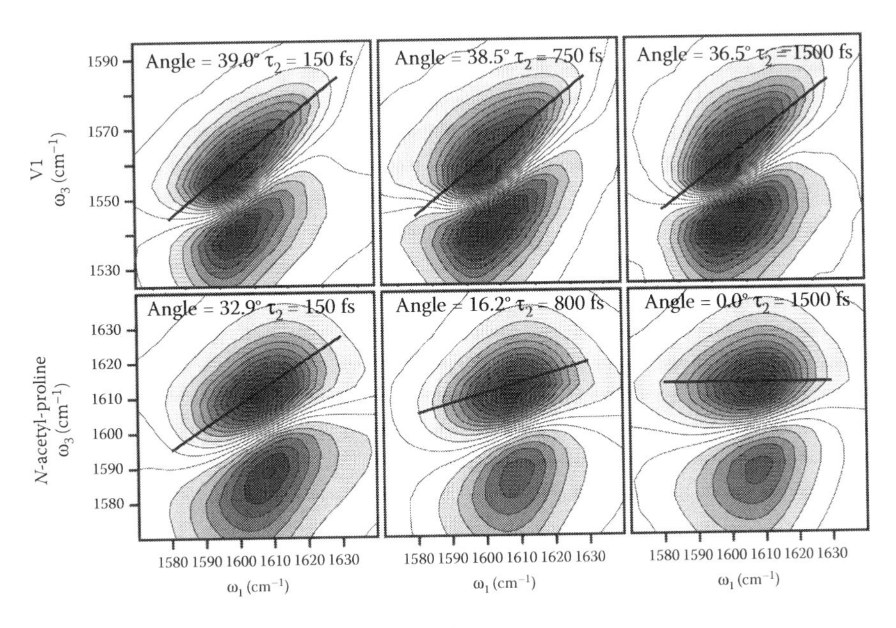

FIGURE 12.12 Waiting time 2D IR spectra for a β-turn peptide GVGVP*GVG and the single amino acid *N*-acetyl-proline in D_2O. The star denotes a $^{13}C{=}^{18}O$ label.

while elongation along the antidiagonal is called *homogeneous* broadening. The example shows how residues that are locked into rigid intramolecular hydrogen bonding configurations have slower dynamics (V1 peak), which lead to smaller frequency fluctuations, and a slower rotation of the nodal line, whereas residues exposed to the solvent (*N*-acetyl-proline) quickly lose the correlation between excitation and detection.

12.3.2 Transient Temperature-Jump 2D IR Spectroscopy

While waiting-time 2D IR spectroscopy offers structural information with subpicosecond time resolution, a powerful experimental toolbox is developed by combining 2D IR spectroscopy with nonequilibrium methods to study longer timescale protein folding, dynamics, and function. Since IR spectroscopy produces ensemble measurements, a rapid trigger event is required to synchronize the time evolution of the ensemble. The time resolution of the pump along with the timescales accessible by the probe often dictates the type of dynamics that can be studied. The "probe" method, such as 2D IR, must have high time resolution and high structure sensitivity, but most importantly, the reaction coordinate must project favorably onto the selected spectroscopic coordinate. In the case of proteins, isotope labels often provide a method to select a set of spectroscopic variables of interest.

Common triggers include the following [79] (approximate time resolution indicated in parenthesis): (1) Electronic excitation (10–100 fs) to probe ultrafast excited-state dynamics and photo-induced reactions, or to gate photorelease of caged compounds [80]. (2) Temperature-jump spectroscopy (1–10 ns) to study thermal denaturation, protein folding, association, and nonequilibrium dynamics [81,82]. (3) Pressure-jump (1 μs) to study pressure-induced conformational changes of proteins such as partial denaturation [83]. (4) pH-jump methods (100 μs) to study pH-induced conformational dynamics or rapid protonation of different protein sites [84]. (5) Stopped-flow rapid mixing techniques (1–3 ms) to probe protein unfolding and conformational changes induced by denaturants or mixed solvents [85]. Among these triggered methods, only electronic excitation and temperature-jump triggering have been demonstrated within the context of ultrafast 2D IR spectroscopy [80,86,87]. In principle, any of these techniques can be combined with a 2D IR probe; one practical limitation, however, is related to the weak nonlinear signals observed in biological samples that inevitably lead to long data collection times. Triggering methods must therefore cause large changes in the signal and have repetition rates commensurate with the timescales of interest so as to maximize the duty time of the measurement.

Here, we focus on nonequilibrium 2D IR spectroscopy initiated by a temperature jump. Figure 12.13 shows a schematic representation of the temperature profile of the sample following a T-jump. A 2D IR pulse sequence is able to probe structural rearrangements from the nanosecond to millisecond timescales. Protein dynamics occur on various timescales, but in general, nanosecond to

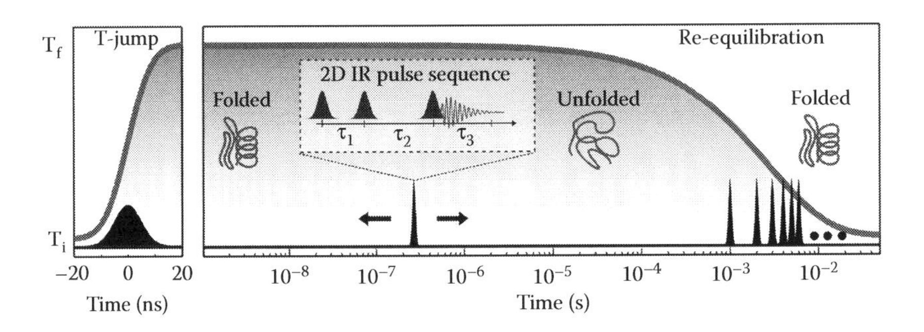

FIGURE 12.13 Schematic representation of the T-jump pulse sequences and temperature profile. The delay between the T-jump pulse and the 2D IR pulse sequence is electronically controlled. The 1 ms repetition rate of the 2D IR pulses enables us to use subsequent pulses in the train to collect time delays from 1 to 50 ms.

microsecond response times correspond to barrierless rearrangements whereas transitions occurring from 100 μs to 1 ms are attributed to barrier-crossing events. Delay times are limited by the repetition rate of the T-jump laser, and the relaxation of the sample. Within our current implementation, the maximum delay is 50 ms [81].

A T-jump pulse must carry sufficient energy to raise the temperature of the solvent by a few degrees. In our implementation, 20 mJ pulses centered around 2 μm are generated by a frequency-doubled neodymium-doped yttrium aluminum garnet (Nd:YAG) laser with a repetition rate of 20 Hz, coupled to an optical parametric oscillator (OPO). The T-jump laser is electronically timed to the 1 kHz femtosecond laser source. The shortest delay is given by the T-jump pulse width (10 ns) and longest delay by the repetition rate of the T-jump laser to (50 ms). T-jump pulses are resonant with the overtone of the optical density (OD) stretch in the solvent. Following the temperature jump, the solvent returns to equilibrium as heat diffuses away from the focus region. The temperature relaxation profile is described by a stretched exponential with a time constant of approximately 3 ms. To achieve uniform heating of the focus region, the T-jump pulses are focused to an ~1 mm spot at the sample and only about 10% of the light is absorbed. Interferometric measurements are particularly demanding since the phase of all beams must be kept constant before and after the trigger event. Within our implementation, all pulses are focused onto the interaction region, and are thus prone to the same changes in solvent absorption and index of refraction induced by the T-jump. Depending on the initial temperature, a 10-degree temperature jump causes transmission changes of approximately 5% in the amide I region that magnifies the nonlinear signal. The spectrum of the reference pulse is collected along with the signal with every laser shot, and the signal is corrected to account for transmission changes. Corrections are necessary to interpret difference 2D IR spectra as well as to obtain undistorted sample kinetics [88]. In contrast to absorption spectroscopy, an important advantage of nonlinear IR spectroscopy is that solvent signals are greatly suppressed; thus, it is not necessary to separately measure a solvent background to isolate the solute response.

A 2D IR spectrum represents the most complete characterization of the third-order vibrational response, but for a variety of reasons, acquisition of a 2D IR spectrum is relatively slow. Techniques that reduce data collection times are better suited to finely sample the time delay following the T-jump to measure kinetics. Three frequency-resolved nonlinear measurements can be carried out with the 2D IR spectrometer: DPP, DVE, and HDVE spectroscopy [89]. DPP measures transmission changes of a probe pulse following an IR excitation by a "pump" pulse. The DPP signal measured is equivalent to the 2D IR signal integrated over the excitation axis; therefore, DPP cannot clearly separate contributions from diagonal and off-diagonal features of a 2D IR spectrum. Similarly, DVE and HDVE represent the four-wave mixing signal emitted by the sample following the interaction with three IR pulses at fixed time delays. Similar information can be extracted from DVE and HDVE measurements, but HDVE, a more demanding phase-sensitive measurement, is sensitive to changes in the amplitude and phase of the reference induced by the T-jump pulse. Unlike DPP, DVE and HDVE are background-free measurements and thus offer improved signal to noise and sensitivity, and DPP signals can be recovered from an HDVE measurement. Fourier transformation of a series of HDVE spectra at various delays between pulses 1 and 2 produces a 2D IR spectrum. Since HDVE spectra can be collected ~250–500 times faster than equivalent 2D IR spectra, HDVE is used for collecting the kinetic data and 2D IR for extracting the structural information at selected time delays.

12.3.3 Sample Preparation

Compared to standard biophysical techniques, the sample preparation procedures for 2D IR spectroscopy are relatively straightforward. Here, we describe the typical sample preparation for peptides and proteins. For amide I spectroscopy, protein samples are often dissolved in D_2O and gently heated for a few hours to partially denature the protein and exchange the labile protons with deuterons. The sample is then lyophilized and redissolved in pure D_2O. Solutions are placed between two

CaF_2 windows with a spacer. The transparency of D_2O limits the pathlength to 50 μm. In principle, the minimum volume needed to collect a spectrum is 10 pL, but in practice, a volume of at least 200 nL is required for each sample. Since the D_2O bend vibration absorbs strongly below 1500 cm^{-1} (see Figure 12.9), the pathlength must be kept to a minimum; therefore, sample concentrations must be relatively high. The residual H_2O bend vibration overlaps with the amide I band; therefore, the residual H_2O contents should be kept to <3%. In proteins, the amide I absorbance is proportional to the number of residues; if we consider the average mass of a residue to be 120 g/mol, the approximate concentration needed for 2D IR is 80 mM/residue (~10 mg/mL) for most proteins. The concentration can be lower when absorption bands are sharp, such as in the case of aggregates or membrane proteins. This sample concentration regime is comparable to that used in NMR spectroscopy, except that sample volumes used in 2D IR are significantly smaller. Carboxylic acid buffers, such as acetate, or compounds that contain a carbonyl group, such as urea, must be avoided due to the absorption overlap with the amide I band. In the case of synthetic peptides, trifluoroacetic acid left from solid-state synthesis and subsequent purification steps must be thoroughly removed by multiple lyophilizations from acidic solution.

12.3.4 Isotope Labeling

Isotope labels spectroscopically isolate single residues by red-shifting their transition frequencies from the main band. A single ^{13}C replacement is not always sufficient to vibrationally isolate a residue, particularly in larger proteins where the ~1% ^{13}C natural abundance increases the probability of finding other naturally occurring ^{13}C sites within the backbone. In addition, broad lineshapes will hide the residue peak underneath the main band. Proline residues also contribute intensity in the 1630 cm^{-1} region [6]. Because of these contributions, often a $^{13}C{=}^{18}O$ double label is necessary to spectrally isolate a region of interest. ^{13}C-labeled amino acids are commercially available but ^{18}O variants have to be synthesized. Recently, a new method was developed for synthesizing ^{13}C,^{18}O-double-labeled *N*-(9-fluorenylmethoxycarbonyl) (FMOC) amino acids with high levels of ^{18}O incorporation for amino acids without acid-labile side-chain protecting groups (Gly, Ala, Val, Ile, Leu, Phe, Trp, and Pro) [90]. The acid hydrolysis reaction is carried out by refluxing the FMOC amino acid for periods between 3 and 30 h in acidic mixtures of $H_2{}^{18}O$ and organic solvents. The organic solvent must dissolve the FMOC amino acid but not act as a source of ^{16}O. The final product is recovered by lyophilization of the reaction mixture. Enrichment ratios are typically >90%; thus, the recovered product may be used without further purification.

12.4 THEORY

An important feature of the IR spectroscopy of biological systems is that the field rests on a well-developed theoretical basis, allowing in many cases for comparison of experimental data with molecular-level models such as MD simulations. In this way, IR spectra can be used as a direct probe into the structure and dynamics of proteins. In this section, we outline the basic features of the theory developed for amide I vibrational spectroscopy applied to IR absorption spectroscopy and 2D IR.

From an experimental point of view, the information content of protein 2D IR spectra comes in two forms: local structural information such as hydrogen bonding and solvent exposure is provided by site-specific isotope labels, while global secondary structure content is encoded in delocalized excitonic bands such as the β-sheet $\upsilon_{//}$ and $\upsilon_{\perp}$ peaks. Theoretical models for amide I spectroscopy must likewise account both for local effects, particularly the influence of an oscillator's local electrostatic environment on its vibrational frequency, and global effects such as the delocalization of amide excitations—excitons—due to site-to-site coupling. We begin our discussion by focusing on local parameters, particularly relevant to isotope labeling experiments, before introducing the exciton treatment of globally delocalized vibrational modes and computational methods for estimating the relevant parameters for amide I vibrations. Finally, after providing a brief explanation of the theory behind 2D IR spectroscopy itself, we close with a short description of practical approaches to simple 2D IR modeling for amide I spectra.

12.4.1 Oscillator Site Energies

At the simplest level, a theoretical description of amide I spectra must describe the frequency of each vibrational oscillator due to its local environment, also known as the site energy. One can understand these effects intuitively in terms of the influence of local interactions on bond strengths within the amide unit. For example, donation of hydrogen bonds from water molecules to the oxygen of the amide unit stabilizes the C=O bond, lowering its vibrational frequency and causing a red-shift of the corresponding amide I absorption peak. Therefore, this site energy can report on the local hydrogen bonding configuration in which the peptide resides. On a computational level, the correlation of local structure and vibrational frequency can be expressed in terms of electrostatic variables (potential, field, gradient, etc.), which is the basis for most of the site energy maps that will be discussed in more detail below [16–18,53,71,91–96]. Although the details of these methods vary, all seek to predict IR spectroscopic features based on the local electrostatic environment of each oscillator in a given structure (e.g., from an MD simulation).

Within a protein structure, these electrostatic factors produce a broad distribution of site energies for individual amide oscillators in the protein backbone, which reports on protein–protein and protein–solvent interactions. The frequency shifts of an amide I vibration relative to an isolated oscillator varies by 40 cm^{-1} depending on the nature of these interactions. Experimentally, the shift of a particular oscillator can be accessed most easily by introducing ^{13}C and/or ^{18}O isotope labels into specific amide I carbonyl groups on the protein backbone. The higher reduced mass of isotope-labeled units results in a red-shift of the corresponding amide I vibration of ~40–75 cm^{-1}, decoupling the selected oscillator from other amide I vibrations and moving its absorption into a spectral window that allows for direct observation of the site energy shift. These isotope-labeled samples are particularly useful in 2D IR spectroscopy since peak shapes provide insight not only into the total distribution of transition frequencies, but also into the timescales for spectral diffusion and dephasing. Thus, a clear understanding of the relationship between local electrostatic effects and oscillator frequency provides a direct link between spectral characteristics and molecular structure and dynamics, such as hydrogen bond configurations and lifetimes.

12.4.2 Delocalized Vibrations and Excitons

Although local electrostatic effects are sufficient to describe the vibrational frequencies of individual sites, interactions between neighboring residues generally complicate the picture by inducing delocalization of vibrational modes across multiple residues. For this reason, beyond electrostatics, the central feature of most theoretical descriptions of amide I and II vibrations is the concept of the *exciton*—a delocalized excited state produced by the interaction of many individual sites. The concept of the exciton was originally introduced in the context of solid-state physics [23], but has found applications in modeling a wide variety of biophysical processes, including the visible absorption spectra of photosynthetic systems [97] and ultraviolet circular dichroism of protein backbones [98].

The vibrational excitons observed in amide I and II spectra behave in many respects as a network of oscillators—or sites—coupled together into a single system—much as in a bedspring, a large number of individual springs are linked together to produce a single mattress. Just as deformation of one spring in a mattress affects neighboring springs, vibrational coupling between individual amide units in a protein causes excitation of any given amide vibration to induce vibrations in adjacent units, a process known as *delocalization*. The extent to which vibrational motion is delocalized over a system depends on the number of sites, the site-to-site coupling strength, and the variation of site energies. For two coupled vibrations, with site energy ε_1 and ε_2 and coupling strength J, the resulting exciton states have energies $E_{\pm} = \frac{1}{2}(\epsilon_1 + \epsilon_2) \pm \frac{1}{2}\sqrt{(\epsilon_1 - \epsilon_2)^2 + 4J^2}$. Similar site energies produce the greatest delocalization since it allows neighboring sites to oscillate in-phase with each other without interference. The case of amide I vibrations proves to be complicated because the variation in vibrational coupling, typically between –10 and +10 cm^{-1}, is of a similar scale to the variation in site energy, 30–40 cm^{-1}.

These concepts were first applied to amide I and II vibrations by Miyazawa who demonstrated that the characteristic amide I IR absorption spectra observed for α-helix and β-sheet structures could be well explained by a normal mode analysis in which each carbonyl group acts as a single oscillator vibrationally coupled to its nearest neighbors and hydrogen bonding partners [99]. As described by Miyazawa, the similarity of carbonyl group site energies across the protein backbone, together with the regularly repeating patterns of α-helices and β-sheets, leads to strongly delocalized excited states. These ideas were developed by many over the years, notably Krimm [100], who introduced a transition dipole coupling (TDC) model for predicting intersite coupling constants from molecular structures, and Torii and Tasumi [48], who more clearly expressed the concept of an amide I subspace adiabatically separated from the remaining protein vibrations.

12.4.3 Amide I Hamiltonian

These methods were adapted for nonlinear spectroscopy by Hamm, Lim, and Hochstrasser who recast the earlier force-constant matrix calculations into a quantum Hamiltonian including doubly excited states [54]. They described the exciton states as eigenstates of a Hamiltonian operator that describes the interactions between N sites

$$\hat{H} = \sum_{n=1}^{N} \varepsilon_n |n\rangle\langle n| + \sum_{m,n=1}^{N} J_{mn} |m\rangle\langle n| + \sum_{m,n=1}^{N} (\varepsilon_m + \varepsilon_n - \Delta\delta_{mn}) |mn\rangle\langle mn| + \sum_{m,n=1}^{N} \sum_{\substack{j,k=1 \\ (m,n)\neq(j,k)}}^{N} J_{mn,jk} |mn\rangle\langle jk| \quad (12.2)$$

In this expression, ε_n and J_{mn} represent the site energies and coupling constants between singly excited states. The second set of terms represent doubly excited states in which a single oscillator is excited twice ($m = n$) or two different oscillators are each excited ($m \neq n$). The *anharmonicity* value Δ (~16 cm^{-1} for amide I) is the difference in the absorption frequency for the fundamental ($0 \rightarrow 1$) transition compared with its overtone ($1 \rightarrow 2$). This value is particularly important for 2D IR spectroscopy since in a perfectly harmonic system (with $\Delta = 0$), the 2D IR signal vanishes due to interference between the fundamental and overtone transitions. Note that coupling between one- and two-quantum states is neglected, so that the states are separated into zero-, one-, and two-quantum subspaces. The resulting block-diagonal amide I Hamiltonian is illustrated graphically in Figure 12.14 in which site energies occur along the diagonal and site-to-site couplings appear off-diagonal within the one- and two-exciton blocks. The eigenstates of the system are obtained by numerically diagonalizing the resulting matrix, providing absorption frequency values, dipole moments, and oscillator strengths.

The necessary input for the calculation is the site energies and coupling constants for each state. In structure-based modeling, the site energies are dependent on the local structure about a site, and the coupling will depend on the configuration of two sites with respect to one another, for instance, a dipole–dipole interaction. These assignments are aided by the structure/spectroscopy maps described below. Usually, a harmonic approximation is assumed to obtain the two-exciton coupling constants from the one-exciton energies ε_n and coupling constants J_{mn}.

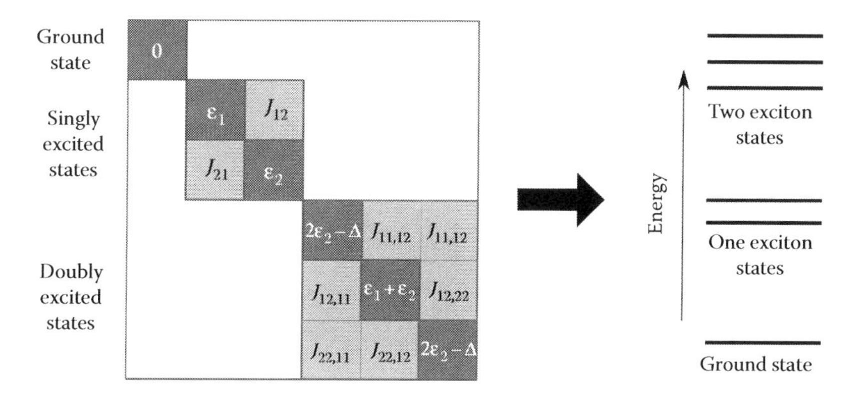

FIGURE 12.14 Schematic representation of the excitonic Hamiltonian matrix and energy levels for a two-oscillator system.

Having outlined the principles of this mixed quantum–classical approach, it is worth stating the assumptions that go into such modeling. The principle assumption is an adiabatic separation of the amide I vibrational motions and the dynamics of the protein. The semiclassical approximation then states that essentially all large-amplitude protein motions such as conformational fluctuations can be treated classically via static protein structures or MD simulations, while the higher-frequency vibrations such as amide I that are interrogated by the IR radiation must be treated quantum-mechanically. The interaction between the amide I vibrations and protein structure is handled with the mapping procedures described below. For dynamic calculations, the classical motion of the protein then determines the parameters of a quantum-excitonic Hamiltonian that describes the high-frequency amide I dynamics and the interaction of the system with the perturbing electromagnetic fields.

12.4.4 Structure-Based Calculations

A number of schemes have been constructed to obtain the one-exciton site energies and coupling constants directly from molecular structures, for example, MD simulations. For small systems, electronic structure methods such as density functional theory can be applied directly to obtain energies, couplings, and transition dipole moments. However, such calculations are limited, at least for the present, to systems not larger than a few amino acids and including at most a few solvent molecules. A variety of parameterized *maps* have been constructed to predict frequencies and couplings in larger systems without the need for electronic structure calculations. The maps take advantage of the correlation that exists between the local structure and electrostatic environment of the amide bond and the frequencies of the associated vibrational modes to greatly reduce the computational expense of calculating site frequencies. As an example, the left frame of Figure 12.15 shows the calculated site-energy shift induced by a +0.5 point charge located around the amide unit using one such map [15]. The right frame shows for comparison the correlation between hydrogen bond donor/acceptor distance and site-energy shift for a small β-turn peptide using another map [16]. Although the maps differ in their details, both reflect the observation that amide I site energies red-shift on hydrogen bond formation. These electrostatic maps are primarily parameterized through electronic structure calculations and, as illustrated in Figure 12.16, generally perform well in reproducing the influence of the local environment on the transition frequencies of small model systems.

Site energy maps may consist of two distinct contributions:

- *Electrostatic maps*: A nonspecific electrostatic frequency shift induced by the electrostatic potential, field, or gradient evaluated at the amide bond. This contribution arises from the net electrostatic effects of *all* atoms in the simulation (excluding the nearest neighbors), regardless of their identity, and accounts for effects such as hydrogen bonding and

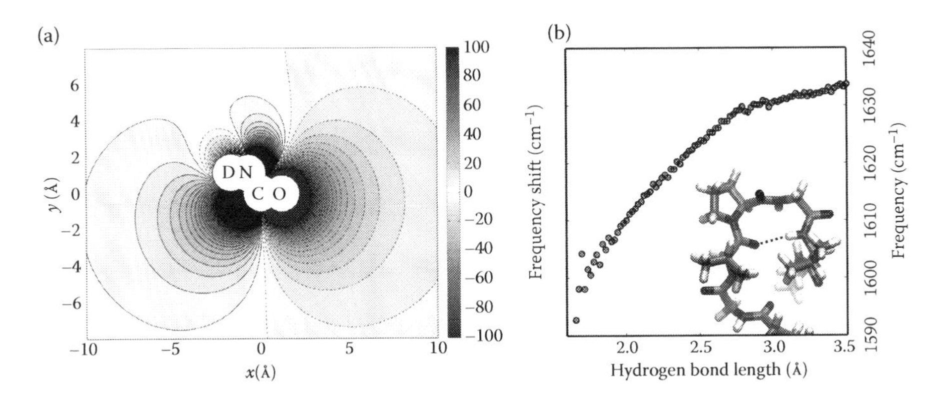

FIGURE 12.15 (a) Frequency shift induced by a 0.5 a.u. test charge around the amide unit calculated using the map of Skinner et al. [12]. (b) Correlation between V_1–V_4 hydrogen bond length and V_1 frequency for the β-turn peptide GVGV$_1$PGV$_4$G. Frequencies were calculated using the parameterization of Jansen and Knoester [17] (see Table 12.2).

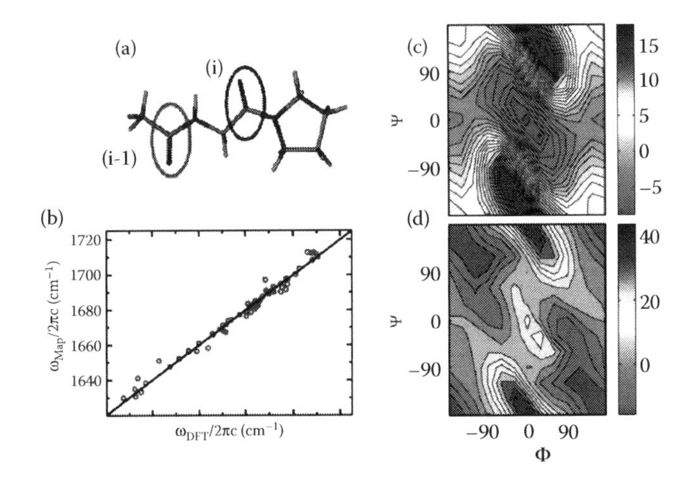

FIGURE 12.16 Electrostatic maps corresponding to a two-residue fragment. (a) Structure of the Pro–Gly fragment the two backbone C=O. (b) Correlation plot between frequencies computed through DFT calculations and the electrostatic map. (c) Coupling constants (cm^{-1}) as a function of phi and psi angles (d) Frequency shift (cm^{-1}) of site i-1 due to the perturbation by site i. (Adapted from Roy S et al. 2011. C. *J Chem Phys* 135:234507.)

solvent-induced shifts. One such example is the map developed by Skinner and coworkers [15], where the site frequency (ω in cm^{-1}) is given by

$$\omega = 1684 + 7729E_C - 3576E_N$$

where E_C and E_N represent the electric field projection along the C=O bond (in atomic units) at the C and N positions. Other common map positions include the O, H, and Cα atoms. The frequency shift from a point charge at different positions in the $x - y$, calculated using the above map, is depicted in Figure 12.15.

- *Nearest-neighbor shifts:* A specific through-bond frequency shift is determined by the dihedral (ψ, φ) angle pair of adjacent amino acids in peptides. This contribution reflects the fact that amide I vibrations are in reality not completely decoupled from other vibrations. It is a function only of the backbone conformation of the polypeptide chain, and accounts for site energy shifts due to nearest-neighbor effects such as steric strain between adjacent residues. Figure 12.16 illustrates one such nearest-neighbor map [96].

Although existing site energy maps are for the most part all based on some combination of these two contributions, they differ considerably in their details. In particular, there is significant variation in the choice of electrostatic parameters considered (generally, the potential, field, or gradient) and the sampling points around the amide bond (usually evaluated at the N, H, C, and O atoms, possibly with the addition of other adjacent points).

Various methods have similarly been proposed to predict the site-to-site coupling constants J_{mn}, including both through-bond (nearest-neighbor) and through-space effects [101].

- *Transition dipole coupling* (TDC) treats through-space coupling by treating each oscillator as a simple dipole vector that interacts with adjacent dipoles through a coupling factor similar to the geometric factor encountered in Förster resonance energy transfer (FRET) studies.
- *Transition charge coupling* (TCC) is an improvement on the TDC model in which a transition charge is assigned to each atom of the amide bond (or other suitably chosen nearby points in space) and an interaction energy is calculated between each atom. The TCC method is accurate to somewhat shorter distances than TDC.
- *Nearest-neighbor coupling* or through-bond coupling is often accounted for in much the same way as the nearest-neighbor frequency shift used for site energy calculations (see Figure 12.16). A set of electronic structure calculations are used to parameterize a dihedral angle map for coupling between adjacent residues, accounting for steric effects and through-bond interaction.

Table 12.2 provides a comparison of the essential features of selected amide vibration maps published in recent years. A comparison of 2D IR experimental data with spectra calculated with various maps is presented in Figure 12.18.

12.4.5 Nonlinear Polarization and Signal

So far, our discussion has focused on a microscopic picture of vibrational modes within an individual protein. Nonlinear spectroscopy, on the other hand, is usually described in terms of a macroscopic *nonlinear polarization* $\mathbf{P}(\mathcal{R},t)$ induced in a sample by a series of perturbing electromagnetic fields (laser pulses) [73,102,106]. This induced polarization in turn gives rise to an oscillating electromagnetic field that is emitted from the sample in the same wavevector-matched directions as the polarization itself and can be measured experimentally. The connection between the two pictures is given by the relation

$$\mathbf{E}(\mathcal{R},t) \propto \mathbf{P}(\mathcal{R},t) = \langle \mu(\mathcal{R},t) \rangle \tag{12.3}$$

that is, the *macroscopic* polarization, and hence the emitted field is proportional to the expectation value of the *microscopic* dipole moment evaluated at different points in the sample [73,102].*

Under the semiclassical approximation that we use for amide I spectroscopy, the resulting expressions for the nonlinear signal from the excitonic system can be written as the sum of two terms. These are known as the *rephasing* and *nonrephasing* signals, and they differ by the phase

* Although formally, the polarization $\mathbf{P}(\mathcal{R},t)$ acts as a source input in Maxwell's equations, to a good approximation, the electric field looks simply like a phase-shifted $\mathbf{P}(\mathcal{R},t)$, so that for most purposes, it suffices to treat the two as equivalent.

TABLE 12.2
Summary of Selected Electrostatic Maps Developed for Amide Vibrational Frequencies

Model	Electrostatic Variable	Sites	Amide Modes	Site Energies	Coupling	System
Torii and Tasumi [55,133]	N/A	N/A	I	No	Yes	Gly_2 and Gly_3
Hamm and Woutersen [101]	N/A	N/A	I	No	Yes	Gly_2
Cho [53]	Potential	4	I	Yes	Yes	NMA and Gly_2
Bouř and Keiderling [91]	Potential	4	I	Yes	No	NMA and penta-peptide
Skinner [17]	Field	4/2	I	Yes	No	NMA and small peptides
Mukamel [71]	Field, Gradient, 2nd Derivative	19	10 lowest modes	Yes	No	NMA
Hirst [92]	Potential	4/7	I	Yes	Yes	NMA and [Leu]-enkephalin
Jansen and Knoester [16,93,96]	Field and Gradient	4	I	Yes	Yes	NMA
Wang [94]	Potential	4	I	Yes	Yes	Ala_2 and Gly_2
Ge [37]	Potential	4	I and II	Yes	Yes	3–10 peptide

Note: The columns indicate the model system used to parameterize the map, the electrostatic parameters used to describe the vibrations, the number of sites (usually the amide bond N, H, C, and O atom locations), and the vibrational modes considered.

of oscillating coherences in the excitation period τ_1 [103]. If we assume that the perturbing field disturbs a system with a static Hamiltonian, we can express the signal in terms of the corresponding eigenstates as

$$\begin{aligned}
E_{ijkl}^{R}(\tau_1,\tau_2,\tau_3) \propto\ & Im\sum_{ab}\Big\{\langle\tilde{\mu}_i^{0a}\mu_j^{0b}\tilde{\mu}_k^{a0}\mu_l^{b0}\rangle e^{-i\omega_{0a}\tau_1}e^{-i\omega_{ba}\tau_2}e^{-i\omega_{b0}\tau_3} \\
& + \langle\tilde{\mu}_i^{0a}\tilde{\mu}_j^{a0}\mu_k^{0b}\mu_l^{b0}\rangle e^{-i\omega_{0a}\tau_1}e^{-i\omega_{b0}\tau_3}\Big\} \\
& - Im\sum_{abc}\Big\{\langle\tilde{\mu}_i^{0a}\mu_j^{0b}\mu_k^{bc}\mu_l^{ca}\rangle e^{-i\omega_{0a}\tau_1}e^{-i\omega_{ba}\tau_2}e^{-i\omega_{ca}\tau_3}\Big\} \\
E_{ijkl}^{NR}(\tau_1,\tau_2,\tau_3) \propto\ & Im\sum_{ab}\Big\{\langle\mu_i^{0a}\tilde{\mu}_j^{0b}\tilde{\mu}_k^{b0}\mu_l^{a0}\rangle e^{-i\omega_{a0}\tau_1}e^{-i\omega_{ab}\tau_2}e^{-i\omega_{a0}\tau_3} \\
& + \langle\mu_i^{0a}\mu_j^{a0}\mu_k^{0b}\mu_l^{b0}\rangle e^{-i\omega_{a0}\tau_1}e^{-i\omega_{b0}\tau_3}\Big\} \\
& - Im\sum_{abc}\Big\{\langle\mu_i^{0a}\tilde{\mu}_j^{0b}\mu_k^{ac}\mu_l^{cb}\rangle e^{-i\omega_{a0}\tau_1}e^{-i\omega_{ab}\tau_2}e^{-i\omega_{cb}\tau_3}\Big\}
\end{aligned} \tag{12.4}$$

In these expressions, the three sequential time intervals τ_1 τ_2 τ_3 are shown in Figure 12.8. The indices a and b run over all possible one-quantum eigenstates, while the third index c is summed over all possible two-quantum eigenstates. The tilde over some matrix elements indicates a complex conjugate. The subscript indices i, j, k, and l indicate the polarization direction of the three laser pulses (i, j, and k) or of the signal analyzer (l) in the laboratory frame; the angular brackets indicate an ensemble average over the orientation of the molecular reference frame [102,104]. A practical guide for simulating 2D IR spectra is provided in Section 12.4.7.

These expressions are easily rationalized in light of the transition dipole elements at the start of each term: the first three elements, μ_i, μ_j, and μ_k, indicate absorption or emission events induced by the three excitation pulses; the final element μ_l indicates the final emission event that is detected in the experiment. The oscillating exponentials corresponding to the three time intervals, τ_1, τ_2, and τ_3, indicate the creation of *coherences* (oscillating expectation values arising from the superposition of two-quantum states) or *populations* (static contributions to observable values reflecting the occupation of a single-quantum state). In both the rephasing and nonrephasing response functions, coherences give rise to fast oscillations in the polarization as a function of the initial and final time delays τ_1 and τ_3, but much slower oscillations as a function of τ_2 since the difference frequency $\omega_{ab} \approx 0$ is for two nearly degenerate one-quantum states.

12.4.6 Frequency-Domain Spectra

The expressions we have discussed so far describe the polarization (or signal electric field) as a function of time. What we wish to obtain in the experiment is a 2D frequency–frequency correlation spectrum linking the "absorption" frequency during τ_1 with the "emission" frequency during τ_3. Experimentally, one frequency dimension can be obtained directly by dispersing the emitted field through a monochromator onto an array detector, essentially giving a Fourier transform of the field along τ_3 directly. This gives a frequency-domain spectrum as a function of the emission frequency (ω_3) for a single value of τ_1, but provides no information on the interaction events during τ_1. Referring back to Equation 12.4, note that although the oscillation frequencies along τ_1 do not directly affect the (integrated) intensity of the emitted signal, they do alter the phase. This dependence can be monitored experimentally by *heterodyne detection* in which the emitted signal is overlapped with a reference pulse (the local oscillator) with a fixed phase; the interference pattern between the two pulses is then recorded as a function of both delay time (τ_1) and emission frequency (ω_3). Finally, the collected data is postprocessed with a numerical Fourier transformation along the τ_1 axis to obtain a 2D frequency–frequency plot in the two variables ω_1 and ω_3. Note that this gives peaks in the rephasing spectrum negative frequencies in ω_1, so the spectrum must be inverted along the ω_1 axis as $-\omega_1 \rightarrow \omega_1$ for comparison with the nonrephasing spectrum [105].

The expressions above (Equation 12.4) describe nonlinear response for a system with a discrete number of energy levels and no population relaxation, giving rise to a discrete set of delta functions in the frequency-domain spectrum. For a real system, of course, neither of these assumptions is strictly appropriate. Population relaxation gives rise to an overall decay of the signal intensity, without directly affecting the phase, as excited oscillators return to the ground state, while coupling to low-frequency—phonon—modes gives rise to pure dephasing of the signal as the phase is gradually scrambled due to interference between closely spaced modes. In the frequency domain, dephasing and relaxation processes together produce *phase-twisted* 2D spectra in which absorptive (amplitude modulation) features along ω_1 are mixed with dispersive (phase modulation) features, resulting in a complicated lineshape that is often difficult to interpret visually [103,106]. These effects are illustrated in the upper panel of Figure 12.17, where experimental and calculated rephasing and nonrephasing spectra are plotted for the protein ubiquitin. Note the strong "smearing" of the spectrum along the diagonal in rephasing and against the diagonal in the nonrephasing spectrum. This distortion can be removed by adding the rephasing and nonrephasing signals together to obtain the *pure absorptive* spectrum plotted on the left. Note that only in the pure absorptive spectrum can the β-sheet cross peak ridge along $\omega_3 \approx 1680\ \mathrm{cm}^{-1}$ region be clearly distinguished from the dispersive features of either rephasing or nonrephasing individually.

12.4.7 Static Simulations

Finally, by combining the 2D IR calculations just described with a parameterized excitonic Hamiltonian as described in Section 12.4.3 above, one can directly link structural data (MD simulations or NMR/x-ray structures) to spectroscopic observables. In the simplest case, one can construct

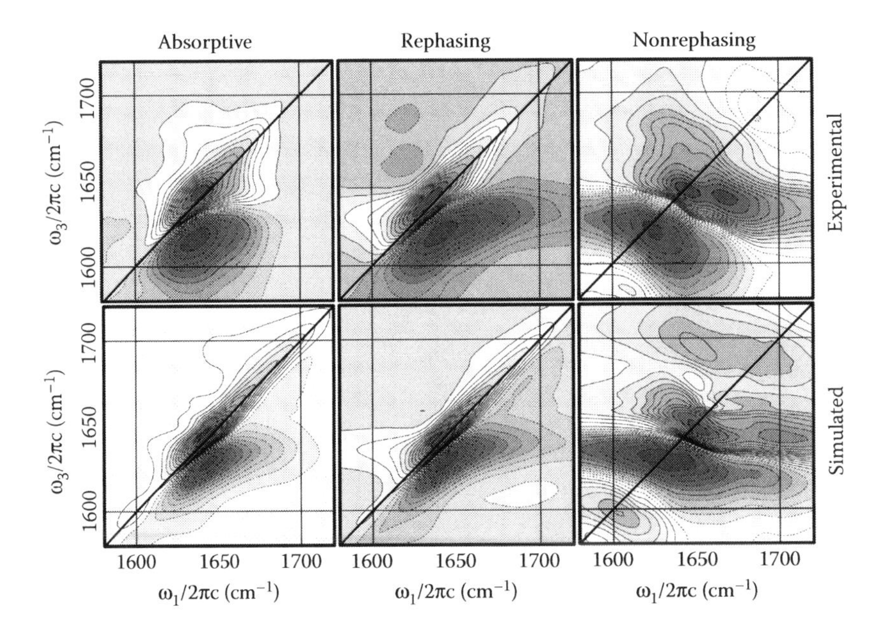

FIGURE 12.17 Experimental and simulated absorptive, rephasing and nonrephasing spectra of ubiquitin. Large rephasing amplitudes give rise to strongly inhomogeneously broadened absorptive spectra, as shown in this example. The amplitudes of the rephasing and nonrephasing signals normalized in the above contour plots. Positive and negative contours are represented by solid and dashed lines, respectively. Nonlinear contour spacing is used to emphasize the low-amplitude features.

a 2D IR spectrum corresponding to a single protein structure by inserting excitonic eigenstate energies and transition dipole vectors obtained from the electrostatic map Hamiltonian into the complex exponentials of Equation 12.4 or their Fourier transform representation as delta functions [31]. Dephasing is incorporated in the time domain by adding exponential decays to each term or in the frequency domain by convolution with a complex Lorentzian (the Fourier transform of an exponential decay). As a reference, we summarize here the steps necessary to perform these calculations along with the necessary numerical factors.

1. *Solvated structures*: To begin, a structural ensemble is generated, usually with the result of fully solvated MD simulations. The inclusion of the explicit solvent is important here to obtain the correct solvent-induced shifts of surface residues.
2. *One-quantum Hamiltonian*: These starting structures are then used as input to generate a Hamiltonian trajectory/ensemble using one of the electrostatic and/or nearest-neighbor maps discussed above. The parameters generated consist of a set of one-quantum site energies ε_n, coupling constants J_{mn}, and zero- to one-quantum transition dipole moments $\mu^{0,m}$.
3. *Two-quantum Hamiltonian*: Using the assumption of a weakly anharmonic system described above, a two-quantum Hamiltonian is generated from the one-quantum Hamiltonian via the relations

$$\begin{aligned} \langle nn \mid \hat{H} \mid nn \rangle &= 2\varepsilon_n - \Delta_n \\ \langle mn \mid \hat{H} \mid mn \rangle &= \varepsilon_m + \varepsilon_n \end{aligned} \tag{12.5}$$

with $\Delta = 16\ \text{cm}^{-1}$ for diagonal entries (site energies) and for off-diagonal elements (couplings)

$$
\begin{aligned}
\langle mm \mid \hat{H} \mid mp \rangle &= \sqrt{2} J_{mp} \\
\langle mn \mid \hat{H} \mid mp \rangle &= J_{np} \\
\langle mn \mid \hat{H} \mid pq \rangle &= 0
\end{aligned}
\tag{12.6}
$$

with n, m, and p, and q as distinct site indices. The dipole moments are similarly obtained via harmonic scaling.

$$
\begin{aligned}
\mu^{m,mm} &= \sqrt{2}\mu^{0,m} \\
\mu^{m,mn} &= \mu^{0,n} \\
\mu^{m,np} &= 0
\end{aligned}
\tag{12.7}
$$

4. *Eigenstates and stick spectra*: For each frame in the trajectory, the Hamiltonian generated above is diagonalized and the eigenvalues and eigenvectors are used to calculate transition frequencies and dipole moments. These values then act as input for stick spectra, that is, the Fourier transform of the static field expressions above

$$
\begin{aligned}
S^{R}_{ijkl}(\omega_1,\omega_3) \propto \sum_{a,b=1}^{N} \delta(\omega_1 + \omega_{a0}) \Big\{ \Big(\langle \mu_i^{0a}\mu_j^{0b}\mu_k^{a0}\mu_l^{b0} \rangle + \langle \mu_i^{0a}\mu_j^{a0}\mu_k^{0b}\mu_l^{b0} \rangle \Big) \delta(\omega_3 - \omega_{b0}) \\
- \sum_{c=1}^{N(N+1)/2} \langle \mu_i^{0a}\mu_j^{0b}\mu_k^{bc}\mu_l^{ca} \rangle \delta(\omega_3 - \omega_{ca}) \Big\} \\
S^{NR}_{ijkl}(\omega_1,\omega_3) \propto \sum_{a,b=1}^{N} \delta(\omega_1 - \omega_{a0}) \Big\{ \Big(\langle \mu_i^{0a}\mu_j^{0b}\mu_k^{b0}\mu_l^{a0} \rangle + \langle \mu_i^{0a}\mu_j^{a0}\mu_k^{0b}\mu_l^{b0} \rangle \Big) \delta(\omega_3 - \omega_{a0}) \\
- \sum_{c=1}^{N(N+1)/2} \langle \mu_i^{0a}\mu_j^{0b}\mu_k^{ac}\mu_l^{cb} \rangle \delta(\omega_3 - \omega_{cb}) \Big\}
\end{aligned}
\tag{12.8}
$$

As above, the indices a and b indicate one-quantum states, while the sum over c is over two-quantum states. Note that the rephasing profile here gives signals at negative frequency in ω_1; the spectrum will be flipped (after convolution) before comparison with the rephasing signal. To obtain the correct peak intensities, the laboratory frame dipole moment components $\mu_i^A\mu_j^B\mu_k^C\mu_l^D$ must be ensemble-averaged to account for the random orientation of molecules in the sample with respect to the laser polarization. For a spherically symmetric distribution of molecular orientations, the results are given analytically as [102,104]

$$
\begin{aligned}
\langle \mu_Z^A\mu_Z^B\mu_Z^C\mu_Z^D \rangle &= \sum_{\alpha} \left[\frac{1}{5} M_\alpha^A M_\alpha^B M_\alpha^C M_\alpha^D + \frac{1}{15} \sum_{\beta \neq \alpha} \left(M_\alpha^A M_\alpha^B M_\beta^C M_\beta^D + M_\alpha^A M_\beta^B M_\alpha^C M_\beta^D + M_\alpha^A M_\beta^B M_\beta^C M_\alpha^D \right) \right] \\
\langle \mu_Z^A\mu_Z^B\mu_Y^C\mu_Y^D \rangle &= \sum_{\alpha} \left[\frac{1}{15} M_\alpha^A M_\alpha^B M_\alpha^C M_\alpha^D + \frac{1}{30} \sum_{\beta \neq \alpha} \left(4M_\alpha^A M_\alpha^B M_\beta^C M_\beta^D - M_\alpha^A M_\beta^B M_\alpha^C M_\beta^D - M_\alpha^A M_\beta^B M_\beta^C M_\alpha^D \right) \right] \\
\langle \mu_Z^A\mu_Y^B\mu_Y^C\mu_Z^D \rangle &= \sum_{\alpha} \left[\frac{1}{15} M_\alpha^A M_\alpha^B M_\alpha^C M_\alpha^D - \frac{1}{30} \sum_{\beta \neq \alpha} \left(M_\alpha^A M_\alpha^B M_\beta^C M_\beta^D - 4M_\alpha^A M_\beta^B M_\alpha^C M_\beta^D + M_\alpha^A M_\beta^B M_\beta^C M_\alpha^D \right) \right]
\end{aligned}
\tag{12.9}
$$

where the indices Z and Y refer to laboratory frame Cartesian axes, the subscripts α and β run over the x, y, and z axes of the molecular frame coordinate system and the quantities M_α^A are molecular frame transition dipole moment components obtained directly from the MD trajectories above. For spherically symmetric systems, only the four laboratory frame polarizations $ZZZZ$, $ZZYY$, $ZYZY$, and $ZYYZ$ give non-vanishing signals, with the last condition $\langle \mu_Z^A \mu_Y^B \mu_Y^C \mu_Z^D \rangle$ obtained via the relation $\langle \mu_Z^A \mu_Z^B \mu_Z^C \mu_Z^D \rangle = \langle \mu_Z^A \mu_Z^B \mu_Y^C \mu_Y^D \rangle + \langle \mu_Z^A \mu_Y^B \mu_Y^C \mu_Z^D \rangle + \langle \mu_Z^A \mu_Y^B \mu_Z^C \mu_Y^D \rangle$.

5. *Convolution*: To simulate dephasing and population decay, the stick spectra are convolved with complex profiles of the form

$$\frac{1}{(i\omega_1 + \gamma)(i\omega_3 + \gamma)} = \frac{\gamma^2 - \omega_1\omega_3}{(\omega_1^2 + \gamma^2)(\omega_3^2 + \gamma^2)} - i\frac{\gamma(\omega_1 + \omega_3)}{(\omega_1^2 + \gamma^2)(\omega_3^2 + \gamma^2)} \tag{12.10}$$

that is the half-sided Fourier transform of a 2D exponential decay.

6. *Combined spectra*: Finally, the rephasing spectrum is flipped horizontally by exchanging ω_1 with $-\omega_1$. The complex spectra can then be plotted in a variety of ways for comparison to experiment. The real part of the sum of rephasing and nonrephasing signals gives the so-called purely absorptive (or correlation) surface, while the real part of the difference gives the dispersive contribution. Absolute value surfaces (of the complex signal—not the real part) are sometimes plotted instead, avoiding the appearance of negative peaks in the spectrum, but significantly increasing the width of the peaks.

A comparison of static ensemble spectra simulated using three different maps shows how the features observed in experimental spectra are all qualitatively reproduced by the different maps (see Figure 12.18) [16,31,91,92]. The two-peak spectra associated with β-sheets, and the single symmetric peak of α-helices is reproduced by the three maps; however, some of the details, such as overall shift of the spectra, the ratio of intensities corresponding to the $\upsilon_{//}$ and $\upsilon_{\perp}$ peaks along with the ridge structure associated with β-sheet spectra not well captured by the simulations. The parameterization does not include effects from other protein residues, which may or may not be captured in the simulations. Finally, the broadness of the calculated peaks in comparison with experiment is, in part, due to motional narrowing effects that are not captured by the static simulations [16].

12.4.8 Dynamic Simulations

The methods described so far model 2D IR spectra by combining an ensemble of "snapshot" 2D spectra calculated for a single static structure. In reality, although the global protein structure is static on the picosecond timescale relevant to 2D IR measurements, local solvent environments do fluctuate on a subpicosecond timescale giving rise to complex dynamic effects in the 2D spectrum, in particular, as a function of the *waiting time*. Such dephasing and relaxation processes can be modeled at a higher level using a numerical integration approach originally described for amide I vibrations by Torii [107] and applied to a wide variety of systems by Jansen and Knoester [72]. In this approach, the time-dependent excitonic Hamiltonian generated from an MD trajectory (step 3 in the list above) is broken down into small time intervals over which it can be assumed to be static. An exponential time-evolution operator is then calculated for each individual time step, allowing the time-dependent Schrödinger equation for the unperturbed system to be integrated numerically. This approach is particularly useful for providing realistic lineshapes for 2D surfaces as well as providing a numerically straightforward description of energy transfer between sites or bands, such as amide I and II [18,108]. The accuracy of this approach depends strongly on the accuracy of the mapping used to parameterize the excitonic Hamiltonian.

More recently, a variety of flavors of mixed quantum mechanics/molecular mechanics (QM/MM)-type calculations have begun to be applied to linear and 2D amide I spectra [109–111]. These

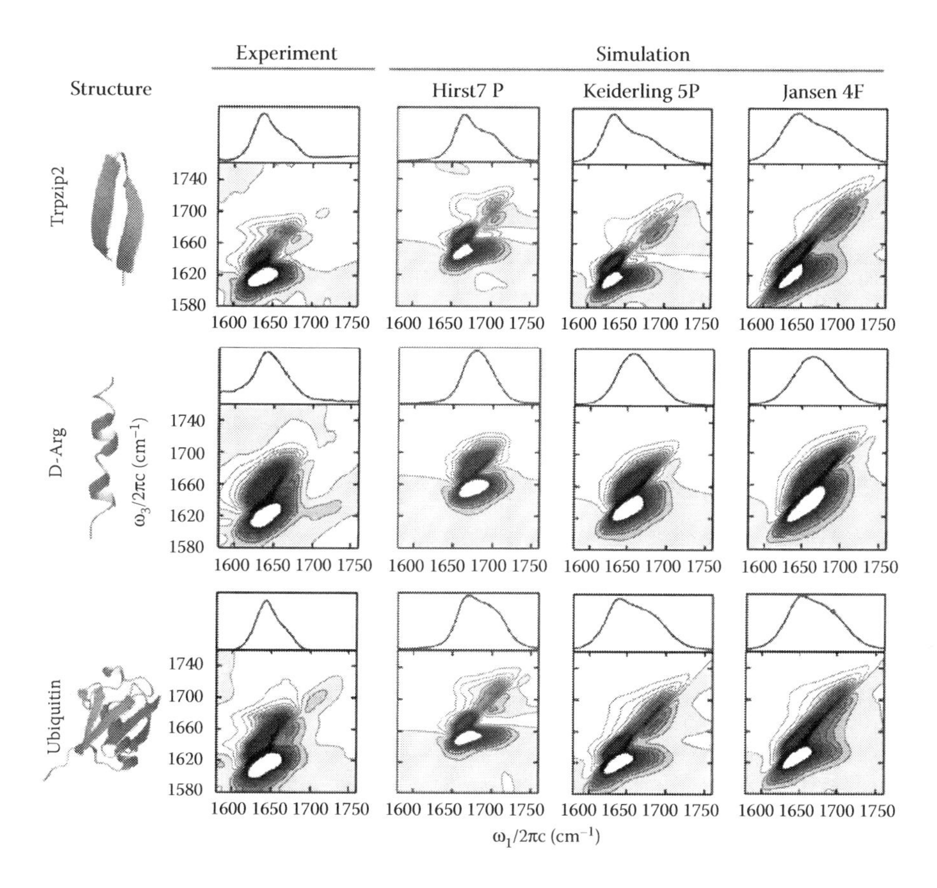

FIGURE 12.18 Comparison between calculated and experimental spectra for three different site maps. (Adapted from Ganim Z, Tokmakoff A. 2006. *Biophys J* 91:2636–2646.)

approaches are appealing in that they in principle allow spectroscopic features to be obtained directly from *ab initio* calculations, although at least for the present, computational requirements limit both the accuracy of the calculations and the size of the system to which they can be applied. In the coming years, an important challenge for all of these approaches will be to bridge the gap between qualitative descriptions of amide I spectroscopy and a quantitative map between IR spectroscopy and protein structure and dynamics in solution.

12.4.9 Summary and Outlook

Structure-spectroscopy maps provide the essential connection between structure and spectra. Though the agreement is for the most part qualitative, the simulations provide an intuitive interpretation of the spectral features to help interpret the experimental data and inform the design of new experiments. For example, simulations can be used to explore the effects of isotope labels at different positions to strategically select labeling sites. It is important to emphasize that experimental spectra provide structural constraints, and the quality of the map determines the degree to which the experimental data can be interpreted and how much structural information can be extracted from the measured spectra. Since the simulations are relatively inexpensive and trivially parallelizable, hundreds or thousands of trial structures can be simulated to extract structures that best describe the experimental data. Importantly, the maps can be easily interfaced with Markov state models to explore the heterogeneity of folding mechanisms in small proteins [12].

It is worth noting a number of possibilities for future work that may prove useful in broadening the scope of amide I spectral modeling capabilities. Although, as illustrated above, the existing models provide a good qualitative description of the experimental amide I spectra, quantitative precision is often lacking. Several opportunities for improvement stem from protein side-chain considerations. For example, although several side-chain groups absorb in the amide I spectral region (especially Asp, Glu, Asn, Gln, and Arg), these moieties are usually neglected for spectral calculations. An accurately parameterized site energy map for these groups would be of interest for many applications, particularly since their vibrations are expected to be largely localized, making them potentially useful as probes of local structure. Similarly, a side-chain-specific study of amide I and II vibrations would be useful to more closely examine the effects of individual side chains on the site energies of adjacent peptide groups [96,112]. More generally, it should be noted that the existing spectral maps have been parameterized and tested against a relatively small number of model compounds (notably NMA and alanine and glycine dipeptides), raising questions about their quantitative accuracy when transferred to new systems such as large proteins. The basic assumption involved in these calculations is that site energy shifts are essentially nonspecific, so that, for example, the electrostatic influence of solvating water can be treated identically to that of adjacent side chains and backbone atoms. It would be of great interest to examine these assumptions on a case-by-case basis, potentially allowing for greater accuracy through individual parameterizations for different amino acids or functional groups. On the experimental front, residue-specific isotope labels could assist this process, since they provide nonperturbative, spectrally isolated local probes of the site energy shifts associated with individual residues.

12.5 EXAMPLES OF PROTEIN 2D IR SPECTROSCOPY

The capabilities of multidimensional spectroscopy as a technique for measuring protein structure, conformational flexibility, and transient peptide unfolding mechanisms are demonstrated by the examples in this section. The first example shows how protein secondary structure is extracted from amide I 2D IR spectra. The second example highlights how cross peaks provide a direct structural view of the solvent exposure of proteins. The third example describes the unfolding mechanism of a small peptide, TrpZip2, investigated using nonequilibrium temperature-jump 2D IR spectroscopy.

12.5.1 Determination of Protein Structure Composition with 2D IR Spectroscopy

This example illustrates the analytical aspect of 2D IR spectroscopy and its application to the quantitative measurement of secondary structure composition based on amide I 2D spectra. As described in previous sections, secondary protein structures exhibit particular signatures in the amide I spectrum. However, to date, most IR studies of protein structure have provided qualitative structural information, or have been carried out on systems with known equilibrium structures. Previous attempts to quantitate secondary structure from FTIR spectra have been based on spectral deconvolution combined with lineshape fitting methods that rely on a number of assumptions that limit the usefulness and applicability of the technique [113]. Here, we provide an example of how the enhanced structural sensitivity of 2D IR spectroscopy can be harnessed to develop a quantitative assay of protein secondary structure composition without *a priori* spectral assignments [14].

In brief, a library is constructed by collecting spectra of proteins with known crystal structures, and the signatures of individual secondary structures are extracted through singular value decomposition (SVD), a form of the principal component analysis. For the purpose of the analysis, proteins are treated as mixtures of α-helices, β-sheets, and unassigned conformations. The analysis makes two important assumptions: (1) All α-helices, or β-sheets, have the same spectral signatures regardless of the structural details. For example, parallel and antiparallel β-sheets are assumed to have the same spectrum. As described previously, peak center frequencies and intensity ratios depend on the size and number of strands in a β-sheet. (2) Spectra are not sensitive to supersecondary structure, namely, the coupling between secondary structures is negligible.

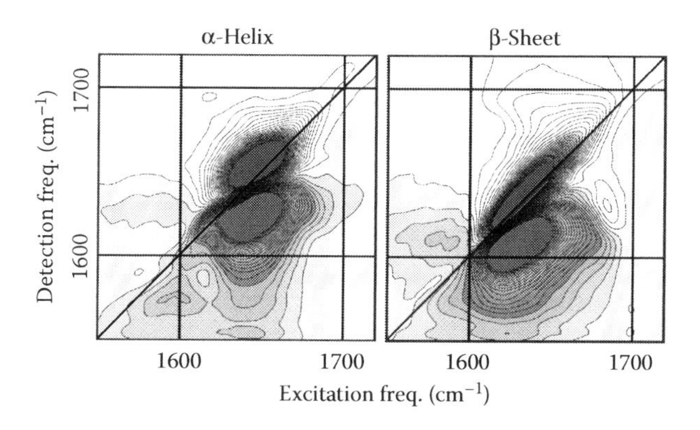

FIGURE 12.19 Singular value components extracted from a library containing sixteen protein spectra with mixed α/β components. (Adapted from Baiz CR et al. 2012. *Analyst* 137:1793–1799.)

Figure 12.19 shows component 2D IR spectra decomposed from the protein set. The helix spectrum shows a typical "figure-8 shape" represented by a single peak centered near 1650 cm^{-1} and the β-sheet shows a diagonally elongated band, corresponding to the $\upsilon_\perp$, and $\upsilon_{//}$ bands centered around 1630 and 1670 cm^{-1}, respectively, along with the cross ridges that give the spectrum its characteristic "Z-shape." The projection of an "unknown" spectrum onto the α-helix and β-sheet component spectra is proportional to the percentage of residues in α-helix and β-sheet conformations in the unknown protein, respectively. Figure 12.20 shows the fraction of residues in different secondary structures predicted from 2D IR spectra along with the respective fractions extracted from analysis of the crystal structures. The root-mean-squared errors are: 6.7% for α-helices, 7.7% for β-sheets, and 8.2% for unstructured conformations. These results represent a significant improvement over linear IR absorption spectroscopy, for which the same singular value decomposition analysis produces an average error of 19.5% for α-helices, 8.3% for β-sheets, and 21.5% for unassigned conformations. The comparison highlights the enhanced structural sensitivity afforded by nonlinear spectroscopies, which are inherently sensitive to the anharmonic site energies and coupling contributions to the amide I vibrations.

Although the data analysis overlooks the finer spectral details associated with different protein structures, the accuracy of the results justifies the assumptions. In principle, the conformational

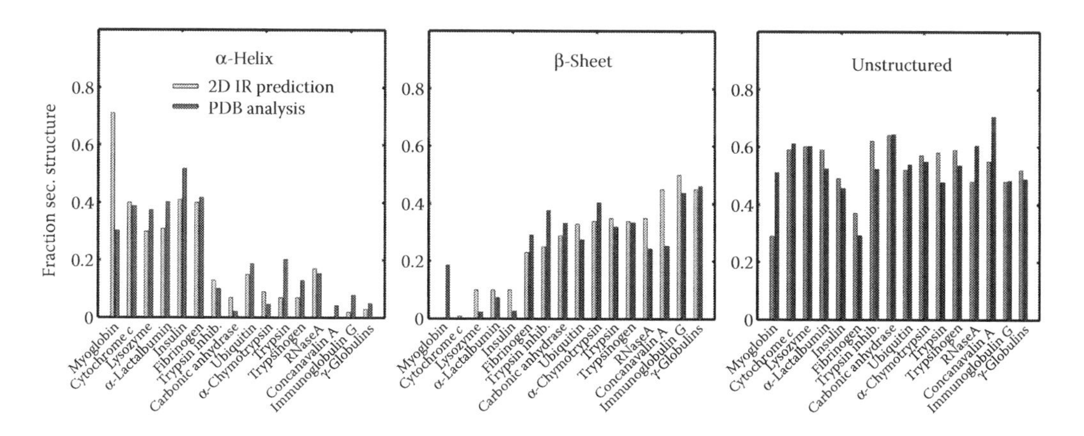

FIGURE 12.20 Percent of residues in α-helix, β-sheet, and unstructured configurations for 16 proteins with known crystal structures. The dark bars indicate the values extracted from the crystal structures and the light bars indicate the values predicted from analysis of experimental 2D IR spectra. (Adapted from Baiz CR et al. 2012. *Analyst* 137:1793–1799.)

selectivity can be further refined by expanding the protein set and including more varied secondary and supersecondary structural motifs. Finally, the example demonstrated here highlights particular strengths of IR spectroscopy: (1) The analysis is performed on stock proteins without the need for complex sample preparation techniques. (2) Protein spectra can be collected in solution under biological conditions.* (3) The analysis is not limited to fully solvated globular proteins: membrane proteins, protein complexes, aggregates, or other samples, which remain difficult to characterize with traditional techniques, can be analyzed with amide I 2D IR spectroscopy.

12.5.2 Measuring Solvent Exposure of Secondary Structures through Multimode Amide I/II 2D IR Spectroscopy

While the amide I band has received a great deal of attention as a structural probe, amide II vibrations are excellent markers of conformational flexibility and solvent exposure. The predominant N–H bending character is responsible for the large shift observed upon hydrogen–deuterium (H–D) exchange. Two-dimensional spectroscopy has the advantage of being able to measure the couplings between amide I and amide II modes, combining the structural sensitivity of amide I with the solvent-exposure sensitivity of amide II [46,114]. The two bands are coupled primarily through shared motions of the backbone atoms such that modes that share common residues are more strongly coupled, than modes that are delocalized over different sections of the protein backbone. The number of shared residues determines the strength of the coupling. During an H/D exchange experiment, residues that are solvent exposed exchange faster than those buried within the protein core. The 100 cm^{-1} frequency difference between amide II and II′ modes allows for selective excitation of deuterated—solvent exposed—or protonated–buried—residues. On-site couplings allow for efficient energy transfer to the amide I band. As described previously, amide I peaks can be related to protein secondary structure; thus, the 2D IR cross peaks between amide II/II′ and amide I/I′ contain a *structural* view of solvent exposure. In simpler terms, the amide II excitation selectively "tags" either buried or exposed residues, and the amide I detection extracts the spectral information related to secondary structure.

Figure 12.21 shows a series of multimode 2D IR spectra of concanavalin A, myoglobin, RNase A, and ubiquitin, together with projections onto the detection axis of the main amide I band and amide I/II cross peaks. The difference between the features observed in the diagonal peak projection and the cross peaks provides a qualitative view of the secondary structures. The absence of cross peaks indicates rapid H–D exchange for that protein. First, in the case of concanavalin A and ubiquitin, the amide II/I slices show significantly less intensity in the center of the band (1660–1680 cm^{-1}), compared to the amide I/I diagonal peak slices, indicating that turn and coil regions exchange faster than α-helices or β-sheets amide I 2D IR spectra and crystal structures are shown in Figure 12.3. Similarly, the cross-peak projection of myoglobin is red-shifted with respect to the diagonal slice, indicating exchange primarily in the coil regions. RNase A contains features corresponding to β-sheet and α-helices, indicating a high degree of structural stability under the experimental conditions. In the case of ubiquitin, the lack of intensity around 1680 cm^{-1} suggests that residues in β-sheet conformations are less stable and more prone to undergo H–D exchange.

In summary, multimode H–D exchange spectroscopy gives a direct view of structural stability and solvent exposure of secondary structures within the protein. The technique is general, does not require labeling, and has high time resolution, which in principle allows for the combination of H–D exchange with temperature jump to probe structural changes during protein unfolding. Millisecond time resolution can be attained through rapid mixing experiments. Combined with isotope labeling, multimode 2D IR spectroscopy has the ability to measure the exchange kinetics of individual residues and thus provide a much more detailed view of the backbone solvent exposure.

---

* In this example, spectra were collected at pH* = 1 so as to shift side-chain vibrations outside of the amide I window to simplify the data analysis.

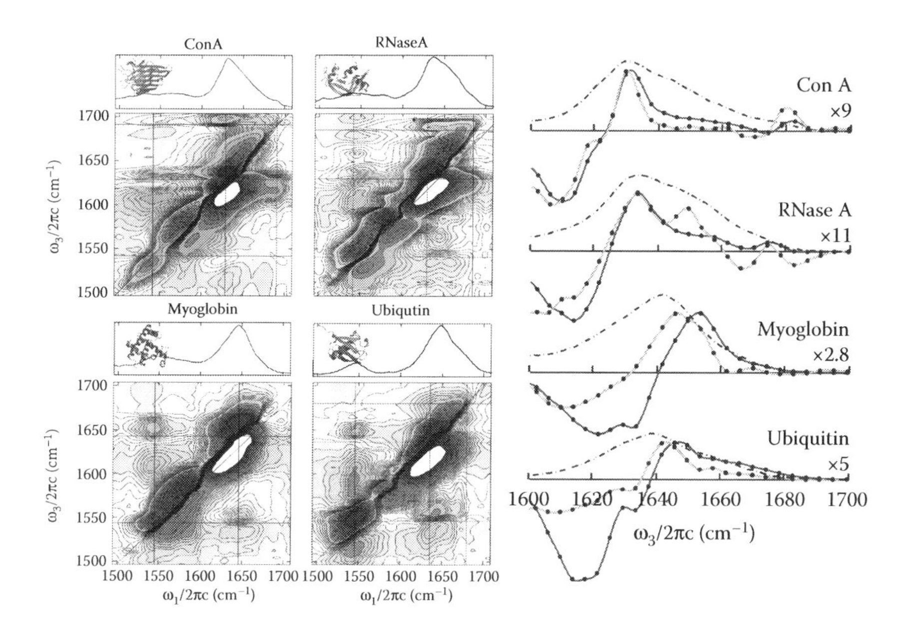

FIGURE 12.21 Amide I/II 2D IR spectra of concanavalin A (Con A), RNaseA, ubiquitin, and myoglobin. The spectra show two main diagonal bands corresponding to amide I/I′ (~1650 cm^{-1}), amide II (1550 cm^{-1}). (Adapted from DeFlores LP et al. 2009. *J Am Chem Soc* 131:3385–3391.)

12.5.3 Transient T-Jump Spectroscopy as a Probe of Unfolding Dynamics of TrpZip2

Understanding protein folding is an outstanding challenge in modern biophysics and biology. The three-dimensional structure of proteins is encoded in the sequence, but it is not understood how proteins adopt a stable conformation, how sequence determines structure, or what microscopic interactions, among the thousands of degrees of freedom, are relevant to the folding process. Conformational dynamics dictated by the subtle energetic interactions between contact formation and protein–solvent enthalpy and entropy remain extremely difficult to study with current experimental techniques. Advances in computer technologies have recently enabled atomistic MD simulations of protein folding processes [115–117]. Although the simulations provide very detailed mechanisms for protein folding, there has been virtually no experimental data to validate the predictions. Temperature-jump (T-jump) 2D IR spectroscopy is a promising new technique that leverages the structural resolution power of 2D IR spectroscopy and applies it to the study of thermally induced changes in peptides and proteins [81,88,118]. As illustrated in Section 12.2, isotope-edited amide I 2D IR spectroscopy provides a localized view of structural disorder and heterogeneity of peptides undergoing denaturation. In addition to ultrafast picosecond dynamics of protein–protein or protein–solvent interactions, ultrafast measurements enable differentiation of conformations interconverting on the micro- to nanosecond timescales.

The tryptophan zipper 2 (TrpZip2), shown in Figure 12.22, forms a type-I′ β-turn stabilized by cross-strand hydrophobic interactions involving the tryptophan indole rings [119]. Fast folding rates and small number of residues make it an attractive target for experimental and computational folding studies. As described in Section 12.2, temperature-dependent 2D IR spectroscopy has been used to characterize the residual structure and disorder of TrpZip2 [12]. The spectra suggest the presence of at least two stable turn conformations: a native type-I β-turn, and a disordered bulged loop.

T-jump spectroscopy is able to assess the rate of interconversion between the states by examining the kinetics of the isotope peaks as a function of delay following the T-jump. Here, we present a combination of T-jump HDVE and 2D IR spectroscopy. Since HDVE spectra can be collected ~300× faster than 2D IR, the standard approach involves obtaining kinetic rates by finely sampling the T-jump delays using HDVE spectroscopy, and extracting structural information at selected times

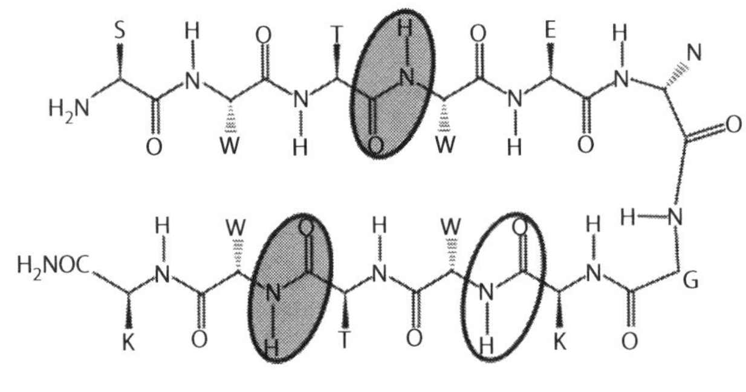

FIGURE 12.22 Tryptophan Zipper 2 (TrpZip2). Residues selected for isotope labels T3, K8, and T10 are highlighted for reference.

using T-jump 2D IR spectroscopy. Figure 12.23 shows the measured sample response times. The curves are extracted from a singular value decomposition (SVD) analysis of the kinetic traces. Three distinct timescales are observed: a small-amplitude nanosecond rise, a stronger microsecond rise, and a millisecond decay. The millisecond decay is due to reequilibration as heat is dissipated from the laser focus region. The first two components are attributed to the response of the peptide. To interpret the rates from a structural perspective, 2D IR spectra are collected at nanosecond and microsecond delays following the T-jump. T-jump difference (transient–equilibrium) spectra are shown in

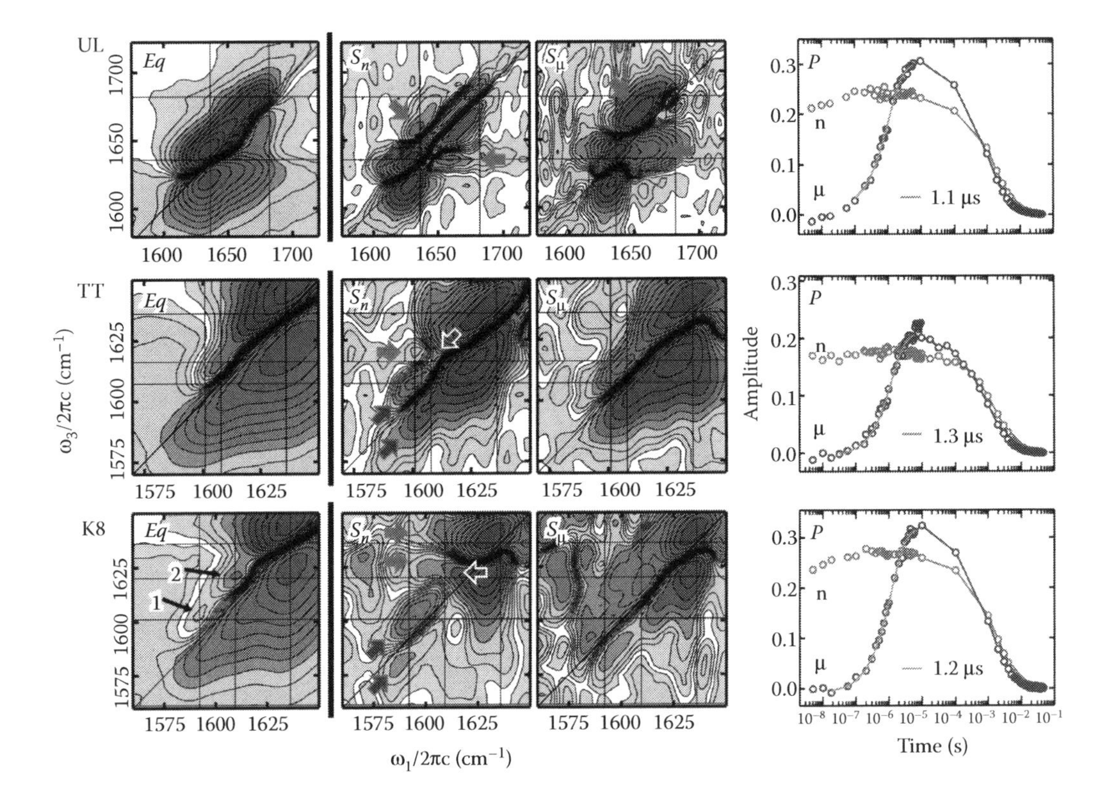

FIGURE 12.23 TrpZip2 transient and equilibrium 2D IR spectra and HDVE kinetic curves. The left column shows equilibrium 2D IR spectra of unlabeled (UL), T3T10 (TT), and K8 $^{13}C{=}^{18}O$ isotope-labeled samples. The center columns show difference spectra associated with the nanosecond (right) and microsecond (left) responses. Solid and dashed contours represent positive and negative peaks, respectively. The right column shows the response curves measured with HDVE spectroscopy.

Figure 12.23. Peaks above the diagonal are positive whereas peaks below the diagonal are negative in the equilibrium spectra. In difference spectra, negative peaks above the diagonal represent loss of signal, whereas negative peaks below the diagonal represent gain of signal and vice versa.

12.5.3.1 Transient 2D IR: Nanosecond Response

The nanosecond component shows positive/negative doublet peaks with narrow antidiagonal widths spanning the entire diagonal region. The bleach is most intense in the low- and high-frequency regions ~1635 and 1670 cm^{-1} corresponding to the equilibrium loss of the parallel and antiparallel β-sheet modes. The off-diagonal ridges are attributed to a broadening of the peak due to increased conformational flexibility and weaker interstrand hydrogen bonds. The K8 and TT double labels show features consistent with the loss of hydrogen bonding: loss of the equilibrium peaks, indicated by diagonal arrows, along with appearance of blue-shifted peaks, indicated by horizontal arrows. Spectral shifts can result from changes in hydrogen bond strengths and do not necessarily reflect large conformational changes within the system.

12.5.3.2 Transient 2D IR: Microsecond Response

Changes observed in the microsecond spectra (Figure 12.23, third column) are indicative of disordering and potential fraying of the β-strands. Two loss features at 1640 and 1680 cm^{-1} along with the corresponding off-diagonal ridges are associated with the loss of β-sheet structure. The gain features in the 1660 cm^{-1} region, indicated by the arrows, are attributed to frayed or partially disordered peptide conformations. The interpretation is consistent with results from Markov state modeling, which indicate that disordered conformations show an intensity gain in the 1650 cm^{-1} region. The double-label, TT, spectrum shows loss features above and below the diagonal indicative of an overall loss of intensity due to weakening of contacts in the mid-strand region. A ~15 cm^{-1} blue-shift is expected for the TT peak upon loss of coupling, which has the effect of hiding the nascent peak within the main amide I bleach feature.

12.5.3.3 Assignment of Markov States

Interpretation of the T-jump 2D IR spectra in the context of Markov-state simulations (see Ref. [12]) suggest the presence of three main states: a *folded* state characterized by a type-I′ turn stabilized by interstrand hydrogen bonds and side-chain packing, a *misfolded* conformation exhibiting a bulged turn conformation with a solvent-exposed K8, and misregistered hydrogen bond contacts, a *disordered* state representing conformations that lack the standard interstrand hydrogen bonds. The equilibrium spectra indicate that a significant portion of the misfolded state exists in equilibrium with the folded state. The pulse-width limited response (<5 ns) is attributed to loosening of the interstrand contacts; the microsecond response indicates a partial shift of population from the folded and misfolded into the disordered state. Since the refolding kinetics are significantly faster than the temperature relaxation, there is no trigger event to synchronize the refolding process; thus, the peptide response follows the solvent relaxation kinetics, making it difficult to separate the spectral changes associated with refolding from the solvent response.

12.5.4 Additional Examples

The examples above are vignettes that illustrate some capabilities that are enabled by amide 2D IR spectroscopy of proteins. Recent applications of 2D IR spectroscopy in the study of the structure and dynamics of proteins are numerous, and we here briefly point the reader to some of the many recent examples of this method. Recent studies of amyloid formation by amylin have been used to explain the mechanism of fibril nucleation and growth [51] and the binding of a fibril inhibitor [4]. Structural studies using isotope labeling have described conformationally heterogeneous states of peptides [6,12]. A variety of implementations have been used to investigate structure, conformational changes, and fast fluctuations in membrane proteins including the CD3ζ transmembrane peptide

[120], the influenza virus M2 proton channel [7,8], and transmembrane helix dimers [121]. Transient 2D IR has been used in protein folding studies aimed at understanding photo-initiated structural rearrangements in small peptides [122], and temperature-jump-induced unfolding dynamics of ubiquitin [123,124]. Transient and steady-state methods have revealed fast timescale fluctuations and site-specific protein dynamics in the active sites of enzymes, such as myoglobin [125], horseradish peroxidase [126], and formate dehydrogenase [127], and bulk solution, villin headpiece [128,129], and human immunodeficiency virus (HIV) reverse transcriptase [130]. Protein–protein interactions and the folding associated with binding have been observed in insulin dimer [131]. A broader perspective can be gained from a number of recent reviews on protein 2D IR [3,30,55].

12.6 CONCLUSION AND OUTLOOK

Over the past decade, protein 2D IR spectroscopy has evolved from a novel optical technique into a more mature research tool that provides the structural sensitivity and ultrafast time resolution required to shed light on problems related to protein structure and conformational dynamics. With the ability to probe structural information in systems that are not readily accessed by traditional methods, we imagine that the discipline will continue to find a variety of applications. It will find increasing use in the study of membrane proteins, where it can be applied to structure and folding of single proteins and complexes, the mechanism of transduction of ions and small molecules through channels [132], and signal transduction processes in receptors. The combination of isotope-edited 2D IR and structural modeling will not only provide restraints to solve structures for systems not amenable to crystallography or NMR, but also enable studies of disordered proteins and unstructured states. We imagine that these studies will provide structural ensembles that characterize intrinsically disordered chains, encountered in coupled folding and binding problems in protein–protein and protein–DNA interactions. The level of structural detail needed for various protein structural problems varies from atomistic to nanometer scale. 2D IR can be fashioned to reveal contacts in binding sites, gross features in protein complexes, and oligomerization processes. To enable such studies, isotope labeling need not only be at specific sites, but it can be applied to an entire secondary structure, domain, or protein to facilitate studies of mesoscopic structures.

The truly unique nature of 2D IR lies in its combined structural and temporal resolution, which enables kinetics and dynamics studies and opens the door to direct observation of the function of proteins. Conformational dynamics in folding, binding, and signal transduction are processes that can potentially be visualized at high time resolution with triggered transient 2D IR methods. Similarly, proton and ion transport processes in enzymes and membrane proteins can also be triggered and followed with ultrafast time resolution.

Technological advances have improved and simplified the experimental methods, so that 2D IR instruments can be commercialized. Nonetheless, the need for innovation and development of the method remains higher than ever. Further developments in instrumentation, theory, and synthesis are needed to make them more widely available. An increased library of protein vibrations available for detailed structural interpretation and the new pulsed IR light sources that can simultaneously excite and probe many vibrational modes are both avenues that will greatly increase the number of applications. Significant efforts have to be made in order to develop structure-based models for 2D IR constraints that are quantitative and have accuracy to meaningfully distinguish conformational variation and solvent environment on the angstrom scale. Close ties with the protein simulation community are needed for experimentalists to atomistically interpret data, and will also provide the new data that can be used to validate the force fields used in simulation. Similarly, methods for incorporating isotope and molecular labels into proteins will become increasingly important for these studies. Fortunately, this area of research is of wide interest to many spectroscopies, and is closely related to problems that the protein NMR community has worked on for many years.

In conclusion, the ultrafast time resolution and structural sensitivity of 2D IR spectroscopy offers a broadly applicable approach to study protein structure and fast conformational motions that remain virtually inaccessible by conventional techniques. Given the possibilities and the applicability to complex samples such as membrane proteins, complexes, fibers, oligomers and aggregates, and disordered systems, we believe there will be extraordinary future growth in information content and applications of protein 2D IR.

ACKNOWLEDGMENTS

We acknowledge members of the Tokmakoff group whose accomplishments over the past decade are summarized in this chapter: Hoi Sung Chung, Matthew Decamp, Lauren DeFlores, Nuri Demirdöven, Ziad Ganim, Munira Khalil, Poul Petersen, and Adam Smith. We acknowledge Adam Squires and Ziad Ganim for providing the unpublished insulin data, Krupa Ramasesha and Luigi DeMarco for providing the BBIR spectrum in Figure 12.9, Kevin Jones for the TrpZip2 T-jump data, and Joshua Lessing for the waiting time data on the elastin-like peptide. We thank Kevin Jones and Chunte Sam Peng for helpful comments on the manuscript. The work is supported by the National Science Foundation (CHE-0911107) and Agilent Technologies. Some research described in this chapter was partially funded by the Department of Energy (DE-FG02–99ER14988).

REFERENCES

1. Hamm P, Zanni M. 2011. *Concepts and Methods of 2D Infrared Spectroscopy* (Cambridge University Press, Cambridge, New York).
2. Cho M. 2009. *Two-Dimensional Optical Spectroscopy* (CRC Press, Boca Raton, FL).
3. Fayer MD. 2009. Dynamics of liquids, molecules, and proteins measured with ultrafast 2D IR vibrational echo chemical exchange spectroscopy. *Annu Rev Phys Chem* 60:21.
4. Middleton CT et al. 2012. Two-dimensional infrared spectroscopy reveals the complex behaviour of an amyloid fibril inhibitor. *Nat Chem* 4:355–360.
5. Kim YS, Liu L, Axelsen PH, Hochstrasser RM. 2009. 2D IR provides evidence for mobile water molecules in beta-amyloid fibrils. *Proc Natl Acad Sci USA* 106:17751–17756.
6. Lessing J et al. 2012. Identifying residual structure in intrinsically disordered systems: A 2D IR spectroscopic study of the GVGXPGVG peptide. *J Am Chem Soc* 134:5032–5035.
7. Ghosh A, Qiu J, DeGrado WF, Hochstrasser RM. 2011. Tidal surge in the M2 proton channel, sensed by 2D IR spectroscopy. *Proc Natl Acad Sci USA* 108:6115–6120.
8. Manor J et al. 2009. Gating mechanism of the influenza A M2 channel revealed by 1D and 2D IR spectroscopies. *Structure* 17:247–254.
9. Munoz V. 2007. Conformational dynamics and ensembles in protein folding. *Annu Rev Bioph Biom* 36: 395–412.
10. Kubelka J, Hofrichter J, Eaton WA. 2004. The protein folding "speed limit". *Curr Opin Struc Biol* 14: 76–88.
11. Chung HS, Shandiz A, Sosnick TR, Tokmakoff A. 2008. Probing the folding transition state of ubiquitin mutants by temperature-jump-induced downhill unfolding. *Biochemistry USA* 47:13870–13877.
12. Smith AW et al. 2010. Melting of a beta-Hairpin peptide using isotope-edited 2D IR spectroscopy and simulations. *J Phys Chem B* 114:10913–10924.
13. Woutersen S, Mu Y, Stock G, Hamm P. 2001. Hydrogen-bond lifetime measured by time-resolved 2D IR spectroscopy: N-methylacetamide in methanol. *Chem Phys* 266:137–147.
14. Baiz CR, Peng CS, Reppert ME, Jones KC, Tokmakoff A. 2012. Coherent two-dimensional infrared spectroscopy: Quantitative analysis of protein secondary structure in solution. *Analyst* 137:1793–1799.
15. Wang L, Middleton CT, Zanni MT, Skinner JL. 2011. Development and validation of transferable amide I vibrational frequency maps for peptides. *J Phys Chem B* 115:3713–3724.
16. Jansen TL, Knoester J. 2006. A transferable electrostatic map for solvation effects on amide I vibrations and its application to linear and two-dimensional spectroscopy. *J Chem Phys* 124:044502.
17. Schmidt JR, Corcelli SA, Skinner JL. 2004. Ultrafast vibrational spectroscopy of water and aqueous N-methylacetamide: Comparison of different electronic structure/molecular dynamics approaches. *J Chem Phys* 121:8887–8896.
18. Bloem R, Dijkstra AG, Jansen TLC, Knoester J. 2008. Simulation of vibrational energy transfer in two-dimensional infrared spectroscopy of amide I and amide II modes in solution. *J Chem Phys* 129:055101.

19. Barth A. 2007. Infrared spectroscopy of proteins. *Bba-Bioenergetics* 1767:1073–1101.
20. Barth A, Zscherp C. 2002. What vibrations tell us about proteins. *Q Rev Biophys* 35:369–430.
21. Jackson M, Mantsch HH. 1995. The use and misuse of FTIR spectroscopy in the determination of protein-structure. *Crit Rev Biochem Mol* 30:95–120.
22. Krimm S, Bandekar J. 1986. Vibrational spectroscopy and conformation of peptides, polypeptides, and proteins. *Adv Protein Chem* 38:181–364.
23. Davydov AS. 1971. *Theory of Molecular Excitons* (Plenum Press, New York).
24. Chung HS, Tokmakoff A. 2006. Visualization and characterization of the infrared active amide I vibrations of proteins. *J Phys Chem B* 110:2888–2898.
25. Cheatum CM, Tokmakoff A, Knoester J. 2004. Signatures of beta-sheet secondary structures in linear and two-dimensional infrared spectroscopy. *J Chem Phys* 120:8201–8215.
26. Torii H, Tasumi M. 1992. 3-Dimensional doorway-state theory for analyses of absorption-bands of many-oscillator systems. *J Chem Phys* 97:86–91.
27. Torii H, Tasumi M. 1992. Application of the 3-dimensional doorway-state theory to analyses of the amide-I infrared bands of globular-proteins. *J Chem Phys* 97:92–98.
28. Abramavicius D, Zhuang W, Mukamel S. 2004. Peptide secondary structure determination by three-pulse coherent vibrational spectroscopies: A simulation study. *J Phys Chem B* 108:18034–18045.
29. Demirdoven N et al. 2004. Two-dimensional infrared spectroscopy of antiparallel beta-sheet secondary structure. *J Am Chem Soc* 126:7981–7990.
30. Ganim Z et al. 2008. Amide I two-dimensional infrared spectroscopy of proteins. *Acc Chem Res* 41: 432–441.
31. Ganim Z, Tokmakoff A. 2006. Spectral signatures of heterogeneous protein ensembles revealed by MD simulations of 2D IR spectra. *Biophys J* 91:2636–2646.
32. Nevskaya NA, Chirgadze YN. 1976. Infrared-spectra and resonance interactions of amide-one and amide-2 vibrations of alpha-helix. *Biopolymers* 15:637–648.
33. Woutersen S, Hamm P. 2002. Nonlinear two-dimensional vibrational spectroscopy of peptides. *J Phys.: Condens Matter* 14:1035–1062–1035–1062.
34. Wang JP, Hochstrasser RM. 2004. Characteristics of the two-dimensional infrared spectroscopy of helices from approximate simulations and analytic models. *Chem Phys* 297:195–219.
35. Kubelka J, Silva RAGD, Keiderling TA. 2002. Discrimination between peptide 3(10)- and alpha-helices. Theoretical analysis of the impact of alpha-methyl substitution on experimental spectra. *J Am Chem Soc* 124:5325–5332.
36. Silva RAGD et al. 2002. Discriminating 3(10)- from alpha helices: Vibrational and electronic CD and IR absorption study of related Aib-containing oligopeptides. *Biopolymers* 65:229–243.
37. Maekawa H, Toniolo C, Moretto A, Broxterman QB, Ge NH. 2006. Different spectral signatures of octapeptide 3(10) and alpha-helices revealed by two-dimensional infrared spectroscopy. *J Phys Chem B* 110:5834–5837.
38. Maekawa H, Toniolo C, Broxterman QB, Ge NH. 2007. Two-dimensional infrared spectral signatures of 3(10)- and alpha-helical peptides. *J Phys Chem B* 111:3222–3235.
39. Wang JP. 2008. Conformational dependence of anharmonic vibrations in peptides: Amide-I modes in model dipeptide. *J Phys Chem B* 112:4790–4800.
40. Karjalainen EL, Ravi HK, Barth A. 2011. Simulation of the amide I absorption of stacked beta-sheets. *J Phys Chem B* 115:749–757.
41. Kubelka J, Keiderling TA. 2001. Differentiation of beta-sheet-forming structures: Ab initio-based simulations of IR absorption and vibrational CD for model peptide and protein beta-sheets. *J Am Chem Soc* 123:12048–12058.
42. Lee C, Cho MH. 2004. Local amide I mode frequencies and coupling constants in multiple-stranded antiparallel beta-sheet polypeptides. *J Phys Chem B* 108:20397–20407.
43. Moore WH, Krimm S. 1975. Transition dipole coupling in amide I modes of beta polypeptides. *Proc Natl Acad Sci USA* 72:4933–4935.
44. Maekawa H, Ge NH. 2010. Comparative study of electrostatic models for the amide-I and -II modes: Linear and two-dimensional infrared spectra. *J Phys Chem B* 114:1434–1446.
45. Choi JH, Kim JS, Cho MH. 2005. Amide I vibrational circular dichroism of polypeptides: Generalized fragmentation approximation method. *J Chem Phys* 122:174903.
46. DeFlores LP, Ganim Z, Nicodemus RA, Tokmakoff A. 2009. Amide I "-II" 2D IR spectroscopy provides enhanced protein secondary structural sensitivity. *J Am Chem Soc* 131:3385–3391.
47. Middleton CT, Buchanan LE, Dunkelberger EB, Zanni MT. 2011. Utilizing lifetimes to suppress random coil features in 2D IR spectra of peptides. *J Phys Chem Lett* 2:2357–2361.

48. Torii H, Tasumi M. 1992. Model-calculations on the amide-I infrared bands of globular-proteins. *J Chem Phys* 96:3379–3387.
49. Wang L et al. 2011. 2D IR spectroscopy of human amylin fibrils reflects stable beta-sheet structure. *J Am Chem Soc* 133:16062–16071.
50. Middleton CT, Woys AM, Mukherjee SS, Zanni MT. 2010. Residue-specific structural kinetics of proteins through the union of isotope labeling, mid-IR pulse shaping, and coherent 2D IR spectroscopy. *Methods* 52:12–22.
51. Shim SH et al. 2009. Two-dimensional IR spectroscopy and isotope labeling defines the pathway of amyloid formation with residue-specific resolution. *Proc Natl Acad Sci USA* 106:6614–6619.
52. Gnanakaran S, Hochstrasser RM. 2001. Conformational preferences and vibrational frequency distributions of short peptides in relation to multidimensional infrared spectroscopy. *J Am Chem Soc* 123:12886–12898.
53. Ham S, Kim JH, Lee H, Cho MH. 2003. Correlation between electronic and molecular structure distortions and vibrational properties. II. Amide I modes of NMA-nD(2)O complexes. *J Chem Phys* 118:3491–3498.
54. Hamm P, Lim MH, Hochstrasser RM. 1998. Structure of the amide I band of peptides measured by femtosecond nonlinear-infrared spectroscopy. *J Phys Chem B* 102:6123–6138.
55. Kim YS, Hochstrasser RM. 2009. Applications of 2D IR spectroscopy to peptides, proteins, and hydrogen-bond dynamics. *J Phys Chem B* 113:8231–8251.
56. Decatur SM. 2006. Elucidation of residue-level structure and dynamics of polypeptides via isotope-edited infrared spectroscopy. *Acc Chem Res* 39:169–175.
57. Barber-Armstrong W, Donaldson T, Wijesooriya H, Silva RAGD, Decatur SM. 2004. Empirical relationships between isotope-edited IR spectra and helix geometry in model peptides. *J Am Chem Soc* 126:2339–2345.
58. Fang C, Hochstrasser RM. 2005. Two-dimensional infrared spectra of the C-13=O-18 isotopomers of alanine residues in an alpha-helix. *J Phys Chem B* 109:18652–18663.
59. Bandekar J. 1992. Amide modes and protein conformation. *Biochim Biophys Acta* 1120:123–143.
60. Barth A. 2000. The infrared absorption of amino acid side chains. *Prog Biophys Mol Bio* 74:141–173.
61. Nagarajan S et al. 2011. Differential ordering of the protein backbone and side chains during protein folding revealed by site-specific recombinant infrared probes. *J Am Chem Soc* 133:20335–20340.
62. Lindquist BA, Furse KE, Corcelli SA. 2009. Nitrile groups as vibrational probes of biomolecular structure and dynamics: An overview. *Phys Chem Chem Phys* 11:8119–8132.
63. Oh KI et al. 2008. Nitrile and thiocyanate IR probes: Molecular dynamics simulation studies. *J Chem Phys* 128:154504.
64. Choi JH, Oh KI, Cho MH. 2008. Azido-derivatized compounds as IR probes of local electrostatic environment: Theoretical studies. *J Chem Phys* 129:174512.
65. Thielges MC et al. 2011. Two-dimensional IR spectroscopy of protein dynamics using two vibrational labels: A site-specific genetically encoded unnatural amino acid and an active site ligand. *J Phys Chem B* 115:11294–11304.
66. Hendrickson WA, Horton JR, Lemaster DM. 1990. Selenomethionyl proteins produced for analysis by multiwavelength anomalous diffraction (Mad)—A vehicle for direct determination of 3-dimensional structure. *Embo J* 9:1665–1672.
67. Wang L, Xie J, Schultz PG. 2006. Expanding the genetic code. *Annu Rev Bioph Biom* 35:225–249.
68. Hill JR et al. 1994. Vibrational dynamics of carbon-monoxide at the active-site of myoglobin—Picosecond infrared free-electron laser pump-probe experiments. *J Phys Chem USA* 98:11213–11219.
69. Fayer MD. 2001. Fast protein dynamics probed with infrared vibrational echo experiments. *Annu Rev Phys Chem* 52:315–356.
70. King JT, Arthur EJ, Brooks CL, Kubarych KJ. 2012. Site-specific hydration dynamics of globular proteins and the role of constrained water in solvent exchange with amphiphilic cosolvents. *J Phys Chem B* 116:5604–5611.
71. Hayashi T, Zhuang W, Mukamel S. 2005. Electrostatic DFT map for the complete vibrational amide band of NMA. *J Phys Chem A* 109:9747–9759.
72. Jansen TLC, Knoester J. 2009. Waiting time dynamics in two-dimensional infrared spectroscopy. *Acc Chem Res* 42:1405–1411.
73. Mukamel S. 1999. *Principles of Nonlinear Optical Spectroscopy* (Oxford University Press, USA).
74. Petersen PB, Tokmakoff A. 2010. Source for ultrafast continuum infrared and terahertz radiation. *Opt Lett* 35:1962–1964.
75. Baiz CR, Kubarych KJ. 2011. Ultrabroadband detection of a mid-IR continuum by chirped-pulse upconversion. *Opt Lett* 36:187–189.

76. Zanni MT, Ge NH, Kim YS, Hochstrasser RM. 2001. Two-dimensional IR spectroscopy can be designed to eliminate the diagonal peaks and expose only the crosspeaks needed for structure determination. *Proc Natl Acad Sci* 98:11265–11265.
77. Khalil M, Demirdoven N, Tokmakoff A. 2003. Coherent 2D IR spectroscopy: Molecular structure and dynamics in solution. *J Phys Chem A* 107:5258–5279.
78. Roberts ST, Loparo JJ, Tokmakoff A. 2006. Characterization of spectral diffusion from two-dimensional line shapes. *J Chem Phys* 125:084502.
79. Gruebele M. 1999. The fast protein folding problem. *Annu Rev Phys Chem* 50:485–516.
80. Bredenbeck J, Hamm P. 2007. Transient 2D IR spectroscopy: Towards a molecular movie. *Chimia* 61:45–46.
81. Chung HS, Khalil M, Smith AW, Tokmakoff A. 2007. Transient two-dimensional IR spectrometer for probing nanosecond temperature-jump kinetics. *Rev Sci Instrum* 78:063101.
82. Callender R, Dyer RB. 2002. Probing protein dynamics using temperature jump relaxation spectroscopy. *Curr Opin Struc Biol* 12:628–633.
83. Dumont C, Emilsson T, Gruebele M. 2009. Reaching the protein folding speed limit with large, sub-microsecond pressure jumps. *Nat Methods* 6:515–U570.
84. Gutman M, Nachliel E. 1990. The dynamic aspects of proton-transfer processes. *Biochim Biophys Acta* 1015:391–414.
85. Fabian H, Naumann D. 2004. Methods to study protein folding by stopped-flow FT-IR. *Methods* 34: 28–40.
86. Baiz C, Nee M, McCanne R, Kubarych K. 2008. Ultrafast nonequilibrium Fourier-transform two-dimensional infrared spectroscopy. *Opt Lett* 33:2533–2535.
87. Xiong W et al. 2009. Transient 2D IR spectroscopy of charge injection in dye-sensitized nanocrystalline thin films. *J Am Chem Soc* 131:18040
88. Jones KC, Ganim Z, Peng CS, Tokmakoff A. 2012. Transient two-dimensional spectroscopy with linear absorption corrections applied to temperature-jump two-dimensional infrared. *J Opt Soc Am B* 29: 118–129.
89. Jones KC, Ganim Z, Tokmakoff A. 2009. Heterodyne-detected dispersed vibrational echo spectroscopy. *J Phys Chem A* 113:14060–14066.
90. Marecek J et al. 2007. A simple and economical method for the production of C-13, O-18-labeled Fmoc-amino acids with high levels of enrichment: Applications to isotope-edited IR studies of proteins. *Org Lett* 9:4935–4937.
91. Bouř P, Keiderling TA. 2003. Empirical modeling of the peptide amide I band IR intensity in water solution. *J Chem Phys* 119:11253–11262.
92. Watson TM, Hirst JD. 2005. Theoretical studies of the amide I vibrational frequencies of [Leu]-enkephalin. *Mol Phys* 103:1531–1546.
93. Jansen TL, Dijkstra AG, Watson TM, Hirst JD, Knoester J. 2006. Modeling the amide I bands of small peptides. *J Chem Phys* 125:044312.
94. Cai KC, Han C, Wang JP. 2009. Molecular mechanics force field-based map for peptide amide-I mode in solution and its application to alanine di- and tripeptides. *Phys Chem Chem Phys* 11:9149–9159.
95. Lin YS, Shorb JM, Mukherjee P, Zanni MT, Skinner JL. 2009. Empirical amide I vibrational frequency map: Application to 2D IR line shapes for isotope-edited membrane peptide bundles. *J Phys Chem B* 113:592–602.
96. Roy S et al. 2011. Solvent and conformation dependence of amide I vibrations in peptides and proteins containing proline. *J Chem Phys* 135:234507.
97. Van Amerongen H, Valkunas L, Van Grondelle R. 2000. *Photosynthetic Excitons* (World Scientific), ISBN 9810232802.
98. Moffitt W. 1956. Optical rotatory dispersion of helical polymers. *J Chem Phys* 25:467–478.
99. Miyazawa T. 1960. Perturbation treatment of the characteristic vibrations of polypeptide chains in various configurations. *J Chem Phys* 32:1647–1652.
100. Krimm S, Abe Y. 1972. Intermolecular interaction effects in amide I vibrations of beta polypeptides. *Proc Natl Acad Sci USA* 69:2788–2792.
101. Hamm P, Woutersen S. 2002. Coupling of the amide I modes of the glycine dipeptide. *B Chem Soc Jpn* 75:985–988.
102. Sung J, Silbey RJ. 2001. Four wave mixing spectroscopy for a multilevel system. *J Chem Phys* 115: 9266–9287.
103. Khalil M, Demirdoven N, Tokmakoff A. 2003. Obtaining absorptive line shapes in two-dimensional infrared vibrational correlation spectra. *Phys Rev Lett* 90: 047401.
104. Golonzka O, Tokmakoff A. 2001. Polarization-selective third-order spectroscopy of coupled vibronic states. *J Chem Phys* 115:297–309.

105. Jonas DM. 2003. Two-dimensional femtosecond spectroscopy. *Annu Rev Phys Chem* 54:425–463.
106. Faeder SMG, Jonas DM. 1999. Two-dimensional electronic correlation and relaxation spectra: Theory and model calculations. *J Phys Chem A* 103:10489–10505.
107. Torii H. 2006. Effects of intermolecular vibrational coupling and liquid dynamics on the polarized Raman and two-dimensional infrared spectral profiles of liquid N,N-dimethylformamide analyzed with a time-domain computational method. *J Phys Chem A* 110:4822–4832.
108. Jansen TL, Knoester J. 2006. Nonadiabatic effects in the two-dimensional infrared spectra of peptides: Application to alanine dipeptide. *J Phys Chem B* 110:22910–22916.
109. Gaigeot MP. 2010. Theoretical spectroscopy of floppy peptides at room temperature. A DFTMD perspective: Gas and aqueous phase. *Phys Chem Chem Phys* 12:3336–3359.
110. Ingrosso F, Monard G, Farag MH, Bastida A, Ruiz-Lopez MF. 2011. Importance of polarization and charge transfer effects to model the infrared spectra of peptides in solution. *J Chem Theory Comput* 7: 1840–1849.
111. Jeon J, Cho M. 2010. Direct quantum mechanical/molecular mechanical simulations of two-dimensional vibrational responses: N-methylacetamide in water. *New J Phys* 12:065001.
112. Gorbunov RD, Kosov DS, Stock G. 2005. Ab initio-based exciton model of amide I vibrations in peptides: Definition, conformational dependence, and transferability. *J Chem Phys* 122:224904.
113. Byler DM, Susi H. 1986. Examination of the secondary structure of proteins by deconvolved FTIR spectra. *Biopolymers* 25:469–487.
114. DeFlores LP, Ganim Z, Ackley SF, Chung HS, Tokmakoff A. 2006. The anharmonic vibrational potential and relaxation pathways of the amide I and II modes of N-methylacetamide. *J Phys Chem B* 110: 18973–18980.
115. Pande V. 2011. Folding@home: Sustained petaflops for production calculations today, exaflops soon? *Abstr Pap Am Chem S* 242.
116. Pande V. 2010. Simulating protein folding *in vitro* and in vivo. *Biochem Cell Biol* 88:409–409.
117. Lindorff-Larsen K, Piana S, Dror RO, Shaw DE. 2011. How fast-folding proteins fold. *Science* 334: 517–520.
118. Chung HS, Ganim Z, Jones KC, Tokmakoff A. 2007. Transient 2D IR spectroscopy of ubiquitin unfolding dynamics. *Proc Natl Acad Sci USA* 104:14237–14242.
119. Cochran AG, Skelton NJ, Starovasnik MA. 2001. Tryptophan zippers: Stable, monomeric beta-hairpins. *Proc Natl Acad Sci USA* 98:5578–5583.
120. Mukherjee P, Kass I, Arkin IT, Zanni MT. 2006. Structural disorder of the CD3 xi transmembrane domain studied with 2D IR spectroscopy and molecular dynamics simulations. *J Phys Chem B* 110:24740–24749.
121. Remorino A, Korendovych IV, Wu YB, DeGrado WF, Hochstrasser RM. 2011. Residue-specific vibrational echoes yield 3D structures of a transmembrane helix dimer. *Science* 332:1206–1209.
122. Hamm P, Helbing J, Bredenbeck J. 2008. Two-dimensional infrared spectroscopy of photoswitchable peptides. *Annu Rev Phys Chem* 59:291–317.
123. Chung HS, Tokmakoff A. 2008. Temperature-dependent downhill unfolding of ubiquitin. II. Modeling the free energy surface. *Proteins* 72:488–497.
124. Chung HS, Tokmakoff A. 2008. Temperature-dependent downhill unfolding of ubiquitin. I. Nanosecond-to-millisecond resolved nonlinear infrared spectroscopy. *Proteins* 72:474–487.
125. Bredenbeck J, Helbing J, Nienhaus K, Nienhaus GU, Hamm P. 2007. Protein ligand migration mapped by nonequilibrium 2D IR exchange spectroscopy. *Proc Natl Acad Sci USA* 104:14243–14248.
126. Finkelstein IJ, Ishikawa H, Kim S, Massari AM, Fayer MD. 2007. Substrate binding and protein conformational dynamics measured by 2D IR vibrational echo spectroscopy. *Proc Natl Acad Sci USA* 104: 2637–2642.
127. Bandaria JN et al. 2010. Characterizing the dynamics of functionally relevant complexes of formate dehydrogenase. *Proc Natl Acad Sci USA* 107:17974–17979.
128. Chung JK, Thielges MC, Fayer MD. 2011. Dynamics of the folded and unfolded villin headpiece (HP35) measured with ultrafast 2D IR vibrational echo spectroscopy. *Proc Natl Acad Sci USA* 108:3578–3583.
129. Urbanek DC, Vorobyev DY, Serrano AL, Gai F, Hochstrasser RM. 2010. The two-dimensional vibrational echo of a nitrile probe of the Villin HP35 protein. *J Phys Chem Lett* 1:3311–3315.
130. Fang C et al. 2008. Two-dimensional infrared spectra reveal relaxation of the nonnucleoside inhibitor TMC278 complexed with HIV-1 reverse transcriptase. *Proc Natl Acad Sci USA* 105:1472–1477.
131. Ganim Z, Jones KC, Tokmakoff A. 2010. Insulin dimer dissociation and unfolding revealed by amide I two-dimensional infrared spectroscopy. *Phys Chem Chem Phys* 12:3579–3588.
132. Ganim Z, Tokmakoff A, Vaziri A. 2011. Vibrational excitons in ionophores: Experimental probes for quantum coherence-assisted ion transport and selectivity in ion channels. *New J Phys* 13:113030.
133. Torii H, Tasumi M. 1998. Ab initio molecular orbital study of the amide I vibrational interactions between the peptide groups in di- and tripeptides and considerations on the conformation of the extended helix. *J Raman Spectrosc* 29:81–86.

13 Quasi-Particle Approach to 2D IR Spectra of Vibrational Excitons in Biomolecules

Molecular Dynamics versus Stochastic Simulation Protocols

Cyril Falvo, František Šanda, and Shaul Mukamel

CONTENTS

13.1 INTRODUCTION

The idea that nearly degenerate coupled vibrations can be described by a delocalized excitation within the system (an exciton) first emerged in the context of molecular crystals [1]. A similar picture applies to the amide vibrations in proteins where delocalization among C=O stretching modes of nearby amide-I groups results in disordered exciton dynamics [2,3]. This approach was used to link the molecular conformation with infrared absorption and Raman spectroscopy [4,5]; the band frequencies are given by the eigenvalues of the exciton Hamiltonian and intensities are determined by the transition dipole moments (absorption) or polarizabilities (Raman).

Exciton delocalization in biomolecules depends primarily on two factors: the hydrogen-bonding structure and the interaction between the transition dipoles [2,3,6]. The first directly modifies the frequencies of the local vibrations (diagonal disorder) and the second modifies the excitonic coupling [7] (off-diagonal disorder). Early studies emphasized the importance of static disorder for the dynamics of vibrational excitons in proteins. In these studies, the time variation of the disorder was neglected.

The advent of novel femtosecond spectroscopies had made it possible to probe the dynamics of excitons in fluctuating environments. Pump-probe and two-dimensional infrared (2D IR) spectroscopy measurements have been performed on systems such as peptides, proteins, membrane systems, and hydrogen-bonded liquids [8–14]. Bath-induced perturbations have long been studied in electronic excitons. They were extended to vibrations and applied to study solvation and hydrogen bond dynamics [15–17] as well as intramolecular dynamics of water in the condensed phase [18–20]. Two theoretical approaches have been used to simulate the effects of classical bath: a microscopic description based on molecular dynamics (MD) and phenomenological stochastic dynamics.

Early models considered small Gaussian adiabatic perturbations and nonadiabatic couplings between vibrational states (curve crossing) were neglected. These fluctuations were treated either as an extension of Kubo's stochastic model through the cumulant Gaussian fluctuation (CGF) or via decoherence rates incorporated in the nonlinear exciton equations (NEE) [21,22]. CGF is used in the sum-over-states (SOS) picture where the signals are expressed in terms of transitions between the global eigenstates of the exciton Hamiltonian. NEE, a quasi-particle approach (QP), expresses the signals as scattering of excitons, thereby avoiding the explicit calculation of multiexciton eigenstates. The QP approach has several advantages. First, it does not suffer from cancellations of terms (Liouville space pathways) occurring in the SOS approach. For N chromophores, each pathway contributes to the third-order signal scales as $\sim N^2$ whereas the total signal only scales as $\sim N$, for large N [23]. This stems from the fact that only interactions with the laser fields occurring within a coherence vibrational size contribute to the signal. Instead, in the quasi-particle picture, this cancellation is naturally built in and the signal is computed directly within the coherence size, which greatly simplifies its calculation and interpretation. A second advantage of the QP picture is that it suggests new approximations to treat large system such as the mean-field approximation [22].

Conformational or hydrogen-bonding dynamics or strongly fluctuating molecular vibrations cannot be modeled with the assumptions of the CGF–SOS or NEE models. A more complete description to treat these systems must include finite-timescale fluctuations of eigenstates and non-Gaussian types of perturbation. The time-average approximation (TAA) introduces a free parameter to separate slow and fast bath fluctuations and creates an interpolation between uncoupled chromophore with fast fluctuations and coupled chromophore in the static fluctuation limit [24,25]. An arbitrary fluctuation timescale has been incorporated into the SOS approach through direct numerical integration of the Schrödinger equation (NISE) [26], or by combining quantum (system) and stochastic (bath) dynamics into the stochastic Liouville equations (SLE) [27,28].

While the two methods are formally equivalent, NISE requires the generation of Hamiltonian trajectories by MD. The SLE, in contrast, requires identifying a few relevant Markovian collective coordinates as the main source of fluctuations. It is thus linked to phenomenological descriptions of bath.

Here, we survey the recent advances in the quasi-particle approach. Both the NISE [29] and the SLE [30] strategies can be adapted into the QP picture and applied to the microscopic and stochastic bath dynamics. In Section 13.2, we describe how the contributions of various Liouville space pathways for the third-order signals are combined in the quasi-particle formalism. In Section 13.3, we apply this formalism to a microscopic description of the bath. In Section 13.4, we apply the results of Section 13.2 to Markovian fluctuations. This is equivalent to the stochastic nonlinear exciton equations (SNEE), derived originally by adapting NEE to finite timescales [30]. These formal developments are illustrated by simulations of 2D IR spectra of water, amyloid fibrils, and solvent dynamics.

13.2 VIBRATIONAL HAMILTONIAN AND 2D IR SIGNALS

13.2.1 VIBRATIONAL EXCITON HAMILTONIAN

Assuming an adiabatic decoupling between the primary optically active vibrational modes and the remaining modes that act as a bath, the vibrational dynamics can be described by a fluctuating excitonic Hamiltonian.

$$H = \sum_{nm} h_{nm}(\mathbf{q}) b_n^{\dagger} b_m + \sum_{nmn'm'} U_{nmn'm'}(\mathbf{q}) b_n^{\dagger} b_m^{\dagger} b_{n'} b_{m'} + H_{\mathrm{B}}(\mathbf{q},\mathbf{p}), \tag{13.1}$$

where $b_n^{\dagger}$ and b_n are respectively the boson creation and annihilation operator of the n-th primary vibrational mode with the commutation relation $[b_n, b_m^{\dagger}] = \delta_{nm}$. $\mathbf{q} = \{q_i\}$ and $\mathbf{p} = \{p_i\}$ are the coordinates and momenta of the bath degrees of freedom. h_{nm} is the one-exciton Hamiltonian, $U_{nmn'm'}$ denotes the exciton–exciton anharmonic interactions, and H_{B} is the bath Hamiltonian. Both h_{nm} and $U_{nmn'm'}$ depend parameterically on the bath coordinates q_i. The diagonal and the off-diagonal parts of the one-exciton Hamiltonian represent the vibrational frequencies $h_{nn} = \hbar\omega_n$ and their vibrational couplings. The Hamiltonian (Equation 13.1) commutes with the exciton number operator $v = \sum_m b_m^{\dagger} b_m$, $[H,v] = 0$. Therefore, the exciton number is conserved, and the Hamiltonian is block diagonal into n-exciton blocks. Three blocks are relevant for third-order nonlinear spectroscopy: the ground, the one-exciton, and the two-exciton states. Let us first ignore the bath and separately diagonalize each block of the Hamiltonian for a system of N vibrational modes. The three blocks contain 1, N, and $N(N+1)/2$ levels and are depicted in Figure 13.1. The bath has several effects. First, the vibrational eigenvalues are modified, and level splittings turn each level into a continuum (or multiple continua if some bath coordinates jump between several wells). Second, if we describe the excitons by a fixed basis set (e.g., defined as the eigenstates for a given set of bath coordinates), then bath fluctuations will induce transport in that basis.

The interaction Hamiltonian between the vibrations and the optical field is

$$H'(t) = -\mathbf{E}(\mathbf{r},t) \cdot \mathbf{V}(\mathbf{q}), \tag{13.2}$$

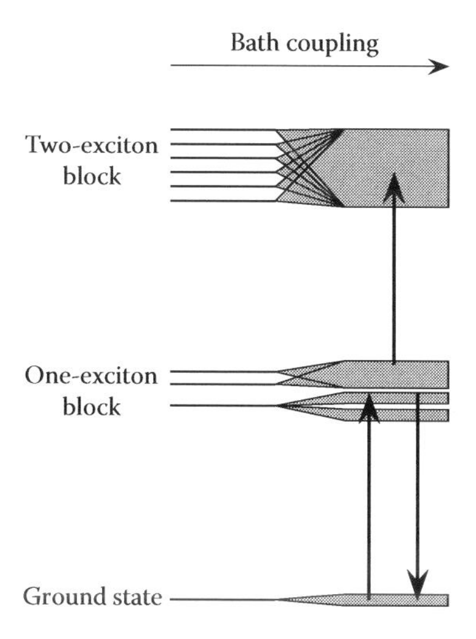

FIGURE 13.1 Energy level scheme for the vibrational exciton Hamiltonian coupled to a bath.

with $\mathbf{E}(\mathbf{r}, t)$ the optical field at the coordinate $\mathbf{r}$ and $\mathbf{V}(\mathbf{q})$ the dipole operator defined as

$$\mathbf{V}(\mathbf{q}) = \sum_m \boldsymbol{\mu}_m(\mathbf{q})\left(b_m + b_m^\dagger\right), \tag{13.3}$$

where $\boldsymbol{\mu}_m(\mathbf{q})$ are the transition dipoles. H' induces transitions between the different blocks. It can change the number of excitons one at a time $\Delta v = \pm 1$ as shown in Figure 13.1.

Calculating the eigenstates of the Hamiltonian H (Equation 13.1) for systems with a large number of bath coordinates is a difficult task, even when the bath dynamics are harmonic. Several approximate schemes have been used to solve the exciton dynamics for a harmonic bath [22,31–33]. Here, we do not assume a harmonic bath but make two key approximations:

- The bath is classical
- Bath dynamics are not affected by the state of the system

These conditions are expected to generally hold provided the bath is large and at high temperature. Mixed classical/quantum dynamics (QM/MM) algorithms where some degrees of freedom are quantum and the others are classical have been widely used. Examples are the Ehrenfest approach, hopping surface method, and the Bohmian trajectories [34–36]. By neglecting the effect of the quantum system on the classical bath, no further treatment of the classical/quantum interaction is necessary and the classical bath dynamics is simply given by Hamilton's equations

$$\dot{q}_i = \frac{\partial H_{\mathrm{B}}}{\partial p_i}, \tag{13.4}$$

$$\dot{p}_i = -\frac{\partial H_{\mathrm{B}}}{\partial q_i}. \tag{13.5}$$

The system's dynamics is described by the Schrödinger equation

$$i\hbar \frac{\mathrm{d}}{\mathrm{d}t} |\Psi\rangle = H_{\mathrm{S}}(t)\, |\Psi\rangle, \tag{13.6}$$

with the fluctuating Hamiltonian $H_{\mathrm{S}}(t)$

$$H_{\mathrm{S}}(t) = \sum_{nm} h_{nm}(\mathbf{q}(t)) b_n^\dagger b_m + \sum_{nmn'm'} U_{nmn'm'}(\mathbf{q}(t)) b_n^\dagger b_m^\dagger b_{n'} b_{m'}, \tag{13.7}$$

where $q_i(t)$ and $p_i(t)$ are the trajectories obtained from Hamilton's Equations 13.4 and 13.5. The bath coordinates can be viewed as external sources that act on the excitonic system. Since $H_{\mathrm{S}}(t)$ is explicitly time dependent, the system energy is not conserved. The model can account for the key effects of the bath: level splittings are modeled by the fluctuations of the eigenvalues of H_{S} and population transport is caused by fluctuations of the eigenstates. Equation 13.2 is similarly given by

$$H'(t) = -\sum_m \mathbf{E}(\mathbf{r},t) \cdot \boldsymbol{\mu}_m(\mathbf{q}(t))\left(b_m + b_m^\dagger\right), \tag{13.8}$$

where $\boldsymbol{\mu}_m$ are now explicitly time dependent.

13.2.2 Nonlinear Optical Response

In this section, we apply the vibrational exciton Hamiltonian described above to calculate four wave mixing signals. The experiment is depicted in Figure 13.2. Three laser pulses with wavevectors $\mathbf{k}_1$, $\mathbf{k}_2$, and $\mathbf{k}_3$ interact with the system to produce a nonlinear polarization. Heterodyne detection of this polarization is performed using a fourth pulse in the direction $\mathbf{k}_4$. The formalism can also describe homodyne experiments. These will not be considered here.

The third-order nonlinear polarization $\mathbf{P}^{(3)}(\mathbf{r},t)$ induced by the interaction between the system and the laser pulses can be expressed as

$$P^{(3)}_{\nu_4}(\mathbf{r},\tau_4) = \int_{-\infty}^{\tau_4} d\tau_3 \int_{-\infty}^{\tau_3} d\tau_2 \int_{-\infty}^{\tau_2} d\tau_1\, R^{(3)}_{\nu_4\nu_3\nu_2\nu_1}(\tau_4,\tau_3,\tau_2,\tau_1) E_{\nu_3}(\mathbf{r},\tau_3) E_{\nu_2}(\mathbf{r},\tau_2) E_{\nu_1}(\mathbf{r},\tau_1), \quad (13.9)$$

where ν_i are the Cartesian polarization indices and the third-order response function [31] is a fourth-order tensor given by

$$R^{(3)}_{\nu_4\nu_3\nu_2\nu_1}(\tau_4,\tau_3,\tau_2,\tau_1) = \left(\frac{i}{\hbar}\right)^3 \left\langle [[[V_{\nu_4}(\tau_4), V_{\nu_3}(\tau_3)], V_{\nu_2}(\tau_2)], V_{\nu_1}(\tau_1)] \right\rangle. \quad (13.10)$$

In Equation 13.10, $\langle \cdots \rangle$ denotes an ensemble average over the quantum states of the Hamiltonian H_S as well as the classical bath trajectories.

We consider vibrations with frequencies much higher than the temperature (e.g., 1600 cm^{-1} for the amide-I vibrations) so that the system is initially in the vibrational ground state $|g\rangle$. Averaging over bath trajectories involves path integration where the average of an operator $A(t)$ is written as

$$\langle A(t) \rangle = \int \mathcal{D}\mathbf{q}(t) \langle g | A(\mathbf{q}(t)) | g \rangle. \quad (13.11)$$

The path integral can be calculated by dividing the time variable into small segments

$$\int \mathcal{D}\mathbf{q}(t) = \lim_{n\to\infty} \prod_{p=1}^{n} \left(\int d\mathbf{q}(t_p) \right) P(\mathbf{q}(t_n), \ldots, \mathbf{q}(t_1)), \quad (13.12)$$

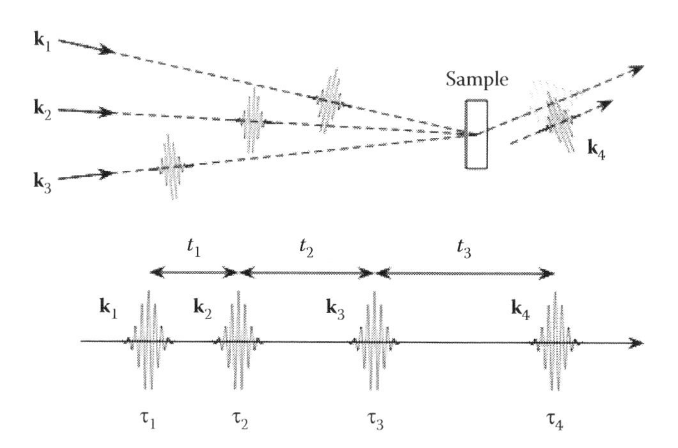

FIGURE 13.2 Pulse configuration for a coherent four wave mixing experiment.

where $P(\mathbf{q}(t_n), \ldots, \mathbf{q}(t_1))$ correspond to the probabilities of trajectories. The three commutators in Equation 13.10 yield eight Liouville space pathways and each nonlinear technique that corresponds to the measurement of the nonlinear polarization in a specific direction selects a subgroup of pathways [22]. The optical field is decomposed into three incoming pulses with wavevectors $\mathbf{k}_1$, $\mathbf{k}_2$, and $\mathbf{k}_3$

$$\mathbf{E}(\mathbf{r},t) = \sum_{\ell=1}^{3} \mathbf{E}_\ell(t - t_\ell)e^{i\mathbf{k}_\ell \mathbf{r} - i\omega_\ell(t-t_\ell)} + \text{c.c.}, \tag{13.13}$$

where $\mathbf{E}_\ell$ and ω_ℓ are respectively the envelope and the central frequency of the ℓ-th pulse and t_ℓ is its arrival time. We assume that the pulses envelopes are short compared to the vibrational dynamics. The signal is generated in three phase-matching directions: $\mathbf{k}_\text{I} = -\mathbf{k}_1 + \mathbf{k}_2 + \mathbf{k}_3$, $\mathbf{k}_\text{II} = \mathbf{k}_1 - \mathbf{k}_2 + \mathbf{k}_3$, and $\mathbf{k}_\text{III} = \mathbf{k}_1 + \mathbf{k}_2 - \mathbf{k}_3$. Each constitutes a distinct technique with a different window into the vibrational dynamics. In particular, the $\mathbf{k}_\text{I}$ technique is also known as photon-echo spectroscopy.

To better represent the response, we separate the dipole into its raising and lowering components

$$\mu_{n_1}^-(t) = \mu_{n_1}(\mathbf{q}(t))b_{n_1}, \tag{13.14}$$

$$\mu_{n_1}^+(t) = \mu_{n_1}(\mathbf{q}(t))b_{n_1}^\dagger. \tag{13.15}$$

Since the system is initially in the vibrational ground state $|g\rangle$, and using the rotating wave approximation, we find that three pathways contribute to the photon-echo signal $\mathbf{k}_\text{I}$ (Figure 13.3).

$$\begin{aligned} R^{\mathbf{k}_\text{I}}_{\nu_4\nu_3\nu_2\nu_1}(\tau_4,\tau_3,\tau_2,\tau_1) &= \left(\frac{i}{\hbar}\right)^3 \\ &\times \sum_{n_1n_2n_3n_4} \Big\{ \langle \mu^-_{n_1;\nu_1}(\tau_1)\mathcal{U}(\tau_1,\tau_2)\mu^+_{n_2;\nu_2}(\tau_2)\mathcal{U}(\tau_2,\tau_4)\mu^-_{n_4;\nu_4}(\tau_4)\mathcal{U}(\tau_4,\tau_3)\mu^+_{n_3;\nu_3}(\tau_3)\rangle \\ &+ \langle \mu^-_{n_1;\nu_1}(\tau_1)\mathcal{U}(\tau_1,\tau_3)\mu^+_{n_3;\nu_3}(\tau_3)\mathcal{U}(\tau_3,\tau_4)\mu^-_{n_4;\nu_4}(\tau_4)\mathcal{U}(\tau_4,\tau_2)\mu^+_{n_2;\nu_2}(\tau_2)\rangle \\ &- \langle \mu^-_{n_1;\nu_1}(\tau_1)\mathcal{U}(\tau_1,\tau_4)\mu^-_{n_4;\nu_4}(\tau_4)\mathcal{U}(\tau_4,\tau_3)\mu^+_{n_3;\nu_3}(\tau_3)\mathcal{U}(\tau_3,\tau_2)\mu^+_{n_2;\nu_2}(\tau_2)\rangle \Big\}, \end{aligned} \tag{13.16}$$

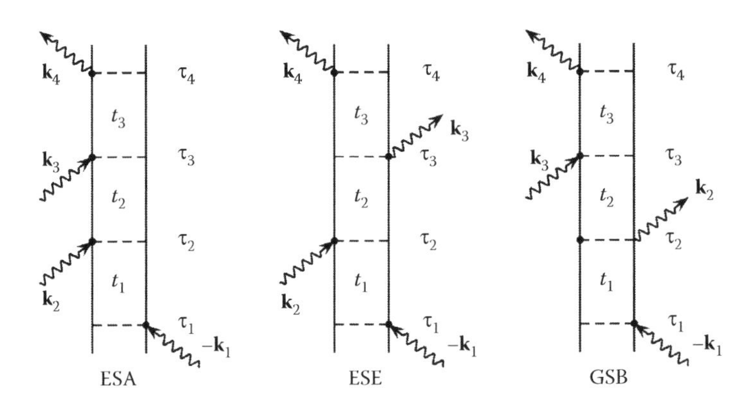

FIGURE 13.3 Ladder diagrams for the three elementary contributions (Liouville space pathways) to the $\mathbf{k}_\text{I}$ signal: the excited-state absorption (ESA), the excited-state emission (ESE), and the ground-state bleaching (GSB).

where $\mathcal{U}(\tau_2,\tau_1)$ is the evolution operator

$$\mathcal{U}(\tau_2,\tau_1) = \exp_+\left(-\frac{i}{\hbar}\int_{\tau_1}^{\tau_2} H_S(\tau)\mathrm{d}\tau\right), \tag{13.17}$$

and $\exp_+$ is the time-ordered exponential. Because $H_S(t)$ conserves the number of excitons v, the system remains in a given block (ground-state, one-exciton block, and two-exciton block) during the evolution periods between interactions with the laser field. To describe the forward evolution in time, we introduce the one-exciton Green's functions

$$G_{n_2,n_1}(\tau_2,\tau_1) = \theta(\tau_2-\tau_1)\langle g\,|\,b_{n_2}\mathcal{U}(\tau_2,\tau_1)b^\dagger_{n_1}\,|\,g\rangle, \tag{13.18}$$

and the two-exciton Green's functions

$$\mathcal{G}_{n_2m_2,n_1m_1}(\tau_2,\tau_1) = \theta(\tau_2-\tau_1)\langle g|b_{n_2}b_{m_2}\mathcal{U}(\tau_2,\tau_1)b^\dagger_{n_1}b^\dagger_{m_1}|g\rangle. \tag{13.19}$$

The signal is then given as

$$\begin{aligned} R^{\mathbf{k}_\mathrm{I}}_{\nu_4\nu_3\nu_2\nu_1}(\tau_4,\tau_3,\tau_2,\tau_1) &= \left(\frac{i}{\hbar}\right)^3 \sum_{n_1n_2n_3n_4} \mu_{n_1;\nu_1}(\tau_1)\mu_{n_2;\nu_2}(\tau_2)\mu_{n_3;\nu_3}(\tau_3)\mu_{n_4;\nu_4}(\tau_4) \\ &\times \begin{bmatrix} G_{n_4,n_3}(\tau_4,\tau_3)G^*_{n_2,n_1}(\tau_2,\tau_1) + G_{n_4,n_2}(\tau_4,\tau_2)G^*_{n_3,n_1}(\tau_3,\tau_1) \\ -\sum_{m_1m_2}\mathcal{G}_{n_4m_1,n_3m_2}(\tau_4,\tau_3)G_{m_2,n_2}(\tau_3,\tau_2)G^*_{m_1,n_1}(\tau_4,\tau_1) \end{bmatrix}. \end{aligned} \tag{13.20}$$

To simplify the notation, we have written $\mu_{n;\nu}(\tau) = \mu_{n;\nu}(\mathbf{q}(\tau))$ and omitted the integration over the bath trajectories. The three contributions to the signal in Equation 13.20 are known respectively as ground-state bleaching (GSB), excited-state emission (ESE), and excited-state absorption (ESA). They are represented by ladder diagrams as shown in Figure 13.3. Similarly, the other signals can be represented by a sum of three ($\mathbf{k}_\mathrm{II}$) and two diagrams ($\mathbf{k}_\mathrm{III}$) [22].

Let us first analyze the simple case of a harmonic vibrational system by neglecting the anharmonic couplings $U_{nmn'm'} = 0$. The evolution operator is then written as

$$\mathcal{U}(\tau_2,\tau_1) = \mathcal{U}^{(0)}(\tau_2,\tau_1) = \exp_+\left(-\frac{i}{\hbar}\int_{\tau_1}^{\tau_2} H_S^{(0)}(\tau)\mathrm{d}\tau\right), \tag{13.21}$$

with the harmonic Hamiltonian $H_S^{(0)}(\tau) = \sum_{nm} h_{nm}(\mathbf{q}(t))b^\dagger_n b_m$. In this situation, the two-exciton Green's function can be factorized into a symmetrized product of one-exciton Green's functions since the two excitons are free bosons and evolve independently

$$\mathcal{G}^{(0)}_{n_2m_2,n_1m_1}(\tau_2,\tau_1) = G_{n_2,n_1}(\tau_2,\tau_1)G_{m_2,m_1}(\tau_2,\tau_1) + G_{n_2,m_1}(\tau_2,\tau_1)G_{m_2,n_1}(\tau_2,\tau_1). \tag{13.22}$$

By inserting $\mathcal{G}^{(0)}_{n_2m_2,n_1m_1}(\tau_2,\tau_1)$ and using the relation

$$G_{ij}(\tau_i,\tau_j) = \sum_k G_{ik}(\tau_i,s)G_{kj}(s,\tau_j), \tag{13.23}$$

where s is an arbitrary time between τ_1 and τ_2, we find that the $\mathbf{k}_\mathrm{I}$ response function vanishes. This should not be a surprise since the response of a harmonic system linearly coupled to the field is known to be linear. This shows that harmonic vibrations, even when they are anharmonically coupled to external coordinates (through fluctuations of the harmonic Hamiltonian) do not show a nonlinear response. This is due to the fact that the bath is not affected by the system. Otherwise, an effective anharmonic interaction between the excitons will result in a nonvanishing nonlinear optical response [37,38]. This effect is quite general and a similar cancellation occurs for the $\mathbf{k}_\mathrm{II}$ and $\mathbf{k}_\mathrm{III}$ signals.

Including the anharmonicity is thus essential for creating the nonlinear optical response. We now turn to the general anharmonic case $U_{nmn'm'} \neq 0$. We shall write the two-exciton Green's function using the two-particle Dyson equation that is exact and includes the anharmonicity explicitly

$$\mathcal{G}_{n_2m_2,n_1m_1}(\tau_2,\tau_1) = \mathcal{G}^{(0)}_{n_2m_2,n_1m_1}(\tau_2,\tau_1) - \frac{i}{\hbar}\sum_{n_3m_3n_4m_4}\int_{\tau_1}^{\tau_2} ds\, \mathcal{G}^{(0)}_{n_2m_2,n_4m_4}(\tau_2,s)U_{n_4m_4n_3m_3}(s)\mathcal{G}_{n_3m_3,n_1m_1}(s,\tau_1). \quad (13.24)$$

Substituting Equation 13.24 in Equation 13.20 and noting the cancellation of the $\mathcal{G}^{(0)}$ term, we find that the $\mathbf{k}_\mathrm{I}$ response function is given by a single term

$$R^{\mathbf{k}_\mathrm{I}}_{\nu_4\nu_3\nu_2\nu_1}(\tau_4,\tau_3,\tau_2,\tau_1) = 2\left(\frac{i}{\hbar}\right)^4 \sum_{n_1n_2n_3n_4}\sum_{m_1m_2m_3m_4p_2} \mu_{n_1;\nu_1}(\tau_1)\mu_{n_2;\nu_2}(\tau_2)\mu_{n_3;\nu_3}(\tau_3)\mu_{n_4;\nu_4}(\tau_4)$$
$$\times \int_{\tau_3}^{\tau_4} ds\, G_{n_4m_4}(\tau_4,s)\,U_{m_4m_1m_3m_2}(s)\mathcal{G}_{m_3m_2,n_3p_2}(s,\tau_3)G_{p_2,n_2}(\tau_3,\tau_2)G^*_{m_1,n_1}(s,\tau_1). \quad (13.25)$$

Expressions for the $\mathbf{k}_\mathrm{II}$ and $\mathbf{k}_\mathrm{III}$ signals may be derived by proceeding along similar lines [29].

$$R^{\mathbf{k}_\mathrm{II}}_{\nu_4\nu_3\nu_2\nu_1}(\tau_4,\tau_3,\tau_2,\tau_1) = 2\left(\frac{i}{\hbar}\right)^4 \sum_{n_1n_2n_3n_4}\sum_{m_1m_2m_3m_4p_1} \mu_{n_1;\nu_1}(\tau_1)\mu_{n_2;\nu_2}(\tau_2)\mu_{n_3;\nu_3}(\tau_3)\mu_{n_4;\nu_4}(\tau_4)$$
$$\times \int_{\tau_3}^{\tau_4} ds\, G_{n_4m_4}(\tau_4,s)U_{m_4m_2m_3m_1}(s)\mathcal{G}_{m_3m_1,n_3p_1}(s,\tau_3)G_{p_1,n_1}(\tau_3,\tau_1)G^*_{m_2,n_2}(s,\tau_2), \quad (13.26)$$

and

$$R^{\mathbf{k}_\mathrm{III}}_{\nu_4\nu_3\nu_2\nu_1}(\tau_4,\tau_3,\tau_2,\tau_1) = 2\left(\frac{i}{\hbar}\right)^4 \sum_{n_1n_2n_3n_4}\sum_{m_1m_2m_3m_4} \mu_{n_1;\nu_1}(\tau_1)\mu_{n_2;\nu_2}(\tau_2)\mu_{n_3;\nu_3}(\tau_3)\mu_{n_4;\nu_4}(\tau_4)$$
$$\times \int_{\tau_3}^{\tau_4} ds\, G_{n_4m_4}(\tau_4,s)U_{m_4m_3m_2m_1}(s)G^*_{m_3,n_3}(s,\tau_3)\mathcal{G}_{m_2m_1,n_2p_1}(s,\tau_2)G_{p_1,n_1}(\tau_2,\tau_1). \quad (13.27)$$

The nonlinear response functions Equations 13.25 through 13.27 are now given by a time integral over the interval s between interactions with the $\mathbf{k}_3$ and $\mathbf{k}_4$ pulses. The exact cancellation of the harmonic part in Equation 13.20 for the $\mathbf{k}_\mathrm{I}$, $\mathbf{k}_\mathrm{II}$, and $\mathbf{k}_\mathrm{III}$ signals has now been accounted for. These expressions depend explicitly on the anharmonicity $U_{nmn'm'}$ to first order; higher orders enter through the two-exciton Green's function $\mathcal{G}(s,\tau_3)$. Each of the signals (Equations 13.25 through 13.27) can be represented by the single diagram as shown in Figure 13.4. Time evolves from bottom to top. A wavy line represents an interaction with the laser field. A solid line represents a one-exciton Green's function

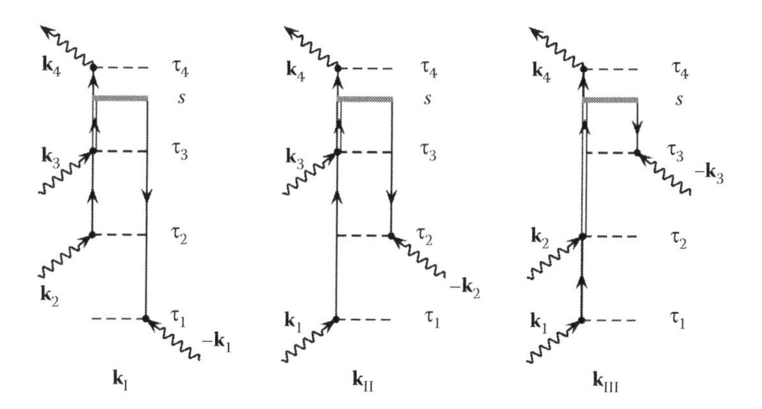

FIGURE 13.4 Diagrams representing the third-order signals $\mathbf{k}_\text{I}$, $\mathbf{k}_\text{II}$, and $\mathbf{k}_\text{III}$ in the quasi-particle representation.

propagating forward (upward arrow) or backward (downward arrow). The two-exciton Green's function $\mathcal{G}$ is represented by a double line. Finally, the gray band represents the region between times τ_3 and τ_4 where exciton scattering takes place. This scattering stems from the interaction $U_{nmn'm'}$ that splits the two-exciton Green's function into an exciton propagating forward from s to τ_4 and a second exciton propagating backward from s to τ_1 ($\mathbf{k}_\text{I}$ signal), τ_2 ($\mathbf{k}_\text{II}$ signal), or τ_3 ($\mathbf{k}_\text{III}$ signal).

Upon averaging over the bath trajectories, the nonlinear signals become invariant to translation of time and depends only on the three time intervals between the four interactions

$$\mathcal{R}_{\nu_4\nu_3\nu_2\nu_1}(t_3,t_2,t_1) = R_{\nu_4\nu_3\nu_2\nu_1}(t_3+t_2+t_1,t_2+t_1,t_1,0). \tag{13.28}$$

Multidimensional signals are more conveniently represented by frequency–frequency correlation plots. The $\mathbf{k}_\text{I}$ and $\mathbf{k}_\text{II}$ signals will be represented by a Fourier transform with respect to the times t_1 and t_3

$$\mathcal{R}^{\mathbf{k}_\text{I};\mathbf{k}_\text{II}}_{\nu_4\nu_3\nu_2\nu_1}(\Omega_1,t_2,\Omega_3) = \int_0^\infty\int_0^\infty \mathcal{R}^{\mathbf{k}_\text{I};\mathbf{k}_\text{II}}_{\nu_4\nu_3\nu_2\nu_1}(t_1,t_2,t_3)e^{i\Omega_1 t_1 + i\Omega_3 t_3}\mathrm{d}t_1\mathrm{d}t_3. \tag{13.29}$$

For $\mathbf{k}_\text{III}$, we perform a Fourier transform along t_2 and t_3

$$\mathcal{R}^{\mathbf{k}_\text{III}}_{\nu_4\nu_3\nu_2\nu_1}(t_1,\Omega_2,\Omega_3) = \int_0^\infty\int_0^\infty \mathcal{R}^{\mathbf{k}_\text{III}}_{\nu_4\nu_3\nu_2\nu_1}(t_1,t_2,t_3)e^{i\Omega_2 t_2 + i\Omega_3 t_3}\mathrm{d}t_2\mathrm{d}t_3. \tag{13.30}$$

Most systems probed with third-order coherent spectroscopy are centrosymmetric and a proper orientational averaging of the signal must be performed. If the system is slowly rotating compared to the timescale of the experiment, the average can be computed by a sum of different contributions, each corresponding to a specific laser polarization configuration [39,40]. If the system rotates on a fast timescale, a direct average over random orientations must be performed.

The present QP approach suggests new approximations. One example is the mean-field approximation. In that level of theory, we replace the two-exciton Green's function $\mathcal{G}_{n_2m_2,n_1m_1}(\tau_2,\tau_1)$ in Equations 13.25 through 13.27 by the harmonic Green's function $\mathcal{G}^{(0)}_{n_2m_2,n_1m_1}(\tau_2,\tau_1)$ (Equation 13.22). This greatly simplifies the calculation of nonlinear signals since it avoids the time-consuming calculation of the two-exciton Green's function and replaces it by a product of one-exciton Green's functions. The mean-field approximation is equivalent to expanding the nonlinear signal to first order in the anharmonicities.

Below, we present simulations of the third-order signals based on Equations 13.25 through 13.27. Two algorithms will be employed to describe the bath dynamics. The first uses MD simulations to compute bath trajectories. The second is based on the SLE that describe the evolution of the distributions of collective bath coordinates and allow an analytical integration of bath trajectories.

13.3 MOLECULAR DYNAMICS SIMULATIONS

We model the environment by directly solving Hamilton's Equations 13.4 and 13.5 for a set of coordinates q_i using a microscopic description of the bath. Assuming that the bath is at equilibrium, averaging over bath trajectories is replaced by an average over initial conditions, which, in practice, ultimately corresponds to a sum over a finite number N_p of initial conditions $(q_i^{(p)}(\tau_1), p_i^{(p)}(\tau_1))$,

$$\int \mathcal{D}\mathbf{q}(t) \rightarrow \lim_{T\to\infty} \frac{1}{2T} \int_{-T}^{T} d\tau_1 \rightarrow \frac{1}{N_p} \sum_{p=1}^{N_p}. \tag{13.31}$$

MD simulations are commonly used to model the vibrational dynamics of complex systems, including liquids and proteins [41,42]. In these simulations, molecular interactions are parameterized through force fields such as CHARMM [43] or AMBER [44] designed to reproduce the equilibrium properties as well as the slow motions. An ensemble of trajectories $q_i(t)$ is then obtained. The fluctuating Hamiltonian $H_S(t)$ and the transition dipole $\boldsymbol{\mu}_n(t)$ that govern the interaction with the laser pulses must be parameterized in terms of these coordinates $q_i(t)$.

13.3.1 Hamiltonian Parameterization

Many approaches have been used to model the influence of the environment on the frequency ω_n of an isolated vibrational mode. It appears that the effect of the local electric field is dominant and the vibrational frequency may often be modeled by a simple relation with some electrostatic parameters, through "electrostatic maps." Several parameterization schemes have been employed. For the amide-I mode, Cho and coworkers parameterized the amide-I vibration by identifying the electrostatic potential at four coordinates corresponding to the atoms C, O, N, and H of the amide bond [15–17]. A similar approach was used by Bour and Keiderling [45]. An anharmonic vibrational Hamiltonian for the amide I, II, III, and A modes has been recast in terms of 19 components of an external electric field and its first and second derivative tensors evaluated at a single point [46]. Other parameterizations were introduced to model amide-I and amide-II based on the electrostatic field and its gradients at several points [47–50]. A similar map has been introduced for modeling the fluctuations of carboxylate side chain in proteins [51]. Corrections to the amide-I electrostatic maps were introduced to take into account more accurately nearest-neighbor residues in proteins [52]. Electrostatic maps have been developed for the OH and OD stretch of liquid H_2O, HOD, and D_2O [53–57] and for ice and water clusters [18,19,58]. All these approaches relate the vibrational fluctuations to the electrostatic environment sampled in the neighborhood of the atoms involved in the relevant vibrational mode.

Transition dipole fluctuations can be mostly accounted for by considering the effect of molecular orientations. An early model introduced by Torii et al. [7] proposed a fixed transition dipole in the local frame associated with the amide group. More advanced models of the amide-I transition dipole based on an electrostatic parameterization were introduced [46,47]. A similar parameterization has been proposed for the OD stretch of liquid water [53]. The transition dipole coupling (TDC) is the most popular model for the interaction between local vibrational modes. In this model, the vibrational couplings are given by

$$h_{nm} = J(\boldsymbol{\mu}_m, \boldsymbol{\mu}_n) = \frac{1}{4\pi\varepsilon r_{nm}^3}\left(\boldsymbol{\mu}_m \cdot \boldsymbol{\mu}_n - 3\frac{(\boldsymbol{\mu}_m \cdot \mathbf{r}_{nm})(\boldsymbol{\mu}_n \cdot \mathbf{r}_{nm})}{r_{nm}^2}\right), \tag{13.32}$$

where r_{nm} is the distance between the vibrations n and m. Different models were used to describe the anharmonicities. Coherent exciton transport is usually modeled through the transition dipole model with the harmonic part of the Hamiltonian. The small contribution of anharmonicities to this transport is neglected and only diagonal anharmonicities are included

$$U_{nmn'm'} = \frac{\Delta_{nm}}{4}\left(\delta_{nn'}\delta_{mm'} + \delta_{nm'}\delta_{mn'}\right). \tag{13.33}$$

When the anharmonicity is weak, a fixed (not fluctuating) anharmonicity is commonly used. This is the case, for example, of the amide-I vibration [50,59–61]. For systems with larger anharmonicity, fluctuating anharmonicities models were introduced, for example, in the case of water [53].

13.3.2 Quantum Propagation

For computational efficiency, we do not simulate the Green's function in Equations 13.25 through 13.27. Instead, we directly propagate the one-exciton and the two-exciton wavefunctions. Propagating a vector rather than a matrix reduces memory cost and computational time. This strategy is illustrated here for the $\mathbf{k}_I$ signal; the other signals may be calculated similarly. For each pulse, we choose a direction of polarization ε_ℓ^ν and calculate the signal corresponding to this set of polarization vectors

$$R^{\mathbf{k}_I}(\tau_4,\tau_3,\tau_2,\tau_1) = \sum_{\nu_1\nu_2\nu_3\nu_4} \varepsilon_1^{\nu_1}\varepsilon_2^{\nu_2}\varepsilon_3^{\nu_3}\varepsilon_4^{\nu_4} R^{\mathbf{k}_I}_{\nu_4\nu_3\nu_2\nu_1}(\tau_4,\tau_3,\tau_2,\tau_1). \tag{13.34}$$

The interaction with the first and second laser pulses at time τ_1 and τ_2 creates a population in Liouville space or alternatively two one-exciton wave packets in Hilbert space defined as

$$\psi^{(1)}_{m_1;1}(\tau_1;\tau_1) = \boldsymbol{\varepsilon}_1 \cdot \boldsymbol{\mu}_{m_1}(\tau_1), \tag{13.35}$$

$$\psi^{(1)}_{m_2;2}(\tau_2;\tau_2) = \boldsymbol{\varepsilon}_2 \cdot \boldsymbol{\mu}_{m_2}(\tau_2). \tag{13.36}$$

Its time evolution is described by the Green's function

$$\psi^{(1)}_{m_i;i}(s;\tau_i) = \sum_{n_i} G_{m_i,n_i}(s,\tau_i)\psi^{(1)}_{n_i;i}(\tau_i), \tag{13.37}$$

where $i = 1,2$. The interaction between the system and the third laser pulse at time τ_3 creates a two-exciton wave packet defined as a symmetrized product of the one-exciton wavefunction $\psi^{(1)}_{m;2}(\tau_3;\tau_2)$ and the transition dipole $\boldsymbol{\varepsilon}_3 \cdot \boldsymbol{\mu}_{m_3}(\tau_3)$,

$$\psi^{(2)}_{m_2m_3;2,3}(\tau_3;\tau_3;\tau_2) = \boldsymbol{\varepsilon}_3 \cdot \boldsymbol{\mu}_{m_2}(\tau_3)\psi^{(1)}_{m_3;2}(\tau_3;\tau_2) + \boldsymbol{\varepsilon}_3 \cdot \boldsymbol{\mu}_{m_3}(\tau_3)\psi^{(1)}_{m_2;2}(\tau_3;\tau_2). \tag{13.38}$$

The time evolution is now given by the two-exciton Green's function

$$\psi^{(2)}_{m_2m_3;2,3}(s;\tau_3;\tau_2) = \frac{1}{2}\sum_{n_2n_3} \mathcal{G}_{m_2m_3,n_2n_3}(s,\tau_3)\psi^{(2)}_{n_2n_3;2,3}(\tau_3;\tau_3;\tau_2). \tag{13.39}$$

The first exciton is created at time τ_2 and propagates until time τ_3 where a second exciton is created and propagates until time s. Using these definitions, we can recast Equation 13.16 in the form

$$R^{\mathbf{k}_\mathrm{I}}(\tau_4,\tau_3,\tau_2,\tau_1) = 2\left(\frac{i}{\hbar}\right)^4 \sum_{n_4} \boldsymbol{\varepsilon}_4 \cdot \boldsymbol{\mu}_{n_4}(\tau_4) S_{n_4}^{\mathbf{k}_\mathrm{I}}(\tau_4,\tau_3,\tau_2,\tau_1), \tag{13.40}$$

with

$$S_{n_4}^{\mathbf{k}_\mathrm{I}}(\tau_4,\tau_3,\tau_2,\tau_1) = \sum_{m_4} \int_{\tau_3}^{\tau_4} \mathrm{d}s\, G_{n_4,m_4}(\tau_4,s) X_{m_4}^{\mathbf{k}_\mathrm{I}}(s;\tau_3,\tau_2,\tau_1), \tag{13.41}$$

and

$$X_{m_4}^{\mathbf{k}_\mathrm{I}}(s;\tau_3,\tau_2,\tau_1) = \sum_{m_1 m_2 m_3} U_{m_4 m_1 m_3 m_2}(s) \psi_{m_3 m_2;2,3}^{(2)}(s;\tau_3;\tau_2) \psi_{m_1;1}^{(1)*}(s;\tau_1). \tag{13.42}$$

The one- and the two-exciton wavefunctions are computed by direct integration of the Schrödinger equation. For the one-exciton wavefunction, we have

$$i\hbar \frac{\mathrm{d}}{\mathrm{d}t} |\psi^{(1)}(t;\tau_1)\rangle = H(t)|\psi^{(1)}(t;\tau_1)\rangle, \tag{13.43}$$

where $|\psi^{(1)}(t;t_0)\rangle = \sum_n \psi_n^{(1)}(t;t_0) b_n^\dagger |g\rangle$. A similar equation holds for the two-exciton wavefunction.

$$i\hbar \frac{\mathrm{d}}{\mathrm{d}t} |\psi^{(2)}(t;\tau_2;\tau_1)\rangle = H(t)|\psi^{(2)}(t;\tau_2;\tau_1)\rangle, \tag{13.44}$$

where $|\psi^{(2)}(t;\tau_2;\tau_1)\rangle = 1/2\sum_{n_1 n_2} \psi_{n_1 n_2}^{(2)}(t;\tau_2;\tau_1) b_{n_1}^\dagger b_{n_2}^\dagger |g\rangle$. The response function $\mathcal{R}(t_3,t_2,t_1)$ is computed by repeating the calculation and varying the time intervals between the pulses $t_1 = \tau_2 - \tau_1$, $t_2 = \tau_3 - \tau_2$, and $t_3 = \tau_4 - \tau_3$.

Our simulation protocol for a $\mathbf{k}_\mathrm{I}$ signal is based on a fluctuating Hamiltonian trajectory and can be summarized as follows:

1. Choose an initial time τ_1 along the Hamiltonian trajectory.
2. The first one-exciton wavefunction is created at time τ_1 (Equation 13.35) and propagated until time $\tau_1 + t_1 + t_2 + t_3$ using Equation 13.43.
3. A second one-exciton wavefunction is created at time $\tau_1 + t_1$ and propagated until time $\tau_1 + t_1 + t_2$ using Equation 13.43.
4. At time $\tau_1 + t_1 + t_2$, the second exciton is used to create a two-exciton wavefunction (Equation 13.38) that is propagated until time $\tau_1 + t_1 + t_2 + t_3$ using Equation 13.44.
5. Using Equations 13.41 and 13.42, the function $S_{n_4}(s,\tau_1 + t_1 + t_2,\tau_1 + t_1,\tau_1)$ is computed between $s = \tau_1 + t_1 + t_2$, where S_{n_4} is set to zero and the time $s = \tau_1 + t_1 + t_2 + t_3$. The response function is finally given by Equation 13.40.

Ensemble averaging is performed by repeating these steps for several initial conditions and orientations. A similar algorithm may be used for the other two techniques ($\mathbf{k}_\mathrm{II}$ and $\mathbf{k}_\mathrm{III}$), where the single- and two-exciton wavefunctions are created at different times, as follows:

$$X_{m_4}^{\mathbf{k}_\mathrm{II}}(s;\tau_3,\tau_2,\tau_1) = \sum_{m_1 m_2 m_3} U_{m_4 m_2 m_3 m_1}(s) \psi_{m_3 m_1;1,3}^{(2)}(s;\tau_3;\tau_1) \psi_{m_2;2}^{(1)*}(s;\tau_2), \tag{13.45}$$

$$X_{m_4}^{\mathbf{k}_{\mathrm{III}}}(s;\tau_3,\tau_2,\tau_1) = \sum_{m_1 m_2 m_3} U_{m_4 m_3 m_2 m_1}(s)\psi^{(2)}_{m_2 m_1;1,2}(s;\tau_2;\tau_1)\psi^{(1)*}_{m_3;3}(s;\tau_3). \quad (13.46)$$

13.3.3 Quantum Exciton Dynamics in Liquid Water

The following simulations of coherent third-order nonlinear spectra of liquid water demonstrate the power of our methodology. Our simulations are based on the fluctuating Hamiltonian of Ref. [57]. An MD simulation of 64 water molecules at 300 K using periodic boundary conditions and the SPC/E water model [62] was performed. An electrostatic map based on *ab initio* calculations at the MP2/6-31+G(d,p) level was used to parameterize the Hamiltonian [57,63]. The exciton simulation included $N = 128$ modes corresponding to two OH stretching (symmetric and asymmetric) modes per water molecules. The signal was averaged over 50 trajectories of 1 ps long and for each over 20 random orientations.

Figure 13.5 depicts the imaginary part of the $\mathbf{k}_{\mathrm{I}}$ and $\mathbf{k}_{\mathrm{II}}$ signals for two delay times: $t_2 = 0$ and $t_2 = 500$ fs. The signal shows two peaks of opposite signs but similar magnitude. The high-frequency (positive) peak along Ω_3 corresponds to the GSB and ESE processes. The lower-frequency (negative) peak along Ω_3 corresponds to the ESA process. This peak is red-shifted along the Ω_3 axis by the strong OH stretch anharmonicity. Note that the peaks are not aligned along Ω_1 due to large frequency fluctuations of the same order as the anharmonicity. Both peaks are elongated along the diagonal. This is characteristic of the $\mathbf{k}_{\mathrm{I}}$ technique for which the photon-echo process eliminates the inhomogeneous broadening in the antidiagonal direction. For this reason, this is often called the rephasing signal. The $\mathbf{k}_{\mathrm{II}}$ signal also exhibits two peaks of opposite signs but with a very different shape compared to the $\mathbf{k}_{\mathrm{I}}$ signal. This is because for this technique, inhomogeneous broadening is present in both diagonal and antidiagonal directions. Therefore, this signal is often denoted as non-rephasing. Note that for $t_2 = 0$, the maximum amplitude of the $\mathbf{k}_{\mathrm{II}}$ signal is about a third of that of

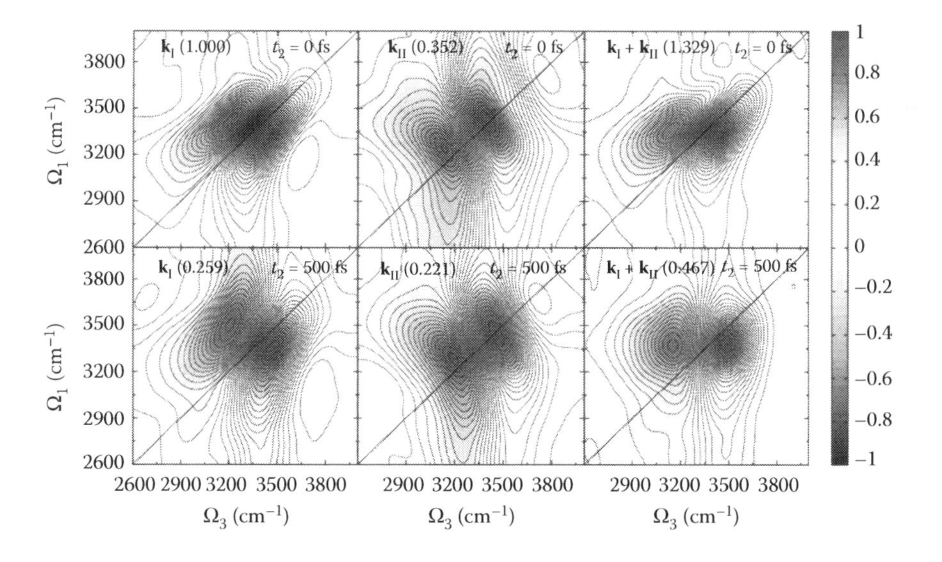

FIGURE 13.5 The $\mathbf{k}_{\mathrm{I}}$, $\mathbf{k}_{\mathrm{II}}$, and $\mathbf{k}_{\mathrm{I}} + \mathbf{k}_{\mathrm{II}}$ signals of liquid water (imaginary part) with parallel polarization for $t_2 = 0$ (upper row) and $t_2 = 500$ fs (lower row). Each panel is normalized to its maximum. The relative maximum with respect to the $\mathbf{k}_{\mathrm{I}}$ signal at time $t_2 = 0$ is indicated in parenthesis. The $\mathbf{k}_{\mathrm{I}}$ signal is displayed for negative Ω_1 frequencies. (Reprinted with permission from Falvo C, Palmieri B, Mukamel S. 2009. Coherent infrared multidimensional spectra of the OH stretching band in liquid water simulated by direct nonlinear exciton propagation. *J. Chem. Phys.* 130:184501. Copyright 2009, American Institute of Physics.)

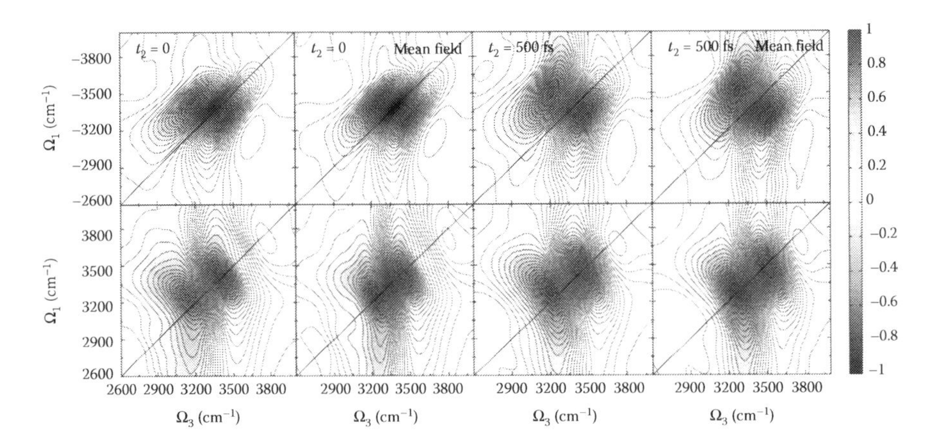

FIGURE 13.6 Upper row: full simulation of $\mathbf{k}_\mathrm{I}$ signal of liquid water for $t_2 = 0$ and $t_2 = 500$ fs compared with the mean-field approximation. Lower row: same for the $\mathbf{k}_\mathrm{II}$ technique. (Reprinted with permission from Falvo C, Palmieri B, Mukamel S. 2009. Coherent infrared multidimensional spectra of the OH stretching band in liquid water simulated by direct nonlinear exciton propagation. *J. Chem. Phys.* 130:184501. Copyright 2009, American Institute of Physics.)

the $\mathbf{k}_\mathrm{I}$ signal. The sum of the rephasing and nonrephasing spectra shows sharper absorption peaks and is referred to as 2D IR spectrum

$$\mathcal{R}^{\mathbf{k}_\mathrm{I}+\mathbf{k}_\mathrm{II}}(\Omega_1,t_2,\Omega_3) = \mathcal{R}^{\mathbf{k}_\mathrm{I}}(-\Omega_1,t_2,\Omega_3) + \mathcal{R}^{\mathbf{k}_\mathrm{II}}(\Omega_1,t_2,\Omega_3). \tag{13.47}$$

As shown in Figure 13.5, since for $t_2 = 0$, $\mathbf{k}_\mathrm{I}$ is much stronger than $\mathbf{k}_\mathrm{II}$, the 2D IR spectrum is dominated by the shape of the $\mathbf{k}_\mathrm{I}$ signal and appears strongly elongated along the diagonal. Both peaks now appear aligned along the Ω_1 axis at a frequency corresponding to the absorption of the band. The splitting of the two peaks directly reveals the anharmonicity. Upon increasing the time t_2 to 500 fs, the $\mathbf{k}_\mathrm{I}$ signal amplitude is reduced by a factor of 4 compared to its $t_2 = 0$ value. The $\mathbf{k}_\mathrm{II}$ signal is hardly affected and its maximum decreases only to 0.221 at $t_2 = 500$ fs, starting from 0.352 at $t_2 = 0$. Both $\mathbf{k}_\mathrm{I}$ and $\mathbf{k}_\mathrm{II}$ now contribute equally to the 2D IR spectrum. Both peaks in the 2D IR spectrum at $t_2 = 500$ fs have lost their elongated shape. This is characteristic of vibrational dephasing.

The $\mathbf{k}_\mathrm{I}$ and $\mathbf{k}_\mathrm{II}$ spectra at $t_2 = 0$ and $t_2 = 500$ fs are compared with the mean-field simulations in Figure 13.6. In all cases, the mean-field approximation appears almost identical to the full calculation. However, this calculation is much faster. The quasi-particle protocol shows a great potential for the simulation of large systems. Systems containing thousands of vibrational modes can be readily computed at the mean-field approximation level.

13.3.4 2D IR Spectroscopy of Amyloid Fibrils

We show a second example of our simulation protocol applied to amyloid fibrils. Amyloid fibrils are self-assembled filaments and their formation and deposition are associated with more than 20 neurodegenerative diseases, including Alzheimer's, Parkinson's, Huntington's diseases, the transmissible spongiform encephalopathies, and type II diabetes [64–68]. In the case of Alzheimer's disease, fibrils are composed of β-amyloid (Aβ) peptides ranging from 39 to 42 residues rich in β-sheet secondary structure. Based on solid-state nuclear magnetic resonance (NMR) data, Tycko and coworkers have built detailed molecular models of 40-residues β-amyloid ($A\beta_{1-40}$) [69–73], in particular a twofold symmetry structure as shown in Figure 13.7.

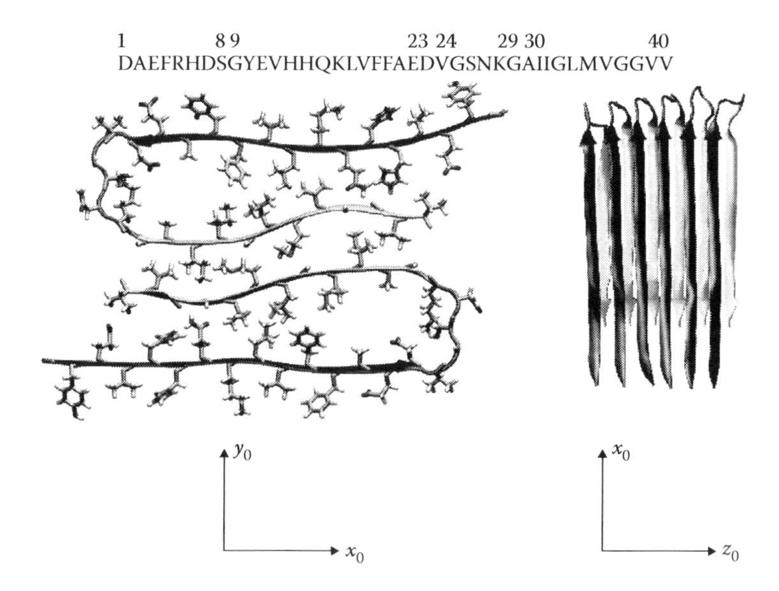

FIGURE 13.7 Sequence of Aβ_{1-40} peptide and the Aβ_{9-40} molecular model described in Ref. [71]. z_0 is the fibril axis. (Reprinted with permission from Falvo C et al. 2012. Frequency distribution of the amide I vibration sorted by residues in amyloid fibrils revealed by 2D IR measurements and simulations. *J. Phys. Chem. B* 116:3322–3330. Copyright 2012, American Chemical Society.)

Detailed isotope-edited 2D IR photon echo of Aβ_{1-40} mature fibrils were reported [9,74,75]. Isotope labeling of a specific residue shifts the corresponding amide-I transitions for all strands. The assembly of strands forms a linear exciton chain of labeled amide units, which has an absorption frequency that is red-shifted by the intermolecular coupling [74]. The 2D IR spectra of these linear chains have been measured for 18 residues located between Val12 and Val39 [9]. This extensive data set provides local information on the amide-I vibrational dynamics inside the amyloid fibrils. We have used MD based on the twofold symmetry molecular model of Tycko and coworkers [71] and compared the simulated nonlinear spectroscopy to the experiment. Simulation details are given in Ref. [59]. The experiment was performed on dry amyloid fibrils; however, evidence of trapped water molecules inside the fibrils was deduced from the ultrafast decay of the frequency correlation function [75]. Simulations were aimed at understanding the effect of water molecules on the 2D IR lineshape. The MD molecular model was completely embedded in water molecules. We have performed two series of 2D IR simulations. The first included the effect of water molecules in the amide-I vibrational dynamics and the other series did not include this effect. The QP formalism is particularly useful because it fully captures the exciton dynamics (strong coupling between the neighboring units in the β-sheets and all the fluctuation timescales).

Figure 13.8 compares the experimental 2D IR (left column) and simulated signals for various isotopomers. By comparing simulations without (central column) and with the water contribution (right column), we find that interaction with the water molecules strongly modify the 2D IR spectra for certain residues. In the simulations, the most homogeneous transitions (residues L17, I32, G33, and L34) have the smallest effect of water. The splitting of some isotopologue spectra is seen in both theory and experiment. For residues G25, G29, A30, I31, and G37, multiple peaks appear in both the experiment and the simulation and these peaks appear whether or not water is present in the simulation. The simulated 2D IR spectra for residues L17, V18, L34, and V36 show homogeneous broadened peaks while the experiment shows inhomogeneously broadened peaks. In the simulations, water molecules were unable to penetrate in the regions near these residues while experiments suggest the existence of trapped water molecules [75]. Our MD simulation did not allow water molecules near these residues,

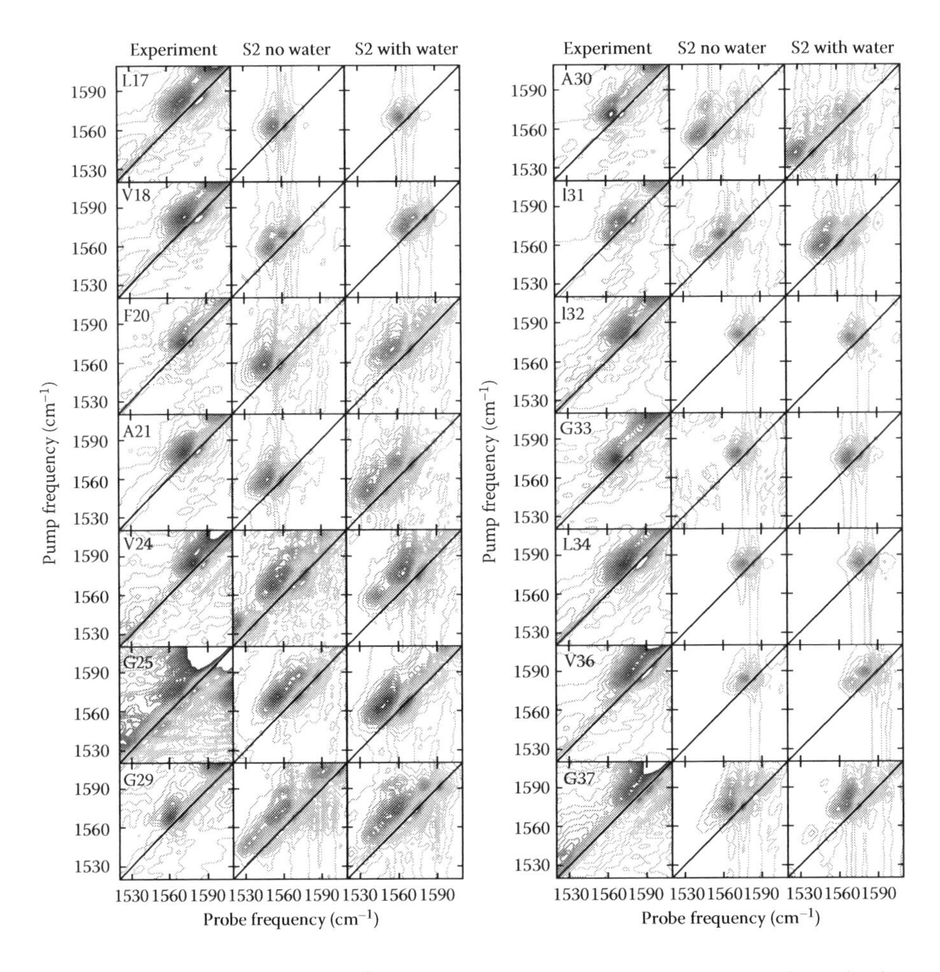

FIGURE 13.8 2D IR spectra of various $A\beta_{1-40}$ isotopomers. Left column: experiment. Central column: simulation without the water contribution in the frequency shift. Right column: simulation including the water contribution. (Reprinted with permission from Falvo C et al. 2012. Frequency distribution of the amide I vibration sorted by residues in amyloid fibrils revealed by 2D IR measurements and simulations. *J. Phys. Chem. B* 116:3322–3330. Copyright 2012, American Chemical Society.)

but the strong difference in shape between the simulated and the measured 2D IR reinforces the conjecture that water molecules have somehow penetrated these regions.

13.4 STOCHASTIC DYNAMICS OF HAMILTONIAN FLUCTUATIONS

13.4.1 Stochastic Formalism

The explicit molecular-dynamical treatment of large number of molecular coordinates is not always feasible. In an alternative approach, few classical collective coordinates $\mathbf{q} \equiv \{q_i\}$ are included explicitly in theoretical description, and their evolution is viewed as a stochastic process, rather than the result of deterministic Hamiltonian evolution subjected to a force from other "irrelevant" molecular coordinates. The general form of the path integration $\mathcal{D}\mathbf{q}(t)$ can often be anticipated from physical arguments, leaving only few free physically meaningful parameters to be adjusted.

For instance, according to the Donsker's theorem [76] (also known as functional central limit theorem), independent stochastic processes may be composed into Gaussian macroscopic collective coordinate. Elongated elliptical two-dimensional (2D) photon-echo peak shapes characteristic of slow Gaussian spectral diffusion [77] have been observed in many cases, when the vibrational (or electronic) transitions are well resolved in the absorption spectrum. In a different class of models, the Hamiltonian parameters (Equation 13.7) can only assume a few discrete values. Examples are hydrogen bonding in water or organic solvents, conformational dynamics, and so on, where the molecular system fluctuates between distinct structures [14]. The vibrational frequencies depend on the structure, and transitions between these structures can be approximated as jump processes. We *a priori* assume some statistical properties of $\mathcal{D}\mathbf{q}(t)$ subjected to two constraints: they must be physically sound and numerically feasible.

It has been shown that certain types of memory erasing along stochastic paths $\mathbf{q}(t)$ substantially reduce computational cost. Path integration has been transformed into matrix algebra for a wide class of continuous-time spectral random walks [78,79]. Even more effective descriptions are at hand for memory-less (Markovian) processes [80,81]. The path integration can be carried out analytically using the method of marginal averages [82] for quantum evolution with Markovian fluctuations of the Hamiltonian (Equation 13.7). Gaussian–Markovian relaxation is of particular interest, since it leaves only two free parameters—the amplitude and relaxation (autocorrelation) time. This *Ornstein–Uhlenbeck process* [83] is thus widely used to model spectral diffusion. More importantly, the method of marginal averages is not limited to Gaussian models. The implementation of spectral jumps parameterized by few kinetic rates for transitions is particularly simple [84].

Hereafter, we focus on Markovian processes. The marginal average method when applied to Liouville–von Neumann evolution equation results in the SLE [85]. These were employed to simulate spectra of water [63], organic solutions [28], peptides [27], or small model Frenkel exciton aggregates [86]. These applications do not involve strong interferences of Liouville space pathways. Such interference and cancellations are massive in collections of many bosons, and usually reduce the accuracy of simulations of larger systems based on Liouville equations. The QP NEE approach avoids these problems, since it does not dissect the response into pathways and seems to work better in many cases [22].

We show how the description of Markovian processes can be combined with the QP approach developed in Section 13.2. The following results were originally derived using the SNEE [30]. Here, we recover the same results by the direct integration of Equations 13.25 through 13.27. We first employ some elementary properties of Markovian processes to simplify the path integrations (Equation 13.12). Markovian evolution in different (noncoinciding) time intervals can be separated out. We divide the time axis into intervals (τ_1,τ_2), (τ_2,τ_3), (τ_3,s), and (s,τ_4). The path integration over trajectories of Markovian processes on interval (τ_1,τ_4) can be recast according to the Chapman–Kolmogorov theorem:

$$\mathcal{D}\mathbf{q}(t) = \int d\mathbf{q}_4 \int d\mathbf{q}_s \int d\mathbf{q}_3 \int d\mathbf{q}_2 \int d\mathbf{q}_1 \mathcal{D}^{\tau_4,s}_{\mathbf{q}_4,\mathbf{q}_s}\mathbf{q}(t)\, \mathcal{D}^{s,\tau_3}_{\mathbf{q}_s,\mathbf{q}_3}\mathbf{q}(t)\, \mathcal{D}^{\tau_3,\tau_2}_{\mathbf{q}_3,\mathbf{q}_2}\mathbf{q}(t)\, \mathcal{D}^{\tau_2,\tau_1}_{\mathbf{q}_2,\mathbf{q}_1}\mathbf{q}(t)\;\; P(\mathbf{q}_1), \quad (13.48)$$

where we denoted path integrations over trajectories on the interval (τ_1,τ_2) with fixed initial $\mathbf{q}_1 = \mathbf{q}(\tau_1)$ and final $\mathbf{q}_2 = \mathbf{q}(\tau_2)$ conditions by

$$\mathcal{D}^{\tau_2,\tau_1}_{\mathbf{q}_2,\mathbf{q}_1}\mathbf{q}(t) \equiv \lim_{n\to\infty} \prod_{p=1;\tau_2>t_p>\tau_1}^{n} \left(\int d\mathbf{q}(t_p)\right) P(\mathbf{q}_2,\mathbf{q}(t_n),\ldots,\mathbf{q}(t_1),\mathbf{q}_1)/P(\mathbf{q}_1). \quad (13.49)$$

We next turn to the factors of Equation 13.16 that originate from the quantum evolution. The evolution of bra vectors (Green's functions of the left-hand side of diagram in Figure 13.4) is always accompanied by the evolution of ket vectors (Green's functions of the right-hand side of diagram). These evolutions along a given trajectory $\mathbf{q}(t)$ share the same Hamiltonian evolution $H_S(\mathbf{q}(t))$. *The evolution of both Green's functions must thus be averaged jointly.*

The Green's functions $G^*_{m_1n_1}(s,\tau_1)$ in the right-hand side of Equation 13.25 overlap during three time delay periods. In the various intervals, it is accompanied by the evolution of different bra vectors, and its averaging must thus proceed in a different way. We decompose $G^*_{m_1n_1}(s,\tau_1)$ into intervals according to Equation 13.23

$$G^*_{m_1n_1}(s,\tau_1) = \sum_{m_1'm_1''} G^*_{m_1,m_1'}(s,\tau_3)G^*_{m_1',m_1''}(\tau_3,\tau_2)G^*_{m_1'',n_1}(\tau_2,\tau_1). \tag{13.50}$$

We next average the Green's functions in the various intervals. In the (τ_1,τ_2) and (s,τ_4) intervals, one-particle Green's functions G,G^* are accompanied with trivial evolution of ground state. We thus define

$$\hat{G}_{m_1''\mathbf{q}_2,n\mathbf{q}_1}(\tau_2,\tau_1) \equiv \int G_{m_1'',n}(\tau_2,\tau_1;\mathbf{q}(t))\mathcal{D}^{\tau_2,\tau_1}_{\mathbf{q}_2,\mathbf{q}_1}\mathbf{q}(t), \tag{13.51}$$

where we added $\mathbf{q}(t)$ to parameters of Green's function in the right-hand side to emphasize that the Green's functions of the previous sections depend on the entire trajectory. In contrast, as a consequence of the Markovian property, the left-hand side of Equation 13.51 only depends on boundary initial and final values.

During the other intervals (τ_2,τ_3), and (τ_3,s), we consider both evolutions of the ket and bra sides of diagram and jointly average the relevant Green's functions

$$\hat{G}^{(N)}_{m_1'p_2\mathbf{q}_3,m_1''n_2\mathbf{q}_2}(\tau_3,\tau_2) \equiv \int G^*_{m_1',m_1''}(\tau_3,\tau_2;\mathbf{q}(t))G_{p_2,n_2}(\tau_3,\tau_2;\mathbf{q}(t))\mathcal{D}^{\tau_3,\tau_2}_{\mathbf{q}_3,\mathbf{q}_2}\mathbf{q}(t), \tag{13.52}$$

$$\hat{\mathcal{G}}^{(Z)}_{m_1''m_3m_2\mathbf{q}_s,m_1'n_3p_2\mathbf{q}_3}(s,\tau_3) \equiv \int G^*_{m_1'',m_1'}(s,\tau_3;\mathbf{q}(t))\mathcal{G}_{m_3n_2,n_3p_2}(s,\tau_3;\mathbf{q}(t))\mathcal{D}^{s,\tau_3}_{\mathbf{q}_s,\mathbf{q}_3}\mathbf{q}(t). \tag{13.53}$$

With definitions 13.49 through 13.53 and the Chapman–Kolmogorov decomposition (Equation 13.48), we can integrate Equation 13.25 to obtain the final expression for photon-echo signal

$$R^{\mathbf{k}_1}_{\nu_4\nu_3\nu_2\nu_1}(\tau_4,\tau_3,\tau_2,\tau_1) = 2\left(\frac{i}{\hbar}\right)^4 \sum_{n_1n_2n_3n_4}\ \sum_{m_1m_2m_3m_4m_{1'}m_{1''}p_2} \int d\mathbf{q}_1 d\mathbf{q}_2 d\mathbf{q}_3 d\mathbf{q}_s d\mathbf{q}_4$$

$$\times \mu_{n_4\mathbf{q}_4;\nu_4}\mu_{n_3\mathbf{q}_3;\nu_3}\mu_{n_2\mathbf{q}_2;\nu_2}\mu_{n_1\mathbf{q}_1;\nu_1}\int_{\tau_3}^{\tau_4} ds\, \hat{G}_{n_4\mathbf{q}_4,m_4\mathbf{q}_s}(\tau_4,s)U_{m_4m_1m_3m_2;\mathbf{q}_s}$$

$$\times \hat{\mathcal{G}}^{(Z)}_{m_1m_3m_2\mathbf{q}_s,m_1'n_3p_2\mathbf{q}_3}(s,\tau_3)\hat{G}^{(N)}_{m_1'p_2\mathbf{q}_3,m_1''n_2\mathbf{q}_2}(\tau_3,\tau_2)\,\hat{G}^*_{m_1''\mathbf{q}_2,n_1\mathbf{q}_1}(\tau_2,\tau_1)P(\mathbf{q}_1)\,. \tag{13.54}$$

Equation 13.26 can be integrated in an analogous way. Here, we first factorize the one-exciton Green's functions G into intervals as follows:

$$G_{p_1,n_1}(\tau_3,\tau_1) = \sum_{p_1'} G_{p_1,p_1'}(\tau_3,\tau_2)G_{p_1',n_1}(\tau_2,\tau_1), \tag{13.55}$$

$$G^*_{m_2,n_2}(s,\tau_2) = \sum_{m_2'} G^*_{m_2,m_2'}(s,\tau_3)G^*_{m_2',n_2}(\tau_3,\tau_2). \tag{13.56}$$

Using Equations 13.48 and 13.49, we obtain for the $\mathbf{k}_{\text{II}}$ signal

$$R^{\mathbf{k}_{\text{II}}}_{\nu_4\nu_3\nu_2\nu_1}(\tau_4,\tau_3,\tau_2,\tau_1) = 2\left(\frac{i}{\hbar}\right)^4 \sum_{m_1m_2m_3m_4p_1'm_2'p_1} \int d\mathbf{q}_1 d\mathbf{q}_2 d\mathbf{q}_3 d\mathbf{q}_s d\mathbf{q}_4$$
$$\times \mu_{n_4\mathbf{q}_4;\nu_4}\mu_{n_3\mathbf{q}_3;\nu_3}\mu_{n_2\mathbf{q}_2;\nu_2}\mu_{n_1\mathbf{q}_1;\nu_1}\int_{\tau_3}^{\tau_4} ds\, \hat{G}_{n_4\mathbf{q}_4,m_4\mathbf{q}_s}(\tau_4,s)U_{m_4m_2m_3m_1;\mathbf{q}_s}$$
$$\times \hat{\mathcal{G}}^{(Z)}_{m_2m_3m_1\mathbf{q}_s,m_2'n_3p_1\mathbf{q}_3}(s,\tau_3)\hat{G}^{(N)}_{m_2'p_1\mathbf{q}_3,n_2p_1'\mathbf{q}_2}(\tau_3,\tau_2)\hat{G}_{p_1'\mathbf{q}_2n_1\mathbf{q}_1}(\tau_2,\tau_1)P(\mathbf{q}_1). \tag{13.57}$$

Finally, the integration of Equation 13.27 requires an additional factorization of the two-exciton Green's function

$$\mathcal{G}_{m_2m_1,n_2p_1}(s,\tau_2) = \frac{1}{2}\sum_{m_2'm_1'} \mathcal{G}_{m_2m_1,m_2'm_1'}(s,\tau_3)\mathcal{G}_{m_2'm_1',n_2p_1}(\tau_3,\tau_2), \tag{13.58}$$

and its averaging over Markovian stochastic trajectories on interval (τ_2,τ_3). We thus define

$$\hat{\mathcal{G}}_{m_2'm_1'\mathbf{q}_3,n_2p_1\mathbf{q}_2}(\tau_3,\tau_2) \equiv \int \mathcal{G}_{m_2'm_1',n_2p_1}(\tau_3,\tau_2)\mathcal{D}^{\tau_3,\tau_2}_{\mathbf{q}_3,\mathbf{q}_2}\mathbf{q}(t). \tag{13.59}$$

Inserting definitions Equations 13.48, 13.49, and 13.59 into Equation 13.27, we finally obtain for the $\mathbf{k}_{\text{III}}$ signal

$$R^{\mathbf{k}_{\text{III}}}_{\nu_4\nu_3\nu_2\nu_1}(\tau_4,\tau_3,\tau_2,\tau_1) = \left(\frac{i}{\hbar}\right)^4 \sum_{m_1m_2m_3m_4m_1'm_2'p_1} \int d\mathbf{q}_1 d\mathbf{q}_2 d\mathbf{q}_3 d\mathbf{q}_s d\mathbf{q}_4$$
$$\times \mu_{n_4\mathbf{q}_4;\nu_4}\mu_{n_3\mathbf{q}_3;\nu_3}\mu_{n_2\mathbf{q}_2;\nu_2}\mu_{n_1\mathbf{q}_1;\nu_1}\int_{\tau_3}^{\tau_4} ds\hat{G}_{n_4\mathbf{q}_4,m_4\mathbf{q}_s}(\tau_4,s)U_{m_4m_3m_2m_1;\mathbf{q}_s}$$
$$\times \hat{\mathcal{G}}^{(Z)}_{m_3m_2m_1\mathbf{q}_s,n_3m_2'm_1'n_3\mathbf{q}_3}(s,\tau_3)\hat{\mathcal{G}}_{m_2'm_1'\mathbf{q}_3,n_2p_1\mathbf{q}_2}(\tau_3,\tau_2)\hat{G}_{p_1\mathbf{q}_2,n_1\mathbf{q}_1}(\tau_2,\tau_1)P(\mathbf{q}_1). \tag{13.60}$$

Recasting Equations 13.25 through 13.27 in the form of Equations 13.54, 13.57, and 13.60 simplifies the simulations of response because the Markovian stochastic subtrajectories on the various intervals can be readily combined into complete trajectories, which properly sample the entire trajectory space.

The full power of the present formalism hinges upon the development of practical algorithm for calculating the averaged Green's functions $\hat{G}$, $\hat{G}^{(N)}$, $\hat{G}^{(Z)}$, and $\hat{\mathcal{G}}$. To this end, let us first recall that Markovian processes are semigroups that can be represented by master (evolution) equations for probability densities $P(\mathbf{q})$ of collective coordinate $\mathbf{q}$

$$\frac{dP(\mathbf{q})}{dt} = (T^{\mathbf{q}}P)(\mathbf{q}). \tag{13.61}$$

The linear operator T contains the complete information about the stochastic process and generates the Markovian dynamics. Path integration $\mathcal{D}\mathbf{q}(t)$ can eventually be built up by dissecting the time interval to infinitesimal segments Δt and prescribing convolution kernel of the master equation (Green's function solution to Equation 13.61 with boundary condition $\mathcal{K}(\mathbf{q},\mathbf{q}',t=0)=\delta(\mathbf{q}-\mathbf{q}')$)

$$\mathcal{D}^{n\Delta t,0}_{\mathbf{q}_n\mathbf{q}_0}\mathbf{q}(t)=\int \mathrm{d}\mathbf{q}_{n-1}\mathcal{K}(\mathbf{q}_n,\mathbf{q}_{n-1},\Delta t)\int \mathrm{d}\mathbf{q}_{n-2}\mathcal{K}(\mathbf{q}_{n-1},\mathbf{q}_{n-2},\Delta t)\ldots\mathcal{K}(\mathbf{q}_1,\mathbf{q}_0,\Delta t). \tag{13.62}$$

Below, we present the generator T for some common examples of stochastic processes. The Ornstein–Uhlenbeck coordinates $\mathbf{q}=\{q_i\}$ satisfy the Fokker–Planck (Smoluchowski) equation [87]

$$T^{\mathbf{q}}=\sum_i \Lambda_i \frac{\partial}{\partial q_j}\left(q_i+\sigma_i^2\frac{\partial}{\partial q_i}\right). \tag{13.63}$$

Equation 13.63 represents a diffusion of a particle in harmonic potential [87]. Here, Λ_i is inverse autocorrelation time and σ_i is the equilibrium width of the distribution of the coordinate q_i. The solution of Equation 13.61 is given by the Gaussian kernel

$$\mathcal{K}(\mathbf{q},\mathbf{q}',t)=\prod_i\sqrt{\frac{1}{2\pi\sigma_i^2(1-\mathrm{e}^{-2\Lambda_i t})}}\exp\left[-\frac{(q_i-\mathrm{e}^{-\Lambda_i t}q_i')^2}{2\sigma_i^2(1-\mathrm{e}^{-2\Lambda_i t})}\right]. \tag{13.64}$$

Multistate processes are described by a matrix of rate constants $T_{jj'}$ for jumps from state j' to j, $j\neq j'$. Diagonal elements are defined by $T_{jj}\equiv-\sum_{j';j'\neq j}T_{jj'}$. We also assign to each state j some value ξ_j of the coordinate q. The evolution of probability densities can be then expressed along the same lines for both continuous and discrete processes in terms of master equation (Equation 13.61). For a multistate jump process, the operator T is given by

$$T^{\mathbf{q}}=\sum_{jj'}\delta(q-\xi_j)T_{jj'}\int_{\xi_{j'}-\varepsilon}^{\xi_{j'}+\varepsilon}\mathrm{d}q, \tag{13.65}$$

where $\varepsilon\to 0$.

The master equation description of the Markovian stochastic process (Equation 13.61) provides a simple algorithm for computing the averaged Green's function. Using the marginal averages method, the averaged one-exciton Green's functions satisfy the following equations:

$$\begin{aligned}\frac{\mathrm{d}}{\mathrm{d}\tau}\hat{G}_{m\mathbf{q},m'\mathbf{q}'}(\tau,\tau')=&-\frac{i}{\hbar}\sum_{m''}h_{mm'';\mathbf{q}}G_{m''\mathbf{q},m'\mathbf{q}'}(\tau,\tau')\\&+T^{\mathbf{q}}(G_{m\mathbf{q},m'\mathbf{q}'}(\tau,\tau'))+\delta_{mm'}\delta(\mathbf{q}-\mathbf{q}')\delta(\tau-\tau').\end{aligned} \tag{13.66}$$

Equation 13.66 is in fact the Green's function solution to the SNEE derived in Ref. [30] for $\langle b_m\rangle_{\mathbf{q}}(t)\equiv\{\langle\psi(t)|b_m|\psi(t)\rangle|\,\mathbf{q}(t)=\mathbf{q}\}$. The connection with the present derivation may be verified by checking that the resulting Green's functions are the same.

The other averaged Green's functions can be obtained by solving the following equations of motion:

$$\begin{aligned}\frac{\mathrm{d}}{\mathrm{d}\tau}\hat{\mathcal{G}}_{mn\mathbf{q},m'n'\mathbf{q}'}(\tau,\tau')=&-\frac{i}{\hbar}\sum_{m''n''}h^{(Y)}_{mn,m''n'';\mathbf{q}}\hat{\mathcal{G}}_{m''n''\mathbf{q},m'n'\mathbf{q}'}(\tau,\tau')\\&+T^{\mathbf{q}}(\hat{\mathcal{G}}_{mn\mathbf{q},m'n'\mathbf{q}'}(\tau,\tau'))+(\delta_{mm'}\delta_{nn'}+\delta_{nm'}\delta_{mn'})\delta(\mathbf{q}-\mathbf{q}')\delta(\tau-\tau'),\end{aligned} \tag{13.67}$$

$$\frac{d}{d\tau}\hat{G}^{(N)}_{mn\mathbf{q},m'n'\mathbf{q}'}(\tau,\tau') = \frac{i}{\hbar}\sum_k\left[h_{km;\mathbf{q}}\hat{G}^{(N)}_{kn\mathbf{q},m'n'\mathbf{q}'}(\tau,\tau') - h_{nk;\mathbf{q}}\hat{G}^{(N)}_{mk\mathbf{q},m'n'\mathbf{q}'}(\tau,\tau')\right]$$

$$+ T^{\mathbf{q}}(\hat{G}^{(N)}_{mn\mathbf{q},m'n'\mathbf{q}'}(\tau,\tau')) + \delta_{mm'}\delta_{nn'}\delta(\mathbf{q}-\mathbf{q}')\delta(\tau-\tau'), \quad (13.68)$$

$$\frac{d}{d\tau}\hat{\mathcal{G}}^{(Z)}_{kmn\mathbf{q},k'm'n'\mathbf{q}'}(\tau,\tau') = -\frac{i}{\hbar}\sum_{m''n''} h^{(Y)}_{mn,m''n'';\mathbf{q}}\hat{\mathcal{G}}^{(Z)}_{km''n''\mathbf{q},k'm'n'\mathbf{q}'}(\tau,\tau')$$

$$+\frac{i}{\hbar}\sum_{k''} h_{k''k;\mathbf{q}}\hat{\mathcal{G}}^{(Z)}_{k''mn\mathbf{q},k'm'n'\mathbf{q}'}(\tau,\tau') + T^{\mathbf{q}}(\mathcal{G}^{(Z)}_{kmn\mathbf{q},k'm'n'\mathbf{q}'}(\tau,\tau'))$$

$$+ \delta_{kk'}(\delta_{mm'}\delta_{nn'} + \delta_{nm'}\delta_{mn'})\delta(\mathbf{q}-\mathbf{q}')\delta(\tau'-\tau), \quad (13.69)$$

where $h^{(Y)}_{mn,m'n';\mathbf{q}} \equiv h_{mm';\mathbf{q}}\delta_{nn'} + h_{nn';\mathbf{q}}\delta_{mm'} + U_{mnm'n';\mathbf{q}} + U_{nmm'n';\mathbf{q}}$ is the Hamiltonian for the two-exciton manifold. These are Green's function solutions to the other SNEE of Ref. [30], in particular for quantities $\langle b_m b_n\rangle_{\mathbf{q}}(t) \equiv \{\langle\psi(t) \mid b_m b_n \mid \psi(t)\rangle \mid \mathbf{q}(t) = \mathbf{q}\}$, $\langle b_n^\dagger b_m\rangle_{\mathbf{q}}(t) \equiv \{\langle\psi(t) \mid b_m^\dagger b_n \mid \psi(t)\rangle \mid \mathbf{q}(t) = \mathbf{q}\}$, and $\langle b_k^\dagger b_m b_n\rangle_{\mathbf{q}}(t) \equiv \{\langle\psi(t) \mid b_k^\dagger b_m b_n \mid \psi(t)\rangle \mid \mathbf{q}(t) = \mathbf{q}\}$, respectively.

Equations 13.41 through 13.66 considerably simplify the calculations of stochastic averages, since they do not require the generation of the individual Markovian trajectories. The Hamiltonian coefficients $h_{mm'';\mathbf{q}}$, $\mu_{n;\mathbf{q}}$, and $U_{mnm'n';\mathbf{q}}$ are no longer stochastic time-dependent quantities, but constants in an extended joint space made of the excitonic system and a limited number of collective bath degrees of freedom $\mathbf{q}$. The resulting set of linear partial differential equations with constant coefficients can be solved by standard methods. The Green's functions $\hat{G}(\tau_2,\tau_1)$, $\hat{G}^{(N)}(\tau_2,\tau_1)$, $\hat{\mathcal{G}}(\tau_2,\tau_1)$, and $\hat{\mathcal{G}}^{(Z)}(\tau_2,\tau_1)$ form a semigroup in the extended space. As such, they can be easily calculated in the frequency domain $\hat{G}_{m\mathbf{q},m'\mathbf{q}'}(\Omega) \equiv \int_0^\infty e^{i\Omega\tau}\hat{G}_{m\mathbf{q},m'\mathbf{q}'}(\tau,0)d\tau$ using operator–matrix inversion $^{-1}$. For instance

$$\hat{G}_{m\mathbf{q},m'\mathbf{q}'}(\Omega) = \left[\left(i\Omega - \frac{i}{\hbar}\hat{h} + T\right)^{-1}\right]_{m\mathbf{q},m'\mathbf{q}'}. \quad (13.70)$$

Here, $\hat{h}$ should be interpreted as the h operator in the extended space $[\hat{h}]_{m\mathbf{q},m'\mathbf{q}'} = h_{mm';\mathbf{q}}\delta(\mathbf{q}-\mathbf{q}')$. Similarly, $T_{m\mathbf{q},m'\mathbf{q}'} = T_{\mathbf{q}\mathbf{q}'}\delta_{mm'}$. The other Green's functions $\hat{G}^{(N)}$, $\hat{\mathcal{G}}$, $\hat{\mathcal{G}}^{(Z)}$ can be obtained similarly. These Green's functions are defined in spaces with larger number of exciton indices; which one is acted upon is distinguished by lower index [...], for example, $[\hat{h}_{[1]}]_{mn\mathbf{q},m'n'\mathbf{q}'} = h_{mm';\mathbf{q}}\delta_{nn'}\delta(\mathbf{q}-\mathbf{q}')$, $[\hat{h}_{[2]}]_{mn\mathbf{q},m'n'\mathbf{q}'} = \delta_{mm'}h_{nn';\mathbf{q}}\delta(\mathbf{q}-\mathbf{q}')$, # stands for the transpose $[\hat{h}^{\#}]_{m\mathbf{q},m'\mathbf{q}'} = h_{m'm;\mathbf{q}}\delta(\mathbf{q}-\mathbf{q}')$. In this notation, the solutions to Equation 13.70 are

$$\hat{\mathcal{G}}_{mn\mathbf{q},m'n'\mathbf{q}'}(\Omega) = \left[\left(i\Omega - \frac{i}{\hbar}\hat{h}^{(Y)} + T\right)^{-1}\right]_{mn\mathbf{q},m'n'\mathbf{q}'} + \left[\left(i\Omega - \frac{i}{\hbar}\hat{h}^{(Y)} + T\right)^{-1}\right]_{mn\mathbf{q},n'm'\mathbf{q}'} \quad (13.71)$$

$$\hat{G}^{(N)}_{mn\mathbf{q},m'n'\mathbf{q}'}(\Omega) = \left[\left(i\Omega - \frac{i}{\hbar}\hat{h}_{[2]} + \frac{i}{\hbar}\hat{h}^{\#}_{[1]} + T\right)^{-1}\right]_{mn\mathbf{q},m'n'\mathbf{q}'} \quad (13.72)$$

$$\hat{\mathcal{G}}^{(Z)}_{kmn\mathbf{q},k'm'n'\mathbf{q}'}(\Omega) = \left[\left(i\Omega - \frac{i}{\hbar}\hat{h}^{(Y)}_{[23]} + \frac{i}{\hbar}\hat{h}^{\#}_{[1]} + T\right)^{-1}\right]_{kmn\mathbf{q},k'm'n'\mathbf{q}'}$$

$$+ \left[\left(i\Omega - \frac{i}{\hbar}\hat{h}^{(Y)}_{[23]} + \frac{i}{\hbar}\hat{h}^{\#}_{[1]} + T\right)^{-1}\right]_{kmn\mathbf{q},k'n'm'\mathbf{q}'}. \quad (13.73)$$

Nonlinear optical signals may be calculated directly in the frequency domain

$$\mathcal{R}_{\nu_4\nu_3\nu_2\nu_1}(\Omega_3,\Omega_2,\Omega_1) \equiv \int_0^\infty \mathrm{d}t_1 \int_0^\infty \mathrm{d}t_2 \int_0^\infty \mathrm{d}t_3 e^{i\Omega_1 t_1 + i\Omega_2 t_2 + i\Omega_3 t_3} \mathcal{R}_{\nu_4\nu_3\nu_2\nu_1}(t_3,t_2,t_1). \tag{13.74}$$

A numerical Fourier transform of Equation 13.28 is not necessary. We obtain

$$\begin{aligned}\mathcal{R}^{\mathbf{k}_\mathrm{I}}_{\nu_4\nu_3\nu_2\nu_1}(\Omega_3,\Omega_2,\Omega_1) &= 2\left(\frac{i}{\hbar}\right)^4 \sum_{n_1n_2n_3n_4} \sum_{m_1m_2m_3m_4m_1'm_1''p_2} \int \mathrm{d}\mathbf{q}_1\mathrm{d}\mathbf{q}_2\mathrm{d}\mathbf{q}_3\mathrm{d}\mathbf{q}_s\mathrm{d}\mathbf{q}_4 \\ &\times \mu_{n_4\mathbf{q}_4;\nu_4}\mu_{n_3\mathbf{q}_3;\nu_3}\mu_{n_2\mathbf{q}_2;\nu_2}\mu_{n_1\mathbf{q}_1;\nu_1}\hat{G}_{n_4\mathbf{q}_4,m_4\mathbf{q}_s}(\Omega_3)U_{m_4m_1m_3m_2;\mathbf{q}_s} \\ &\times \hat{\mathcal{G}}^{(Z)}_{m_1m_3m_2\mathbf{q}_s,m_1'n_3p_2\mathbf{q}_3}(\Omega_3)\hat{G}^{(N)}_{m_1'p_2\mathbf{q}_3,m_1''n_2\mathbf{q}_2}(\Omega_2)\hat{G}^{*}_{m_1''\mathbf{q}_2,n_1\mathbf{q}_1}(\Omega_1)P(\mathbf{q}_1),\end{aligned} \tag{13.75}$$

$$\begin{aligned}\mathcal{R}^{\mathbf{k}_\mathrm{II}}_{\nu_4\nu_3\nu_2\nu_1}(\Omega_3,\Omega_2,\Omega_1) &= 2\left(\frac{i}{\hbar}\right)^4 \sum_{m_1m_2m_3m_4p_1'm_2'p_1} \int \mathrm{d}\mathbf{q}_1\mathrm{d}\mathbf{q}_2\mathrm{d}\mathbf{q}_3\mathrm{d}\mathbf{q}_s\mathrm{d}\mathbf{q}_4 \\ &\times \mu_{n_4\mathbf{q}_4;\nu_4}\mu_{n_3\mathbf{q}_3;\nu_3}\mu_{n_2\mathbf{q}_2;\nu_2}\mu_{n_1\mathbf{q}_1;\nu_1}\hat{G}_{n_4\mathbf{q}_4,m_4\mathbf{q}_s}(\Omega_3)U_{m_4m_2m_3m_1;\mathbf{q}_s} \\ &\times \hat{\mathcal{G}}^{(Z)}_{m_2m_3m_1\mathbf{q}_s,m_2'n_3p_1\mathbf{q}_3}(\Omega_3)\hat{G}^{(N)}_{m_2'p_1\mathbf{q}_3,n_2p_1'\mathbf{q}_2}(\Omega_2)\hat{G}_{p_1'\mathbf{q}_2n_1\mathbf{q}_1}(\Omega_1)P(\mathbf{q}_1),\end{aligned} \tag{13.76}$$

$$\begin{aligned}\mathcal{R}^{\mathbf{k}_\mathrm{III}}_{\nu_4\nu_3\nu_2\nu_1}(\Omega_3,\Omega_2,\Omega_1) &= \left(\frac{i}{\hbar}\right)^4 \sum_{m_1m_2m_3m_4m_1'm_2'p_1} \int \mathrm{d}\mathbf{q}_1\mathrm{d}\mathbf{q}_2\mathrm{d}\mathbf{q}_3\mathrm{d}\mathbf{q}_s\mathrm{d}\mathbf{q}_4 \\ &\times \mu_{n_4\mathbf{q}_4;\nu_4}\mu_{n_3\mathbf{q}_3;\nu_3}\mu_{n_2\mathbf{q}_2;\nu_2}\mu_{n_1\mathbf{q}_1;\nu_1}\hat{G}_{n_4\mathbf{q}_4,m_4\mathbf{q}_s}(\Omega_3)U_{m_4m_3m_2m_1;\mathbf{q}_s} \\ &\times \hat{\mathcal{G}}^{(Z)}_{m_3m_2m_1\mathbf{q}_s,n_3m_2'm_1'n_3\mathbf{q}_3}(\Omega_3)\hat{\mathcal{G}}_{m_2'm_1'\mathbf{q}_3,n_2p_1\mathbf{q}_2}(\Omega_2)\hat{G}_{p_1\mathbf{q}_2,n_1\mathbf{q}_1}(\Omega_1)P(\mathbf{q}_1).\end{aligned} \tag{13.77}$$

A similar strategy may be used for signals in a mixed time–frequency domain. For instance, $\mathbf{k}_\mathrm{I}$ and $\mathbf{k}_\mathrm{II}$ are commonly displayed as the 2D frequency–frequency correlation plots (Equation 13.29)

$$\begin{aligned}\mathcal{R}^{\mathbf{k}_\mathrm{I}}_{\nu_4\nu_3\nu_2\nu_1}(\Omega_3,t_2,\Omega_1) &= 2\left(\frac{i}{\hbar}\right)^4 \sum_{n_1n_2n_3n_4} \sum_{m_1m_2m_3m_4m_1'm_1''p_2} \int \mathrm{d}\mathbf{q}_1\mathrm{d}\mathbf{q}_2\mathrm{d}\mathbf{q}_3\mathrm{d}\mathbf{q}_s\mathrm{d}\mathbf{q}_4 \\ &\times \mu_{n_4\mathbf{q}_4;\nu_4}\mu_{n_3\mathbf{q}_3;\nu_3}\mu_{n_2\mathbf{q}_2;\nu_2}\mu_{n_1\mathbf{q}_1;\nu_1}\hat{G}_{n_4\mathbf{q}_4,m_4\mathbf{q}_s}(\Omega_3)U_{m_4m_1m_3m_2;\mathbf{q}_s} \\ &\times \hat{\mathcal{G}}^{(Z)}_{m_1m_3m_2\mathbf{q}_s,m_1'n_3p_2\mathbf{q}_3}(\Omega_3)\hat{G}^{(N)}_{m_1'p_2\mathbf{q}_3,m_1''n_2\mathbf{q}_2}(t_2,0)\hat{G}^{*}_{m_1''\mathbf{q}_2,n_1\mathbf{q}_1}(\Omega_1)P(\mathbf{q}_1)\end{aligned} \tag{13.78}$$

$$\begin{aligned}\mathcal{R}^{\mathbf{k}_\mathrm{II}}_{\nu_4\nu_3\nu_2\nu_1}(\Omega_3,t_2,\Omega_1) &= 2\left(\frac{i}{\hbar}\right)^4 \sum_{m_1m_2m_3m_4p_1'm_2'p_1} \int \mathrm{d}\mathbf{q}_1\mathrm{d}\mathbf{q}_2\mathrm{d}\mathbf{q}_3\mathrm{d}\mathbf{q}_s\mathrm{d}\mathbf{q}_4 \\ &\times \mu_{n_4\mathbf{q}_4;\nu_4}\mu_{n_3\mathbf{q}_3;\nu_3}\mu_{n_2\mathbf{q}_2;\nu_2}\mu_{n_1\mathbf{q}_1;\nu_1}\hat{G}_{n_4\mathbf{q}_4,m_4\mathbf{q}_s}(\Omega_3)U_{m_4m_2m_3m_1;\mathbf{q}_s} \\ &\times \hat{\mathcal{G}}^{(Z)}_{m_2m_3m_1\mathbf{q}_s,m_2'n_3p_1\mathbf{q}_3}(\Omega_3)\hat{G}^{(N)}_{m_2'p_1\mathbf{q}_3,n_2p_1'\mathbf{q}_2}(t_2,0)\hat{G}_{p_1'\ \mathbf{q}_2n_1\mathbf{q}_1}(\Omega_1)P(\mathbf{q}_1).\end{aligned} \tag{13.79}$$

So far, we have used a compact functional notation to represent bath variables $\mathbf{q}$ of arbitrary nature. Equations 13.48 through 13.60 and 13.66 through 13.79 may be readily applied to continuous variables, as well as to discrete multistate jump processes, when represented along the lines of Equation 13.65. However, the discrete case may be described using a much simpler notation.

Let us connect the probability densities $P(q)$ and occupation probabilities of j-th state p_j by setting $P(q,t) = \sum_j p_j(t)\delta(q - \xi_j)$. Equation 13.61 is then equivalent to the master equation

$$\frac{dp_j}{dt} = \sum_k T_{jj'} p_{j'}(t). \tag{13.80}$$

The Green's function calculation can be reduced from operator to a matrix algebra by prescribing $\hat{G}_{mq,m'q'=\xi_{j'}} = \sum_j \delta(q - \xi_j)\hat{g}_{mj,m'j'}$.* Equation 13.66 then becomes

$$\frac{d}{d\tau}\hat{g}_{mj,m'j'}(\tau,\tau') = -\frac{i}{\hbar}\sum_{m''} h_{mm'';j}\hat{g}_{m''j,m'j'}(\tau,\tau') + \sum_{j''} T_{jj''}\hat{g}_{mj'',m'j'}(\tau,\tau') + \delta_{mm'}\delta_{jj'}\delta(\tau - \tau'). \tag{13.81}$$

Other Green's functions can be reduced similarly by defining $\hat{\mathcal{G}}_{mnq,m'n'q'=\xi_{j'}} = \sum_j \delta(q - \xi_j)\hat{\gamma}_{mnj,m'n'j'}$, $\hat{G}^{(N)}_{mnq,m'n'q'=\xi_{j'}} = \sum_j \delta(q - \xi_j)\hat{g}^{(N)}_{mnj,m'n'j'}$, and $\hat{\mathcal{G}}^{(Z)}_{kmnq,k'm'n'q'=\xi_{j'}} = \sum_j \delta(q - \xi_j)\hat{\gamma}^{(Z)}_{kmnj,k'm'n'j'}$. The evolution equations (Equations 13.67 through 13.69) are thus transformed by using the obvious scheme: the coordinate $\mathbf{q}$ is replaced by a discrete index j, $T^{\mathbf{q}}(G_{\mathbf{qq}'\ldots}) \to \sum_{j''} T_{jj''} g_{j''j'\ldots}$ and $\delta(\mathbf{q} - \mathbf{q}') \to \delta_{jj'}$

$$\begin{aligned}\frac{d}{d\tau}\hat{\gamma}_{mnq,m'n'j'}(\tau,\tau') = &-\frac{i}{\hbar}\sum_{m''n''} h^{(Y)}_{mn,m''n'';j}\hat{\gamma}_{m''n''j,m'n'j'}(\tau,\tau') \\ &+ \sum_{j''} T_{jj''}\hat{\gamma}_{mnj'',m'n'j'}(\tau,\tau') + (\delta_{mm'}\delta_{nn'} + \delta_{nm'}\delta_{mn'})\delta_{jj'}\delta(\tau - \tau')\end{aligned} \tag{13.82}$$

$$\begin{aligned}\frac{d}{d\tau}\hat{g}^{(N)}_{mnj,m'n'j'}(\tau,\tau') = &\frac{i}{\hbar}\sum_k \left[h_{km;j}\hat{g}^{(N)}_{knj,m'n'j'}(\tau,\tau') - h_{nk;j}\hat{g}^{(N)}_{mkj,m'n'j'}(\tau,\tau')\right] \\ &+ \sum_{j''} T_{jj''}\hat{g}^{(N)}_{mnj'',m'n'j'}(\tau,\tau') + \delta_{mm'}\delta_{nn'}\delta_{jj'}\delta(\tau - \tau')\end{aligned} \tag{13.83}$$

$$\begin{aligned}\frac{d}{d\tau}\hat{\gamma}^{(Z)}_{kmnj,k'm'n'j'}(\tau,\tau') = &-\frac{i}{\hbar}\sum_{m''n''} h^{(Y)}_{mn,m''n'';j}\hat{\gamma}^{(Z)}_{km''n''j,k'm'n'j'}(\tau,\tau') + \frac{i}{\hbar}\sum_{k''} h_{k''k;j}\hat{\gamma}^{(Z)}_{k''mnj,k'm'n'j'}(\tau,\tau') \\ &+ \sum_{j''} T_{jj''}\hat{\gamma}^{(Z)}_{kmnj'',k'm'n'j'}(\tau,\tau') + \delta_{kk'}(\delta_{mm'}\delta_{nn'} + \delta_{nm'}\delta_{mn'})\delta_{jj'}\delta(\tau' - \tau).\end{aligned} \tag{13.84}$$

To rewrite Equations 13.54 through 13.60, we further need to replace the integrations $\int d\mathbf{q}$ by summations $\sum_j$. The nonlinear signals are finally given by

$$\begin{aligned}R^{\mathbf{k}_\mathrm{I}}_{\nu_4\nu_3\nu_2\nu_1}(\tau_4,\tau_3,\tau_2,\tau_1) = &2\left(\frac{i}{\hbar}\right)^4 \sum_{n_1n_2n_3n_4}\ \sum_{m_1m_2m_3m_4m_1'm_1''p_2}\ \sum_{j_1j_2j_3j_sj_4} \mu_{n_4j_4;\nu_4}\mu_{n_3j_3;\nu_3}\mu_{n_2j_2;\nu_2}\mu_{n_1j_1;\nu_1} \\ &\times \int_{\tau_3}^{\tau_4} ds\, \hat{g}_{n_4j_4,m_4j_s}(\tau_4,s)U_{m_4m_1m_3m_2;j_s}\hat{\gamma}^{(Z)}_{m_1m_3m_2j_s,m_1'n_3p_2j_3}(s,\tau_3)\hat{g}^{(N)}_{m_1'p_2j_3,m_1''n_2j_2}(\tau_3,\tau_2)\hat{g}^{*}_{m_1''j_2,n_1j_1}(\tau_2,\tau_1)p_{j_1},\end{aligned} \tag{13.85}$$

* Values outside $q' = \xi_j$ have no physical meaning. The formal solution of Equation 13.66 yields $\hat{G}_{mq,m'q'}(\tau,\tau') = \theta(\tau - \tau')\delta_{mm'}\delta(q - q')$, but it does not enter in the final formulas of Equations 13.54, 13.57, or 13.60.

$$R^{\mathbf{k}_{\mathrm{II}}}_{\nu_4\nu_3\nu_2\nu_1}(\tau_4,\tau_3,\tau_2,\tau_1) = 2\left(\frac{i}{\hbar}\right)^4 \sum_{n_1n_2n_3n_4} \sum_{m_1m_2m_3m_4p_1'm_2'p_1} \sum_{j_1j_2j_3j_sj_4} \mu_{n_4j_4;\nu_4}\mu_{n_3j_3;\nu_3}\mu_{n_2j_2;\nu_2}\mu_{n_1j_1;\nu_1}$$
$$\times \int_{\tau_3}^{\tau_4} \mathrm{d}s\, \hat{g}_{n_4j_4,m_4j_s}(\tau_4,s)\, U_{m_4m_2m_3m_1;j_s} \hat{\gamma}^{(Z)}_{m_2m_3m_1j_s,m_2'n_3p_1j_3}(s,\tau_3)\hat{g}^{(N)}_{m_2'p_1j_3,n_2p_1'j_2}(\tau_3,\tau_2)\hat{g}_{p_1'j_2n_1j_1}(\tau_2,\tau_1)p_{j_1}, \tag{13.86}$$

$$R^{\mathbf{k}_{\mathrm{III}}}_{\nu_4\nu_3\nu_2\nu_1}(\tau_4,\tau_3,\tau_2,\tau_1) = \left(\frac{i}{\hbar}\right)^4 \sum_{n_1n_2n_3n_4} \sum_{m_1m_2m_3m_4m_1'm_2'p_1} \sum_{j_1j_2j_3j_sj_4} \mu_{n_4j_4;\nu_4}\mu_{n_3j_3;\nu_3}\mu_{n_2j_2;\nu_2}\mu_{n_1j_1;\nu_1}$$
$$\times \int_{\tau_3}^{\tau_4} \mathrm{d}s\hat{g}_{n_4j_4,m_4j_s}(\tau_4,s)U_{m_4m_3m_2m_1;j_s}\; \hat{\gamma}^{(Z)}_{m_3m_2m_1j_s,n_3m_2'm_1'n_3j_3}(s,\tau_3)\hat{\gamma}_{m_2'm_1'j_3,n_2p_1j_2}(\tau_3,\tau_2)\,\hat{g}_{p_1j_2,n_1j_1}(\tau_2,\tau_1)p_{j_1}. \tag{13.87}$$

To connect the present formalism to the NEE approach [22], we assume that the bath is fast $\Lambda_j \gg |h_{mn,\sigma_j} - h_{mn,-\sigma_j}|$, so that the stochastic process can explore the entire $\mathbf{q}$ space before quantum evolution takes place. The Green's functions may then be factorized as follows:

$$\hat{G}_{m\mathbf{q},m'\mathbf{q}'}(\tau,\tau') = \bar{G}_{mm'}(\tau,\tau')P(\mathbf{q}), \tag{13.88}$$

$$\hat{\mathcal{G}}_{mn\mathbf{q},m'n'\mathbf{q}'}(\tau,\tau') = \bar{\mathcal{G}}_{mn,m'n'}(\tau,\tau')P(\mathbf{q}), \tag{13.89}$$

$$\hat{G}^{(N)}_{mn\mathbf{q},m'n'\mathbf{q}'}(\tau,\tau') = \bar{G}^{(N)}_{mn,m'n'}(\tau,\tau')P(\mathbf{q}), \tag{13.90}$$

$$\hat{\mathcal{G}}^{(Z)}_{kmn\mathbf{q},k'm'n'\mathbf{q}'}(\tau,\tau') = \bar{\mathcal{G}}^{(Z)}_{kmn,k'm'n'}(\tau,\tau')P(\mathbf{q}). \tag{13.91}$$

Bath variables may be entirely integrated out of Equations 13.54 through 13.60, and we recover the NEE results of Ref. [22]:

$$R^{\mathbf{k}_{\mathrm{I}}}_{\nu_4\nu_3\nu_2\nu_1}(\tau_4,\tau_3,\tau_2,\tau_1) = 2\left(\frac{i}{\hbar}\right)^4 \sum_{n_1n_2n_3n_4} \sum_{m_1m_2m_3m_4m_1'm_1''p_2} \bar{\mu}_{n_4;\nu_4}\bar{\mu}_{n_3;\nu_3}\bar{\mu}_{n_2;\nu_2}\bar{\mu}_{n_1;\nu_1}$$
$$\times \int_{\tau_3}^{\tau_4} \mathrm{d}s\, \bar{G}_{n_4,m_4}(\tau_4,s)\bar{U}_{m_4m_1m_3m_2}\bar{\mathcal{G}}^{(Z)}_{m_1m_3m_2,m_1'n_3p_2}(s,\tau_3)\bar{G}^{(N)}_{m_1'p_2,m_1''n_2}(\tau_3,\tau_2)\bar{G}^{*}_{m_1'',n_1}(\tau_2,\tau_1), \tag{13.92}$$

$$R^{\mathbf{k}_{\mathrm{II}}}_{\nu_4\nu_3\nu_2\nu_1}(\tau_4,\tau_3,\tau_2,\tau_1) = 2\left(\frac{i}{\hbar}\right)^4 \sum_{n_1n_2n_3n_4} \sum_{m_1m_2m_3m_4p_1'm_2'p_1} \bar{\mu}_{n_4;\nu_4}\bar{\mu}_{n_3;\nu_3}\bar{\mu}_{n_2;\nu_2}\bar{\mu}_{n_1;\nu_1}$$
$$\times \int_{\tau_3}^{\tau_4} \mathrm{d}s\, \bar{G}_{n_4\mathbf{q}_4,m_4}(\tau_4,s)\bar{U}_{m_4m_2m_3m_1}\bar{\mathcal{G}}^{(Z)}_{m_2m_3m_1,m_2'n_3p_1}(s,\tau_3)\bar{G}^{(N)}_{m_2'p_1,n_2p_1'}(\tau_3,\tau_2)\bar{G}_{p_1',n_1}(\tau_2,\tau_1), \tag{13.93}$$

$$R^{\mathbf{k}_{\mathrm{III}}}_{\nu_4\nu_3\nu_2\nu_1}(\tau_4,\tau_3,\tau_2,\tau_1) = \left(\frac{i}{\hbar}\right)^4 \sum_{n_1n_2n_3n_4} \sum_{m_1m_2m_3m_4m_1'm_2'p_1} \bar{\mu}_{n_4;\nu_4}\bar{\mu}_{n_3;\nu_3}\bar{\mu}_{n_2;\nu_2}\bar{\mu}_{n_1;\nu_1}$$
$$\times \int_{\tau_3}^{\tau_4} \mathrm{d}s\bar{G}_{n_4,m_4}(\tau_4,s)\bar{U}_{m_4m_3m_2m_1}\bar{\mathcal{G}}^{(Z)}_{m_3m_2m_1,n_3m_2'm_1'n_3}(s,\tau_3)\bar{\mathcal{G}}_{m_2'm_1',n_2p_1}(\tau_3,\tau_2)\bar{G}_{p_1,n_1}(\tau_2,\tau_1). \tag{13.94}$$

where we averaged $\bar{U}_{m_4m_3m_2m_1} \equiv \int U_{m_4m_3m_2m_1;\mathbf{q}}P(\mathbf{q})\mathrm{d}\mathbf{q}$, and $\bar{\mu}_{n_1;\nu_1} \equiv \int \mu_{n_1\mathbf{q};\nu_1}P(\mathbf{q})\mathrm{d}\mathbf{q}$.

The averaged Green's functions $\bar{G}$, $\bar{\mathcal{G}}$, $\bar{G}^{(N)}$, $\bar{\mathcal{G}}^{(Z)}$ are semigroups in the excitonic space, and represent Redfield type dynamics [88]. Their calculation involves averaging the Hamiltonian coefficients $\bar{h}_{mm'} \equiv \int h_{mm';\mathbf{q}} P(\mathbf{q})\mathrm{d}\mathbf{q}$ and adding relaxation. The relaxation terms depend on details of the bath. For linearly coupled Ornstein–Uhlenbeck coordinates $h = \bar{h} + \sum_i q_i \Delta^i$, the one-exciton Green's function can be calculated by solving [89]

$$\frac{d}{d\tau}\bar{G}_{m,m'}(\tau,\tau') = -\frac{i}{\hbar}\sum_{m''}\bar{h}_{mm''}\bar{G}_{m'',m'}(\tau,\tau') - \sum_{im''m'''}\frac{\sigma_i^2}{\Lambda_i}\Delta^i_{mm'''}\Delta^i_{m'''m''}\bar{G}_{m'',m'}(\tau,\tau') + \delta_{mm'}\delta(\tau-\tau'). \tag{13.95}$$

Similar equations can be developed in the fast bath limit for the other averaged Green's functions. This level of theory reduces the effects of Hamiltonian fluctuations to a few transport and decoherence rates. It has been successfully used to simulate electronic spectra in the visible and to describe signatures of electron transfer in photosynthetic complexes and organic dyes. However, since the fluctuation timescales are completely neglected in this approach, it is less adequate for simulations of infrared spectra of protein vibrations.

13.4.2 Numerical Simulations

We first apply the stochastic approach to hydrogen-bonding dynamics in organic solvents. In the experiments of Ref. [14], phenol with deuterated hydroxyl group was dissolved in mixture of benzene and tetrachloromethane. The phenolic OD group can form a hydrogen bond with the surrounding benzene molecules. Such binding is not possible in tetrachloromethane; in their vicinity, the phenolic group remains free. Complexation and dissociation of hydrogen bond follow the changes of phenol surroundings in the course of time. This exchange dynamics has been monitored by 2D IR spectra of the OD stretch [14]. The absorption spectrum shows two peaks attributed to complexed and dissociated benzene molecules, respectively. We associated the free $q_1 = 1$ and complexed phenol $q_1 = -1$ with two states of random-telegraph jump process $q_1(t)$. The state of the process alters vibrational frequency h of the monitored OD stretch. A number of smaller environmental fluctuations are expected to compose additional Gaussian collective coordinate $q_2(t)$. The fluctuating vibrational frequency was modeled [28] by multilinear form linear of both stochastic processes $h = \zeta_0 + \zeta_1 q_1(t) + \zeta_2 q_2(t) + \zeta_3 q_1(t) q_2(t)$. The coefficients $\zeta_0, \ldots, \zeta_3$ and the anharmonicity $2U$ (here taken as constant) may be adjusted to experiment ($H_S = h(\mathbf{q}(t))b^\dagger b + Ub^\dagger b^\dagger bb$; $\mathbf{q} \equiv \{q_1, q_2\}$).

The dynamics of $\mathbf{q}(t)$ is taken to be Markovian. Master (Smoluchowski) Equation 13.63 with $\sigma_2 = 1$ and adjustable relaxation rate Λ_2 describe Ornstein–Uhlenbeck dynamics of q_2 coordinate. The kinetic equation (Equation 13.65) describes the probability evolution of random-telegraph process q_1 with dissociation rates $T_{-11} = -T_{11} = k_{\mathrm{com}}$ and complexation rates $T_{1-1} = -T_{-1-1} = k_{\mathrm{dis}}$.

This simple model [28] provides a good agreement with experiment [14] using the parameters given in the caption of Figure 13.9. The linear spectrum (top panel) shows two peaks at $\zeta_0 \pm \zeta_1$ frequencies (free and complexed phenol) with correct peak widths. The various relaxation regimes are correctly reproduced in delay time evolution of 2D spectrum. The memory of the q_2 coordinate is lost after 2 ps due to spectral diffusion and peak shapes are changed from diagonally elongated to circular on this timescale. Cross peaks appearing at ~10 ps signify the formation and dissociation of the hydrogen bond.

In the 2D spectra of phenol OD stretch, the positive and negative peaks are well separated. The ESA pathway (Figure 13.3) does not interfere with GSB and ESE pathways, and the QP picture thus does not provide substantial advantages over the SLE used in Ref. [28]. The power of

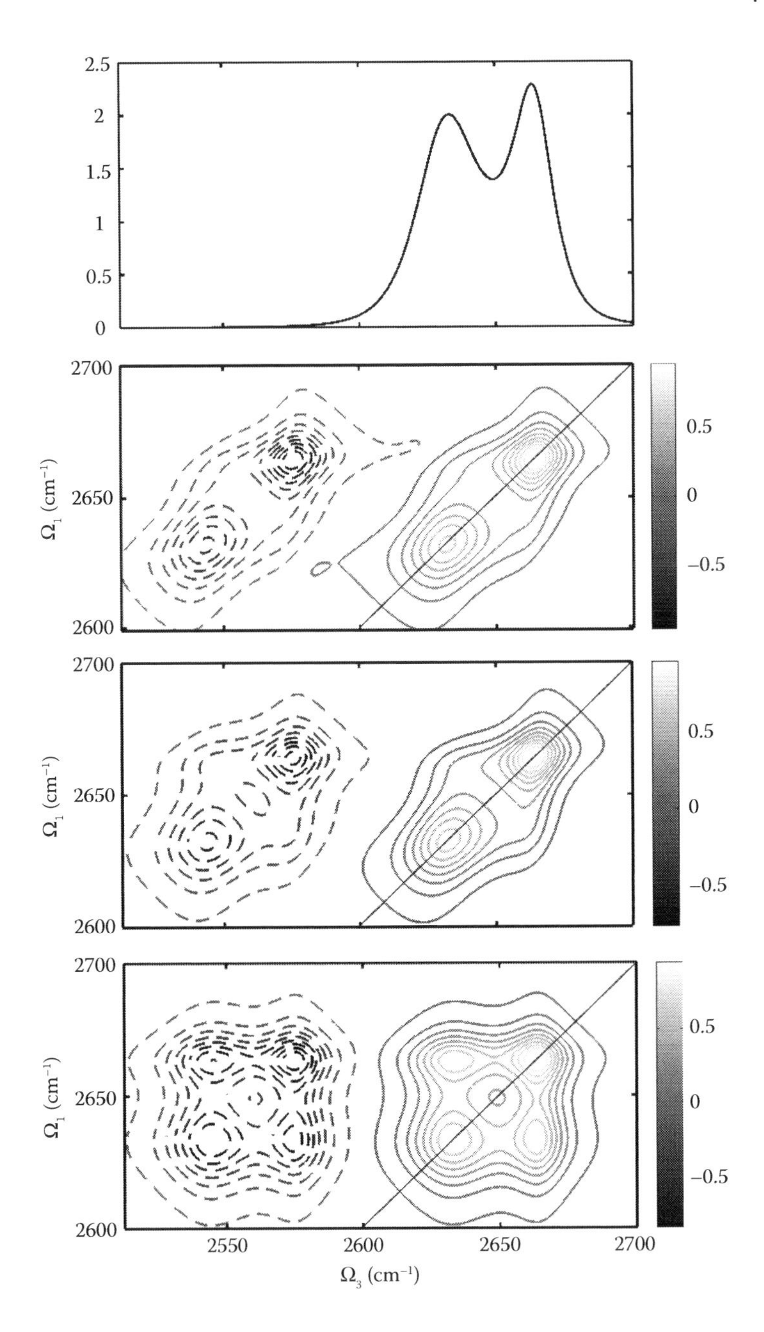

FIGURE 13.9 The absorption lineshape (top panel), and the $\mathbf{k}_{\rm I} + \mathbf{k}_{\rm II}$ signal (Equation 13.47) of a stochastic exchange model for delay times (from top to bottom) $t_2 = 0$, $t_2 = 2$ ps, and $t_2 = 10$ ps of the OD stretch [28]. Parameters used: $\zeta_0 = 2648$ cm^{-1}, $\zeta_1 = 17$ cm^{-1}, $\zeta_2 = 11$ cm^{-1}, $\zeta_3 = -2.4$ cm^{-1}, $U = -45$ cm^{-1}, $k_{\rm dis} = 0.125$ ps^{-1}, $k_{\rm com} = 0.1$ ps, $\Lambda_1 = 0.4$ ps^{-1}. Solid (dashed) peaks are positive (negative).

the QP approach is demonstrated for the following model of linear chain of vibrations with stochastic orientation jumps. The spectra of linear tetramer in which each ($i = 1, \ldots, 4$) vibration can assume two spatial orientations (modeled by discrete stochastic coordinate $j_i = \pm 1$) are shown in Figure 13.10. Four discrete Markovian coordinates $\mathbf{j} = \{j_i\}$ form the relevant configuration space of geometries.

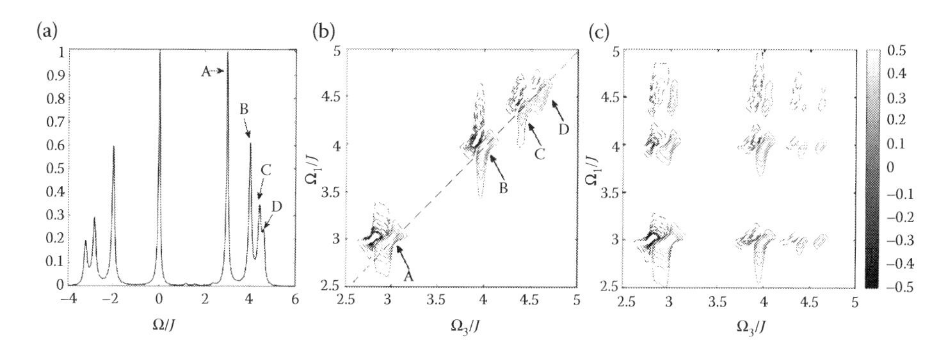

FIGURE 13.10 The calculations of Ref. [30] for Frenkel excitons adapted for collections of bosons by taking finite anharmonicities. (a) Absorption spectrum of a tetramer. Vibrations jump between perpendicular $\mu_n = (1,0,0)$ and parallel $\mu_n = (0,1,0)$ orientation to chain. Frequencies are parameterized by the coupling between two dipoles perpendicular to chain $J \equiv J(\mu_i(q_i = 1), \mu_{i+1}(q_{i+1} = 1))$. Parameters: local field $\boldsymbol{\varepsilon} = (3J,0,0)$, transition rate $k = 0.02J$, and anharmonicity $2U = 0.2J$. Peaks A, B, C, and D correspond to size 1, 2, 3, and 4 domain of dipoles perpendicular to the chain. (b) Nonlinear $\mathbf{k}_\text{I}$ signal at zero delay times $t_2 = 0$. (A–D) region of spectrum is presented. Solid (dashed) peaks are positive (negative). (c) The same after relaxation $kt_2 = 1$.

The primary effect of reorientation is to change the dipole moment $\mu_m(\mathbf{j}) = \mu(j_m)$. Orientational dependence of the transition frequency is described by local field $\boldsymbol{\varepsilon}$, and the coupling is approximated by the dipole–dipole interaction (Equation 13.32) of nearest neighbors

$$h_{mn;\mathbf{j}} = [\bar{\varepsilon} + \boldsymbol{\varepsilon} \cdot \mu(j_m)]\delta_{mn} + J(\mu(j_m); \mu(j_n))(\delta_{mn+1} + \delta_{m+1n}). \tag{13.96}$$

The anharmonicity $2U$ is taken as fixed (no fluctuation) $U_{mnm'n'} = U\delta_{mn}\delta_{m'n'}\delta_{m'n}$. The stochastic Markovian dynamics was described by high-temperature master equation (Equation 13.80) with

$$T_{\mathbf{jj}'} = \sum_{i=1}^{4} \tilde{T}^{(i)}_{j_i j_{i'}} \tag{13.97}$$

representing the composition of symmetric random-telegraph processes on each vibration $\tilde{T}^{(i)}_{1-1} = \tilde{T}^{(i)}_{-11} = -\tilde{T}^{(i)}_{11} = -\tilde{T}^{(i)}_{-1-1} = k$.

Equations 13.82 through 13.87 have been used to calculate linear and the third-order response. We oriented the chain along the y axis and considered flips between perpendicular $\mu(j = 1) = (1,0,0)$ and parallel $\mu(j = -1) = (0,1,0)$ orientation of the dipole.

The peaks of linear spectra (Figure 13.10a) represent delocalized excitons. Delocalization between oscillators of different orientation is less effective, since the coupling between perpendicular dipoles vanishes. Delocalized excitons can thus be associated with domains of ordered dipoles (see caption) that represent excitons of different delocalization length.

The 2D spectra shown at the central and right panels of Figure 13.10 correspond to dynamics of the "perpendicular domains" over various delay times. At the zero delay time (central), diagonal peaks show up accompanied with overtone peaks. These correspond to A, B, C, and D peaks of linear spectrum, thus with delocalized excitons over ordered domain. Cross peaks at larger delays (right) represent ultrafast domain transformations. The stochastic simulations are inexpensive once the relevant collective coordinates have been identified; calculations shown in Figures 13.9 and 13.10 took just a few minutes on the standard personal computer (PC).

13.5 CONCLUSIONS AND FUTURE PERSPECTIVES

We have presented a QP approach to describe nonlinear optical response of vibrational excitons. Exciton fluctuations are treated using either MD or stochastic simulations. The QP picture avoids the separation of the response into interfering (and almost canceling) Liouville space pathways. It provides useful insights into the structure of nonlinear response, and a practical alternative to the more common SOS simulation protocols. Applications to vibrational excitons in liquid–water, amyloid fibrils, solute, or phenol in benzene–tetrachloromethane demonstrate the power of the approach.

The QP approach provides a complete simulation protocol that description of exciton dynamics with environmental fluctuations for arbitrary timescales. The classical bath used here has several limitations. Our mixed classical-quantum approach neglects back-reaction of the quantum evolution to the classical bath. Some effects, such as effective anharmonic couplings between excitons, or relaxation within a band (at fixed quantum number), are not taken into account by this adiabatic decoupling. For Gaussian modulation (Equation 13.63), the Stokes shift associated with the classical coordinate has been studied in Refs. [90,91]. A different, but equivalent, algebraic formulation, known as the Kubo–Tanimura hierarchy [85] was originally motivated by computing the temperature corrections to stochastic quantum dynamics. Non-Gaussian extensions and the connections to MD remains an open issue.

The present formalism neglects population relaxation. We assume that the vibrational wavefunction of each local oscillator does not vary with the bath. Population relaxation is only included phenomenologically by adding a decay rate. For proteins, it has been shown that the amide-I vibrational population relaxation is of the order of 1.2 ps [8]. In liquid water, the population relaxation occurs at even lower timescale around 200 fs [92]. Most approaches to population relaxation rely on either perturbation theory or classical dynamics [93,94]. The development of more rigorous quantum simulation algorithms that account for population relaxation in nonlinear spectroscopy is an important future challenge.

ACKNOWLEDGMENTS

The support of the Agence Nationale de la Recherche (Grant No. ANR-2011-BS04-027-03), the Grant Agency of the Czech Republic (Grant No. 205/10/0989), the National Institutes of Health (Grant No. GM59230), and the National Science Foundation (Grant No. CHE-0745892), is gratefully acknowledged. František Šanda acknowledges Václav Perlík for his contribution to the work reviewed here.

REFERENCES

1. Davydov AS. 1962. *Theory of Molecular Excitons* (McGraw-Hill, New York).
2. Krimm S, Abe Y. 1972. Intermolecular interaction effects in the amide I vibrations of β-polypeptides. *Proc. Natl. Acad. Sci. U.S.A.* 69:2788–2792.
3. Miyazawa T. 1960. Perturbation treatment of the characteristic vibrations of polypeptide chains in various configurations. *J. Chem. Phys.* 32:1647–1652.
4. Elliott AA, Ambrose EJ, Robinson C. 1950. Chain configurations in natured and denatured insulin: Evidence from infra-red spectra. *Nature* 166:194–194.
5. Elliott A. 1954. Infra-red spectra of polypeptides with small side chains. *Proc. Roy. Soc. (London) A* 226:408–421.
6. Torii H, Tasumi M. 1992. Model calculations on the amide-I infrared bands of globular proteins. *J. Chem. Phys.* 96:3379–3387.
7. Torii H, Tasumi M. 1998. Ab initio molecular orbital study of the amide-I vibrational interactions between the peptide groups in di- and tripeptides and considerations on the conformation of the extended helix. *J. Raman Spectrosc.* 29:81–86.
8. Hamm P, Lim M, Hochstrasser RM. 1998. Structure of the amide I band of peptides measured by femtosecond nonlinear-infrared spectroscopy. *J. Phys. Chem. B* 102:6123–6138.
9. Kim YS, Hochstrasser RM. 2009. Applications of 2D IR spectroscopy to peptides, proteins, and hydrogen-bond dynamics. *J. Phys. Chem. B* 113:8231–8251.

10. Lim M, Hamm P, Hochstrasser RM. 1998. Protein fluctuations are sensed by stimulated infrared echoes of the vibrations of carbon monoxide and azide probes. *Proc. Natl. Acad. Chem. Soc. U.S.A.* 95: 15315–15320.
11. Owrutsky JC, Li M, Locke B, Hochstrasser RM. 1995. Vibrational relaxation of the CO stretch vibration in hemoglobin-CO, myoglobin-CO, and protoheme-CO. *J. Phys. Chem.* 99:4842–4846.
12. Park J, Ha JH, Hochstrasser RM. 2004. Multidimensional infrared spectroscopy of the N-H bond motions in formamide. *J. Chem. Phys.* 121:7281–7292.
13. Zanni MT, Hochstrasser RM. 2001. Two-dimensional infrared spectroscopy: A promising new method for the time resolution of structures. *Curr. Opin. Struc. Biol.* 11:516–522.
14. Zheng J et al. 2005. Ultrafast dynamics of solute-solvent complexation observed at thermal equilibrium in real time. *Science* 309:1338–1343.
15. Choi JH, Ham S, Cho M. 2003. Local amide I mode frequencies and coupling constants in polypeptides. *J. Phys. Chem. B* 107:9132–9138.
16. Ham S, Kim JH, Lee H, Cho M. 2003. Correlation between electronic and molecular structure distortions and vibrational properties. II. amide I modes of NMA-nD_2O complexes. *J. Chem. Phys.* 118:3491–3498.
17. Ham S, Cha S, Choi JH, Cho M. 2003. Amide I modes of tripeptides: Hessian matrix reconstruction and isotope effects. *J. Chem. Phys.* 119:1451–1461.
18. Buch V, Devlin JP. 1999. A new interpretation of the OH-stretch spectrum of ice. *J. Chem. Phys.* 110:3437–3443.
19. Buch V, Bauerecker S, Devlin JP, Buck U, Kazimirski JK. 2004. Solid water clusters in the size range of tens-thousands of H_2O: A combined computational/spectroscopic outlook. *Int. Rev. Phys. Chem.* 23:375–433.
20. Lawrence CP, Skinner JL. 2002. Vibrational spectroscopy of HOD in liquid D_2O. II. Infrared line shapes and vibrational stokes shift. *J. Chem. Phys.* 117:8847–8854.
21. Mukamel S, Abramavicius D. 2004. Many-body approaches for simulating coherent nonlinear spectroscopies of electronic and vibrational excitons. *Chem. Rev.* 104:2073–2098.
22. Abramavicius D, Palmieri B, Voronine DV, Šanda F, Mukamel S. 2009. Coherent multidimensional optical spectroscopy of excitons in molecular aggregates: Quasiparticle versus supermolecule perspectives. *Chem. Rev.* 109:2350–2408.
23. Spano FC, Mukamel S. 1989. Nonlinear susceptibilities of molecular aggregates: Enhancement of $\chi^{(3)}$ by size. *Phys. Rev. A* 40:5783–5801.
24. Auer BM, Skinner JL. 2007. Dynamical effects in line shapes for coupled chromophores: Time-averaging approximation. *J. Chem. Phys.* 127:104105.
25. Jansen TLC, Ruszel WM. 2008. Motional narrowing in the time-averaging approximation for simulating two-dimensional nonlinear infrared spectra. *J. Chem. Phys.* 128:214501.
26. Jansen TLC, Knoester J. 2006. Nonadiabatic effects in the two-dimensional infrared spectra of peptides: Application to alanine dipeptide. *J. Phys. Chem. B* 110:22910–22916.
27. Jansen TLC, Zhuang W, Mukamel S. 2004. Stochastic Liouville equation simulation of multidimensional vibrational line shapes of trialanine. *J. Chem. Phys.* 121:10577.
28. Šanda F, Mukamel S. 2006. Stochastic simulation of chemical exchange in two dimensional infrared spectroscopy. *J. Chem. Phys.* 125:014507.
29. Falvo C, Palmieri B, Mukamel S. 2009. Coherent infrared multidimensional spectra of the OH stretching band in liquid water simulated by direct nonlinear exciton propagation. *J. Chem. Phys.* 130:184501.
30. Šanda F, Perlk V, Mukamel S. 2010. Exciton coherence length fluctuations in chromophore aggregates probed by multidimensional optical spectroscopy. *J. Chem. Phys.* 133:014102.
31. Mukamel S. 1995. *Principles of Nonlinear Optical Spectroscopy* (Oxford University Press, Oxford).
32. Pouthier V. 2011. Quantum decoherence in finite size exciton-phonon systems. *J. Chem. Phys.* 134:114516.
33. Barišic O, Barišic S. 2008. Phase diagram of the Holstein polaron in one dimension. *Eur. Phys. J. B* 64:1–18.
34. Gindensperger E, Meier C, Beswick JA. 2000. Mixing quantum and classical dynamics using Bohmian trajectories. *J. Chem. Phys.* 113:9369–9372.
35. Tully JC. 1998. In *Classical and Quantum Dynamics in Condensed Phase Simulations*, eds Berne B, Cicotti G, Cokes D (World Scientific, Singapore), pp. 489–514.
36. Billing GD. 1994. Classical path method in inelastic and reactive scattering. *Intern. Rev. Phys. Chem.* 13:309–335.
37. Pouthier V. 2003. Two-vibron bound states in alpha-helix proteins: The interplay between the intramolecular an harmonicity and the strong vibron-phonon coupling. *Phys. Rev. E* 68:021909.

38. Edler J, Pfister R, Pouthier V, Falvo C, Hamm P. 2004. Direct observation of self-trapped vibrational states in α-helices. *Phys. Rev. Lett.* 93:106405.
39. Andrews DL, Thirunamachandran T. 1977. On three-dimensional rotational averages. *J. Chem. Phys.* 67:5026.
40. Hochstrasser RM. 2001. Two-dimensional IR-spectroscopy: Polarization anisotropy effects. *Chem. Phys.* 266:273–284.
41. Allen MP, Tildesley DJ. 1989. *Computer Simulation of Liquids* (Oxford University Press, Oxford).
42. Becker OM, Mackerell, Jr. AD, Roux B, Watanabe M. 2001. *Computational Biochemistry and Biophysics* (Marcel Dekker, New York).
43. MacKerell AD et al. 1998. All-atom empirical potential for molecular modeling and dynamics studies of proteins. *J. Phys. Chem. B* 102:3586–3616.
44. Cornell WD et al. 1995. A second generation force field for the simulation of proteins, nucleic acids, and organic molecules. *J. Am. Chem. Soc.* 117:5179–5197.
45. Bour P, Keiderling TA. 2003. Empirical modeling of the peptide amide I band IR intensity in water solution. *J. Chem. Phys.* 119:11253–11262.
46. Hayashi T, Zhuang W, Mukamel S. 2005. Electrostatic DFT map for the complete vibrational amide band of NMA. *J. Phys. Chem. A* 109:9747–9759.
47. Jansen TLC, Knoester J. 2006. A transferable electrostatic map for solvation effects on amide I vibrations and its application to linear and two-dimensional spectroscopy. *J. Chem. Phys.* 124:044502.
48. Bloem R, Dijkstra AG, Jansen TLC, Knoester J. 2008. Simulation of vibrational energy transfer in two-dimensional infrared spectroscopy of amide I and amide II modes in solution. *J. Chem. Phys.* 129:055101.
49. Schmidt JR, Corcelli SA, Skinner JL. 2004. Ultrafast vibrational spectroscopy of water and aqueous *N*-methylacetamide: Comparison of different electronic structure/molecular dynamics approaches. *J. Chem. Phys.* 121:8887–8896.
50. Lin YS, Shorb JM, Mukherjee P, Zanni MT, Skinner JL. 2009. Empirical amide I vibrational frequency map: Application to 2D IR line shapes for isotope-edited membrane peptide bundles. *J. Phys. Chem. B* 113:592–602.
51. Bagchi S, Falvo C, Mukamel S, Hochstrasser RM. 2009. 2D IR experiments and simulations of the coupling between amide I and ionizable side chains in proteins: Application to the villin headpiece. *J. Phys. Chem. B* 113:11260–11273.
52. Jansen TLC, Dijkstra AG, Watson TM, Hirst JD, Knoester J. 2006. Modeling the amide I bands of small peptides. *J. Chem. Phys.* 125:044312.
53. Auer B, Kumar R, Schmidt JR, Skinner JL. 2007. Hydrogen bonding and Raman, IR, and 2D IR spectroscopy of dilute HOD in liquid D_2O. *Proc. Natl. Acad. Sci. U.S.A.* 104:14215–14220.
54. Auer BM, Skinner JL. 2008. IR and Raman spectra of liquid water: Theory and interpretation. *J. Chem. Phys.* 128:224511.
55. Auer BM, Skinner JL. 2008. Vibrational sum-frequency spectroscopy of the water liquid/vapor interface. *J. Phys. Chem. B* 113:4125–4130.
56. Corcelli SA, Skinner JL. 2005. Infrared and Raman line shapes of dilute HOD in liquid H_2O and D_2O from 10 to 90°C. *J. Phys. Chem. A* 109:6154–6165.
57. Paarmann A, Hayashi T, Mukamel S, Miller RJD. 2008. Probing intermolecular couplings in liquid water with two-dimensional infrared photon echo spectroscopy. *J. Chem. Phys.* 128:191103.
58. Buch V. 2005. Molecular structure and OH-stretch spectra of liquid water surface. *J. Phys. Chem. B* 109:17771–17774.
59. Falvo C et al. 2012. Frequency distribution of the amide I vibration sorted by residues in amyloid fibrils revealed by 2D IR measurements and simulations. *J. Phys. Chem. B* 116:3322–3330.
60. Hahn S, Ham S, Cho M. 2005. Simulation studies of amide I IR absorption and two-dimensional IR spectra of β hairpins in liquid water. *J. Phys. Chem. B* 109:11789–11801.
61. Wang J, Chen J, Hochstrasser RM. 2006. Local structure of β-hairpin isotopomers by FTIR, 2D IR, and ab initio theory. *J. Phys. Chem. B* 110:7545–7555.
62. Berendsen HJC, Grigera JR, Straatsma TP. 1987. The missing term in effective pair potentials. *J. Phys. Chem.* 91:6269–6271.
63. Jansen TLC, Hayashi T, Zhuang W, Mukamel S. 2005. Stochastic Liouville equations for hydrogen-bonding fluctuations and their signatures in two-dimensional vibrational spectroscopy of water. *J. Chem. Phys.* 123:114504.
64. Selkoe DJ. 1994. Cell biology of the amyloid beta-protein precursor and the mechanism of Alzheimer's disease. *Annu. Rev. Cell Biol.* 10:373–403.

65. Tycko R. 2004. Progress towards a molecular-level structural understanding of amyloid fibrils. *Curr. Opin. Struct. Biol.* 14:96–103.
66. Caughey B, Lansbury PT. 2003. Protofibrils, pores, fibrils, and neurodegeneration: Separating the responsible protein aggregates from the innocent bystanders. *Annu. Rev. Neurosci.* 26:267–298.
67. Sunde M, Blake CCF. 1998. From the globular to the fibrous state: Protein structure and structural conversion in amyloid formation. *Q. Rev. Biophys.* 31:1.
68. Lester-Coll N et al. 2006. Intracerebral streptozotocin model of type 3 diabetes: Relevance to sporadic alzheimer's disease. *J. Alzheimer's Dis.* 9:13–33.
69. Petkova AT et al. 2002. A structural model for alzheimer's β-amyloid fibrils based on experimental constraints from solid state NMR. *Proc. Natl. Acad. Sci. U.S.A.* 99:16742–16747.
70. Petkova AT et al. 2005. Self-propagating, molecular-level polymorphism in alzheimer's β-amyloid fibrils. *Science* 307:262–265.
71. Petkova AT, Yau WM, Tycko R. 2006. Experimental constraints on quaternary structure in alzheimer's β-amyloid fibrils. *Biochemistry* 45:498–512.
72. Paravastu AK, Petkova AT, Tycko R. 2006. Polymorphic fibril formation by residues 10–40 of the alzheimer's β-amyloid peptide. *Biophys. J.* 90:4618–4629.
73. Paravastu AK, Leapman RD, Yau WM, Tycko R. 2008. Molecular structural basis for polymorphism in alzheimer's β-amyloid fibrils. *Proc. Natl. Acad. Sci. U.S.A.* 105:18349–18354.
74. Kim YS, Liu L, Axelsen PH, Hochstrasser RM. 2008. Two-dimensional infrared spectra of isotopically diluted amyloid fibrils from Aβ40. *Proc. Natl. Acad. Sci. U.S.A.* 105:7720–7725.
75. Kim YS, Liu L, Axelsen PH, Hochstrasser RM. 2009. 2D IR provides evidence for mobile water molecules in β-amyloid fibrils. *Proc. Natl. Acad. Sci. U.S.A.* 106:17751–17756.
76. Donsker MD. 1952. Justification and extension of Doob's heuristic approach to the Kolmogorov-Smirnov theorems. *Ann. Math. Statist.* 23:277–281.
77. Okumura K, Tokmakoff A, Tanimura Y. 1999. Two-dimensional line-shape analysis of photon-echo signal. *Chem. Phys. Lett.* 314:488–495.
78. Šanda F, Mukamel S. 2006. Anomalous continuous-time random-walk spectral diffusion in coherent third-order optical response. *Phys. Rev. E* 73:011103.
79. Šanda F, Mukamel S. 2007. Anomalous lineshapes and aging effects in two-dimensional correlation spectroscopy. *J. Chem. Phys.* 127:154107.
80. van Kampen NG. 1992. *Stochastic Processes in Physics and Chemistry* (North-Holland, Amsterdam).
81. Gamliel D, Levanon H. 1995. *Stochastic Processes in Magnetic Resonance* (World Scientific, Singapore).
82. Burshtein AI. 1966. Kinetics of relaxation induced by a sudden potential change. *Sov. Phys. JETP* 22:939–947.
83. Uhlenbeck GE, Ornstein LS. 1930. On the theory of the Brownian motion. *Phys. Rev.* 36:823–841.
84. Anderson PW. 1954. A mathematical model for the narrowing of spectral lines by exchange or motion. *J. Phys. Soc. Jpn.* 9:316–339.
85. Tanimura Y. 2006. Stochastic Liouville, Langevin, Fokker-Planck, and master equation approaches to quantum dissipative systems. *J. Phys. Soc. Jpn.* 75:082001.
86. Šanda F, Mukamel S. 2008. Stochastic Liouville equations for coherent multidimensional spectroscopy of excitons. *J. Phys. Chem. B* 112:14212–14220.
87. Risken H. 1989. *The Fokker-Plank Equation* (Springer, Berlin).
88. Redfield AG. 1957. On the theory of relaxation processes. *J. Phys. Chem. B* 1:19–31.
89. Haken H, Strobl G. 1973. An exactly solvable model for coherent and incoherent exciton motion. *Z. Physik* 262:135–148.
90. Garg A, Onuchic JN, Ambegaokar V. 1985. Effect of friction on electron transfer in biomolecules. *J. Chem. Phys.* 83:4491–4503.
91. Zusman LD. 1980. Outer-sphere electron transfer in polar solvents. *Chem. Phys.* 110:295–304.
92. Cowan ML et al. 2005. Ultrafast memory loss and energy redistribution in the hydrogen bond network of liquid H_2O. *Nature* 434:199–202.
93. Botan V et al. 2007. Energy transport in peptide helices. *Proc. Natl. Acad. Chem. Soc. U.S.A.* 104: 12749–12754.
94. Stock G. 2009. Classical simulation of quantum energy flow in biomolecules. *Phys. Rev. Lett.* 102:118301.

14 Ultrafast Infrared Spectroscopy of Amylin Solution and Fibrils

L. Wang, L. E. Buchanan, E. B. Dunkelberger, J. J. de Pablo, M. T. Zanni, and J. L. Skinner

CONTENTS

14.1 INTRODUCTION

The aggregation of proteins from their native functional state to highly organized amyloid fibrils is associated with a wide range of diseases, including Alzheimer's and Huntington's disease and type 2 diabetes [1,2]. Despite the fact that each disease is associated with a particular protein, the amyloid fibrils have many characteristics in common. For example, they typically contain straight and unbranched fibrils, with diameters on the order of 10 nm and varying lengths ranging from 100 nm to many microns. They are rich in β-sheet content, and peptides in the fibril usually adopt a "cross-β" configuration, in which continuous β-strands are arranged perpendicular to the fibril axis and are hydrogen bonded along the fibril axis [2,3]. For detection, the fibrils can be stained with Congo red and produce a characteristic green birefringence under polarized light [1,3].

The peptide associated with type 2 diabetes is amylin, also known as islet amyloid polypeptide or IAPP, which is a 37-residue peptide hormone cosecreted with insulin by pancreatic β-cells. The amyloid deposits formed from human IAPP (hIAPP) are commonly found in the islet cells of patients with type 2 diabetes [4–7]. Although the aggregation of hIAPP has been linked to the disease, the cytotoxic structure of the peptide oligomer, the mechanism of amyloid formation, and the molecular basis of cell disruption are poorly understood, mainly due to the extremely fast aggregation kinetics.

Increasing evidence suggests that intermediates populated during amyloid formation, instead of the mature amyloid fibrils, may be the key toxic species [8–12]. Lipid membranes are postulated to play an important role as they greatly enhance the rate of hIAPP fibril formation in *in vitro* experiments [10,13,14]. Concomitantly, membrane damage is observed during the aggregation process, indicating that hIAPP-induced membrane disruption might be the origin of its cytotoxicity [8,10–12,15]. Therefore, understanding the aggregation of hIAPP both in the absence and in the presence of lipid membranes is of crucial importance to the understanding of its pathological roles.

The aggregation of proteins can be probed by a variety of experimental techniques. X-ray diffraction and nuclear magnetic resonance (NMR) experiments are powerful methods that provide the protein three-dimensional (3D) structure at high resolution [2,16–20]. Electron paramagnetic resonance (EPR) spectroscopy, in conjunction with site-specific spin labeling techniques, has revealed structural information about the aggregation process [21,22]. Other techniques such as circular dichroism (CD) and fluorescence spectroscopy are also commonly used for structure and kinetics analysis [23–26].

Linear and two-dimensional infrared (IR) spectroscopy are sensitive and versatile tools that complement other techniques. They can be readily applied to small amounts of proteins in a variety of sample conditions, such as in aqueous solution or lipid membranes [27,28]. The IR experiments typically involve the amide I vibrational mode, which is primarily associated with the peptide bond carbonyl stretch [27]. Each local amide I mode is called a chromophore, and the many chromophores in a protein interact with each other to form the amide I band in a frequency range of 1600–1700 cm^{-1}. The frequencies at which the amide I band occurs, along with the band widths and intensities, depend on different patterns of intra- and intermolecular couplings and therefore are sensitive to protein secondary structures [29–31]. For instance, α-helices absorb near 1650 cm^{-1}, while antiparallel β-sheets have two bands at ca. 1620 and 1675 cm^{-1}. Moreover, residue-specific information can be revealed by the isotope-labeling technique. For example, a $^{13}C{=}^{18}O$ isotope label lowers the amide I frequency by about 70 cm^{-1} [32–37]. Carbonyls with isotope labels are essentially shifted out of the main amide I band and can be resolved individually. Since isotope labeling simply replaces an atom with its isotope, it produces a minimal perturbation to the system.

Two-dimensional IR (2D IR) spectroscopy spreads the IR absorption spectra in two dimensions and greatly enhances the spectral resolution. By using ultrafast laser pulses, time-resolved 2D IR spectroscopy has enabled the measurement of molecular dynamics over subpicosecond timescales. Cross peaks in 2D IR spectra directly reveal the interactions, such as mode coupling or energy transfer, between chromophores [38–44]. 2D IR experiments, coupled with $^{13}C{=}^{18}O$ isotope labeling techniques, have been applied to many biologically important systems such as proton channels, antimicrobial peptides, and amyloidogenic polypeptides [45–54].

In particular, 2D IR has been utilized to investigate the aggregation and inhibition of hIAPP in aqueous solution and in lipid vesicles [47,48,51,52]. For example, by isotope labeling six residues that span the length of the hIAPP peptide, Shim and coworkers measured the 2D IR spectroscopic changes during the aggregation process of hIAPP and monitored the aggregation kinetics at the level of individual residues [47]. They proposed that the fibril formation, which involves the assembly of soluble hIAPP monomers to form β-sheet-rich aggregates, follows a multistep pathway. Multiple hIAPP monomers first associate to develop well-ordered structure in the loop region of the fibril, which serves as the aggregation nucleus. This step is followed by the formation of two parallel β-sheet regions, with the N-terminal β-sheet likely forming before the C-terminal sheet [47].

Experimental spectra usually have complex features, the interpretation of which is facilitated by theoretical calculations. Theoretical modeling of IR spectra of proteins requires an accurate and efficient description of the amide I frequencies. We have developed frequency "maps" that generate the amide I frequencies directly from molecular dynamics (MD) simulations [55]. These maps are developed for both protein backbone and side-chain chromophores, from model compounds in solvents that represent distinct electrostatic environments. As the frequency maps are designed to be transferable to different solvent conditions, they are potentially applicable to heterogeneous systems such as membrane proteins.

We combine the frequency maps with the existing nearest-neighbor frequency shift method and coupling schemes [56–58], and a mixed quantum/classical framework to form a theoretical strategy for calculating protein linear and 2D IR spectra in the amide I region directly from MD simulations [55]. The theoretical method is validated by applying it to peptides with various secondary structures in aqueous solution [55]. It is then applied to the study of hIAPP, both in solution phase and in the fibril form, in combination with atomistic MD simulations and IR experiments [59,60]. This method bridges the gap between computer simulations and spectroscopic experiments, which allows a critical assessment of MD simulations by comparison to experiment, and enables the interpretation of experimental spectra at the molecular level.

14.2 DEVELOPMENT OF THE FREQUENCY MAPS

For a protein with N atoms and m amide chromophores, it is impractical to model the full 3N-dimensional Hamiltonian when dynamics must be considered. One common approach is to treat the modes of interest, in this case the amide stretches, quantum mechanically and all other nuclear degrees of freedom classically [57,61–74]. Within such a mixed quantum/classical approach, a central quantity is the exciton Hamiltonian in the amide I subspace. At any time t, the amide I Hamiltonian (divided by $\hbar$), $\kappa(t)$, is an m × m matrix, with

$$\kappa_{ij}(t) = \omega_i(t)\delta_{ij} + \omega_{ij}(t)(1 - \delta_{ij}). \tag{14.1}$$

Its diagonal elements $\omega_i(t)$ are the transition frequencies of the local amide chromophores, and the off-diagonal terms $\omega_{ij}(t)$ correspond to the interactions (or couplings) between pairs of chromophores.

The time-dependent fluctuations of the exciton Hamiltonian depend on the classical coordinates, which evolve in time according to a classical MD simulation. For each configuration of the classical coordinates (or each time step in the MD simulation), *ab initio* electronic structure calculations can in principle be used to determine $\kappa(t)$ [75,76]. However, in practice, it is extremely difficult to perform calculations for large proteins in condensed phase with high accuracy. This has led to the development of frequency and coupling "maps," which correlate the Hamiltonian elements with certain collective coordinates of the system and thus allow the calculation of the Hamiltonian relatively accurately and efficiently.

It has been widely accepted that electrostatic interactions play an important role in modulating molecular vibrations, and researchers have chosen electrostatic properties as the collective coordinates. Various frequency maps for protein backbone chromophores have been developed that describe how the electrostatic interactions between the amide groups and the environment (solvents, lipids, ions, etc.) affect the amide I local frequencies [62,64–66,67,70,73,74]. Couplings can be understood as interactions between transition charges. At long distances, those interactions are mainly between transition dipole moments. This assumption leads to the transition dipole coupling (TDC) scheme [27,61,77–79].

Although this electrostatic picture of the Hamiltonian has been tested extensively [74,80,81], interactions between amide groups that are covalently linked should be treated differently as they are quantum mechanical in nature [56–58,82,83]. To account for these effects on local frequencies, *ab initio*-based nearest-neighbor frequency shift (NNFS) maps have been developed to show explicitly how $\omega_i(t)$ is affected by the neighboring (ϕ,ψ) angles [57,58,66]. Similarly, nearest-neighbor coupling (NNC) maps have been developed for couplings between adjacent backbone amide groups [57,58,66,82,84].

We aim to provide a reliable scheme to form the exciton Hamiltonian, which will allow the calculation of protein linear and 2D IR spectra in the amide I region based on MD simulations. We focus on the development of frequency maps in this section, and we will discuss the construction

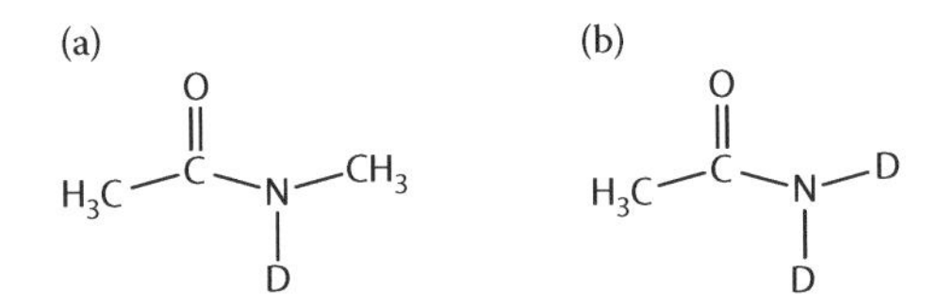

FIGURE 14.1 Molecular structures of (a) NMAD and (b) ACED.

of the full Hamiltonian in the next section. We start by considering the local frequencies of protein backbone chromophores. For this purpose, *N*-methylacetamide (NMA), a model compound with a single backbone chromophore, is employed. The IR experiments on proteins often use heavy water (D_2O) instead of water as solvent, to avoid the interference of the water bend mode (ca. 1640 cm^{-1}) with the amide I band. The major isotopic form of NMA in D_2O is N-deuterated NMA (NMAD), as shown in Figure 14.1a.

NMAD contains a single chromophore, for which the frequency perturbations entirely come from the environment (in this case, the solvent molecules). Based on previous studies in the Skinner group, electric fields from solvent molecules on atoms of NMAD have proven to be good collective coordinates to describe local frequencies [65,73]. We would like to develop a frequency map that relates local amide I frequencies with these electric fields and that applies equally well in different solvent conditions. We have chosen the three solvents D_2O, DMSO, and $CHCl_3$, each representing a different electrostatic environment, and developed the frequency map from the experimental IR line shapes and vibrational lifetimes of NMAD in all three solvents [55].

Besides the backbone chromophores, amide groups in the side chains of asparagine (Asn) and glutamine (Gln) also contribute to the amide I band. It is therefore important to include them in the exciton Hamiltonian for the modeling of Asn- or Gln-containing proteins. Moreover, as side chains often play crucial roles in biological processes such as protein folding and aggregation, the ability to track their spectroscopic changes is particularly meaningful. We have chosen N-deuterated acetamide (ACED, as shown in Figure 14.1b) as a model system, and developed a side-chain frequency map in a similar way to the backbone map.

The frequency maps are developed using the GROMOS96 53a6 force field [85–87], which has been used extensively in biological simulations [55,59,60,88]. The optimized backbone and side-chain maps are [55]

$$\omega_i^0 = 1684 + 7729E_{Ci} - 3576E_{Ni}, \tag{14.2}$$

$$\omega_i^s = 1714 + 2154E_{Ci} + 3071E_{Ni}. \tag{14.3}$$

ω_i^0 and ω_i^s are the instantaneous backbone and side-chain frequencies, respectively, for the *i*th chromophore, and they are in cm^{-1} units. E_{Ci} and E_{Ni} (in atomic units) are the electric fields on the C and N atoms in the *i*th chromophore along the C=O bond direction. The intercepts for the backbone and side-chain maps are 1684 and 1714 cm^{-1}, representing the frequency of NMAD and ACED, respectively, in a nonpolar environment where E_{Ci} and E_{Ni} are zero. These two values are in reasonable agreement with experiments [55,89–91].

To quantify the line shapes of NMAD and ACED in the three solvents, their peak frequencies ω and full-width-half-maxima Γ are extracted from the calculated IR spectra and compared with experiments in Table 14.1. Γ values for ACED in $CHCl_3$ are not listed because the experimental spectra contains two overlapping peaks in the amide I region due to the self-association of ACED, and as a result the line width is not well resolved. From Table 14.1 in all cases, the deviations between theory and experiment are, at most, 3 and 10 cm^{-1} for ω and Γ, respectively. This good

TABLE 14.1
Summary of Simulated and Experimental IR Line Shape Parameters for NMAD and ACED in Different Solvents

	ω_{D_2O}	Γ_{D_2O}	ω_{DMSO}	Γ_{DMSO}	ω_{CHCl_3}	Γ_{CHCl_3}
Sim (NMAD)	1621	34	1656	22	1668	16
Exp (NMAD)	1623 [92]	28 [92]	1659	23	1665	26
Sim (ACED)	1634	40	1666	22	1697	
Exp (ACED)	1633	37	1664	22	1700	

Source: Reprinted with permission from Wang L. et al., *J. Phys. Chem. B*, 115, 3713. Copyright 2011, American Chemical Society.
Note: All quantities are in units of cm^{-1}.

agreement suggests that the frequency maps are transferable and are potentially applicable to more heterogeneous, including hydrophobic, environments [55].

14.3 THEORETICAL PROTOCOL FOR THE CALCULATION OF PROTEIN IR SPECTRA IN THE AMIDE I REGION

14.3.1 Modeling the Amide I Hamiltonian

In this section, we construct the exciton Hamiltonian, from which the linear and 2D IR spectra can be calculated using line shape theory. For this purpose, at any instant in time ω_i and ω_{ij} are needed. ω_i could be a backbone frequency ω_i^b or a side-chain frequency ω_i^s. Frequency maps developed in the last section are for model systems with a single chromophore. Reasonable extension is required to apply them to proteins that contain multiple chromophores.

For a side-chain chromophore, all atoms in the protein and all molecules from the environment can be treated as charged sites, and their electric fields are used to calculate ω_i^s using Equation 14.3. We include all electric fields within a cutoff of 20 Å. A backbone chromophore, on the other hand, feels the interactions from nearby amide linkages as well as the environment. Effects due to the adjacent peptide units are treated using the NNFS maps developed by Jansen and coworkers [57,58]. For the ith backbone chromophore, the NNFS maps relate the contributions from the $(i-1)$th and $(i+1)$th residues (termed as $\Delta\omega_N$ and $\Delta\omega_C$) to the corresponding (ϕ,ψ) dihedral angles. Note that the NNFS maps for the N- and C-sites were swapped in Ref. 57 [58]. All the other peptide units as well as all side-chain atoms and the environment within 20 Å of the ith chromophore contribute to the electric fields that are used to calculate ω_i^0 (Equation 14.2). Therefore, the backbone frequency ω_i^b is

$$\omega_i^b = \omega_i^0 + \Delta\omega_N\left(\phi_{i-1},\psi_{i-1}\right) + \Delta\omega_C\left(\phi_{i+1},\psi_{i+1}\right). \tag{14.4}$$

Note that when the C-terminus of the peptide is amidated, this creates an additional chromophore with the same form as the Asn and Gln side chains. Thus, for these chromophores, we use Equation 14.3, but also add the NNFS contribution $\Delta\omega_N$.

Couplings between adjacent peptide units, as discussed above, are due to the through-bond effect [57,82] and are determined from an NNC map [57]. Couplings between all other peptide units are represented by the TDC scheme [27,56,61,77,78],

$$\omega_{ij} = \frac{A}{\epsilon}\left\{\frac{\vec{m}_i \cdot \vec{m}_j}{r_{ij}^3} - 3\frac{\left(\vec{m}_i \cdot \vec{r}_{ij}\right)\left(\vec{m}_j \cdot \vec{r}_{ij}\right)}{r_{ij}^5}\right\}. \tag{14.5}$$

In the above equation, the vector $\vec{r}_{ij}$ (in Å) connects the two transition dipoles $\vec{m}_i$ and $\vec{m}_j$, which are in the units of $DÅ^{-1}u^{-1/2}$ (u is the atomic mass unit). ϵ is the dielectric constant and is taken to be 1. The conversion factor $A = 0.1 \times 848619/1650$ gives the couplings in cm^{-1} [27,68]. The TDC parameters are taken from *ab initio* calculations by Torii and Tasumi [56]. Specifically, the transition dipoles have the magnitude of $2.73\ DÅ^{-1}u^{-1/2}$ and are oriented 10.0 away from the C=O bond, pointing toward the N atom. Their origins are at $\vec{r}_C + 0.665\hat{n}_{CO} + 0.258\hat{n}_{CN}$ (in units of Å), where $\vec{r}_C$ is the location of the carbonyl carbon atom, $\hat{n}_{CO} = (\vec{r}_O - \vec{r}_C)/|\vec{r}_O - \vec{r}_C|$ and $\hat{n}_{CN} = (\vec{r}_N - \vec{r}_C)/|\vec{r}_N - \vec{r}_C|$.

14.3.2 Line Shape Theory

Line shape theory states that the absorption line shape can be calculated from the Fourier transform of the quantum dipole time-correlation function [93]. If the electric field of the excitation light is polarized in the $\hat{\epsilon}$ direction, the IR absorption line shape is

$$I(\omega) \sim Re\int_0^\infty dt\, e^{-i\omega t}\langle \hat{\epsilon}\cdot\vec{\mu}(0)\vec{\mu}(t)\cdot\hat{\epsilon}\rangle, \tag{14.6}$$

where $\vec{\mu}$ is the dipole operator of the system. The brackets indicate a quantum equilibrium statistical mechanical average, which is impossible to evaluate for proteins in condensed phase.

As discussed in the last section, one practical approximation is to treat the amide I subspace quantum mechanically, ignore other high-frequency modes, and treat the low-frequency degrees of freedom (translations, rotations, and torsions) classically. Within such a mixed quantum/classical approach, the IR line shape is [55,57,94–96]

$$I(\omega) \sim Re\int_0^\infty dt\, e^{-i\omega t}\sum_{ij}\langle m_i(0)F_{ij}(t)m_j(t)\rangle e^{-t/2T_1}, \tag{14.7}$$

where i and j index the amide I vibrational chromophores. The matrix $F(t)$ describes the time propagation of the Hamiltonian $\kappa(t)$,

$$\dot{F}(t) = iF(t)\kappa(t), \tag{14.8}$$

with the initial condition $F_{ij}(0) = \delta_{ij}$. $m_i(t)$ is the projection of the transition dipole moment $\vec{m}_i(t)$ on the polarization unit vector of the excitation light, that is, $m_i(t) = \hat{\epsilon}\cdot\vec{m}_i(t)$. Since in experiment the protein is not oriented, we average over all possible orientations by averaging over the three polarizations $\hat{\epsilon} = \hat{\imath}, \hat{\jmath}, \hat{k}$. The angular brackets in Equation 14.7 indicate a classical equilibrium statistical mechanical average. T_1 is the lifetime of the first excited state of an isolated amide I vibration, and the term $e^{-t/2T_1}$ is added phenomenologically to incorporate lifetime broadening. T_1 is set to be 600 fs [34].

14.3.3 General Protocol for Protein IR Spectra Calculation

A theoretical protocol for the calculation of protein IR absorption spectra in the amide I region is formed. This protocol is readily extendable to 2D IR calculations, which we will briefly discuss in Section 14.6.

Details of the protocol are as follows:

1. Generate configuration trajectories from MD simulations. At each time step, the configuration can be used to calculate backbone (ϕ,ψ) angles and electric fields on each chromophore.

2. Calculate ω_i^0 and ω_i^s from Equations 14.2 and 14.3, respectively.
3. Determine NNFS and NNC from the corresponding (ϕ,ψ) angles [57,58].
4. ω_i^b is then calculated by adding ω_i^0 and NNFS, as in Equation 14.4.
5. If isotope labels are placed in the protein, shift the local frequencies of the labeled residues by the appropriate amount. For example, the isotope shift for $^{13}C{=}^{18}O$ used in this work is -70 cm^{-1}, consistent with previous experiments and *ab initio* calculations [32,33,97–99].
6. Couplings between all non-nearest-neighbor amide groups are calculated from TDC [56].
7. Construct the fluctuating Hamiltonian matrix κ and propagate the F matrix [100,101].
8. Calculate the absorption line shape using Equation 14.7.

14.4 VALIDATION OF THE THEORETICAL PROTOCOL

Validations of the theoretical protocol have been carried out [55]. In our original validation, model peptides AKA and Trpzip2 were chosen to represent α-helical and β-hairpin secondary structures. The swapped NNFS maps from Ref. [57] were used in the prediction of their IR absorption spectra [55]. Correction of the NNFS maps leads to very minor changes to the AKA spectra. That is because for a residue in an α-helix, its N- and C-site (ϕ,ψ) angles are almost identical. The total nearest-neighbor effect, which is the sum of $\Delta\omega_N$ and $\Delta\omega_C$, is therefore not altered much by the swapping of the NNFS maps. For Trpzip2, however, its spectra have more significant changes after correcting the NNFS maps. Trpzip2 adopts multiple configurations at room temperature [102,103], and thus we are not certain whether our theoretical spectra calculated from a single conformation (its PDB structure) is directly comparable to experiment. To avoid this issue, more recently, we have considered a model cyclic peptide [104]. It adopts a stable parallel β-sheet conformation, which is enforced by the constraint of its cyclic configuration. By $^{13}C{=}^{18}O$ labeling six pairs of residues, we are able to test the theoretical local frequencies and couplings by comparison to 2D IR experiments. Results for the cyclic parallel β-sheet peptide will be shown in a future publication [104]. Below we will present the calculation on AKA, both with and without isotope labeling.

The sequence of AKA is shown in Figure 14.2. Its special sequence motif of (AAAAK)$_n$ has been shown to form α-helical structures, especially in the central region [105–109]. The *N*-acetyl capping further increases the helical stability at the N-terminus [107]. MD simulations of AKA in water were performed for a total of 10 ns using the GROMACS simulation package [110–112]. The initial structure of AKA was constructed by assuming a perfect α-helix with all (ϕ,ψ) angles set to ($-57°$, $-47°$). The simulation temperature was set to be 275 K, in accordance with experiment [99,113,114]. The GROMACS source code was modified to report local frequencies on the fly using Equations 14.2 and 14.3. Frequencies and configurations were saved every 10 fs for spectral calculations. Throughout the 10 ns simulation, the central part of the peptide stays α-helical, while the C-terminus frays, in agreement with previous experimental studies [106,107]. A representative configuration of AKA is shown in Figure 14.3.

Using the theoretical protocol, IR absorption spectra of unlabeled AKA are calculated and compared with experiment [113] in Figure 14.4a. The theoretical spectrum has a peak frequency of 1627 cm^{-1}, differing from experiment by only 4 cm^{-1}. This peak position is determined by both the diagonal and off-diagonal elements of the Hamiltonian [115], and we analyze these terms individually. The average local frequencies of individual chromophores, at least in the central region of the helix, are around 1640 cm^{-1}, which provides a central frequency that will be modified by couplings. Coupling constants between peptide units that are adjacent to each other (β_{12}),

AKA Ac-AAAAKAAAAKAAAAKAAAAKAAAAY-NH_2

FIGURE 14.2 Sequence of the AKA peptide.

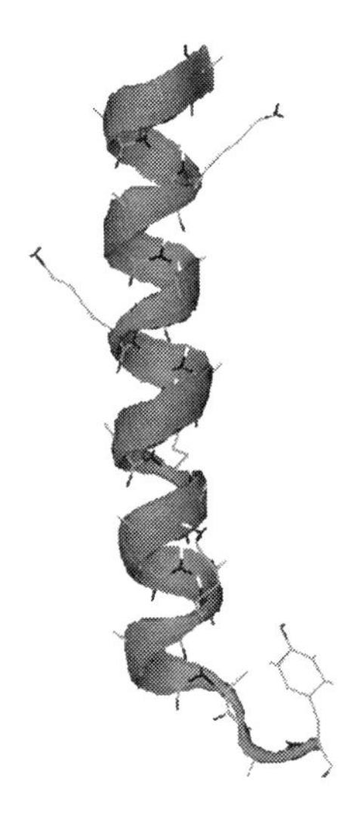

FIGURE 14.3 A representative snapshot of AKA.

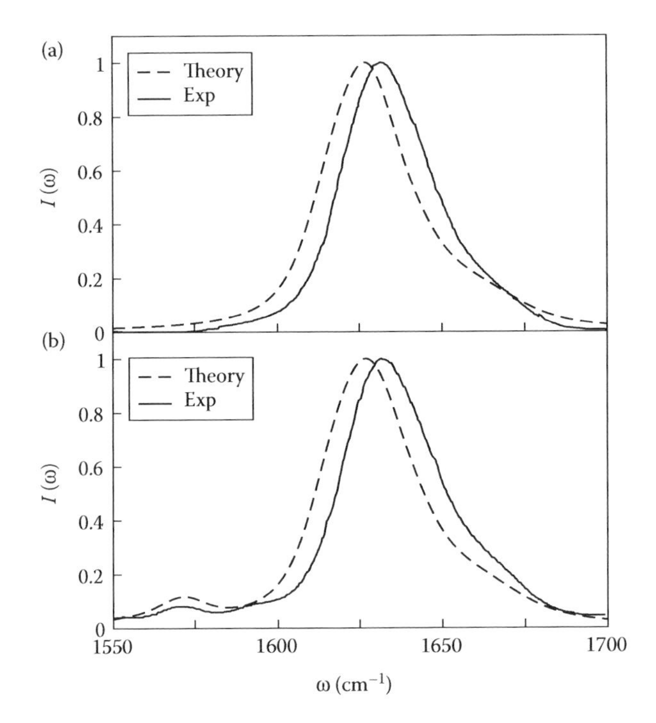

FIGURE 14.4 Theoretical and experimental IR absorption spectra of (a) unlabeled [113], and (b) [12] labeled [99] AKA peptide.

one residue apart (β_{13}) and two residues apart (β_{14}) have the largest magnitude, and thus have the largest impact on the overall spectrum. Their values, averaged over residues 2–20, are shown in Table 14.2. β_{12}'s are positive for α-helices, as predicted by the NNC map and Ham et al. [98], which lead to an overall blue shift of the peak frequency. On the other hand, β_{13}, β_{14}, and other couplings between nonadjacent neighbors are negative, which counteract the effect of β_{12} and cause a final red shift of the peak position. The experimental coupling constants, extracted by

TABLE 14.2
Theoretical and Experimental [114] Coupling Constants for the AKA Peptide

	β_{12}	β_{13}	β_{14}
Theory	5.2 ± 0.3	−1.5 ± 0.2	−5.6 ± 0.2
Experiment	8.5 ± 1.8	−5.4 ± 1.0	−6.6 ± 0.8

Source: Reprinted with permission from Wang L et al., *J. Phys. Chem. B*, 115, 3713. Copyright 2011, American Chemical Society.
Note: Theoretical values are averaged over residues 2–20. All coupling constants are in cm^{-1}.

systematically isotope labeling two residues zero, one and two residues apart, are also listed in Table 14.2 [114]. Compared to experiment, the signs and relative magnitude of the three coupling constants are correct.

We also carried out calculations on the isotope-edited AKA peptide. A single $^{13}C{=}^{18}O$ label was placed on residue Ala12 (denoted as [12]), and the full spectrum was calculated by shifting the frequency of Ala12 and keeping all the other local frequencies and couplings unchanged. The [12] spectrum is compared with experiment [99] in Figure 14.4b. Note that the overall line shape and the isotope feature in the theoretical spectrum are in good agreement with experiment. To study the labeled features more systematically, we assume *each* residue is $^{13}C{=}^{18}O$ labeled, one at a time, and treat each one of them as an isolated chromophore and neglect couplings. The theoretical peak frequency of each labeled residue is shown in Figure 14.5 as a function of residue numbers, along with the experimental values [99] for residues 12–15. Peak frequencies in the central region stay more or less constant, consistent with a homogeneous α-helical configuration. The good agreement between theory and experiment (differing by only a few cm^{-1}) confirms our combined electrostatic and NNFS [57,58] scheme for calculating the backbone frequencies.

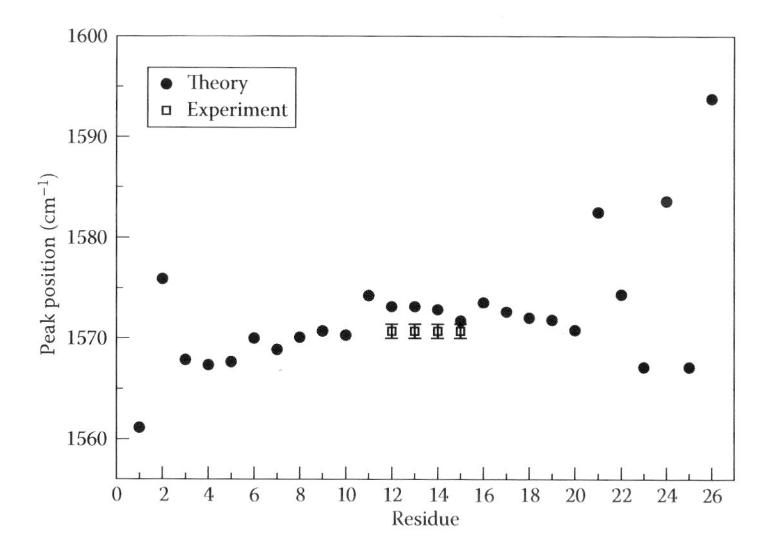

FIGURE 14.5 $^{13}C{=}^{18}O$-labeled peak frequencies as a function of residue number for the AKA peptide. Experimental isotope-labeled peak frequencies are shown for residues 12–15 [99].

14.5 SOLUTION STRUCTURE OF hIAPP

The sequence of hIAPP is shown in Figure 14.6. Due to the fast aggregation kinetics, detailed solution structures of monomeric hIAPP are not well understood. *In vitro* CD experiments indicate that soluble hIAPP is mostly disordered in aqueous solution, although CD spectroscopy is low in structural resolution and hIAPP is not necessarily in the monomeric state in the measurement [116,117]. Recent NMR experiments, combined with a better aggregate removal method to ensure a monomeric peptide, suggest that the N-terminal part of hIAPP preferentially adopts an α-helical conformation, but the peptide does not adopt a unique 3D structure or fold [19]. Using a combination of ion mobility mass spectrometry experiment and replica-exchange MD (REMD) simulations, Dupuis and coworkers found two distinct solution structural families of hIAPP monomer [118]. Their data is consistent with one conformation that contains extended β-hairpin and the other one with compact helix-coil structure [118].

As a first step toward understanding the aggregation of hIAPP, in this section we investigate its solution structure. All-atom REMD simulations with explicit water were performed to obtain stable conformers of monomeric hIAPP [59]. Based on these atomistic-level models, IR absorption spectra are calculated [59]. This work presented a first atomistic description of the full-length hIAPP peptide in explicit water.

Atomistic REMD simulations were performed for 100 ns, which reveal that hIAPP adopts α-helical, β-hairpin, and random-coil conformations in aqueous solution [59]. The α-helical conformer has an α-helical segment from residues 9 to 17 and a short antiparallel β-sheet comprising residues 24–28 and 31–35. The β-hairpin conformer adopts an extended antiparallel β-hairpin configuration with residues 20–23 being the turn region. The random coil conformer is completely disordered. Representative snapshots of the three conformers are shown in Figure 14.7. Their relative stability was determined using the thermodynamic integration method. The α-helical, β-hairpin, and random coil conformers have relative populations of 31%, 40%, and 29%, respectively [59].

hIAPP $^{+}H_3N$-KCNTATCATQRLANFLVHSSNNFGAILSSTNVGSNTY-NH_2

FIGURE 14.6 The primary sequences of hIAPP. The peptides have an amidated C-terminus and a disulfide bond between residues 2 and 7.

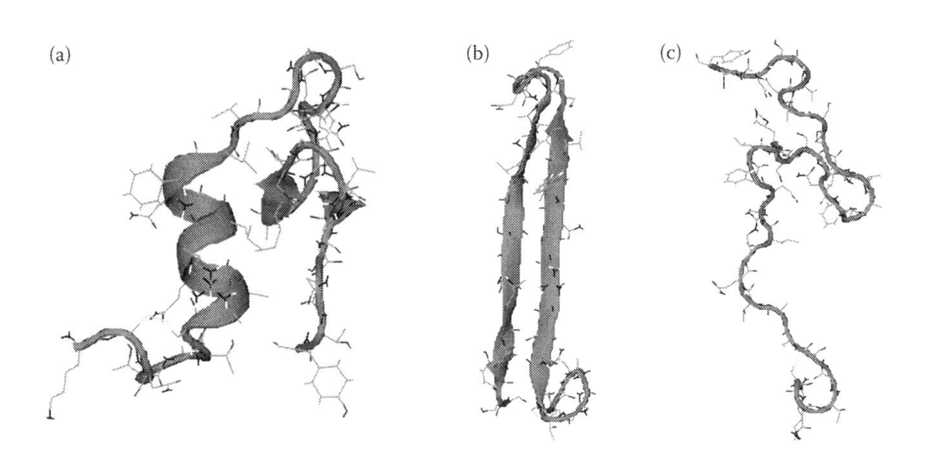

FIGURE 14.7 Representative snapshots of hIAPP monomer in (a) α-helical, (b) β-hairpin, and (c) random coil conformation [59].

IR absorption spectra of hIAPP are measured at pH 7 at the early stage of aggregation, as shown in Figure 14.8 [59]. The experimental spectrum shows a minor change when the peptide concentration is doubled [59], and it agrees well with a previous experiment with very low peptide concentration (0.01 μM) [119], in part confirming that the spectrum is for the monomer state of hIAPP.

Theoretical IR absorption spectra for each conformer are calculated based on MD simulations, as shown in Figure 14.8. The α-helical and random coil conformers have peak frequencies of 1645 and 1656 cm^{-1}, respectively. The misfolded conformer has two peaks, in agreement with the general spectral features of an antiparallel β-sheet [120–122]. Note that the main peak is at 1644 cm^{-1}, instead of 1620 cm^{-1} (which is characteristic of antiparallel β-sheet). It is due to the fact that in a β-hairpin the vibrational excitons can only delocalize over two strands rather than three or more, which is necessary to obtain a lower β-sheet vibrational frequency. Spectra of the three conformers are added, according to their relative populations, to give the total spectrum, which is also shown in Figure 14.8. The good agreement between the theoretical total spectrum and experiment is consistent with our contention that there is a substantial population of the β-hairpin conformer in solution.

This study illustrates the power of our combined MD simulation, theoretical and experimental spectroscopy approach. It enables a quantitative cycle of structure prediction and validation for systems, such as aggregation-prone proteins, that are challenging for other techniques. The predicted conformer structures have important biological implications, as discussed below.

It is postulated that when interacting with lipid membranes, hIAPP initially folds into an α-helical structure at the N-terminus [10,14,25]. Assembly of the α-helical N-termini possibly facilitates the formation of β-sheet oligomers by increasing the local concentration of the highly amyloidogenic regions between residues 20 and 29 [10,14,25,123]. This N-terminal α-helical segment, as predicted in this study and in previous NMR experiments [19], might therefore play an important role in the association of hIAPP with membranes, and might be related to the membrane-catalyzed aggregation.

The amyloid fibrils formed from hIAPP are rich in β-sheet content [16,17,21]. A structural model for the fibril has been proposed by Luca and coworkers, as shown in Figure 14.9 [17]. We note that the turn region in the β-hairpin conformer predicted in this study coincides with the turn region of the fibril model [17], indicating that this conformer might be a precursor for aggregation [59,118].

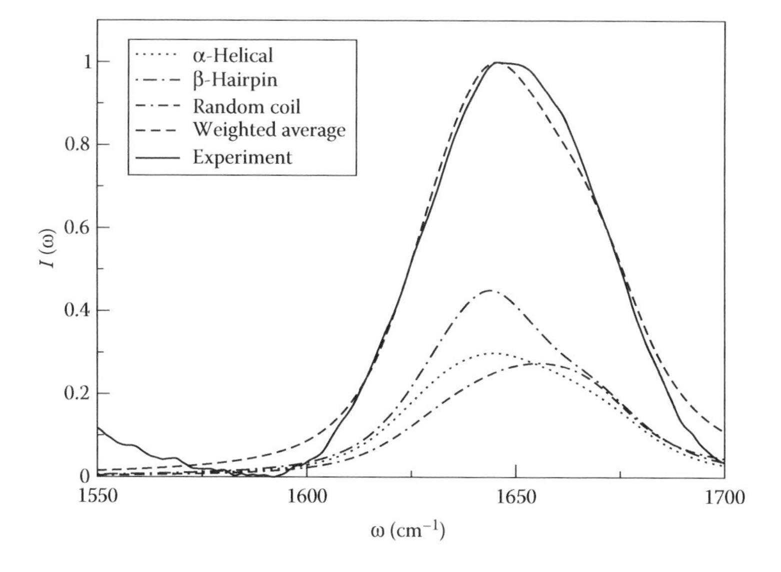

FIGURE 14.8 Theoretical (weighted average) and experimental IR line shapes in the amide I stretch region of hIAPP in D_2O. Also shown are line shapes for the α-helical, β-hairpin, and random coil states, weighted by their relative probabilities [59].

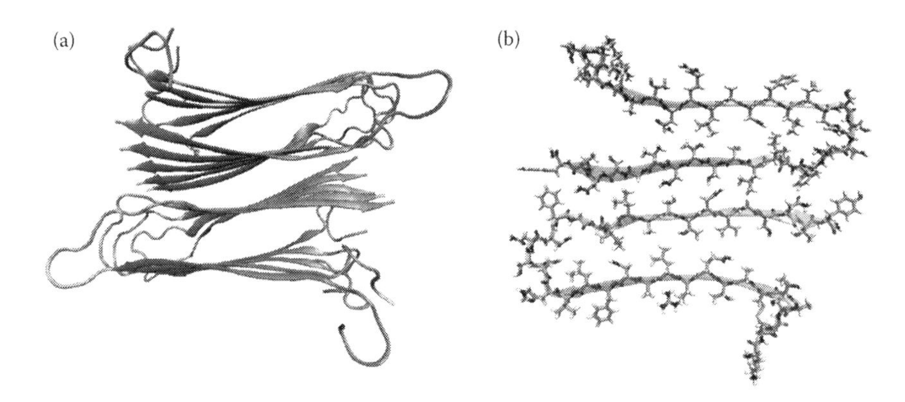

FIGURE 14.9 (a) The overall backbone arrangement of the protofilament structural models [17]. (b) The cross section of the structural model that corresponds to the structure depicted in Figure 11C of Ref. [17].

We concur with the proposal by Dupuis et al. [118] that a possible aggregation pathway involves the formation and aggregation of monomeric β-hairpin hIAPP peptides [59].

14.6 2D IR SPECTROSCOPY REFLECTS STABLE β-SHEET STRUCTURE IN hIAPP FIBRILS

Amyloid fibrils formed from hIAPP are under extensive experimental study [16,17,21]. The fibrils adopt the classic cross-β structure [16]. EPR spectroscopy combined with site-directed spin labeling reveals that in the fibril, hIAPP forms in-register parallel β-sheets [21]. Based on scanning transmission electron microscopy and NMR measurements, Luca and coworkers propose a structural model of the protofilament, which is the basic structural unit of the fibril [17]. As shown in Figure 14.9a, this protofilament model has a twofold symmetry, containing four layers of parallel β-strands with two symmetric columns of hIAPP molecules [17]. In this model, the N-terminal half of each hIAPP peptide faces water, and the C-terminal half forms the fibril core [17]. The cross section of the protofilament is shown in Figure 14.9b, which is composed of two hIAPP peptides, one from each column [17]. Each hIAPP peptide has the structure of two β-strands, one in the N-region and one in the C-region of the chain, connected by a turn. Its N-terminus (residues 1–7) is disordered due to the constraint of the Cys2–Cys7 disulfide bond.

In this section, we study the structure and dynamics of hIAPP fibrils using theoretical and experimental 2D IR spectroscopy, combined with extensive MD simulations. Residue-specific information is obtained through a series of isotope labels of the fibril. Theoretical 2D IR spectra are predicted for each residue based on atomistic MD simulations. Both theoretical and experimental 2D IR line widths suggest a “W” pattern, as a function of residue number. Small line widths in the two β-sheet regions indicate that they are very stable secondary structures. This work also provides a theoretical strategy for bridging MD simulations and 2D IR experiments for future aggregation studies.

14.6.1 Materials and Methods

14.6.1.1 Calculation of the 2D IR Spectra

In a typical 2D IR experiment, three pulses with time intervals t_1 and t_2 are incident into the sample. A photon echo is generated and detected after time t_3. The 2D IR spectrum is obtained by double Fourier transforming with respect to t_1 and t_3.

Experimentally, for each sample, a $^{13}C{=}^{18}O$ isotope label is placed at a specific residue, and the singly labeled hIAPP monomers are diluted by the unlabeled monomers. The labeled residue is modeled as an isolated chromophore that is decoupled from the rest of the amide I band. The instantaneous frequency of each labeled chromophore is calculated along the MD trajectory. Specifically, the frequencies for residues 1 through 36 are calculated using Equations 14.2 and 14.4. For residue 37, which is a C-terminal amidation, Equation 14.3 is used for its frequencies, and $\Delta\omega_N$ is added to correct its local frequency.

The frequencies of each chromophore are fluctuating around an average value $\langle\omega\rangle$. At time t, set $\delta\omega(t) = \omega(t) - \langle\omega\rangle$. The frequency time-correlation function (FTCF) is calculated from

$$C(t) = \langle\delta\omega(t)\delta\omega(0)\rangle. \tag{14.9}$$

The line shape function $g(t)$ is then calculated as

$$g(t) \equiv \int_0^t d\tau\,(t-\tau)C(\tau). \tag{14.10}$$

We set the "waiting time" t_2 to zero, consistent with experiment, and assume infinitely short pulses. Within the second-order cumulant expansion, the rephasing and nonrephasing response functions become [50,73,96,124]

$$R_R(t_1,0,t_3) \sim e^{-2g(t_1)-2g(t_3)+g(t_1+t_3)}e^{-(t_1+t_3)/2T_1}\left[e^{i\langle\omega_{10}\rangle(t_1-t_3)} - e^{i\langle\omega_{10}\rangle t_1 - i\langle\omega_{21}\rangle t_3}e^{-t_3/T_1}\right], \tag{14.11}$$

$$R_{NR}(t_1,0,t_3) \sim e^{-g(t_1+t_3)}e^{-(t_1+t_3)/2T_1}\left[e^{-i\langle\omega_{10}\rangle(t_1+t_3)} - e^{-i\langle\omega_{10}\rangle t_1 - i\langle\omega_{21}\rangle t_3}e^{-t_3/T_1}\right]. \tag{14.12}$$

$\langle\omega_{10}\rangle$ and $\langle\omega_{21}\rangle$ are the average transition frequencies between the ground and the first-excited vibrational state, and between the first and the second excited states, respectively. They differ by the vibrational anharmonicity, which is taken to be 14 cm^{-1} [37]. The terms $e^{-(t_1+t_3)/2T_1}$ and e^{-t_3/T_1} are added phenomenologically to include lifetime broadening [124]. T_1 is set to be 600 fs for all residues [34], although it has recently been discovered that residues in β-sheet conformations have longer lifetimes [51].

The rephasing and nonrephasing spectra are the double Fourier transforms of their corresponding response functions

$$S_R(\omega_1,0,\omega_3) \sim \int_0^\infty dt_1 e^{-i\omega_1 t_1}\int_0^\infty dt_3 e^{i\omega_3 t_3}\mathrm{Re}\{R_R(t_1,0,t_3)\} \tag{14.13}$$

and

$$S_{NR}(\omega_1,0,\omega_3) \sim \int_0^\infty dt_1 e^{i\omega_1 t_1}\int_0^\infty dt_3 e^{i\omega_3 t_3}\,\mathrm{Re}\{R_{NR}(t_1,0,t_3)\}. \tag{14.14}$$

The 2D IR absorptive spectrum is obtained by summing the rephasing and nonrephsing spectra

$$I(\omega_1,0,\omega_3) \sim \mathrm{Re}\{S_R(\omega_1,0,\omega_3) + S_{NR}(\omega_1,0,\omega_3)\}. \tag{14.15}$$

MD simulations were performed for 110 ns [60]. A total of 12 fibril configurations, 10 ns apart from each other, are extracted from the simulation and are used for the 2D IR calculations. Each configuration serves as a starting structure to provide better sampling of the fibril structure. Note that our simulation involves only one protofilament with 10 hIAPP peptides. An ideal fibril involves an infinite number of hIAPP peptides, each of which forms hydrogen bonds with its neighboring peptides. Such an environment is well represented by the two central peptides (one in each column) in the structural models, while all the other hIAPP peptides are affected by edge effects. 2D IR spectra are calculated by averaging over the central two peptides for each simulation and over the 12 simulations. The diagonal line widths (Γ_d) are extracted from the spectra to quantify the comparison with experiment. Error bars are calculated as twice the standard deviation of the mean from the 12 simulations.

14.6.1.2 Sample Preparation and 2D IR Measurements

Methods for synthesizing $^{13}C{=}^{18}O$ isotope-labeled amino acids and incorporating the labeled amino acids into hIAPP have been published elsewhere [32,125,126]. Briefly, commercially available FMOC and side-chain-protected amino acids are used for solid-phase peptide synthesis. The hydrophobicity and high aggregation tendency of hIAPP lead to difficulties in its synthesis. We have developed an efficient synthesis method using two pseudoprolines in combination with microwave technology [126]. To integrate isotope labels, amino acids with ^{13}C at the backbone carbonyl position are purchased with protecting groups attached. The ^{18}O label is incorporated easily in amino acids with nonreactive side chains (Gly, Ala, Leu, Ile, Val, Phe) using an acid-catalyzed exchange mechanism with ^{18}O-enriched water [32,125]. As the FMOC group is base labile, it is retained during the synthesis. Side-chain-protecting groups, however, are acid labile and thus this method will not work for amino acids with reactive side chains. Instead, a new method has been demonstrated that allows the oxygen to be exchanged at pH ~ 5 and all the protecting groups to be retained [126].

The peptides are purified using reversed-phase HPLC and are denatured and stored in d-HFIP at 1 mM. To prepare samples for 2D IR, 5 μL of peptide solution is lyophilized and then redissolved in 20 mM deuterated potassium phosphate buffer at pH 7.4 to initiate aggregation. Final peptide concentration is between 500 μM and 1 mM. The peptide solution is placed in an IR cell between CaF_2 windows with a 56 μm spacer.

2D IR spectra are collected in a pump-probe beam geometry using a mid-IR pulse shaper; we have published several reviews on collecting spectra in this manner [45,115,125,127–129]. Briefly, femtosecond pulses of mid-IR light are generated in an optical parametric amplifier with difference frequency mixing. Generally, around 3 μJ pulses are produced. Roughly 10% of each pulse is split off to serve as the probe (E_3) while the rest is sent into a mid-IR pulse shaper. The pulses are dispersed into the frequency domain, focused in the Fourier plane onto a Ge acousto-optic modulator, then recombined. An arbitrary waveform generator is used to create a sinusoidal amplitude mask that amplifies and phase modulates the dispersed pulse and creates two pump pulses in the time domain, E_1 and E_2, with a variable delay between them. This delay is scanned over 2560 fs in 24 fs steps to generate a 2D IR spectrum. The pump and probe pulses are focused onto the sample, which emits a third-order electric field in the same direction as the probe beam. The E_3 probe beam also serves as the local oscillator (LO), which allows us to implement heterodyne detection easily. The combined signal field and LO are dispersed through a monochromator and detected with an MCT array (Figure 14.10).

2D IR absorptive spectra were collected continuously to achieve satisfactory signal-to-noise ratio. Average spectral amplitude was calculated as a function of pump and probe frequencies to give the final 2D IR spectra. The standard deviation of the spectral amplitude was calculated as well. Using 2D interpolation, diagonal slices through the average contours were obtained. Each slice was fit to a Gaussian for the label peak and one for the unlabeled β-sheet peak, sometimes incorporating a small baseline offset (held at experimental values) in the fit. The standard deviations were used as weighting factors in the nonlinear least-squares fitting procedure. The fit line widths and uncertainties (twice the standard deviation) are reported.

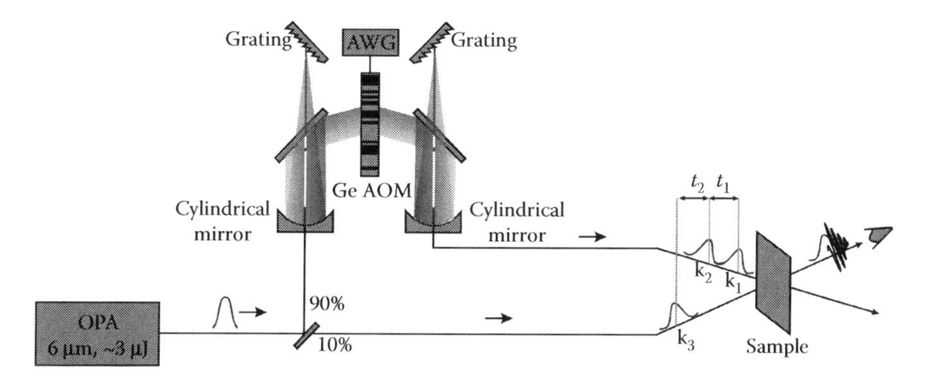

FIGURE 14.10 Diagram of the pulse shaper used in the 2D IR experiment.

14.6.2 Results and Discussion

Experimentally, a single $^{13}C{=}^{18}O$ label was placed at residues Ala13, Leu16, Ser19, Ser20, Ala25, Val32, or Gly33, one at a time, and the 2D IR spectra are measured. Selective spectra for Ala13, Ser19, Ala25, and Val32, along with the diagonal slices along their 2D contour, are presented in Figure 14.11. Three features are most prominent in the experimental 2D IR spectra. (1) Peaks along the diagonal appear in pairs, which are characteristic for all 2D vibrational spectroscopy. The negative peak, or the fundamental peak, is generated by the transition from the vibrational ground state to the first excited state. The positive peak corresponds to a transition from the first to the second vibrational excited states. (2) A peak pair appears at pump and probe frequencies ~1620 cm^{-1}, which is due to the unlabeled chromophores and is characteristic of parallel β-sheets of the mature fibril [47,49]. (3) Another peak pair appears at pump and probe frequencies ~1595 cm^{-1} due to the carbonyl stretch of the $^{13}C{=}^{18}O$-labeled residue. Since we label only one of the many chromophores in the fibril, the labeled signal has a much weaker intensity compared to the unlabeled feature. The scale of Figure 14.11 is chosen to emphasize the labeled feature, so the unlabeled peak around 1620 cm^{-1} is too strong to be fully visible. Note that cross peaks appear at a pump frequency of ~1595 cm^{-1} and a probe frequency of ~1620 cm^{-1}, which indicate the interaction of the labeled and unlabeled chromophores [47,49,60].

It can be observed directly from Figure 14.11 that Γ_d of the fundamental peak varies from residue to residue. Γ_d are extracted and plotted in Figure 14.12 as squares. It shows a trend that residues in the β-sheet regions have narrow line widths. For example, Γ_d for Ala13 and Ala25 are 22 ± 5 and 20 ± 6 cm^{-1}, respectively. On the other hand, residues near the turn and the terminal regions have broader line widths. Examples are Ser19 and Val32, which have a Γ_d of 33 ± 7 and 29 ± 5 cm^{-1}, respectively.

While the isotope-edited 2D IR experiments are (at the current stage) limited by the amino acids that can be labeled, theoretically we can do a more systematic study by assuming that every residue is labeled, one at a time. Theoretical values of Γ_d are also presented in Figure 14.12 as a function of residue number, which are in good agreement with experiment. A characteristic *W* pattern appears for both theory and experiment. Theoretical spectral analyses suggest that Γ_d is dominated by inhomogeneous frequency distributions, which are ultimately determined by the structural inhomogeneity [60].

We then directly illustrate the structural fluctuations of the fibril. We have chosen the distances between the *i*th C_α in the central peptide chain and the *i*th C_α's in the nearest two chains to represent the backbone structure. Root-mean-square deviations (RMSDs) of the distances, averaged over the two columns in each simulation and over the 12 simulations, are shown in Figure 14.13. From the RMSD values, the β-sheet regions of residues 8–16 and 27–36 have small fluctuations (<0.5 Å) throughout the simulations, consistent with their small Γ_d values. Note that discrepancies between the Γ_d and RMSD patterns might occur for the residues right at the edge of the β-sheets, for example, Ala8, mainly due to the modulation of the surrounding water molecules.

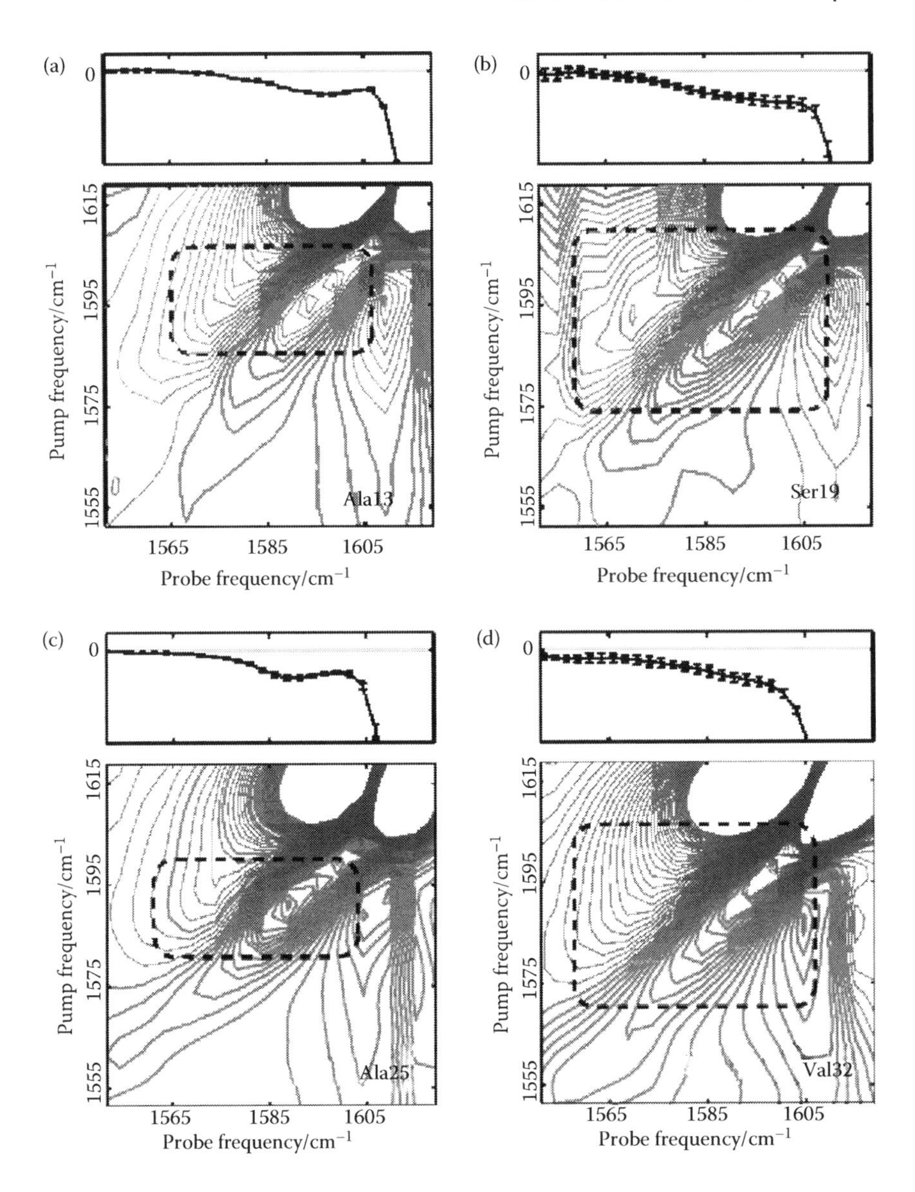

FIGURE 14.11 Experimental 2D IR spectra for (a) Ala13, (b) Ser19, (c) Ala25, and (d) Val32. The label peaks are identified with boxes. The top panels show diagonal slices (with error bars) of the 2D IR spectra and the fit to the label peaks. (Reprinted with permission from Wang L et al., *J. Am. Chem. Soc.*, 133, 16062. Copyright 2011, American Chemical Society.)

By comparing the spectral and RMSD analyses, a spectrum–structure relation is proposed to explain the W pattern in Γ_d. Structurally, each hIAPP monomer contains two β-strands connected by a flexible turn. The two β-sheet regions are well ordered and do not exhibit significant static or dynamic disorder, while the terminal and the turn regions are particularly disordered and are in close contact with water molecules. These structural features manifest themselves in the 2D IR spectra, with the β-sheet regions showing smaller line widths and the turn and terminal regions having enhanced diagonal line widths.

In this work, our combined MD simulation, theoretical and experimental isotope-edited 2D IR spectroscopy approach is used for the study of hIAPP fibril with residue-specific resolution.

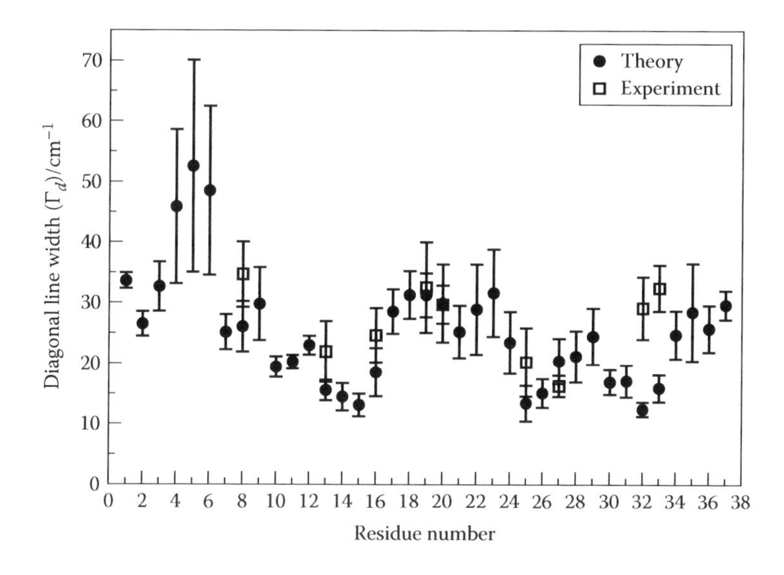

FIGURE 14.12 Comparison of the theoretical and experimental Γ_d (in cm^{-1}) as a function of residue number [60].

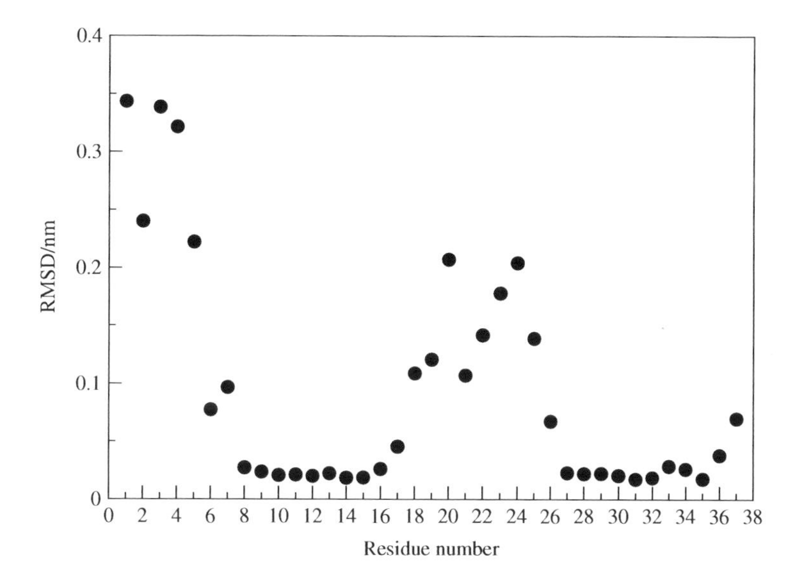

FIGURE 14.13 RMSD of the C_α distance between the central peptide chain and nearest two chains. (Reprinted with permission Wang L et al., from *J. Am. Chem. Soc.*, 133, 16062. Copyright 2011, American Chemical Society.)

Previous experiments have identified that the β-sheet regions are particularly amyloidogenic, and mutations in these regions often reduce the aggregation propensity [117,130,131]. Their structural stability, as revealed in this work, indicate that they are possible drug-binding sites for the purpose of inhibition of amyloid formation.

14.7 CONCLUSIONS

In this chapter, we have developed frequency maps for protein backbone and side chain amide I absorptions, from experimental line shapes of model compounds in different solvents. We have combined the maps with previously developed NNFS [57,58] and coupling schemes [56,57] and a mixed quantum/classical framework to form a theoretical strategy for the calculation of protein linear and 2D IR spectra in the amide I region. After validating the theoretical method, we apply it to study the structure and dynamics of hIAPP, in conjunction with extensive MD simulations and IR experiments. In all cases, we are able to reproduce experimental IR spectra well, without any *ad hoc* frequency shifts or further adjustments. This and other theoretical frameworks that can accurately and efficiently calculate linear and 2D IR spectra provide a way to bridge computer simulations and spectroscopic experiments, and will enable the understanding of protein structure and dynamics at the molecular level.

2D IR spectroscopy, in conjunction with isotope labeling, has the capability to detect *in situ* kinetics of highly dynamical processes such as protein folding and aggregation, both in aqueous solution and in lipid membranes. From a theoretical perspective, since the frequency maps are designed to be transferable to different electrostatic environments, the theoretical protocol we have developed is potentially applicable to heterogeneous systems such as membrane proteins. A combined theoretical and experimental spectroscopy approach, coupled with atomistic MD simulations, is a powerful technique for the study of biological problems.

For example, this combined approach is promising in the study of the assembly of hIAPP in the absence and presence of lipid membranes. Using 2D IR spectroscopy, Ling and coworkers investigated the aggregation of hIAPP in aqueous solutions and in lipid vesicles, and were able to probe directly the nature of the intermediates in real time [48]. The synergy between theory, simulation, and experiment will allow us to understand the spectroscopic changes along the aggregation pathway, provide detailed structural information on oligomers and aggregates, unravel the mechanism of hIAPP-induced membrane damage, and ultimately shed light on the medical treatment of type 2 diabetes [52].

ACKNOWLEDGMENTS

This work was supported in part by a grant from the NSF (CHE-0832584) to JLS and MTZ. MTZ also thanks NIH for support through grant DK79895. JLS also thanks NSF through grant CHE-1058752. JLS and JJdP thank NIH through 1R01DK088184-01A1. LW thanks Dr. Chris Middleton for preparing the 2D IR figure. LEB is supported by the NSF Graduate Research Fellowship Program (DGE-0718123).

REFERENCES

1. Dobson CM. 1999. Protein misfolding, evolution and disease. *Trends. Biochem. Sci.* 24:329.
2. Tycko R. 2006. Molecular structure of amyloid fibrils: Insights from solid-state NMR. *Quart. Rev. Biophys.* 39:1.
3. Sipe JD, ed. 2005. *Amyloid Proteins: The Beta Sheet Conformation and Disease* (WILEY-VCH, Weinheim, Germany).
4. Clark A et al. 1987. Islet amyloid formed from diabetes-associated peptide may be pathogenic in type-2 diabetes. *Lancet* 2:231.
5. Lorenzo A, Razzaboni B, Weir GC, Yankner BA. 1994. Pancreatic islet cell toxicity of amylin associated with type-2 diabetes mellitus. *Nature* 368:756.
6. Kahn SE, Andrikopoulos S, Verchere CB. 1999. Islet amyloid: A long-recognized but underappreciated pathological feature of type 2 diabetes. *Diabetes* 48:241.
7. Chiti F, Dobson CM. 2006. Protein misfolding, functional amyloid, and human disease. *Annu. Rev. Biochem.* 75:333.
8. Kayed R et al. 2004. Permeabilization of lipid bilayers is a common conformation-dependent activity of soluble amyloid oligomers in protein misfolding diseases. *J. Biol. Chem.* 279:46363.
9. Sparr E et al. 2004. Islet amyloid polypeptide-induced membrane leakage involves uptake of lipids by forming amyloid fibers. *FEBS Lett.* 577:117.
10. Jayasinghe SA, Langen R. 2007. Membrane interaction of islet amyloid polypeptide. *Biochim. Biophys. Acta* 1768:2002.

11. Engel MFM et al. 2008. Membrane damage by human islet amyloid polypeptide through fibril growth at the membrane. *Proc. Natl. Acad. Sci. USA* 105:6033.
12. Smith PES, Brender JR, Ramamoorthy A. 2009. Induction of negative curvature as a mechanism of cell toxicity by amyloidogenic peptides: The case of islet amyloid polypeptide. *J. Am. Chem. Soc.* 131:4470.
13. Knight JD, Miranker AD. 2004. Phospholipid catalysis of diabetic amyloid assembly. *J. Mol. Biol.* 341:1175.
14. Knight JD, Hebda JA, Miranker AD. 2006. Conserved and cooperative assembly of membrane-bound α-helical states of islet amyloid polypeptide. *Biochemistry* 45:9496.
15. Mirzabekov TA, Lin M, Kagan BL. 1996. Pore formation by the cytotoxic islet amyloid peptide amylin. *J. Biol. Chem.* 271:1988.
16. Makin OS, Serpell LC. 2004. Structural characterization of islet amyloid polypeptide fibrils. *J. Mol. Biol.* 335:1279.
17. Luca S, Yau WM, Leapman R, Tycko R. 2007. Peptide conformation and supramolecular organization in amylin fibrils: Constraints from solid-state NMR. *Biochemistry* 46:13505.
18. Nanga RPR, Brender JR, Xu J, Veglia G, Ramamoorthy A. 2008. Structures of rat and human islet amyloid polypeptide $IAPP_{1-19}$ in micelles by NMR spectroscopy. *Biochemistry* 47:12689.
19. Yonemoto IT, Kroon GJA, Dyson HJ, Balch WE, Kelly JW. 2008. Amylin proprotein processing generates progressively more amyloidogenic peptides that initially sample the helical state. *Biochemistry* 47:9900.
20. Cort JR et al. 2009. Solution state structures of human pancreatic amylin and pramlintide. *Protein Eng. Des. Sel.* 22:497.
21. Jayasinghe SA, Langen R. 2004. Identifying structural features of fibrillar islet amyloid polypeptide using site-directed spin labeling. *J. Biol. Chem.* 279:48420.
22. Apostolidou M, Jayasinghe SA, Langen R. 2008. Structure of α-helical membrane-bound human islet amyloid polypeptide and its implications for membrane-mediated misfolding. *J. Biol. Chem.* 283:17205.
23. Goldsbury C et al. 2000. Amyloid fibril formation from full-length and fragments of amylin. *J. Struct. Biol.* 130:352.
24. Padrick SB, Miranker AD. 2001. Islet amyloid polypeptide: Identification of long-range contacts and local order on the fibrillogenesis pathway. *J. Mol. Biol.* 308:783.
25. Patil SM, Xu S, Sheftic SR, Alexandrescu AT. 2009. Dynamic α-helix structure of micelle-bound human amylin. *J. Biol. Chem.* 284:11982.
26. Sasahara K, Hall D, Hamada D. 2010. Effect of lipid type on the binding of lipid vesicles to islet amyloid polypeptide amyloid fibrils. *Biochemistry* 49:3040.
27. Krimm S, Bandekar J. 1986. Vibrational spectroscopy and conformation of peptides, polypeptides, and proteins. *Adv. Protein Chem.* 38:181.
28. Barth A, Zscherp C. 2002. What vibrations tell us about proteins. *Quart. Rev. Biophys.* 35:369.
29. Susi H, Byler DM. 1986. Resolution-enhanced Fourier transform infrared spectroscopy of enzymes. *Methods Enzymol.* 130:290.
30. Haris PI, Chapman D. 1992. Does Fourier-transform infrared spectroscopy provide useful information on protein structures? *Trends. Biochem. Sci.* 17:328.
31. Surewicz WK, Mantsch HH, Chapman D. 1993. Determination of protein secondary structure by Fourier transform infrared spectroscopy: A critical assessment. *Biochemistry* 32:389.
32. Torres J, Adams PD, Arkin IT. 2000. Use of a new label, $^{13}C{=}^{18}O$, in the determination of a structural model of phospholamban in a lipid bilayer. Spatial restraints resolve the ambiguity arising from interpretations of mutagenesis data. *J. Mol. Biol.* 300:677.
33. Fang C et al. 2003. Two-dimensional infrared measurements of the coupling between amide modes of an α-helix. *Chem. Phys. Lett.* 382:586.
34. Mukherjee P et al. 2004. Site-specific vibrational dynamics of the CD3ζ membrane peptide using heterodyned two-dimensional infrared photon echo spectroscopy. *J. Chem. Phys.* 120:10215.
35. Arkin IT. 2006. Isotope-edited IR spectroscopy for the study of membrane proteins. *Curr. Opin. Chem. Biol.* 10:394.
36. Decatur SM. 2006. Elucidation of residue-level structure and dynamics of polypeptides via isotope-edited infrared spectroscopy. *Acc. Chem. Res.* 39:169.
37. Mukherjee P, Kass I, Arkin I, Zanni MT. 2006. Picosecond dynamics of a membrane protein revealed by 2D IR. *Proc. Natl. Acad. Sci. USA* 103:3528.
38. Hamm P, Lim M, Hochstrasser RM. 1998. Structure of the amide I band of peptides measured by femtosecond nonlinear-infrared spectroscopy. *J. Phys. Chem. B* 102:6123.

39. Zanni MT, Hochstrasser RM. 2001. Two-dimensional infrared spectroscopy: A promising new method for the time resolution of structures. *Curr. Opin. Struct. Biol.* 11:516.
40. Hochstrasser RM. 2007. Two-dimensional spectroscopy at infrared and optical frequencies. *Proc. Natl. Acad. Sci. USA* 104:14190.
41. Park S, Kwak K, Fayer MD. 2007. Ultrafast 2D IR vibrational echo spectroscopy: A probe of molecular dynamics. *Laser Phys. Lett.* 4:704.
42. Cho M. 2008. Coherent two-dimensional optical spectroscopy. *Chem. Rev.* 108:1331.
43. Ganim Z et al. 2008. Amide I two-dimensional infrared spectroscopy of proteins. *Acc. Chem. Res.* 41:432.
44. Strasfeld DB, Ling YL, Shim SH, Zanni MT. 2008. Tracking fiber formation in human islet amyloid polypeptide with automated 2D IR spectroscopy. *J. Am. Chem. Soc.* 130:6698.
45. Shim SH, Strasfeld DB, Ling YL, Zanni MT. 2007. Automated 2D IR spectroscopy using a mid-IR pulse shaper and application of this technology to the human islet amyloid polypeptide. *Proc. Natl. Acad. Sci. USA* 104:14197.
46. Manor J et al. 2009. Gating mechanism of the influenza A M2 channel revealed by 1D and 2D spectroscopies. *Structure* 17:247.
47. Shim SH et al. 2009. Two-dimensional IR spectroscopy and isotope labeling defines the pathway of amyloid formation with residue specific resolution. *Proc. Natl. Acad. Sci. USA* 106:6614.
48. Ling YL, Strasfeld DB, Shim SH, Raleigh DP, Zanni MT. 2009. Two-dimensional infrared spectroscopy provides evidence of an intermediate in the membrane-catalyzed assembly of diabetic amyloid. *J. Phys. Chem. B* 113:2498.
49. Strasfeld DB, Ling YL, Gupta R, Raleigh DP, Zanni MT. 2009. Strategies for extracting structural information from 2D IR spectroscopy of amyloid: Application to islet amyloid polypeptide. *J. Phys. Chem. B* 113:15679.
50. Woys AM et al. 2010. 2D IR line shapes probe ovispirin peptide conformation and depth in lipid bilayers. *J. Am. Chem. Soc.* 132:2832.
51. Middleton CT, Buchanan LE, Dunkelberger EB, Zanni MT. 2011. Utilizing lifetimes to suppress random coil features in 2D IR spectra of peptides. *J. Phys. Chem. Lett.* 2:2357.
52. Middleton CT et al. 2012. Two-dimensional infrared spectroscopy reveals the complex behaviour of an amyloid fibril inhibitor. *Nature Chem.* 4:355.
53. Moran SD et al. 2012. Two-dimensional IR spectroscopy and segmental ^{13}C labeling reveals the domain structure of human γD-crystallin amyloid fibrils. *Proc. Natl. Acad. Sci. USA* 109:3329.
54. Kim YS, Liu L, Axelsen PH, Hochstrasser RM. 2009. 2D IR provides evidence for mobile water molecules in β-amyloid fibrils. *Proc. Natl. Acad. Sci. USA* 106:17751.
55. Wang L, Middleton CT, Zanni MT, Skinner JL. 2011. Development and validation of transferable amide I vibrational frequency maps for peptides. *J. Phys. Chem. B* 115:3713.
56. Torii H, Tasumi M. 1998. *Ab initio* molecular orbital study of the amide I vibrational interactions between the peptide groups in di- and tripeptides and considerations on the conformation of the extended helix. *J. Raman Spectrosc.* 29:81.
57. Jansen TLC, Dijkstra AG, Watson TM, Hirst JD, Knoester J. 2006. Modeling the amide I bands of small peptides. *J. Chem. Phys.* 125:044312.
58. Jansen TLC, Dijkstra AG, Watson TM, Hirst JD, Knoester J. 2012. Erratum: Modeling the amide I bands of small peptides. [*J. Chem. Phys.* 125, 044312 (2006)]. *J. Chem. Phys.* 136:209901.
59. Reddy AS et al. 2010. Stable and metastable states of human amylin in solution. *Biophys. J.* 99:2208.
60. Wang L et al. 2011. 2D IR spectroscopy of human amylin fibrils reflects stable β-sheet structure. *J. Am. Chem. Soc.* 133:16062.
61. Torii H, Tasumi M. 1992. Model calculations on the amide-I infrared bands of globular proteins. *J. Chem. Phys.* 96:3379.
62. Ham S, Kim JH, Lee H, Cho M. 2003. Correlation between electronic and molecular structure distortions and vibrational properties. II. Amide I modes of NMA-nD_2O complexes. *J. Chem. Phys.* 118:3491.
63. Choi JH, Ham S, Cho M. 2003. Local amide I mode frequencies and coupling constants in polypeptides. *J. Phys. Chem. B* 107:9132.
64. Bour P, Keiderling T. 2003. Empirical modeling of the peptide amide I band IR intensity in water solution. *J. Chem. Phys.* 119:11253.
65. Schmidt JR, Corcelli SA, Skinner JL. 2004. Ultrafast vibrational spectroscopy of water and aqueous *N*-methylacetamide: Comparison of different electronic structure/molecular dynamics approaches. *J. Chem. Phys.* 121:8887.
66. Gorbunov RD, Kosov DS, Stock G. 2005. *Ab initio*-based exciton model of amide I vibrations in peptides: Definition, conformational dependence, and transferability. *J. Chem. Phys.* 122:224904.

67. Hayashi T, Zhuang W, Mukamel S. 2005. Electrostatic DFT map for the complete vibrational amide band of NMA. *J. Phys. Chem. A* 109:9747.
68. Zhuang W, Abramavicius D, Hayashi T, Mukamel S. 2006. Simulation protocols for coherent femtosecond vibrational spectra of peptides. *J. Phys. Chem. B* 110:3362.
69. Hochstrasser RM. 2006. Dynamical models for two-dimensional infrared spectroscopy of peptides. *Adv. Chem. Phys.* 132:1.
70. Jansen TLC, Knoester J. 2006. A transferable electrostatic map for solvation effects on amide I vibrations and its application to linear and two-dimensional spectroscopy. *J. Chem. Phys.* 124:044502.
71. Torii H. 2006. In *Atoms, Molecules and Clusters in Electric Fields. Theoretical Approaches to the Calculation of Electric Polarizability*, ed Maroulis G (Imperial College Press, London), p. 179.
72. Bloem R, Dijkstra AG, Jansen TLC, Knoester J. 2008. Simulation of vibrational energy transfer in two-dimensional infrared spectroscopy of amide I and amide II modes in solution. *J. Chem. Phys.* 129:055101.
73. Lin YS, Shorb JM, Mukherjee P, Zanni MT, Skinner JL. 2009. Empirical amide I vibrational frequency map: Application to isotope-edited membrane peptide bundles. *J. Phys. Chem. B* 113:592.
74. Maekawa H, Ge NH. 2010. Comparative study of electrostatic models for the amide-I and -II modes: Linear and two-dimensional infrared spectra. *J. Phys. Chem. B* 114:1434.
75. Kim J, Huang R, Kubelka J, Bour P, Keiderling TA. 2006. Simulation of infrared spectra for β-hairpin peptides stabilized by an Aib-Gly turn sequence: Correlation between conformational fluctuation and vibrational coupling. *J. Phys. Chem. B* 110:23590.
76. Grahnen JA, Amunson KE, Kubelka J. 2010. DFT-based simulations of IR amide *I′* spectra for a small protein in solution. Comparison of explicit and empirical solvent models. *J. Phys. Chem. B* 114:13011.
77. Krimm S, Abe Y. 1972. Intermolecular interaction effects in the amide I vibrations of β peptides. *Proc. Natl. Acad. Sci. USA* 69:2788.
78. Moore WH, Krimm S. 1975. Transition dipole coupling in amide I modes of β polypeptides. *Proc. Natl. Acad. Sci. USA* 72:4933.
79. Torii H, Tatsumi T, Kanazawa T, Tasumi M. 1998. Effects of intermolecular hydrogen-bonding interactions on the amide I mode of N-methyl acetamide: Matrix-isolation infrared studies and *ab initio* molecular orbital calculations. *J. Phys. Chem. B* 102:309.
80. Gorbunov RD, Nguyen PH, Kobus M, Stock G. 2007. Quantum-classical description of the amide I vibrational spectrum of trialanine. *J. Chem. Phys.* 126:054509.
81. Sengupta N et al. 2009. Sensitivity of 2D IR spectra to peptide helicity: A concerted experimental and simulation study of an octapeptide. *J. Phys. Chem. B* 113:12037.
82. Hamm P, Woutersen S. 2002. Coupling of the amide I modes of the glycine dipeptide. *Bull. Chem. Soc. Jpn.* 75:985.
83. Gorbunov RD, Stock G. 2007. *Ab initio* based building block model of amide I vibrations in peptides. *Chem. Phys. Lett.* 437:272.
84. Ham S, Cho M. 2003. Amide I modes in the *N*-methylacetamide dimer and glycine dipeptide analog: Diagonal force constants. *J. Chem. Phys.* 118:6915.
85. van Gunsteren WF et al. 1996. *Biomolecular Simulation: The GROMOS96 Manual and User Guide* (Hochschuleverlag AG an der ETH Zürich, Zürich, Switzerland).
86. Scott WRP et al. 1999. The GROMOS biomolecular simulation program package. *J. Phys. Chem. A* 103:3596.
87. Oostenbrink C, Villa A, Mark AE, van Gunsteren WF. 2004. A biomolecular force field based on the free enthalpy of hydration and solvation: The GROMOS force-field parameter sets 53a5 and 53a6. *J. Comput. Chem.* 25:1656.
88. Reddy AS et al. 2010. Solution structures of rat amylin peptide: Simulation, theory and experiment. *Biophys. J.* 98:443.
89. Eaton G, Symons MCR, Rastogi PP. 1989. Spectroscopic studies of the solvation of amides with N-H groups. Part 1. The carbonyl group. *J. Chem. Soc., Faraday Trans.* 85:3257.
90. Kubelka J, Keiderling TA. 2001. *Ab initio* calculations of amide carbonyl stretch vibrational frequencies in solution with modified basis sets. 1. N-methyl acetamide. *J. Phys. Chem. A* 105:10922.
91. Nyquist RA. 1963. The structural configuration of some α-substituted secondary acetamides in dilute CCl_4 solution. *Spectrochim. Acta* 19:509.
92. DeCamp MF et al. 2005. Amide I vibrational dynamics of N-methylacetamide in polar solvents: The role of electrostatic interactions. *J. Phys. Chem. B* 109:11016.
93. McQuarrie DA. 1976. *Statistical Mechanics* (Harper and Row, New York).
94. Auer BM, Skinner JL. 2007. Dynamical effects in line shapes for coupled chromophores: Time-averaging approach. *J. Chem. Phys.* 127:104105.

95. Auer BM, Skinner JL. 2008. IR and Raman spectra of liquid water: Theory and interpretation. *J. Chem. Phys.* 128:224511.
96. Mukamel S. 1995. *Principles of Nonlinear Optical Spectroscopy* (Oxford, New York).
97. Torres J, Kukol A, Goodman JM, Arkin IT. 2001. Site-specific examination of secondary structure and orientation determination in membrane proteins: The peptidic $^{13}C{=}^{18}O$ group as a novel infrared probe. *Biopolymers* 59:396.
98. Ham S, Cha S, Choi JH, Cho M. 2003. Amide I modes of tripeptides: Hessian matrix reconstruction and isotope effects. *J. Chem. Phys.* 119:1451.
99. Fang C, Hochstrasser RM. 2005. Two-dimensional infrared spectra of the $^{13}C{=}^{18}O$ isotopomers of alanine residues in an α-helix. *J. Phys. Chem. B* 109:18652.
100. Jansen TLC, Knoester J. 2006. Nonadiabatic effects in the two-dimensional infrared spectra of proteins: Application to alanine dipeptide. *J. Phys. Chem. B* 110:22910.
101. Yang M, Skinner JL. 2010. Signatures of coherent energy transfer in IR and Raman line shapes for liquid water. *Phys. Chem. Chem. Phys.* 12:982.
102. Chodera JD, Singhal N, Pande VS, Dill KA, Swope WC. 2007. Automatic discovery of metastable states for the construction of Markov models of macromolecular conformational dynamics. *J. Chem. Phys.* 126: 155101.
103. Smith AW et al. 2010. Melting of a β-hairpin peptide using isotope-edited 2D IR spectroscopy and simulations. *J. Phys. Chem. B* 114:10913.
104. Woys AM et al. 2012. Parallel β-sheet vibrational couplings revealed by 2D IR spectroscopy of an isotopically labeled macrocycle: Quantitative benchmark for the interpretation of amyloid and protein infrared spectra. *J. Am. Chem. Soc.*, submitted.
105. Marqusee S, Robbins VH, Baldwin RL. 1989. Unusually stable helix formation in short alanine-based peptides. *Proc. Natl. Acad. Sci. USA* 86:5286.
106. Decatur SM, Antonic J. 1999. Isotope-edited infrared spectroscopy of helical peptides. *J. Am. Chem. Soc.* 121:11914.
107. Decatur SM. 2000. IR spectroscopy of isotope-labeled helical peptides: Probing the effect on N-acetylation on helix stability. *Biopolymers* 54:180.
108. Silva RAGD, Nguyen JY, Decatur SM. 2002. Probing the effect of side chains on the conformation and stability of helical peptides via isotope-edited infrared spectroscopy. *Biochemistry* 41:15296.
109. Huang R et al. 2004. Nature of vibrational coupling in helical peptides: An isotopic labeling study. *J. Am. Chem. Soc.* 126:2346.
110. Berendsen HJC, van der Spoel D, van Drunen R. 1995. GROMACS: A message-passing parallel molecular dynamics implementation. *Comput. Phys. Commun.* 91:43.
111. van der Spoel D et al. 2005. *GROMACS User Manual Version 3.3* (www.gromacs.org).
112. van der Spoel D et al. 2005. GROMACS: Fast, flexible and free. *J. Comput. Chem.* 26:1701.
113. Barber-Armstrong W, Donaldson T, Wijesooriya H, Silva RAGD, Decatur SM. 2004. Empirical relationships between isotope-edited IR spectra and helix geometry in model peptides. *J. Am. Chem. Soc.* 126:2339.
114. Fang C et al. 2004. Two-dimensional infrared spectroscopy of isotopomers of an alanine rich α-helix. *J. Phys. Chem. B* 108:10415.
115. Hamm P, Zanni M. 2011. *Concepts and Methods of 2D Infrared Spectroscopy* (Cambridge University Press, Cambridge, UK).
116. Kayed R et al. 1999. Conformational transitions of islet amyloid polypeptide (IAPP) in amyloid formation in vitro. *J. Mol. Biol.* 287:781.
117. Jaikaran ETAS, Clark A. 2001. Islet amyloid and type 2 diabetes: From molecular misfolding to islet pathophysiology. *Biochim. Biophys. Acta* 1537:179.
118. Dupuis NF, Wu C, Shea JE, Bowers MT. 2009. Human islet amyloid polypeptide monomers form ordered β-hairpins: A possible direct amyloidogenic precursor. *J. Am. Chem. Soc.* 131:18283.
119. Jha S, Sellin D, Seidel R, Winter R. 2009. Amyloidogenic properties and conformational properties of proIAPP and IAPP in the presence of lipid bilayer membranes. *J. Mol. Biol.* 389:907.
120. Dijkstra AG, Knoester J. 2005. Collective oscillations and the linear and two-dimensional infrared spectra of inhomogeneous β-sheets. *J. Phys. Chem. B* 109:9787.
121. Smith AW, Tokmakoff A. 2007. Amide I two-dimensional infrared spectroscopy of β-hairpin peptides. *J. Chem. Phys.* 126:045109.
122. Jansen TLC, Knoester J. 2008. Two-dimensional infrared population transfer spectroscopy for enhancing structural markers of proteins. *Biophys. J.* 94:1818.
123. Abedini A, Raleigh DP. 2009. A role for helical intermediates in amyloid formation by natively unfolded polypeptides? *Phys. Biol.* 6:015005.

124. Schmidt JR et al. 2007. Are water simulation models consistent with steady-state and ultrafast vibrational spectroscopy experiments? *Chem. Phys.* 341:143.
125. Middleton CT, Woys AM, Mukherjee SS, Zanni MT. 2010. Residue-specific structural kinetics of proteins through the union of isotope labeling, mid-IR pulse shaping, and coherent 2D IR spectroscopy. *Methods* 52:12.
126. Marek P, Woys AM, Sutton K, Zanni MT, Raleigh DP. 2010. Efficient microwave-assisted synthesis of human islet amyloid polypeptide designed to facilitate the specific incorporation of labeled amino acids. *Org. Lett.* 12:4848.
127. Grumstrup EM, Shim SH, Montgomery MA, Damrauer NH, Zanni MT. 2007. Facile collection of two-dimensional electronic spectra using femtosecond pulse-shaping technology. *Opt. Express* 15:16681.
128. Shim SH, Zanni MT. 2009. How to turn your pump-probe instrument into a multidimensional spectrometer: 2D IR and Vis spectroscopies via pulse shaping. *Phys. Chem. Chem. Phys.* 11:748.
129. Buchanan LE, Dunkelberger EB, Zanni MT. 2011. In *Protein Folding and Misfolding: Shining Light by Infrared Spectroscopy*, eds Fabian H, Naumann D (Springer, Heidelberg).
130. Abedini A, Raleigh DP. 2006. Destabilization of human IAPP amyloid fibrils by proline mutations outside of the putative amyloidogenic domain: Is there a critical amyloidogenic domain in human IAPP? *J. Mol. Biol.* 355:274.
131. Fox A et al. 2010. Selection for nonamyloidogenic mutants of islet amyloid polypeptide (IAPP) identifies an extended region for amyloidogenicity. *Biochemistry* 49:7783.

Index

E

F

G

H

J

K

L

M

T

U

W

X

Z